MW00837520

Second Edition

The Metabolism and Molecular Physiology of *Saccharomyces cerevisiae*

Second Edition

The Metabolism and Molecular Physiology of *Saccharomyces cerevisiae*

Edited by

J. Richard Dickinson

Michael Schweizer

CRC PRESS

Boca Raton London New York Washington, D.C.

First edition published 1999
by Taylor & Francis Ltd
Second edition published 2004
by Taylor & Francis Ltd
11 New Fetter Lane, London EC4P 4EE

Simultaneously published in the USA and Canada
by Taylor & Francis Inc,
325 Chestnut Street, 8th Floor, Philadelphia PA 19106

Taylor & Francis is an imprint of the Taylor & Francis Group

Typeset in Baskerville MT by
Newgen Imaging Systems (P) Ltd, Chennai, India
Cover images courtesy of Jeremy Brown

British Library Cataloguing in Publication Data
A catalogue record for this book is available from the British Library

Library of Congress Cataloging in Publication Data
A catalog record for this book has been requested

ISBN 0–415–29900–4

Contents

Figures

Tables

Contributors

Michael Breitenbach
Institut für Genetik und Allgemeine Biologie
Universität Salzburg
Hellbrunnerstrasse 34
A-5020 Salzburg
Austria

Jeremy D. Brown
School of Cell and Molecular Biosciences
The Medical School
University of Newcastle
Newcastle
NE2 4HH
United Kingdom

Stanley Brul
Swammerdam Institute for Life Sciences-Microbiology
University of Amsterdam
Nieuwe Achtergracht 166
1018 WV Amsterdam
The Netherlands

Ian W. Dawes
School of Biotechnology and Biomolecular Sciences
University of New South Wales
Sydney
New South Wales 2052
Australia

J. Richard Dickinson
Cardiff School of Biosciences
Cardiff University
PO Box 915
Cardiff
CF10 3TL
United Kingdom

Piet de Groot
Swammerdam Institute for Life Sciences-Microbiology
University of Amsterdam
Nieuwe Achtergracht 166
1018 WV Amsterdam
The Netherlands

Gino Heeren
Institut für Genetik und Allgemeine Biologie
Universität Salzburg
Hellbrunnerstrasse 34
A-5020 Salzburg
Austria

Klaas Hellingwerf
Swammerdam Institute for Life Sciences-Microbiology
University of Amsterdam
Nieuwe Achtergracht 166
1018 WV Amsterdam
The Netherlands

Stephi Jarolim
Institut für Genetik und Allgemeine Biologie
Universität Salzburg
Hellbrunnerstrasse 34
A-5020 Salzburg
Austria

Frans M. Klis
Swammerdam Institute for Life Sciences-Microbiology
University of Amsterdam
Nieuwe Achtergracht 166
1018 WV Amsterdam
The Netherlands

Arthur L. Kruckeberg
Gothia Yeast Solutions AB
Terrassgatan 7
41133 Göteborg
Sweden

Peter Laun
Institut für Genetik und Allgemeine Biologie
Universität Salzburg
Hellbrunnerstrasse 34
A-5020 Salzburg
Austria

Alena Pichova
Institute of Microbiology of the Czech Academy of Sciences
Prague
Czech Republic

Michael Schweizer
School of Life Sciences
Heriot-Watt University
Edinburgh
EH14 4AS
United Kingdom

Michael J. R. Stark
Division of Gene Regulation and Expression
School of Life Sciences
MSI/WTB Complex
University of Dundee
Dundee
DD1 5EH
United Kingdom

Preface

An enormous amount has been discovered since the publication (at the end of 1999) of the first edition of *The Metabolism and Molecular Physiology of Saccharomyces cerevisiae.* To take account of this, all of the previous chapters have been completely rewritten, updated and expanded. On the whole, we have tried not to repeat information given in the first edition but to focus on new results, except where we felt it was essential. Two areas of particular growth of our knowledge are ageing and the molecular organization and biogenesis of the cell wall; this second edition has new chapters on each. As before, we hope that the book will be suitable for newcomer and expert alike.

Chapter 1

Life cycle and morphogenesis

J. Richard Dickinson

1.1 Introduction

The purpose of this chapter is to give a brief, general introduction to the budding yeast *Saccharomyces cerevisiae*. The various components of yeast's life cycle are described in subsequent sections of this chapter. Details of the molecular events which govern and comprise (e.g.) the cell cycle, ageing, stress resistance, etc., are cross-referenced and can be found in later chapters.

1.2 Life cycle

As shown in Figure 1.1, *S. cerevisiae* can exist both as a haploid and as a diploid. Given adequate nutrients both haploids and diploids can undergo repeated rounds of vegetative growth and mitosis. Yeast has a considerable number of alternative developmental options; the signals for all of these (except mating) are nutritional. Haploids exist in one of two mating types called **a** and α. Haploids of mating type **a** produce a pheromone ('**a** factor') and haploids of α mating type produce a pheromone ('α factor'). Each cell type has a cell surface receptor for pheromone produced by cells of the opposite mating type. **a** factor causes α haploids to arrest in the G1 phase of the cell cycle and α factor has the same effect on **a** cells. Consequently, when in each other's presence, haploids of opposite mating type cease proliferation and commence the development of protuberances towards each other. The resultant shape is called a 'shmoo'. Eventually, there is cell contact and subsequent fusion, culminating in the formation of a diploid.

When nutrients become depleted, both haploids and diploids arrest as stationary phase cells. These are morphologically and biochemically distinct from proliferating yeast cells. They are unbudded, round, phase-bright and contain much higher levels of the storage carbohydrates trehalose and glycogen than proliferating cells. Compared with proliferating cells, stationary phase cells also have increased resistance to a large number of stresses and adverse environmental conditions (Werner-Washburne *et al.*, 1993).

Diploid cells starved of nitrogen and in the presence of a poor carbon source such as acetate will undergo meiosis and spore formation. Four haploid spores are formed which are contained within an ascus. The spores have even greater resistance to environmental extremes than stationary phase cells. If the spores are returned to rich nutrient conditions they will germinate and commence growth as haploids.

The life cycle of *S. cerevisiae* with its alternation between haplophase and diplophase is exploited in conventional strain construction. Haploids of opposite mating type, each

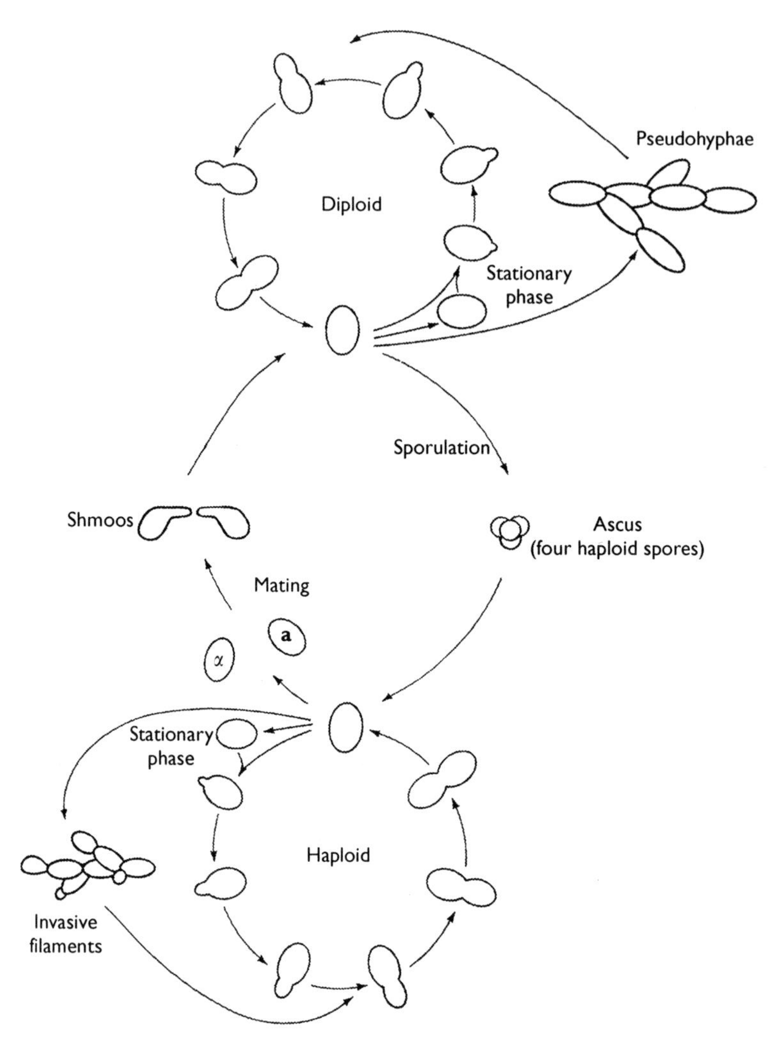

Figure 1.1 The life cycle of *Saccharomyces cerevisiae*. (Redrawn from Dickinson (1999) with the permission of Taylor & Francis.)

having a certain desirable genotype are allowed to mate on a medium permissive to the growth of both and the diploid is duly formed. The mating mixture, containing the diploid is then replica-plated to a selective medium on which only the new diploid (but neither parental haploid) can grow. After a further day or two, the diploid is then replica-plated to a sporulation medium to induce sporulation. Because the formation of spores involves meiosis, some of the haploid progeny will have different combinations of genes and mutations from those present in the original parents. The ascus wall is digested away using glucanase and then the four individual spores from each ascus are transferred separately to fresh medium (using a micromanipulator) to grow as haploid clones. The precise combination of genes and mutations in each spore clone can be determined by analysing its phenotype and/or by a genetic screen (e.g. diagnostic polymerase chain reactions using specific primers).

Pseudohyphae and haploid filaments are easily recognized by their greater cell length and the fact that they form chains. They are both distinct developmental forms and are not merely 'clumpy' or sticky cells. It is generally believed that this filamentation allows foraging for nutrients away from the site where a colony initially resided. *Saccharomyces cerevisiae* has also been described as being able to form biofilm (defined as the ability to adhere to plastic) (Reynolds and Fink, 2001), an activity not normally associated with this organism. Key regulators of haploid invasive growth and diploid pseudohyphal formation are also required for biofilm formation (Kuchin *et al.*, 2002).

1.3 Cell cycle

Yeast's vegetative proliferation via budding (Figure 1.2) is one of the most well-known features in biology. The use of even an unsophisticated microscope will reveal bud initiation at the start of S-phase. The underlying molecular machinery started in G1 by the combination of the cyclin-dependent kinase (Cdk) Cdc28 with the G1 cyclins (Cln1, Cln2 and Cln3). ('Cyclins' were so-called because of their regular appearance and disappearance in each cell cycle.) The earliest activation event is from the association of Cln3 and Cdc28. This enables the transcriptional activators SBF (which comprises Swi4 and Swi6) and MBF (Swi4 and Mbp1) to start-off the transcription of *CLN1* and *CLN2*. Consequently, the levels

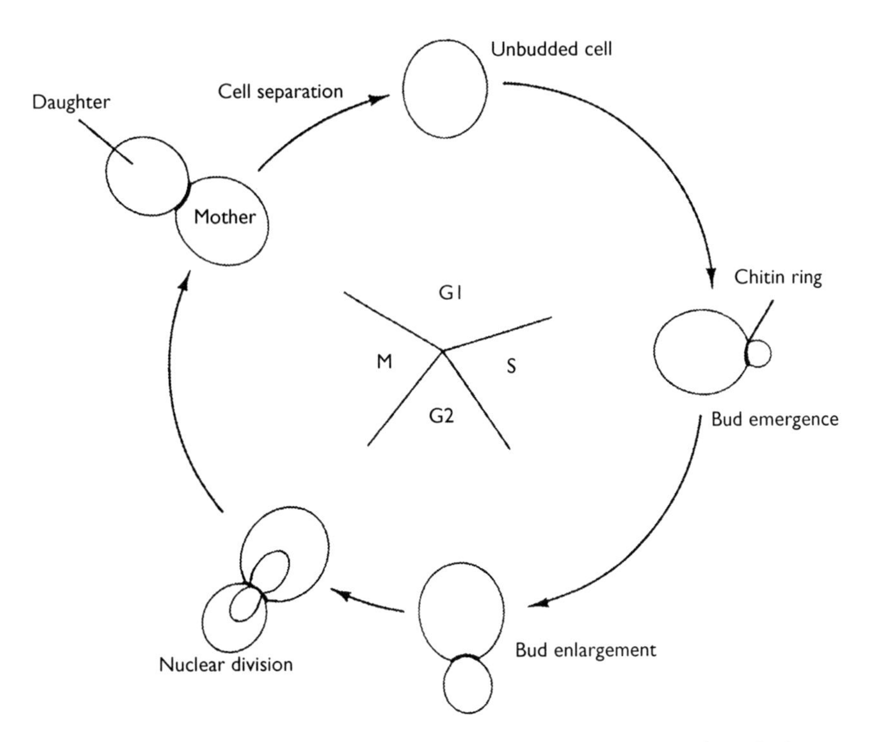

Figure 1.2 Simplified representation of the cell cycle of *S. cerevisiae*. (Redrawn from Dickinson (1999) with the permission of Taylor & Francis.)

of Cln1 and Cln2 build up in late G1 thereby leading to the formation of Cln1–Cdc28 and Cln2–Cdc28 complexes which trigger the 'Start' commitment point to the cell cycle seen by bud emergence. The majority of genes induced at 'Start' have multiple binding sites for SBF and MBF in their promoters (Spellman *et al.*, 1998). SBF and MBF bind to these genes (Iyer *et al.*, 2001; Simon *et al.*, 2001). There is a size control which operates over 'Start'. Cells cannot pass 'Start' unless a minimum size (which varies according to the growth medium) has been reached. cAMP levels play a key role in controlling this step via cAMP-dependent protein kinase (PKA) which simultaneously inhibits the transcription of *CLN1* and *CLN2* (Baroni *et al.*, 1994; Toikwa *et al.*, 1994) and promotes growth (probably by a direct stimulation of protein synthesis (Ashe *et al.*, 2000)). Since Cln3 levels are mainly regulated at the level of translation, the increased protein synthesis leads to an increase in the amount of Cln3 in the cell and the induction of *CLN1* and *CLN2* as already described.

A number of other events are also triggered at 'Start' including the initiation of DNA synthesis, polarized growth and the duplication of spindle pole bodies. The Cln1–Cdc28 and Cln2–Cdc28 complexes lead to the expression of other cyclins (Clb1–6) which control the activity of the Cdc28 protein kinase at subsequent stages of the cell cycle. Clb5–Cdc28 and Clb6–Cdc28 promote DNA replication whilst the association of Clb1–4 with Cdc28 causes the cell to be propelled towards mitosis. Clb–Cdc28 complexes also serve to inhibit re-budding later in the cell cycle (Lew and Reed, 1993, 1995; Amon *et al.*, 1994; Shimada *et al.*, 2000). The functioning of Cdc28 in the cell cycle (as far as it is understood) is described in full in Chapter 8 and Figure 8.2.

The attachment of chromosomes to the mitotic spindle triggers the anaphase–metaphase transition in which the anaphase promoting complex (APC) causes the ubiquitination of proteins which are proteolytically cleaved so as to allow anaphase and exit from mitosis. Cdc20 and Cdh1 activate APC towards its targets. Three subunits of APC (Cdc16, Cdc23 and Cdc27) are phosphorylated by Cdc28, which results in the activation of Cdc20. Cdc28 also phosphorylates Cdh1, which inactivates it and thereby renders it unable to bind APC. APC^{Cdc20} causes the degradation of Pds1 that starts a chain of events culminating in the separation of sister chromatids (see Chapter 8, pp. 292–294). For the next stage (mitotic exit) Clbs are destroyed so that Cdc28 protein kinase activity is rendered low once more. APC^{Cdh1} destroys the Clbs. Cdh1 is activated by dephosphorylation, which is principally accomplished by Cdc14. Until recently, it was thought that degradation of Clb5 was necessary for the dephosphorylation and activation of Cdh1 and (the Cdk inhibitor) Sic1 (Morgan, 1999; Zachariae and Nasmyth, 1999). However, Wäsch and Cross (2002) have shown that spindle disassembly and cell division occurred without significant APC^{Cdc20}-mediated Clb5 degradation and in the absence of both Cdh1 and Sic1. These authors report that destruction box-dependent degradation of Clb2 was essential for mitotic exit and state that 'APC^{Cdc20} may be required for an essential early phase of Clb2 degradation, and this phase may be sufficient for most aspects of mitotic exit. Cdh1 and Sic1 may be required for further inactivation of Clb2–Cdk1, regulating cell size and the length of G1' (Wäsch and Cross, 2002).

Cyclic AMP is also involved in the metaphase–anaphase transition and in the exit from mitosis by inhibiting the onset of anaphase and exit from mitosis in response to growth rate (Spevak *et al.*, 1993; Anghileri *et al.*, 1999). Yeast mutants defective in their APC are inviable on media with high concentrations of glucose but can be rescued by *ras2Δ*, or *cdc25* mutations, or elevated levels of the high *Km* cAMP phosphodiesterase Pde2, all of which reduce cAMP levels in the cell (Irniger *et al.*, 2000). Hence, cAMP levels influence the timing of the G1–S transition, and exit from mitosis, and consequently the cell size at budding and at division.

The yeast cell cycle comprises an array of 'checkpoint controls' whereby cell cycle progression is prevented if certain necessary processes have not taken place for example, mitosis is delayed if DNA replication has not been completed or if damage has occurred to DNA, the actin cytoskeleton or the mitotic spindle. Abnormality in the actin cytoskeleton prevents bud formation with a consequent delay to mitosis through negative regulation of Clb–Cdc28 by Swe1 (McMillan *et al.*, 1999). In response to bud formation, Swe1 is negatively regulated by Hsl1 (Ma *et al.*, 1996; McMillan *et al.*, 1999; Lew, 2000; Longtine *et al.*, 2000). Swe1 is targeted to the bud neck after formation. The interacting proteins Hsl1 and Hsl7 are required for neck localization (Shulewitz *et al.*, 1999; Longtine *et al.*, 2000). Thus, morphogenesis is coupled to proliferation.

The morphogenic aspects of the yeast cell cycle are considerable. These include bud site selection, polarity, pattern and rate of growth of the daughter cell and organellar distribution. The budding pattern of haploids and diploids is different. In rich media haploids bud in an axial pattern whereas normal *MAT***a**/*MAT*α diploids show polar budding (Figure 1.3). (Homozygous *MAT***a**/*MAT***a** and *MAT*α/*MAT*α diploids also bud axially.) The complete story is more complicated than this however, because environmental conditions also affect the budding patterns of both haploids and diploids. Since yeast cells are rarely in constant conditions (in batch cultivation and on Petri dishes the composition of the medium changes as a consequence of yeast metabolism and proliferation), these differences have implications not only for normal vegetative 'yeast form' proliferation but also for the various developmental alternatives such as the formation of invasive filaments by haploids and pseudohyphae by diploids (see Section 1.5).

Two fundamentally different patterns of growth occur during different portions of the yeast cell cycle. The emergence of a bud and its initial growth are accomplished by polarized growth. Later in the cell cycle (approximately early in G2), there is a switch to isotropic growth, which brings about uniform swelling of the bud. Finally, for cytokinesis, secretion refocuses to the neck between mother and daughter cells. The establishment of cell polarity and bud emergence depend upon the Rho-like GTPase Cdc42 (Adams *et al.*, 1990). During the initial apical growth, Cdc42 localizes to the bud tip, where it recruits various polarity

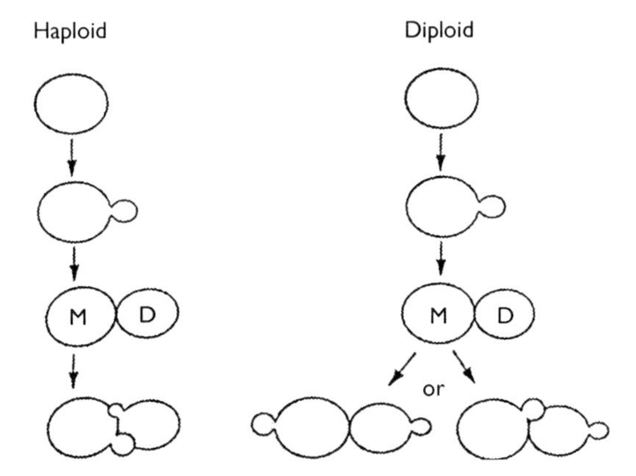

Figure 1.3 The budding patterns in haploids and diploids. 'M' denotes the mother cell and 'D' the daughter. (Redrawn from Dickinson (1999) with the permission of Taylor & Francis.)

factors including actin. During the subsequent isotropic growth, Cdc42 is not required which shows that the establishment of polarity and its maintenance are separate processes. The septins Cdc3, Cdc10, Cdc11 and Cdc12 copolymerize to form a ring, which is localized to the plasma membrane of the mother-bud neck. The septins are required for correct chitin localization into the chitin ring that forms the neck between the mother and daughter cells. Mutants defective in any of *CDC3, CDC10, CDC11* or *CDC12* have delocalized chitin (Longtine *et al.*, 1996; DeMarini *et al.*, 1997) and do not switch to isotropic growth, instead they continue growing apically and form elongated buds. The septins are also very important for actin compartmentalization during isotropic growth and for the correct operation of the morphogenesis checkpoint via Swe1 (mentioned before).

Since yeast has a cell wall (see Chapter 5 for a complete description of the cell wall), growth can only be accomplished at sites of exocytosis (i.e. where new cell wall material and cell wall remodelling enzymes are delivered). Exocytosis comprises vesicle delivery and the docking and fusion of vesicles with the plasma membrane. The actin cytoskeleton (actin cables running between the mother and bud) are envisaged to serve as tracks for the myosin-dependent delivery of vesicles to the bud (Faty *et al.*, 2002). Polarization of the actin cables towards the bud requires a protein complex called the polarisome which comprises Bni1, Spa2, Pea2 and Bud6. Vesicle docking and fusion to the cell membrane is accomplished by the exocyst (which is comprised of Sec3, Sec5, Sec6, Sec8, Sec10, Sec15, Exo70 and Exo84). Cdc42 is believed to coordinate the vesicle docking machinery and the actin cytoskeleton for polarized secretion (Zhang *et al.*, 2001). Hence, during apical growth, the polarisome and the exocyst localize to the tip of the bud. When the switch to isotropic growth occurs, they are relocated to the whole of the surface of the bud and are excluded from the mother cell by the septins, which act as a passive barrier to diffusion (TerBush *et al.*, 1996; Finger and Novick, 1997; Barral *et al.*, 2000).

The genes required for bud site selection comprise three different classes: (1) genes required for axial budding (*BUD3, BUD4, AXL1* and *AXL2*) which do not affect bipolar budding; (2) genes that control bipolar budding of diploids but are not required for axial budding in haploids; (3) genes required for both axial and bipolar budding (*BUD1/RSR1, BUD2* and *BUD5*) (Madden and Snyder, 1998 and references therein). Both axial and bipolar budding are thought to rely upon information from a cytokinesis tag, which is used to direct where the next bud site shall be. In haploid cells the filaments of the neck serve as templates for the assembly of the axial bud site components Bud3, Bud4 and Axl2. This complex directs the formation of a new neck filament ring adjacent to the pre-existing one. The general bud site proteins Bud1, Bud2 and Bud3 (i.e. those which are used in both types of budding) signal this position to the polarity establishment protein Cdc42, its nucleotide exchange factor Cdc24, and Bem1 (Chant and Pringle, 1995) which, in the next cell cycle, will establish polarized growth adjacent to the current bud as already described. The localization of Cdc24 is dependent upon Bud1. The lysine repeat in the hypervariable region and a CAAX motif in Bud1 are both important for its localization to the plasma membrane and sites of polarized growth (Park *et al.*, 2002). In a diploid daughter cell, there is 'positional information' in two locations: adjacent to the site of attachment to its mother and distal (i.e. at the tip where it was previously a bud). The distal site is used for formation of its bud. A diploid new mother cell has two sites: one being where it had been attached to its own mother and the other where its recently born daughter was attached. The new mother cell initiates fresh bud formation adjacent to the site where its daughter had been attached. With each successive generation, older mothers show an increased use of distal sites. It has been deduced that haploids express an extra level of control over bud site selection compared with diploids because a *bud3, bud4* or *bud10* mutation will change the axial budding pattern of haploids to polar, whereas these mutations have no effect in diploids (Madden and Snyder, 1992; Chant and Pringle, 1995). Thus, the default budding pattern is polar but this is overridden in haploids. Bud8 and Bud9 are predicted to be transmembrane proteins which establish cortical tags for bud site selection. Bud8 is thought to be involved in establishing distal tags in daughter cells and Bud9 is said to function by establishing proximal sites during cell divisions of daughter cells (Zahner *et al.*, 1996; Madden and Snyder, 1998).

Even if it were to be kept in the presence of an unlimited supply of nutrients, an individual yeast cell could not continue to produce new daughters indefinitely. Every strain has a characteristic finite lifetime, which is usually measured as the mean or median number of generations produced before it is no longer able to proliferate. Chapter 2 is devoted entirely to yeast ageing, which will not be described here. However, as outlined in Chapter 2, it is currently widely held that the accumulation of oxidative damage is the cause of ageing. Reactive oxygen species, which cause oxidative damage also trigger apoptosis in a variety of organisms. Apoptosis, a form of programmed cell death which involves the flipping of phosphatidylserine to the outer leaflet of the plasma membrane, chromatin condensation, nuclear fragmentation and the formation of apoptotic bodies, had always been considered as a metazoan phenomenon and not a response to be expected of a unicellular organism such as yeast. However, despite the fact that the *S. cerevisiae* genome has no obvious relatives of apoptotic proteins, there is growing evidence of apoptotic machinery within the yeast cell. The first evidence came from Madeo *et al.* (1997) who isolated an unusual *cdc48* mutant which died with a typical apoptotic phenotype. The same group have also reported that a caspase-related protease (Yca1) mediates apoptosis in yeast (Madeo *et al.*, 2002). (Caspases are a family of cysteine proteases that destroy the cell in a well-defined sequence.) Other evidence comes from the observation that expression of heterologous pro-apoptotic genes in yeast causes death of the yeast in an apoptotic fashion (Greenhalf *et al.*, 1996; Kang *et al.*, 1999). None of these findings prove that apoptosis occurs naturally in yeast. Other observations concerning the role of mitochondria are either contradicted by opposite results or open to alternative explanations. Hence, many are still not persuaded that apoptosis exists in *S. cerevisiae*. As this book went to press Qi *et al.* (2003) reported that inactivation of yeast's telomere binding protein Cdc13 resulted in abnormal telomeres and a variety of apoptotic signals including caspase activation, and flipping of phosphatidyl serine. Mitochondrial functions were required because in a ρ^0 petite (lacking mitochondrial DNA) the apoptotic signals were suppressed (Qi *et al.*, 2003). This would seem to be incontrovertible evidence that apoptosis is a genuine physiological response in yeast.

1.4 Pheromone response and mating

The response to pheromone secreted by a cell(s) of opposite mating type begins with the binding of pheromone to the cognate G-coupled receptor on the cell surface. This binding stimulates a variety of responses including global transcriptional changes, arrest of the cell cycle in G1 and initiation of highly polarized morphogenesis resulting in the formation of a 'shmoo' (Figure 1.1). Only the mating type-specific cell surface receptors vary between **a** cells and α cells. In **a** cells the surface receptor for α-factor is Ste2, in α cells the surface receptor for **a**-factor is Ste3. The remainder is accomplished via a MAP kinase signalling system, which is common to both cell types. This MAP kinase signalling system (the pheromone response pathway) is described in full detail in Chapter 8 and Figure 8.9. The sole purpose of this section is to highlight the cell physiology. When two cells of opposite mating type are located in close proximity to each other on a surface, both experience a gradient of pheromone. In the presence of such a gradient, each cell grows toward the pheromone source (the potential mating partner) by chemotropism. The binding of pheromone to the appropriate cell surface receptor triggers dissociation of the heterotrimeric G protein into G_α and $G_{\beta\gamma}$ subunits (Jackson *et al.*, 1991). $G_{\beta\gamma}$ then recruits a complex comprising Far1, Cdc42 and other proteins involved in establishment of polarity to the location in the cell where the receptor was activated (Butty *et al.*, 1998; Nern and Arkowitz, 1998, 1999). Far1 is required for both the growth arrest and chemotropism (Valtz *et al.*, 1995). Hence, growth site selection uses different G proteins for bud formation (Bud1) and for response to mating pheromone ($G_{\alpha\beta\gamma}$).

As explained in Section 1.3 each bud is sited at a predictable location (depending upon the ploidy of the cell and, possibly, whether it is a mother or daughter cell). In contrast, a pheromone signal could potentially require polarized growth in any direction (depending upon the location of the mating partner). Usually, the pheromone signal overrides bud site selection (Madden and Snyder, 1992). If a cell were in a homogeneous liquid culture to which mating pheromone had been added, then that cell would not experience a pheromone gradient but a uniform concentration over the whole of its surface. In this instance, the bud site is used to produce the mating protuberance (Madden and Snyder, 1992; Dorer *et al.*, 1995). The same is true of cells with mating-specific mutations (Dorer *et al.*, 1995; Valtz *et al.*, 1995; Nern and Arkowitz, 1998). Strains carrying *bud* mutations have defects in bud site selection (see Section 1.3), but they mate normally.

Whilst the early stages of mating, especially response to pheromone, have been studied intensively, the later stages have received much less attention. Once the haploid mating partners have made contact, the subsequent stages are cell fusion and then nuclear fusion to yield the diploid. Cell fusion requires the removal of cell wall material in the region where the two haploids have made contact. Like so many other parts of the yeast life- and cell cycle, this also requires correct spatial regulation. Removal of cell wall material is intrinsically dangerous and removal of cell wall components from the wrong part of the cell would be catastrophic in most cases. Cell wall degradation starts from the centre of the contact region and proceeds outwards (Osumi *et al.*, 1974; Gammie *et al.*, 1998). The plasma membranes fuse shortly thereafter followed quickly by fusion of the nuclei and mitochondria.

The cell surface flocculins Fig2 and Aga1 are induced during mating (Roy *et al.*, 1991; Erdman *et al.*, 1998). These proteins have a domain structure very similar to that in the better known Flo1, Flo5, Flo9 and Flo10, but utilize a completely unrelated amino acid sequence (Guo *et al.*, 2000). Fusion of the mating partners' plasma membranes is not simply a consequence of proximity. Prm1 is a pheromone-regulated multispanning membrane protein, which facilitates plasma membrane fusion during mating (Heiman and Walter, 2000). The 'fusase' has to be present in at least one of the maters for efficient mating to occur. Prm1 seems to assist membrane fusion over the final 8 nm between **a** and α partners (Heiman and Walter, 2000). As the mating partners form shmoos, their nuclei and spindle pole bodies become orientated and fixed to the tips of the shmoos by a microtubular 'search and capture mechanism' (Maddox *et al.*, 1999). A bar of microtubules is formed between the shmoo tip bundles, which slowly shortens as the two nuclei and their spindle pole bodies approach and fuse (Maddox *et al.*, 1999). Nuclear fusion (karyogamy) has long been known to rely upon the spindle pole body protein Kar1. The γ-tubulin complex-binding protein Spc72 interacts directly with the N-terminal domain of Kar1. As a consequence of this interaction, the γ-tubulin complex is targeted to the substructure of the spindle pole body known as the 'half bridge' (Pereira *et al.*, 1999). The binding of Spc72 to Kar1 is of minor importance during the normal cell cycle but is essential for karyogamy (Pereira *et al.*, 1999). It appears that the first mitosis of the newly formed diploid lacks the checkpoints that are present in normal vegetative cell cycling (Maddox *et al.*, 1999).

1.5 Pseudohyphae and invasive haploid filaments

As indicated in Figure 1.1, diploids can form pseudohyphae and haploids can form invasive filaments. Although originally considered to be completely different, it is now clear that, as well as their visual similarities, the two forms have much in common. The ability of haploid filaments to invade agar was once thought to be distinctive, but some diploid strains (dependent upon genetic background) can do this too (Gimeno *et al.*, 1992). Filamentation by haploids and diploids has attracted a great deal of attention, not least because this may form part of the pathogenicity of certain other yeast species (e.g. *Candida albicans*) which do not share the genetic tractability of *S. cerevisiae*. Thus, much of the motivation has been the expectation that the control systems discovered in the generally safe *S. cerevisiae*, would then also be found in the pathogens (e.g. see Lorenz and Fink, 2001).

The two most obvious aspects of filamentation are that the cells are longer and that they grow in chains. Both aspects involve changes to many aspects of cell cycle control. In yeast-form proliferation, the daughter is smaller than the mother which gave rise to it. As explained in Section 1.3, cells cannot pass 'Start' until they exceed a certain size, and since yeast-form daughter cells are smaller than the critical size, they undergo a period of growth in G1 before initiating a new round of bud formation. Mother cells are large enough, so they pass through 'Start' before their new-born daughters. In filament formation, the G2 phase of the cell cycle is extended resulting in the mother and daughter being the same size at mitosis. In this case, the mother and daughter start bud formation at the same time. These features were succinctly summarized by the statement that yeast-form proliferation is asymmetric and asynchronous whilst pseudohyphal proliferation is symmetric and synchronous (Kron and Gow, 1995). Growth as filaments requires a change in the budding patterns of both diploids and haploids. In diploids it becomes unipolar, in haploids distal; retention of the yeast-form budding patterns would not yield filaments. Additionally, mothers and daughters no longer separate but remain attached to each other, indicating important alterations to the cell surface.

There are many conditions which will elicit the formation of pseudohyphae or invasive haploid filaments. These include nitrogen starvation of diploid or polyploid strains on fermentable carbon sources, carbon source limitation of haploids on rich media, a poor fermentable carbon source supplied to diploids, and oxygen limitation of polyploids in the presence of both ammonia and glucose (Gimeno *et al.*, 1992; Roberts and Fink, 1994; Kuriyama and Slaughter, 1995; Lambrechts *et al.*, 1996). From such a varied set of conditions it might seem improbable that there could be a common regulatory pathway(s). At least part of the explanation is that the cells are not responding primarily to the specific conditions but to their own metabolic by-products which are indicators of all of those conditions. Evidence for this comes from the observation that 'fusel' alcohols stimulate the formation of filamentous forms in all sorts of conditions, including rich complex media where no nutrients are limiting (Dickinson, 1996; Lorenz *et al.*, 2000a). Fusel alcohols are the products of amino acid catabolism, which accumulate when nutrients become limiting (Dickinson *et al.*, 1997; 1998; 2000; 2003).

Two signalling pathways regulating filamentous growth have been recognized: a MAP kinase pathway (comprising elements of the mating signal transduction pathway) and a cAMP signalling pathway (Liu *et al.*, 1993; Roberts and Fink, 1994; Cook *et al.*, 1997; Kűbler *et al.*, 1997; Lorenz and Heitman, 1997; Madhani *et al.*, 1997; Robertson and Fink, 1998; Pan and Heitman, 1999; Rupp *et al.*, 1999; Lorenz *et al.*, 2000a,b). Everything that is known about these two signal transduction pathways, the way they co-operate and their interactions with other regulatory systems is fully described in Chapter 8 (pp. 326–329) and summarized in Figure 8.9. The key feature is that both signalling pathways converge to control expression of the cell wall flocculin Flo11 (Pan and Heitman, 1999; Rupp *et al.*, 1999), which is a glycerol phosphoinositol-anchored cell surface protein required for the adhesion of mothers to daughters, and for substrate-binding and penetration (Lambrechts *et al.*, 1996; Lo and Dranginis, 1998). The MAP kinase pathway involves Ste20, Ste11, Ste7 and Ste12 (Liu *et al.*, 1993; Roberts and Fink, 1994). The cAMP pathway controls the expression of *FLO11* via Flo8 and Sfl1 which act antagonistically (Pan and Heitman, 2002). The transcription of *FLO11* is also controlled by Snf1 (Kuchin *et al.*, 2002), which is better known as a global regulator of carbon metabolism (see Chapter 3). The repressors Nrg1 and Nrg2 which interact with Snf1, serve as negative regulators of invasive growth and repressors of *FLO11*. Snf1, Nrg1 and Nrg2 also affect biofilm formation (Kuchin *et al.*, 2002).

Current notions on the dual (MAP kinase and cAMP) signalling pathways do not explain the way in which fusel alcohols act. Very recently it has been shown that isoamyl alcohol induces the rapid formation of linear chains of anucleate buds, each of which is accompanied by the formation of a septin ring at its neck. Swe1, and Slt2 (Mpk1) are required. Cdc28 was phosphorylated on tyrosine 19 in a Swe1-dependent manner, while Slt2 became activated by dual tyrosine threonine/phosphorylation. A defective response in a *slt2Δ* mutant was rescued by a *mih1Δ* mutation, suggesting that Slt2 acts as a negative regulator of Mih1 (Martinez-Anaya *et al.*, 2003). Taken together, these observations show that isoamyl alcohol induces the morphogenesis checkpoint. Hence, the resulting pseudohyphae arise in an entirely different way from the routes previously identified.

1.6 Meiosis and sporulation

Normal diploids which are heterozygous at the mating type locus (*MAT***a**/*MATα*) and respiratorily complete can undergo meiosis and spore formation. Haploids, *MAT***a**/*MAT***a** or *MATα*/*MATα* diploids and petites cannot. Sporulation-competent cells must be in the G1 phase of the cell cycle and have reached a minimum size to be able to switch to this alternative developmental pathway. Experimentally, sporulation is usually induced by transferring the cells to potassium acetate. Sporulation can take place on any non-fermentable carbon source, but for most strains acetate is by far the best. Fermentable carbon sources are not permissive and glucose, even at low concentration in the presence of acetate, is strongly repressing. Nitrogen sources must also be absent. As with glucose (yeast's preferred carbon source), ammonia (its preferred nitrogen source) represses sporulation.

When yeast cells sporulate in acetate, some of the acetate is catabolized to carbon dioxide which gives rise to bicarbonate in the medium. Consequently, the pH of the medium increases (Esposito *et al.*, 1969). It has long been known that if the sporulation medium is buffered, sporulation is reduced or completely abolished. Furthermore, it has also been known for nearly 30 years, that there are minimal and optimal cell densities required for efficient, synchronous sporulation (Fast, 1973). The reasons for both of these empirical observations were discovered by Ohkuni *et al.* (1998) who showed that yeast cells sporulating on acetate use the bicarbonate produced to signal to other cells to sporulate. The experimental addition of bicarbonate to the medium enhances sporulation of cells at low densities and even overcomes the asporogeny of malate dehydrogenase-defective mutants (Ohkuni *et al.*, 1998). It is not known at this time whether bicarbonate can restore sporulation in any other respiratory deficient (petite) mutants. However, the addition of bicarbonate does elevate the levels of *IME1* and *IME2* mRNA (Ohkuni *et al.*, 1998). *IME1* and *IME2* are key regulators of sporulation (see below). These results and others from the laboratory of Yamashita (Hayashi *et al.* 1998a,b), along with considerations of apoptosis (Section 1.3), suggest that *S. cerevisiae* is not a free-living unicellular eukaryote but a social, colonial organism.

Many genetic investigations have attempted to identify sporulation-specific genes by specifying that such genes are only expressed in (or required in) *MAT***a**/*MAT*α diploids and not in haploids, *MAT***a**/*MAT***a** or *MAT*α/*MAT*α diploids. Sporulation-specific genes have also been defined and envisaged as required during sporulation but not required for vegetative growth. It has always seemed to this author that these criteria were too strict because they exclude genes which, though not sporulation specific, are nevertheless *essential* for sporulation (Dickinson, 1988). Recent systematic analyses using single-gene deletion mutant banks have tended to confirm this view (Briza *et al.*, 2002; Enyenihi and Saunders, 2003). It is very informative to compare these functional analyses with expression-based analyses such as the transcriptome data of Chu *et al.* (1998). Whilst there is a correlation between requirement and increased expression of individual genes, there is little overlap between those genes shown to be essential for sporulation and those with the greatest fold increases in expression (Enyenihi and Saunders, 2003). Deletion of highly expressed genes in most cases did not yield a sporulation phenotype. This led Enyenihi and Saunders (2003) to comment 'This comparison indicates that expression-based profiling and phenotypic analyses are yielding different sets of sporulation genes.'

A MAP kinase signal transduction cascade for sporulation is supposed to exist but to date, only two components have been positively identified (see Figure 8.7 in Chapter 8.) The MAP kinase Sps1 is required for the expression of late sporulation genes. (Sporulation genes are assigned as being 'early', 'mid', 'mid-late' and 'late'.) The initiation of meiosis is dependent upon the transcriptional activator Ime1 (Smith and Mitchell, 1989; Smith *et al.*, 1993; Mandel *et al.*, 1994). Due to its important role and the fact that it is highly expressed during meiosis (Kupiec *et al.*, 1997), it is often assumed that *IME1* is a sporulation-specific gene, but this is not the case. Ime1 is tethered to the promoters of early sporulation genes specifically by Ume6 (Yoshida *et al.*, 1990). A *ume6*Δ / *ume6*Δ mutant is viable but expresses its sporulation genes during vegetative growth and fails to sufficiently derepress these genes during meiosis and is thus asporogenous (Bowdish *et al.*, 1995). The mutant demonstrates that Ume6 functions both as a repressor and as an activator. Ume6 functions as a repressor by interacting with two separate corepressor complexes (one containing Sin3 and Rpd3; the other containing Isw2). When Ume6 interacts with Ime1 it brings about activation of early sporulation genes by relieving the repression of Sin3. The combination of Ume6/Ime1 also induces Ndt80 which is a transcription factor required for the expression of mid-sporulation genes (see Williams *et al.*, 2002 and references therein.)

One of the early genes induced by Ime1 is *IME2* (a true, early sporulation gene) which encodes a serine/threonine protein kinase. *ime2*/*ime2* homozygous diploids have delayed expression of early sporulation genes and reduced expression of early, mid, mid-late and late genes (Mitchell *et al.*, 1990; Yoshida *et al.*, 1990). Ime2 acts as both a positive and negative regulator of the various stages of sporulation. Ime2 associates with Ime1 and has been shown to be capable of the *in vitro* phosphorylation of the C-terminal tail of Ime1 (Guttmann-Raviv *et al.*, 2002). The *in vivo* stability of Ime1 depends upon the protein kinase activity of Ime2. It appears that this is accomplished by Ime2 phosphorylating Ime1 and thereby rendering it available for rapid degradation via the 26S proteasome (Guttmann-Raviv *et al.*, 2002). It has been proposed that there is a feedback loop involving Ime1, Ime2 and the 26S proteasome for the control of meiosis. *IME1* transcription (which is autoregulated) leads to the transcription of sporulation-specific genes including *IME2*. Ime2 then phosphorylates Ime1 leading to its destruction by the proteasome. Since Ime1 and Ime2 are mutually antagonistic, the functioning of Ime2 is delayed until after Ime1 has promoted the transcription of other early genes. This is achieved by *IME2* being transcribed later than *IME1* and by Ime2 having a very short half-life (Guttmann-Raviv *et al.*, 2002). It is not clear at this time what controls the half-life of Ime2.

The initiation of meiosis and sporulation involves far more regulators than just Ime1 and Ime2. A number of *RIM* (regulator of *IME2*) genes have also been identified (Su and Mitchell, 1993). Prominent among these is *RIM4* which is believed to be unique because it is required for both Ime1 and Ime2 to fully activate gene expression (Soushko and Mitchell, 2000). Rim4 is a RNA binding protein of the RRM (RNA recognition motifs) (Nagai *et al.*, 1995) class that is expressed at high level early in meiosis. Rim4 is required for the expression of early and mid-sporulation genes and may act upstream of Ime2 in one or more signalling pathways (Soushko and Mitchell, 2000).

Metabolism (see Chapter 3 for full details of enzymes and pathways mentioned here) varies throughout the course of sporulation. Upon initiation of sporulation in acetate there is a rapid change in operation of the tricarboxylic acid (TCA) and glyoxylate cycles such that most of the acetate taken up is converted to glutamate (Dickinson *et al.*, 1983). This occurs because the activity of α-ketoglutarate dehydrogenase is reduced (Dickinson *et al.*, 1985). At about 4 hours, net glutamate synthesis ceases and storage carbohydrate (trehalose) synthesis starts using intermediates derived mainly from the glyoxylate cycle. There is an absolute requirement for an intact sequence of gluconeogenic enzymes to accomplish this stage of the process (Dickinson and Williams, 1986). Subsequently fatty acid synthesis begins at the same time as the onset of operation of the hexose monophosphate pathway. The hexose monophosphate pathway is used to generate the NADPH required for fatty acid biosynthesis (Dickinson and Hewlins, 1988). The incorporation of trehalose into the spores gives resistance to heat, dehydration, salt and solvents. Additional resistance to the spore wall is afforded by a highly cross-linked polymer of LL-dityrosine and DL-dityrosine (Briza *et al.*, 1986). This dityrosine-containing polymer is external to a layer of chitosan which surrounds an innermost layer of glucan and mannan (Briza *et al.*, 1986; 1990) organized similarly to the wall of vegetative cells (see Chapter 5).

As cells progress through sporulation metabolic and genetic changes can be as dramatic as the accompanying morphological changes. Yeast can sporulate both in (adequately aerated) liquid media and on the surface of agar but the regulation is not always identical. For example, Purnapatre and Honigberg (2002) showed that *CLN3* is required to repress meiosis and sporulation in colonies but not in liquid cultures. There are, doubtless, many other instances where it has been erroneously assumed that results obtained with cells on solid medium apply equally to liquid media (or *vice versa*).

1.7 Stationary phase

When nutrients become depleted then the orderly cessation of cell cycling and arrest in stationary phase is an important and necessary action. The biochemical changes which accompany stationary phase arrest have been known for many years. Cells originally growing on glucose switch from a fermentative metabolism using mainly glycolysis and forming ethanol, to a respiratory metabolism in which the ethanol formed in the earlier stages of growth is consumed using the TCA and glyoxylate cycles and the mitochondrial electron transport chain (see Chapter 3 for a more complete description). The changes in enzyme activity are well characterized (Polakis and Bartley, 1965; Polakis *et al.*, 1965). As stationary phase is approached the cells accumulate trehalose and glycogen (Lillie and Pringle, 1980). More recently, the accompanying genome-wide changes in transcription have been characterized (DeRisi *et al.*, 1997).

Currently, interest has become focused on the question of how long yeast cells can survive in stationary phase before being returned to fresh nutrients. This period of time over which the non-dividing, stationary phase cells maintain viability has been named the 'chronological lifespan'. It is measured using the circadian clock or calendar and differs fundamentally from 'replicative ageing' (also known as 'budding lifespan' or 'mother cell-specific ageing') which is described in generations and forms the substance of Chapter 2. It is argued that the factors affecting chronological lifespan in yeast might be the same as those which influence ageing in the postmitotic tissues of higher organisms (MacLean *et al.*, 2001). Since it was known that incubating stationary phase cells in water instead of spent growth medium prolongs their survival (Granot and Snyder, 1991, 1993), MacLean *et al.* (2001) compared the survival of stationary phase yeast which had been grown on a variety of different media. They found that the most important factor for survival was that the cells should be grown on carbon sources which are metabolized respiratorily rather than fermentatively. In glycerol containing media, cells grown to early stationary phase survived longer in water than cells grown to early- or mid-exponential phase (MacLean *et al.*, 2001). There is currently no proper explanation for this. It may indicate that a protein(s) present in stationary phase (but not in exponential phase) is required for growth when the cells are returned to fresh media. Alternatively, there may be a molecule (large or small) produced in stationary phase that preserves cell structure during starvation in water in the same way that trehalose enhances the resistance properties of spores.

Chronological lifespan is extended by mutations in *CYR1* (adenylate cyclase) and *SCH9* (protein kinase of the fermentable growth medium (FGM) pathway) (Fabrizio *et al.*, 2001). As described throughout this chapter, cAMP has a role in many aspects of yeast's life cycle. It is also involved (with Sch9) in the regulation of carbon metabolism (see Chapter 3). Hence, there may be many reasons (direct and indirect) for the lifespan extension. However, the stress resistance transcription factors Msn2 and Msn4 and the protein kinase Rim15 were required, which led Fabrizio *et al.* (2001) to suggest that increased protection and repair may slow the ageing process.

When cells are returned to complete glucose medium they quickly resume proliferation. Before budding is detected, *CLN3, BCK2* and *CDC28* mRNAs increase rapidly. The induction of these genes requires the transport and metabolism of glucose but does not require the cAMP, Tor-, *RGT2/SNF3*-, or *HXK2*-mediated signalling pathways (Newcomb *et al.*, 2002; 2003). The fascinating conclusion to this work is that somehow, glycolytic rate is linked to expression of the aforementioned genes that are required for passage through 'Start'.

References

Adams, A. E. M., Johnson, D. I., Longnecker, R. M., Sloat, B. F. and Pringle, J. R. (1990) *CDC42* and *CDC43*, two additional genes involved in budding and the establishment of cell polarity in the yeast *Saccharomyces cerevisiae*. *Journal of Cell Science* **111**, 131–142.

Amon, A., Irniger, S. and Nasmyth, K. (1994) Closing the cell cycle circle in yeast: G2 cyclin proteolysis initiated at mitosis persists until the activation of G1 cyclins in the next cycle. *Cell* **77**, 1037–1050.

Anghileri, P., Branduardi, P., Sternieri, F., Monti, P., Visintin, R., Bevilacqua, A., Alberghina, L., Martegani, E. and Baroni, M. D. (1999) Chromosome separation and exit from mitosis in budding yeast: dependence on growth rate revealed by cAMP-mediated inhibition. *Experimental Cell Research* **250**, 510–523.

Ashe, M. P., De Long, S. K. and Sachs, A. B. (2000) Glucose depletion rapidly inhibits translation in yeast. *Molecular Biology of the Cell* **11**, 833–848.

Baroni, M. D., Monti, P. and Alberghina, L. (1994) Repression of growth regulated G1 cyclin expression by cyclic AMP in budding yeast. *Nature* **371**, 339–342.

Barral, Y., Mermall, V., Mooseker, M. S. and Snyder, M. (2000) Compartmentalization of the cell cortex by septins is required for maintenance of cell polarity in yeast. *Genes and Development* **13**, 176–187.

Bowdish, K. S., Yuan, H. E. and Mitchell, A. P. (1995) Positive control of yeast meiotic genes by the negative regulator UME6. *Molecular and Cellular Biology* **15**, 2955–2961.

Briza, P., Winkler, G., Kalchhauser, H. and Breitenbach, M. (1986) Dityrosine is a prominent component of the yeast ascospore wall. A proof of its structure. *Journal of Biological Chemistry* **261**, 4288–4294.

Briza, P., Ellinger, A., Winkler, G. and Breitenbach, M. (1990) Characterization of a DL-dityrosine-containing macromolecule from yeast ascospore walls. *Journal of Biological Chemistry* **265**, 15118–15123.

Briza, P., Bogengruber, E., Thür, A., Rützler, M., Münsterkötter, M., Dawes, I. W. and Breitenbach, M. (2002) Systematic analysis of sporulation phenotypes in 624 non-lethal homozygous deletion strains of *Saccharomyces cerevisiae*. *Yeast* **19**, 402–422.

Butty, A. C., Pryciak, P. M., Huang. L. S., Herskowitz, I. and Peter, M. (1998) The role of Far1 in linking the heterotrimeric G protein to polarity establishment proteins during yeast mating. *Science* **282**, 1511–1516.

Chant, J. and Pringle, J. R. (1995) Patterns of bud site selection in the yeast *Saccharomyces cerevisiae*. *Journal of Cell Biology* **129**, 751–756.

Chu, S., DeRisi, J., Eisen, M., Mulholland, J., Botstein, D., Brown, P. O. and Herskowitz, I. (1998) The transcriptional program of sporulation in budding yeast. *Science* **282**, 699–705 and *erratum: Science* **282**, 1421.

Cook, J. G., Bardwell, L. and Thorner, J. (1997) Inhibitory and activating functions for MAPK Kss1 in the *S. cerevisiae* filamentous-growth signalling pathway. *Nature* **390**, 85–88.

DeMarini, D. J., Adams, A. E. M., Faras, H., Virgilio, C. D., Valle, G., Chuang, J. S. and Pringle, J. R. (1997) A septin-based hierarchy of proteins required for localized deposition of chitin in the *Saccharomyces cerevisiae* cell wall. *Journal of Cell Biology* **139**, 75–93.

DeRisi, J. L., Iyer, V. R. and Brown, P. O. (1997) Exploring the metabolic and genetic control of gene expression on a genomic scale. *Science* **278**, 680–686.

Dickinson, J. R. (1988) The metabolism of sporulation in yeast. *Microbiological Sciences* **5**, 121–123.

Dickinson, J. R. (1996) 'Fusel' alcohols induce hyphal-like extensions and pseudohyphal formation in yeasts. *Microbiology* **142**, 1391–1397.

Dickinson, J. R. (1999) Life cycle and morphogenesis. In: J. R. Dickinson and M. Schweizer (eds) *The Metabolism and Molecular Physiology of Saccharomyces cerevisiae*, pp. 1–22. London: Taylor & Francis.

Dickinson, J. R. and Williams, A. S. (1986) A genetic and biochemical analysis of the role of gluconeogenesis in sporulation of *Saccharomyces cerevisiae*. *Journal of General Microbiology* **132**, 2605–2610.

Dickinson, J. R. and Hewlins, M. J. E. (1988) A study of the role of the hexose monophosphate pathway with respect to fatty acid biosynthesis in sporulation of *Saccharomyces cerevisiae*. *Journal of General Microbiology* **134**, 333–337.

Dickinson, J. R., Ambler, R. P. and Dawes, I. W. (1985) Abnormal amino acid metabolism in mutants of *Saccharomyces cerevisiae* affected in the initiation of sporulation. *European Journal of Biochemistry* **148**, 405–406.

Dickinson, J. R., Harrison, S. J. and Hewlins, M. J. E. (1998) An investigation of the metabolism of valine to isobutyl alcohol in *Saccharomyces cerevisiae*. *Journal of Biological Chemistry* **273**, 25751–25756.

Dickinson, J. R., Salgado, L. E. J. and Hewlins, M. J. E. (2003) The catabolism of amino acids to long chain and complex alcohols in *Saccharomyces cerevisiae*. *Journal of Biological Chemistry* **278**, 8028–8034.

Dickinson, J. R., Dawes, I. W., Boyd, A. S. F. and Baxter, R. L. (1983) ^{13}C NMR studies of acetate metabolism during sporulation of *Saccharomyces cerevisiae*. *Proceedings of the National Academy of Sciences USA* **80**, 5847–5851.

Dickinson, J. R., Lantermann, M. M., Danner, D. J., Pearson, B. M., Sanz, P., Harrison, S. J. and Hewlins, M. J. E. (1997) A ^{13}C nuclear magnetic resonance investigation of the metabolism of leucine to isoamyl alcohol in *Saccharomyces cerevisiae*. *Journal of Biological Chemsitry* **272**, 26871–26878.

Dickinson, J. R., Harrison, S. J., Dickinson, J. A. and Hewlins, M. J. E. (2000) An investigation of the metabolism of isoleucine to active amyl alcohol in *Saccharomyces cerevisiae*. *Journal of Biological Chemistry* **275**, 10937–10942.

Dorer, R., Pryciak, P. M. and Hartwell, L. H. (1995) *Saccharomyces cerevisiae* cells execute a default pathway to select a mate in the absence of pheromone gradients. *Journal of Cell Biology* **131**, 845–861.

Enyenihi, A. H. and Saunders, W. S. (2003) Large-scale functional genomic analysis of sporulation and meiosis in *Saccharomyces cerevisiae*. *Genetics* **163**, 47–54.

Erdman, S., Lin, L., Malczynski, M. and Snyder, M. (1998) Pheromone-regulated genes required for yeast mating differentiation. *Journal of Cell Biology* **140**, 461–483.

Esposito, M. S., Esposito, R. E., Arnaud, M. and Halvorson, H. O. (1969) Acetate utilization and macromolecular synthesis during sporulation of yeast. *Journal of Bacteriology* **100**, 180–186.

Fabrizio, P., Pozza, F., Pletcher, S. D., Gendron, C. M. and Longo, V. D. (2001) Regulation of longevity and stress resistance by Sch9 in yeast. *Science* **292**, 288–290.

Fast, D. (1973) Sporulation synchrony in yeast. *Journal of Bacteriology* **116**, 925–930.

Faty, M., Fink, M. and Barral, Y. (2002) Septins: a ring to part mother and daughter. *Current Genetics* **41**, 123–131.

Finger, F. P. and Novick, P. (1997) Sec3 is involved in secretion and morphogenesis in *Saccharomyces cerevisiae*. *Molecular Biology of the Cell* **8**, 647–662.

Gammie, A. E., Brizzo, V. and Rose, M. D. (1998) Distinct morphological phenotypes of cell fusion mutants. *Molecular Biology of the Cell* **6**, 1395–1410.

Granot, D. and Snyder, M. (1991) Glucose induces cAMP-independent growth-related changes in stationary phase cells of *Saccharomyces cerevisiae*. *Proceedings of the National Academy of Sciences USA* **88**, 5724–5728.

Granot, D. and Snyder, M. (1993) Carbon source induces growth of stationary phase yeast cells, independent of carbon source metabolism. *Yeast* **9**, 465–479.

Gimeno, C. J., Ljungdahl, P. O., Styles, C. A. and Fink, G. R. (1992) Unipolar cell divisions in the yeast *S. cerevisiae* lead to filamentous growth: regulation by starvation and *RAS*. *Cell* **68**, 1077–1090.

Greenhalf, W., Stephan, C. and Chaudhuri, B. (1996) Role of mitochondria and C-terminal membrane anchor of Bcl-2 in Bax induced growth arrest and mortality in *Saccharomyces cerevisiae*. *FEBS Letters* **380**, 169–175.

Guo, B., Styles, C. A., Feng, Q. and Fink, G. R. (2000) A *Saccharomyces* gene family involved in invasive growth, cell–cell adhesion, and mating. *Proceedings of the National Academy of Sciences USA* **97**, 12158–12163.

Guttmann-Raviv, N., Martin, S. and Kassir, Y. (2002) Ime2, a meiosis-specific kinase in yeast, is required for destabilization of its transcriptional activator, Ime1. *Molecular and Cellular Biology* **22**, 2047–2056.

Hayashi, M., Ohkuni, K. and Yamashita, I. (1998a) An extracellular meiosis-promoting factor in *Saccharomyces cerevisiae*. *Yeast* **14**, 617–622.

Hayashi, M., Ohkuni, K. and Yamashita, I. (1998b) Control of division arrest and entry into meiosis by extracellular alkalisation in *Saccharomyces cerevisiae*. *Yeast* **14**, 905–913.

Heiman, M. G. and Walter, P. (2000) Prm1p, a pheromone-regulated multispanning membrane protein, facilitates plasma membrane fusion during yeast mating. *Journal of Cell Biology* **151**, 719–730.

Irniger, S., Bäumer, M. and Braus, G. H. (2000) Glucose and Ras activity influence the ubiquitin ligases APC/C and SCF in *Saccharomyces cerevisiae*. *Genetics* **154**, 1509–1521.

Iyer, V. R., Horak, C. E., Scafe, C. S., Botstein, D., Snyder, M. and Brown, P. O. (2001) Genomic binding sites of the yeast cell-cycle transcription factors SBF and MBF. *Nature* **409**, 533–538.

Jackson, C. L., Konopka, J. B. and Hartwell, L. H. (1991) *S. cerevisiae* alpha pheromone receptors activate a novel signal transduction pathway for mating partner discrimination. *Cell* **67**, 389–402.

Kang, J. J., Schaber, M. D., Srinivasula, S., Alnmeri, E. S., Litwak, G., Hall, D. J. and Bjornsti, M. A. (1999) Cascades of mammalian caspase activation in the yeast *Saccharomyces cerevisiae*. *Journal of Biological Chemsitry* **274**, 3189–3198.

Kron, S. J. and Gow, N. A. R. (1995) Budding yeast morphogenesis: signalling, cytoskeleton and cell cycle. *Current Opinion in Cell Biology* **5**, 845–855.

Kűbler, E., Mősch, H. U., Rupp, S. and Lisanti, M. P. (1997) Gpa2p a G-protein α-subunit, regulates growth and pseudohyphal development in *Saccharomyces cerevisiae* via a cAMP-dependent mechanism. *Journal of Biological Chemistry* **272**, 20321–20323.

Kuchin, S., Vyas, V. K. and Carlson, M. (2002) Snf1 protein kinase and the repressors Nrg1 and Nrg2 regulate *FLO11*, haploid invasive growth, and diploid pseudohyphal differentiation. *Molecular and Cellular Biology* **22**, 3994–4000.

Kupiec, M., Byers, B., Esposito, R. E. and Mitchell, A. P. (1997) Meiosis and sporulation in *Saccharomyces cerevisiae*. In: J. R. Pringle, J. R. Broach and E. W. Jones (eds) *The Molecular and Cellular Biology of Saccharomyces. Vol. 3*, pp.889–1036. Cold Spring Harbor: Cold Spring Harbor Laboratory.

Kuriyama, H. and Slaughter, J. C. (1995) Control of cell morphology of the yeast *Saccharomyces cerevisiae* by nutrient limitation in continuous culture. *Letters in Applied Microbiology* **20**, 37–40.

Lambrechts, M. G., Bauer, F. F., Marmur, J. and Pretorius, I. S. (1996) Muc1, a mucin-like protein that is regulated by Mss10, is critical for pseudohyphal differentiation in yeast. *Proceedings of the National Academy of Sciences USA* **93**, 8419–8424.

Lew, D. J. (2000) Cell-cycle checkpoints that ensure coordination between nuclear and cytoplasmic events in *Saccharomyces cerevisiae*. *Current Opinion in Genetics and Development* **10**, 47–53.

Lew, D. J. and Reed, S. I. (1993) Morphogenesis in the yeast cell cycle: regulation by Cdc28 and cyclins. *Journal of Cell Biology* **120**, 1305–1320.

Lew, D. J. and Reed, S. I. (1995) Cell cycle control of morphogenesis in budding yeast. *Current Opinion in Genetics and Development* **5**, 17–23.

Lillie, S. H. and Pringle, J. R. (1980) Reserve carbohydrate metabolism in *Saccharomyces cerevisiae*: response to nutrient limitation. *Journal of Bacteriology* **143**, 1384–1394.

Liu, H., Styles, C. A. and Fink, G. R. (1993) Elements of the yeast pheromone response pathway required for filamentous growth of diploids. *Science* **262**, 1741–1744.

Lo, W.-S. and Dranginis, A. M. (1998) The cell surface flocculin Flo11 is required for pseudohyphae formation and invasion by *Saccharomyces cerevisiae*. *Molecular and Cellular Biology* **9**, 161–171.

Longtine, M. S., DeMarini, D. J., Valenick, M. L., Al-Awar, O. S., Fares, H., DeVirgilio, C. and Pringle, J. R. (1996) The septins: roles in cytokinesis and other processes. *Current Opinion in Cell Biology* **8**, 106–119.

Longtine, M. S., Theesfeld, C. L., McMillan, J. N., Weaver, E., Pringle, J. R. and Lew, D. J. (2000) Septin-dependent assemble of a cell cycle-regulatory module in *Saccharomyces cerevisiae*. *Molecular and Cellular Biology* **20**, 4049–4061.

Lorenz, M. C. and Heitman, J. (1997) Yeast pseudohyphal growth is regulated by *GPA2*, a G protein α homolog. *EMBO Journal* **16**, 7008–7018.
Lorenz, M. C. and Fink, G. R. (2001) The glyoxylate cycle is required for fungal virulence. *Nature* **412**, 83–86.
Lorenz, M. C., Cutler, N. S. and Heitman, J. (2000a) Characterization of alcohol-induced filamentation in *Saccharomyces cerevisiae. Molecular Biology of the Cell* **11**, 183–199.
Lorenz, M. C., Pan, X., Harashima, T., Cardenas, M. E., Xue, Y., Hirsch, J. P. and Heitman, J. (2000b) The G protein-coupled receptor GPR1 is a nutrient sensor that regulates pseudohyphal differentiation in *Saccharomyces cerevisiae. Genetics* **154**, 609–622.
Ma, X. J., Lu, Q. and Grunstein, M. (1996) A search for proteins that interact genetically with histone H3 and H4 amino termini uncovers novel regulators of the swe1 kinase in *Saccharomyces cerevisiae. Genes and Development* **10**, 1327–1340.
MacLean. M., Harris, N. and Piper, P. W. (2001) Chronological lifespan of stationary phase yeast cells; a model for investigating the factors that might influence the ageing of postmitotic tissues in higher organisms. *Yeast* **18**, 499–509.
McMillan, J. N., Longtine, M. S., Sia, R. A., Theesfeld, C. L., Bardes, E. S., Pringle, J. R. and Lew, D. J. (1999) The morphogenesis checkpoint in *Saccharomyces cerevisiae*: cell cycle control of Swe1p degradation by Hsl1p and Hsl7p. *Molecular and Cellular Biology* **19**, 6929–6939.
Madden, K. and Snyder, M. (1992) Specification of sites of polarized growth in *Saccharomyces cerevisiae* and the influence of external factors on site selection. *Molecular Biology of the Cell* **3**, 1025–1035.
Madden, K. and Snyder, M. (1998) Cell polarity and morphogenesis in budding yeast. *Annual Reviews of Microbiology* **52**, 687–744.
Maddox, P., Chin, E., Mallavarapu, A., Yeh, E., Salmon, E. D. and Bloom, K. (1999) Microtubule dynamics from mating through the first zygotic division in the budding yeast *Saccharomyces cerevisiae. Journal of Cell Biology* **144**, 977–987.
Madeo, F., Fröhlich, E. and Fröhlich, K.-U. (1997) A yeast mutant showing diagnostic markers of early and late apoptosis. *Journal of Cell Science* **139**, 729–734.
Madeo, F., Herker, E., Maldena, C., Wissing, S., Lächelt, S., Herlan, M., Fehr, M., Lauber, K., Sigrist, S. J., Wesselborg, S. and Fröhlich, K.-U. (2002) A caspase-related protein regulates apoptosis in yeast. *Molecular Cell* **9**, 911–917.
Madhani, H. D., Styles, C. A. and Fink, G. R. (1997) MAP kinases with distinct inhibitory functions impart signalling specificity during yeast differentiation. *Cell* **91**, 673–684.
Mandel, S., Robzyk, K. and Kassir, Y. (1994) IME1 gene encodes a transcriptional factor which is required to induce meiosis in *Saccharomyces cerevisiae. Developmental Genetics* **15**, 139–147.
Martinez-Anaya, C., Dickinson, J. R. and Sudbery, P. E. (2003) In yeast, the pseudohyphal phenotype induced by isoamyl alcohol results from the operation of the morphogenesis checkpoint. *Journal of Cell Science* **116**. 3423–3431.
Mitchell, A. P., Driscoll, S. E. and Smith, H. E. (1990) Positive control of sporulation-specific genes by *IME1* and *IME2* products in *Saccharomyces cerevisiae. Molecular and Cellular Biology* **10**, 2104–2110.
Morgan, D. (1999) Regulation of the APC and the exit from mitosis. *Nature Cell Biology* **1**, E47–E53.
Nagai, K., Oubridge, C., Ito, N., Avis, J. and Evans, P. (1995) The RNP domain: a sequence-specific RNA-binding domain involved in processing and transport of RNA. *Trends in Biochemical Sciences* **20**, 235–240.
Nern, A. and Arkowitz, R. A. (1998) A GTP-exchange factor required for cell orientation. *Nature* **391**, 195–198.
Nern, A. and Arkowitz, R. A. (1999) A Cdc24p, Far1p, $G_{\beta\gamma}$ protein complex required for yeast orientation during mating. *Journal of Cell Biology* **144**, 1187–1202.
Newcomb, L. L., Hall, D. D. and Heideman, W. (2002) AZF1 is a glucose-dependent positive regulator of *CLN3* transcription in *Saccharomyces cerevisiae. Molecular and Cellular Biology* **22**, 1607–1614.
Newcomb, L. L., Diderich, J. A., Slattery, M. G. and Heideman, W. (2003) Glucose regulation of *Saccharomyces cerevisiae* cell cycle genes. *Eukaryotic Cell* **2**, 143–149.

Ohkuni, K., Hayashi, M. and Yamashita, I. (1998) Bicarbonate-mediated social communication stimulates meiosis and sporulation of *Saccharomyces cerevisiae. Yeast,* **14**, 623–631.

Osumi, M., Shimoda, C. and Yanagishima, N. (1974) Mating reaction in *Saccharomyces cerevisiae.* V. Changes in fine structure during the mating reaction. *Archives of Microbiology* **97**, 27–38.

Pan, X. and Heitman, J. (1999) Cyclic AMP-dependent protein kinase regulates pseudohyphal differentiation in *Saccharomyces cerevisiae. Molecular and Cellular Biology* **19**, 4874–4887.

Pan, X. and Heitman, J. (2002) Protein kinase A operates a molecular switch that governs yeast pseudohyphal differentiation. *Molecular and Cellular Biology* **22**, 3981–3993.

Park, H.-O., Kang, P. J. and Rachfal, A. W. (2002) Localization of the Rsr1/Bud1 GTPase involved in selection of a proper growth site in yeast. *Journal of Biological Chemistry* **277**, 26721–26724.

Pereira, G., Grueneberg, U., Knop, M. and Schiebel, E. (1999) Interaction of the yeast gamma-tubulin complex-binding protein Spc72 with Kar1 is essential for microtubule function during karyogamy. *EMBO Journal* **18**, 4180–4195.

Polakis, E. S. and Bartley, W. (1965) Changes in the enzyme activities of *Saccharomyces cerevisiae* during aerobic growth on different carbon sources. *Biochemical Journal* **97**, 284–297.

Polakis, E. S., Bartley, W. and Meek, G. A. (1965) Changes in the activities of respiratory enzymes during the aerobic growth of yeast on different carbon sources. *Biochemical Journal* **97**, 298–302.

Purnapatre, K. and Honigberg, S. M. (2002) Meiotic differentiation during colony maturation in *Saccharomyces cerevisiae. Current Genetics* **42**, 1–8.

Qi, H., Li T.-K, Kuo, D., Kamal, A. N.-E. and Liu, L. F. (2003) Inactivation of Cdc13p triggers MEC1-dependent apoptotic signals in yeast. *Journal of Biological Chemistry* **278**, 15136–15141.

Reynolds, T. B. and Fink, G. R. (2001) Bakers' yeast, a model for fungal biofilm formation. *Science* **291**, 878–881.

Roberts, R. L. and Fink, G. R. (1994) Elements of a single MAP kinase cascade in *Saccharomyces cerevisiae* mediate two developmental programs in the same cell type: mating and invasive growth. *Genes and Development* **8**, 2974–2985.

Robertson, L. S. and Fink, G. R. (1998) The three yeast A kinases have specific signalling functions in pseudohyphal growth. *Proceedings of the National Academy of Sciences USA* **95**, 13783–13787.

Roy, A., Lu, C. F., Marykwas, D. L., Lipke, P. N. and Kurjan, J. (1991) The *AGA1* product is involved in cell surface attachment of the *Saccharomyces cerevisiae* cell adhesion glycoprotein **a**-agglutinin. *Molecular and Cellular Biology* **11**, 4196–4206.

Rupp, S., Summers, E., Lo, H., Madhani, H. and Fink, G. R. (1999) MAP kinase and cAMP filamentation signalling pathways converge on the unusually large promoter of the yeast *FLO11* gene. *EMBO Journal* **18**, 1257–1269.

Shimada, Y., Gulli, M. and Peter, M. (2000) Nuclear sequestration of the exchange factor Cdc24 by Far1 regulates cell polarity during mating. *Nature Cell Biology* **2**, 117–124.

Shulewitz, M. J., Inouye, C. J. and Thorner, J. (1999) Hsl7 localizes to a septin ring and serves as an adapter in a regulatory pathway that relieves tyrosine phosphorylation of Cdc28 protein kinase in *Saccharomyces cerevisiae. Molecular and Cellular Biology* **19**, 7123–7137.

Simon, I., Barnett, J., Hannet, N., Harbison, C. T., Rinaldi, N. J., Volkert, T. L., Wyrick, J. J., Zeitlinger, J., Gifford, D. K., Jaakkola, T. S. and Young, R. A. (2001) Serial regulation of transcriptional regulators in the yeast cell cycle. *Cell* **106**, 697–708.

Soushko, M. and Mitchell, A. P. (2000) An RNA-binding protein homologue that promotes sporulation-specific gene expression in *Saccharomyces cerevisiae. Yeast* **16**, 631–639.

Smith, H. E. and Mitchell, A. P. (1989) A transcriptional cascade governs entry into meiosis in *Saccharomyces cerevisiae. Molecular and Cellular Biology* **9**, 2142–2152.

Smith, H. E., Driscoll, S. E., Sia, R. A., Yuan, H. E. and Mitchell, A. P. (1993) Genetic evidence for transcriptional activation by the yeast *IME1* gene product. *Genetics* **133**, 775–784.

Spevak, W., Keiper, B. D., Stratowa, C. and Castañón, M. J. (1993) *Saccharomyces cerevisiae cdc15* mutants arrested in a late stage in anaphase are rescued by *Xenopus* cDNAs encoding N-ras or a protein with β-transducin repeats. *Molecular and Cellular Biology* **13**, 4953–4966.

Spellman, P. T., Sherlock, G., Zhang, M. Q., Iyer, V. R., Anders, K., Eisen, M. B., Brown, P. O., Botstein, D. and Futcher, B. (1998) Comprehensive identification of cell cycle-regulated genes of the yeast *Saccharomyces cerevisiae* by microarray hybridisation. *Molecular Biology of the Cell* **9**, 3273–3297.

Su, S. S. and Mitchell, A. P. (1993) Identification of functionally related genes that stimulate early meiotic gene expression in yeast. *Genetics* **133**, 67–77.

TerBush, D. R., Maurice, T., Roth, D. and Novick, P. (1996) The exocyst is a multprotein complex required for exocytosis in *Saccharomyces cerevisiae. EMBO Journal* **15**, 6483–6494.

Toikwa, G., Tyers, M., Volpe, T. and Futcher, B. (1994) Inhibition of G1 cyclin activity by the Rascyclic AMP pathway in yeast. *Nature* **371**, 343–345.

Valtz, N., Peter, M. and Herskowitz, I. (1995) *FAR1* is required for oriented polarization of yeast cells in response to mating pheromones. *Journal of Cell Biology* **131**, 863–873.

Wäsch, R. and Cross, F. (2002) APC-dependent proteolysis of the mitotic cyclin Clb2 is essential for mitotic exit. *Nature* **418**, 556–562.

Werner-Washburn, M., Braun, E., Johnston, G. C. and Singer, R. A. (1993) Stationary phase in the yeast *Saccharomyces cerevisiae. Microbiological Reviews* **57**, 383–391.

Williams, R. M., Primig, M., Washburn, B. K., Winzeler, E. A., Bellis, M., Sarrauste de Menthière, C., Davis, R. W. and Esposito, R. E. (2002) The Ume6 regulon coordinates metabolic and meiotic gene expression in yeast. *Proceedings of the National Academy of Sciences USA* **99**, 13431–13436.

Yoshida, M., Kawaguchi, H., Sakata, Y., Kominami, K., Hirano, M. and Shima, H. (1990) Initiation of meiosis and sporulation in *Saccharomyces cerevisiae* requires a novel protein kinase homologue. *Molecular and General Genetics* **221**, 176–186.

Zachariae, W. and Nasmyth, K. (1999) Whose end is destruction: cell division and the anaphase-promoting complex. *Genes and Development* **13**, 2039–2058.

Zahner, J. A., Harkins, H. A. and Pringle, J. R. (1996) Genetic analysis of the bipolar pattern of bud site selection in the yeast *Saccharomyces cerevisiae. Molecular and Cellular Biology* **16**, 1857–1870.

Zhang, X., Bi, E., Novick, P., Du, L., Kozminski, K. G., Lipschtz, J. H. and Guo, W. (2001) Cdc42 interacts with the exocyst and regulates polarized secretion. *Journal of Biological Chemistry* **276**, 46745–46750.

Chapter 2

Mother cell-specific ageing

Michael Breitenbach, Peter Laun, Gino Heeren, Stephi Jarolim and Alena Pichova

2.1 Introduction

This chapter reviews experimental results and hypotheses pertaining to the physiology of mother cell-specific ageing in yeast. There is currently a worldwide renaissance of ageing research. This has to do with the fact that human life expectancy at birth has increased considerably over the last 50 years resulting in an increasing amount of the lifetime of people spent during retirement. This has not only produced considerable socio-economic problems but has also resulted in a growing awareness of how little we know about the basic biology of the ageing process and has intensified clinical, biomedical and biological ageing research. We start by pointing out what in our opinion are the most relevant open questions and proceed then to the more specific topic, that is, the metabolic changes and the influence of manipulation of metabolism on the ageing process in yeast. This treatment of yeast mother cell-specific ageing is, therefore, not comprehensive.

2.2 Model systems for ageing research

We need simple model systems for the ageing process because human genetics has not resulted in sufficient evidence to pinpoint a generally recognized mechanism for the ageing process of the organism and because, as outlined later, the *in vitro* ageing process of human cells also does not tell us the most important and interesting facts about ageing of the organism. Family studies of longevity reveal that the heritability of lifespan is less than 50% (Finch and Ruvkun, 2001). The so-called premature ageing syndromes (progerias) only illuminate 'segmental' aspects of ageing (Martin and Oshima, 2000). Certain tissues or organs of those patients give the impression of an old age phenotype (skin, connective tissue) but others do not. The most prominent of the premature ageing syndromes (Werner's disease) is caused by a recessive mutation in a gene coding for a DNA helicase which is probably involved in DNA repair. However, as there are a relatively large number of helicases in the genome and helicases are needed also for transcription, recombination, ribosome biogenesis and many other processes, we can presently not be absolutely sure about the importance of DNA repair or which aspects of the repair pathways are critical for the ageing process.

Hayflick (1965) discovered that human fibroblasts have a finite lifespan and age clonally *in vitro*. In this context 'clonally' means that all cells derived from a single cell stop dividing and reach the so-called Hayflick limit after a certain number of cell divisions. What was really intriguing about this finding was that fibroblasts derived from older individuals had a progressively shorter lifespan (Hayflick, 1998). Fibroblasts derived from short-lived species displayed a shorter lifespan *in vitro* (Hayflick, 1998). However, it is by no means clear whether the Hayflick limit observed in cell culture is functionally relevant to *in vivo* ageing of the organism. Cells can escape the Hayflick limit by oncogenic point mutations. They will then grow indefinitely, but when transferred back into animals (mice) they give rise to tumours and display markers of tumour cells, such as the accumulation of chromosome mutations (inversions, translocations, aneuploidy).

One of the more prominent ageing theories states that the Hayflick limit is determined by progressive shortening of telomeres in cultured fibroblasts. Indeed, overexpressing telomerase in those cells extends the Hayflick limit (Bodnar *et al.*, 1998; for review see Hayflick, 1998). However, shortened telomeres were not observed in fibroblasts obtained through biopsies of centenarians (Mondello *et al.*, 1999) and in the mouse, a homozygous telomerase deletion mutant did not show premature ageing but, quite unexpectedly, infertility and a number of other defects (Rudolph *et al.*, 1999). This example shows that there are serious difficulties in trying to relate findings in cell culture to the organismic ageing process. It was, however, found in muscle satellite cells of patients suffering from muscle dystrophy, that the telomeres in those cells were shortened (discussed later).

2.3 Determination of replicative (mother cell-specific) lifespan in yeast

In yeast, two different physiological processes could be called ageing. For both processes, parallels with (organismic) ageing processes of higher organisms have been pointed out. First, there is postmitotic ageing, senescence and ultimately death, which is equivalent to the changes occurring in stationary cells (see Section 1.7 of Chapter 1) over relatively long periods of time. Stationary phase cells, depending on the exact conditions under which they arrest, lose viability over a time period of weeks to months. Postmitotic ageing is also observed, for instance, in the brain, where neurons cannot be replaced (or hardly so) and therefore ageing of the brain depends on the survival and function of postmitotic neurons of the brain.

The second ageing process of yeast is the one which will be discussed in detail here; it is the mother cell-specific ageing of growing cells (Mortimer and Johnston, 1959). It occurs in the presence of nutrients and does not depend on starvation. Parallels to the ageing process of higher organisms are less obvious but will be discussed later (muscle ageing). Mother cell specificity means that only the mother cell ages in the asymmetric cell division process of yeast (Figure 2.1). Eventually, the mother cell reaches a state of senescence, defined here as the terminal stage at which no further cell divisions are performed. On the contrary, the daughter cell resets the clock to zero and its lifespan is the same regardless of whether it is the first or the tenth daughter of an individual mother cell. Only the very last daughters of old mothers have a somewhat shortened lifespan (Kennedy *et al.*, 1994). The lifespan characteristic for a strain is determined by micromanipulating and counting all daughters of a set ('cohort') of at least 60 virgin cells that were themselves isolated by micromanipulation (Figure 2.3). Lifespan depends only on the number of generations, not on calendar time (Jazwinski, 1993). It takes about a week for a skilled worker to determine the lifespan curve of a strain. The median lifespan is a convenient characterization of a given strain and is well reproducible. On the other hand, the maximum lifespan of the strain is less well defined, because, of necessity, the statistics are poor at the end of the lifespan curve when only a few cells survive.

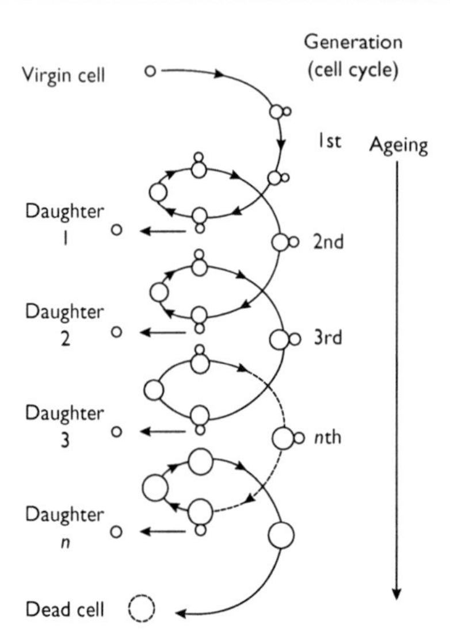

Figure 2.1 Schematic representation of mother cell-specific ageing. In every consecutive cell division cycle the mother cell (on the right) gains one additional bud scar and ages due to accumulation of an unknown 'death factor'. The daughters produced in every cell division cycle reset their clock to zero (with the exception of the daughters of very old mothers).

Source: Jazwinski *et al.*, *Experimental Gerontology* **24**, 423–436 (1989), with permission.

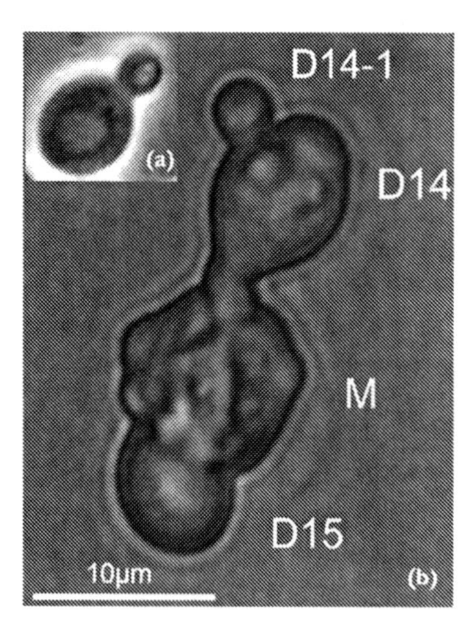

Figure 2.2 A terminal stage of an old yeast mother cell isolated by micromanipulation and photographed in phase contrast (b). For comparison, a young cell is shown at the same magnification (a). M: mother cell, D15: the fifteenth (last) daughter; note that this cell cycle was not completed. D14: the fourteenth daughter; D14-1: the first granddaughter of D14; also this cell cycle was not completed. Note the surface changes of M.

Source: Nestelbacher *et al.*, *Experimental Gerontology* **34**, 895–896 (1999), with permission of the publisher.

2.4 Mother cell specificity – public or private phenomenon?

It is important to explain that the mother cell-specific ageing process of yeast cells which was some time ago viewed as an exotic biological phenomenon, might be very common indeed. Even fission yeast which produces two daughters of equal size, in reality displays an asymmetric division and undergoes mother cell-specific ageing (Barker and Walmsley, 1999). More importantly, the various stem cell populations of the human body do undergo asymmetric divisions resulting in one new stem cell and one cell that made the first step on the way to differentiation. The two daughters resulting from such stem cell divisions may look morphologically similar, but they certainly differ in gene expression. In striated muscle, but not in heart muscle, a limited supply of satellite cells which are equivalent to muscle stem cells are used for muscle growth and repair during our lifetime. This phenomenon is dramatically visible in patients suffering from muscular dystrophies where the muscles undergo continuous cycles of degeneration and regeneration but can also be observed in high level sportspersons suffering from Fatigued Athlete's Muscle Syndrome (FAMS) who have used up their supply through excessive training that damages many muscle fibres and consequently, the continual growth and repair uses up the pool of stem cells. The lesson it teaches us is that even in the human body mother cell-specific ageing might occur. Otherwise, the supply of stem cells would not be exhaustible. Shortening of telomeres was observed in the satellite cells of patients suffering from muscle dystrophies, but not in muscle biopsies derived from old patients. (Decary *et al.*, 2000; Renault *et al.*, 2002; Butler-Browne, personal communication).

2.5 Statistics of lifespan distributions

In this section we briefly discuss the statistical measures and methods used to describe the lifespan distribution of a cohort of yeast cells. The curve shown in Figure 2.3 depicts the lifespan distribution of a cohort of yeast cells. This curve theoretically and practically follows the Gompertz law, the same law that very well describes the lifespan distribution of humans! The important characteristic of the distribution is that the probability to survive the next fraction of time exponentially decreases with age. The curve is not represented by a normal distribution and therefore the arithmetical mean is not the best measure to describe the lifespan of a strain (a cohort of individuals that are genetically identical). Rather, the median age (in generations, at which 50% of the cells survive) is used to best describe the lifespan of the strain. The standard deviation of that median (at the 95% level) can be interpreted as a measure of the deviation of the curve from the Gompertz law. This standard deviation is given as an error bar in the graphs. We determined by systematic experimentation and data analysis how many cells we need in a cohort to get a reasonably small error of the median lifespan. This is shown in Figure 2.3 and the number of cells generally used in our laboratory for lifespan determination is 60.

In practice, it is important to determine whether two lifespan curves are significantly different. This is done by applying the non-parametric Wilcoxon test (Wilcoxon, 1945), compare the lifespans shown in Figure 2.5. Ideally, the Gompertz law would yield a symmetric distribution, although it is not a normal distribution. However, in practice, the distribution is biased or tailed towards the right, in other words, there is a fraction of cells that lives longer than expected. This is true both for yeast mother cells and for human individuals. However, the deviation is small, and generally the use of the Gompertz law is justified.

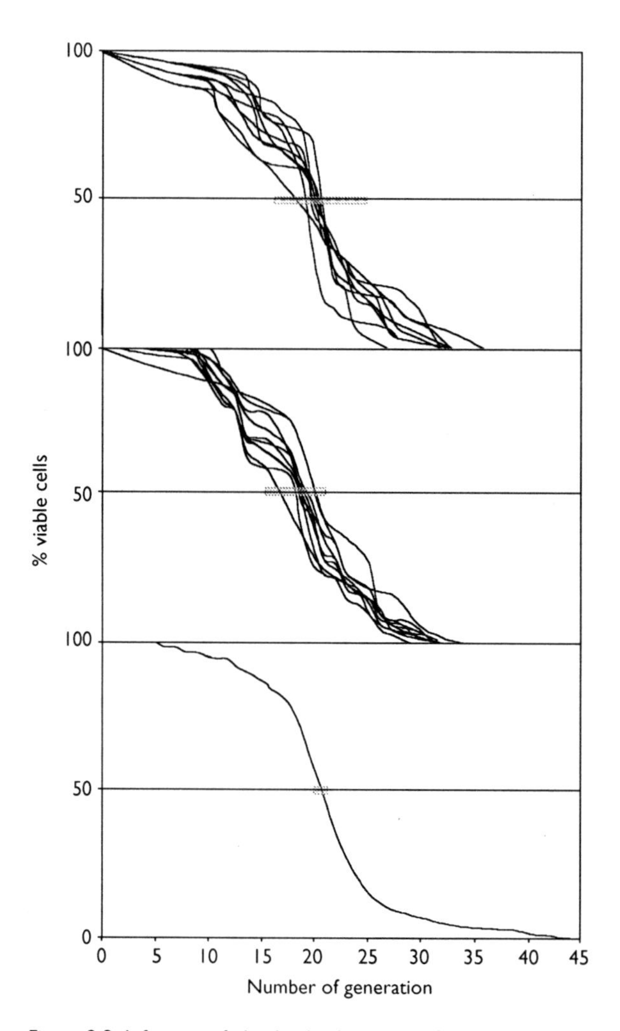

Figure 2.3 Lifespan of the haploid strain W303 *MAT***a** (bottom panel) obtained from 280 individual cells on SC (synthetic complete) media. The standard deviation of the median is shown as a horizontal bar. In the middle panel, 10 different randomly chosen data sets of 60 individual cells taken from the 280 are shown. The upper panel shows 10 different randomly chosen data sets of 30 individual cells taken from the 280. The horizontal bars in the middle and upper panel show the maximum standard deviation of the 10 curves in each panel. Note the large variance in the upper panel.

Source: Unpublished data from our laboratory.

2.6 Technical considerations: preparation of old cells

Biochemical investigation of the senescent phenotype requires an enrichment procedure that leads to sufficiently pure terminally senescent yeast mother cells. Purification of these cells which occur in a batch grown yeast culture as a very minute fraction of all cells, has proven to be a major difficulty in yeast ageing research. All old cell preparations are contaminated with young cells, because in the cell cycle of budding yeast, the daughter adheres to the mother and is not separated immediately. Such mother/daughter aggregates behave like a large single cell during physical separation of large and small cells. The separation of daughters from their mothers at least in some favourable strains is more nearly complete in stationary phase. However, reaching the stationary phase should be avoided in order not to mix up the effects of mother cell-specific ageing, which is independent of nutrient availability, with the physiological effects of starvation. Egilmez *et al.* (1990) separated young and old yeast cells in sucrose density gradients. Their fractions appear very pure and have led to the first phenotypic characterization of old cells based on a bulk cell preparation. Next, the magnetic bead technology was employed to immobilize yeast cells. The daughters are budded off and washed away, fresh growth medium is supplied either continuously or after certain times and, eventually, a population of yeast cells still immobilized on the beads is obtained that have a rather equal age structure (Smeal *et al.*, 1996; Laun, 1999). As we have shown (Laun, 1999), the method is limited by the fact that the extensive cell surface alterations of very old (senescent) cells leads to the loss of these cells from the beads. Therefore, the method is excellent for enriching small amounts of cells of the 10th or even 20th generation in a strain with a median lifespan of 25 generations. It was not possible however, to isolate significant numbers of senescent (terminally aged) cells on magnetic beads.

We now turn to the enrichment of senescent cells by elutriation centrifugation. This method was described in detail earlier (Laun *et al.*, 2001) and is based in part on previous work by Woldringh *et al.* (1995). The strain to be investigated is batch grown for 20 h (about 10 generations) without reaching stationary phase and the cells are separated into six fractions according to size by elutriation centrifugation. Fractions III and IV (large cells that are not damaged or broken) are re-inoculated into fresh medium and again grown for 20 h. The elutriation procedure is repeated and fraction V (largest undamaged cells) and fraction II (virgin daughter cells) are washed and used for cytological and biochemical analysis. Fraction II cells are a convenient control because they have been shown to be mostly virgin cells and first generation cells and have undergone the same purification procedure as the old cells, thus excluding artefacts of the purification procedure. Fraction V cells are to about 30% final stage senescent cells. About one-third of these cells stop dividing within the first cell cycle when tested in a standard lifespan test and show the markers of yeast apoptosis to be discussed next. As the lifespan curve shows, the preparation contains a considerable amount of young cells (represented by the tail of the curve) which originated from daughters adhering to their mothers.

2.7 The senescent phenotype

We are using the term ‘ageing’ for a gradual process over many cell generations, while the term ‘senescence’ is used for the terminal state of aged cells in their last one or two cell cycles when a sharp decline in vitality is observed. Terminally senescent mother cells of yeast (Figure 2.2(b) and Nestelbacher *et al.*, 1999) display dramatic changes in appearance and physiology, compared to young cells. Some of these changes are strikingly reminiscent of those seen in senescent endothelial cells (or the more commonly used fibroblasts) in culture. Both the clonally aged human umbilical vein endothelial cells (‘HUVEC’) and the senescent yeast mother cells are much larger than young cells. Also, cytoskeletal abnormalities, especially in the actin cytoskeleton, are observed in both types of senescent cells. Senescent yeast cells create internal oxidative stress (Laun *et al.*, 2001, see Section 2.8) and consequently display a patchy actin cytoskeleton instead of the usual dots and cables picture (Figure 2.4(a,b)). Senescent HUVEC are not spindle shaped but rather flat and spread out (Pfister, personal communication). A fraction of them display markers of apoptosis including markers of internal oxidative stress (Wagner *et al.*, 2001). Those cells also display patchy actin cytoskeleton and start to detach from the glass surface (Figure 2.4(c,d)). Nuclear chromatin stained with DAPI appears more diffuse than in young cells (Pichova *et al.*, 1997; Laun *et al.*, 2001). Frequently we see cells without nucleus or with more than one nucleus due to nuclear fragmentation or, possibly, endomitosis. Perhaps as a consequence of the irregular cytoskeleton, the cell surface (plasma membrane) of human umbilical vein endothelial cells looks 'folded' or shrunken and a similar picture is seen in yeast cells (transmission electron

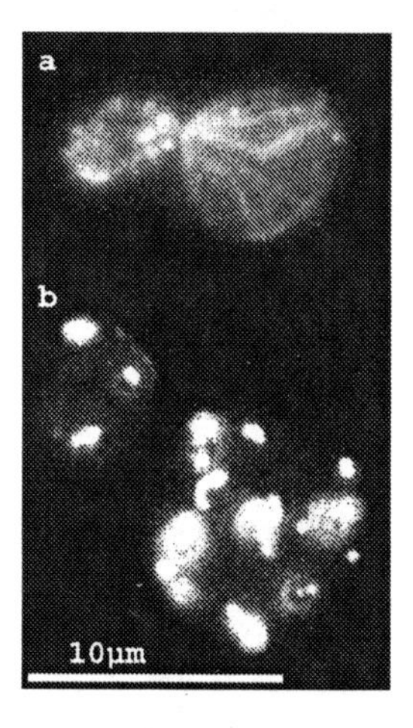

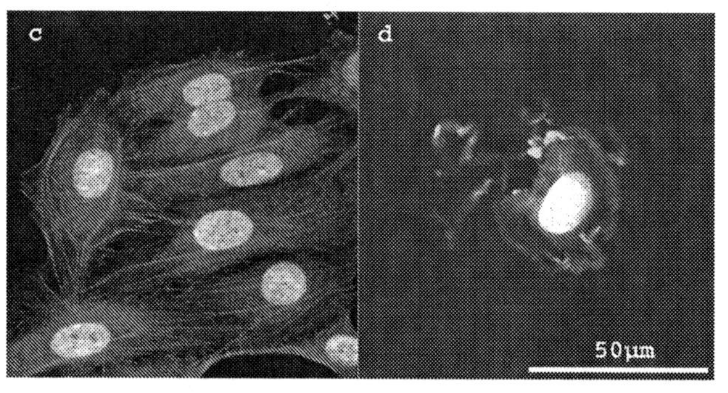

Figure 2.4 (a,b) Yeast cells strain JC482 (unpublished data of A. Pichova); (c,d) HUVEC the fluorescent stain is specific for actin (unpublished data, courtesy: G. Pfister and G. Wick). (a) The typical dots and cables appearance of a young yeast cell with a small daughter cell. (b) An old mother cell with her daughter in the upper right. The actin cytoskeleton is collapsed to a number of cytoplasmic patches. The two yeast cells are shown at the same magnification. (c) Spindle shaped adherent young HUVEC. The actin cytoskeleton shows typical fibres. (d) An old HUVEC in the process of detaching from the substrate. The larger part with the well-stained nucleus is rounded and detached and the smaller part of the cell (upper right) is still adherent. The cytoplasmic actin cytoskeleton looks patchy. (c) and (d) are shown at the same magnification.

microscopy, unpublished data), and even the cell wall of senescent yeast cells looks folded (Figure 2.2(b)). When lifespans are determined by micromanipulation, it is consistently observed that cells that approach senescence need much longer to perform a cell division cycle. The same is true for cultured human cells.

In order to also observe biochemical and physiological changes in senescent cells, a method for enriching these old cells is needed. This is no problem with cultured human cells (all of them senesce when the Hayflick limit is reached), but is a major problem with yeast cells (see Section 2.6). Using gradient purified old yeast mother cells, a slowing down of protein synthesis was observed in these cells (Egilmez *et al.*, 1990). The biochemical markers of oxidative stress-induced damage and of apoptosis is discussed in the next section.

2.8 Ageing and oxidative stress, respiration and the retrograde response

For a general discussion of oxidative stress in yeast, please consult Chapter 9. The involvement of the toxic side products of respiratory oxygen metabolism in ageing and in a number of human diseases was considered by Harman (Harman, 1956); for review see (Harman, 1998) when he formulated the 'free radical theory of ageing' that also became known as the 'oxygen theory of ageing' (Ames *et al.*, 1993; Halliwell and Gutteridge, 1999). Essentially, the theory states that one electron reduction of oxygen occurs relatively frequently in the respiratory chain (at the bc1 segment) as a side product. This process yields superoxide radical which is not itself extremely reactive, but gives rise to follow-up products, so-called ROS (reactive oxygen species) that are toxic and mutagenic. One of the most important ROS is the OH radical that avidly reacts with DNA (producing 8-oxo-guanine, which is mutagenic), proteins (producing protein carbonyls), lipids and other bio-molecules. All these reactions have been implicated in ageing (Halliwell and Gutteridge, 1999). Another important follow-up product of ROS, 4-hydroxy-2-nonenal, a product of peroxidation of multiply unsaturated fatty acids, was studied intensively in higher organisms, but apparently does not occur naturally in yeast cells (although it is toxic for yeast cells.). The eukaryotic cell has evolved a number of parallel and partly overlapping detoxification pathways for ROS. The superoxide dismutase (SOD)/catalase system that removes superoxide and hydrogen peroxide and the glutathione cycle (Chapter 9) using reducing equivalents originating from NADPH to detoxify hydrogen peroxide and organic peroxides.

The oxygen theory of ageing is the only one of a large number of ageing theories (not all of them mentioned here) for which experimental evidence has been supplied in all major model systems of ageing (Halliwell and Gutteridge, 1999). The most important facts are outlined here without going into detail, followed by a detailed discussion of yeast. ROS production in mitochondria increases with age and so do biochemical markers of oxidative damage in rodents, in flies and in human cells. This is true at the organismic level and also at the cellular level in cultured cells reaching the Hayflick limit (Halliwell and Gutteridge, 1999; Sitte *et al.*, 2000). Especially in rodents, antioxidants in the diet improve survival and cognitive performance (Floyd, 1991; Harman, 1998). Overexpression of catalase and superoxide dismutase increases lifespan, and mutants selected for longevity without selecting for oxidative stress resistance, nevertheless show oxidative stress resistance in Drosophila (Orr and Sohal, 1994; Sohal and Weindruch, 1996; Arking *et al.*, 2000). Perhaps the most striking fact in this connection is that it was found relatively recently that the most important long-lived mutants in *Caenorhabditis elegans* (affected in the insulin-like growth factor signalling pathway) depend completely on functional defence against oxidative stress (Gems, 1999; Taub *et al.*, 1999). Moreover, expression of catalase in neurons was sufficient to restore longevity, pointing to the fact that in higher organisms, the ageing *zeitgeber* might be located in certain differentiated tissues, most likely in neurons, dictating ageing of the organism (Wolkow *et al.*, 2000).

In yeast, conditions that increase the load of the cells with ROS invariably decrease the reproductive lifespan. This was shown by disrupting the two yeast catalase genes (Nestelbacher *et al.*, 2000; Figure 2.5), by disrupting Cu/Zn-SOD (Barker *et al.*, 1999) and by disrupting both yeast *SOD* genes, the mitochondrial superoxide dismutase Mn-SOD and the cytoplasmic Cu/Zn-SOD (Wawryn *et al.*, 1999). Both *SOD*-disruption mutations reduced the lifespan and the effect was additive (Wawryn *et al.*, 1999). Increasing the partial pressure of oxygen decreased the lifespan (Nestelbacher *et al.*, 2000). Conversely, the biological antioxidant, reduced glutathione (GSH, 1 mM) in the growth medium restored the lifespan of catalase mutant strains to the normal lifespan observed under standard conditions (Nestelbacher *et al.*, 2000; Figure 2.5). Glutathione is the major 'redox buffer' of the cell (Schafer and Buettner, 2001) and is known to be taken up efficiently by yeast cells.

Using a method developed in our laboratory for the enrichment of senescent (terminal) yeast mother cells (see Section 2.6), we investigated cytological markers of oxidative stress and of apoptosis in senescent cells and, for comparison, in virgin daughter cells that had undergone the same elutriation procedure (Laun *et al.*, 2001). Care was taken not to apply any external oxidative or starvation stress to these cells. The remaining median lifespan of this fraction was only 3 generations as compared to the normal median lifespan of 22 generations in the virgin cells. About one-third of the cells in the large cell fraction (fraction V) did enter a new cell cycle in the lifespan experiment, but did not complete it. Also, one-third of these cells showed inversion of the plasma membrane (Annexin V test),

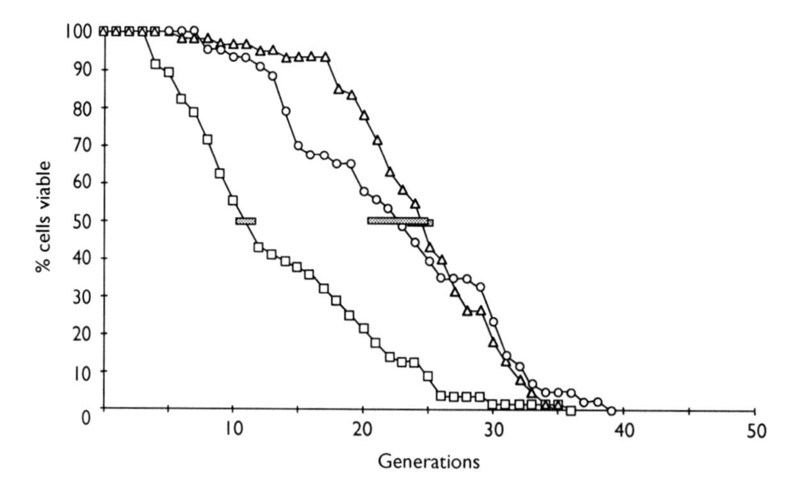

Figure 2.5 Yeast strain W303 *MAT***a**, *ctt1::URA3, cta1::URA3*. Circles: ambient air, squares: 55% oxygen, triangles: 55% oxygen, 1 mM GSH. The complete lack of catalase causes a large decrease in lifespan (in 55% oxygen) which is reversed to a normal lifespan by adding 1 mM in the medium. The two curves for ambient air and for 55% oxygen plus 1mM GSH are not significantly different by applying the Wilcoxon non-parametric test.

Source: Nestelbacher *et al.*, *Experimental Gerontology* **35**, 63–70 (2000), with permission of the publisher.

diffuse nuclear chromatin, accumulation of DNA strand breaks (TdT-mediated dUTP nick end labelling 'TUNEL' test; Figure 2.6), and intense staining with dihydrorhodamine 123 (DHR) indicating a sufficiently positive redox potential in the cell to oxidize DHR. We conclude that these old wild-type cells developed internal oxidative stress in the absence of any external stressors, simply as the consequence of ageing. The morphology of the subcellular structures staining positively with DHR was clearly mitochondrial and no such staining was seen in young cells. As a control the same cells were also stained with 2-(4-(dimethylamino)styryl)-1-methylpyridinium iodide (DASPMI), a mitochondrial-specific fluorescent stain that does not depend on the redox potential (Figure 2.7). Although these experiments are still not conclusive to prove a causal role for mitochondrial ROS in ageing, we hypothesize that the build-up of the markers of oxidative stress and apoptosis occurs only in senescence and during the last unfinished cell cycle and is mediated through an active genetically controlled mechanism in response to still unknown signals. These signals could be described as the accumulation of a 'death factor' in the mother cell, but not in the daughter cell. As to the nature of the death factor, only speculation is currently possible. It could be one or more radical species (superoxide) or simply the damaged cellular material. Whatever it is, it must have the capacity to create a signal in the cell and to be inherited very asymmetrically between mother and daughter. One example was given recently, showing that such

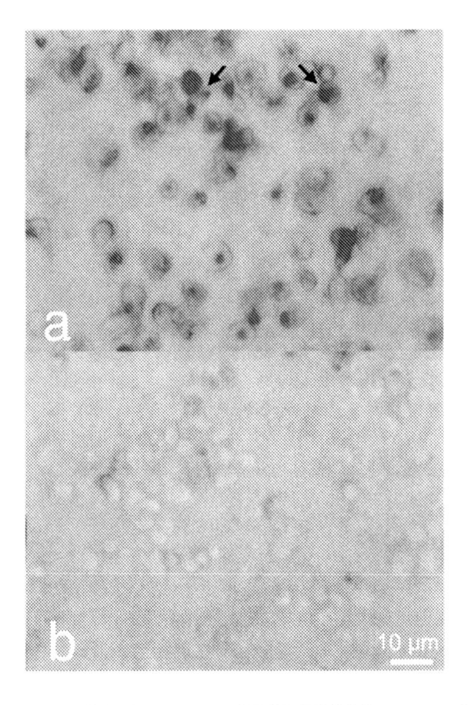

Figure 2.6 Yeast strain: JC482. TUNEL staining for DNA strand breaks. (a) Fraction V old mother cells that were to about 30% in their final cell cycle and had entered apoptosis. Arrows mark mother daughter pairs that were both TUNEL-positive. (b) Fraction II virgin cells of the same strain. TUNEL-staining is practically absent.

Source: P. Laun *et al.*, *Molecular Microbiology* **39**, 1166–1173 (2001), with permission of the publisher.

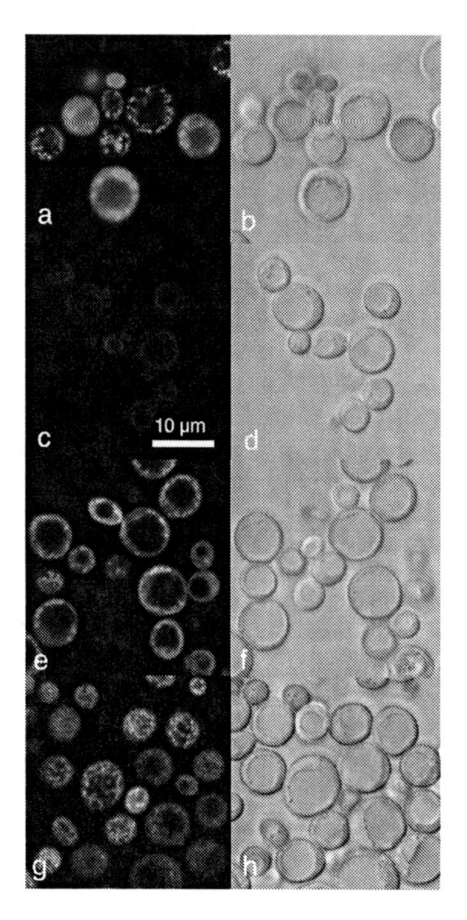

Figure 2.7 Yeast strain: JC482. (a,c) fluorescent stain with DHR; (e,g) fluorescent stain with DASPMI; (b,d,f,h) phase contrast. (a,b) fraction V old yeast mother cells. Positive DHR stain indicates a positive internal redox potential. The stain is mitochondrial as shown by comparison with (e) and (g). (c,d) fraction II virgin cells. DHR-positive cells are absent. (e,f) fraction V old mother cells. (e) stained with DASPMI to show mitochondria. (g,h) young virgin cells. (g) stained with DASPMI to show mitochondria.

Source: Laun *et al.*, *Molecular Microbiology* **39**, 1166–1173 (2001), with permission of the publisher.

asymmetric inheritance of damaged material is indeed possible. Immune fluorescence staining of mother/daughter cell pairs for protein carbonyls showed that the oxidized (carbonylated) proteins stay in the mother and are not transmitted to the daughter cell (Hugo Aguilaniu, personal communication). A cautionary note should be struck at this point: it is possible that the DHR in our experiments is oxidized in the cytoplasm of the cells and then moves into the mitochondria. Even staining with rhodamine which does not need oxidation in the cell, stains mitochondria. Experiments addressing this important question are under way. Some recent experiments with yeast *RAS2* mutants would point to the fact that a non-mitochondrial mechanism is feasible.

Experiments performed with respiratory-deficient rho-zero mutants and with anaerobiosis can also clarify the role of respiration in ageing. Naively, one would assume that eliminating the respiratory chain completely, as is the case in rho-zero mutants which depend on ATP produced by glycolysis, would also eliminate oxygen toxicity. However, this is not true. There are also cytoplasmic oxygen-dependent reactions that could give rise to oxygen toxicity (Rosenfeld *et al.*, 2002). Rho-zero cells are DHR positive under conditions where rho-plus cells are not, and the GSH equilibrium of rho-zero cells is shifted towards a positive intracellular redox potential. Rho-zero strains are hypersensitive to chemicals inducing oxidative stress (our own unpublished observations, collaboration with the group of Ian Dawes). A likely explanation of these results is that oxygen toxicity arising from non-mitochondrial reactions plays a role in determining oxidative stress defence and lifespan in these strains.

The lifespan of rho-zero cells has been measured and compared to isogenic wild-type cells by a number of laboratories (Berger and Yaffe, 1998; Kirchman *et al.*, 1999; Powell *et al.*, 2000 and our own unpublished observations). The lifespan of rho-zero cells was found to be shorter than that of the corresponding wild-type cells. In a typical strain, W303, the median lifespan is 22 generations for wild type and 17 generations for the corresponding rho-zero strain (our own unpublished observations). Results with other strains are similar. However, in one case (Kirchman *et al.*, 1999) a strain was described in which the lifespan is longer for the isogenic rho-zero derivative. This result is discussed in terms of the retrograde response (Small *et al.*, 1995; Jazwinski, 2002). The retrograde response in non-respiring yeast cells leads to a global metabolic shift, characterized, for instance, by a shift from the mitochondrial tricarboxylic acid cycle to the peroxisomal glyoxylate cycle which is conveniently monitored by measuring expression of Cit2p, the citrate synthase isoenzyme characteristic for the peroxisome (see Section 3.7 of Chapter 3). The authors maintain that in those strains where the retrograde response is sufficiently strong, the rho-zero strain has a longer lifespan compared to its rho-plus counterpart.

2.9 Ageing and the *ras* genes

In yeast, the two homologues of the ras proto-oncogenes of higher organisms are *RAS1* and *RAS2*. Deletion of both genes is lethal, while deletion of *RAS1* has little effect and deletion of *RAS2* leads to inability to grow efficiently on non-fermentable carbon sources (Tatchell, 1986; Thevelein and de Winde, 1999). The two *RAS* genes were found to be differentially expressed during replicative ageing of yeast cells. Deletion and overexpression experiments showed antagonistic effects of the two genes on the replicative lifespan (Sun *et al.*, 1994). Overexpressing *RAS2* increased the lifespan, while a deletion of *RAS2* slightly decreased the lifespan. The major and well-documented signal transduction pathway involving the yeast *RAS* genes is the well-known RAS/cAMP/PKA pathway (fully described in Section 8.3 of Chapter 8) which controls entry into a new cell division cycle in response to nutrient availability and also the cell's response to various stress conditions (Thevelein and de Winde, 1999). For instance, expression of the yeast catalase T gene, *CTT1*, is under strict control by this pathway, in response to heat stress or osmotic stress (Bissinger *et al.*, 1989). Stress response is probably the link between the *RAS2* gene and ageing (Jazwinski, 2002). Further experiments (using a version of *RAS2* defective in its effector domain) led Jazwinski to postulate that the effects of *RAS2* to increase lifespan and to protect yeast cells from the effect of chronic heat shock on lifespan were independent of cAMP (Sun *et al.*, 1994), thus possibly defining a new regulatory pathway in which the Ras proteins might be involved.

In yeast, the dominant 'oncogenic' point mutation *RAS2-ala18,val19* is homologous to the human *Ha-ras-val12* mutation. Despite the fact that this mutation is frequently found in tumours, introducing it in isolation into human cultured cells leads to premature ageing and apoptosis (Serrano *et al.*, 1997). This effect depends on oxygen; lowering the atmospheric oxygen completely prevents apoptosis and ageing under these conditions (Lee *et al.*, 1999). This points to the fact that the effect of the dominant *ras* mutation might be through cell–internal creation of oxidative stress. We found (Pichova *et al.*, 1997) that yeast *RAS2-ala18,val19* cells have a dramatically shortened lifespan which is partly restored by 1 mM GSH (unpublished). The elutriated fraction V cells of this strain display the same markers of apoptosis and oxidative stress as wild-type cells, however, senescence is reached much earlier than in wild type (Pichova, unpublished). Young cells of *RAS2-ala18,val19*, of wild type, of the *ras2::LEU2* deletion and of the corresponding rho-zero strains were investigated (Heeren, unpublished). rho^+ and rho-zero cells of only the *RAS2-ala18,val19* strain showed a clear signal for superoxide radical anion in EPR (electron paramagnetic resonance) experiments. Both the rho+ and the rho-zero strain had the same extremely short lifespan and both showed a strong signal with DHR in their mitochondria indicating a positive redox potential. Thus we hypothesize that extramitochondrial sources of oxidants are probably responsible for shifting the redox potential (unpublished data from our laboratory, collaboration with the group of Hans Nohl; Figure. 2.8).

2.10 Caloric restriction and nutritional control of ageing

The elongation of lifespan by eating less (without serious undernourishment, which, of course, shortens lifespan) has been studied intensively in rodents (Finch and Ruvkun, 2001). It is debatable whether the manipulation of yeast growth media is comparable with caloric restriction in mammals. It should be noted that yeast lifespans should be determined on synthetic media, because complex media contain known and unknown substances that influence lifespan (for instance, GSH) in varying concentrations. The effects of low glucose, of different carbon sources, and of the reduction of amino acids in synthetic media was found to increase the lifespan. Also in this case, the phenomenon is strain dependent.

Glucose repression of respiration and other metabolic pathways is a well-studied phenomenon in yeast that is highly strain dependent. Several laboratories have found moderate effects of reducing the glucose concentration from the usual 2% (repressing in many strains) to the non-repressing 0.1%, resulted in an increase of lifespan (Jiang *et al.*, 2000; Lin *et al.*, 2000; our own unpublished results). The role that glucose repression might play in limiting lifespan was not studied in detail. In the same vein, the lifespan determined for a particular strain on 2% glucose was compared to the lifespan determined on non-fermentable carbon sources. These carbon sources (ethanol, glycerol) led to an increase in respiration. Their influence on lifespan was controversial and strongly depended on the strain used (Muller *et al.*, 1980; Barker and Walmsley, 1999; Van Zandycke *et al.*, 2002; our own unpublished results).

Finally, a very strong influence on lifespan was found (Jiang *et al.*, 2000) when in a strain prototrophic for amino acids, these amino acids which are normally contained in synthetic complete media, were left out. The elongation of lifespan measured under these conditions is big (two-fold). Here it could be argued that relieving the cells from nitrogen repression could be the mechanism, but further experiments addressed to this question have not been performed.

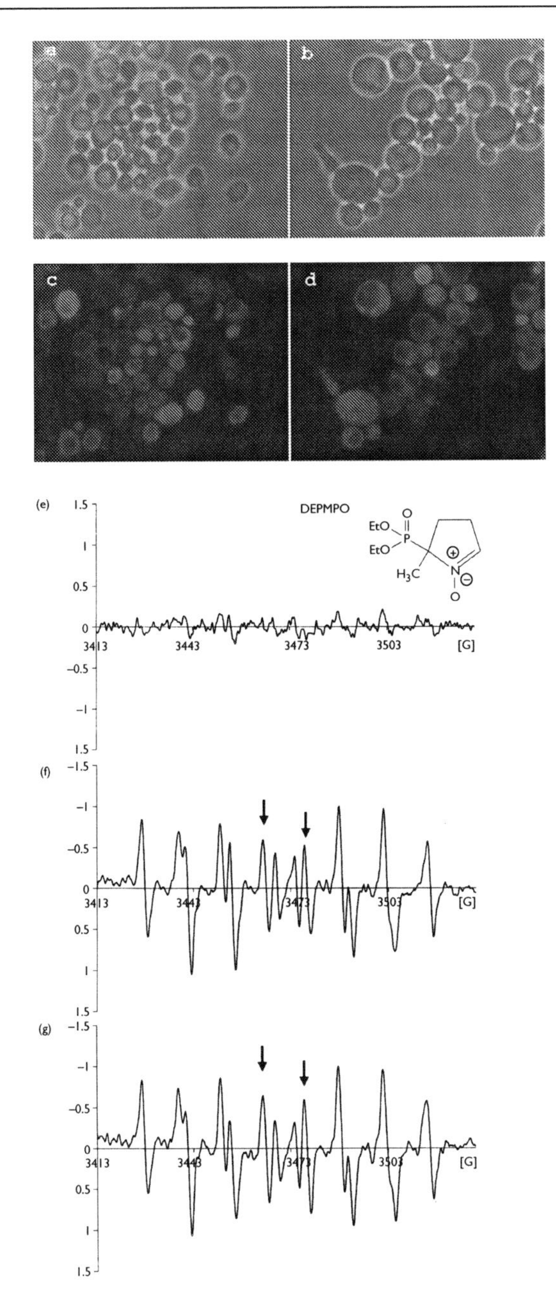

Figure 2.8 Yeast strain JC488 *RAS2-ala18 val19* (Pichova *et al.*, 1997): panels (a,c,f) Yeast strain JC488 *RAS2-ala18 val19* rho-zero (unpublished): panels (b,d,g) Yeast strain JC482 corresponding wild type: panel (e) (a,b) phase contrast (c,d) fluorescence microscopy, DHR staining, note that the respiratory deficient strain (d) is DHR-positive. (e,f,g) EPR spectra, field strength in Gauss, abscissa: relative signal intensity. In (e) the spectrum obtained with the respiring wild-type strain JC482 is shown. In (f) the spectrum obtained with the respiring strain JC488 *RAS2-ala18 val19* is shown. In (g) the spectrum obtained with the non-respiring rho-zero strain JC488 *RAS2-ala18 val19* is shown. The lines marked with arrows are indicative of superoxide. In (e) the structure of the spin-trap that was used for obtaining the EPR spectra is shown.

Source: EPR spectra, courtesy: K. Stolze and H. Nohl.

2.11 Ageing and accumulation of mutations, ERCs, silencing, telomeres and DNA metabolism

The accumulation of chromosomal mutations has been considered as a cause of postmitotic ageing in higher organisms (Halliwell and Gutteridge, 1999; Finch and Ruvkun, 2001). These mutations could be caused by the slow accumulation of ROS-induced damage. This kind of process is logically excluded for mother cell-specific ageing, since DNA replication is semi-conservative and the mother and daughter cells inherit the same chromosomal DNA sequence, and therefore, mutations. However, this kind of process is feasible for mitochondrial DNA (mDNA). Of course, mDNA is also replicated semi-conservatively. However, a population of circular mDNA exists in the cell, and only part of it could be subject to the accumulation of mutations. Moreover, it has been hypothesized that the process of mitochondrial mutation accumulation would lead to a vicious circle, in which mDNA would be particularly prone to mutation because ROS are believed to originate in the mitochondria, and mutations in the mDNA would lead to mutations in mitochondrially encoded components of the respiratory chain, which in turn would lead to further increases in mitochondrial mutations. Such a situation would necessarily include the selective asymmetric distribution of damaged and undamaged mitochondria between mother and daughter cell. Preliminary evidence exists for such an asymmetric distribution of oxidatively damaged material between mother and daughter cell (Hugo Aguilaniu, personal communication). Loss of mitochondrial function in ageing cells of mammals (Ames *et al.*, 1993) and yeast cells (Lai *et al.*, 2002) has been demonstrated. However, the hypothesis on the whole has never been rigorously tested.

A large body of work deals with the role of extrachromosomal ribosomal DNA circles (ERCs) in yeast mother cell-specific ageing (Sinclair and Guarente, 1997; Defossez *et al.*, 1998; Guarente, 2000). These authors originally described ERCs as the cause of yeast ageing, but they have become much more cautious today. ERCs arise as a product of recombination between the rDNA repeats on chromosome XII of yeast. The resulting circular DNAs range from one to several repeats each of approximately 9 kb. They contain a functional origin of replication and in the cell cycle behave like autonomously replicating plasmids without a centromere sequence. They, therefore, accumulate in the mother cell (as was shown previously for non-centromeric plasmids) and lead to enlarged nucleoli. Eventually, according to the theory put forward by the authors, the numerous ERCs titrate out the Sir2p (which binds to rDNA), leading to a deprivation of the sub-telomeric SIR complex, which is shown by the fact that sub-telomeric genes in these cells are not transcriptionally repressed like in young cells, but rather derepressed. This results in expression of the *HML* and *HMR* loci and in sterility of haploid aged cells. However, the titrating out of Sir2p must also cause other more severe metabolic changes (defects in chromosomal DNA replication) to really lead to senescence. Several experimental results support the role of ERCs in ageing. For instance, deletion of *FOB1* is a mutation which strongly suppresses recombination of rDNA (Defossez *et al.*, 1999) and increases the median replicative lifespan of the strain by about six generations. Likewise, overexpressing *SIR2* increases the median lifespan (Kaeberlein *et al.*, 1999). However, there are also serious difficulties with the ERC theory. The cells used for measuring the amount of ERCs were isolated by the magnetic bead method and were thus presumably still far from senescence. No adequate statistics were presented concerning the correlation of enlarged nucleoli with an increase of ERC DNA in young and old cells (Sinclair *et al.*, 1997). Subsequent work (Kirchman *et al.*, 1999) showed that under certain conditions (activation of the retrograde response) ERCs are massively accumulated although the lifespan is increased. It is important to note that in spite of intensive searching, ERCs were never found to be involved in the ageing of higher cells or organisms (Guarente, 2000; Grummt, personal communication).

The silencing protein, Sir2p, enhances silencing through its NAD-dependent histone deacetylase activity, by the same mechanism it prevents ERC formation by repressing recombination of rDNA and promotes longevity. *SIR2* is the founding member of a gene family now called the 'sirtuin' family. It is highly conserved in eukaryotic cells and its involvement in ageing (and cancer) is being actively investigated presently (Bitterman *et al.*, 2002; Rogina *et al.*, 2002). It has been hypothesized that *SIR2* might be the link between the nutritional status of the cell (represented by the available NAD) and chromatin silencing, not only in yeast (Guarente, 2000). In mammalian cells, seven members of the *SIR2* family were identified, a function in ageing (but not in ERC formation) was found (Langley *et al.*, 2002) and one member, *SIRT3*, was found to be located in the mitochondria (Onyango *et al.*, 2002) which could be significant given the involvement of mitochondria in ROS production and in programmed cell death.

From about 1996 onwards the protein-coding genes coding for the catalytic (*EST2*) and accessory (other *EST* genes, and additional genes) subunits and the gene *TLC1* (which codes for the template RNA of yeast telomerase) were identified. This in turn enabled the identification of the corresponding genes in higher organisms, using the genetic system of yeast (Shore, 2001). These findings also enabled workers in the field to rigorously test the telomere hypothesis of ageing, both in yeast and in higher organisms. In higher organisms (not discussed in detail here), the Hayflick limit of fibroblasts (in the admittedly artificial situation of cell culture), can be increased by ectopically expressing telomerase and maintaining telomere length. These cells normally display progressively smaller telomeres when they clonally reach the Hayflick limit. The situation is, however, complicated by the fact that there is also at least one additional mechanism for maintaining telomere length by a process of recombination, and many immortalized cell lines, having escaped the crisis at the Hayflick limit, do not show telomerase activity. The mouse knock-out mutation for telomerase leads to the progressive shortening of telomeres but does not lead to organismic premature ageing. It leads, quite unexpectedly, after six generations with normal lifespan, to sterility and increased malignancies (Rudolph *et al.*, 1999). In yeast, the loss of telomerase by deletion of *EST2* (or other essential subunits) leads to delayed mortality after about 70 cell generations (Lendvay *et al.*, 1996). This kind of 'senescence' is clonal (like the senescence of fibroblasts in culture) and clearly shows the essential function of telomeres for the yeast cell. However, this process of clonal ageing is not observed in wild-type yeast and is, very probably, unrelated to the mother cell-specific ageing process of yeast cells. Old yeast mother cells isolated by the gradient method, displayed a normal length of the telomeres (D'Mello and Jazwinski, 1991).

SGS1 (slow growth suppressor 1) is an interesting yeast gene in several respects for the topic discussed here. It was discovered by Rothstein and co-workers (Gangloff *et al.*, 1994) as a suppressor of a slowly growing topoisomerase mutant and subsequently shown to be needed for maintaining recombinational repair and the integrity of the rDNA repeats on chromosome XII of yeast. It also has a function in telomere maintenance in the absence of telomerase, that is, it is involved in telomerase-independent telomere maintenance in certain yeast mutants that escape the 'slow death' of the *est2* deletion mutant. The gene sequence shows that *SGS1* is the only clear member of the RecQ family of ATP-dependent DNA helicases in the yeast genome. Its closest homologues in the human genome are the *wrn* and *blm* DNA helicases. *wrn* is involved in a hereditary progeria (Werner's syndrome) while Bloom's syndrome (mutant *blm*) presents with neurological symptoms and cancer predisposition, but not with progeria. Both genes were shown to be DNA repair helicases. The loss of Sgs1p function leads to a drastically reduced mother cell-specific lifespan in yeast cells, which can be complemented by expressing the human *blm* (but not *wrn*) cDNA in yeast cells deleted for *SGS1* (Heo *et al.*, 1999). For all of the reasons stated here, in particular because *SGS1* has a proven role in suppressing excessive recombination of rDNA, the *sgs1* mutant is an ideal test system for the ERC hypothesis of ageing. However, the authors did not observe an increase of ERCs in the very short-lived *sgs1* mutant nor a decrease of ERCs after restoring the normal lifespan by expressing BLM. This shows that, at least in the background and in the system tested by Heo *et al.* (1999), ERCs are not causative for mother cell-specific ageing. In another study, an increase in ERCs was found in an *sgs1* deletion strain in a different genetic background (Sinclair *et al.*, 1997). All this underlines the notion that yeast mother cell specific ageing may be caused by a number of different and independent biochemical defects, depending on the strain background and the experimental parameters. Heo *et al.* (1999) hypothesize that the reason for the short lifespan of the *sgs1* mutant cells might be a genome-wide increase in double strand breaks and fragmentation of the nucleolus/redistribution of Sir2p (Sinclair *et al.*, 1997), which could be mother cell-specific, if the nucleolus is inherited asymmetrically by the mother cell. These explanations are hypothetical and raise more questions than they answer. For example, for this hypothesis to be true, the DNA damage checkpoint would have to be inactivated in order to allow accumulation of DNA strand breaks over several generations. One has to realize that these authors investigated 7th generation cells, which are not senescent, but still dividing. Replication rounds in the presence of DNA strand breaks would lead to aneuploidy and loss of chromosome fragments within one or two generations, inconsistent with actual observations of strand breaks in senescent but not in aged presenescent cells (Laun *et al.*, 2001).

2.12 Genetic screens for ageing mutants

In recent reviews (Jazwinski, 2001, 2002), genes are listed that have been published as being specifically involved in the ageing process. Many different biochemical functions are implicated showing that genetic studies in yeast have, up to now, not led to a simple picture for the causes or mechanisms of ageing. Similarly, genome-wide transcription studies have led to a very heterogeneous set of genes that are over- or underexpressed in old cells. This situation is somewhat different from the situation in *C. elegans*, where the genes constituting a main signalling pathway (the insulin-like growth factor 1 pathway) give rise to the majority of long-lived mutants.

It makes sense to screen primarily for long-lived mutants, because as long as the field of yeast ageing is wide open (and it is!) we have to discover new and unknown cellular functions that will clarify ageing physiology. A long-lived mutant is *a priori* much more likely to shed light on the true ageing physiology, while short-lived mutants could reduce fitness in many different ways, which, as a secondary effect, could also shorten the lifespan of the cell. It has proved to be extremely difficult to set up a direct genetic selection for lifespan mutants. For such a system to work, one would have to purify very old mother cells to homogeneity. These cells after many generations could be plated out and only long-lived mutants would survive. However, because of the technical difficulties with these purifications (see Section 2.6 on the preparation of old cells), there is presently no way for such a procedure. Therefore, indirect methods have been used. Jazwinski and co-workers (Egilmez *et al.*, 1989; D'Mello *et al.*, 1994) investigated genes corresponding to mRNA overexpressed in old cells and analysed by differential hybridization. The *LAG1* (longevity assurance gene 1) was isolated using this method.

Another method that has been widely used, is pre-selection of mutants under stress conditions that kill the wild-type cells and screening the stress-resistant mutants one by one for their lifespan. The stresses employed were starvation, heat, a combination of starvation and heat stress, and also oxidative stress using either one of a number of oxidants (menadione, hydrogen peroxide, *tert*-butyl hydroperoxide, diamide, etc.). A prominent example is the isolation of a mutant in the *SIR4* gene (Kennedy *et al.*, 1995), coding for a member of the Sir complex of yeast. The mutant was semi-dominant, while the deletion of the gene caused a short lifespan. The phenotype of the mutant was pleiotropic, giving rise also to a co-segregating mating defect, which could be used to clone the gene. The implication of *SIR4* in ageing is intriguing, because also another member of the Sir complex (*SIR2*) plays a prominent role in ageing.

Finally, we present a method (our own work in progress) which could, in principle, improve previous methods because a large number of pre-selected mutants can be screened without doing individual lifespan determinations by micromanipulation. The method is based on a strain (Bobola *et al.*, 1996) from the laboratory of K. Nasmyth. The strain is engineered to express a conditional mutant phenotype only on glucose. An essential gene (*CDC6*) is disrupted and another copy of *CDC6* is placed under the control of the mother cell-specific *HO*-promoter. Under these conditions the essential *CDC6* gene is expressed only in the mother, not in the daughter cell. In addition, *CDC6* is present under control of the *GAL1/10* promoter. On glucose, the strain grows linearly (not exponentially) as long as the mother cell specific lifespan allows. All daughters produced under these conditions stop growing at the G2 phase of the cell cycle (as a mother cell with a large daughter). The final optical density of the culture is therefore directly proportional to the number of cell division cycles that the strain has undergone. Placing 96 pre-selected mutants in a microtiter plate and monitoring optical density with an ELISA reader resulted in a simple lifespan measurement. The mutants were pre-selected on the oxidant molecules mentioned above and several hundred were tested for their lifespan on glucose, resulting in some promising mutants which, at least on glucose, display about twofold increase in lifespan. Direct selection without pre-selecting resistance mutations proved to be not practicable, because the frequency of generation of mutants influencing the regulation of the HO-promoter was too high. These mutants grew exponentially on glucose and were therefore easily recognized.

2.13 Genomic approaches

The only paper published at the time of writing covering genome-wide transcriptional analysis of ageing yeast cells is by Lin *et al.* (2001). These authors compared transcript levels of wild-type virgin cells and 7th generation mother cells sorted on magnetic beads and find a metabolic shift away from glycolysis and towards gluconeogenesis in cells of the 7th generation. This is further corroborated by investigating two mutant strains, *sip2Δ* and *snf4Δ* under similar conditions. The mutation which increases lifespan, *snf4Δ*, shows a less pronounced shift towards gluconeogenesis, while *sip2Δ* shows increased expression of gluconeogenetic genes and a shorter lifespan.

Work in progress in our laboratory aims at the genome-wide transcriptional analysis of senescent yeast mother cells. The large cell fraction V from our elutriation experiment that was characterized in detail previously (Laun *et al.*, 2001) was used for mRNA preparation. For comparison, the small, virgin, fraction II cells were used. Single stranded cDNA was prepared from these cells using established methods and the transcript levels corresponding to each one of about 6000 yeast genes were determined and compared between old and young cells by two different methods. In the first method, genomic filters supplied by J. Hoheisel (Hauser *et al.*, 1998) and ^{33}P labelling were used. In the second approach, using a different strain, the DNA-microarrays of Ontario Cancer Research corporation were used and equimolar mixtures of young and old cDNA labelled with Cy5-CTP and Cy3-CTP, respectively, were applied. In each case, we obtained about 200 genes that were differentially expressed with a high quality measure (sum of those over- and underexpressed). More than 50% of the genes found with the filter method were again found with the microarray method. The fact that the two sets of genes are not identical is for the most part explained by genes that could not be analysed on the filters because of lower data quality and also by strain differences. The gene set contains several genes with a known function in oxidative stress response, some known housekeeping genes and a large fraction of 'orphan' genes (i.e. those of unknown function). Testing oxidative stress resistance/hypersensitivity in deletion mutants corresponding to all of these genes resulted in a number of new genes involved in this, and determining lifespans in some promising cases, is underway.

Acknowledgements

The authors are extremely grateful to Gerald Pfister and Georg Wick for the fluorescence micrographs showing human cells and to Klaus Stolze and Hans Nohl for the EPR spectra. This work was financially supported by grant number P14574-MOB of the FWF (Austria) to M.B. and by grant number 301/03/0289 of the GACR (Czech Republic) to A.P.

References

Ames, B. N., Shigenaga, M. K. and Hagen, T. M. (1993) Oxidants, antioxidants, and the degenerative diseases of aging. *Proceedings of the National Academy of Sciences USA* **90**, 7915–7922.

Arking, R., Burde, V., Graves, K., Hari, R., Feldman, E., Zeevi, A., Soliman, S., Saraiya, A., Buck, S., Vettraino, J., Sathrasala, K., Wehr, N. and Levine, R. L. (2000) Forward and reverse selection for longevity in Drosophila is characterized by alteration of antioxidant gene expression and oxidative damage patterns. *Experimental Gerontology* **35**, 167–185.

Barker, M. G. and Walmsley, R. M. (1999) Replicative ageing in the fission yeast *Schizosaccharomyces pombe. Yeast* **15**, 1511–1518.

Barker, M. G., Brimage, L. J. and Smart, K. A. (1999) Effect of Cu, Zn superoxide dismutase disruption mutation on replicative senescence in *Saccharomyces cerevisiae. FEMS Microbiology Letters* **177**, 199–204.

Berger, K. H. and Yaffe, M. P. (1998) Prohibitin family members interact genetically with mitochondrial inheritance components in *Saccharomyces cerevisiae. Molecular and Cellular Biology* **18**, 4043–4052.

Bissinger, P. H., Wieser, R., Hamilton, B. and Ruis, H. (1989) Control of *Saccharomyces cerevisiae* catalase T gene (*CTT1*) expression by nutrient supply via the RAS-cyclic AMP pathway. *Molecular and Cellular Biology* **9**, 1309–1315.

Bitterman, K. J., Anderson, R. M., Cohen, H. Y., Latorre-Esteves, M. and Sinclair, D. A. (2002) Inhibition of silencing and accelerated aging by nicotinamide, a putative negative regulator of yeast *sir2* and human *SIRT1*. *Journal of Biological Chemistry* **277**, 45099–45107.

Bobola, N., Jansen, R. P., Shin, T. H. and Nasmyth, K. (1996) Asymmetric accumulation of Ash1p in postanaphase nuclei depends on a myosin and restricts yeast mating-type switching to mother cells. *Cell* **84**, 699–709.
Bodnar, A. G., Ouellette, M., Frolkis, M., Holt, S. E., Chiu, C. P., Morin, G. B., Harley, C. B., Shay, J. W., Lichtsteiner, S. and Wright, W. E. (1998) Extension of life-span by introduction of telomerase into normal human cells. *Science* **279**, 349–352.
D'Mello, N. P. and Jazwinski, S. M. (1991) Telomere length constancy during aging of *Saccharomyces cerevisiae*, *Journal of Bacteriology* **173**, 6709–6713.
D'Mello, N. P., Childress, A. M., Franklin, D. S., Kale, S. P., Pinswasdi, C. and Jazwinski, S. M. (1994) Cloning and characterization of *LAG1*, a longevity-assurance gene in yeast. *Journal of Biological Chemistry* **269**, 15451–15459.
Decary, S., Hamida, C. B., Mouly, V., Barbet, J. P., Hentati, F. and Butler-Browne, G. S. (2000) Shorter telomeres in dystrophic muscle consistent with extensive regeneration in young children. *Neuromuscular Disorders* **10**, 113–120.
Defossez, P. A., Park, P. U. and Guarente, L. (1998) Vicious circles: a mechanism for yeast aging. *Current Opinion in Microbiology* **1**, 707–711.
Defossez, P. A., Prusty, R., Kaeberlein, M., Lin, S. J., Ferrigno, P., Silver, P. A., Keil, R. L. and Guarente, L. (1999) Elimination of replication block protein Fob1 extends the lifespan of yeast mother cells, *Molecular Cell* **3**, 447–455.
Egilmez, N. K., Chen, J. B. and Jazwinski, S. M. (1989) Specific alterations in transcript prevalence during the yeast lifespan, *Journal of Biological Chemistry* **264**, 14312–14317.
Egilmez, N. K., Chen, J. B. and Jazwinski, S. M. (1990) Preparation and partial characterization of old yeast cells. *Journal of Gerontology* **45**, B9–17.
Finch, C. E. and Ruvkun, G. (2001) The genetics of aging. *Annual Reviews of Genomics and Human Genetics* **2**, 435–462.
Floyd, R. A. (1991) Oxidative damage to behavior during aging. *Science* **254**, 1597.
Gangloff, S., McDonald, J. P., Bendixen, C., Arthur, L. and Rothstein, R. (1994) The yeast type I topoisomerase Top3 interacts with Sgs1, a DNA helicase homolog: a potential eukaryotic reverse gyrase. *Molecular and Cellular Biology* **14**, 8391–8398.
Gems, D. (1999) Putting metabolic theories to the test. *Current Biology* **9**, R614–R616.
Guarente, L. (2000) Sir2 links chromatin silencing, metabolism and aging. *Genes and Development* **14**, 1021–1026.
Halliwell, B. and Gutteridge, J. M. C. (1999) *Free Radicals in Biology and Medicine*. New York: Oxford University Press.
Harman, D. (1956) Free radical involvement in aging. *Drugs and Aging* **3**, 60.
Harman, D. (1998) Extending functional lifespan. *Experimental Gerontology* **33**, 95–112.
Hauser, N. C., Vingron, M., Scheideler, M., Krems, B., Hellmuth, K., Entian, K. D. and Hoheisel, J. D. (1998) Transcriptional profiling on all open reading frames of *Saccharomyces cerevisiae*. *Yeast* **14**, 1209–1221.
Hayflick, L. (1965) The limited *in vitro* lifespan of human diploid cell strains. *Experimental Cell Research* **37**, 614–636.
Hayflick, L. (1998) How and why we age. *Experimental Gerontology* **33**, 639–653.
Heo, S. J., Tatebayashi, K., Ohsugi, I., Shimamoto, A., Furuichi, Y. and Ikeda, H. (1999) Bloom's syndrome gene suppresses premature ageing caused by Sgs1 deficiency in yeast. *Genes to Cells* **4**, 619–625.
Jazwinski, S. M. (1993) The genetics of aging in the yeast *Saccharomyces cerevisiae*. *Genetica* **91**, 35–51.
Jazwinski, S. M. (2001) New clues to old yeast. *Mechanisms of Ageing and Development* **122**, 865–882.
Jazwinski, S. M. (2002) Growing old: metabolic control and yeast aging. *Annual Reviews of Microbiology* **56**, 769–792.
Jiang, J. C., Jaruga, E., Repnevskaya, M. V. and Jazwinski, S. M. (2000) An intervention resembling caloric restriction prolongs lifespan and retards aging in yeast. *FASEB Journal* **14**, 2135–2137.

Kaeberlein, M., McVey, M. and Guarente, L. (1999) The SIR2/3/4 complex and SIR2 alone promote longevity in *Saccharomyces cerevisiae* by two different mechanisms. *Genes and Development* **13**, 2570–2580.

Kennedy, B. K., N. R. Austriaco, J. and Guarente, L. (1994) Daughter cells of *Saccharomyces cerevisiae* from old mothers display a reduced lifespan. *Journal of Cell Biology* **127**, 1985–1993.

Kennedy, B. K., Austriaco, N. R., Jr, Zhang, J. and Guarente, L. (1995) Mutation in the silencing gene *SIR4* can delay aging in *S. cerevisiae*. *Cell* **80**, 485–496.

Kirchman, P. A., Kim, S., Lai, C. Y. and Jazwinski, S. M. (1999) Interorganelle signaling is a determinant of longevity in *Saccharomyces cerevisiae*. *Genetics* **152**, 179–190.

Lai, C. Y., Jaruga, E., Borghouts, C. and Jazwinski, S. M. (2002) A mutation in the *ATP2* gene abrogates the age asymmetry between mother and daughter cells of the yeast *Saccharomyces cerevisiae*. *Genetics* **162**, 73–87.

Langley, E., Pearson, M., Faretta, M., Bauer, U. M., Frye, R. A., Minucci, S., Pelicci, P. G. and Kouzarides, T. (2002) Human SIR2 deacetylates p53 and antagonizes PML/p53-induced cellular senescence. *EMBO Journal* **21**, 2383–2396.

Laun, P. (1999) Immobilisierung von Hefezellen durch genetische Derivatisierung der Zelloberfläche. Diploma, University of Salzburg, Salzburg, Austria.

Laun, P., Pichova, A., Madeo, F., Fuchs, J., Ellinger, A., Kohlwein, S., Dawes, I., Frohlich, K. U. and Breitenbach, M. (2001) Aged mother cells of *Saccharomyces cerevisiae* show markers of oxidative stress and apoptosis. *Molecular Microbiology* **39**, 1166–1173.

Lee, A. C., Fenster, B. E., Ito, H., Takeda, K., Bae, N. S., Hirai, T., Yu, Z. X., Ferrans, V. J., Howard, B. H. and Finkel, T. (1999) Ras proteins induce senescence by altering the intracellular levels of reactive oxygen species, *Journal of Biological Chemistry* **274**, 7936–7940.

Lendvay, T. S., Morris, D. K., Sah, J., Balasubramanian, B. and Lundblad, V. (1996) Senescence mutants of *Saccharomyces cerevisiae* with a defect in telomere replication identify three additional EST genes. *Genetics* **144**, 1399–1412.

Lin, S. J., Defossez, P. A. and Guarente, L. (2000) Requirement of NAD and *SIR2* for Life-Span Extension by Calorie Restriction in *Saccharomyces cerevisiae*. *Science* **289**, 2126–2128.

Lin, S. S., Manchester, J. K. and Gordon, J. I. (2001) Enhanced gluconeogenesis and increased energy storage as hallmarks of aging in *Saccharomyces cerevisiae*. *Journal of Biological Chemistry* **276**, 36000–36007.

Martin, G. M. and Oshima, J. (2000) Lessons from human progeroid syndromes. *Nature* **408**, 263–266.

Mondello, C., Petropoulou, C., Monti, D., Gonos, E. S., Franceschi, C. and Nuzzo, F. (1999) Telomere length in fibroblasts and blood cells from healthy centenarians. *Experimental Cell Research* **248**, 234–242.

Mortimer, R. K. and Johnston, J. R. (1959) Lifespan of individual yeast cells. *Nature* **183**, 1751–1752.

Muller, I., Zimmermann, M., Becker, D. and Flomer, M. (1980) Calendar lifespan versus budding lifespan of *Saccharomyces cerevisiae*. *Mechanisms of Ageing and Development* **12**, 47–52.

Nestelbacher, R., Laun, P. and Breitenbach, M. (1999) Images in experimental gerontology. A senescent yeast mother cell. *Experimental Gerontology* **34**, 895–896.

Nestelbacher, R., Laun, P., Vondrakova, D., Pichova, A., Schuller, C. and Breitenbach, M. (2000) The influence of oxygen toxicity on yeast mother cell-specific aging. *Experimental Gerontology* **35**, 63–70.

Onyango, P., Celic, I., McCaffery, J. M., Boeke, J. D. and Feinberg, A. P. (2002) SIRT3, a human SIR2 homologue, is an NAD-dependent deacetylase localized to mitochondria. *Proceedings of the National Academy of Sciences USA* **99**, 13653–13658.

Orr, W. C. and Sohal, R. S. (1994) Extension of life-span by overexpression of superoxide dismutase and catalase in *Drosophila melanogaster*. *Science* **263**, 1128–1130.

Pichova, A., Vondrakova, D. and Breitenbach, M. (1997) Mutants in the *Saccharomyces cerevisiae RAS2* gene influence lifespan, cytoskeleton and regulation of mitosis. *Canadian Journal of Microbiology* **43**, 774–781.

Powell, C. D., Quain, D. E. and Smart, K. A. (2000) The impact of media composition and petite mutation on the longevity of a polyploid brewing yeast strain. *Letters in Applied Microbiology* **31**, 46–51.

Renault, V., Thornell, L. E., Butler-Browne, G. and Mouly, V. (2002) Human skeletal muscle satellite cells: aging, oxidative stress and the mitotic clock. *Experimental Gerontology* **37**, 1229–1236.

Rogina, B., Helfand, S. L. and Frankel, S. (2002) Longevity regulation by Drosophila Rpd3 deacetylase and caloric restriction. *Science* **298**, 1745.

Rosenfeld, E., Beauvoit, B., Rigoulet, M. and Salmon, J. M. (2002) Non-respiratory oxygen consumption pathways in anaerobically-grown *Saccharomyces cerevisiae*: evidence and partial characterization. *Yeast* **19**, 1299–1321.

Rudolph, K. L., Chang, S., Lee, H. W., Blasco, M., Gottlieb, G. J., Greider, C. and DePinho, R. A. (1999) Longevity, stress response and cancer in aging telomerase-deficient mice. *Cell* **96**, 701–712.

Schafer, F. Q. and Buettner, G. R. (2001) Redox environment of the cell as viewed through the redox state of the glutathione disulfide/glutathione couple. *Free Radicals in Biology and Medicine* **30**, 1191–1212.

Serrano, M., Lin, A. W., McCurrach, M. E., Beach, D. and Lowe, S. W. (1997) Oncogenic *ras* provokes premature cell senescence associated with accumulation of p53 and p16INK4a. *Cell* **88**, 593–602.

Shore, D. (2001) Telomeric chromatin: replicating and wrapping up chromosome ends. *Current Opinion in Genetics and Development* **11**, 189–198.

Sinclair, D. A. and Guarente, L. (1997) Extrachromosomal rDNA circles – a cause of aging in yeast. *Cell* **91**, 1033–1042.

Sinclair, D. A., Mills, K. and Guarente, L. (1997) Accelerated aging and nucleolar fragmentation in yeast *sgs1* mutants. *Science* **277**, 1313–1316.

Sitte, N., Merker, K., von Zglinicki, T. and Grune, T. (2000) Protein oxidation and degradation during proliferative senescence of human MRC-5 fibroblasts. *Free Radicals in Biology and Medicine* **28**, 701–708.

Small, W. C., Brodeur, R. D., Sandor, A., Fedorova, N., Li, G., Butow, R. A. and Srere, P. A. (1995) Enzymatic and metabolic studies on retrograde regulation mutants of yeast. *Biochemistry* **34**, 5569–5576.

Smeal, T., Claus, J., Kennedy, B., Cole, F. and Guarente, L. (1996) Loss of transcriptional silencing causes sterility in old mother cells of *S. cerevisiae*. *Cell* **84**, 633–642.

Sohal, R. S. and Weindruch, R. (1996) Oxidative stress, caloric restriction and aging. *Science* **273**, 59–63.

Sun, J., Kale, S. P., Childress, A. M., Pinswasdi, C. and Jazwinski, S. M. (1994) Divergent roles of *RAS1* and *RAS2* in yeast longevity. *Journal of Biological Chemistry* **269**, 18638–18645.

Tatchell, K. (1986) *RAS* genes and growth control in *Saccharomyces cerevisiae*. *Journal of Bacteriology* **166**, 364–367.

Taub, J., Lau, J. F., Ma, C., Hahn, J. H., Hoque, R., Rothblatt, J. and Chalfie, M. (1999) A cytosolic catalase is needed to extend adult lifespan in *C. elegans daf-C* and *clk-1* mutants. *Nature* **399**, 162–166.

Thevelein, J. M. and de Winde, J. H. (1999) Novel sensing mechanisms and targets for the cAMP–protein kinase A pathway in the yeast *Saccharomyces cerevisiae*. *Molecular Microbiology* **33**, 904–918.

Van Zandycke, S. M., Sohier, P. J. and Smart, K. A. (2002) The impact of catalase expression on the replicative lifespan of *Saccharomyces cerevisiae*. *Mechanisms of Ageing and Development* **123**, 365–373.

Wagner, M., Hampel, B., Bernhard, D., Hala, M., Zwerschke, W. and Jansen-Durr, P. (2001) Replicative senescence of human endothelial cells in vitro involves G1 arrest, polyploidization and senescence-associated apoptosis. *Experimental Gerontology* **36**, 1327–1347.

Wawryn, J., Krzepilko, A., Myszka, A. and Bilinski, T. (1999) Deficiency in superoxide dismutases shortens lifespan of yeast cells. *Acta Biochimica Polonica* **46**, 249–253.

Wilcoxon, F. (1945) Individual comparisons by ranking methods. *Biometrics* **1**, 80–83.

Woldringh, C. L., Fluiter, K. and Huls, P. G. (1995) Production of senescent cells of *Saccharomyces cerevisiae* by centrifugal elutriation. *Yeast* **11**, 361–369.

Wolkow, C. A., Kimura, K. D., Lee, M. S. and Ruvkun, G. (2000) Regulation of *C. elegans* life-span by insulinlike signaling in the nervous system. *Science* **290**, 147–150.

Chapter 3

Carbon metabolism

Arthur L. Kruckeberg and J. Richard Dickinson

3.1 Introductory remarks

In this chapter we have sought to emphasize new discoveries and insights which have been made since the first edition of the book was published. This is most apparent in the section on transporters, especially hexose transporters which are now recognized as being of major importance in setting the overall catabolic rate. However, we also felt obliged to re-state the basic elements and cellular *raisons d'etre* of the individual pathways for the benefit of those readers who are new to yeast and/or its carbon metabolism. *Saccharomyces cerevisiae* is able to grow on a wide range of carbon compounds but glucose is the preferred substrate and a truly amazing set of controls exists to ensure the efficient metabolism of this compound.

3.2 Transport of carbon compounds into the cell and nutrient sensing

Assimilatory and dissimilatory metabolism of carbon compounds necessarily begins with their transport into the cell. Most compounds utilized by yeast are sufficiently hydrophilic that they are impermeable to the phospholipid bilayer of the plasma membrane, and hence pass from the periplasm to the cytoplasm via specific transport proteins. Important exceptions are ethanol and glycerol, which are able to diffuse freely across the plasma membrane (although a specific channel, *FPS1*, also contributes to flux of glycerol across the membrane; see later). Some of the transporters mediate facilitated diffusion, that is, transport only occurs down a concentration gradient. Examples include the hexose transporters. Other transporters mediate active transport, usually as proton symporters. These transporters can utilize the membrane potential to accumulate their substrate inside the cell. Examples include the maltose and lactate transporters (Eddy, 1982; van der Rest *et al.*, 1995b). The underlying basis for facilitated diffusion of some solutes and concentrative transport of others may relate to the substrate affinities of the subsequent metabolic steps. For example, after transport, hexoses are phosphorylated by hexokinase, which has a Km of approximately 0.1 mM (Entian, 1997), while maltose is hydrolyzed by maltase (α-glucosidase) which has a Km of approximately 20 mM (Matsusaka *et al.*, 1977; Needleman *et al.*, 1978; Tabata *et al.*, 1984).

Yeast is also able to metabolize some carbohydrates after their extracellular hydrolysis. For example, sucrose (and raffinose) are metabolized after extracellular hydrolysis by invertase (β-fructofuranosidase) (and melibiase, also known as α-galactosidase), followed by uptake of the resulting hexoses. Transport of sucrose into the cell followed by intracellular hydrolysis has been reported as well, but the existence of sucrose transporters seems to be confined to specific strains. Alternatively, sucrose could be transported by maltose transporters and hydrolyzed by maltase. Some strains also possess functional *STA* genes encoding secreted glucoamylase (glucan 1,4-α-glucosidase), and are thus able to hydrolyze starch to glucose (Pretorius *et al.*, 1991); the hexose is subsequently taken up and metabolized. Maltotriose is also fermented (Zastrow *et al.*, 2000) which is especially important in brewing strains.

Thirty-eight genes have been identified in the yeast genome that encode transporters of non-nitrogenous carbon compounds across the plasma membrane (excluding transporters of vitamins, drugs, waste products, etc. (Paulsen *et al.*, 1998; Table 3.1). They can be broadly classified with respect to substrate specificity as transporters of alcohols (glycerol, inositol, glycerophosphoinositol), carbohydrates (hexoses, maltose, trehalose, α-glucosides), and organic acids (lactate, pyruvate). The substrates of some of these transporters have been determined by biochemical and genetic analysis. In other cases, the substrate specificity of the putative transport proteins is inferred by their sequence homology with a known transporter. Genes encoding transporters of carbon compounds fall into two broad families: *FPS1* and its three homologues encode channel-type proteins, while the remainder encode members of the major facilitator superfamily (Nelissen *et al.*, 1997). Of the facilitator-type proteins, Jen1, the lactate transporter, has no close relatives in yeast. The remainder are members of the sugar porter family; this family shares significant sequence similarity despite the wide range of substrate specificities displayed by its members (from hexoses to inositol to inorganic phosphate) (Paulsen *et al.*, 1998; Van Belle and Andre, 2001). The sugar porter family is also conserved among bacteria, plants and animals, and has numerous members in the genomes of each of these groups of organisms (Baldwin and Henderson, 1989; Kruckeberg, 1996). The evolutionary history of the sugar porter family in fungi, and in particular in yeasts such as *S. cerevisiae* and its relatives, remains to be elucidated in detail. However, it is clear that some of the many members of the family have arisen due to duplication of genes and chromosomal blocks quite recently in *Saccharomyces* evolution (Van Belle and Andre, 2001). The sugar porter family in *S. cerevisiae* is more numerous (0.6% of predicted genes) than the 20 members of the carbohydrate transporter family recognized in *Schizosaccharomyces pombe* (0.4% of predicted genes; Wood *et al.*, 2002; http://www.genedb.org/genedb/pombe/).

Hexose transport

The carbohydrate transporter family is the largest family of transporters in the yeast genome (Paulsen *et al.*, 1998). Among these are the twenty genes of the *HXT* gene family which encode proteins with hexoses as substrate (Kruckeberg, 1996). The *HXT* family includes bona fide transporters (based on genetic analysis), transporter homologues, and two hexose sensors (*SNF3* and *RGT2*; see later). A yeast strain with null mutations in *HXT1–HXT7* does not grow on glucose media (Reifenberger *et al.*, 1995). Expression of each of these seven genes individually (with the exception of *HXT5*) in the null strain restores growth. The growth phenotypes conferred by the six *HXT* genes vary depending on the concentration of glucose or fructose provided. This suggests that the genes are differentially regulated by the hexose concentration in the medium. When the glucose transport kinetics of strains containing only one *HXT* gene (*HXT1, HXT2, HXT3, HXT4,* or *HXT7*) were determined, it was found that *HXT1, HXT3,* and *HXT4* confer low-affinity transport on the null strain, while *HXT7* confers high-affinity transport (*HXT6* and *HXT7* differ by only two conserved amino acid changes, and are assumed to have similar transport kinetics). The *HXT2*-only

Table 3.1 Yeast carbon compound transporters

Gene	*ORF*	*Protein encoded*
Transporters of carbohydrates		
AGT1	YGR289C	α-glucoside and trehalose permease
GAL2	YLR081W	Galactose permease
HXT1	YHR094C	Low-affinity hexose facilitator
HXT2	YMR011W	Hexose facilitator of moderately low affinity for glucose
HXT3	YDR345C	Low-affinity hexose facilitator
HXT4	YHR092C	Hexose facilitator of moderately low affinity
HXT5	YHR096C	Hxt family protein with intrinsic hexose transport activity
HXT6	YDR343C	High-affinity hexose facilitator
HXT7	YDR342C	High-affinity hexose facilitator
HXT8	YJL214W	Hxt family protein with intrinsic hexose transport activity – unknown physiological function
HXT9	YJL219W	Hxt family protein with intrinsic hexose transport activity – unknown physiological function
HXT10	YFL011W	Hxt family protein with intrinsic hexose transport activity – unknown physiological function
HXT11	YOL156W	Hxt family protein with intrinsic hexose transport activity – unknown physiological function
HXT13	YEL069C	Hxt family protein with intrinsic hexose transport activity – unknown physiological function
HXT14	YNL318C	Protein of the sugar transporter superfamily – able to sustain slow growth on galactose
HXT15	YDL245C	Hxt family protein with intrinsic hexose transport activity – unknown physiological function
HXT16	YJR158W	Hxt family protein with intrinsic hexose transport activity – unknown physiological function
HXT17	YNR072W	Hxt family protein with intrinsic hexose transport activity – unknown physiological function
MAL61	YBR298C	Maltose permease (from the *MAL6* locus)
RGT2	YDL138W	Transporter-like sensor of high glucose concentrations
SNF3	YDL194W	Transporter-like sensor of low glucose concentrations
MPH2	YDL247W	α-glucoside permease
MPH3	YJR160C	α-glucoside permease
Transporters of organic acids		
JEN1	YKL217W	Lactate and pyruvate permease
Transporters of alcohols		
FPS1	YLL043W	Glycerol channel of the plasma membrane – also involved in arsenite and antimonite uptake (MIP family)
GIT1	YCR098C	Glycerophosphoinositol transporter
ITR1	YDR497C	Inositol permease
ITR2	YOL103W	Inositol permease

strain displayed more complex transport kinetics, with intermediate glucose affinity after growth in high glucose medium, and biphasic kinetics (both low and high affinity for glucose) after growth in low glucose medium (Reifenberger *et al.*, 1997).

Glucose transport and *HXT* gene expression have been studied in a wild-type strain after batch or chemostat cultivation. The chemostats allowed the investigation of hexose transport kinetics and *HXT* mRNA levels in cells cultivated under steady-state aerobic growth conditions with glucose limitation at growth rates ranging from 0.05 to $0.38\,h^{-1}$ (which includes respiro-fermentative growth, comparable to exponential growth in a batch culture, and fully respiratory growth, comparable to the diauxic shift of a batch culture). Cells cultivated in aerobic chemostats with fructose, galactose, ethanol, or ammonium limitation, and in anaerobic chemostats with glucose or ammonium limitation (all at a growth rate of $0.1\,h^{-1}$) were analysed as well. The levels of *HXT1–HXT7* transcripts were often high in these various conditions. With high glucose concentrations in the medium (exponential phase of batch culture, respiro-fermentative growth in aerobic glucose-limited chemostats, anaerobic nitrogen-limited chemostat cultivation) the low-affinity transporter genes *HXT1, HXT3,* and *HXT4* were highly expressed. Their expression was associated with the expression of low-affinity transport kinetics. At lower external glucose concentrations (e.g. diauxic shift of batch culture, respiratory growth in aerobic glucose-limited chemostats) the high-affinity transporter genes *HXT2, HXT6,* and *HXT7* were expressed. Cells from these conditions displayed high-affinity transport kinetics. At the extremes of glucose limitation (post-diauxic shift in batch culture, low dilution rates in glucose-limited chemostats, ethanol-limited chemostat cultivation) *HXT5* expression was abundant as well. This study emphasized the relationship between *HXT* gene expression, hexose transport kinetics, and nutrient (particularly carbon source) availability (Diderich *et al.*, 1999a).

Initial studies on *HXT5* did not identify a growth or transport phenotype for this gene. However, a strain expressing only *HXT5* (i.e. with null mutations in *HXT1–HXT4, HXT6,* and *HXT7*) displayed vigorous glucose fermentation after prolonged growth on maltose (Klaassen and Raamsdonk, 1998). This fact, combined with the high levels of expression of *HXT5* observed in chemostat cultivations, has prompted renewed interest in this gene. The *HXT5* promoter is active in cells grown on galactose, ethanol, and glycerol (Diderich *et al.*, 1999a; Özcan and Johnston, 1999; Diderich *et al.*, 2001a), and is up-regulated under various stress conditions such as osmotic shock, heat shock, nitrogen starvation, starvation-induced sporulation, the diauxic shift, and prolonged maintenance in stationary phase (DeRisi *et al.*, 1997; Chu *et al.*, 1998; Gasch *et al.*, 2000; Rep *et al.*, 2000; Diderich *et al.*, 2001a; Verwaal *et al.*, 2002). When cells are pre-grown on ethanol, Hxt5 is highly expressed; *HXT5*-only cells grown under these conditions display moderate-affinity glucose transport activity (Km ~10 mM) and can carry out significant glycolytic flux. The inability of *HXT5*-only cells to grow on glucose is due to the stringent glucose repression of the gene (Diderich *et al.*, 2001b). The glucose transporter encoded by *HXT5* might provide glucose-starved cells with sufficient transport capacity to initiate glucose metabolism should the sugar become available. Indeed, the lag phase of *hxt5* null mutant cells is detectably longer than that of wild-type cells when glucose is added to stationary-phase cultures (Diderich *et al.*, 2001b).

Early research in yeast genetics included analysis of mutations that impair growth on galactose. In 1954 it was recognized that mutations in *GAL2* 'influenc[e] the rate of penetration of galactose into the cell' (Douglas and Condie, 1954). Subsequent studies showed that *GAL2* is necessary for galactose transport activity, and encodes a protein with homology to other transport proteins. Expression of *GAL2* is strongly induced by provision of galactose, and is also subject to glucose repression; this pattern of regulation is shared with the genes encoding the Leloir pathway enzymes of galactose metabolism (see later) (Tschopp *et al.*, 1986; Nehlin *et al.*, 1989; Szkutnicka *et al.*, 1989). The Gal2 transporter has a Km for galactose of ~1 mM. The protein is also able to transport glucose with approximately the same affinity (Nishizawa *et al.*, 1995; Liang and Gaber, 1996; Reifenberger *et al.*, 1997); however, in the presence of glucose, strong transcriptional repression combined with rapid inactivation of transport activity and degradation of the Gal2 protein ensure that it does not contribute to glucose transport under normal conditions (Tschopp *et al.*, 1986; Horak and Wolf, 1997).

Intracellular galactose is required for induction of *GAL2* and the Leloir pathway genes. The sugar stabilizes the interaction of Gal80 (the repressor of the Gal4 transcriptional activator) with Gal1 or Gal3 (Zenke *et al.*, 1996; Platt and Reece, 1998). The intracellular galactose is probably provided by Gal2, which would be expressed at basal levels in order to prevent inducer exclusion.

Transport

The cellular abundance and transport activities of Hxt2 and Hxt7 have been quantified. The transporters were fused with the green fluorescent protein (GFP), and the fusion proteins were shown to be functional transporters with kinetics similar to the wild-type proteins. Under conditions of high expression, the Hxt2–GFP fusion protein was expressed at 1.4×10^5 copies per cell, and had a catalytic centre activity of $53\,s^{-1}$ (Kruckeberg *et al.*, 1999). Similarly, the Hxt7–GFP fusion protein was expressed at 1.8×10^4 copies per cell, and had a catalytic centre activity of approximately $200\,s^{-1}$ (Ye *et al.*, 2001). These values are in agreement with predictions for hexose transporter abundance and catalytic centre activity that are based on rates of glucose consumption by yeast and estimated levels of transporter protein expression (Serrano, 1991).

The effect of the phospholipid composition of the plasma membrane on hexose transport is less well characterized than its effect on amino acid transport. This is unfortunate given the significant changes in phospholipid composition that occur during batch growth, or in response to nutrient availability and strain characteristics (Homann *et al.*, 1987; Alexandre *et al.*, 1994; Janssen *et al.*, 2000). Particularly relevant to transport activity are the annular phospholipids, those which are in direct contact with transport proteins. For example, when the HUP1 glucose transporter (an *HXT* homologue) from the unicellular alga *Chlorella* was expressed and purified from yeast, functional protein was tightly associated with phosphatidyl choline, phosphatidyl ethanolamine and ergosterol. Phosphatidyl choline was essential for reconstitution of HUP1-dependent hexose transport activity by delipidated protein (Robl *et al.*, 2000). The solute transport characteristics of a mutant strain unable to synthesize phosphatidyl ethanolamine have been compared to a wild-type strain; it was found that the amino acid transport systems of the mutant were significantly inhibited, whereas transport of hexoses was unaffected (Robl *et al.*, 2001).

Metabolic control

It has been proposed that transport of glucose into the cell across the plasma membrane is the rate-limiting step of glycolysis. Evidence for this proposal includes the observations that (1) during steady-state glycolytic flux the transport step is far from equilibrium, with internal glucose concentrations of approximately 2–4 mM at external concentrations of around 200 mM. This indicates that metabolic capacity is in excess of transport capacity (Pedler *et al.*, 1997; Teusink *et al.*, 1998); (2) conditions that lead to inactivation of glucose transport activity (e.g. nitrogen starvation) lead to a concomitant decline in glycolytic flux and the activities of glycolytic enzymes are relatively unaffected by these conditions (Lagunas *et al.*, 1982; Schulze *et al.*, 1996; van Hoek *et al.*, 2000; Nilsson *et al.*, 2001); (3) the maximal activity of the glucose transport system in the absence of further metabolism (e.g. when assayed with non-metabolizable glucose analogues or hexose kinase mutants) is comparable to the measured flux through glycolysis (Serrano and Delafuente, 1974); and (4) over-expression of individual or multiple glycolytic enzymes does not increase the glycolytic flux or the rate of growth on glucose (Heinisch, 1986; Schaaff *et al.*, 1989; Rosenzweig, 1992; Hauf *et al.*, 2000; Smits *et al.*, 2000).

Metabolic Control Analysis (MCA) is a valuable tool for describing the control of metabolic pathways in biochemically and mathematically precise terms. MCA posits that the control of flux through a metabolic pathway is distributed among all enzymatic steps in that pathway. Each step has a control coefficient which describes that step's contribution to steady-state flux through the pathway. The control coefficient is defined as the relative change in flux per relative change in the activity of the step. The sum of control coefficients for a linear metabolic pathway is one. Each step has a control coefficient ranging from 0 (or infinitesimal) to nearly 1, that is, it will range from having virtually no control to being truly 'rate-limiting' (Kacser and Burns, 1981; Fell, 1997). A related term in MCA is the response coefficient, defined as the relative change in flux per relative change in a parameter. The parameter can be the concentration of the substrate or an enzyme inhibitor (external response coefficient) or the enzyme itself (internal response coefficient). Thus, the response coefficient is directly amenable to experimental manipulation. The response and control coefficients are related by the elasticity coefficient, which is defined as the relative change in the activity of a pathway step per relative change in a parameter that affects the activity of that step in isolation. The elasticity is also directly measurable by experimentation. In situations where the elasticity equals one (e.g. where the concentration of an enzyme is proportional to its activity) the response and control coefficients are equal (Kacser and Burns, 1981; Fell, 1997).

MCA has been used to assess the proportion of the control on glycolytic flux and growth on glucose that is exerted by glucose transport. In one approach, the transport activity was modulated by varying the concentrations of maltose or glucose provided during transport assays. Maltose is a competitive inhibitor of glucose transport; therefore both it and glucose were used to determine external response coefficients of glucose transport with respect to glycolytic flux. The response coefficients varied from approximately 0.1 to 0.3. The elasticity coefficients were derived from the steady-state glucose transport activities at different extracellular glucose concentrations using a rate equation, and were then used to calculate the control coefficients. The control coefficients of glucose transport on glycolytic flux were high (>0.5) over a wide range of extracellular glucose concentrations (1–200 mM) (Diderich *et al.*, 1999b).

In a second approach, glucose transport activity was varied by altering the level of expression of *HXT7* in strains which expressed only that transporter (i.e. had null mutations in *HXT1–HXT6*). The glucose transport activity of these strains, and the parental wild-type strain, ranged from 27 to 364 $\text{nmol}\cdot\text{min}^{-1}\cdot(\text{mg protein})^{-1}$. The rates of growth and glycolytic flux varied proportionally with transport activity. Under the conditions of the experiment (and assuming an elasticity of 1) the control coefficient of transport with respect to growth was 0.54, and with respect to glycolytic flux was 0.90 (Ye *et al.*, 1999).

Taken together, these two studies clearly demonstrate that a high proportion of control over glycolytic flux and fermentative growth resides in the glucose transport step. This conclusion is tempered by observations on the range over which the intracellular glucose concentration can vary during steady-state glucose consumption, from <0.1 to 3 mM or higher. The intracellular glucose concentration is affected by the extracellular glucose concentration, the affinity for glucose of the transport system expressed by the cells under consideration, and the capacity of the glycolytic pathway. Since glucose transport is a reversible process, and the inward-facing binding site of the transporters is assumed to have a substrate affinity comparable to the outward-facing site, the factors that influence the intracellular glucose concentration will also influence the magnitude of the control coefficient of the transport step (Teusink *et al.*, 1998).

Transcriptional regulation

Transcription of *HXT1–HXT4, HXT6,* and *HXT7* is regulated according to glucose availability by two main pathways: glucose repression and glucose induction (Özcan and Johnston, 1995, 1996; Liang and Gaber, 1996; Petit *et al.*, 2000; Ye *et al.*, 2001). Glucose repression acts on the high-affinity transporters *HXT2, HXT6,* and *HXT7,* as well as on a wide variety of genes involved in respiration, mitochondrial biogenesis, and metabolism of other carbon sources (see later). This pathway involves hexokinase II, the Snf1 protein kinase, and the Mig1 transcriptional repressor. The glucose induction pathway involves the glucose sensors Snf3 and Rgt2, the Grr1 protein (a component of an SCF ubiquitin ligase complex) and the Rgt1 transcription factor. Rgt1 is a bifunctional factor: it represses transcription of target genes in the absence of glucose and activates transcription at high glucose concentrations (Özcan *et al.*, 1996a,b, 1998). The interplay of the glucose repression and induction pathways results in differential transcription of the *HXT* genes. For example, in the absence of glucose, Rgt1 represses *HXT1* and *HXT2*. At high concentrations of glucose, *HXT2* is repressed by the Mig1 repressor, while *HXT1* is activated by Rgt1 (and is unaffected by Mig1, which has no binding sites in the *HXT1* promoter). At low glucose concentrations, both the Rgt1 and Mig1 repressors are inactive, resulting in *HXT2* transcription; *HXT1* is neither repressed nor induced by Rgt1, and its transcription is low (Özcan and Johnston, 1995). Similar regulatory circuits operate on the *HXT3, HXT4, HXT6,* and *HXT7* genes (Liang and Gaber, 1996; Özcan and Johnston, 1999).

HXT4 transcription is apparently unique among hexose transporter genes in being regulated by the Gcr1 and Gcr2 transcription factors. These transcription factors are known to up-regulate the genes encoding glycolytic enzymes during growth on glucose (see later). Superimposing regulation by Gcr1 and Gcr2 onto the other regulatory pathways that impinge on *HXT4* transcription ensures that low-affinity transport activity is expressed in concert with glycolytic enzyme activity during fermentative growth (Turkel and Bisson, 1999).

Genetic studies on *HXT5* transcriptional regulation have not been carried out. However, the high level of *HXT5* transcription in conditions of extreme glucose limitation is remarkable (Diderich *et al.*, 1999a, 2001b). Two consensus binding sites for the Hap transcriptional activator complex upstream of *HXT5* suggest that this gene might be positively co-regulated with other genes by the absence of glucose (see later) (Diderich *et al.*, 1999a, 2001b).

Transcription of *HXT8–HXT17* is generally quite low under various laboratory growth conditions, although the *HXT13* promoter was fairly active under derepressing conditions

(Diderich *et al.*, 1999a; Özcan and Johnston, 1999). Considering the low codon adaptation indices of their open reading frames (ORF) (Kruckeberg, 1996), it is apparent that these transporters are not expressed at levels comparable to the proteins encoded by *HXT1–HXT7*, which is consistent with their inability to sustain growth on glucose in the *hxt1–hxt7* null mutant strain (Reifenberger *et al.*, 1995). Indeed, a strain with null mutations in *HXT1, HXT2, HXT4, HXT5,* and *HXT8–HXT17* (i.e. possessing only *HXT3, HXT6,* and *HXT7*) has a wild-type growth phenotype on glucose (Wieczorke *et al.*, 1999). Further deletion of *HXT3, HXT6,* and *HXT7* abolishes growth on glucose, fructose and mannose; subsequent deletion of *GAL2* markedly reduces growth on galactose. Individual over-expression in the *hxt1–hxt17 gal2* null strain of the *HXT8–17* genes (with the exception of *HXT12*) restores growth on various hexoses: over-expression of *HXT8, HXT13, HXT15, HXT16,* and *HXT17* restores growth on glucose, fructose, and mannose; over-expression of *HXT9, HXT10,* and *HXT11* restores growth on these hexoses and galactose as well. Over-expression of *HXT14* restores growth only on galactose. When the *hxt1–hxt17 gal2* null strain was pre-grown on maltose it was able to consume glucose, albeit at low rates. This residual uptake capacity was in part due to members of the maltose transporter family; it was abolished by deletion of the *AGT1, MPH2,* and *MPH3* genes (but not *MAL31*), and over-expression of these genes in the *hxt1–hxt17 gal2* null strain restores growth on glucose. Surprisingly, deletion of the *SNF3* glucose sensor gene in the *hxt1–hxt17 gal2 agt1 mph2 mph3* null strain restores growth on glucose, suggesting that another protein(s) is able to transport glucose as well, and that its expression is repressed by Snf3 (Wieczorke *et al.*, 1999). However, conditions in which the native *HXT8–HXT17* genes contribute significantly to glycolytic flux have not been identified. It is not clear if they play this role under conditions that have not yet been tested, if they play a different metabolic role, or if they are evolutionary vestiges.

Expression of the *SNF3* and *RGT2* genes is low, albeit readily detectable; the two glucose sensors are generally co-expressed except at the extremes of the available glucose concentrations: in anaerobic nitrogen-limited chemostats (high residual glucose) *RGT2* expression is increased relative to *SNF3*, and in ethanol-limited chemostats (no exogenous carbohydrate) *SNF3* expression is increased relative to *RGT2* (Diderich *et al.*, 1999a). In batch cultivation, *SNF3* expression is repressed by glucose, while *RGT2* expression is constitutive (Celenza *et al.*, 1988; Gamo *et al.*, 1994; Özcan *et al.*, 1996a).

Glucose sensing

As mentioned above, two members of the *HXT* family, *SNF3* and *RGT2*, encode glucose sensors (Özcan *et al.*, 1996a, 1998). The sensors regulate expression of other *HXT* family members in response to the extracellular glucose concentration. The regulatory roles of the Snf3 and Rgt2 proteins are a remarkable evolutionary story. Virtually all organisms have sugar transporters of the *HXT* type, including bacteria, fungi, plants, and animals (Baldwin and Henderson, 1989). The proteins they encode are similar in sequence and (predicted) structure, which attests to the efficiency of this protein architecture in mediating solute transport. In yeast this architecture has been modified in Snf3 and Rgt2, resulting in proteins that do not transport a solute, but instead transduce a signal into the cell about the presence of the solute in the extracellular environment. The signal is presumed to be initiated by binding of glucose to the outward-facing binding site of the proteins, which causes a change in conformation. The conformational change is then communicated across the membrane to a signal transduction pathway that regulates the expression of various homologues of the sensors (Özcan *et al.*, 1996a, 1998).

Snf3 induces transcription of high-affinity transporter genes such as *HXT2* and *HXT7* when cells are exposed to low glucose concentrations, and Rgt2 induces transcription of low-affinity transporter genes such as *HXT1* when cells are exposed to high glucose concentrations. This suggests that the two proteins differ in their affinity for glucose, with Rgt2 acting as a sensor for high glucose concentrations and Snf3 acting as a sensor for low glucose concentrations. Differences in affinity for their ligand do not fully account for their different signalling properties, however. The *SNF3* and *RGT2* genes are differentially expressed only at extremes of glucose availability: as mentioned earlier, *SNF3* is repressed at low glucose concentrations, whereas *RGT2* is expressed under a wide variety of conditions (although it is down-regulated in ethanol-limited chemostats). The greater responsiveness of *SNF3* expression to glucose is a good adaptation, since signalling by the Snf3 sensor would be fully stimulated at glucose concentrations at which Rgt2 signalling is appropriate. Conversely, the low affinity of Rgt2 would make this sensor insensitive to the low glucose concentrations at which Snf3 signalling is appropriate.

Signalling by Snf3 and Rgt2 is dependent upon long C-terminal tails of the proteins (approximately 300 and 200 amino acids, respectively) that are located on the cytoplasmic face of the plasma membrane. In particular, three repeat sequences of 25 amino acids are necessary for signalling; the tail of Snf3 has two of these repeats and the tail of Rgt2 has one (Coons *et al.*, 1997; Özcan *et al.*, 1998; Vagnoli *et al.*, 1998). The other members of the *HXT* family in *S. cerevisiae* do not have long C-terminal tails or sequences similar to the repeat sequences. These motifs presumably interact with the subsequent components of the signal transduction pathway to convey the signal that glucose is bound to the sensor. The cytoplasmic components that interact with the sensor tails are Std1 and Mth1. These homologous proteins are required for proper sensor-dependent regulation of *HXT* gene transcription. They display dual localization at the cell surface and in the nucleus, and have been shown by two-hybrid analysis to physically interact with the tails of Snf3 and Rgt2, but their detailed mechanism of action remains unclear (Özcan *et al.*, 1993; Gamo *et al.*, 1994; Schmidt *et al.*, 1999; Lafuente *et al.*, 2000; Schulte *et al.*, 2000).

Another important component in the Snf3/Rgt2 signalling pathway is the Grr1 protein (Özcan *et al.*, 1994; Vallier *et al.*, 1994). Grr1 is a substrate recognition subunit that characterizes one class of SCF complex. SCF complexes ubiquitinate specific proteins, which targets them for proteolysis (Li and Johnston, 1997). The ultimate target of Grr1 in the Snf3/Rgt2 pathway is the transcription factor Rgt1, though it is not clear whether the Grr1-type SCF complex ubiquitinates Rgt1 directly (Özcan *et al.*, 1996b). Rgt1 is a zinc-finger DNA binding protein with binding sites in a number of *HXT* gene promoters. It acts in concert with the transcriptional co-repressors Ssn6 and Tup1 to repress high-affinity glucose transporters such as *HXT2* and *HXT7* when glucose is absent from the environment. At high glucose concentrations Rgt1 is converted into an activator of transcription, and induces expression of *HXT1* under these conditions. At low glucose concentrations the regulatory activities of Rgt1 are neutralized. Grr1 is involved in interconverting Rgt1 among these three states, suggesting that Grr1 somehow integrates signals from Snf3 about low glucose concentrations and from Rgt2 about high glucose concentrations to differentially regulate *HXT* transcription via a single species of DNA-binding protein (Özcan and Johnston, 1996a,b).

The evolution of transporter-like sensors extends beyond *S. cerevisiae* and glucose sensing. Nutrient sensors related to Snf3 and Rgt2 have been identified in other ascomycetes, but have not been found outside the fungal kingdom (Madi *et al.*, 1997; Kruckeberg *et al.*, 1998;

Betina *et al.*, 2001). Moreover, a similar sensor system for amino acids has been recognized in *S. cerevisiae*; in this case, an amino acid transporter orthologue, Ssy1, acts as a broad-specificity sensor of amino acids and requires an extended N-terminal tail for signal transduction (Didion *et al.*, 1998; Jorgensen *et al.*, 1998; Klasson *et al.*, 1999) (see Chapter 4). Remarkably, Grr1 is also a component of the Ssy1 signalling pathway, and affects the transactivation activity of the Uga35 zinc-finger DNA-binding protein (Iraqui *et al.*, 1999). Ssy1 regulates transcription of other genes involved in nitrogen metabolism as well, including genes encoding peptide transporters (Hauser *et al.*, 2001) and genes involved in methionine biosynthesis (Kodama *et al.*, 2002).

Transporter trafficking

Integral membrane proteins are delivered to the plasma membrane by the secretory pathway. *HXT* mRNAs are translated into transporter proteins by endoplasmic reticulum (ER)-associated ribosomes, with concomitant insertion of the hydrophobic domains of the polypeptides into the ER membrane. The transporters are included in the cargo of vesicles that deliver the proteins to the Golgi apparatus, then through the compartments of the Golgi, and finally to the plasma membrane. Exocytosis results in incorporation of the membrane protein cargo of the vesicles into the plasma membrane (Tschopp *et al.*, 1984; Brada and Schekman, 1988). Expression of galactose and high-affinity glucose transport activity by yeast cells uses this pathway, as shown by analysis of transport activity in secretory pathway mutants (Tschopp *et al.*, 1986; Bisson, 1988): temperature-sensitive mutants in *SEC1* (encoding a SNARE-binding protein required for exocytosis), *SEC4* (encoding a Ras-like GTP-binding protein required for exocytosis), *SEC7* (encoding a guanine-nucleotide exchange factor involved in ER-to-Golgi and intra-Golgi vesicle transport), *SEC14* (encoding a phosphatidyl inositol transfer protein required for vesicle budding from the Golgi), and *SEC17* (encoding a SNAP protein involved in vesicle traffic) were unable to express these transport activities under inducing conditions at the restrictive temperature. Furthermore, expression of high-affinity glucose transport activity was impaired in a *sec18* mutant (encoding an ATPase involved in vesicle fusion). Surveillance of Hxt2::GFP fusion proteins in a temperature-sensitive *sec6* mutant (impaired in fusion of post-Golgi vesicles with the plasma membrane) revealed that the protein accumulated in intracellular structures at the restrictive temperature (Kruckeberg *et al.*, 1999).

The *GSF2* gene was identified as being required for proper glucose repression of the *SUC2* gene encoding invertase; further analysis has shown that it is required for efficient release of some sugar transporters (e.g. Hxt1 and Gal2) but not others (e.g. Hxt2 and maltose transporters) from the ER; *GSF2* encodes a 46 kDa protein localized in the ER membrane (Sherwood and Carlson, 1999). The role of Gsf2 in protein trafficking, as well as the basis for its transport protein specificity, remain unclear.

Hexose transport activity is inactivated by various nutritional conditions. In particular, nitrogen starvation or the arrest of protein synthesis (e.g. by treatment with cycloheximide) lead to an accelerated loss of hexose transport capacity. Inactivation is further accelerated by provision of fermentable carbon sources (Busturia and Lagunas, 1986). The inactivation process is initiated by ubiquitination of the transport proteins, followed by their removal from the membrane by endocytosis. The fate of the ubiquitinated proteins is proteolysis in the vacuole, not the proteasome as is usual for many cellular proteins after ubiquitination (Riballo and Lagunas, 1994a,b; Krampe *et al.*, 1998; Kruckeberg *et al.*, 1999). Vacuolar degradation after ubiquitination has been observed for a number of other plasma membrane proteins as well (Hicke, 1997, 1999; Rotin *et al.*, 2000). The details of this pathway have been characterized by analysis of the fate of Gal2, Hxt2, and Hxt7 in various mutant strains (Horak and Wolf, 1997, 2001; Krampe *et al.*, 1998; Kruckeberg *et al.*, 1999; Horak *et al.*, 2002; Krampe and Boles, 2002). Thus, the role of ubiquitination has been revealed by a prolonged half-life of hexose transport proteins in strains with mutations in the *NPI1* ubiquitin-protein ligase and the *NPI2* ubiquitin-protein hydrolase genes. Mutations in endocytic factors encoded by *END3, END4, ACT1*, and *DID4* (*REN1*) also prolong the half-life of hexose transporters in the plasma membrane. The role of vacuolar rather than proteasomal proteolysis has been demonstrated with *pep4* mutants (encoding the central vacuolar protease) that show extended transporter half-lives, and with *pre1, pre2, cim3*, and *cim5* mutants (encoding subunits of the proteasome) that show wild-type (i.e. short) half-lives of transporter proteins and transport activity upon imposition of inactivating conditions.

Later steps in the autophagy pathway, encoded by *APG1* (a protein kinase) and *AUT2* (a cysteine peptidase), are also involved in targeting hexose transporters to the vacuole in response to starvation (Krampe and Boles, 2002). Autophagy is well known to be induced by starvation to bring about non-specific degradation of cytoplasmic components (soluble proteins, RNAs, etc.) in the vacuole (Abeliovich and Klionsky, 2001). The fact that starvation-induced trafficking of hexose transporters involves the autophagy pathway suggests that inactivation of transporters under these conditions may differ mechanistically from the normal turnover of the proteins during balanced growth when the autophagy pathway is not active.

The signal that initiates hexose transporter inactivation has been explored for Gal2, using various mutant strains as well as sugar analogues. Hexokinase II (encoded by the *HXK2* gene) is involved (while the other hexose-phosphorylating enzymes, hexokinase I and glucokinase, are not). Importantly, inactivation requires hexose phosphorylation rather than mere presence of Hxk2 protein, as shown by the inability of non-phosphorylatable sugars such as fucose or 6-deoxyglucose to stimulate inactivation. The inactivation signal does not seem to require the cyclic AMP protein kinase pathway, or the Snf3/Rgt2/Rgt1 glucose-sensing pathway. Surprisingly, however, Grr1, which is a downstream component of the glucose sensing pathway, and Reg1, the regulatory subunit of the Glc7 protein phosphatase of the glucose repression pathway (see later), are required for proper inactivation of Gal2 (Horak *et al.*, 2002).

Maltose transport and utilization

Maltose is an important carbon source for yeast during brewing and bread-making, and strain selection for these industries has apparently contributed to the diversity of maltose utilizing genotypes found in *S. cerevisiae* (ten Berge *et al.*, 1974; Needleman and Michels, 1983; Jespersen *et al.*, 1999). Metabolism of maltose is initiated by transport of the carbohydrate across the plasma membrane, followed by hydrolysis by maltase (α-glucosidase). The resulting glucose is catabolized by the glycolytic pathway. The structural genes for maltose utilization are clustered in multi-genic loci. Five such loci have been described (*MAL1, MAL2, MAL3, MAL4*, and *MAL6*; in general, *MALX*); the loci include genes for a maltose transporter (*MALX1*), maltase (*MALX2*), and a transcriptional activator of the transporter and maltase genes (*MALX3*). Different strains have one or more (active) loci (Needleman and Michels, 1983; Cohen *et al.*, 1984, 1985; Michels and Needleman, 1984; Needleman *et al.*,

1984; Charron *et al.*, 1986). These loci occur at the telomeres of their respective chromosomes, and telomeric duplication and translocation have presumably played a role in the diversification of the *MAL* loci (Charron *et al.*, 1989; Chow *et al.*, 1989; Naumov *et al.*, 1994).

The transport proteins encoded by *MALX1* (typified by *MAL61* and *MAL11*, the first members of the *MALX1* family to be cloned and sequenced (Cheng and Michels, 1989, 1991; Yao *et al.*, 1989)) are orthologous to other members of the sugar porter family. Maltose transport exploits the electrochemical gradient of the plasma membrane, and is co-transported with protons in a 1:1 stoichiometry (Serrano, 1977; Van Leeuwen *et al.*, 1992); the Km for maltose is 1–5 mM (Görts, 1969; Cheng and Michels, 1991; Van Leeuwen *et al.*, 1992). A low-affinity system with a Km of 70–80 mM has been described as well (Cheng and Michels, 1991; Crumplen *et al.*, 1996) but may be artefactual (Benito and Lagunas, 1992). Maltose–proton symport allows accumulation of the carbohydrate inside the cell, and this may be necessary for its efficient metabolism, since the Km of maltase for maltose is ~20 mM (Matsusaka *et al.*, 1977; Needleman *et al.*, 1978; Tabata *et al.*, 1984). Studies with isolated membrane vesicles containing maltose transporters have shown that non-concentrative maltose transport is also possible in the absence of an electrochemical gradient, and that maltose transport is in principle reversible (van der Rest *et al.*, 1995a).

Maltose transport is inactivated by various treatments such as addition of glucose to the growth medium or inhibition of protein synthesis by nitrogen starvation or treatment with protein synthesis inhibitors (Robertson and Halvorson, 1957; Görts, 1969; Cheng and Michels, 1991). Inactivation by inhibition of protein synthesis generally requires the availability of a fermentable carbon source; indeed, maltose itself can promote the inactivation of maltose transport activity under nitrogen starvation conditions, if the strain is able to hydrolyze the maltose to glucose (Penalver *et al.*, 1998; Jiang *et al.*, 2000a; Brondijk *et al.*, 2001). Non-fermentable carbon sources such as ethanol do not promote inactivation of transport activity in resting cells (Lucero *et al.*, 1993), although both maltose transport activity and maltase are inactivated during sporulation (which is induced by nitrogen starvation and provision of the non-fermentable carbon source acetate) (Ferreira *et al.*, 2000). Inactivation of transport activity occurs concomitantly with loss of transporter protein (Lucero *et al.*, 1993; Medintz *et al.*, 1996). The inactivation process can be divided into two components: loss of transport activity and proteolytic breakdown of the transporter (Peinado and Loureiro, 1986; Medintz *et al.*, 1996; Brondijk *et al.*, 2001). These two processes are under separate, but overlapping, genetic control. Both require the Glc7-Reg1/Reg2 protein phosphatase type-1, while proteolysis *per se* is also promoted via the action of Rgt2 and Grr1 (Jiang *et al.*, 1997, 2000a,b). Inactivation of transport activity requires glucose transport and hexokinase activity, and may be signalled by the levels of intracellular trehalose or trehalose-6-phosphate (Jiang *et al.*, 2000a; Brondijk *et al.*, 2001). Studies with mutant strains have shown that proteolysis occurs in the vacuole following mono-ubiquitination and endocytosis of the transporter (Lucero *et al.*, 1993; Riballo *et al.*, 1995; Medintz *et al.*, 1996, 1998; Lucero and Lagunas, 1997). Phosphorylation of the transporter and targeting of a PEST sequence (Rechsteiner and Rogers, 1996) are required for proteolysis and for inactivation of transport activity (Brondijk *et al.*, 1998; Medintz *et al.*, 2000).

AGT1 encodes an α-glucoside transporter with a broader substrate range than the Malx1 proteins. In addition to maltose and turanose, Agt1 is able to transport isomaltose, α-methylglucoside, maltotriose, trehalose, sucrose, palatinose, and melezitose. It carries out proton-coupled active transport of its substrates. *AGT1* occurs at the *MAL1* locus of some strains, apparently replacing *MAL11* via a recombinational event (Han *et al.*, 1995; Jespersen *et al.*, 1999; Stambuk *et al.*, 1999). The open reading frames (ORFs) YDL247W and YJR160C have also recently been characterized as α-glucoside permeases capable of transporting maltose, maltotriose, α-methylglucoside, and turanose (Day *et al.*, 2002). YDL247W has been named '*MPH2*' and YJR160C has been named '*MPH3*' (Day *et al.*, 2002). Hence, yeast has five maltose permeases with strongly overlapping, but distinct functions.

Maltase (α-glucosidase, EC 3.2.1.20), encoded by *MALX2* genes, is a cytoplasmic enzyme that catalyses hydrolysis of terminal, non-reducing 1,4-linked D-glucose residues with release of D-glucose. In practice, it is able to hydrolyze various naturally occurring saccharides such as maltose, maltotriose, α-methylglucoside and sucrose. It is required for fermentation of most of these sugars, and can permit growth on sucrose in strains lacking invertase, if those strains constitutively express maltase and the Agt1 α-glucoside transporter (Khan *et al.*, 1973; Stambuk *et al.*, 1999).

Basal levels of *MALX1* and *MALX2* transcription occur during growth on non-repressing carbon sources such as ethanol and galactose; transcription is induced by maltose, and is repressed in the presence of glucose (Federoff *et al.*, 1983; Needleman *et al.*, 1984). Induction is mediated by the transcriptional activator protein Malx3 (Chang *et al.*, 1988). Mal63 is a zinc-finger protein (Kim and Michels, 1988; Sollitti and Marmur, 1988) that binds to an upstream activation sequence (UAS) between the divergently transcribed *MAL61* and *MAL62* genes (Hong and Marmur, 1987; Ni and Needleman, 1990; Levine *et al.*, 1992; Yao *et al.*, 1994). Maltose transport is required for Mal63-dependent transcriptional activation (Charron *et al.*, 1986). The basal level of *MALX1* expression provides sufficient transport activity for uptake of the inducer (Hu *et al.*, 2000). Various constitutive (i.e. maltose-independent) alleles of *MALX3* have been identified, both in laboratory and industrial strains. The mutations are clustered in the carboxy-terminal portion of the protein which is presumed to have a maltose-responsive negative regulatory role (Gibson *et al.*, 1997; Higgins *et al.*, 1999). Regulation of *MAL* gene expression is also affected by naturally occurring mutations in the shared promoter of the *MALX1* and *MALX2* genes (Wang and Needleman, 1996; Bell *et al.*, 1997). Glucose represses transcription of *MAL61, MAL62*, and *MAL63* via the canonical glucose repression pathway (see later), due to the action of the Mig1 transcriptional repressor which is able to bind to the promoters of all three genes (Hu *et al.*, 1995; Klein *et al.*, 1996; Wang and Needleman, 1996). Glucose also inhibits *MAL* gene expression in a Mig1-independent manner, by interfering with the ability of the cell to sense maltose (Hu *et al.*, 2000).

Sucrose utilization

Invertase (β-fructofuranosidase, EC 3.2.1.26) catalyses the hydrolysis of sucrose to fructose and glucose (and of raffinose to fructose and melibiose; melibiose can be further hydrolyzed to glucose and galactose by melibiase (see later)). Yeast expresses two forms of invertase (Moreno *et al.*, 1975); one form is localized in the cytoplasm and requires uptake of sucrose (e.g. via the Agt1 α-glucoside transporter; Stambuk *et al.*, 1999) to function. The other form is localized in the periplasmic space between the plasma membrane and the cell wall. The hexoses produced by the activity of the periplasmic form are taken up by hexose transporters (Bisson *et al.*, 1987). Net expression of invertase is repressed by glucose (Elorza *et al.*, 1977a,b), but this is specific for the periplasmic form. The two forms result from different sites of transcriptional initiation at the *SUC* genes (Carlson and Botstein, 1982; Carlson *et al.*,

1983). The periplasmic form includes an amino-terminal leader peptide that directs its entry into the secretory pathway (Taussig and Carlson, 1983). During secretion the signal sequence is proteolytically removed and the mature protein becomes glycosylated (Perlman and Halvorson, 1981). Glycosylated forms of the enzyme are more stable to environmental insults than the non-glycosylated cytoplasmic form, but they have otherwise similar catalytic properties (Chu *et al.*, 1985; Williams *et al.*, 1985; Kern *et al.*, 1992). Transcription of the two forms is regulated independently (Perlman *et al.*, 1984): in general, the cytoplasmic form is expressed constitutively, while transcription of the periplasmic form is repressed by high concentrations of glucose (Carlson and Botstein, 1982, 1983). This glucose repression is mediated by an UAS (Sarokin and Carlson, 1984, 1985; Hohmann and Gozalbo, 1988) and depends on the transcriptional repressors Mig1 (Nehlin and Ronne, 1990), Mig2 (Lutfiyya and Johnston, 1996), and Nrg1 (Vyas *et al.*, 2001; Zhou and Winston, 2001). Transcription of the secreted form of invertase is also induced by low levels of glucose (Özcan *et al.*, 1997). In addition to transcriptional control, invertase mRNA degradation is also accelerated at high glucose concentration (Cereghino and Scheffler, 1996).

Invertase is encoded by the *SUC* gene family, made up of six members (*SUC1–SUC5* and *SUC7*). Strains typically have only a subset of these genes (with *SUC2* being the most common among those strains tested), and strains lacking sucrose fermentation ability may have *suc* pseudogenes or silent genes, or lack *SUC* genetic information altogether (Carlson and Botstein, 1983; Naumov *et al.*, 1996a). Most *SUC* genes (with the exception of *SUC2*) occur adjacent to chromosomal telomeres (Carlson *et al.*, 1985). All *SUC* genes are similar in their sequence and structure (Carlson *et al.*, 1985), though they differ in their level of expression (Hohmann and Zimmermann, 1986; del Castillo Agudo *et al.*, 1992).

Galactose utilization

After uptake of galactose via Gal2, it is converted into glucose-6-phosphate by the Leloir pathway enzymes. First, galactose is phosphorylated to galactose-1-phosphate by galactokinase (encoded by *GAL1*); then the glucose group of UDP-glucose is exchanged with the galactose group by galactose-1-P uridylyltransferase (encoded by *GAL7*) to yield UDP-galactose and glucose-1-phosphate. UDP-galactose 4-epimerase (encoded by *GAL10*) converts UDP-galactose to UDP-glucose, which is then available for modification of another galactose-1-phosphate molecule. Glucose-1-phosphate produced in the cycle is converted by phosphoglucomutase (the major isozyme of which is encoded by *PGM2*) to glucose-6-phosphate, an intermediate in the glycolytic pathway (Leloir, 1964). Many yeast strains growing on galactose display a higher respiratory capacity than cells growing on glucose. As a result, aerobic galactose cultures yield more biomass and less fermentative products than glucose cultures (Lagunas, 1976; Ostergaard *et al.*, 2000). This is a consequence of increased expression of respiratory enzymes in galactose-grown cells (Herrero *et al.*, 1985; Lodi *et al.*, 1991; Rodriguez and Gancedo, 1999).

GAL1, *GAL7*, and *GAL10* are tightly linked on chromosome II, while *GAL2* is on chromosome XII. *GAL1* and *GAL10* are transcribed divergently from a common promoter and *GAL7* is downstream of *GAL10* (Douglas and Condie, 1954; Douglas and Hawthorne, 1964). All four of these *GAL* genes are induced by galactose, and repressed by glucose, and all four have binding sites for the Gal4 transcriptional activator in their promoter regions (Bram *et al.*, 1986; Huibregtse *et al.*, 1993). Transcription of *PGM2* and some of the genes encoding other regulatory factors in the galactose-response pathway (e.g. *GAL3* and *GAL80*) is also activated by Gal4 (Bajwa *et al.*, 1988; Oh and Hopper, 1990), whereas *GAL4* transcription is only affected by glucose repression (see later; Griggs and Johnston, 1991; Nehlin *et al.*, 1991). Galactose induction is mediated by the Gal4 transcriptional activator in the following manner. The Gal4 zinc-finger protein binds to UAS sequences in the promoters of its target genes in cells growing on both non-inducing and inducing carbon sources (Giniger *et al.*, 1985; Selleck and Majors, 1987). In the absence of intracellular galactose, the Gal80 protein binds to and inactivates Gal4 (Lue *et al.*, 1987; Ma and Ptashne, 1987). When galactose is present it binds with ATP to the Gal3 protein, which then interacts with Gal80 and overcomes its inactivation of Gal4, resulting in transcription of the target genes (Platt and Reece, 1998; Sil *et al.*, 1999). Gal3 is homologous to the Gal1 galactokinase, and indeed Gal1 is able to partially complement the defect in galactose induction of *gal3* mutants (Bajwa *et al.*, 1988; Bhat and Hopper, 1992; Platt *et al.*, 2000). In the presence of glucose, galactose induction is impaired, due in part to diminished levels of Gal4 protein and in part to Mig1-dependent repression of *GAL* target genes (Lamphier and Ptashne, 1992; Johnston *et al.*, 1994; Frolova *et al.*, 1999).

Melibiose utilization

Melibiase (α-galactosidase, EC 3.2.1.22) hydrolyzes melibiose to glucose and galactose. The enzyme is secreted into the periplasm as a glycoprotein (Lazo *et al.*, 1977). The products of the reaction it catalyses are substrates for hexose transport proteins. Melibiase is encoded by the *MEL* gene family, of which *MEL1* is the prototype (Buckholz and Adams, 1981; Liljestrom, 1985; Sumner-Smith *et al.*, 1985). *MEL1* is co-ordinately regulated at the transcriptional level with the *GAL* genes under control of Gal3, Gal4, and Gal80, being induced by galactose and repressed by glucose (Kew and Douglas, 1976; Lazo *et al.*, 1981; Post-Beittenmiller *et al.*, 1984). The ability to utilize melibiose varies among yeast strains; *mel(0)* strains lack sequences coding for melibiase (Post-Beittenmiller *et al.*, 1984; Lyness *et al.*, 1993), while Mel^+ strains have one or more of the eight *MEL* loci (*MEL1–MEL3, MEL5,* and *MEL8–MEL11*) at chromosomal telomeres (Naumov *et al.*, 1990, 1991, 1996b; Turakainen *et al.*, 1994).

3.3 Glycolysis

In glycolysis glucose and other hexoses are catabolized to pyruvate with the concomitant formation of ATP and NADH. No net oxidation occurs in the process, since the oxidation of some pathway intermediates is balanced by reduction of NAD^+, which is restored by other metabolic reactions such as reduction of acetaldehyde to ethanol. Glycolysis consists of ten enzymatic reactions. In the first five, ATP is consumed in the phosphorylation of the hexose, with subsequent hydrolysis to yield two triose phosphates. In the subsequent five steps, the potential energy of these intermediates is used to synthesize ATP by substrate-level phosphorylation. The pyruvate formed can be fermented to ethanol and carbon dioxide. Alternatively, it can be oxidized via the citric acid cycle and electron transport chain to carbon dioxide and water with the formation of more ATP. Glycolysis plays an anabolic role as well: a number of glycolytic intermediates are utilized by biosynthetic pathways for production of amino acids, nucleotides, and lipids.

In the first step of glycolysis, hexokinase transfers a phosphate group from Mg-ATP to a hexose molecule; glucose, fructose, and mannose are substrates of hexokinase I and hexokinase II (EC 2.7.1.1, encoded by *HXK1* and *HXK2*, respectively), while glucokinase (EC 2.7.1.2, encoded by *GLK1*) only phosphorylates glucose and mannose (Lobo and Maitra, 1977; Walsh *et al.*, 1983). The sequences of *HXK1* and *HXK2* are highly similar, while the sequence of *GLK1* is divergent except in the region of the active site (Fröhlich *et al.*, 1985; Kopetzki *et al.*, 1985; Stachelek *et al.*, 1986; Albig and Entian, 1988). The affinities of the enzymes for glucose are ~0.1 mM (hexokinase I), 0.25 mM (hexokinase 2), and 0.03 mM (glucokinase) (reviewed in Entian, 1997). Hexokinase activity is inhibited by high physiological concentrations of ATP (Kopetzki and Entian, 1985; Moreno *et al.*, 1986; Larsson *et al.*, 2000) and trehalose-6-phosphate (Blazquez *et al.*, 1993; Noubhani *et al.*, 2000), and these effects may influence the rate of entry of glucose into the glycolytic pathway (see later). The *HXK1* and *GLK1* genes are transcriptionally repressed in high-glucose media, while *HXK2* expression is induced by glucose (Herrero *et al.*, 1995, 1999; De Winde *et al.*, 1996; Martinez-Campa *et al.*, 1996; Rodriguez *et al.*, 2001). In addition to their catalytic roles, the hexose kinases (particularly hexokinase II) have regulatory effects (see later).

Glucose-6-phosphate and mannose-6-phosphate are isomerized to fructose-6-phosphate by glucose-6-phosphate isomerase (EC 5.3.1.9, encoded by *PGI1*; Maitra and Lobo, 1977a; Aguilera and Zimmermann, 1986; Green *et al.*, 1988; Tekamp-Olson *et al.*, 1988) and mannose-6-phosphate isomerase (EC 5.3.1.8, encoded by *PMI40*, Herrera *et al.*, 1976), respectively. *PGI1* expression is essentially constitutive with respect to carbon source (Green *et al.*, 1988; Tekamp-Olson *et al.*, 1988).

Fructose-6-phosphate is phosphorylated to fructose-1,6-bisphosphate by phosphofructokinase (EC 2.7.1.11) at the expense of ATP; this reaction is exergonic and essentially irreversible. The yeast enzyme is an octamer composed of four molecules each of an α and a β subunit (encoded by *PFK1* and *PFK2*, respectively; the two subunits are homologous to one another; Clifton *et al.*, 1978; Heinisch, 1986; Heinisch, *et al.*, 1989). Enzyme activity is under complex allosteric control: it is inhibited by ATP and activated by AMP and fructose-2,6-bisphosphate (Avigad, 1981; Bartrons *et al.*, 1982; reviewed in Kopperschläger and Heinisch, 1997). According to Heinisch *et al.*, (1991a) the genes encoding the two subunits are expressed constitutively with respect to carbon source.

Fructose-1,6-bisphosphate is cleaved to dihydroxyacetone phosphate and glyceraldehyde-3-phosphate by aldolase (EC 4.1.2.13, encoded by *FBA1*; Lobo, 1984; Schwelberger *et al.*, 1989). *FBA1* is transcribed constitutively with respect to carbon source (Compagno *et al.*, 1991). The products of the aldolase reaction are interconverted by triose-phosphate isomerase (EC 5.3.1.1, encoded by *TPI1*; Ciriacy and Breitenbach, 1979; Alber and Kawasaki, 1982). Transcription of the *TPI1* gene is essentially constitutive (Scott *et al.*, 1990; Scott and Baker, 1993). The isomerization of the triose phosphates is important because only glyceraldehyde-3-phosphate is used by the subsequent step of glycolysis, whereas dihydroxyacetone phosphate is a precursor of glycerol. Mutation of *TPI1* results in accumulation of dihydroxyacetone phosphate and increased glycerol production at the expense of ethanol production (Compagno *et al.*, 2001). The *tpi1* mutant cannot grow on glucose as sole carbon source, but the addition of ethanol or acetate to glucose-containing media, or the use of raffinose or galactose, permit growth (Compagno *et al.*, 2001).

Glycerol formation from dihydroxyacetone phosphate occurs in two steps. The first step (reduction of dihydroxyacetone phosphate to glycerol-3-phosphate) is catalysed by NAD-dependent glycerol-3-phosphate dehydrogenase. Two isoenzymes exist. *GPD1* encodes the major form which is used in aerobic conditions and is induced by high osmolarity to provide glycerol as a compatible solute (the 'HOG' pathway). Gpd2 is used anaerobically. The second step (dephosphorylation of glycerol-3-phosphate) also involves two isoenzymes: Gpp1 and Gpp2. *GPP1* and *GPP2* are both induced by hyperosmotic stress but are differently regulated in other respects. For example, the expression of *GPP2* is unaffected by transfer to anaerobiosis whereas *GPP1* is transiently induced and *gpp1* mutants grow poorly without oxygen (Påhlman *et al.*, 2001). In addition, *GPP1* expression is much reduced in strains lacking protein kinase A (Påhlman *et al.*, 2001).

The sixth reaction of glycolysis is catalysed by glyceraldehyde-3-phosphate dehydrogenase (EC 1.2.1.12), encoded by the *TDH1, TDH2,* and *TDH3* genes (Holland and Holland, 1979, 1980). The genes share high sequence homology; the major isozyme is encoded by *TDH3* (McAlister and Holland, 1985). The three genes are coordinately expressed, and show a twofold stimulation in cells growing in glucose compared to ethanol (McAlister and Holland, 1985). The oxidation of glyceraldehyde-3-phosphate results in reduction of NAD^+ to NADH, and in the formation of 1,3-bisphosphoglycerate, which contains an energy-rich acyl-phosphate group. The potential energy of this group is used to drive the synthesis of ATP in the next step of glycolysis, catalysed by phosphoglycerate kinase (EC 2.7.2.3, encoded by *PGK1*; Hitzeman *et al.*, 1980, 1982). The *PGK1* gene is well known as one that is expressed on all carbon sources but is up-regulated approximately fivefold at the transcriptional level in glucose compared to gluconeogenic carbon sources (Chambers *et al.*, 1989). The other product of the phosphoglycerate kinase reaction, 3-phosphoglycerate, is the substrate for phosphoglycerate mutase (EC 5.4.2.1 formerly EC 2.7.5.3), encoded by *GPM1* (Rodicio and Heinisch, 1987; Heinisch *et al.*, 1991b); *GPM2* and *GPM3* are non-functional homologues (Heinisch *et al.*, 1998). A cofactor of the reaction is 2,3-bisphosphoglycerate; its synthesis might also be catalysed by phosphoglycerate mutase (Fothergill-Gilmore and Watson, 1990). The *GPM1* gene is slightly induced in glucose medium (Moore *et al.*, 1991; Rodicio *et al.*, 1993). The product of the phosphoglycerate mutase reaction, 2-phosphoglycerate, is subsequently converted to phospho*enol*pyruvate, which contains a high-energy phosphoryl bond. This conversion is carried out by the enzyme enolase (EC 4.2.1.11, encoded by *ENO1* and *ENO2*; Holland *et al.*, 1981). *ENO1* is expressed constitutively, while *ENO2* is induced by glucose (McAlister and Holland, 1982). As a result, Eno1 is the predominant isozyme under gluconeogenic conditions, while Eno2 is the major fermentative isozyme. The two isozymes have corresponding differences in kinetic regulation: Eno1 is inhibited by the fermentative substrate 2-phosphoglycerate, and Eno2 is inhibited by the gluconeogenic substrate phospho*enol*pyruvate (Entian *et al.*, 1987).

The energy in phospho*enol*pyruvate is used to transfer the phosphate group to ADP; this second substrate-level phosphorylation, catalysed by pyruvate kinase (EC 2.7.1.40, encoded by *PYK1*; Maitra and Lobo, 1977b; Kawasaki and Fraenkel, 1982; Burke *et al.*, 1983), yields pyruvate as well as ATP. The reaction is exergonic and essentially irreversible; the enzyme is under complex allosteric regulation, with fructose-1,6-bisphosphate, phospho*enol*pyruvate, and ADP as the important activators and citric acid and ATP as the important inhibitors (Hess *et al.*, 1966; Haeckel *et al.*, 1968; Morris *et al.*, 1986). Pyruvate kinase activity is elevated in glucose-grown cells due to transcriptional activation of the *PYK1* gene (Maitra and Lobo, 1971; Nishizawa *et al.*, 1989; Moore *et al.*, 1991). The other isozyme of pyruvate kinase (Pyk2) is insensitive to fructose-1,6-bisphosphate and its expression is repressed by glucose (Boles *et al.*, 1997). Recently, both Pyk1 and Pyk2 have been shown to be *in vitro* and *in vivo* substrates of (cAMP dependent) protein kinase A ('PKA') (Portela *et al.*, 2002). Thus it seems that PKA regulates glycolysis by activating pyruvate kinase by phosphorylation and its consequent increased sensitivity to small changes in the concentration of the substrate phospho*enol*pyruvate or the activator fructose-1,6,-bisphosphate (Portela *et al.*, 2002).

Pyruvate is at a central branch-point in metabolism; it can either be fermented to ethanol or oxidized via the citric acid cycle. Both routes re-oxidize the NADH that is formed by glyceraldehyde-3-phosphate dehydrogenase, thus replenishing the NAD^+ pool. The oxidative route yields more ATP per mole of pyruvate, but it requires oxygen. Moreover, *S. cerevisiae* displays the Crabtree effect: even in the presence of oxygen, most of the pyruvate formed by glycolysis enters the fermentative route (van Dijken *et al.*, 1993). The fermentative route consists of two enzymes. The first enyzme, pyruvate decarboxylase (EC 4.1.1.1, isozymes of which are encoded by *PDC1, PDC5,* and *PDC6*; Schmitt and Zimmermann, 1982; Schmitt *et al.*, 1983; Seeboth *et al.*, 1990; Hohmann, 1991), converts pyruvate into acetaldehyde and CO_2; this is an irreversible reaction. The enzyme is allosterically activated by its substrate pyruvate (Schellenberger and Hubner, 1968; Hubner *et al.*, 1978; Lu *et al.*, 2000; Sergienko and Jordan, 2002). Pdc1 is the predominant isozyme form (80–90% of activity in wild-type cells), and Pdc5 the minor form expressed on glucose. The *PDC6* gene has been said to be generally silent (Hohmann and Cederberg, 1990; Hohmann, 1991; Flikweert *et al.*, 1996) however, its true role is not in glycolysis but in amino acid catabolism because Pdc1, Pdc5 and Pdc6 will accept the α-ketoacids which are intermediates in amino acid catabolism to 'fusel' alcohols (see Chapter 4 and Dickinson *et al.*, 1997; 1998; 2000; ter Schure *et al.*, 1998). Transcription of both *PDC1* and *PDC5* is higher in cells grown on glucose than on ethanol (Schmitt *et al.*, 1983; Entian *et al.*, 1984; Kellermann *et al.*, 1986; Seeboth *et al.*, 1990). This is due to induction mediated by the transcription factors Rap1 and Gcr1 (Kellermann and Hollenberg, 1988; Butler *et al.*, 1990; Liesen *et al.*, 1996), and to autoregulation of *PDC* transcription by pyruvate decarboxylase protein (Hohmann and Cederberg, 1990; Eberhardt *et al.*, 1999; Muller *et al.*, 1999).

In the second step of the fermentative route, the acetaldehyde formed by pyruvate decarboxylase is reduced to ethanol by ethanol dehydrogenase (EC 1.1.1.1) with the concomitant oxidation of NADH. The ethanol formed diffuses across the plasma membrane into the growth medium, and the NAD^+ formed is available for further glycolytic metabolism. *Saccharomyces cerevisiae* has 20 alcohol dehydrogenases. There are the putative aryl alcohol dehydrogenases encoded by the seven-member family of *AAD* genes all of which are of unknown function (Delneri *et al.*, 1999a,b) and 13 other alcohol dehydrogenases (see Table 3.2). BDH1 (YAL060w) has recently been shown to be involved in the formation of 2,3-butanediol (González *et al.*, 2000). On the basis of similarity, it is safe to assume that YAL061W can be called '*BDH2*'and that it has a similar activity. *SOR1* (YJR159W), '*SOR2*' (YDL246C) and *XDH1* (YLR070C) encode sugar alcohol dehydrogenases (Sarthy *et al.*, 1994; Richard *et al.*, 1999). '*CDH1*' (YCR105W) and *ADH6* (YMR318C) encode NADPH-dependent enzymes (Larroy *et al.*, 2000). This leaves six other alcohol dehydrogenases. Adh1 is the main cytosolic enzyme involved in the formation of ethanol during glycolysis. Adh2, which is also cytosolic, is the glucose-repressed isozyme which is needed for growth on ethanol. Adh3 is mitochondrial, it is induced on glucose and its role is not really understood. Recent work from Pronk's group suggests it is involved in a redox shuttle (Bakker *et al.*, 2000). Adh4 is present at only very low levels in most laboratory strains, but is plentiful in brewing strains. Adh5 was discovered by genome sequencing and its role is unexplained. Sfa1 is part of a bi-functional enzyme which has both glutathione-dependent formaldehyde dehydrogenase activity (the gene name comes from sensitive to formaldehyde, which is the phenotype of mutants affected in this gene); it is also described as being capable of catalysing the

Table 3.2 The alcohol dehydrogenases of *Saccharomyces cerevisiae*

Gene	*ORF*	*Functions*
ADH1	YOL086c	Main cytosolic Adh, forms ethanol in glycolysis
ADH2	YMR303c	Cytosolic, glucose repressed, used for growth on EtOH
ADH3	YMR083w	Mitochondrial, role as redox shuttle?
ADH4	YGL256w	Present at very low levels in most lab strains
ADH5	YBR145w	Unknown role
SFA1	YDL168w	Part of bifunctional formaldehyde dehydrogenase
BDH1	YAL060w	2,3-butanediol dehydrogenase
'*BDH2*'	YAL061w	51% identical to Bdh1
SOR1	YJR159w	Sorbitol dehydrogenase
'*SOR2*'	YDL246c	99% identical to Sor1
XDH1	YLR070c	xylitol dehydrogenase
'*CDH1*'	YCR105w	NADP(H)-dependent cinnamyl alcohol dHase
'*ADH6*'	YMR318c	NADP(H)-dependent cinnamyl alcohol dHase

breakdown of long-chain alcohols (Wehner *et al.*, 1993). *ADH1* transcription is up-regulated during growth on glucose (Denis *et al.*, 1983) due to activation by the Rap1 and Gcr1 transcription factors (Santangelo and Tornow, 1990). The expression of *ADH2*, and hence the utilization of ethanol, is controlled by the transcriptional acivator Adr1 (which also controls the expression of genes required for the utilization of glycerol and fatty acids). Adr1 binds directly to the promoter of *ADH2*. Snf1 (protein kinase) promotes the binding of Adr1 to chromatin in the absence of glucose and Glc7-Reg1 (protein phosphatase) represses chromatin binding in the presence of glucose (Young *et al.*, 2002).

Acetaldehyde formed by pyruvate decarboxylase is also essential during growth on glucose for synthesis of cytosolic acetyl-CoA by the so-called pyruvate dehydrogenase bypass (Flikweert *et al.*, 1996, 1997). In this bypass, acetaldehyde is oxidized to acetate by acetaldehyde dehydrogenase. [The yeast genome contains seven potential aldehyde dehydrogenases (Wang *et al.*, 1998) and the nomenclature of these has been confusing. However, only Ald4 (mitochondrial, encoded by YOR374W) and Ald6 (cytosolic, encoded by YPL061W) seem to have a role in acetaldehyde metabolism (Meaden *et al.*, 1997; Tessier *et al.*, 1999). Ald2 (encoded by YMR170C) and Ald3 (encoded by YMR169C), both of which are cytosolic, are involved in the conversion of 3-aminopropanal to β-alanine and are thus involved in the biosynthesis of CoA (White *et al.*, 2003). Ald5 (encoded by YER073W) is mitochondrial and of uncertain function. Msc7 (encoded by YHR039C) has a role in meiotic recombination (Thompson and Stahl, 1999). The cellular location and function of Ymr110Cp are both unknown.] Acetate is then condensed to acetyl-CoA at the expense of ATP by acetyl-CoA synthetase (EC 6.2.1.1). Two isozymes have been identified: that encoded by *ACS1* is repressed by glucose, while that encoded by *ACS2* is constitutively transcribed (Steensma *et al.*, 1993; Van den Berg and Steensma, 1995; Van den Berg *et al.*, 1996). The regulation of *ACS1* is complex. A carbon source responsive element (CSRE) and a binding site for Adr1 combine to mediate about 80% of derepression. The balance between repression by Ume6 and induction by Abf1 is also important (Kratzer and Schüller, 1997). Acs2 is the isoenzyme required for growth in anaerobic conditions as an *acs2* mutant is unable to grow on glucose but grows on ethanol or acetate. Acs1 is regarded as the isoenzyme required for growth in aerobic conditions as an *acs1Δ* mutant can grow on all carbon sources. *acs1 acs2* double mutants are inviable (Van den Berg and Steensma, 1995). The cytosolic acetyl-CoA produced by this bypass pathway is available for lipid (Chapter 6) and amino acid biosynthesis, and can also be transported into the mitochondria by the carnitine-acetyl transferase shuttle (reviewed in Pronk *et al.*, 1996). Acetyl CoA synthetase is, of course, essential for vegetative growth on acetate and for sporulation which is usually induced on acetate (Chapter 1).

The major oxidative route from pyruvate begins with its transport into the mitochondria, followed by conversion to acetyl-CoA and CO_2 with the reduction of NAD^+. The reaction is catalysed by the pyruvate dehydrogenase complex (EC 1.2.4.1), a large multi-enzyme complex that is located in the mitochondria. The complex is composed of five different subunits encoded by *PDA1* (E1α), *PDB1* (E1β), *LAT1* (E2), *LPD1* (E3) and *PDX1* (protein X) (reviewed in Steensma, 1997). The 'E3' component (lipoamide dehydrogenase) is also common to the related multi-enzyme complexes α-ketoglutarate dehydrogenase, branched-chain ketoacid dehydrogenase and glycine decarboxylase (Dickinson *et al.*, 1986; Ross *et al.*, 1988; Dickinson and Dawes, 1992; Sinclair *et al.*, 1993; Sinclair and Dawes, 1995). Hence, whilst *PDA1, PDB1, LAT1* and *PDX1* appear to be constitutively expressed (Wenzel *et al.*, 1995; Steensma, 1997), the transcription of *LPD1* is repressed on glucose and activated by the Hap transcriptional activation complex (Bowman *et al.*, 1992), and is regulated by Gcn4 in response to the quality of the nitrogen source (Zaman *et al.*, 1999).

The distribution of pyruvate between the fermentative and oxidative branches is influenced in part by the kinetic properties of pyruvate dehydrogenase and pyruvate decarboxylase. Pyruvate dehydrogenase has a tenfold higher affinity for pyruvate than pyruvate decarboxylase; however, its activity in the cell is relatively low. Therefore, low concentrations of pyruvate will favour its oxidation, whereas high concentrations of pyruvate (which occur when the flux through glycolysis is high, such as during growth on high glucose concentrations) will favour its fermentation. Further control of the distribution of pyruvate between fermentation and respiration is effected at the level of gene expression, since glucose induces expression of pyruvate decarboxylase, and represses expression of acetyl-CoA synthetase and the Lpd1 subunit of pyruvate dehydrogenase.

Pyruvate has a third possible fate since it is required for replenishment of the oxaloacetate pool of the citric acid cycle. This is achieved by the anaplerotic reaction catalysed by pyruvate carboxylase (EC 6.4.1.1, encoded by the homologous *PYC1* and *PYC2* genes (Morris *et al.*, 1987; Lim *et al.*, 1988; Stucka *et al.*, 1991; Walker *et al.*, 1991)) which converts pyruvate to oxaloacetate at the expense of ATP. The pyruvate carboxylase isozyme encoded by *PYC1* is the predominant form. Its expression is essentially constitutive with respect to carbon source whereas *PYC2* transcription is activated by glucose (Brewster *et al.*, 1994). In contrast, *PYC1* transcription is regulated according to the quality of the nitrogen source, a response not observed with *PYC2* (Huet *et al.*, 2000).

The glycolytic pathway is highly active in yeast: one yeast cell is able to consume approximately five times its weight in glucose during one hour of fermentation. However the capacity of the pathway (and the gluconeogenic reactions) needs to adapt to the dynamic nutritional status of the environment, and to be coordinated with other cellular functions such as anabolic pathways and the cell cycle. Therefore, a number of mechanisms have evolved that regulate the levels of enzyme activity and functional protein, which in turn participate in the control of flux through the pathway. The major mechanisms include transcriptional activation and repression, post-transcriptional effects on mRNA and protein abundance, post-translational covalent modifications of proteins, and reversible regulation of enzyme activities. The result of their concerted action affects the magnitude and direction of net flux, as well as the size of the pools of intermediate metabolites.

Regulation of enzyme activity by competitive inhibition or allosteric effects is most prominent for enzymes that catalyse reactions which are essentially irreversible under physiological conditions, that is, hexokinase, phosphofructokinase, and pyruvate kinase. The inhibition of hexokinase by trehalose-6-phosphate (Blazquez *et al.*, 1993) has been examined as a candidate regulator of glycolytic flux. Mutants in *TPS1* (which encodes the trehalose-6-phosphate synthase component of the trehalose synthase complex, EC 2.4.1.15; van de Poll *et al.*, 1974; Bell *et al.*, 1992, 1998; Van Aelst *et al.*, 1993), accumulate high levels of hexose phosphates and are unable to grow on glucose or fructose. This suggests that hexokinase activity must be restricted at the onset of glycolysis in order to prevent depletion of the ATP pool (Ernandes *et al.*, 1998; Bonini *et al.*, 2000; Noubhani *et al.*, 2000). However, regulatory effects of Tps1 and trehalose-6-phosphate on influx into glycolysis that are not simply due to hexokinase inhibition have been proposed (Hohmann *et al.*, 1993, 1996; Noubhani *et al.*, 2000).

Fructose-2,6-bisphosphate is a potent allosteric effector *in vitro* of a pair of yeast glycolytic and gluconeogenic enzymes: it is an activator of phosphofructokinase (Avigad, 1981; Bartrons *et al.*, 1982) and an inhibitor of fructose-1,6-bisphosphatase (Gancedo *et al.*, 1982). Fructose-2,6-bisphosphate is synthesized by 6-phosphofructo-2-kinase (phosphofructokinase-2; EC 2.7.1.105, encoded by *PFK26* and *PFK27*; Kretschmer and Fraenkel, 1991; Kretschmer *et al.*, 1991; Boles *et al.*, 1996) and hydrolyzed to fructose-6-phosphate by fructose-2,6-bisphosphatase (fructose-2,6-bisphosphate 2-phosphatase, EC 3.1.3.46, encoded by *FBP26* (Paravicini and Kretschmer, 1992) as well as non-specific phosphatases (e.g. Pho8; Francois *et al.*, 1988a; Plankert *et al.*, 1991). The levels of fructose-2,6-bisphosphate in the cell are regulated by the relative levels of the kinase and phosphatase activities. Pfk26 and Pfk27 activities are stimulated, and Fbp26 activity is inhibited, by cyclic AMP (Francois *et al.*, 1984, 1987, 1988b; Hofmann *et al.*, 1989) (as described later, cyclic AMP levels increase when glucose is added to non-fermenting cells). Furthermore, expression of the less-abundant Pfk27 enzyme is activated in glucose-grown cells. The intracellular concentration of fructose-2,6-bisphosphate is generally correlated with the rates of glycolytic flux (and inversely correlated with the occurrence of gluconeogenesis), although high concentrations can be found in cells growing on gluconeogenic carbon sources (Clifton and Fraenkel, 1983; Francois *et al.*, 1988b). Analysis of mutations in the *PFK26, PFK27*, and/or *FBP26* genes indicate that fructose-2,6-bisphosphate affects the pool size of glycolytic intermediates rather than the growth rate or glycolytic flux. Apparently fructose-2,6-bisphosphate is most important during the initiation of glycolysis when glucose is added to resting cells (Boles *et al.*, 1996; Muller *et al.*, 1997; Raamsdonk *et al.*, 2001). As such, regulation of the counter-poised phosphofructokinase and fructose bisphosphatase activities by fructose-2,6-bisphosphate can function to prevent futile cycling if both activities are present in the cell (Wilson and Bhattacharjee, 1986; Navas *et al.*, 1993; de Jong-Gubbels *et al.*, 1995).

The third mode of regulation of enzymatic activity considered here is the strong allosteric activation of pyruvate kinase by fructose-1,6-bisphosphate *in vitro*. This regulation can also function to prevent futile cycling, in this case between phospho*enol*pyruvate and pyruvate (Wilson and Bhattacharjee, 1986; Navas *et al.*, 1993; de Jong-Gubbels *et al.*, 1995). However, the importance of this regulation on metabolic fluxes *in vivo* has not been tested, despite the availability of pyruvate kinase mutants that are respectively insensitive to, or absolutely dependent on, fructose-1,6-bisphosphate (Maitra and Lobo, 1977b; Collins *et al.*, 1995).

Textbooks of biochemistry generally propose that, since they catalyse exergonic reactions that are subject to regulation by metabolic effectors, phosphofructokinase or pyruvate kinase are the rate-limiting or controlling steps of glycolysis (Lehninger *et al.*, 1993; Voet and Voet, 1995). However, over-expression of these enzymes does not increase steady-state flux from glucose to ethanol (Heinisch, 1986; Schaaff *et al.*, 1989; Rosenzweig, 1992; Hauf *et al.*, 2000; Smits *et al.*, 2000; Pearce *et al.*, 2001). As discussed earlier, MCA points out that the control of flux through a metabolic pathway is distributed over the various steps of that pathway; in the case of glycolysis, regulation of 'rate-limiting' enzyme activities will not result in proportional changes in flux through the pathway, since the control coefficients of these steps is less than 1 (Kacser and Burns, 1981; Fell, 1997; Teusink *et al.*, 2000). On the other hand, Pearce *et al.* (2001) observed that a strain in which pyruvate kinase activity was reduced to about 65% of wild type, glycolytic flux was reduced. They also observed that ethanol and glycerol production were decreased. Furthermore, the decrease in Pyk1 caused increased cycling of the tricarboxylic acid (TCA) cycle. This leads to the conclusion that Pyk1 exerts a significant level of control over both the rate and direction of carbon flux in yeast growing on glucose!

3.4 Gluconeogenesis

When yeast is growing on non-fermentable carbon sources it uses gluconeogenesis to synthesize glucose-6-phosphate, which is required in various biosynthetic reactions including the formation of ribose-5-phosphate (via the hexose phosphate pathway, see Section 3.5) for the synthesis of histidine and purine ribonucleotides and deoxyribonucleotides; and erythrose-4-phosphate (also via the hexose monophosphate pathway) for the synthesis of aromatic amino acids. The carbon skeletons to be built into glucose-6-phosphate originate from the TCA and glyoxylate cycles in the form of oxaloacetate which is converted into pyruvate by phospho*enol*pyruvate carboxykinase (the first enzyme unique to gluconeogenesis). The pyruvate is then converted by (the predominantly gluconeogenic) enolase 1 (see earlier text) to 2-phosphoglycerate and subsequently by 'reverse glycolysis' as far as fructose-1,6-bisphosphate when the second enzyme unique to gluconeogenesis (fructose-1, 6-bisphosphatase) converts the fructose-1,6-bisphosphate to fructose-6-phosphate. The fructose-6-phosphate is subsequently converted to glucose-6-phosphate by phosphoglucose isomerase. The regulatory controls over phopshofructokinase and pyruvate kinase, which do not operate during gluconeogenesis, have also been described earlier. Two other enzymes (isocitrate lyase and malate synthase) are required to make up for the extraction of oxaloacetate from the TCA cycle which would otherwise cease due to depletion of oxaloacetate. These latter two activities comprise the glyoxylate bypass and are described in Section 3.7.

Glycolysis by cells growing on a non-fermentable carbon source and gluconeogenesis when glucose is plentiful are both metabolically futile and need to be avoided. Phospho*enol*pyruvate carboxykinase is encoded by *PCK1* (Valdes-Hevia *et al.*, 1989). This gene's expression is strongly up-regulated in cells growing on gluconeogenic carbon sources (Mercado *et al.*, 1994). *FBP1* which encodes fructose-1,6-bisphosphatase (Sedivy and Fraenkel, 1985; Entian *et al.*, 1988), and *PCK1* are both repressed in cells growing on glucose (Sedivy and Fraenkel, 1985; Mercado *et al.*, 1991). This glucose repression is extremely sensitive as it is triggered at glucose concentrations of only 0.005% glucose (Yin *et al.*, 1996) and gluconeogenically grown cells rapidly inhibit and proteolytically degrade phospho*enol*pyruvate carboxykinase and fructose-1,6-bisphosphatase if glucose becomes available (Gancedo, 1971; Haarasilta and Oura, 1975; Müller and Holzer, 1981). The expression of *FBP1* and *PCK1* when glucose is present resulted in an increase in generation time of approximately

20% (Navas *et al.*, 1993): evidence (if any were needed) that yeast has evolved extremely tight controls over its central metabolic pathways.

3.5 Hexose monophosphate pathway

The hexose monophosphate pathway (or pentose phosphate pathway) is studied very little, a situation which belies its central importance to many aspects of metabolism (Figure 3.1). The pathway can be considered to start with the dual glycolytic/gluconeogenic intermediate glucose-6-phosphate which is converted to 6-phosphogluconolactone by glucose-6-phosphate dehydrogenase (Zwf1) (Nogae and Johnston, 1990; Thomas *et al.*, 1991). It was once believed that this enzyme was the limiting step of the pathway, but this has been shown to be not the case (Dickinson *et al.*, 1995) This reaction is an important source of NADPH which is required for a variety of reductive biosyntheses. The subsequent conversion of 6-phosphogluconate to ribulose-5-phosphate catalysed by 6-phosphogluconate dehydrogenase (major isoenzyme Gnd1, minor isoenzyme Gnd2; Lobo and Maitra, 1982) also yields NADPH. Ribulose-5-phosphate can be epimerized to xylulose-5-phosphate (catalysed by ribulose-5-phosphate 3-epimerase, Rpe1) or isomerized to ribose-5-phosphate by ribose-5-phosphate ketolisomerase, Rki1 (Miosga and Zimmermann, 1996). Xylulose-5-phosphate and ribose-5-phosphate can then react together in a reaction catalysed by transketolase to produce glyceraldehyde-3-phosphate and sedoheptulose-7-phosphate. There are two isoenzymes of transketolase: Tkl1 and Tkl2; the former appears to be the major isoenzyme (Fletcher *et al.*, 1992; Schaaff-Gerstenschläger *et al.*, 1993; Sundström *et al.*, 1993). The glyceraldehyde-3-phosphate can re-enter glycolysis/gluconeogenesis or can undergo further reaction with sedoheptulose-7-phosphate to yield fructose-6-phosphate and erythrose-4-phosphate; the latter reaction being catalysed by transaldolase (Tal1). The fructose-6-phosphate formed in the transaldolase reaction can also re-enter glycolysis/gluconeogenesis or can undergo a transketolase-catalysed reaction with erythrose-4-phosphate to produce glyceraldehyde-3-phosphate and xylulose-5-phosphate.

The hexose monophosphate pathway fulfils many metabolic requirements: the provision of NADPH, ribose skeletons (needed for the synthesis of histidine, tryptophan and purine ribo- and deoxyribonucleotides) and erythrose-4-phosphate for the synthesis of the aromatic amino acids phenylalanine, tyrosine, and tryptophan.

3.6 Trehalose and glycogen

Trehalose is produced from glucose-6-phosphate in two steps. First, UDP-glucose (or ADP-glucose) and glucose-6-phosphate are converted into trehalose-6-phosphate and UDP (or ADP) by trehalose-6-phosphate synthase (Tps1) (Vuorio *et al.*, 1993). Then the phosphate group of trehalose-6-phosphate is removed by trehalose-6-phosphate phosphatase (Tps2) (De Viriglio *et al.*, 1993). Both enzymes are present in a complex along with Tsl1 (or Tps3) (Thevelein and Hohmann, 1995; Ferreira *et al.*, 1996). Trehalose is important in resistance against many adverse conditions including heat, cold, dehydration, osmotic stress, solvents, and free radicals (see Chapter 9). Hence, its synthesis and degradation are highly regulated. Trehalose can also serve as a carbon source if supplied externally and is transported into the cell for this purpose (see Section 3.2). There are two trehalases which hydrolyze trehalose into two molecules of glucose. The neutral trehalase (Nth1) is cytosolic; the acid trehalase (Ath1) is vacuolar (Kopp *et al.*, 1993; Destruelle *et al.*, 1995).

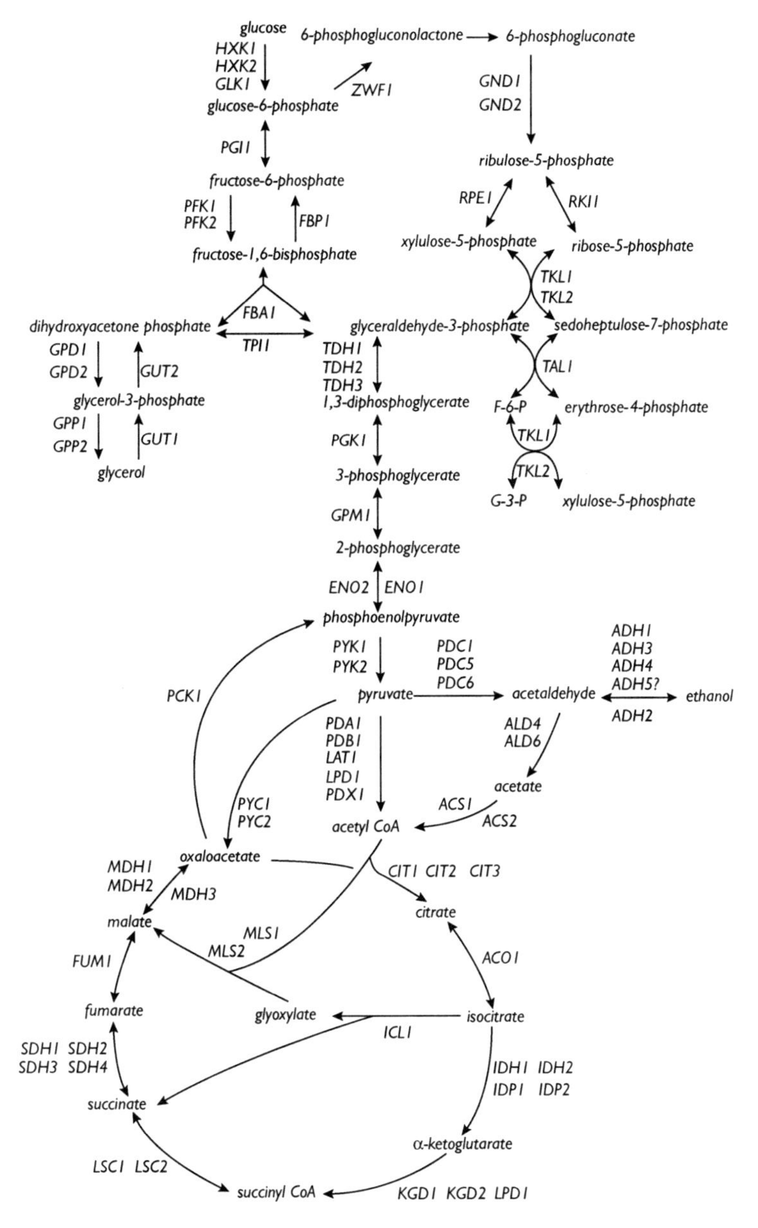

Figure 3.1 The principal pathways of carbon metabolism in *S. cerevisiae*. Only intermediates are shown; other reactants, co-enzymes, and products in individual reactions (e.g. ATP, NADH, Pi, etc.) are omitted for the sake of clarity but are described in the relevant section of text. Double-headed arrows denote reversible reactions; single arrows irreversible reactions. Separate arrows between two intermediates indicate that completely different types of enzymes are used in opposite directions. The italicized three-letter acronyms are the genes which encode the appropriate enzymes. In the hexose monophosphate pathway the abbreviations 'F-6-P' and 'G-3-P' denote fructose-6-phosphate and glyceraldehyde-3-phosphate; these are not in separate pools but are fully exchangeable with the F-6-P and G-3-P of glycolysis.

Glycogen synthesis also uses UDP-glucose but results in a polymer not simply a disaccharide as with trehalose. An α-1,4 glucosyl 'primer' is required initially; this is formed from UDP-glucose by Glg1 and Glg2 (Cheng *et al.*, 1995). Glycogen synthase performs the subsequent elongation. Gsy1 and Gsy2 represent the minor and major isoforms respectively (Farkas *et al.*, 1991). Gsy3 forms α-1,6 branch points. The breakdown of glycogen into glucose and glucose-1-phosphate is accomplished by glycogen phosphorylase (Gph1) (Hwang *et al.*, 1989). Phosphorylation inactivates glycogen synthase and activates glycogen phosphorylase; dephosphorylation has the opposite effect on both activities (the details are fully described in Chapter 8). Dual reciprocal control is obviously efficient as phosphorylation will increase the availability of glucose units by stopping glycogen synthesis and starting glycogen breakdown; whilst dephosphorylation stops glycogen breakdown and simultaneously starts glycogen synthesis.

3.7 Tricarboxylic acid cycle and glyoxylate bypass

S. cerevisiae can utilize certain non-fermentable compounds as its sole source of carbon and energy. Some of these compounds (e.g. lactate, oleate) may be available in the environment due to the metabolism of other organisms (microbes, plants, or animals). In addition, a number of non-fermentable compounds (e.g. ethanol, glycerol, pyruvate, and acetate) are produced by yeast and excreted into the medium during fermentative growth. After sugar fermentation is completed, a lag occurs during which the genes required for utilization of non-fermentable compounds are expressed. Subsequently the non-fermentable compounds are metabolized. This is an example of diauxic growth (Monod, 1942; Jones and Kompala, 1999).

Ethanol, glycerol, and the undissociated forms of organic acids are able to diffuse freely across the yeast plasma membrane. Glycerol and organic acids also cross the plasma membrane via protein-mediated transport systems. Two glycerol transport systems have been identified. One system is a channel protein of the Major Intrinsic Protein (MIP) family, encoded by *FPS1*. The glycerol transport facilitators of this family allow passive transport of glycerol across the membrane; transport occurs only down a concentration gradient (Van Aelst *et al.*, 1991; Reizer *et al.*, 1993; Luyten *et al.*, 1995; Sutherland *et al.*, 1997). One cellular role of glycerol is as a compatible solute, protecting cellular functions from hyperosmotic conditions. The physiological role of Fps1 is in glycerol export rather than uptake, acting to maintain an intracellular glycerol concentration appropriate to the osmotic stress (Tamas *et al.*, 1999). The amino terminus of the Fps1 protein regulates channel opening. Deletion of this domain results in constitutive glycerol efflux, which stimulates over-production of glycerol (Tamás *et al.*, 1999; Remize *et al.*, 2001). Conversely, loss of function mutations of Fps1 result in accumulation of glycerol, which activates the Sln1 histidine kinase of the high-osmolarity glycerol (HOG) mitogen-activated protein kinase pathway involved in osmotic stress responses (Tao *et al.*, 1999; Hohmann, 2002). Fps1 also mediates the uptake of arsenite and antimonite into *S. cerevisiae* (Wysocki *et al.*, 2001).

Yeast also displays high affinity (Km = 2 mM) H^+–glycerol co-transport. This transport is mediated by the Gup1 and Gup2 proteins: Gup1 is involved in glycerol uptake for its metabolism under normal growth conditions, and Gup2 is involved in glycerol uptake under conditions of hyperosmotic salt stress (Lages and Lucas, 1997; Sutherland *et al.*, 1997; Holst *et al.*, 2000). Glycerol kinase (Gut1) which is cytosolic, converts glycerol to glycerol-3-phosphate which is then converted to dihydroxyacetone phosphate by glycerol-3-phosphate dehydrogenase (Gut2). Some of the dihydoxyacetone phosphate is used in gluconeogenesis and some is catabolized using the glycolysis pathway via pyruvate and pyruvate dehydrogenase to acetyl-CoA which then enters the TCA cycle (Figure 3.1). The *GUT2* gene is repressed by glucose and derepressed by non-fermentable carbon sources. This derepression works via Snf1 and the Hap2/3/4/5 complex (Grauslund and Rønnow, 2000).

Yeast has two transport systems for short-chain monocarboxylic acids. One system recognizes lactate, pyruvate, propionate, and acetate, and the second recognizes acetate, propionate, and formate (Cássio *et al.*, 1987). The latter system has not been well characterized genetically (Paiva *et al.*, 1999). Transport of lactate and pyruvate is dependent on the *JEN1* gene product (Casal *et al.*, 1999; Akita *et al.*, 2000). This gene encodes a substrate–H^+ transporter that has Km values of ~1 and ~4 mM, respectively, for lactate and pyruvate. Transcription of *JEN1* is repressed by glucose in a Mig1- and Mig2-dependent fashion, and is activated by Cat8 and the Hap2/3/4/5 complex. This results in coordinate expression of *JEN1* and genes encoding enzymes of the glyoxylate cycle and gluconeogenesis (Bojunga and Entian, 1999; Andrade and Casal, 2001; Lodi *et al.*, 2002). Lactate results in higher expression than ethanol, and this requires intracellular lactate metabolism (Andrade and Casal, 2001); similarly, induction of L- and D-lactate dehydrogenase by lactate requires *JEN1* function (Lodi *et al.*, 2002). This suggests that a lactate induction pathway exists that requires a low, basal expression of *JEN1* and lactate-metabolizing enzymes. The Jen1 protein is targeted to the plasma membrane via the secretory pathway (Paiva *et al.*, 2002). *JEN1* transcription ceases rapidly after addition of glucose to derepressed cells, and *JEN1* mRNA is quickly degraded (it has a half-life of 15 min, compared to >2 h in the absence of glucose) (Andrade and Casal, 2001). Glucose also stimulates a rapid inactivation of the Jen1 protein; within 10 min most of the protein has been endocytosed and transported to the vacuole. This internalization process depends on ubiquitination of Jen1 (Paiva *et al.*, 2002).

Two isoenzymes of lactate dehydrogenase encoded by *DLD1* (YDL174c) and *DLD3* (YEL071w) have been identified. The former is mitochondrial (Lodi and Ferrero, 1993), the latter cytoplasmic (Chelstowska *et al.*, 1999). *DLD3* expression is dependent on the functional state of the mitochondria and upon Rtg1, Rtg2, and Rtg3 because the expression of Dld3 is 'retrograde-responsive' (the phenomenon in which the expression of nuclear genes is regulated by the mitochondria). *CIT2* which encodes an isoenzyme of citrate synthase (see later), is similarly regulated. The 5' flanking region of *DLD3* and *CIT2* contain two R box binding sites for the Rtg1/Rtg3 heterodimeric (basic helix–loop–helix/leucine zipper) transcription complex. Both boxes are required for full expression of *DLD3* (Chelstowska *et al.*, 1999). Rtg2, a 68 kDa protein with an N-terminal ATP binding domain, acts upstream of Rtg1 and Rtg3. A related protein encoded by *AIP2* (YDL178w) (actin interacting protein 2) has been shown to possess lactate dehydrogenase activity (Chelstowska *et al.*, 1999) and has, accordingly, been nominated as 'Dld2' – the name has now been adopted. The physiological role of this third lactate dehydrogenase activity, which is located in the mitochondrial matrix, is not clear. The *dld1* mutants are unable to grow on a medium in which D-lactate is the sole carbon source, whereas there is no known phenotype for *dld2, dld3,* or *dld2 dld3* double mutants (Lodi and Ferrero, 1993; Chelstowska *et al.*, 1999).

Acetyl-CoA (derived from ethanol or acetate via acetyl-CoA synthetase, from pyruvate via pyruvate dehydrogenase, or from fatty acid oxidation) reacts with oxaloacetate to form citrate, the first intermediate in the TCA cycle. The reaction is catalysed by citrate synthase. There are three isoenzymes of citrate synthase encoded by *CIT1* (mitochondrial matrix), *CIT2* (peroxisome), and *CIT3* (mitochondrial matrix) (McAlister-Henn and Small, 1977). As noted earlier, *CIT2* is subject to retrograde response. In wild-type cells the expression of *CIT2* is low, but in cells which are mitochondrially compromised the transcription of *CIT2* is greatly increased (Liao and Butow, 1993; Chelstowska and Butow, 1995). Metabolites produced in the peroxisomes can also be used in the TCA cycle (in the mitochondria). Thus, the retrograde regulation of *CIT2* can regulate the efficiency by which the cells use two-carbon compounds in anaplerotic pathways especially the glyoxylate cycle.

Aconitase encoded by *ACO1* is located in both the cytosol and mitochondrial matrix. There is no genuine *ACO2* gene but Yjl200c has considerable sequence similarity to Aco1. The expression of aconitase is repressed by glucose and glutamate where its presence would be unnecessary. As one would predict, *aco1* mutants are unable to grow on non-fermentable carbon sources and are glutamate auxotrophs on glucose because α-ketoglutarate (the precursor to glutamate) is not made (Crocker and Bhattacharjee, 1973). It would be interesting to see whether over-expression of YJL200c could complement an *aco1* mutation.

The next step of the TCA cycle is the oxidative decarboxylation of isocitrate to α-ketoglutarate. There are three isoenzymes of isocitrate dehydrogenase. NAD-specific isocitrate dehydrogenase is an allosterically regulated octamer composed of four subunits of Idh1 and four of Idh2. It is mitochondrially located. Both subunits can bind the substrate: Idh1 for allosteric activation by AMP and Idh2 for catalysis (Lin *et al.*, 2001). Homologous serine residues in both subunits (92 in Idh1 and 98 in Idh2) are required for this binding (Lin and McAlister-Henn, 2002). A strong correlation has been noted between the level of isocitrate dehydrogenase activity and the ability of cells to grow on acetate or glycerol (Lin *et al.*, 2001). There are two NADP-dependent isocitrate dehydrogenases: Idp1 (mitochondrial) and Idp2 (cytosolic). A mutation in either *IDH1* or *IDH2* renders the cell unable to grow on acetate or pyruvate but it can still grow on ethanol. In contrast, *idp1* or *idp2* mutants as well as *idp1 idp2* double mutants, can grow on either acetate or ethanol. NAD-specific isocitrate dehydrogenase binds specifically and with high affinity to the 5′ untranslated leader sequences of all mitochondrial mRNAs. It has been suggested that this binding is to suppress inappropriate translation because cells disrupted for NAD-specific isocitrate dehydrogenase display increased mitochondrial translation (de Jong *et al.*, 2000). However, despite the increased rate of synthesis, subunits 1, 2, and 3 of cytochrome c oxidase, and cytochrome b (all encoded by mitochondrial genes) are all reduced in the absence of NAD-specific isocitrate dehydrogenase due to more rapid degradation (de Jong *et al.*, 2000). Clearly, a number of lines of evidence point to NAD-specific isocitrate dehydrogenase having a role in regulating the rate of mitochondrial assembly besides its role in the TCA cycle.

α-ketoglutarate dehydrogenase catalyses the oxidative decarboxylation of α-ketoglutarate via succinyl-CoA to succinate. The multi-enzyme complex comprises a dehydrogenase ('E1') encoded by *KGD1*, succinyl transferase ('E2') encoded by *KGD2* and lipoamide dehydrogenase ('E3') encoded by *LPD1*. Mutants in E1, E2, or E3 components are able to grow on glucose but not on acetate or glycerol (Dickinson *et al.*, 1986; Repetto and Tzagoloff, 1989, 1990). The complex assembles spontaneously *in vivo*. Succinyl-CoA ligase α and β subunits are encoded by *LSC1* (YOR142w) and *LSC2* (YGR244c) respectively. All of the aforementioned are subject to catabolite repression.

Succinate dehydrogenase catalyses the interconversion of succinate and fumarate. The flavoprotein precursor of this enzyme is encoded by *SDH1*. The mature Sdh1 binds to Sdh2 (the iron–sulphur protein) to form an active dimer. Two hydrophobic proteins (Sdh3 and Sdh4) anchor the active dimer to the mitochondrial inner membrane. The turnover of *SDH2* mRNA (described in Section 3.8) is crucial to the control of succinate dehydrogenase activity by carbon source.

Fumarase which catalyses the conversion of fumarate to malate is encoded by the single gene *FUM1* but exists as separate cytosolic and mitochondrial forms. A number of explanations have been advanced to explain this situation including different transcription initiation sites to produce two different mRNAs (a longer one encoding the mitochondrial isoenzymes and targeting sequence and a shorter one for the cytosolic form) (Wu and Tzagoloff, 1987), dual translational initiation and selective splicing; the latter two having been established for other proteins. However, the localization and distribution of fumarase appears to be unique because there is only one translation product which is targeted to the mitochondria by an N-terminal presequence which is then removed by the mitochondrial processing peptidase. Some of the fully mature fumarase molecules are then released back into the cytosol. *In vivo* translocation into the mitochondria occurs during translation and *in vitro* translation of the *FUM1* mRNA requires mitochondria (Sass *et al.*, 2001).

Malate dehydrogenase forms the last step in the TCA cycle by catalysing the oxidation of malate to oxaloacetate. The three isoenzymes, encoded by *MDH1* (mitochondrial), *MDH2* (cytoplasmic), and *MDH3* (peroxisomal) are all subject to catabolite repression. The majority (90%) of malate dehydrogenase activity is due to Mdh1 except when acetate or ethanol is the carbon source in which case Mdh2 constitutes 65% of total malate dehydrogenase (Steffan and McAlister-Henn, 1992). Oxaloacetate levels are crucial. If oxaloacetate levels were to become insufficient then further turns of the TCA cycle would not be possible. This could arise due to the consumption of TCA cycle intermediates for example, α-ketoglutarate and oxaloacetate in the formation of glutamate and aspartate (respectively), both of which contribute to the synthesis of other amino acids. The metabolic requirement is ensured in two ways: the anaplerotic (filling-up) glyoxylate bypass and Mdh2. Mdh2, like phospho*enol*pyruvate carboxykinase and fructose-1,6-bisphosphatase is critical in gluconeogenesis (see Section 3.4) and is similarly rapidly inactivated and then proteolytically degraded if glucose is supplied to cells which had been growing gluconeogenically.

The glyoxylate cycle comprises isocitrate lyase and malate synthase. Isocitrate lyase is encoded by *ICL1* (Fernández *et al.*, 1992). Until recently, it was thought that the closely related *ICL2* also encoded an isocitrate lyase that is inactive. However, it is now known that *ICL2* encodes a mitochondrial 2-methylisocitrate lyase involved in propionyl-CoA metabolism (Luttik *et al.*, 2000). Isocitrate lyase catalyses the conversion of isocitrate into glyoxylate and succinate. The enzyme is a homotetramer. Its synthesis is induced by ethanol and repressed by glucose. Transcription of the gene is controlled by Snf1. The enzyme is rapidly inactivated and then proteolytically degraded if glucose is added to gluconeogenically grown cells (Ordiz *et al.*, 1996). As with the other enzymes, mentioned already, whose activity is similarly controlled, phosphorylation by cAMP-dependent protein kinase is the trigger.

Malate synthase catalyses the formation of malate from glyoxylate and acetyl-CoA. There are two malate synthases. *MLS1* encodes an enzyme with both peroxisomal and cytosolic locations. Mls1 is abundant in the peroxisomes of cells grown on oleic acid but in ethanol-grown cells it is mostly cytosolic (Kunze *et al.*, 2002). *MLS1* is repressed by glucose. *MLS2* (better known as *DAL7*) encodes an enzyme required when allantoin is the sole source of nitrogen; it is subject to repression by NH_4^+ (Hartig *et al.*, 1992). The different regulatory controls over the expression and activity of the two forms ensure that yeast can metabolize C-2 compounds, allantoin or both. It has been observed that *S. cerevisiae* which have been phagocytosed have up-regulated both *ICL1* and *MLS1* and that *Candida albicans* lacking *CaICL1* have reduced virulence, leading to the suggestion that the glyoxylate cycle is an important component of fungal virulence (Lorenz and Fink, 2001).

3.8 Regulation and integration of carbon metabolism with other metabolic and physiological processes

Carbon metabolism does not operate in total isolation from other parts of metabolism (e.g. nitrogen metabolism) or other cellular events and processes (e.g. the cell cycle, developmental state, etc.). Many aspects of the regulation of carbon metabolism have been described previously in the relevant sections in this chapter and details of its integration with other processes are described in other chapters. In this section we have attempted to summarize some major examples of regulation and integration.

Covalent modifications: the cAMP/protein kinase A pathway

The signal transduction pathways acting on the cAMP-dependent protein kinase (protein kinase A) regulate many aspects of cell physiology, including the metabolism of storage carbohydrates (trehalose and glycogen), coupling of cell cycle progression to nutrient availability, development of stress resistance in resting cells, and morphogenesis. In glucose-repressed cells the pathway responds to environmental signals by rapidly increasing the intracellular cyclic AMP (cAMP) concentration. The elevated cAMP levels activate protein kinase A by liberating the catalytic subunits from the regulatory subunits (encoded by *BCY1*; Toda *et al.*, 1987a). The catalytic subunits (encoded by *TPK1*, *TPK2*, and *TPK3*; Toda *et al.*, 1987b) then phosphorylate various target proteins. The three catalytic subunits are similar in sequence, and have overlapping but not identical cellular effects (Robertson and Fink, 1998; Pan and Heitman, 1999; Robertson *et al.*, 2000).

Cyclic AMP is synthesized by adenylate cyclase in response to environmental signals. In yeast, this enzyme (Cyr1) is activated by at least two pathways, each responding to a different signal and each involving a different type of G protein (Colombo *et al.*, 1998). Intracellular acidification activates Cyr1 via the Ras pathway. This pathway involves two Ras proteins (encoded by *RAS1* and *RAS2*). Ras is activated upon exchange of GDP for GTP catalysed by the Cdc25 guanine nucleotide exchange factor. Activated Ras-GTP then stimulates adenylate cyclase activity, resulting in cAMP synthesis. Subsequently, GTP-Ras is down-regulated upon hydrolysis of GTP by the intrinsic GTPase activity of Ras; the GTPase activity is stimulated by the Ira1 and Ira2 proteins (Thevelein *et al.*, 2000).

Addition of glucose to glucose-derepressed cells stimulates the cAMP/protein kinase A pathway, which leads to preparation of the cell for fermentative growth and metabolism (Jiang *et al.*, 1998). Glucose activates adenylate cyclase via a pathway comprising Gpr1, a seven-transmembrane domain receptor protein, and Gpa2, a Gα protein (the proteins Gpb1/Gpb2 and Gpg1 may be Gβ and Gγ surrogates that form a heterotrimeric G protein analogue with Gpa2; Harashima and Heitman, 2002). Gpr1 is apparently a receptor for extracellular glucose; in the presence of glucose, Gpr1 activates Gpa2 by promoting exchange of GTP for GDP. Activated GTP-Gpa2 in turn activates adenylate cyclase activity, resulting in cAMP synthesis. GTP-Gpa2 is subsequently down-regulated by the GTPase activator Rgs2 (Versele *et al.*, 1999). Full activation of adenylate cyclase by this pathway also requires transport and phosphorylation of glucose (Rolland *et al.*, 2000); thus this pathway assesses both the availability of fermentable carbon and the capacity of cellular metabolism to utilize it (D'Souza and Heitman, 2001).

Two phosphodiesterase enzymes attenuate protein kinase A activity by catalysing cAMP hydrolysis: Pde1 is a low-affinity enzyme, and Pde2 has a high affinity for cAMP (Sass *et al.*,

1986; Nikawa *et al.*, 1987). Each has a role in regulating intracellular cAMP concentrations; the low-affinity Pde1 enzyme is particularly prominent in attenuating glucose-induced cAMP signalling. Pde1 is phosphorylated and activated by protein kinase A, and functions in a feedback loop regulating cAMP accumulation after stimulation by the agonist glucose (Ma *et al.*, 1999).

Evidence has also been presented that protein kinase A is also activated by a cAMP-independent pathway in response to the nutritional status of the environment (Hirimburegama *et al.*, 1992); this pathway, known as the Fermentable Growth Medium (FGM) pathway, integrates information about the availability of a fermentable carbon source, the quality and quantity of the nitrogen source, and the availability of other essential nutrients such as phosphate and sulphate (Thevelein, 1994; Thevelein *et al.*, 2000). The FGM pathway is proposed to lead to sustained activation of protein kinase A, in contrast to the transient activation which results from cAMP synthesis. The Sch9 protein kinase acts as an antagonist of protein kinase A in the FGM pathway (Crauwels *et al.*, 1997); the upstream factors and downstream targets of the pathway are currently being elucidated.

Glucose repression

Glucose is considered the preferred carbon source of yeast, because the presence of this hexose inhibits the utilization of other carbon sources. Yeast exercises this preference in part by repressing the transcription of genes required for the utilization of the alternative carbon sources. Target genes of this glucose repression pathway include the *SUC* genes encoding invertase, *GAL* and *MAL* genes involved in utilization of galactose and maltose, and the *FBP1* gene encoding fructose bisphosphatase. Genes required for utilization of non-fermentable carbon sources (encoding proteins involved in mitochondrial biogenesis, respiration, the citric acid cycle, etc.) are also repressed by glucose. Other fermentable carbon sources (e.g. fructose, galactose, and maltose) can exert repressive effects on members of this set of target genes as well, though they are generally less potent than glucose.

The canonical glucose repression pathway involves binding of transcriptional repressors to the promoters of affected genes; these repressors, the Mig1 (Nehlin and Ronne, 1990) and Mig2 (Lutfiyya and Johnston, 1996) zinc-finger proteins, require the Ssn6 and Tup1 co-repressors for activity (Keleher *et al.*, 1992; Vallier and Carlson, 1994). Mig1 is localized in the nucleus in glucose-grown cells, and upon removal of glucose it is rapidly phosphorylated and translocated from the nucleus to the cytoplasm (De Vit *et al.*, 1997). Derepression requires the Snf1 protein kinase complex (Carlson *et al.*, 1981; Carlson, 1999). This heterotrimeric complex includes the Snf1 catalytic subunit, which has a catalytic domain and an autoregulatory domain. The other components of the complex are a regulatory subunit (Snf4) involved in responding to glucose, and an oligomerization factor (Sip1, Sip2, or Gal83) that influences the target specificity of the complex (Schmidt and McCartney, 2000). The Snf1 kinase phosphorylates Mig1 (Treitel *et al.*, 1998; Smith *et al.*, 1999), resulting in its dissociation from promoter DNA and thus permitting transcription of target genes. The activity of the Snf1 kinase is regulated by glucose; in the presence of the hexose the catalytic domain is inactive due to intramolecular interaction with the autoregulatory domain. As glucose concentrations decline, the Snf4 protein liberates the Snf1 catalytic domain, which is then able to phosphorylate Mig1. The process is reversed by the Glc7-Reg1 protein phosphatase, which restores the inactive conformation of the Snf1 catalytic and autoregulatory domains at high glucose concentrations (see Section 8.4 of Chapter 8 and Figure 8.5).

The signal initiated by glucose is not yet understood. Two models are currently favoured. The first model is based on the homology between Snf1 and the AMP-activated protein kinase (AMPK), which regulates energy metabolism in mammalian cells. AMPK is activated by high AMP concentrations and low ATP concentrations, and it is proposed that changes in adenylate concentrations in response to glucose availability could affect Snf1 in a similar way (Hardie and Carling, 1997). The second model is based on two types of mutations that result in expression in the presence of glucose of normally glucose-repressible genes. Mutations in *HXK2* encoding the predominant hexokinase in glucose-grown cells lead to derepression, as do mutations that restrict cellular glucose transport activity. Therefore, the intracellular concentrations of glucose, glucose-6-phosphate, or a related non-glycolytic metabolite are implicated in signalling the extracellular glucose concentration to the repression machinery.

The role of Hxk2 is particularly interesting as it has been suggested to have intrinsic regulatory functions as well as a catalytic function in glycolysis. *HXK2* mutations have been identified that separate catalytic activity from the protein's role in glucose repression (Hohmann *et al.*, 1999; Mayordomo and Sanz, 2001a), and Hxk2 has been shown to translocate into the nucleus and act directly at the *SUC2* promoter (Herrero *et al.*, 1998; Randez-Gil *et al.*, 1998a,b). These observations are complicated by observations that hexokinases from many other organisms, from yeast to man, are able to replace Hxk2 in exerting glucose repression (Rose, 1995; Petit and Gancedo, 1999; Mayordomo and Sanz, 2001b), suggesting that the catalytic activity of the enzyme is sufficient for its regulatory role.

Activation

Most glycolytic enzyme genes are expressed at high and essentially constitutive levels. The high level of expression results from strong transcriptional activation due to the concerted action of potent transcriptional factors, including Rap1 (Shore, 1994), Abf1 (Packham *et al.*, 1996), Gcr1 (Holland *et al.*, 1987), Gcr2 (Uemura and Fraenkel, 1990), and Reb1 (Scott and Baker, 1993). The promoters of different glycolytic genes recruit different combinations and configurations of these and other transcription factors, resulting in differences in the details of their expression patterns.

A number of mechanisms bring about transcriptional activation of genes in conditions where glucose is low or absent. Two important mechanisms will be considered here. The first mechanism involves a complex of the Hap2, Hap3, Hap4, and Hap5 proteins, which binds to the promoters of target genes in the absence of glucose and activates their transcription. The target genes are mostly involved in respiration and the utilization of non-fermentable carbon sources. Regulation of the Hap complex takes place at the level of transcription of its components (Pinkham and Guarente, 1985; Forsburg and Guarente, 1989), although the rapidity of its action suggests that other levels of control may exist as well.

The second activation mechanism results in expression of genes involved in utilization of non-fermentable carbon sources, such as *FBP1*, *PCK1*, *ICL1* (encoding isocitrate lyase), *MDH2* (encoding cytosolic malate dehydrogenase), and *JEN1* (encoding the lactate transporter). These genes are under control of Cat8, a zinc-finger DNA-binding protein that activates transcription of target genes (Hedges *et al.*, 1995; Randez-Gil *et al.*, 1997; Bojunga and Entian, 1999; Haurie *et al.*, 2001; Roth and Schuller, 2001). The *CAT8* gene itself is repressed by the canonical glucose repression pathway (Hedges *et al.*, 1995).

RNA stability

The stability of the mRNAs encoding enzymes involved in carbon metabolism has been shown to vary significantly in response to environmental signals, especially availability of glucose. In most of the cases in which mRNA degradation is accelerated by glucose, transcription of the genes is also subject to glucose repression, and in some cases the protein products are subject to glucose-induced degradation as well. In combination, these mechanisms provide rapid and thorough down-regulation of metabolic activities that are not required for growth on glucose.

In one example, the half-life of *SDH2* mRNA (encoding the iron–sulphur subunit of succinate dehydrogenase) declines from >60 min in glycerol-grown cells to ~5 min after addition of glucose (Scheffler *et al.*, 1998). Glucose initiates a process that interferes with translational initiation on these mRNAs at the level of various eIF initiation factors. The RNA molecules with reduced ribosome loading are subject to decapping and rapid nucleolytic degradation (Scheffler *et al.*, 1998). The process occurs when high glucose concentrations are added to gluconeogenically growing cells. Other examples include the observations that addition of glucose to cells that are expressing invertase from one of the *SUC* genes results in interference with translation of invertase mRNA and an increased rate of mRNA degradation (Elorza *et al.*, 1977b; Cereghino and Scheffler, 1996), and that low levels of glucose decrease the half-lives of mRNAs of the gluconeogenic genes *PCK1* and *FBP1*, as well as block their transcription (Mercado *et al.*, 1994; Yin *et al.*, 2000). In the latter case, the glucose signal requires hexose phosphorylation, Snf3, and Ras-cAMP signalling. A further example is the mRNA encoding the Jen1 lactate permease, which is destabilized by adding glucose to cells growing on lactate. The half-life of the mRNA declines to ~15 min in the presence of glucose (Andrade and Casal, 2001).

Complex and cross-pathway regulation

The regulatory pathways discussed here reflect some of the complexity of the regulatory networks that act on carbon metabolism in the yeast cell. Additional layers of control stem from the occurrence of forked regulatory pathways and cross-talk between pathways. For example, Ras2 not only signals to the cAMP pathway but also to the MAP kinase pathway in the context of morphogenesis (filamentous growth) (Mosch *et al.*, 1999). Some genes encoding enzymes of carbon metabolism (e.g. *PYC1, LPD1*) are regulated by the nitrogen source and glucose limitation induces *GCN4* (the transcriptional activator controlling depression of amino acid biosynthetic pathways – see Chapter 4) translation by activation of the Gcn2 protein kinase (Yang *et al.*, 2000). In some cases the links are neither obvious nor currently explained. For example, it has been shown recently that cellular Ca^{2+} homeostasis is coupled to the relative levels of glucose-6-phosphate and glucose-1-phosphate (Aiello *et al.*, 2002).

3.9 Metabolomics as relevant to carbon metabolism: metabolic modelling

The activity of enzymes and metabolic pathways *in vivo* can be determined by measuring the concentrations of metabolic substrates, pathway intermediates, and products, and analysing how they change with time. Metabolic flux will result in a decline in the substrate concentrations and an increase in the product concentrations, although the concentrations of pathway intermediates may be in a steady state. The measurements of these metabolites are important for describing the physiological capabilities of the cell, and for comparing the properties of individual enzymes *in vitro* with their performance when embedded in a metabolic pathway *in vivo*.

Traditional methods for measuring metabolites and metabolism include manometric determination of oxygen consumption and carbon dioxide production, and colorimetric or enzyme-linked assays of soluble metabolites. For example, Pedler *et al.* (1997), determined the fate of glucose in non-growing cells of two yeast strains, displaying either fermentative or respiratory metabolism. They measured the consumption of glucose and oxygen, and the production of carbon dioxide, ethanol, acetate, glycerol, glycogen, and trehalose, over 30 minutes. The sum of the metabolic products they measured accounted for 88–98% of the glucose consumed, demonstrating the efficiency of these analytical methods. In another example, Diderich *et al.* (2001a), compared the physiology of a wild-type strain with a *hxk2* deletion strain that lacked glucose repression. They measured the consumption of glucose and oxygen, and the production of carbon dioxide, ethanol, glycerol, acetate, and pyruvate. They also measured the intracellular metabolites glucose-6-phosphate, fructose-6-phosphate, fructose-1,6-bisphosphate, ATP, ADP, and AMP. The activities of hexokinase, pyruvate decarboxylase, and mitochondrial H^+-ATPase were assayed, and growth was monitored by assessing optical density and total cellular protein. The results give a detailed picture of the changes in carbon metabolism that result from loss of glucose repression.

High performance liquid chromatography (HPLC; alternatively known as high pressure liquid chromatography) can be used to determine the concentrations of multiple metabolites simultaneously (Bhattacharya *et al.*, 1995; Groussac *et al.*, 2000). The set of metabolites detected by this method depends on the chromatography matrix (e.g. reversed phase, anion exchange, or cation exchange chromatography) and the detection method (e.g. amperometry, conductivity, refractometry, UV, or fluorescence). HPLC is also frequently linked in tandem with mass spectrometry (see later).

Nuclear magnetic resonance spectroscopy (NMR) has also been used to assay multiple metabolites simultaneously. NMR measurements can be made on intact cells, which allow a single cell suspension to be assayed repeatedly over time, and which minimize artefacts due to cell harvesting and extraction. The chemical shifts of the nuclei of ^{1}H, ^{2}H, ^{13}C, and ^{31}P isotopes produce characteristic spectra in each chemical compound (e.g. metabolite) in which they occur. Complex mixtures of metabolites (e.g. whole cells or cell extracts) produce spectra from which the (relative) concentrations of each metabolite can be determined (Campbell-Burk and Shulman, 1987; Brindle *et al.*, 1997; Grivet, 2001). For example, ^{13}C NMR has been used to monitor the metabolism of glucose under anaerobic and aerobic conditions in cells displaying fermentative or respiratory physiology (den Hollander *et al.*, 1986), and to determine the distribution of ethanol in fermenting cells (Yuan *et al.*, 2000). The ^{31}P NMR has been used to monitor the intracellular pH and the metabolism of phosphate during the cell cycle (Gillies *et al.*, 1981), and to assess the effect of phosphite on phosphate metabolism and signalling (McDonald *et al.*, 2001). The ^{2}H NMR has been used to monitor glycerol formation during fermentation (Pionnier and Zhang, 2002), and combined ^{1}H and ^{13}C NMR spectroscopy was used to assess the effects on metabolite pools and glycolytic product formation of mutations that stimulate flux through glycolysis (Herve *et al.*, 2000).

The complete sequencing of the yeast genome has led to the development of techniques to assess the full complement of mRNAs (the transcriptome) and proteins (the proteome) expressed by the cell under given conditions (Kumar and Snyder, 2001; Oliver *et al.*, 2002). This has stimulated the development of analogous techniques for assessing the complement of metabolites in the cell, referred to as the metabolome. The metabolome is in principle the easiest of these '-omes' to measure, since the yeast genome is made up of ~6200 genes (Goffeau *et al.*, 1996) yet the cell contains only 600–800 metabolites (Oliver *et al.*, 1998; Palsson, 2000). However, techniques for comprehensive metabolite analysis are faced with the challenge of detecting a chemically diverse set of targets, many of which are labile and present at low concentrations. The new analytical methods that are being employed for comprehensive metabolite determination include Raman and Fourier-transform infrared spectroscopy (Sivakesava *et al.*, 2001a,b; Rhiel *et al.*, 2002), electrospray mass spectroscopy (Buchholz *et al.*, 2001; Wittmann, 2002), and pyrolysis mass spectroscopy (McGovern *et al.*, 1999). Mass spectrometers can be linked in tandem with gas or liquid chromatographs to increase their resolving power (Fiehn *et al.*, 2000; Buchholz *et al.*, 2001; Gombert *et al.*, 2001).

These techniques produce large data sets, with the spectra of many metabolites superimposed on one another. Computational approaches are required to identify individual metabolites in these complex spectra. Alternatively, spectra from global metabolite analyses can be used as 'fingerprints' to compare different metabolic conditions or genotypes without resolving and quantitating individual metabolites (Forster *et al.*, 2002; Oliver *et al.*, 2002; Phelps *et al.*, 2002). For example, Raamsdonk *et al.* (2001) collected metabolite spectra from a series of isogenic yeast strains using ^{1}H NMR. The spectra were analysed statistically, using Principle Components Analysis (to reduce the large number of variables in the raw spectra to a smaller number of correlated variables) followed by Discriminant Function Analysis (to determine which of these variables discriminate among the strains). This approach distinguished the wild-type strain from two classes of mutant strains; the first class of mutants was made up of respiratory-deficient strains, and the other class was made up of strains with mutations in fructose-6-phosphate 2-kinase genes. The authors propose that metabolite fingerprints and statistical analysis can be used to reveal the cellular role of genes of unknown function. Other statistical approaches in metabolomics include Artificial Neural Network and Least Squares Regression analyses; the various statistical approaches are known collectively as chemometrics (Schuster, 2000; Shaw *et al.*, 2000).

Metabolic pathways and networks are complex, dynamic systems. Quantitative mathematical models have been constructed to describe their behaviour. Metabolic models have a number of uses. They assist in the organization and interpretation of experimental data (Edwards *et al.*, 2001), aid in the identification of unrecognized steps and pathways (Weckwerth and Fiehn, 2002), and guide the engineering of improved or novel metabolic pathways for biotechnological applications (Carlson *et al.*, 2002). Along with chemometrics, metabolic modelling is an invaluable application of bio-informatics to metabolomic studies (Aimar-Beurton *et al.*, 2002; Forster *et al.*, 2002; Weckwerth and Fiehn, 2002).

Many different types of metabolic models have been constructed. Some consider only the fluxes and metabolite concentrations of a pathway (e.g. Flux Balance Analysis, Metabolic Control Analysis, Metabolic Network Modelling), while others incorporate the *in vitro* kinetics of the pathway enzymes into their structure (e.g. Biochemical Systems Theory, Kinetic Parameter Modelling). Yeast carbon metabolism has been the subject of numerous models (e.g. Castrillo and Ugalde, 1994; Cascante *et al.*, 1995; Dantigny, 1995; Nissen *et al.*, 1997; Shi and Shimizu, 1998; Jones and Kompala, 1999; Giuseppin and van Riel, 2000; Teusink *et al.*, 2000; Visser *et al.*, 2000; Gombert *et al.*, 2001; Hynne *et al.*, 2001; Lei *et al.*, 2001; Pritchard and Kell, 2002; Stuckrath *et al.*, 2002). These models generally treat glycolysis in isolation during steady-state conditions, although some also consider branches from the pathway (e.g. glycerol, glycogen, and trehalose synthesis) or the response of the modelled pathway to perturbations, for example, a pulse of glucose. The models have predictive value which supports experimental results and can suggest further experimentation. For example, the stoichiometric model of Nissen *et al.* (1997) predicts the existence of a system to shuttle redox equivalents between the cytoplasm and the mitochondrial matrix; the kinetic parameter model of Teusink *et al.* (2000) has pointed out discrepancies between the activities of glycolytic enzymes *in vitro* and their *in vivo* behaviour when embedded in metabolism; and the model of Pritchard and Kell (2002) has identified glucose transport as exerting high control over glycolytic flux, in agreement with experimental results (see earlier text).

Metabolomic studies and metabolic models can also be fruitfully integrated with bioinformatic approaches to genomic and proteomic analysis. For example, Voit and Radivoyevitch (2000) identified the changes in gene expression after administering a heat shock to yeast cells. They also modelled the metabolic response of the cells to the heat shock, and found a correlation between the activities of the enzymes induced by the heat shock and the changes in trehalose, ATP, and NADPH concentrations predicted by the model.

Changes in cellular gene expression and protein composition underlie the changes in metabolic networks and metabolite levels within the cell. Pioneering work by Hood and co-workers has integrated transcriptome and proteome analyses in studies on yeast carbon metabolism and its control (Ideker *et al.*, 2001; Griffin *et al.*, 2002). The results have elucidated some of the regulatory networks influencing the expression of metabolic pathways. Including metabolome analysis in this type of research as a third level of cellular complexity will bring systems biology a step closer to its goal of providing an integrative description of cellular function.

References

Abeliovich, H. and Klionsky, D. J. (2001) Autophagy in yeast: mechanistic insights and physiological function. *Microbiology and Molecular Biology Reviews* **65**, 463–479.

Aguilera, A. and Zimmermann, F. K. (1986) Isolation and molecular analysis of the phosphoglucose isomerase structural gene of *Saccharomyces cerevisiae. Molecular and General Genetics* **202**, 83–89.

Aiello, D. P., Fu, L., Miseta, A. and Bedwell, D. M. (2002) Intracellular glucose-1-phosphate levels modulate Ca^{2+} homeostasis in *Saccharomyces cerevisiae. Journal of Biological Chemistry* **277**, 45751–45758.

Aimar-Beurton, M., Korzeniewski, B., Letellier, T., Ludinard, S., Mazat, J. P. and Nazaret, C. (2002) Virtual mitochondria: metabolic modelling and control. *Molecular Biology Reports* **29**, 227–232.

Akita, O., Nishimori, C., Shimamoto, T., Fujii, T. and Iefuji, H. (2000) Transport of pyruvate in *Saccharomyces cerevisiae* and cloning of the gene encoded pyruvate permease. *Bioscience Biotechnology and Biochemistry* **64**, 980–984.

Alber, T. and Kawasaki, G. (1982) Nucleotide sequence of the triose phosphate isomerase gene of *Saccharomyces cerevisiae. Journal of Molecular and Appied Genetics* **1**, 419–434.

Albig, W. and Entian, K. D. (1988) Structure of yeast glucokinase, a strongly diverged specific aldohexose-phosphorylating isoenzyme. *Gene* **73**, 141–152.

Alexandre, H., Rousseaux, I. and Charpentier, C. (1994) Ethanol adaptation mechanisms in *Saccharomyces cerevisiae. Biotechnology and Applied Biochemistry* **20**, 173–183.

Andrade, R. P. and Casal, M. (2001) Expression of the lactate permease gene *JEN1* from the yeast *Saccharomyces cerevisiae. Fungal Genetics and Biology* **32**, 105–111.

Avigad, G. (1981) Stimulation of yeast phosphofructokinase activity by fructose 2,6-bisphosphate. *Biochemical and Biophysical Research Communications* **102**, 985–991.

Bajwa, W., Torchia, T. E. and Hopper, J. E. (1988) Yeast regulatory gene *GAL3*: carbon regulation; UASGal elements in common with *GAL1, GAL2, GAL7, GAL10, GAL80*, and *MEL1*; encoded protein strikingly similar to yeast and *Escherichia coli* galactokinases. *Molecular and Cellular Biology* **8**, 3439–3447.
Bakker, B., Bro, C., Kotter, P., Luttik, M. A. H., van Dijken, J. P. and Pronk, J. T. (2000) The mitochondrial alcohol dehydrogenase Adh3p is involved in a redox shuttle in *Saccharomyces cerevisiae. Journal of Bacteriology* **182**, 4730–4737.
Baldwin, S. A. and Henderson, P. J. F. (1989) Homologies between sugar transporters from eukaryotes and prokaryotes. *Annual Review of Physiology* **51**, 459–471.
Bartrons, R., Van Schaftingen, E., Vissers, S. and Hers, H. G. (1982) The stimulation of yeast phosphofructokinase by fructose 2,6-bisphosphate. *FEBS Letters* **143**, 137–140.
Bell, W., Klaassen, P., Ohnacker, M., Boller, T., Herweijer, M., Schoppink, P., Van der Zee, P. and Wiemken, A. (1992) Characterization of the 56-kDa subunit of yeast trehalose-6-phosphate synthase and cloning of its gene reveal its identity with the product of *CIF1*, a regulator of carbon catabolite inactivation. *European Journal of Biochemistry* **209**, 951–959.
Bell, P. J., Higgins, V. J., Dawes, I. W. and Bissinger, P. H. (1997) Tandemly repeated 147 bp elements cause structural and functional variation in divergent *MAL* promoters of *Saccharomyces cerevisiae. Yeast* **13**, 1135–1144.
Bell, W., Sun, W., Hohmann, S., Wera, S., Reinders, A., De Virgilio, C., Wiemken, A. and Thevelein, J. M. (1998) Composition and functional analysis of the *Saccharomyces cerevisiae* trehalose synthase complex. *Journal of Biological Chemistry* **273**, 33311–33319.
Benito, B. and Lagunas, R. (1992) The low-affinity component of *Saccharomyces cerevisiae* maltose transport is an artifact. *Journal of Bacteriology* **174**, 3065–3069.
Betina, S., Goffrini, P., Ferrero, I. and Wesolowski-Louvel, M. (2001) *RAG4* gene encodes a glucose sensor in *Kluyveromyces lactis. Genetics* **158**, 541–548.
Bhat, P. J. and Hopper, J. E. (1992) Overproduction of the GAL1 or GAL3 protein causes galactose-independent activation of the GAL4 protein: evidence for a new model of induction for the yeast GAL/MEL regulon. *Molecular and Cellular Biology* **12**, 2701–2707.
Bhattacharya, M., Fuhrman, L., Ingram, A., Nickerson, K. W. and Conway, T. (1995) Single-run separation and detection of multiple metabolic intermediates by anion-exchange high-performance liquid chromatography and application to cell pool extracts prepared from *Escherichia coli. Analytical Biochemistry* **232**, 98–106.
Bisson, L. F. (1988) Derepression of high-affinity glucose uptake requires a functional secretory system in *Saccharomyces cerevisiae. Journal of Bacteriology* **170**, 2654–2658.
Bisson, L. F., Neigeborn, L., Carlson, M. and Fraenkel, D. G. (1987) The *SNF3* gene is required for high-affinity glucose transport in *Saccharomyces cerevisiae. Journal of Bacteriology* **169**, 1656–1662.
Blazquez, M. A., Lagunas, R., Gancedo, C. and Gancedo, J. M. (1993) Trehalose-6-phosphate, a new regulator of yeast glycolysis that inhibits hexokinases. *FEBS Letters* **329**, 51–54.
Bojunga, N. and Entian, K. D. (1999) Cat8p, the activator of gluconeogenic genes in *Saccharomyces cerevisiae*, regulates carbon source-dependent expression of NADP-dependent cytosolic isocitrate dehydrogenase (Idp2p) and lactate permease (Jen1p). *Molecular and General Genetics* **262**, 869–875.
Boles, E., Gohlmann, H. W. and Zimmermann, F. K. (1996) Cloning of a second gene encoding 5-phosphofructo-2-kinase in yeast, and characterization of mutant strains without fructose-2,6-bisphosphate. *Molecular Microbiology* **20**, 65–76.
Boles, E. Schulte, F., Miosga, T., Freidel, K., Schlüter, E., Zimmermann, F. K., Hollenberg, C. P. and Heinisch, J. J. (1997) Characterization of a glucose-repressed pyruvate kinase (Pyk2p) in *Saccharomyces cerevisiae* that is catalytically insensitive to fructose-1,6-bisphosphate. *Journal of Bacteriology* **179**, 2987–2993.
Bonini, B. M., Van Vaeck, C., Larsson, C., Gustafsson, L., Ma, P., Winderickx, J., Van Dijck, P. and Thevelein, J. M. (2000) Expression of *Escherichia coli otsA* in a *Saccharomyces cerevisiae tps1* mutant restores trehalose 6-phosphate levels and partly restores growth and fermentation with glucose and control of glucose influx into glycolysis. *Biochemical Journal* **350**, 261–268.

Bowman, S. B., Zaman, Z., Collinson, L. P., Brown, A. J. and Dawes, I. W. (1992) Positive regulation of the *LPD1* gene of *Saccharomyces cerevisiae* by the HAP2/HAP3/HAP4 activation system. *Molecular and General Genetics* **231**, 296–303.

Brada, D. and Schekman, R. (1988) Coincident localization of secretory and plasma membrane proteins in organelles of the yeast secretory pathway. *Journal of Bacteriology* **170**, 2775–2783.

Bram, R. J., Lue, N. F. and Kornberg, R. D. (1986) A *GAL* family of upstream activating sequences in yeast: roles in both induction and repression of transcription. *EMBO Journal* **5**, 603–608.

Brewster, N. K., Val, D. L., Walker, M. E. and Wallace, J. C. (1994) Regulation of pyruvate carboxylase isozyme (PYC1, PYC2) gene expression in *Saccharomyces cerevisiae* during fermentative and non-fermentative growth. *Archives of Biochemistry and Biophysics* **311**, 62–71.

Brindle, K. M., Fulton, S. M., Gillham, H. and Williams, S. P. (1997) Studies of metabolic control using NMR and molecular genetics. *Journal of Moecularl Recognition* **10**, 182–187.

Brondijk, T. H., van der Rest, M. E., Pluim, D., de Vries, Y., Stingl, K., Poolman, B. and Konings, W. N. (1998) Catabolite inactivation of wild-type and mutant maltose transport proteins in *Saccharomyces cerevisiae*. *Journal of Biological Chemistry* **273**, 15352–15357.

Brondijk, T. H., Konings, W. N. and Poolman, B. (2001) Regulation of maltose transport in *Saccharomyces cerevisiae*. *Archives of Microbiology* **176**, 96–105.

Buckholz, R. G. and Adams, B. G. (1981) Induction and genetics of two alpha-galactosidase activities in *Saccharomyces cerevisiae*. *Molecular and General Genetics* **182**, 77–81.

Buchholz, A., Takors, R. and Wandrey, C. (2001) Quantification of intracellular metabolites in *Escherichia coli* K12 using liquid chromatographic-electrospray ionization tandem mass spectrometric techniques. *Analytical Biochemistry* **295**, 129–137.

Burke, R. L., Tekamp-Olson, P. and Najarian, R. (1983) The isolation, characterization, and sequence of the pyruvate kinase gene of *Saccharomyces cerevisiae*. *Journal of Biological Chemistry* **258**, 2193–2201.

Busturia, A. and Lagunas, R. (1986) Catabolite inactivation of the glucose transport system in *Saccharomyces cerevisiae*. *Journal of General Microbiology* **132**, 379–385.

Butler, G., Dawes, I. W. and McConnell, D. J. (1990) TUF factor binds to the upstream region of the pyruvate decarboxylase structural gene (*PDC1*) of *Saccharomyces cerevisiae*. *Molecular and General Genetics* **223**, 449–456.

Campbell-Burk, S. L. and Shulman, R. G. (1987) High-resolution NMR studies of *Saccharomyces cerevisiae*. *Annual Review of Microbiology* **41**, 595–616.

Carlson, M. (1999) Glucose repression in yeast. *Current Opinion in Micriobiology* **2**, 202–207.

Carlson, M. and Botstein, D. (1982) Two differentially regulated mRNAs with different 5′ ends encode secreted with intracellular forms of yeast invertase. *Cell* **28**, 145–154.

Carlson, M. and Botstein, D. (1983) Organization of the *SUC* gene family in *Saccharomyces*. *Molecular and Cellular Biology* **3**, 351–359.

Carlson, M., Osmond, B. C. and Botstein, D. (1981) Mutants of yeast defective in sucrose utilization. *Genetics* **98**, 25–40.

Carlson, M., Celenza, J. L. and Eng, F. J. (1985) Evolution of the dispersed *SUC* gene family of *Saccharomyces* by rearrangements of chromosome telomeres. *Molecular and Cellular Biology* **5**, 2894–2902.

Carlson, M., Taussig, R., Kustu, S. and Botstein, D. (1983) The secreted form of invertase in *Saccharomyces cerevisiae* is synthesized from mRNA encoding a signal sequence. *Molecular and Cellular Biology* **3**, 439–447.

Carlson, R., Fell, D. and Srienc, F. (2002) Metabolic pathway analysis of a recombinant yeast for rational strain development. *Biotechnology and Bioengineering* **79**, 121–134.

Casal, M., Paiva, S. Andrade, R. P., Gancedo, C. and Leao, C. (1999) The lactate-proton symport of *Saccharomyces cerevisiae* is encoded by *JEN1*. *Journal of Bacteriology* **181**, 2620–2623.

Cascante, M., Curto, R. and Sorribas, A. (1995) Comparative characterization of the fermentation pathway of *Saccharomyces cerevisiae* using biochemical systems theory and metabolic control analysis: steady-state analysis. *Mathematical Biosciences* **130**, 51–69.

Cássio, F., Leão, C. and van Uden, N. (1987) Transport of lactate and other short-chain monocarboxylates in the yeast *Saccharomyces cerevisiae*. *Applied and Environmental Microbiology* **53**, 509–513.

Castrillo, J. I. and Ugalde, U. O. (1994) A general model of yeast energy metabolism in aerobic chemostat culture. *Yeast* **10**, 185–197.

Celenza, J. L., Marshall, C. L. and Carlson, M. (1988) The yeast *SNF3* gene encodes a glucose transporter homologous to the mammalian protein. *Proceedings of the National Academy of Sciences USA* **85**, 2130–2134.

Cereghino, G. P. and Scheffler, I. E. (1996) Genetic analysis of glucose regulation in *Saccharomyces cerevisiae*: control of transcription versus mRNA turnover. *EMBO Journal* **15**, 363–374.

Chambers, A., Tsang, J. S., Stanway, C., Kingsman, A. J. and Kingsman, S. M. (1989) Transcriptional control of the *Saccharomyces cerevisiae PGK* gene by RAP1. *Molecular and Cellular Biology* **9**, 5516–5524.

Chang, Y. S., Dubin, R. A., Perkins, E., Forrest, D., Michels, C. A. and Needleman, R. B. (1988) *MAL63* codes for a positive regulator of maltose fermentation in *Saccharomyces cerevisiae*. *Current Genetics* **14**, 201–209.

Charron, M. J., Dubin, R. A. and Michels, C. A. (1986) Structural and functional analysis of the *MAL1* locus of *Saccharomyces cerevisiae*. *Molecular and Cellular Biology* **6**, 3891–3899.

Charron, M. J., Read, E., Haut, S. R. and Michels, C. A. (1989) Molecular evolution of the telomere-associated *MAL* loci of *Saccharomyces*. *Genetics* **122**, 307–316.

Chelstowska, A. and Butow, R. A. (1995) *RTG* genes in yeast that function in communication between mitochondria and the nucleus are also required for expression of genes encoding peroxisomal proteins. *Journal of Biological Chemistry* **270**, 18141–18146.

Chelstowska, A., Liu, Z., Jia, Y., Amberg, D. and Butow, R. A. (1999) Signalling between mitochondria and the nucleus regulates the expression of a new D-lactate deydrogenase activity in yeast. *Yeast* **15**, 1377–1391.

Cheng, Q. and Michels, C. A. (1989) The maltose permease encoded by the *MAL61* gene of *Saccharomyces cerevisiae* exhibits both sequence and structural homology to other sugar transporters. *Genetics* **123**, 477–484.

Cheng, Q. and Michels, C. A. (1991) *MAL11* and *MAL61* encode the inducible high-affinity maltose transporter of *Saccharomyces cerevisiae*. *Journal of Bacteriology* **173**, 1817–1820.

Cheng, C., Mu, J., Farkas, I., Huang, D. and Goebl, M. G. (1995) Requirement of self-glycosylating initiator proteins, Glg1p and Glg2p, for glycogen accumulation in *Saccharomyces cerevisiae*. *Molecular and Cellular Biology* **15**, 6632–6640.

Chow, T. H., Sollitti, P. and Marmur, J. (1989) Structure of the multigene family of *MAL* loci in *Saccharomyces*. *Molecular and General Genetics* **217**, 60–69.

Chu, F. K., Takase, K., Guarino, D. and Maley, F. (1985) Diverse properties of external and internal forms of yeast invertase derived from the same gene. *Biochemistry* **24**, 6125–6132.

Chu, S., DeRisi, J., Eisen, M., Mulholland, J., Botstein, D., Brown, P. O. and Herskowitz, I. (1998) The transcriptional program of sporulation in budding yeast. *Science* **282**, 699–705.

Ciriacy, M. and Breitenbach, I. (1979) Physiological effects of seven different blocks in glycolysis in *Saccharomyces cerevisiae*. *Journal of Bacteriology* **139**, 152–160.

Clifton, D. and Fraenkel, D. G. (1983) Fructose 2,6-bisphosphate and fructose-6-P 2-kinase in *Saccharomyces cerevisiae* in relation to metabolic state in wild type and fructose-6-P 1-kinase mutant strains. *Journal of Biological Chemistry* **258**, 9245–9249.

Clifton, D., Weinstock, S. B. and Fraenkel, D. G. (1978) Glycolysis mutants in *Saccharomyces cerevisiae*. *Genetics* **88**, 1–11.

Cohen, J. D., Goldenthal, M. J., Buchferer, B. and Marmur, J. (1984) Mutational analysis of the *MAL1* locus of *Saccharomyces*: identification and functional characterization of three genes. *Molecular and General Genetics* **196**, 208–216.

Cohen, J. D., Goldenthal, M. J., Chow, T., Buchferer, B. and Marmur, J. (1985) Organization of the *MAL* loci of *Saccharomyces*. Physical identification and functional characterization of three genes at the *MAL6* locus. *Molecular and General Genetics* **200**, 1–8.

Collins, R. A., McNally, T., Fothergill-Gilmore, L. A. and Muirhead, H. (1995) A subunit interface mutant of yeast pyruvate kinase requires the allosteric activator fructose 1,6-bisphosphate for activity. *Biochemical Journal* **310**, 117–123.

Colombo, S., Ma, P., Cauwenberg, L., Winderickx, J., Crauwels, M., Teunissen, A., Nauwelaers, D., de Winde, J. H., Gorwa, M. F., Colavizza, D. and Thevelein, J. M. (1998) Involvement of distinct G-proteins, Gpa2 and Ras, in glucose- and intracellular acidification-induced cAMP signalling in the yeast *Saccharomyces cerevisiae*. *EMBO Journal* **17**, 3326–3341.

Compagno, C., Ranzi, B. M. and Martegani, E. (1991) The promoter of *Saccharomyces cerevisiae FBA1* gene contains a single positive upstream regulatory element. *FEBS Letters* **293**, 97–100.

Compagno, C., Brambilla, L., Capitanio, D., Boschi, F., Ranzi, B. M. and Porro, D. (2001) Alterations of the glucose metabolism in a triose phosphate isomerase-negative *Saccharomyces cerevisiae* mutant. *Yeast* **18**, 663–670.

Coons, D. M., Vagnoli, P. and Bisson, L. F. (1997) The C-terminal domain of Snf3p is sufficient to complement the growth defect of *snf3* null mutations in *Saccharomyces cerevisiae: SNF3* functions in glucose recognition. *Yeast* **13**, 9–20.

Crauwels, M., Donaton, M. C., Pernambuco, M. B., Winderickx, J., de Winde, J. H. and Thevelein, J. M. (1997) The Sch9 protein kinase in the yeast *Saccharomyces cerevisiae* controls cAPK activity and is required for nitrogen activation of the fermentable-growth-medium-induced (FGM) pathway. *Microbiology* **143**, 2627–2637.

Crocker, W. H. and Bhattacharjee, J. K. (1973) Biosynthesis of glutamic acid in *Saccharomyces cerevisiae*: accumulation of tricarboxylic acid cycle intermediates in a glutamate auxotroph. *Applied Microbiology* **26**, 303–308.

Crumplen, R. M., Slaughter, J. C. and Stewart, G. G. (1996) Characteristics of maltose transporter activity in an ale and large strain of the yeast *Saccharomyces cerevisiae*. *Letters in Applied Microbiology* **23**, 448–452.

Dantigny, P. (1995) Modeling of the aerobic growth of *Saccharomyces cerevisiae* on mixtures of glucose and ethanol in continuous culture. *Journal of Biotechnology* **43**, 213–220.

Day, R. E., Higgins, V. J., Rogers, P. J. and Dawes, I. W. (2002) Characterization of the putative - maltose transporters encoded by YDL247w and YJR160c. *Yeast* **19**, 1015–1027.

De Jong, L., Elzinga, S. D. J., McCammon, M.T. (2000) Increased synthesis and decreased stability of mitochondrial translation products in yeast as a result of loss of mitochondrial NAD^+-dependent isocitrate dehydrogenase. *FEBS Letters* **483**, 62–66.

de Jong-Gubbels, P., Vanrolleghem, P., Heijnen, S., van Dijken, J. P. and Pronk, J. T. (1995) Regulation of carbon metabolism in chemostat cultures of *Saccharomyces cerevisiae* grown on mixtures of glucose and ethanol. *Yeast* **11**, 407–418.

den Hollander, J., Ugurbil, K., Brown, T. R., Bednar, M., Redfield, C. and Shulman, R. G. (1986) Studies of anaerobic and aerobic glycolysis in *Saccharomyces cerevisiae*. *Biochemistry* **25**, 203–211.

De Virgilio, C., Bűrckert, N., Bell, W., Jenő, T. and Wiemken, A. (1993) Disruption of *TPS2*, the gene encoding the 100-kDa subunit of the trehalose-6-phosphate synthase/phosphatase complex in *Saccharomyces cerevisiae*, causes accumulation of trehalose-6-phosphate and loss of trehalose-6-phosphatase activity. *European Journal of Biochemistry* **212**, 315–323.

De Vit, M. J., Waddle, J. A. and Johnston, M. (1997) Regulated nuclear translocation of the Mig1 glucose repressor. *Molecular Biology of the Cell* **8**, 1603–1618.

De Winde, J. H., Crauwels, M., Hohmann, S., Thevelein, J. M. and Winderickx, J. (1996) Differential requirement of the yeast sugar kinases for sugar sensing in establishing the catabolite-repressed state. *European Journal of Biochemistry* **241**, 633–643.

del Castillo Agudo, L., Nieto Soria, A. and Sentandreu, R. (1992) Differential expression of the invertase-encoding *SUC* genes in *Saccharomyces cerevisiae*. *Gene* **120**, 59–65.

Delneri, D., Gardner, D. C. J. and Oliver, S. G. (1999a) Analysis of the seven-member *AAD* gene set demonstrates that genetic redundancy in yeast may be more apparent than real. *Genetics*, **153**, 1591–1600.

Delneri, D., Gardner, D. C. J., Bruschi, C. V. and Oliver, S. G. (1999b) Disruption of seven hypothetical aryl alcohol dehydrogenase genes from *Saccharomyces cerevisiae* and construction of a multiple knock-out strain *Yeast* **15**, 1681–1698.

Denis, C. L., Ferguson, J. and Young, E. T. (1983) mRNA levels for the fermentative alcohol dehydrogenase of *Saccharomyces cerevisiae* decrease upon growth on a nonfermentable carbon source. *Journal of Biological Chemistry* **258**, 1165–1171.

DeRisi, J. L., Iyer, V. R. and Brown, P. O. (1997) Exploring the metabolic and genetic control of gene expression on a genomic scale. *Science* **278**, 680–686.

Destruelle, M., Holzer, H. and Klionsky, D. J. (1995) Isolation and characterization of a novel yeast gene, *ATH1*, that is required for vacuolar acid trehalase activity. *Yeast* **11**, 1015–1025.

Dickinson, J. R. and Dawes, I. W. (1992) The catabolism of branched-chain amino acids occurs via 2-oxoacid dehydrogenase in *Saccharomyces cerevisiae*. *Journal of General Microbiology* **138**, 2022–2033.

Dickinson, J. R., Roy, D. J. and Dawes, I. W. (1986) A mutation affecting lipoamide dehydrogenase, pyruvate dehydrogenase and 2-oxoglutarate dehydrogenase activities in *Saccharomyces cerevisiae*. *Molecular and General Genetics* **204**, 103–107.

Dickinson, J. R., Sobanski, M. A. and Hewlins, M. J. E. (1995) In *Saccharomyces cerevisiae* deletion of phosphoglucose isomerase can be suppressed by increased activities of enzymes of the hexose monophosphate pathway. *Microbiology* **141**, 385–391.

Dickinson, J. R., Lanterman, M. M., Danner, D. J., Pearson, B. M., Sanz, P., Harrison, S. J. and Hewlins, M. J. (1997) A ^{13}C nuclear magnetic resonance investigation of the metabolism of leucine to isoamyl alcohol in *Saccharomyces cerevisiae*. *Journal of Biological Chemistry* **272**, 26871–26878.

Dickinson, J. R., Harrison, S. J. and Hewlins, M. J. (1998) An investigation of the metabolism of valine to isobutyl alcohol in *Saccharomyces cerevisiae*. *Journal of Biological Chemistry* **273**, 25751–25756.

Dickinson, J. R., Harrison, S. J., Dickinson, J. A. and Hewlins, M. J. (2000) An investigation of the metabolism of isoleucine to active amyl alcohol in *Saccharomyces cerevisiae*. *Journal of Biological Chemistry* **275**, 10937–10942.

Diderich, J. A., Schepper, M., van Hoek, P., Luttik, M. A. H., van Dijken, J. P., Pronk, J. T., Klaassen, P., Boelens, H. F. M., Teixeira de Mattos, M. J., van Dam, K. and Kruckeberg, A. L. (1999a) Glucose uptake kinetics and transcription of *HXT* genes in chemostat cultures of *Saccharomyces cerevisiae*. *Journal of Biological Chemistry* **274**, 15350–15359.

Diderich, J. A., Teusink, B., Valkier, J., Anjos, J., Spencer-Martins, I., van Dam, K. and Walsh, M. C. (1999b) Strategies to determine the extent of control exerted by glucose transport on glycolytic flux in the yeast *Saccharomyces bayanus*. *Microbiology* **145**, 3447–3454.

Diderich, J. A., Raamsdonk, L. M., Kruckeberg, A. L., Berden, J. A. and Van Dam, K. (2001a) Physiological properties of *Saccharomyces cerevisiae* from which hexokinase II has been deleted. *Applied and Environmental Microbiology* **67**, 1587–1593.

Diderich, J. A., Schuurmans, J. M., Van Gaalen, M. C., Kruckeberg, A. L. and Van Dam, K. (2001b) Functional analysis of the hexose transporter homologue *HXT5* in *Saccharomyces cerevisiae*. *Yeast* **18**, 1515–1524.

Didion, T., Regenberg, B., Jorgensen, M. U., Kielland-Brandt, M. C. and Andersen, H. A. (1998) The permease homologue Ssy1p controls the expression of amino acid and peptide transporter genes in *Saccharomyces cerevisiae*. *Molecular Microbiology* **27**, 643–650.

Douglas, H. C. and Condie, F. (1954) The genetic control of galactose utilization in *Saccharomyces*. *Journal of Bacteriology* **68**, 662–670.

Douglas, H. C. and Hawthorne, D. C. (1964) Enzymatic expression and genetic linkage of genes controlling galactose utilization in *Saccharomyces*. *Genetics* **49**, 837–844.

D'Souza, C. A. and Heitman, J. (2001) Conserved cAMP signaling cascades regulate fungal development and virulence. *FEMS Microbiology Reviews* **25**, 349–364.

Eberhardt, I., Cederberg, H., Li, H., Konig, S., Jordan, F. and Hohmann, S. (1999) Autoregulation of yeast pyruvate decarboxylase gene expression requires the enzyme but not its catalytic activity. *European Journal of Biochemistry* **262**, 191–201.

Eddy, A. A. (1982) Mechanisms of solute transport in selected eukaryotic micro-organisms. *Advances in Microbial Physiology* **23**, 2–78.

Edwards, J. S., Ibarra, R. U. and Palsson, B. O. (2001) In silico predictions of *Escherichia coli* metabolic capabilities are consistent with experimental data. *Nature Biotechnology* **19**, 125–130.

Elorza, M. V., Lostau, C. M., Villanueva, J. R. and Sentandreu, R. (1977a) Invertase messenger ribonucleic acid in *Saccharomyces cerevisiae*. Kinetics of formation and decay. *Biochimica et Biophysica Acta* **475**, 638–651.

Elorza, M. V., Villanueva, J. R. and Sentandreu, R. (1977b) The mechanism of catabolite inhibition of invertase by glucose in *Saccharomyces cerevisiae*. *Biochimica et Biophysica Acta* **475**, 103–112.

Entian, K.-D. (1997) Sugar Phosphorylation in Yeast. In: F. K. Zimmermann and K.-D. Entian (eds), *Yeast Sugar Metabolism. Biochemistry, Genetics, Biotechnology, and Applications*, pp. 67–79. Lancaster: Technomic.

Entian, K. D., Frohlich, K. U. and Mecke, D. (1984) Regulation of enzymes and isoenzymes of carbohydrate metabolism in the yeast *Saccharomyces cerevisiae*. *Biochimica et Biophysica Acta* **799**, 181–186.

Entian, K. D., Meurer, B., Kohler, H., Mann, K. H. and Mecke, D. (1987) Studies on the regulation of enolases and compartmentation of cytosolic enzymes in *Saccharomyces cerevisiae*. *Biochimica et Biophysica Acta* **923**, 214–221.

Entian, K. D., Vogel, R. F., Rose, M., Hofmann, L. and Mecke, D. (1988) Isolation and primary structure of the gene encoding fructose-1,6-bisphosphatase from *Saccharomyces cerevisiae*. *FEBS Letters* **236**, 195–200.

Ernandes, J. R., De Meirsman, C., Rolland, F., Winderickx, J., de Winde, J., Brandao, R. L. and Thevelein, J. M. (1998) During the initiation of fermentation overexpression of hexokinase PII in yeast transiently causes a similar deregulation of glycolysis as deletion of Tps1. *Yeast* **14**, 255–269.

Farkas, I., Hardy, T. A., Goebl, M. G. and Roach, P. J. (1991) Two glycogen synthase isoforms in *Saccharomyces cerevisiae* are coded by distinct genes that are differentially controlled. *Journal of Biological Chemistry* **266**, 15602–15607.

Federoff, H. J., Eccleshall, T. R. and Marmur, J. (1983) Regulation of maltase synthesis in *Saccharomyces carlsbergensis*. *Journal of Bacteriology* **154**, 1301–1308.

Fell, D. A. (1997) *Understanding the Control of Metabolism*, 1st edn. London: Portland Press.

Fernández, E., Moreno, F. and Rodicio, R. (1992) The *ICL1* gene from *Saccharomyces cerevisiae*. *European Journal of Biochemistry* **204**, 983–990.

Ferreira, J. C., Silva, J. T. and Panek, A. D. (1996) A regulatory role for *TSL1* on trehalose synthase activity. *Biochemistry and Molecular Biology International* **38**, 259–265.

Ferreira, J. C., Panek, A. D. and de Araujo, P. S. (2000) Inactivation of maltose permease and maltase in sporulating *Saccharomyces cerevisiae*. *Canadian Journal of Microbiology* **46**, 383–386.

Fiehn, O., Kopka, J., Dormann, P., Altmann, T., Trethewey, R. N. and Willmitzer, L. (2000) Metabolite profiling for plant functional genomics. *Nature Biotechnology* **18**, 1157–1161.

Fletcher, T. S., Kwee, I. L., Nakada, T., Largman, C. and Martin, B. M. (1992) DNA sequence of the yeast transketolase gene. *Biochemistry* **31**, 1892–1896.

Flikweert, M. T., Van Der Zanden, L., Janssen, W. M., Steensma, H. Y., Van Dijken, J. P. and Pronk, J. T. (1996) Pyruvate decarboxylase: an indispensable enzyme for growth of *Saccharomyces cerevisiae* on glucose. *Yeast* **12**, 247–257.

Flikweert, M. T., van Dijken, J. P. and Pronk, J. T. (1997) Metabolic responses of pyruvate decarboxylase-negative *Saccharomyces cerevisiae* to glucose excess. *Applied and Environmental Microbiology* **63**, 3399–3404.

Forsburg, S. L. and Guarente, L. (1989) Identification and characterization of HAP4: a third component of the CCAAT-bound HAP2/HAP3 heteromer. *Genes and Development* **3**, 1166–1178.

Forster, J., Gombert, A. K. and Nielsen, J. (2002) A functional genomics approach using metabolomics and in silico pathway analysis. *Biotechnology and Bioengineering* **79**, 703–712.

Fothergill-Gilmore, L. A. and Watson, H. C. (1990) Phosphoglycerate mutases. *Biochemical Society Transactions* **18**, 190–193.

Francois, J., Van Schaftingen, E. and Hers, H. G. (1984) The mechanism by which glucose increases fructose 2,6-bisphosphate concentration in *Saccharomyces cerevisiae*. A cyclic-AMP-dependent activation of phosphofructokinase 2. *European Journal of Biochemistry* **145**, 187–193.
Francois, J., Eraso, P. and Gancedo, C. (1987) Changes in the concentration of cAMP, fructose 2,6-bisphosphate and related metabolites and enzymes in *Saccharomyces cerevisiae* during growth on glucose. *European Journal of Biochemistry* **164**, 369–373.
Francois, J., Van Schaftigen, E. and Hers, H. G. (1988a) Characterization of phosphofructokinase 2 and of enzymes involved in the degradation of fructose 2,6-bisphosphate in yeast. *European Journal of Biochemistry* **171**, 599–608.
Francois, J., Villanueva, M. E. and Hers, H. G. (1988b) The control of glycogen metabolism in yeast. 1. Interconversion *in vivo* of glycogen synthase and glycogen phosphorylase induced by glucose, a nitrogen source or uncouplers. *European Journal of Biochemistry* **174**, 551–559.
Fröhlich, K.-U., Entian, K.-D. and Mecke, D. (1985) The primary structure of the yeast hexokinase PII gene (*HXK2*) which is responsible for glucose repression. *Gene* **36**, 105–111.
Frolova, E., Johnston, M. and Majors, J. (1999) Binding of the glucose-dependent Mig1p repressor to the *GAL1* and *GAL4* promoters in vivo: regulationby glucose and chromatin structure. *Nucleic Acids Research* **27**, 1350–1358.
Gamo, F. J., Lafuente, M. J. and Gancedo, C. (1994) The mutation *DGT1-1* decreases glucose transport and alleviates carbon catabolite repression in *Saccharomyces cerevisiae*. *Journal of Bacteriology* **176**, 7423–7429.
Gancedo, C. (1971) Inactivation of fructose-1,6-diphosphatase by glucose in yeast. *Journal of Bacteriology*, **107**, 401–405.
Gancedo, J. M., Mazon, M. J. and Gancedo, C. (1982) Kinetic differences between two interconvertible forms of fructose-1,6-bisphosphatase from *Saccharomyces cerevisiae*. *Archives of Biochemistry and Biophysics* **218**, 478–482.
Gasch, A. P., Spellman, P. T., Kao, C. M., Carmel-Harel, O., Eisen, M. B., Storz, G., Botstein, D. and Brown, P. O. (2000) Genomic expression programs in the response of yeast cells to environmental changes. *Molecular Biology of the Cell* **11**, 4241–4257.
Gibson, A. W., Wojciechowicz, L. A., Danzi, S. E., Zhang, B., Kim, J. H., Hu, Z. and Michels, C. A. (1997) Constitutive mutations of the *Saccharomyces cerevisiae MAL*-activator genes *MAL23*, *MAL43*, *MAL63*, and *mal64*. *Genetics* **146**, 1287–1298.
Gillies, R. J., Ugurbil, K., Hollander, J. A. D. and Shulman, R. G. (1981) ^{31}P NMR studies of intracellular pH and phsophate metabolism during cell division cycle of *Saccharomyces cerevisiae*. *Proceedings of the National Academy of Sciences USA* **78**, 2125–2129.
Giniger, E., Varnum, S. M. and Ptashne, M. (1985) Specific DNA binding of GAL4, a positive regulatory protein of yeast. *Cell* **40**, 767–774.
Giuseppin, M. L. and van Riel, N. A. (2000) Metabolic modeling of *Saccharomyces cerevisiae* using the optimal control of homeostasis: a cybernetic model definition. *Metabolic Engineering* **2**, 14–33.
Goffeau, A., Barrell, B. G., Bussey, H., Davis, R. W., Dujon, B., Feldmann, H., Galibert, F., Hoheisel, J. D., Jacq, C., Johnston, M., Louis, E. J., Mewes, H. W., Murakami, Y., Philippsen, P., Tettelin, H. and Oliver, S. G. (1996) Life with 6000 genes. *Science* **274**, 546, 563–567.
Gombert, A. K., Moreira dos Santos, M., Christensen, B. and Nielsen, J. (2001) Network identification and flux quantification in the central metabolism of *Saccharomyces cerevisiae* under different conditions of glucose repression. *Journal of Bacteriology* **183**, 1441–1451.
González, E., Fernádez, M. R., Larroy, C., Sola, L., Pericàs, M. A., Parés, X. and Biosca, J. A. (2000) Characterization of a (2R,3R)-2,3-butanediol dehydrogenase as the *Saccharomyces cerevisiae* YAL060W gene product. Disruption and induction of the gene. *Journal of Biological Chemistry* **275**, 35876–35885.
Görts, C. P. M. (1969) Effect of glucose on the activity and the kinetics of the maltose uptake system and of alpha-glucosidase in *Saccharomyces cerevisiae*. *Antonie Van Leeuwenhoek* **35**, 233–234.
Grauslund, M. and Rønnow, B. (2000) Carbon source-dependent transcriptional regulation of the mitochondrial glycerol-3-phosphate dehydrogenase gene, *GUT2*, from *Saccharomyces cerevisiae*. *Canadian Journal of Microbiology* **46**, 1069–1100.

Green, J. B., Wright, A. P., Cheung, W. Y., Lancashire, W. E. and Hartley, B. S. (1988) The structure and regulation of phosphoglucose isomerase in *Saccharomyces cerevisiae*. *Molecular and General Genetics* **215**, 100–106.

Griffin, T. J., Gygi, S. P., Ideker, T., Rist, B., Eng, J., Hood, L. and Aebersold, R. (2002) Complementary profiling of gene expression at the transcriptome and proteome levels in *Saccharomyces cerevisiae*. *Molecular and Cellular Proteomics* **1**, 323–333.

Griggs, D. W. and Johnston, M. (1991) Regulated expression of the *GAL4* activator gene in yeast provides a sensitive genetic switch for glucose repression. *Proceedings of the National Academy of Sciences USA* **88**, 8597–8601.

Grivet, J. P. (2001) NMR and microorganisms. *Current Issues in Molecular Biology* **3**, 7–14.

Groussac, E., Ortiz, M. and Francois, J. (2000) Improved protocols for quantitative determination of metabolites from biological samples using high performance ionic-exchange chromatography with conductimetric and pulsed amperometric detection. *Enzyme and Microbial Technology* **26**, 715–723.

Haaarasilta, S. and Oura, E. (1975) On the activity and regulation of anaplerotic and gluconeogenic enzymes during the growth processes of baker's yeast. *European Journal of Biochemistry* **52**, 1–27.

Haeckel, R., Hess, B., Lauterborn, W. and Wuster, K. H. (1968) Purification and allosteric properties of yeast pyruvate kinase. *Hoppe Seyler's Zeitschrift für Physiologische Chemie* **349**, 699–714.

Han, E. K., Cotty, F., Sottas, C., Jiang, H. and Michels, C. A. (1995) Characterization of *AGT1* encoding a general alpha-glucoside transporter from *Saccharomyces*. *Molecular Microbiology* **17**, 1093–1107.

Harashima, T. and Heitman, J. (2002) The Galpha protein Gpa2 controls yeast differentiation by interacting with kelch repeat proteins that mimic Gbeta subunits. *Molecular Cell* **10**, 163–173.

Hardie, D. G. and Carling, D. (1997) The AMP-activated protein kinase – fuel gauge of the mammalian cell? *European Journal of Biochemistry* **246**, 259–273.

Hartig, A., Simon, M. M., Schuster, T., Daugherty, J. R., Yoo, H-S. and Cooper, T. G. (1992) Differentially regulated malate synthase genes participate in carbon and nitrogen metabolism of *S. cerevisiae*. *Nucleic Acids Research* **20**, 5677–5686.

Hauf, J., Zimmermann, F. K. and Muller, S. (2000) Simultaneous genomic overexpression of seven glycolytic enzymes in the yeast *Saccharomyces cerevisiae*. *Enzyme and Microbial Technology* **26**, 688–698.

Haurie, V., Perrot, M., Mini, T., Jeno, P., Sagliocco, F. and Boucherie, H. (2001) The transcriptional activator Cat8p provides a major contribution to the reprogramming of carbon metabolism during the diauxic shift in *Saccharomyces cerevisiae*. *Journal of Biological Chemistry* **276**, 76–85.

Hauser, M., Narita, V., Donhardt, A. M., Naider, F. and Becker, J. M. (2001) Multiplicity and regulation of genes encoding peptide transporters in *Saccharomyces cerevisiae*. *Molecular Membrane Biology* **18**, 105–112.

Hedges, D., Proft, M. and Entian, K. D. (1995) *CAT8*, a new zinc cluster-encoding gene necessary for derepression of gluconeogenic enzymes in the yeast *Saccharomyces cerevisiae*. *Molecular and Cellular Biology* **15**, 1915–1922.

Heinisch, J. J. (1986) Isolation and characterization of the two structural genes coding for phosphofructokinase in yeast. *Molecular and General Genetics* **202**, 75–82.

Heinisch, J. J., Ritzel, R. G., von, B. R., Aguilera, A., Rodicio, R. and Zimmermann, F. K. (1989) The phosphofructokinase genes of yeast evolved from two duplication events. *Gene* **78**, 309–321.

Heinisch, J. J., Vogelsang, K. and Hollenberg, C. P. (1991a) Transcriptional control of yeast phosphofructokinase gene expression. *FEBS Letters* **289**, 77–82.

Heinisch, J. J., von Borstel, R. C. and Rodicio, R. (1991b) Sequence and localization of the gene encoding yeast phosphoglycerate mutase. *Current Genetics* **20**, 167–171.

Heinisch, J. J., Muller, S., Schluter, E., Jacoby, J. and Rodicio, R. (1998) Investigation of two yeast genes encoding putative isoenzymes of phosphoglycerate mutase. *Yeast* **14**, 203–213.

Herrera, L. S., Pascual, C. and Alvarez, X. (1976) Genetic and biochemical studies of phosphomannose isomerase deficient mutants of *Saccharomyces cerevisiae*. *Molecular and General Genetics* **144**, 223–230.

Herrero, P., Fernández, R. and Moreno, F. (1985) Differential sensitivities to glucose and galactose repression of gluconeogenic and respiratory enzymes from *Saccharomyces cerevisiae*. *Archives of Microbiology* **143**, 216–219.

Herrero, P., Galindez, J., Ruiz, N., Martinez-Campa, C. and Moreno, F. (1995) Transcriptional regulation of the *Saccharomyces cerevisiae HXK1, HXK2* and *GLK1* genes. *Yeast* **11**, 137–144.
Herrero, P., Martinez-Campa, C. and Moreno, F. (1998) The hexokinase 2 protein participates in regulatory DNA-protein complexes necessary for glucose repression of the *SUC2* gene in *Saccharomyces cerevisiae. FEBS Letters* **434**, 71–76.
Herrero, P., Flores, L., de la Cera, T. and Moreno, F. (1999) Functional characterization of transcriptional regulatory elements in the upstream region of the yeast *GLK1* gene. *Biochemical Journal* **343**, 319–325.
Herve, M., Buffin-Meyer, B., Bouet, F. and Son, T. D. (2000) Detection of modifications in the glucose metabolism induced by genetic mutations in *Saccharomyces cerevisiae* by ^{13}C- and ^{1}H-NMR spectroscopy. *European Journal of Biochemistry* **267**, 3337–3344.
Hess, B., Haeckel, R. and Brand, K. (1966) FDP-activation of yeast pyruvate kinase. *Biochemical and Biophysical Research Communications* **24**, 824–831.
Hicke, L. (1997) Ubiquitin-dependent internalization and down-regulation of plasma membrane proteins. *FASEB Journal* **11**, 1215–1226.
Hicke, L. (1999) Getting down with ubiquitin: turning off cell-surface receptors, transporters and channels. *Trends in Cell Biology* **9**, 107–112.
Higgins, V. J., Braidwood, M., Bell, P., Bissinger, P., Dawes, I. W. and Attfield, P. V. (1999) Genetic evidence that high noninduced maltase and maltose permease activities, governed by *MALx3*-encoded transcriptional regulators, determine efficiency of gas production by Baker's yeast in unsugared dough. *Applied and Environmental Microbiology* **65**, 680–685.
Hirimburegama, K., Durnez, P., Keleman, J., Oris, E., Vergauwen, R., Mergelsberg, H. and Thevelein, J. M. (1992) Nutrient-induced activation of trehalase in nutrient-starved cells of the yeast *Saccharomyces cerevisiae*: cAMP is not involved as second messenger. *Journal of General Microbiology* **138**, 2035–2043.
Hitzeman, R. A., Clarke, L. and Carbon, J. (1980) Isolation and characterization of the yeast 3-phosphoglycerokinase gene (*PGK*) by an immunological screening technique. *Journal of Biological Chemistry* **255**, 12073–12080.
Hitzeman, R. A., Hagie, F. E., Hayflick, J. S., Chen, C. Y., Seeburg, P. H. and Derynck, R. (1982) The primary structure of the *Saccharomyces cerevisiae* gene for 3-phosphoglycerate kinase. *Nucleic Acids Research* **10**, 7791–7808.
Hofmann, E., Bedri, A., Kessler, R., Kretschmer, M. and Schellenberger, W. (1989) 6-Phosphofructo-2-kinase and fructose-2,6-bisphosphatase from *Saccharomyces cerevisiae. Advances in Enzyme Regulation* **28**, 283–306.
Hohmann, S. (1991) *PDC6*, a weakly expressed pyruvate decarboxylase gene from yeast, is activated when fused spontaneously under the control of the *PDC1* promoter. *Current Genetics* **20**, 373–378.
Hohmann, S. (2002) Osmotic stress signaling and osmoadaptation in yeasts. *Microbiology and Molecular Biology Reviews* **66**, 300–372.
Hohmann, S. and Zimmermann, F. K. (1986) Cloning and expression on a multicopy vector of five invertase genes of *Saccharomyces cerevisiae. Current Genetics* **11**, 217–225.
Hohmann, S. and Gozalbo, D. (1988) Structural analysis of the 5′ regions of yeast SUC genes revealed analogous palindromes in *SUC, MAL* and *GAL. Molecular and General Genetics* **211**, 446–454.
Hohmann, S. and Cederberg, H. (1990) Autoregulation may control the expression of yeast pyruvate decarboxylase structural genes *PDC1* and *PDC5. European Journal of Biochemistry* **188**, 615–621.
Hohmann, S., Bell, W., Neves, M. J., Valckx, D. and Thevelein, J. M. (1996) Evidence for trehalose-6-phosphate-dependent and -independent mechanisms in the control of sugar influx into yeast glycolysis. *Molecular Microbiology* **20**, 981–991.
Hohmann, S., Neves, M. J., de Koning, W., Alijo, R., Ramos, J. and Thevelein, J. M. (1993) The growth and signalling defects of the *ggs1 (fdp1/byp1)* deletion mutant on glucose are suppressed by a deletion of the gene encoding hexokinase PII. *Current Genetics* **23**, 281–289.
Hohmann, S., Winderickx, J., de Winde, J. H., Valckx, D., Cobbaert, P., Luyten, K., de Meirsman, C., Ramos, J. and Thevelein, J. M. (1999) Novel alleles of yeast hexokinase PII with distinct effects on catalytic activity and catabolite repression of *SUC2. Microbiology* **145**, 703–714.

Holland, J. P. and Holland, M. J. (1979) The primary structure of a glyceraldehyde-3-phosphate dehydrogenase gene from *Saccharomyces cerevisiae*. *Journal of Biological Chemistry* **254**, 9839–9845.

Holland, J. P. and Holland, M. J. (1980) Structural comparison of two nontandemly repeated yeast glyceraldehyde-3-phosphate dehydrogenase genes. *Journal of Biological Chemistry* **255**, 2596–2605.

Holland, M. J., Holland, J. P., Thill, G. P. and Jackson, K. A. (1981) The primary structures of two yeast enolase genes. Homology between the 5′ noncoding flanking regions of yeast enolase and glyceraldehyde-3-phosphate dehydrogenase genes. *Journal of Biological Chemistry* **256**, 1385–1395.

Holland, M. J., Yokoi, T., Holland, J. P., Myambo, K. and Innis, M. A. (1987) The *GCR1* gene encodes a positive transcriptional regulator of the enolase and glyceraldehyde-3-phosphate dehydrogenase gene families in *Saccharomyces cerevisiae*. *Molecular and Cellular Biology* **7**, 813–820.

Holst, B., Lunde, C., Lages, F., Oliveira, R., Lucas, C. and Kielland-Brandt, M. C. (2000) *GUP1* and its close homologue *GUP2*, encoding multimembrane-spanning proteins involved in active glycerol uptake in *Saccharomyces cerevisiae*. *Molecular Microbiology* **37**, 108–124.

Homann, M. J., Poole, M. A., Gaynor, P. M., Ho, C. T. and Carman, G. M. (1987) Effect of growth phase on phospholipid biosynthesis in *Saccharomyces cerevisiae*. *Journal of Bacteriology* **169**, 533–539.

Hong, S. H. and Marmur, J. (1987) Upstream regulatory regions controlling the expression of the yeast maltase gene. *Molecular and Cellular Biology* **7**, 2477–2483.

Horak, J. and Wolf, D. H. (1997) Catabolite inactivation of the galactose transporter in the yeast *Saccharomyces cerevisiae*: ubiquitination, endocytosis, and degradation in the vacuole. *Journal of Bacteriology* **179**, 1541–1549.

Horak, J. and Wolf, D. H. (2001) Glucose-induced monoubiquitination of the *Saccharomyces cerevisiae* galactose transporter is sufficient to signal its internalization. *Journal of Bacteriology* **183**, 3083–3088.

Horak, J., Regelmann, J. and Wolf, D. H. (2002) Two distinct proteolytic systems responsible for glucose-induced degradation of fructose-1,6-bisphosphatase and the Gal2p transporter in the yeast *Saccharomyces cerevisiae* share the same protein components of the glucose signaling pathway. *Journal of Biological Chemistry* **277**, 8248–8254.

Hu, Z., Nehlin, J. O., Ronne, H. and Michels, C. A. (1995) *MIG1*-dependent and *MIG1*-independent glucose regulation of *MAL* gene expression in *Saccharomyces cerevisiae*. *Current Genetics* **28**, 258–266.

Hu, Z., Yue, Y., Jiang, H., Zhang, B., Sherwood, P. W. and Michels, C. A. (2000) Analysis of the mechanism by which glucose inhibits maltose induction of *MAL* gene expression in *Saccharomyces*. *Genetics* **154**, 121–132.

Hubner, G., Weidhase, R. and Schellenberger, A. (1978) The mechanism of substrate activation of pyruvate decarboxylase: a first approach. *European Journal of Biochemistry* **92**, 175–181.

Huet, C., Menendez, J., Gancedo, C. and Francois, J. M. (2000) Regulation of pyc1 encoding pyruvate carboxylase isozyme I by nitrogen sources in *Saccharomyces cerevisiae*. *European Journal of Biochemistry* **267**, 6817–6823.

Huibregtse, J. M., Good, P. D., Marczynski, G. T., Jaehning, J. A. and Engelke, D. R. (1993) Gal4 protein binding is required but not sufficient for derepression and induction of *GAL2* expression. *Journal of Biological Chemistry* **268**, 22219–22222.

Hwang, P. K., Tugendreich, S. and Fletterick, R. J. (1989) Molecular analysis of *GPH1*, the gene encoding glycogen phosphorylase in *Saccharomyces cerevisiae*. *Molecular and Cellular Biology* **9**, 1659–1665.

Hynne, F., Dano, S. and Sorensen, P. G. (2001) Full-scale model of glycolysis in *Saccharomyces cerevisiae*. *Biophysical Chemistry* **94**, 121–163.

Ideker, T., Thorsson, V., Ranish, J. A., Christmas, R., Buhler, J., Eng, J. K., Bumgarner, R., Goodlett, D. R., Aebersold, R. and Hood, L. (2001) Integrated genomic and proteomic analyses of a systematically perturbed metabolic network. *Science* **292**, 929–934.

Iraqui, I., Vissers, S., Bernard, F., de Craene, J. O., Boles, E., Urrestarazu, A. and Andre, B. (1999) Amino acid signaling in *Saccharomyces cerevisiae*: a permease-like sensor of external amino acids and F-Box protein Grr1p are required for transcriptional induction of the *AGP1* gene, which encodes a broad-specificity amino acid permease. *Molecular and Cellular Biology* **19**, 989–1001.

Janssen, M. J., Koorengevel, M. C., de Kruijff, B. and de Kroon, A. I. (2000) The phosphatidylcholine to phosphatidylethanolamine ratio of *Saccharomyces cerevisiae* varies with the growth phase. *Yeast* **16**, 641–650.

Jespersen, L., Cesar, L. B., Meaden, P. G. and Jakobsen, M. (1999) Multiple alpha-glucoside transporter genes in Brewer's yeast. *Applied and Environmental Microbiology* **65**, 450–456.

Jiang, H., Medintz, I. and Michels, C. A. (1997) Two glucose sensing/signaling pathways stimulate glucose-induced inactivation of maltose permease in *Saccharomyces. Molecular Biology of the Cell* **8**, 1293–1304.

Jiang, Y., Davis, C. and Broach, J. R. (1998) Efficient transition to growth on fermentable carbon sources in *Saccharomyces cerevisiae* requires signaling through the Ras pathway. *EMBO Journal* **17**, 6942–6951.

Jiang, H., Medintz, I., Zhang, B. and Michels, C. A. (2000a) Metabolic signals trigger glucose-induced inactivation of maltose permease in *Saccharomyces. Journal of Bacteriology* **182**, 647–654.

Jiang, H., Tatchell, K., Liu, S. and Michels, C. A. (2000b) Protein phosphatase type-1 regulatory subunits Reg1p and Reg2p act as signal transducers in the glucose-induced inactivation of maltose permease in *Saccharomyces cerevisiae. Molecular and General Genetics* **263**, 411–422.

Johnston, M., Flick, J. S. and Pexton, T. (1994) Multiple mechanisms provide rapid and stringent glucose repression of GAL gene expression in *Saccharomyces cerevisiae. Molecular and Cellular Biology* **14**, 3834–3841.

Jorgensen, M. U., Bruun, M. B., Didion, T. and Kielland-Brandt, M. C. (1998) Mutations in five loci affecting *GAP1*-independent uptake of neutral amino acids in yeast. *Yeast* **14**, 103–114.

Jones, K. D. and Kompala, D. S. (1999) Cybernetic model of the growth dynamics of *Saccharomyces cerevisiae* in batch and continuous cultures. *Journal of Biotechnology* **71**, 105–131.

Kacser, H. and Burns, J. A. (1981) The molecular basis of dominance. *Genetics* **97**, 639–666.

Kawasaki, G. and Fraenkel, D. G. (1982) Cloning of yeast glycolysis genes by complementation. *Biochemical and Biophysical Research Communications* **108**, 1107–1122.

Keleher, C. A., Redd, M. J., Schultz, J., Carlson, M. and Johnson, A. D. (1992) Ssn6-Tup1 is a general repressor of transcription in yeast. *Cell* **68**, 709–719.

Kellermann, E. and Hollenberg, C. P. (1988) The glucose- and ethanol-dependent regulation of *PDC1* from *Saccharomyces cerevisiae* are controlled by two distinct promoter regions. *Current Genetics* **14**, 337–344.

Kellermann, E., Seeboth, P. G. and Hollenberg, C. P. (1986) Analysis of the primary structure and promoter function of a pyruvate decarboxylase gene (*PDC1*) from *Saccharomyces cerevisiae. Nucleic Acids Research* **14**, 8963–8977.

Kern, G., Schulke, N., Schmid, F. X. and Jaenicke, R. (1992) Stability, quaternary structure, and folding of internal, external, and core-glycosylated invertase from yeast. *Protein Science* **1**, 120–131.

Kew, O. M. and Douglas, H. C. (1976) Genetic co-regulation of galactose and melibiose utilization in *Saccharomyces. Journal of Bacteriology* **125**, 33–41.

Khan, N. A., Zimmermann, F. K. and Eaton, N. R. (1973) Genetic and biochemical evidence of sucrose fermentation by maltase in yeast. *Molecular and General Genetics* **123**, 43–50.

Kim, J. and Michels, C. A. (1988) The *MAL63* gene of *Saccharomyces* encodes a cysteine-zinc finger protein. *Current Genetics* **14**, 319–323.

Klaassen, P. and Raamsdonk, L. (1998) Contribution of the individual *HXT* gene products to CO_2 production. *Folia Microbiologica* **43**, 197–200.

Klasson, H., Fink, G. R. and Ljungdahl, P. O. (1999) Ssy1p and Ptr3p are plasma membrane components of a yeast system that senses extracellular amino acids. *Molecular and Cellular Biology* **19**, 5405–5416.

Klein, C. J., Olsson, L., Ronnow, B., Mikkelsen, J. D. and Nielsen, J. (1996) Alleviation of glucose repression of maltose metabolism by *MIG1* disruption in *Saccharomyces cerevisiae. Applied and Environmental Microbiology* **62**, 4441–4449.

Kodama, Y., Omura, F., Takahashi, K., Shirahige, K. and Ashikari, T. (2002) Genome-wide expression analysis of genes affected by amino acid sensor Ssy1p in *Saccharomyces cerevisiae. Current Genetics* **41**, 63–72.

Kopetzki, E. and Entian, K. D. (1985) Glucose repression and hexokinase isoenzymes in yeast. Isolation and characterization of a modified hexokinase PII isoenzyme. *European Journal of Biochemistry* **146**, 657–662.

Kopetzki, E., Entian, K.-D. and Mecke, D. (1985) Complete nucleotide sequence of the hexokinase PI gene (*HXK1*) of *Saccharomyces cerevisiae*. *Gene* **39**, 95–102.

Kopp, M., Müller, H. and Holzer, H. (1993) Molecular analysis of the neutral trehalase gene from *Saccharomyces cerevisiae*. *Journal of Biological Chemistry* **268**, 4766–4774.

Kopperschläger, G. and Heinisch, J. J. (1997) 'Phosphofructokinase'. In: F. K. Zimmermann and K.-D. Entian (ed.), *Yeast Sugar Metabolism. Biochemistry, Genetics, Biotechnology, and Applications*. pp. 97–118. Lancaster, PA: Technomic.

Krampe, S. and Boles, E. (2002) Starvation-induced degradation of yeast hexose transporter Hxt7p is dependent on endocytosis, autophagy and the terminal sequences of the permease. *FEBS Letters* **513**, 193–196.

Krampe, S., Stamm, O., Hollenberg, C. P. and Boles, E. (1998) Catabolite inactivation of the high-affinity hexose transporters Hxt6 and Hxt7 of *Saccharomyces cerevisiae* occurs in the vacuole after internalization by endocytosis. *FEBS Letters* **441**, 343–347.

Kratzer, S. and Schüller, H.-J. (1997) Transcriptional control of the yeast acetyl-CoA synthetase gene, *ACS1*, by the positive regulators *CAT8* and *ADR1* and the pleiotropic repressor *UME6*. *Molecular Microbiology* **26**, 631–641.

Kretschmer, M. and Fraenkel, D. G. (1991) Yeast 6-phosphofructo-2-kinase: sequence and mutant. *Biochemistry* **30**, 10663–10672.

Kretschmer, M., Tempst, P. and Fraenkel, D. G. (1991) Identification and cloning of yeast phosphofructokinase 2. *European Journal of Biochemistry* **197**, 367–372.

Kruckeberg, A. L. (1996) The hexose transporter family of *Saccharomyces cerevisiae*. *Archives of Microbiology* **166**, 283–292.

Kruckeberg, A. L., Walsh, M. C. and van Dam, K. (1998) How do yeast cells sense glucose? *Bioessays* **20**, 972–976.

Kruckeberg, A. L., Ye, L., Berden, J. A. and van Dam, K. (1999) Functional expression, quantification and cellular localization of the Hxt2 hexose transporter of *Saccharomyces cerevisiae* tagged with the green fluorescent protein. *Biochemical Journal* **339**, 299–307.

Kumar, A. and Snyder, M. (2001) Emerging technologies in yeast genomics. *Nature Reviews in Genetics* **2**, 302–312.

Kunze, M., Kragler, F., Binder, M., Hartig, A. and Gurvitz, A. (2002) Targeting of malate synthase 1 to the peroxisomes of *Saccharomyces cerevisiae* cells depends on growth on oleic acid medium. *European Journal of Biochemistry* **269**, 915–922.

Lafuente, M. J., Gancedo, C., Jauniaux, J. C. and Gancedo, J. M. (2000) Mth1 receives the signal given by the glucose sensors Snf3 and Rgt2 in *Saccharomyces cerevisiae*. *Molecular Microbiology* **35**, 161–172.

Lages, F. and Lucas, C. (1997) Contribution to the physiological characterization of glycerol active uptake in *Saccharomyces cerevisiae*. *Biochimica et Biophysica Acta* **1322**, 8–18.

Lagunas, R. (1976) Energy metabolism of *Sacharomyces cerevisiae*. Discrepancy between ATP balance and known metabolic functions. *Biochimica et Biophysica Acta* **440**, 661–674.

Lagunas, R., Dominguez, C., Busturia, A. and Saez, M. J. (1982) Mechanisms of appearance of the Pasteur effect in *Saccharomyces cerevisiae*: inactivation of sugar transport systems. *Journal of Bacteriology* **152**, 19–25.

Lamphier, M. S. and Ptashne, M. (1992) Multiple mechanisms mediate glucose repression of the yeast *GAL1* gene. *Proceedings of the National Academy of Sciences USA* **89**, 5922–5926.

Larsson, C., Pahlman, I. L. and Gustafsson, L. (2000) The importance of ATP as a regulator of glycolytic flux in *Saccharomyces cerevisiae*. *Yeast* **16**, 797–809.

Larroy, C., Fernández, M. R., González, E., Parés, X. and Biosca, J. A. (2000) Characterization of the *Saccharomyces cerevisiae* YMR318c (*ADH6*) gene product as a broad specificity NADPH-dependent alcohol dehydrogenase: relevance in aldehyde reduction. *Biochemical Journal* **361**, 163–172.

Lazo, P. S., Ochoa, A. G. and Gascon, S. (1977) alpha-Galactosidase from *Saccharomyces carlsbergensis*. Cellular localization, and purification of the external enzyme. *European Journal of Biochemistry* **77**, 375–382.

Lazo, P. S., Florez, I. G., Ochoa, A. G. and Gascon, S. (1981) Induction and catabolite repression of alpha-galactosidase from *Saccharomyces carlsbergensis*. *Cellular and Molecular Biology* **27**, 615–622.

Lehninger, A. L., Nelson, D. L. and Cox, M. M. (1993) *Principles of Biochemistry*, 2nd edn, New York: Worth.

Lei, F., Rotboll, M. and Jorgensen, S. B. (2001) A biochemically structured model for *Saccharomyces cerevisiae*. *Journal of Biotechnology* **88**, 205–221.

Leloir, L. F. (1964) Nucleoside diphosphate sugars and saccharide synthesis. *Biochemical Journal* **91**, 1–8.

Levine, J., Tanouye, L. and Michels, C. A. (1992) The UAS(*MAL*) is a bidirectional promotor element required for the expression of both the *MAL61* and *MAL62* genes of the *Saccharomyces MAL6* locus. *Current Genetics* **22**, 181–189.

Li, F. N. and Johnston, M. (1997) Grr1 of *Saccharomyces cerevisiae* is connected to the ubiquitin proteolysis machinery through Skp1: coupling glucose sensing to gene expression and the cell cycle. *EMBO Journal* **16**, 5629–5638.

Liao, X. and Butow, R. A. (1993) *RTG1* and *RTG2*: two yeast genes required for a novel path of communication from mitochondria to the nucleus. *Cell* **72**, 61–71.

Liang, H. and Gaber, R. F. (1996) A novel signal transduction pathway in *Saccharomyces cerevisiae* defined by Snf3-regulated expression of *HXT6*. *Molecular Biology of the Cell* **7**, 1953–1966.

Liesen, T., Hollenberg, C. P. and Heinisch, J. J. (1996) ERA, a novel cis-acting element required for autoregulation and ethanol repression of *PDC1* transcription in *Saccharomyces cerevisiae*. *Molecular Microbiology* **21**, 621–632.

Liljestrom, P. L. (1985) The nucleotide sequence of the yeast *MEL1* gene. *Nucleic Acids Research* **13**, 7257–7268.

Lim, F., Morris, C.P., Occhiodoro, F., and Wallace, J.C. (1988) Sequence and domain structure of yeast pyruvate carboxylase. *Journal of Biological Chemistry*, **263**, 11493–11497.

Lin, A.-P. and McAlister-Henn, L. (2002) Isocitrate binding at two functionally distinct sites in Yeast NAD^+-specific isocitrate dehydrogenase. *Journal of Biological Chemistry* **277**, 22475–22483.

Lin, A.-P., McCammon, M. and McAlister-Henn, L. (2001) Kinetic and physiological effects of alterations in homologous isocitrate-binding sites of yeast NAD^+-specific isocitrate dehydrogenase. *Biochemistry* **40**, 14291–14301.

Lobo, Z. (1984) *Saccharomyces cerevisiae* aldolase mutants. *Journal of Bacteriology* **160**, 222–226.

Lobo, Z. and Maitra, P. K. (1977) Resistance to 2-deoxyglucose in yeast: a direct selection of mutants lacking glucose-phosphorylating enzymes. *Molecular and General Genetics* **157**, 297–300.

Lobo, Z. and Maitra, P. K. (1982) Pentose phosphate pathway mutants of yeast. *Molecular and General Genetics* **185**, 367–368

Lodi, T. and Ferrero, I. (1993) Isolation of the *DLD* gene of *Saccharomyces cerevisiae* encoding the mitochondrial enzyme D-lactate ferricytochrome c oxidoreductase. *Molecular and General Genetics* **238**, 315–324.

Lodi, T., Donnini, C. and Ferrero, I. (1991) Catabolite repression by galactose in overexpressed *GAL4* strains of *Saccharomyces cerevisiae*. *Journal of General Microbiology* **137**, 1039–1044.

Lodi, T., Fontanesi, F. and Guiard, B. (2002) Co-ordinate regulation of lactate metabolism genes in yeast: the role of the lactate permease gene *JEN1*. *Molecular Genetics and Genomics* **266**, 838–847.

Lorenz, M. C. and Fink, G. R. (2001) The glyoxylate cycle is required for fungal virulence. *Nature* **412**, 83–86.

Lu, G., Dobritzsch, D., Baumann, S., Schneider, G. and Konig, S. (2000) The structural basis of substrate activation in yeast pyruvate decarboxylase. A crystallographic and kinetic study. *European Journal of Biochemistry* **267**, 861–868.

Lucero, P. and Lagunas, R. (1997) Catabolite inactivation of the yeast maltose transporter requires ubiquitin-ligase npi1/rsp5 and ubiquitin-hydrolase npi2/doa4. *FEMS Microbiology Letters* **147**, 273–277.

Lucero, P., Herweijer, M. and Lagunas, R. (1993) Catabolite inactivation of the yeast maltose transporter is due to proteolysis. *FEBS Letters* **333**, 165–168.

Lue, N. F., Chasman, D. I., Buchman, A. R. and Kornberg, R. D. (1987) Interaction of *GAL4* and *GAL80* gene regulatory proteins *in vitro*. *Molecular and Cellular Biology* **7**, 3446–3451.

Lutfiyya, L. L. and Johnston, M. (1996) Two zinc-finger-containing repressors are responsible for glucose repression of *SUC2* expression. *Molecular and Cellular Biology* **16**, 4790–4797.

Luttik, M. A. H., Kotter, P., Salomons, F. A., van der Klei, I. J., van Dijken, J. P and Pronk, J. T. (2000) The *Saccharomyces cerevisiae ICL2* gene encodes a mitochondrial 2-methylisocitrate lyase involved in propionyl-Coenzyme A metabolism. *Journal of Bacteriology* **182**, 7007–7013.

Luyten, K., Albertyn, J., Skibbe, W. F., Prior, B. A., Ramos, J., Thevelein, J. M. and Hohmann, S. (1995) Fps1, a yeast member of the MIP family of channel proteins, is a facilitator for glycerol uptake and efflux and is inactive under osmotic stress. *EMBO Journal* **14**, 1360–1371.

Lyness, C. A., Jones, C. R. and Meaden, P. G. (1993) The *STA2* and *MEL1* genes of *Saccharomyces cerevisiae* are idiomorphic. *Current Genetics* **23**, 92–94.

Ma, J. and Ptashne, M. (1987) The carboxy-terminal 30 amino acids of GAL4 are recognized by GAL80. *Cell* **50**, 137–142.

Ma, P., Wera, S., Van Dijck, P. and Thevelein, J. M. (1999) The PDE1-encoded low-affinity phosphodiesterase in the yeast *Saccharomyces cerevisiae* has a specific function in controlling agonist-induced cAMP signaling. *Molecular Biology of the Cell* **10**, 91–104.

Madi, L., McBride, S. A., Bailey, L. A. and Ebbole, D. J. (1997) *rco-3*, a gene involved in glucose transport and conidiation in *Neurospora crassa*. *Genetics* **146**, 499–508.

Maitra, P. K. and Lobo, Z. (1971) A kinetic study of glycolytic enzyme synthesis in yeast. *Journal of Biological Chemistry* **246**, 475–488.

Maitra, P. K. and Lobo, Z. (1977a) Genetic studies with a phosphoglucose isomerase mutant of *Saccharomyces cerevisiae*. *Molecular and General Genetics* **156**, 55–60.

Maitra, P. K. and Lobo, Z. (1977b) Yeast pyruvate kinase: a mutant from catalytically insensitive to fructose 1,6-bisphosphate. *European Journal of Biochemistry* **78**, 353–360.

Martinez-Campa, C., Herrero, P., Ramirez, M. and Moreno, F. (1996) Molecular analysis of the promoter region of the hexokinase 2 gene of *Saccharomyces cerevisiae*. *FEMS Microbiology Letters* **137**, 69–74.

Matsusaka, K., Chiba, S. and Shimomura, T. (1977) Purification and substrate specificity of brewer's yeast alpha-glucosidase. *Agricultural and Biological Chemistry* **41**, 1917–1923.

Mayordomo, I. and Sanz, P. (2001a) Hexokinase PII: structural analysis and glucose signalling in the yeast *Saccharomyces cerevisiae*. *Yeast* **18**, 923–930.

Mayordomo, I. and Sanz, P. (2001b) Human pancreatic glucokinase (GlkB) complements the glucose signalling defect of *Saccharomyces cerevisiae hxk2* mutants. *Yeast* **18**, 1309–1316.

McAlister, L. and Holland, M. J. (1982) Targeted deletion of a yeast enolase structural gene. Identification and isolation of yeast enolase isozymes. *Journal of Biological Chemistry* **257**, 7181–7188.

McAlister, L. and Holland, M. J. (1985) Differential expression of the three yeast glyceraldehyde-3-phosphate dehydrogenase genes. *Journal of Biological Chemistry* **260**, 15019–15027.

McAlister-Henn, L. and Small, W. L. (1997) Molecular genetics of yeast TCA cycle isozymes. *Progress in Nucleic Acid Research and Molecular Biology* **57**, 317–339.

McDonald, A. E., Niere, J. O. and Plaxton, W. C. (2001) Phosphite disrupts the acclimation of *Saccharomyces cerevisiae* to phosphate starvation. *Canadian Journal of Microbiology* **47**, 969–978.

McGovern, A. C., Ernill, R., Kara, B. V., Kell, D. B. and Goodacre, R. (1999) Rapid analysis of the expression of heterologous proteins in *Escherichia coli* using pyrolysis mass spectrometry and Fourier transform infrared spectroscopy with chemometrics: application to alpha 2-interferon production. *Journal of Biotechnology* **72**, 157–167.

Meaden, P. G., Dickinson, F. M., Mifsud, A., Tessier, W., Westwater, J., Bussey, H. and Midgley, M. (1997) The *ALD6* gene of *Saccharomyces cerevisiae* encodes a cytosolic, Mg(2+)-activated acetaldehyde dehydrogenase. *Yeast* **13**, 1319–1327.

Medintz, I., Jiang, H., Han, E. K., Cui, W. and Michels, C. A. (1996) Characterization of the glucose-induced inactivation of maltose permease in *Saccharomyces cerevisiae*. *Journal of Bacteriology* **178**, 2245–2254.

Medintz, I., Jiang, H. and Michels, C. A. (1998) The role of ubiquitin conjugation in glucose-induced proteolysis of *Saccharomyces* maltose permease. *Journal of Biological Chemistry* **273**, 34454–34462.

Medintz, I., Wang, X., Hradek, T. and Michels, C. A. (2000) A PEST-like sequence in the N-terminal cytoplasmic domain of *Saccharomyces* maltose permease is required for glucose-induced proteolysis and rapid inactivation of transport activity. *Biochemistry* **39**, 4518–4526.

Mercado, J. J., Vincent, O. and Gancedo, J. M. (1991) Regions in the promoter of the yeast *FBP1* gene implicated in catabolite repression may bind the product of the regulatory gene *MIG1*. *FEBS Letters* **291**, 97–100.

Mercado, J. J., Smith, R., Sagliocco, F. A., Brown, A. J. and Gancedo, J. M. (1994) The levels of yeast gluconeogenic mRNAs respond to environmental factors. *European Journal of Biochemistry* **224**, 473–481.

Michels, C. A. and Needleman, R. B. (1984) The dispersed, repeated family of *MAL* loci in *Saccharomyces* spp. *Journal of Bacteriology* **157**, 949–952.

Miosga, T. and Zimmermann, F. K. (1996) Cloning and characterization of the first two genes of the non-oxidative part of the *Saccharomyces cerevisiae* pentose-phosphate pathway. *Current Genetics* **30**, 404–409.

Monod, J. (1942) *Reserches sur la Croissance des Cultures Bacteriennes*. Paris: Hermann and Cie.

Moore, P. A., Sagliocco, F. A., Wood, R. M. and Brown, A. J. P. (1991) Yeast glycolytic mRNAs are differentially regulated. *Molecular and Cellular Biology* **11**, 5330–5337.

Moreno, F., Ochoa, A. G., Gascon, S. and Villanueva, J. R. (1975) Molecular forms of yeast invertase. *European Journal of Biochemistry* **50**, 571–579.

Moreno, F., Fernandez, T., Fernandez, R. and Herrero, P. (1986) Hexokinase PII from *Saccharomyces cerevisiae* is regulated by changes in the cytosolic Mg2+-free ATP concentration. *European Journal of Biochemistry* **161**, 565–569.

Morris, C. N., Ainsworth, S. and Kinderlerer, J. (1986) The regulatory properties of yeast pyruvate kinase. Effect of fructose 1,6-bisphosphate. *Biochemical Journal* **234**, 691–698.

Morris, C. P., Lim, F. and Wallace, J. C. (1987) Yeast pyruvate carboxylase: gene isolation. *Biochemical and Biophysical Research Communications* **145**, 390–396.

Mosch, H. U., Kubler, E., Krappmann, S., Fink, G. R. and Braus, G. H. (1999) Crosstalk between the Ras2p-controlled mitogen-activated protein kinase and cAMP pathways during invasive growth of *Saccharomyces cerevisiae*. *Molecular Biology of the Cell* **10**, 1325–1335.

Müller, D. and Holzer, H. (1981) Regulation of fructose-1,6-bisphosphatase in yeast by phosphorylation/dephosphorylation. *Biochemical and Biophysical Research Communications* **103**, 926–933

Muller, E. H., Richards, E. J., Norbeck, J., Byrne, K. L., Karlsson, K. A., Pretorius, G. H., Meacock, P. A., Blomberg, A. and Hohmann, S. (1999) Thiamine repression and pyruvate decarboxylase autoregulation independently control the expression of the *Saccharomyces cerevisiae PDC5* gene. *FEBS Letters* **449**, 245–250.

Muller, S., Zimmermann, F. K. and Boles, E. (1997) Mutant studies of phosphofructo-2-kinases do not reveal an essential role of fructose-2,6-bisphosphate in the regulation of carbon fluxes in yeast cells. *Microbiology* **143**, 3055–3061.

Naumov, G., Turakainen, H., Naumova, E., Aho, S. and Korhola, M. (1990) A new family of polymorphic genes in *Saccharomyces cerevisiae*: alpha-galactosidase genes *MEL1–MEL7*. *Molecular and General Genetics* **224**, 119–128.

Naumov, G., Naumova, E., Turakainen, H., Suominen, P. and Korhola, M. (1991) Polymeric genes *MEL8*, *MEL9* and *MEL10* – new members of alpha-galactosidase gene family in *Saccharomyces cerevisiae*. *Current Genetics* **20**, 269–276.

Naumov, G. I., Naumova, E. S. and Michels, C. A. (1994) Genetic variation of the repeated *MAL* loci in natural populations of *Saccharomyces cerevisiae* and *Saccharomyces paradoxus*. *Genetics* **136**, 803–812.

Naumov, G. I., Naumova, E. S., Sancho, E. D. and Korhola, M. P. (1996a) Polymeric *SUC* genes in natural populations of *Saccharomyces cerevisiae*. *FEMS Microbiology Letters* **135**, 31–35.

Naumov, G. I., Naumova, E. S., Turakainen, H. and Korhola, M. (1996b) Identification of the alpha-galactosidase *MEL* genes in some populations of *Saccharomyces cerevisiae*: a new gene *MEL11*. *Genetic Research* **67**, 101–108.

Navas, M. A., Cerdan, S. and Gancedo, J. M. (1993) Futile cycles in *Saccharomyces cerevisiae* strains expressing the gluconeogenic enzymes during growth on glucose. *Proceedings of the National Academy of Sciences USA* **90**, 1290–1294.

Needleman, R. B. and Michels, C. (1983) Repeated family of genes controlling maltose fermentation in *Saccharomyces carlsbergensis*. *Molecular and Cellular Biology* **3**, 796–802.

Needleman, R. B., Federoff, H. J., Eccleshall, T. R., Buchferer, B. and Marmur, J. (1978) Purification and characterization of an alpha-glucosidase from *Saccharomyces carlsbergensis*. *Biochemistry* **17**, 4657–4661.

Needleman, R. B., Kaback, D. B., Dubin, R. A., Perkins, E. L., Rosenberg, N. G., Sutherland, K. A., Forrest, D. B. and Michels, C. A. (1984) *MAL6* of *Saccharomyces*: a complex genetic locus containing three genes required for maltose fermentation. *Proceedings of the National Academy of Sciences USA* **81**, 2811–2815.

Nehlin, J. O. and Ronne, H. (1990) Yeast MIG1 repressor is related to the mammalian early growth response and Wilms' tumour finger proteins. *EMBO Journal* **9**, 2891–2898.

Nehlin, J. O., Carlberg, M. and Ronne, H. (1989) Yeast galactose permease is related to yeast and mammalian glucose transporters. *Gene* **85**, 313–319.

Nehlin, J. O., Carlberg, M. and Ronne, H. (1991) Control of yeast *GAL* genes by MIG1 repressor: a transcriptional cascade in the glucose response. *EMBO Journal* **10**, 3373–3377.

Nelissen, B., De Wachter, R. and Goffeau, A. (1997) Classification of all putative permeases and other membrane plurispanners of the major facilitator superfamily encoded by the complete genome of *Saccharomyces cerevisiae*. *FEMS Microbiology Reviews* **21**, 113–134.

Ni, B. F. and Needleman, R. B. (1990) Identification of the upstream activating sequence of MAL and the binding sites for the MAL63 activator of *Saccharomyces cerevisiae*. *Molecular and Cellular Biology* **10**, 3797–3800.

Nikawa, J., Sass, P. and Wigler, M. (1987) Cloning and characterization of the low-affinity cyclic AMP phosphodiesterase gene of *Saccharomyces cerevisiae*. *Molecular and Cellular Biology* **7**, 3629–3636.

Nilsson, A., Pahlman, I. L., Jovall, P. A., Blomberg, A., Larsson, C. and Gustafsson, L. (2001) The catabolic capacity of *Saccharomyces cerevisiae* is preserved to a higher extent during carbon compared to nitrogen starvation. *Yeast* **18**, 1371–1381.

Nishizawa, M., Araki, R. and Teranishi, Y. (1989) Identification of an upstream activating sequence and an upstream repressible sequence of the pyruvate kinase gene of the yeast *Saccharomyces cerevisiae*. *Molecular and Cellular Biology* **9**, 442–451.

Nishizawa, K., Shimoda, E. and Kasahara, M. (1995) Substrate recognition domain of the Gal2 galactose transporter in yeast *Saccharomyces cerevisiae* as revealed by chimeric galactose–glucose transporters. *Journal of Biological Chemistry* **270**, 2423–2426.

Nissen, T. L., Schulze, U., Nielsen, J. and Villadsen, J. (1997) Flux distributions in anaerobic, glucose-limited continuous cultures of *Saccharomyces cerevisiae*. *Microbiology* **143**, 203–218.

Nogae, I. and Johnston, M. (1990) Isolation and characterization of the *ZWF1* gene of *Saccharomyces cerevisiae*, encoding glucose-6-phosphate dehydrogenase. *Gene* **96**, 161–169.

Noubhani, A., Bunoust, O., Rigoulet, M. and Thevelein, J. M. (2000) Reconstitution of ethanolic fermentation in permeabilized spheroplasts of wild-type and trehalose-6-phosphate synthase mutants of the yeast *Saccharomyces cerevisiae*. *European Journal of Biochemistry* **267**, 4566–4576.

Oh, D. and Hopper, J. E. (1990) Transcription of a yeast phosphoglucomutase isozyme gene is galactose inducible and glucose repressible. *Molecular and Cellular Biology* **10**, 1415–1422.

Oliver, S. G., Winson, M. K., Kell, D. B. and Baganz, F. (1998) Systematic functional analysis of the yeast genome. *Trends in Biotechnology* **16**, 373–378.

Oliver, D. J., Nikolau, B. and Wurtele, E. S. (2002) Functional genomics: high-throughput mRNA, protein, and metabolite analyses. *Metabolic Engineering* **4**, 98–106.

Ordiz, I., Herrero, P., Rodicio, R. and Moreno, F. (1996) Glucose-induced inactivation of isocitrate lyase in *Saccharomyces cerevisiae* is mediated by the cAMP-dependent protein kinase catalytic subunits Tpk1 and Tpk2. *FEBS Letters* **385**, 43–46.

Ostergaard, S., Roca, C., Ronnow, B., Nielsen, J. and Olsson, L. (2000) Physiological studies in aerobic batch cultivations of *Saccharomyces cerevisiae* strains harboring the *MEL1* gene. *Biotechnology and Bioengineering* **68**, 252–259.

Özcan, S. and Johnston, M. (1995) Three different regulatory mechanisms enable yeast hexose transporter (*HXT*) genes to be induced by different levels of glucose. *Molecular and Cellular Biology* **15**, 1564–1572.

Özcan, S. and Johnston, M. (1996) Two different repressors collaborate to restrict expression of the yeast glucose transporter genes *HXT2* and *HXT4* to low levels of glucose. *Molecular and Cellular Biology* **16**, 5536–5545.

Özcan, S. and Johnston, M. (1999) Function and regulation of yeast hexose transporters. *Microbiology and Molecular Biology Reviews* **63**, 554–569.

Özcan, S., Freidel, K., Leuker, A. and Ciriacy, M. (1993) Glucose uptake and catabolite repression in dominant *HTR1* mutants of *Saccharomyces cerevisiae*. *Journal of Bacteriology* **175**, 5520–5528.

Özcan, S., Schulte, F., Freidel, K., Weber, A. and Ciriacy, M. (1994) Glucose uptake and metabolism in *grr1/cat80* mutants of *Saccharomyces cerevisiae*. *European Journal of Biochemistry* **224**, 605–611.

Özcan, S., Dover, J., Rosenwald, A. G., Wölfl, S. and Johnston, M. (1996a) Two glucose transporters in *Saccharomyces cerevisiae* are glucose sensors that generate a signal for induction of gene expression. *Proceedings of the National Academy of Sciences USA* **93**, 12428–12432.

Özcan, S., Leong, T. and Johnston, M. (1996b) Rgt1p of *Saccharomyces cerevisiae*, a key regulator of glucose-induced genes, is both an activator and a repressor of transcription. *Molecular and Cellular Biology* **16**, 6419–6426.

Özcan, S., Vallier, L. G., Flick, J. S., Carlson, M. and Johnston, M. (1997) Expression of the *SUC2* gene of *Saccharomyces cerevisiae* is induced by low levels of glucose. *Yeast* **13**, 127–137.

Özcan, S., Dover, J. and Johnston, M. (1998) Glucose sensing and signaling by two glucose receptors in the yeast *Saccharomyces cerevisiae*. *EMBO Journal* **17**, 2566–2573.

Packham, E. A., Graham, I. R. and Chambers, A. (1996) The multifunctional transcription factors Abf1p, Rap1p and Reb1p are required for full transcriptional activation of the chromosomal *PGK* gene in *Saccharomyces cerevisiae*. *Molecular and General Genetics* **250**, 348–356.

Påhlman, A.-K., Granath, K., Ansell, R., Hohmann, S. and Adler, L. (2001) The yeast glycerol 3-phosphatases Gpp1 and Gpp2 are required for glycerol biosynthesis and differentially involved in the cellular responses to osmotic, anaerobic and oxidative stress. *Journal of Biological Chemistry* **276**, 3555–3563.

Paiva, S., Althoff, S., Casal, M. and Leao, C. (1999) Transport of acetate in mutants of *Saccharomyces cerevisiae* defective in monocarboxylate permeases. *FEMS Microbiology Letters* **170**, 301–306.

Paiva, S., Kruckeberg, A. L. and Casal, M. (2002) Utilization of green fluorescent protein as a marker for studying the expression and turnover of the monocarboxylate permease Jen1p of *Saccharomyces cerevisiae*. *Biochemical Journal* **363**, 737–744.

Palsson, B. O. (2000) Hougen 2000 Lectures. Madison College of Engineering. University of Wisconsin. Online: http://gcrg.ucsd.edu/presentations/hougen/all.pdf (accessed 6 October 2002)

Pan, X. and Heitman, J. (1999) Cyclic AMP-dependent protein kinase regulates pseudohyphal differentiation in *Saccharomyces cerevisiae*. *Molecular and Cellular Biology* **19**, 4874–4887.

Paravicini, G. and Kretschmer, M. (1992) The yeast *FBP26* gene codes for a fructose-2, 6-bisphosphatase. *Biochemistry* **31**, 7126–7133.

Paulsen, I. T., Sliwinski, M. K., Nelissen, B., Goffeau, A. and Saier, M. H., Jr. (1998) Unified inventory of established and putative transporters encoded within the complete genome of *Saccharomyces cerevisiae*. *FEBS Letters* **430**, 116–125.

Pearce, A. K., Crimmins, K., Groussac, E., Hewlins, M. J., Dickinson, J. R., Francois, J., Booth, I. R. and Brown, A. J. P. (2001) Pyruvate kinase (Pyk1) levels influence both the rate and direction of carbon flux in yeast under fermentative conditions. *Microbiology* **147**, 391–401.

Pedler, S. M., Wallace, P. G., Wallace, J. C. and Berry, M. N. (1997) The fate of glucose in strains S288C and S173-6B of the yeast *Saccharomyces cerevisiae*. *Yeast* **13**, 119–125.

Peinado, J. M. and Loureiro, D. M. (1986) Reversible loss of affinity induced by glucose in the maltose–H+ symport of *Saccharomyces cerevisiae*. *Biochimica et Biophysica Acta* **856**, 189–192.

Penalver, E., Lucero, P., Moreno, E. and Lagunas, R. (1998) Catabolite inactivation of the maltose transporter in nitrogen-starved yeast could be due to the stimulation of general protein turnover. *FEMS Microbiology Letters* **166**, 317–324.

Perlman, D. and Halvorson, H. O. (1981) Distinct repressible mRNAs for cytoplasmic and secreted yeast invertase are encoded by a single gene. *Cell* **25**, 525–536.

Perlman, D., Raney, P. and Halvorson, H. O. (1984) Cytoplasmic and secreted *Saccharomyces cerevisiae* invertase mRNAs encoded by one gene can be differentially or coordinately regulated. *Molecular and Cellular Biology* **4**, 1682–1688.

Petit, T. and Gancedo, C. (1999) Molecular cloning and characterization of the gene *HXK1* encoding the hexokinase from *Yarrowia lipolytica*. *Yeast* **15**, 1573–1584.

Petit, T., Diderich, J. A., Kruckeberg, A. L., Gancedo, C. and Van Dam, K. (2000) Hexokinase regulates kinetics of glucose transport and expression of genes encoding hexose transporters in *Saccharomyces cerevisiae*. *Journal of Bacteriology* **182**, 6815–6818.

Phelps, T. J., Palumbo, A. V. and Beliaev, A. S. (2002) Metabolomics and microarrays for improved understanding of phenotypic characteristics controlled by both genomics and environmental constraints. *Current Opinion in Biotechnology* **13**, 20–24.

Pinkham, J. L. and Guarente, L. (1985) Cloning and molecular analysis of the *HAP2* locus: a global regulator of respiratory genes in *Saccharomyces cerevisiae*. *Molecular and Cellular Biology* **5**, 3410–3416.

Pionnier, S. and Zhang, B. L. (2002) Application of ^{2}H NMR to the study of natural site-specific hydrogen isotope transfer among substrate, medium, and glycerol in glucose fermentation with yeast. *Analytical Biochemistry* **307**, 138–146.

Plankert, U., Purwin, C. and Holzer, H. (1991) Yeast fructose-2,6-bisphosphate 6-phosphatase is encoded by *PHO8*, the gene for nonspecific repressible alkaline phosphatase. *European Journal of Biochemistry* **196**, 191–196.

Platt, A. and Reece, R. J. (1998) The yeast galactose genetic switch is mediated by the formation of a Gal4p–Gal80p–Gal3p complex. *EMBO Journal* **17**, 4086–4091.

Platt, A., Ross, H. C., Hankin, S. and Reece, R. J. (2000) The insertion of two amino acids into a transcriptional inducer converts it into a galactokinase. *Proceedings of the National Academy of Sciences USA* **97**, 3154–3159.

Portela, P., Howell, S., Moreno, S. and Rossi, S. (2002) *In vivo* and *in vitro* phosphorylation of two isoforms of yeast pyruvate kinase by protein kinase A. *Journal of Biological Chemistry* **277**, 30477–30487.

Post-Beittenmiller, M. A., Hamilton, R. W. and Hopper, J. E. (1984) Regulation of basal and induced levels of the *MEL1* transcript in *Saccharomyces cerevisiae*. *Molecular and Cellular Biology* **4**, 1238–1245.

Pretorius, I. S., Lambrechts, M. G. and Marmur, J. (1991) The glucoamylase multigene family in *Saccharomyces cerevisiae* var. *diastaticus*: an overview. *Critical Reviews in Biochemistry and Molecular Biology* **26**, 53–76.

Pritchard, L. and Kell, D. B. (2002) Schemes of flux control in a model of *Saccharomyces cerevisiae* glycolysis. *European Journal of Biochemistry* **269**, 3894–3904.

Pronk, J. T., Steensma, Y. H. and Van Dijken, J. P. (1996) Pyruvate metabolism in *Saccharomyces cerevisiae*. *Yeast* **12**, 1607–1633.

Raamsdonk, L. M., Teusink, B., Broadhurst, D., Zhang, N., Hayes, A., Walsh, M. C., Berden, J. A., Brindle, K. M., Kell, D. B., Rowland, J. J., Westerhoff, H. V., van Dam, K. and Oliver, S. G. (2001)

A functional genomics strategy that uses metabolome data to reveal the phenotype of silent mutations. *Nature Biotechnology* **19**, 45–50.

Randez-Gil, F., Bojunga, N., Proft, M. and Entian, K. D. (1997) Glucose derepression of gluconeogenic enzymes in *Saccharomyces cerevisiae* correlates with phosphorylation of the gene activator Cat8p. *Molecular and Cellular Biology* **17**, 2502–2510.

Randez-Gil, F., Herrero, P., Sanz, P., Prieto, J. A. and Moreno, F. (1998a) Hexokinase PII has a double cytosolic-nuclear localisation in *Saccharomyces cerevisiae. FEBS Letters* **425**, 475–478.

Randez-Gil, F., Sanz, P., Entian, K. D. and Prieto, J. A. (1998b) Carbon source-dependent phosphorylation of hexokinase PII and its role in the glucose-signaling response in yeast. *Molecular and Cellular Biology* **18**, 2940–2948.

Rechsteiner, M. and Rogers, S. W. (1996) PEST sequences and regulation by proteolysis. *Trends in Biochemical Sciences* **21**, 267–271.

Reifenberger, E., Freidel, K. and Ciriacy, M. (1995) Identification of novel *HXT* genes in *Saccharomyces cerevisiae* reveals the impact of individual hexose transporters on glycolytic flux. *Molecular Microbiology* **16**, 157–167.

Reifenberger, E., Boles, E. and Ciriacy, M. (1997) Kinetic characterization of individual hexose transporters of *Saccharomyces cerevisiae* and their relation to the triggering mechanisms of glucose repression. *European Journal of Biochemistry* **245**, 324–333.

Reizer, J., Reizer, A. and Saier, M. H., Jr. (1993) The MIP family of integral membrane channel proteins: sequence comparisons, evolutionary relationships, reconstructed pathway of evolution, and proposed functional differentiation of the two repeated halves of the proteins. *Critical Reviews in Biochemistry and Molecular Biology* **28**, 235–257.

Remize, F., Barnavon, L. and Dequin, S. (2001) Glycerol export and glycerol-3-phosphate dehydrogenase, but not glycerol phosphatase, are rate limiting for glycerol production in *Saccharomyces cerevisiae. Metabolic Engineering* **3**, 301–312.

Rep, M., Krantz, M., Thevelein, J. M. and Hohmann, S. (2000) The transcriptional response of *Saccharomyces cerevisiae* to osmotic shock. Hot1p and Msn2p/Msn4p are required for the induction of subsets of high osmolarity glycerol pathway-dependent genes. *Journal of Biological Chemistry* **275**, 8290–8300.

Repetto, B. and Tzagoloff, A. (1989). Structure and regulation of *KGD1*, the structural gene for yeast α-ketoglutarate dehydrogenase. *Molecular and Cellular Biology* **9**, 2695–2705.

Repetto, B. and Tzagoloff, A. (1990). Structure and regulation of *KGD2*, the structural gene for yeast dihydrolipoyl transsuccinylase. *Molecular and Cellular Biology* **10**, 4221–4232.

Rhiel, M., Ducommun, P., Bolzonella, I., Marison, I. and von Stockar, U. (2002) Real-time *in situ* monitoring of freely suspended and immobilized cell cultures based on mid-infrared spectroscopic measurements. *Biotechnology and Bioengineering* **77**, 174–185.

Riballo, E. and Lagunas, R. (1994a) Involvement of endocytosis in catabolite inactivation of the K^+ and glucose transport systems in *Saccharomyces cerevisiae. FEMS Microbiology Letters* **121**, 77–80.

Riballo, E. and Lagunas, R. (1994b) Involvement of endocytosis in catabolite inactivation of the transport systems in *Saccharomyces cerevisiae. Folia Microbiologica (Praha)* **39**, 542.

Riballo, E., Herweijer, M., Wolf, D. H. and Lagunas, R. (1995) Catabolite inactivation of the yeast maltose transporter occurs in the vacuole after internalization by endocytosis. *Journal of Bacteriology* **177**, 5622–5627.

Richard, P., Toivari, M. H. and Penttilä, M. (1999) Evidence that the gene YLR070c of *Saccharomyces cerevisiae* encodes a xylitol dehydrogenase. *FEBS Letters* **457**, 135–138.

Robertson, J. J. and Halvorson, H. O. (1957) The components of maltozymase in yeast, and their behavior during deadaptation. *Journal of Bacteriology* **73**, 186–198.

Robertson, L. S. and Fink, G. R. (1998) The three yeast A kinases have specific signaling functions in pseudohyphal growth. *Proceedings of the National Academy of Sciences USA* **95**, 13783–13787.

Robertson, L. S., Causton, H. C., Young, R. A. and Fink, G. R. (2000) The yeast A kinases differentially regulate iron uptake and respiratory function. *Proceedings of the National Academy of Sciences USA* **97**, 5984–5988.

Robl, I., Grassl, R., Tanner, W. and Opekarova, M. (2000) Properties of a reconstituted eukaryotic hexose/proton symporter solubilized by structurally related non-ionic detergents: specific requirement of phosphatidylcholine for permease stability. *Biochimica et Biophysica Acta* **1463**, 407–418.

Robl, I., Grassl, R., Tanner, W. and Opekarova, M. (2001) Construction of phosphatidylethanolamine-less strain of *Saccharomyces cerevisiae*. Effect on amino acid transport. *Yeast* **18**, 251–260.

Rodicio, R. and Heinisch, J. (1987) Isolation of the yeast phosphoglyceromutase gene and construction of deletion mutants. *Molecular and General Genetics* **206**, 133–140.

Rodicio, R., Heinisch, J. J. and Hollenberg, C. P. (1993) Transcriptional control of yeast phosphoglycerate mutase-encoding gene. *Gene* **125**, 125–133.

Rodriguez, C. and Gancedo, J. M. (1999) Glucose signaling in yeast is partially mimicked by galactose and does not require the Tps1 protein. *Molecular and Cellular Biology Research Communications* **1**, 52–58.

Rodriguez, A., De La Cera, T., Herrero, P. and Moreno, F. (2001) The hexokinase 2 protein regulates the expression of the *GLK1*, *HXK1* and *HXK2* genes of *Saccharomyces cerevisiae*. *Biochemical Journal* **355**, 625–631.

Rolland, F., De Winde, J. H., Lemaire, K., Boles, E., Thevelein, J. M. and Winderickx, J. (2000) Glucose-induced cAMP signalling in yeast requires both a G-protein coupled receptor system for extracellular glucose detection and a separable hexose kinase-dependent sensing process. *Molecular Microbiology* **38**, 348–358.

Rose, M. (1995) Molecular and biochemical characterization of the hexokinase from the starch-utilizing yeast *Schwanniomyces occidentalis*. *Current Genetics* **27**, 330–338.

Rosenzweig, R. F. (1992) Regulation of fitness in yeast overexpressing glycolytic enzymes: parameters of growth and viability. *Genetic Research* **59**, 35–48.

Ross, J., Reid, G. A. and Dawes, I. W. (1988) The nucleotide sequence of the *LPD1* gene encoding lipoamide dehydrogenase in *Saccharomyces cerevisiae*: comparison between eukaryotic and prokaryotic sequences for related enzymes and identification of potential upstream control sites. *Journal of General Microbiology* **134**, 1131–1139.

Roth, S. and Schuller, H. J. (2001) Cat8 and Sip4 mediate regulated transcriptional activation of the yeast malate dehydrogenase gene *MDH2* by three carbon source-responsive promoter elements. *Yeast* **18**, 151–162.

Rotin, D., Staub, O. and Haguenauer-Tsapis, R. (2000) Ubiquitination and endocytosis of plasma membrane proteins: role of Nedd4/Rsp5p family of ubiquitin-protein ligases. *Journal of Membrane Biology* **176**, 1–17.

Santangelo, G. M. and Tornow, J. (1990) Efficient transcription of the glycolytic gene *ADH1* and three translational component genes requires the *GCR1* product, which can act through TUF/GRF/RAP binding sites. *Molecular and Cellular Biology* **10**, 859–862.

Sarokin, L. and Carlson, M. (1984) Upstream region required for regulated expression of the glucose-repressible *SUC2* gene of *Saccharomyces cerevisiae*. *Molecular and Cellular Biology* **4**, 2750–2757.

Sarokin, L. and Carlson, M. (1985) Comparison of two yeast invertase genes: conservation of the upstream regulatory region. *Nucleic Acids Research* **13**, 6089–6103.

Sarthy, A., Schopp, C. and Idler, K. B. (1994) Cloning and sequence determination of the gene encoding sorbitol dehydrogenase from *Saccharomyces cerevisiae*. *Gene* **140**, 121–126.

Sass, P., Field, J., Nikawa, J., Toda, T. and Wigler, M. (1986) Cloning and characterization of the high-affinity cAMP phosphodiesterase of *Saccharomyces cerevisiae*. *Proceedings of the National Academy of Sciences USA* **83**, 9303–9307.

Sass, E., Blachinsky, E., Karniely, S. and Pines, O. (2001) Mitochondrial and cytosolic isoforms of yeast fumarase are derivatives of a single translation product and have identical amino termini. *Journal of Biological Chemistry* **276**, 46111–46117.

Schaaff, I., Heinisch, J. and Zimmermann, F. K. (1989) Overproduction of glycolytic enzymes in yeast. *Yeast* **5**, 285–290.

Schaaff-Gerstenschläger, I., Mannhaupt, G., Vetter, I., Zimmermann, F. K. and Feldmann, H. (1993) *TKL2*, a second transketolase gene of *Saccharomyces cerevisiae.* Cloning, sequence and deletion analysis of the gene. *European Journal of Biochemistry* **217**, 487–492.

Scheffler, I. E., de la Cruz, B. J. and Prieto, S. (1998) Control of mRNA turnover as a mechanism of glucose repression in *Saccharomyces cerevisiae. International Journal of Biochemistry and Cell Biology* **30**, 1175–1193.

Schellenberger, A. and Hubner, G. (1968) Binding of the substrate in yeast pyruvate decarboxylase. *Angewandte Chemie International Edition English* **7**, 68–69.

Schmidt, M. C. and McCartney, R. R. (2000) β-Subunits of Snfl kinase are required for kinase function and substrate definition. *EMBO Journal* **19**, 4936–4943.

Schmidt, M. C., McCartney, R. R., Zhang, X., Tillman, T. S., Solimeo, H., Wolfl, S., Almonte, C. and Watkins, S. C. (1999) Std1 and Mth1 proteins interact with the glucose sensors to control glucose-regulated gene expression in *Saccharomyces cerevisiae. Molecular and Cellular Biology* **19**, 4561–4571.

Schmitt, H. D. and Zimmermann, F. K. (1982) Genetic analysis of the pyruvate decarboxylase reaction in yeast glycolysis. *Journal of Bacteriology* **151**, 1146–1152.

Schmitt, H. D., Ciriacy, M. and Zimmermann, F. K. (1983) The synthesis of yeast pyruvate decarboxylase is regulated by large variations in the messenger RNA level. *Molecular and General Genetics* **192**, 247–252.

Schulte, F., Wieczorke, R., Hollenberg, C. P. and Boles, E. (2000) The *HTR1* gene is a dominant negative mutant allele of MTH1 and blocks Snf3- and Rgt2-dependent glucose signaling in yeast. *Journal of Bacteriology* **182**, 540–542.

Schulze, U., Liden, G., Nielsen, J. and Villadsen, J. (1996) Physiological effects of nitrogen starvation in an anaerobic batch culture of *Saccharomyces cerevisiae. Microbiology* **142**, 2299–2310.

Schuster, K. C. (2000) Monitoring the physiological status in bioprocesses on the cellular level. *Advances in Biochemical Engineering and Biotechnology* **66**, 185–208.

Schwelberger, H. G., Kohlwein, S. D. and Paltauf, F. (1989) Molecular cloning, primary structure and disruption of the structural gene of aldolase from *Saccharomyces cerevisiae. European Journal of Biochemistry* **180**, 301–308.

Scott, E. W. and Baker, H. V. (1993) Concerted action of the transcriptional activators *REB1*, *RAP1*, and *GCR1* in the high-level expression of the glycolytic gene *TPI. Molecular and Cellular Biology* **13**, 543–550.

Scott, E. W., Allison, H. E. and Baker, H. V. (1990) Characterization of *TPI* gene expression in isogeneic wild-type and *gcr1*-deletion mutant strains of *Saccharomyces cerevisiae. Nucleic Acids Research* **18**, 7099–7107.

Sedivy, J. M. and Fraenkel, D. G. (1985) Fructose bisphosphatase of *Saccharomyces cerevisiae.* Cloning, disruption and regulation of the *FBP1* structural gene. *Journal of Molecular Biology* **186**, 307–319.

Seeboth, P. G., Bohnsack, K. and Hollenberg, C. P. (1990) *pdc1(0)* mutants of *Saccharomyces cerevisiae* give evidence for an additional structural *PDC* gene: cloning of *PDC5*, a gene homologous to *PDC1*. *Journal of Bacteriology* **172**, 678–685.

Selleck, S. B. and Majors, J. E. (1987) In vivo DNA-binding properties of a yeast transcription activator protein. *Molecular and Cellular Biology* **7**, 3260–3267.

Sergienko, E. A. and Jordan, F. (2002) New model for activation of yeast pyruvate decarboxylase by substrate consistent with the alternating sites mechanism: demonstration of the existence of two active forms of the enzyme. *Biochemistry* **41**, 3952–3967.

Serrano, R. (1977) Energy requirements for maltose transport in yeast. *European Journal of Biochemistry* **80**, 97–102.

Serrano, R. (1991) Transport across yeast vacuolar and plasma membranes. In: J. R. Broach, J. R. Pringle, and Jones E. W. (eds), *The Molecular and Cellular Biology of the Yeast Saccharomyces: Genome Dynamics, Protein Synthesis, and Energetics*, pp. 523–585. Cold Spring Harbor, NY: Cold Spring Harbor Press.

Serrano, R. and Delafuente, G. (1974) Regulatory properties of the constitutive hexose transport in *Saccharomyces cerevisiae. Molecular and Cellular Biochemistry* **5**, 161–171.

Shaw, A. D., Winson, M. K., Woodward, A. M., McGovern, A. C., Davey, H. M., Kaderbhai, N., Broadhurst, D., Gilbert, R. J., Taylor, J., Timmins, E. M., Goodacre, R., Kell, D. B., Alsberg, B. K. and Rowland, J. J. (2000) Rapid analysis of high-dimensional bioprocesses using multivariate spectroscopies and advanced chemometrics. *Advances in Biochemical Engineering and Biotechnology* **66**, 83–113.

Sherwood, P. W. and Carlson, M. (1999) Efficient export of the glucose transporter Hxt1p from the endoplasmic reticulum requires Gsf2p. *Proceedings of the National Academy of Sciences USA* **96**, 7415–7420.

Shi, H. and Shimizu, K. (1998) On-line metabolic pathway analysis based on metabolic signal flow diagram. *Biotechnology and Bioengineering* **58**, 139–148.

Shore, D. (1994) RAP1: a protean regulator in yeast. *Trends in Genetics* **10**, 408–412.

Sil, A. K., Alam, S., Xin, P., Ma, L., Morgan, M., Lebo, C. M., Woods, M. P. and Hopper, J. E. (1999) The Gal3p–Gal80p–Gal4p transcription switch of yeast: Gal3p destabilizes the Gal80p–Gal4p complex in response to galactose and ATP. *Molecular and Cellular Biology* **19**, 7828–7840.

Sinclair, D. A. and Dawes, I. W. (1995) Genetics of the synthesis of serine from glycine and the utilization of glycine as sole nitrogen source in *Saccharomyces cerevisiae. Genetics* **140**, 1213–1222.

Sinclair, D. A., Dawes, I. W. and Dickinson, J. R. (1993) Purification and characterization of the branched-chain α-ketoacid dehydrogenase complex from *Saccharomyces cerevisiae. Biochemistry and Molecular Biology International* **5**, 911–922.

Sivakesava, S., Irudayaraj, J. and Ali, D. (2001a) Simultaneous determination of multiple components in lactic acid fermentation using FT-MIR, NIR, and FT-Raman spectroscopic techniques. *Process Biochemistry* **37**, 371–378.

Sivakesava, S., Irudayaraj, J. and Demirci, A. (2001b) Monitoring a bioprocess for ethanol production using FT-MIR and FT-Raman spectroscopy. *Journal of Indian Microbiology and Biotechnology* **26**, 185–190.

Smith, F. C., Davies, S. P., Wilson, W. A., Carling, D. and Hardie, D. G. (1999) The SNF1 kinase complex from *Saccharomyces cerevisiae* phosphorylates the transcriptional repressor protein Mig1p in vitro at four sites within or near regulatory domain 1. *FEBS Letters* **453**, 219–223.

Smits, H. P., Hauf, J., Muller, S., Hobley, T. J., Zimmermann, F. K., Hahn-Hagerdal, B., Nielsen, J. and Olsson, L. (2000) Simultaneous overexpression of enzymes of the lower part of glycolysis can enhance the fermentative capacity of *Saccharomyces cerevisiae. Yeast* **16**, 1325–1334.

Sollitti, P. and Marmur, J. (1988) Primary structure of the regulatory gene from the *MAL6* locus of *Saccharomyces carlsbergensis. Molecular and General Genetics* **213**, 56–62.

Stachelek, C., Stachelek, J., Swan, J., Botstein, D. and Konigsberg, W. (1986) Identification, cloning and sequence determination of the genes specifying hexokinase A and B from yeast. *Nucleic Acids Research* **14**, 945–963.

Stambuk, B. U., da Silva, M. A., Panek, A. D. and de Araujo, P. S. (1999) Active alpha-glucoside transport in *Saccharomyces cerevisiae. FEMS Microbiology Letters* **170**, 105–110.

Steensma, H. Y. (1997) From Pyruvate to acetyl-coenzyme A and oxaloacetate. In: F. K. Zimmermann and K.-D. Entian (eds), *Yeast Sugar Metabolism. Biochemistry, Genetics, Biotechnology, and Applications*, pp. 339–357. Lancaster: Technomic.

Steensma, H. Y., Barth, G. and de Virgilio, C. (1993) Genetic and physical localization of the acetyl-coenzyme A synthetase gene *ACS1* on chromosome I of *Saccharomyces cerevisiae. Yeast* **9**, 419–421.

Steffan, J. S. and McAlister-Henn, L. (1992) Isolation and characterization of the yeast gene encoding the MDH3 isozyme of malate dehydrogenase. *Journal of Biological Chemistry* **267**, 24708–24715.

Stucka, R., Dequin, S., Salmon, J. M. and Gancedo, C. (1991) DNA sequences in chromosomes II and VII code for pyruvate carboxylase isoenzymes in *Saccharomyces cerevisiae*: analysis of pyruvate carboxylase-deficient strains. *Molecular and General Genetics* **229**, 307–315.

Stuckrath, I., Lange, H. C., Kotter, P., van Gulik, W. M., Entian, K. D. and Heijnen, J. J. (2002) Characterization of null mutants of the glyoxylate cycle and gluconeogenic enzymes in *S. cerevisiae* through metabolic network modeling verified by chemostat cultivation. *Biotechnology and Bioengineering* **77**, 61–72.

Sumner-Smith, M., Bozzato, R. P., Skipper, N., Davies, R. W. and Hopper, J. E. (1985) Analysis of the inducible *MEL1* gene of *Saccharomyces carlsbergensis* and its secreted product, alpha-galactosidase (melibiase). *Gene* **36**, 333–340.

Sundström, M., Lindqvist, Y., Schneider, G., Hellman, U. and Ronne, H. (1993) Yeast *TKL1* gene encodes a transketolase that is required for efficient glycolysis and biosynthesis of aromatic amino acids. *Journal of Biological Chemistry*, **268**, 24346–24352.

Sutherland, F. C., Lages, F., Lucas, C., Luyten, K., Albertyn, J., Hohmann, S., Prior, B. A. and Kilian, S. G. (1997) Characteristics of Fps1-dependent and -independent glycerol transport in *Saccharomyces cerevisiae*. *Journal of Bacteriology* **179**, 7790–7795.

Szkutnicka, K., Tschopp, J. F., Andrews, L. and Cirillo, V. P. (1989) Sequence and structure of the yeast galactose transporter. *Journal of Bacteriology* **171**, 4486–4493.

Tabata, S., Ide, T., Umemura, Y. and Torii, K. (1984) Purification and characterization of alpha-glucosidases produced by *Saccharomyces* in response to three distinct maltose genes. *Biochimica et Biophysica Acta* **797**, 231–238.

Tamas, M. J., Luyten, K., Sutherland, F. C., Hernandez, A., Albertyn, J., Valadi, H., Li, H., Prior, B. A., Kilian, S. G., Ramos, J., Gustafsson, L., Thevelein, J. M. and Hohmann, S. (1999) Fps1p controls the accumulation and release of the compatible solute glycerol in yeast osmoregulation. *Molecular Microbiology* **31**, 1087–1104.

Tao, W., Deschenes, R. J. and Fassler, J. S. (1999) Intracellular glycerol levels modulate the activity of Sln1p, a *Saccharomyces cerevisiae* two-component regulator. *Journal of Biological Chemistry* **274**, 360–367.

Taussig, R. and Carlson, M. (1983) Nucleotide sequence of the yeast *SUC2* gene for invertase. *Nucleic Acids Research* **11**, 1943–1954.

Tekamp-Olson, P., Najarian, R. and Burke, R. L. (1988) The isolation, characterization and nucleotide sequence of the phosphoglucoisomerase gene of *Saccharomyces cerevisiae*. *Gene* **73**, 153–161.

ten Berge, A. M., Zoutewelle, G. and Needleman, R. B. (1974) Regulation of maltose fermentation in *Saccharomyces carlsbergensis*. 3. Constitutive mutations at the *MAL6*-locus and suppressors changing a constitutive phenotype into a maltose negative phenotype. *Molecular and General Genetics* **131**, 113–121.

ter Schure, E. G., Flikweert, M. T., van Dijken, J. P., Pronk, J. T. and Verrips, C. T. (1998) Pyruvate decarboxylase catalyzes decarboxylation of branched-chain 2-oxo acids but is not essential for fusel alcohol production by *Saccharomyces cerevisiae*. *Applied and Environmental Microbiology* **64**, 1303–1307.

Tessier, W. M., Dickinson, F. M. and Midgley, M. (1999) The roles of acetaldehyde dehydrogenases in *Saccharomyces cerevisiae*. *Enzymology and Molecular Biology of Carbohydrate Metabolism* **7**, 29–34.

Teusink, B., Diderich, J. A., Westerhoff, H. V., van Dam, K. and Walsh, M. C. (1998) Intracellular glucose concentration in derepressed yeast cells consuming glucose is high enough to reduce the glucose transport rate by 50%. *Journal of Bacteriology* **180**, 556–562.

Teusink, B., Passarge, J., Reijenga, C. A., Esgalhado, E., van der Weijden, C. C., Schepper, M., Walsh, M. C., Bakker, B. M., van Dam, K., Westerhoff, H. V. and Snoep, J. L. (2000) Can yeast glycolysis be understood in terms of in vitro kinetics of the constituent enzymes? Testing biochemistry. *European Journal of Biochemistry* **267**, 5313–5329.

Thevelein, J. M. (1994) Signal transduction in yeast. *Yeast* **10**, 1753–1790.

Thevelein, J. M. and Hohmann, S. (1995) Trehalose synthase, guard to the gate of glycolysis in yeast? *Trends in Biochemical Sciences* **20**, 3–10.

Thevelein, J. M., Cauwenberg, L., Colombo, S., De Winde, J. H., Donation, M., Dumortier, F., Kraakman, L., Lemaire, K., Ma, P., Nauwelaers, D., Rolland, F., Teunissen, A., Van Dijck, P., Versele, M., Wera, S. and Winderickx, J. (2000) Nutrient-induced signal transduction through the protein kinase A pathway and its role in the control of metabolism, stress resistance, and growth in yeast. *Enzyme and Microbial Technology* **26**, 819–825.

Thomas, D., Cherest, H. and Surdin-Kerjan, Y. (1991) Identification of the structural gene for glucose-6-phosphate dehydrogenase in yeast. Inactivation leads to a nutritional requirement for organic sulphur. *EMBO Journal*, **10**, 547–553.

Thompson, D. A. and Stahl, F. W. (1999) Genetic control of recombination in yeast meiosis: isolation and characterization of mutants elevated for meiotic unequal sister-chromatid recombination. *Genetics* **153**, 621–641.

Toda, T., Cameron, S., Sass, P., Zoller, M., Scott, J. D., McMullen, B., Hurwitz, M., Krebs, E. G. and Wigler, M. (1987a) Cloning and characterization of *BCY1*, a locus encoding a regulatory subunit of the cyclic AMP-dependent protein kinase in *Saccharomyces cerevisiae*. *Molecular and Cellular Biology* **7**, 1371–1377.

Toda, T., Cameron, S., Sass, P., Zoller, M. and Wigler, M. (1987b) Three different genes in *S. cerevisiae* encode the catalytic subunits of the cAMP-dependent protein kinase. *Cell* **50**, 277–287.

Treitel, M. A., Kuchin, S. and Carlson, M. (1998) Snf1 protein kinase regulates phosphorylation of the Mig1 repressor in *Saccharomyces cerevisiae*. *Molecular and Cellular Biology* **18**, 6273–6280.

Tschopp, J., Esmon, P. C. and Schekman, R. (1984) Defective plasma membrane assembly in yeast secretory mutants. *Journal of Bacteriology* **160**, 966–970.

Tschopp, J. F., Emr, S. D., Field, C. and Schekman, R. (1986) *GAL2* codes for a membrane-bound subunit of the galactose permease in *Saccharomyces cerevisiae*. *Journal of Bacteriology* **166**, 313–318.

Turakainen, H., Kristo, P. and Korhola, M. (1994) Consideration of the evolution of the *Saccharomyces cerevisiae MEL* gene family on the basis of the nucleotide sequences of the genes and their flanking regions. *Yeast* **10**, 1559–1568.

Turkel, S. and Bisson, L. F. (1999) Transcription of the *HXT4* gene is regulated by Gcr1p and Gcr2p in the yeast *S. cerevisiae*. *Yeast* **15**, 1045–1057.

Uemura, H. and Fraenkel, D. G. (1990) *gcr2*, a new mutation affecting glycolytic gene expression in *Saccharomyces cerevisiae*. *Molecular and Cellular Biology* **10**, 6389–6396.

Vagnoli, P., Coons, D. M. and Bisson, L. F. (1998) The C-terminal domain of Snf3p mediates glucose-responsive signal transduction in *Saccharomyces cerevisiae*. *FEMS Microbiology Letters* **160**, 31–36.

Valdes-Hevia, M. D., de la Guerra, R. and Gancedo, C. (1989) Isolation and characterization of the gene encoding phosphoenolpyruvate carboxykinase from *Saccharomyces cerevisiae*. *FEBS Letters* **258**, 313–316.

Vallier, L. G. and Carlson, M. (1994) Synergistic release from glucose repression by *mig1* and *ssn* mutations in *Saccharomyces cerevisiae*. *Genetics* **137**, 49–54.

Vallier, L. G., Coons, D., Bisson, L. F. and Carlson, M. (1994) Altered regulatory responses to glucose are associated with a glucose transport defect in *grr1* mutants of *Saccharomyces cerevisiae*. *Genetics* **136**, 1279–1285.

Van Aelst, L., Hohmann, S., Zimmermann, F. K., Jans, A. W. and Thevelein, J. M. (1991) A yeast homologue of the bovine lens fibre MIP gene family complements the growth defect of a *Saccharomyces cerevisiae* mutant on fermentable sugars but not its defect in glucose-induced RAS-mediated cAMP signalling. *EMBO Journal* **10**, 2095–2104.

Van Aelst, L., Hohmann, S., Bulaya, B., de Koning, W., Sierkstra, L., Neves, M. J., Luyten, K., Alijo, R., Ramos, J. and Coccetti, P. (1993) Molecular cloning of a gene involved in glucose sensing in the yeast *Saccharomyces cerevisiae*. *Molecular Microbiology* **8**, 927–943.

Van Belle, D. and Andre, B. (2001) A genomic view of yeast membrane transporters. *Current Opinion in Cell Biology* **13**, 389–398.

van de Poll, K. W., Kerkenaar, A. and Schamhart, D. H. (1974) Isolation of a regulatory mutant of fructose-1,6-diphosphatase in *Saccharomyces carlsbergensis*. *Journal of Bacteriology* **117**, 965–970.

Van den Berg, M. A., de Jong-Gubbels, P., Kortland, C. J., van Dijken, J. P., Pronk, J. T. and Steensma, H. Y. (1996) The two acetyl-coenzyme A synthetases of *Saccharomyces cerevisiae* differ with respect to kinetic properties and transcriptional regulation. *Journal of Biological Chemistry* **271**, 28953–28959.

Van den Berg, M. A. and Steensma, H. Y. (1995) *ACS2*, a *Saccharomyces cerevisiae* gene encoding acetyl-coenzyme A synthetase, essential for growth on glucose. *European Journal of Biochemistry* **231**, 704–713.

van der Rest, M. E., de Vries, Y., Poolman, B. and Konings, W. N. (1995a) Overexpression of Mal61p in *Saccharomyces cerevisiae* and characterization of maltose transport in artificial membranes. *Journal of Bacteriology* **177**, 5440–5446.

van der Rest, M. E., Kamminga, A. H., Nakano, A., Anraku, Y., Poolman, B. and Konings, W. N. (1995b) The plasma membrane of *Saccharomyces cerevisiae*: structure, function, and biogenesis. *Microbiology Reviews* **59**, 304–322.

van Dijken, J. P., Weusthuis, R. A. and Pronk, J. T. (1993) Kinetics of growth and sugar consumption in yeasts. *Antonie Van Leeuwenhoek* **63**, 343–352.

van Hoek, P., van Dijken, J. P. and Pronk, J. T. (2000) Regulation of fermentative capacity and levels of glycolytic enzymes in chemostat cultures of *Saccharomyces cerevisiae. Enzyme and Microbial Technology* **26**, 724–736.

Van Leeuwen, C. C., Weusthuis, R. A., Postma, E., Van den Broek, P. J. and Van Dijken, J. P. (1992) Maltose/proton co-transport in *Saccharomyces cerevisiae.* Comparative study with cells and plasma membrane vesicles. *Biochemical Journal* **284**, 441–445.

Versele, M., de Winde, J. H. and Thevelein, J. M. (1999) A novel regulator of G protein signalling in yeast, Rgs2, downregulates glucose-activation of the cAMP pathway through direct inhibition of Gpa2. *EMBO Journal* **18**, 5577–5591.

Verwaal, R., Paalman, J. W., Hogenkamp, A., Verkleij, A. J., Verrips, C. T. and Boonstra, J. (2002) *HXT5* expression is determined by growth rates in *Saccharomyces cerevisiae. Yeast* **19**, 1029–1038.

Visser, D., van der Heijden, R., Mauch, K., Reuss, M. and Heijnen, S. (2000) Tendency modeling: a new approach to obtain simplified kinetic models of metabolism applied to *Saccharomyces cerevisiae. Metabolic Engineering* **2**, 252–275.

Voet, E. O. and Voet, J. G. (1995) *Biochemistry*, 2nd edn, New York: Wiley.

Voit, E. O. and Radivoyevitch, T. (2000) Biochemical systems analysis of genome-wide expression data. *Bioinformatics* **16**, 1023–1037.

Vuorio, O., Kalkkinen, N. and Londesborough, J. (1993) Cloning of two related genes encoding the 56-kDa subunits of *Saccharomyces cerevisiae* trehalose synthase. *European Journal of Biochemistry* **216**, 849–861.

Vyas, V. K., Kuchin, S. and Carlson, M. (2001) Interaction of the repressors Nrg1 and Nrg2 with the Snf1 protein kinase in *Saccharomyces cerevisiae. Genetics* **158**, 563–572.

Walker, M. E., Val, D.L., Rhode, M., Devenish, R. J. and Wallace, J. C. (1991) Yeast pyruvate carboxylase: identification of 2 genes encoding isoenzymes. *Biochemical and Biophysical Research Communications* **176**, 1210–1270.

Walsh, R. B., Kawasaki, G. and Fraenkel, D. G. (1983) Cloning of genes that complement yeast hexokinase and glucokinase mutants. *Journal of Bacteriology* **154**, 1002–1004.

Wang, J. and Needleman, R. (1996) Removal of Mig1p binding site converts a *MAL63* constitutive mutant derived by interchromosomal gene conversion to glucose insensitivity. *Genetics* **142**, 51–63.

Wang, X., Mann, C. J., Bai, Y., Ni, L. and Weiner, H. (1998) Molecular cloning, characterization, and potential roles of cytosolic and mitochondrial aldehyde dehydrogenases in ethanol metabolism in *Saccharomyces cerevisiae. Journal of Bacteriology* **180**, 822–830.

Weckwerth, W. and Fiehn, O. (2002) Can we discover novel pathways using metabolomic analysis? *Current Opinion in Biotechnology* **13**, 156–160.

Wehner, E. P., Rao, E. and Brendel, M. (1993) Molecular structure and genetic regulation of *SFA*, a gene responsible for resistance to formaldehyde in *Saccharomyces cerevisiae*, and characterization of its protein product. *Molecular and General Genetics* **237**, 351–358.

Wenzel, T. J., Teunissen, A. W. and Steensma, H. Y. (1995) *PDA1* mRNA: a standard for quantitation of mRNA in *Saccharomyces cerevisiae* superior to *ACT1* mRNA. *Nucleic Acids Research* **23**, 883–884.

White, W. H., Skatrud, P. L., Xue, Z. and Toyn, J. H. (2003) Specialization of function among aldehyde dehydrogenases: the *ALD2* and *ALD3* genes are required for β-alanine biosynthesis in *Saccharomyces cerevisiae. Genetics* **163**, 69–77.

Wieczorke, R., Krampe, S., Weierstall, T., Freidel, K., Hollenberg, C. P. and Boles, E. (1999) Concurrent knock-out of at least 20 transporter genes is required to block uptake of hexoses in *Saccharomyces cerevisiae. FEBS Letters* **464**, 123–128.

Williams, R. S., Trumbly, R. J., MacColl, R., Trimble, R. B. and Maley, F. (1985) Comparative properties of amplified external and internal invertase from the yeast *SUC2* gene. *Journal of Biological Chemistry* **260**, 13334–13341.

Wilson, A. J. and Bhattacharjee, J. K. (1986) Regulation of phospho*enol*pyruvate carboxykinase and pyruvate kinase in *Saccharomyces cerevisiae* grown in the presence of glycolytic and gluconeogenic carbon sources and the role of mitochondrial function on gluconeogenesis. *Canadian Journal of Microbiology* **32**, 969–972.

Wittmann, C. (2002) Metabolic flux analysis using mass spectrometry. *Advances in Biochemical Engineering and Biotechnology* **74**, 39–64.

Wood, V., Gwilliam, R., Rajandream, M. A., Lyne, M., Lyne, R., Stewart, A., Sgouros, J., Peat, N., Hayles, J., Baker, S., Basham, D., Bowman, S., Brooks, K., Brown, D., Brown, S., Chillingworth, T., Churcher, C., Collins, M., Connor, R., Cronin, A., Davis, P., Feltwell, T., Fraser, A., Gentles, S., Goble, A., Hamlin, N., Harris, D., Hidalgo, J., Hodgson, G., Holroyd, S., Hornsby, T., Howarth, S., Huckle, E. J., Hunt, S., Jagels, K., James, K., Jones, L., Jones, M., Leather, S., McDonald, S., McLean, J., Mooney, P., Moule, S., Mungall, K., Murphy, L., Niblett, D., Odell, C., Oliver, K., O'Neil, S., Pearson, D., Quail, M. A., Rabbinowitsch, E., Rutherford, K., Rutter, S., Saunders, D., Seeger, K., Sharp, S., Skelton, J., Simmonds, M., Squares, R., Squares, S., Stevens, K., Taylor, K., Taylor, R. G., Tivey, A., Walsh, S., Warren, T., Whitehead, S., Woodward, J., Volckaert, G., Aert, R., Robben, J., Grymonprez, B., Weltjens, I., Vanstreels, E., Rieger, M., Schafer, M., Muller-Auer, S., Gabel, C., Fuchs, M., Fritzc, C., Holzer, E., Moestl, D., Hilbert, H., Borzym, K., Langer, I., Beck, A., Lehrach, H., Reinhardt, R., Pohl, T. M., Eger, P., Zimmermann, W., Wedler, H., Wambutt, R., Purnelle, B., Goffeau, A., Cadieu, E., Dreano, S., Gloux, S., Lelaure, V. *et al.* (2002) The genome sequence of *Schizosaccharomyces pombe. Nature* **415**, 871–880.

Wu, M. and Tzagoloff, A. (1987) Mitochondrial and cytoplasmic fumarases in *Saccharomyces cerevisiae* are encoded by a single nuclear gene *FUM1*. *Journal of Biological Chemistry* **262**, 12275–12282.

Wysocki, R., Chéry, C. C., Wawrzycka, D., Van Hulle, M., Cornelis, R., Thevelein, J. M. and Tamás, M. J. (2001) The glycerol channel Fpslp mediates the uptake of arsenites and antimonite in *Saccharomyces cerevisiae. Molecular Microbiology* **40**, 1391–1398.

Yang, R., Wek, S. A. and Wek, R. C. (2000) Glucose limitation induces *GCN4* translation initiation by activation of Gcn2 protein kinase. *Molecular and Cellular Biology* **20**, 2706–2717.

Yao, B., Sollitti, P. and Marmur, J. (1989) Primary structure of the maltose-permease-encoding gene of *Saccharomyces carlsbergensis. Gene* **79**, 189–197.

Yao, B., Sollitti, P., Zhang, X. and Marmur, J. (1994) Shared control of maltose induction and catabolite repression of the *MAL* structural genes in *Saccharomyces. Molecular and General Genetics* **243**, 622–630.

Ye, L., Kruckeberg, A. L., Berden, J. A. and van Dam, K. (1999) Growth and glucose repression are controlled by glucose transport in *Saccharomyces cerevisiae* cells containing only one glucose transporter. *Journal of Bacteriology* **181**, 4673–4675.

Ye, L., Berden, J. A., van Dam, K. and Kruckeberg, A. L. (2001) Expression and activity of the Hxt7 high-affinity hexose transporter of *Saccharomyces cerevisiae. Yeast* **18**, 1257–1267.

Yin, Z., Smith, R. J. and Brown, A. J. P. (1996) Multiple signalling pathways trigger the exquisite sensitivity of yeast gluconeogenic mRNAs to glucose. *Molecular Microbiology* **20**, 751–764.

Yin, Z., Hatton, L. and Brown, A. J. P. (2000) Differential post-transcriptional regulation of yeast mRNAs in response to high and low glucose concentrations. *Molecular Microbiology* **35**, 553–565.

Young, E. T., Kacherovsky, N. and Van Riper, K. (2002) Snf1 protein kinase regulates Adr1 binding to chromatin but not transcription activation. *Journal of Biological Chemistry* **277**, 38095–38103.

Yuan, Y. J., Obuchi, K. and Kuriyama, H. (2000) Dynamics of ethanol translocation in *Saccharomyces cerevisiae* as detected by ^{13}C-NMR. *Biochimica et Biophysica Acta* **1474**, 269–272.

Zaman, Z., Bowman, S. B., Kornfeld, G. D., Brown, A. J. P. and Dawes, I. W. (1999) Transcription factor GCN4 for control of amino acid biosynthesis also regulates the expression of the gene for lipoamide dehydrogenase. *Biochemical Journal* **340**, 855–862.

Zastrow, C. R. Mattos, M. A. and Stambuk, B U. (2000) Maltotriose metabolism by *Saccharomyces cerevisiae*. *Biotechnology Letters* **22**, 455–459.

Zenke, F. T., Engles, R., Vollenbroich, V., Meyer, J., Hollenberg, C. P. and Breunig, K. D. (1996) Activation of Gal4p by galactose-dependent interaction of galactokinase and Gal80p. *Science* **272**, 1662–1665.

Zhou, H. and Winston, F. (2001) *NRG1* is required for glucose repression of the *SUC2* and *GAL* genes of *Saccharomyces cerevisiae*. *BioMed Central Genetics* **2**, 5.

Chapter 4

Nitrogen metabolism

J. Richard Dickinson

4.1 Introduction

The aims of this chapter are to focus on the control of nitrogen metabolism, to emphasize how nitrogen metabolism is integrated with other aspects of metabolism and cellular physiology, and to highlight selected important new results. Most of the individual anabolic and catabolic pathways are not described here because they can easily be found elsewhere (Jones and Fink, 1982; Hinnebusch, 1992; Magasanik, 1992; KEGG, 2003).

4.2 Ammonia, glutamate and glutamine: the major interface with carbon metabolism

Ammonia is usually considered to be yeast's preferred nitrogen source, but for many strains glutamine equally permits a maximal rate of growth. The reason for this is clear when one considers the relationship of these two compounds with glutamate, which is derived from the TCA cycle intermediate α-ketoglutarate (Figure 4.1), that all nitrogen sources are converted to ammonia and glutamate, and that this node of metabolism is required for the biosynthesis of 16 amino acids. Glutamate is converted into proline; it is used to produce aspartate which gives rise to threonine, methionine and cysteine; it is used to produce tyrosine, phenylalanine, serine, leucine, isoleucine, valine and histidine and is essential in the biosynthesis of lysine and arginine.

There are three glutamate dehydrogenases: Gdh1 and Gdh3 are $NADP^+$-dependent enzymes and catalyse the synthesis of glutamate from ammonia and α-ketoglutarate; Gdh2 uses NAD^+ and degrades glutamate producing ammonia and α-ketoglutarate. The presence of two $NADP^+$-dependent glutamate dehydrogenases is believed to be unique amongst micro-organisms and has prompted much questioning as to the reason for its presence in *Saccharomyces cerevisiae.* Purification of Gdh1 and Gdh3 and characterization of their individual kinetic properties showed that they had different allosteric properties and rates of α-ketoglutarate utilization (DeLuna *et al.*, 2001). The expression of *GDH1* and *GDH3* is differently modulated by carbon source. The activities and amounts of Gdh1 are similar in glucose and ethanol. In contrast, Gdh3 activity is very low in glucose but is 20-fold higher in ethanol such that it constitutes 25% of $NADP^+$-dependent activity on this carbon source (DeLuna *et al.*, 2001). These relative contributions towards total activity are conserved in rich glucose medium (YPD) where Gdh1 is the sole $NADP^+$-depndent glutamate dehydrogenase activity when the glucose concentration is still high (exponential phase) but changes by late stationary phase when 70% of this activity is now due to Gdh3 (DeLuna *et al.*, 2001). Control

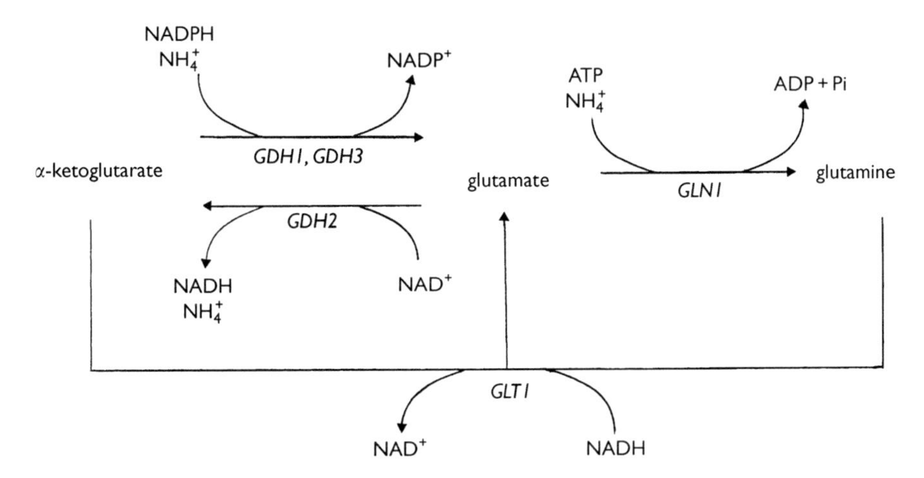

Figure 4.1 The α-ketoglutarate – glutamate – glutamine node of metabolism. The italicized three-letter acronyms are the structural genes for the enzymes involved: *GDH1, GDH3* – $NADP^+$-dependent glutamate dehydrogenase; *GDH2* – NAD^+-dependent glutamate dehydrogenase; *GLN1* – glutamine synthase; *GLT1* – GOGAT.

is achieved by glucose repression of *GDH3* and proteolytic degradation of Gdh1 upon glucose depletion (Mazón and Hemmings, 1979; DeLuna *et al.*, 2001).

As depicted in Figure 4.1, the isoenzymes of glutamate dehydrogenase are not the only route available for the synthesis of glutamate as this compound can also be produced via glutamate synthase ('GOGAT', Glt1) and glutamine synthase (Gln1). GOGAT is another $NADP^+$-dependent enzyme. It catalyses the conversion of one molecule of α-ketoglutarate and one molecule of glutamine to two molecules of glutamate. Glutamine synthase uses ammonium ion and ATP to convert glutamate into glutamine, ADP and Pi. Thus, just as there are two anabolic enzymes (Gdh1 and Gdh3) of glutamate formation from α-ketoglutarate, there are two routes for the catabolic formation of glutamate from glutamine (Gdh2 and Gln1 + Glt1). NAD^+-dependent glutamate dehydrogenase is the low affinity, high activity route; GOGAT and glutamine synthase is the high affinity, low activity route (Holmes *et al.*, 1989, 1991).

4.3 Control of nitrogen metabolism

The genes transcribed and enzymes used in the presence of a 'good' (e.g. glutamine) and a 'poor' (e.g. proline) nitrogen source are quite different. The net effect is to achieve efficient utilization of the particular nitrogen source and prevent the operation of pathways which would be useless or futile under the given set of conditions. The phenomenon is called nitrogen catabolite repression ('NCR') (Magasanik, 1992). For example, in the presence of glutamine, *GLN1* (which encodes glutamine synthase) and *GAP1* (which encodes the general amino acid permease) are repressed because their gene products are not required. In the presence of a poor nitrogen source their expression is derepressed. A considerable number of NCR-sensitive genes are now known including the GATA-type transcription activators Gln3 and Gat1(Nil1), the repressors Dal80 and Gzf3(Nil2) (Courchesne and Magasanik,

1988; Stanbrough *et al.*, 1995; Coffman *et al.*, 1996) and the zinc finger transcription factors Dal81 and Dal82 (Coornaert *et al.*, 1991). Gln3, which controls the expression of the majority of NCR-sensitive genes, has its own activity regulated by Ure2, a protein which has attracted a lot of attention in recent years because it is a pro-prion (Taylor *et al.*, 1999). Ure2 binds to Gln3 and Gat1. When the nitrogen source is poor or limiting, Gln3 and Gat1 are concentrated in the nucleus where they bind GATA sequences upstream of NCR-sensitive genes and thereby activate their transcription. In a good nitrogen source, Gln3 and Gat1 are excluded from the nucleus, hence the upstream GATA sequences are unoccupied and the genes are not expressed. The amino acid sequences required for interaction between Gln3 and Ure2 and for nuclear exclusion have been determined (Kulkarni *et al.*, 2001). Interestingly, Ure2 apparently serves as a glutathione S-transferase and protects *S. cerevisiae* from heavy metal ion and oxidant toxicity (Rai *et al.*, 2003).

Carbon starvation also causes the nuclear localization of Gln3 and increased NCR-sensitive transcription. This has been shown to be due to indirect effects on nitrogen metabolism because Gln3 does not move into the nucleus of carbon-starved cells if glutamine is supplied as the nitrogen source instead of ammonia (Cox *et al.*, 2002). This is an important example of where glutamine and ammonia are not equivalent nitrogen sources. Cox *et al.* (2002) also showed that Gln3 was not distributed uniformly throughout the cytoplasm, but localized to punctate or tubular structures that are believed to be vesicular.

It is also known that the target of rapamycin ('TOR') proteins, Tor1 and Tor2, and Snf1 signalling pathways also regulate the intracellular localization of Gln3 and Gat1 (Beck and Hall, 1999; Cardenas *et al.*, 1999; Bertram *et al.*, 2000, 2002; Carvalho *et al.*, 2001). Tor1 and Tor2 are serine/threonine protein kinases which are involved in nutrient-mediated control of growth and proliferation (Rohde *et al.*, 2001), salt stress (Crespo *et al.*, 2001), the cell integrity pathway (Torres *et al.*, 2002) and (Tor2 only) polarization of the actin cytoskeleton (Kunz *et al.*, 1993; Zheng *et al.*, 1995). The TOR signalling pathway is fully described in Chapter 8 (pp. 308–309); rapamycin-sensitive shared functions of TOR (on pp. 309–312) and Tor2's unique function (on p. 312) are also described in Chapter 8. In addition, the mechanisms which underlie the shared and unique TOR functions in yeast are shown in Figure 8.6(b). Snf1 is a global regulator of carbon metabolism (see Chapter 3; Section 8.4 of Chapter 8 and Figure 8.5). Thus, it is apparent that NCR-sensitive genes are subject to and involved in, a complex regulatory matrix.

As might be expected, *gln3Δ* mutants are partially resistant to rapamycin and *ure2Δ* mutants are hypersensitive to growth inhibition by the drug. *vid30Δ* mutants are more sensitive to rapamycin than wild-type strains, but less sensitive than *ure2Δ* mutants (van der Merwe *et al.*, 2001). *VID30* expression is slightly NCR-sensitive, responsive to deletion of *URE2* and greatly increases when ammonia concentration in the medium is low. Analysis of gene expression patterns in a *vid30Δ* mutant indicates that Vid30 function shifts the balance of nitrogen metabolism towards the production of glutamate when yeast is grown in low ammonia (van der Merwe *et al.*, 2001).

Amino acid biosynthesis

As outlined before, *S. cerevisiae* uses a matrix of regulatory mechanisms to control its nitrogen metabolism. These controls could not operate without systems for sensing the nitrogen source(s) present. Ammonium availability is sensed by the ammonium-specific permease Mep2 (Lorenz and Heitman, 1998). External amino acids are sensed by the amino acid receptor Ssy1 (Iraqui *et al.*, 1999; Klasson *et al.*, 1999) in a similar way to which glucose is sensed by Snf3 and Rgt2 (see Chapter 3). Internal amino acids are sensed by the system termed 'general control of amino acid biosynthesis' (Hinnebusch, 1992).

Fundamental to general amino acid control is the protein kinase, Gcn2 (Wek *et al.*, 1989). The substrate for Gcn2 is the α subunit of translation initiation factor-2 (eIF2α), which is phosphorylated by Gcn2 at serine-51 (Samuel, 1993). eIF2 is a trimeric factor (comprising subunits α, β and γ) that can bind either GDP or GTP. However, only when bound to GTP is it able to execute its physiological function of binding Met-tRNA to the ribosome and transferring it to the 40S ribosomal subunit. Following attachment of the [eIF2·GTP·Met-tRNA] complex to the 40S subunit, the GTP is hydrolyzed to GDP and Pi and eIF2 is released as an inactive [eIF2·GTP] complex. Phosphorylation of eIF2α inhibits the recycling of bound GDP to GTP, decreasing the rate of protein synthesis. Gcn2-mediated phosphorylation of eIF2α under conditions of amino acid deprivation increases expression of amino acid biosynthesis genes through the action of the transcriptional activator Gcn4 (Hinnebusch, 1997). Gcn2 is believed to be activated by interacting with uncharged tRNA (Wek *et al.*, 1989). The region of the protein responsible for this is a C-terminal domain of approximately 400 amino acid residues that shows sequence homology with histidyl-tRNA synthetases (Wek *et al.*, 1989). The presence of adjacent catalytic and His-tRNA synthetase-like domains is indicative of all eukaryotic Gcn2-type protein kinases. Gcn2 is likely to be assisted by Gcn1 and Gcn20, which are associated with the ribosome and are necessary for efficient phosphorylation of eIF2α (Vazques de Aldana *et al.*, 1995; Marton *et al.*, 1993, 1997). The C-terminal segment of Gcn1 binds to the so-called GI domain which occurs at the N-terminus of Gcn2 (Kubota *et al.*, 2001)

Up-regulation of *GCN4* occurs at the translational level; translation initiating from a downstream initiation codon of the *GCN4* gene that is not used in conditions of amino acid sufficiency (Hinnebusch, 1992, 1994). Conversely, when amino acids are available in the growth medium, ribosomes translate the first of four small upstream open reading frames (uORFs) present in the 5′ region of Gcn4, re-initiate at uORF 2–4 and are unable to recognize the later *GCN4* start codon. Gcn4 binds to GCN4 response elements (GCREs) in the promoters of the many genes which it regulates. Some genes have several GCREs, others have merely one. Microarray analysis identified 539 yeast genes that are induced by amino acid starvation (Natarajan *et al.*, 2001). These included genes in every amino acid biosynthetic pathway except cysteine, the supply of amino acid precursors, vitamin biosynthetic enzymes, peroxisomal components, mitochondrial carrier proteins, autophagy proteins and many genes encoding protein kinases and transcription factors. These results certainly confirm Gcn4 as a master regulator during amino acid starvation.

Just as nitrogen catabolite repression is regulated by both nitrogen and carbon sources (see earlier text), glucose limitation induces *GCN4* translation by activation of the Gcn2 protein kinase (Yang *et al.*, 2000). However, unlike the response to amino acid starvation, Gcn20 is not required for the response to glucose limitation, which indicates that yeast is able to use Gcn2 in different ways in response to limitation for these two major nutrients. The Gcn2 induction of *GCN4* translation in response to carbon limitation results in increased storage of amino acids in the vacuole and facilitates entry into exponential growth when cells are shifted from low glucose to a high glucose medium. It also contributes to the maintenance of glycogen levels during prolonged starvation for glucose, showing another important link between carbon and nitrogen metabolism (Yang *et al.*, 2000). Gcn4 is also activated in response to the stimulation of growth by glucose addition or UV radiation in a way that involves Gcn2, but does not increase eIF2α phosphorylation. The phenomenon is mediated through the Ras/cAMP pathway (Marbach *et al.*, 2001) which suggests how the link between nitrogen metabolism and carbon metabolism may operate (see Chapter 3). Gcn2 has also been shown to mediate sodium toxicity (Goossens *et al.*, 2001). *gcn1* or *gcn3* mutations improved salt tolerance and all mutations which activate Gcn4 function caused salt sensitivity indicating that the entire general control of amino acid biosynthesis pathway is involved in the mediation of the toxic effects of growth under salt stress (Goossens *et al.*, 2001). Clearly, this control system is like many others in yeast: it is capable of receiving multiple (environmental) inputs and modulating multiple physiological outputs. As such, it too, should be seen as part of the cell's entire signalling network rather than a discrete separate pathway.

New insights into serine biosynthesis

Our ideas on the genes involved and metabolic pathways used for the biosynthesis of serine in yeast have changed dramatically over the years as a comparison of the review by Jones and Fink (1982) (see figure 5 in that publication) with the recent publication by Albers *et al.* (2003) will reveal. Current thought is that there are four routes of serine biosynthesis (Figure 4.2) although, as explained later, not all operate simultaneously.

The 'threonine pathway' using threonine aldolase, which is encoded by *GLY1*, has been said to be the main source of glycine when glucose is the carbon source (Monschau *et al.*, 1997), but, as Albers *et al.* (2003) have pointed out, this has not been verified. On non-fermentable carbon sources gluconeogenesis operates along with the glyoxylate cycle (see Chapter 3), hence glyoxylate represents another starting point for the biosynthesis of glycine using alanine glyoxylate aminotransferase which is encoded by YFL030w (no gene name has been adopted). Serine is formed from glycine by the isoenzymes of serine hydroxymethyltransferase which are encoded by *SHM1* and *SHM2*. This activity requires N^5-N^{10}-methylene tetrahydrofolate, which is produced from tetrahydrofolate ('THF') by the glycine decarboxylase multi-enzyme complex (encoded by *GCV1, GCV2, GCV3* and *LPD1*) (Sinclair and Dawes, 1995; Piper *et al.*, 2000). Serine hydroxymethyltransferase and glycine decarboxylase are both mitochondrially located. Since the glyoxylate cycle, alanine glyoxylate amino transferase and glycine decarboxylase are all repressed by glucose the gluconeogenic and threonine pathway routes can only supply the cell with serine in the presence of non-fermentable carbon sources, not when it is growing on glucose (Ulane and Ogur, 1972; Takada and Noguchi, 1985; Melcher and Entian, 1992; Melcher *et al.*, 1995). This, of itself would not preclude threonine aldolase from serving as a source of *glycine* when glucose is the carbon source (but see next paragraph).

When glucose is the carbon source, the glycolytic intermediate, 3-phosphoglycerate can provide the carbon skeleton to start the biosynthesis of serine. The pathway via 3-phosphohydroxypyruvate and 3-phopshoserine had been known for years and the *SER1* and *SER2* gene products (3-phosphoserine aminotransferase and 3-phosphoserine phosphatase) identified. An important addition to our knowledge of this pathway was made very recently by Albers *et al.* (2003) who identified the two genes *SER3* (YER081w) and *SER33* (YIL074c) that encode phosphoglycerate dehydrogenases. The two proteins are 92% identical to each other and appear to be the only enzymes with this activity in *S. cerevisiae*. Ser33 is apparently the major enzyme. All of the *SER* genes are coordinately regulated by carbon source (expression is low on ethanol) and all except *SER3* are similarly regulated by nitrogen source. *SER3*

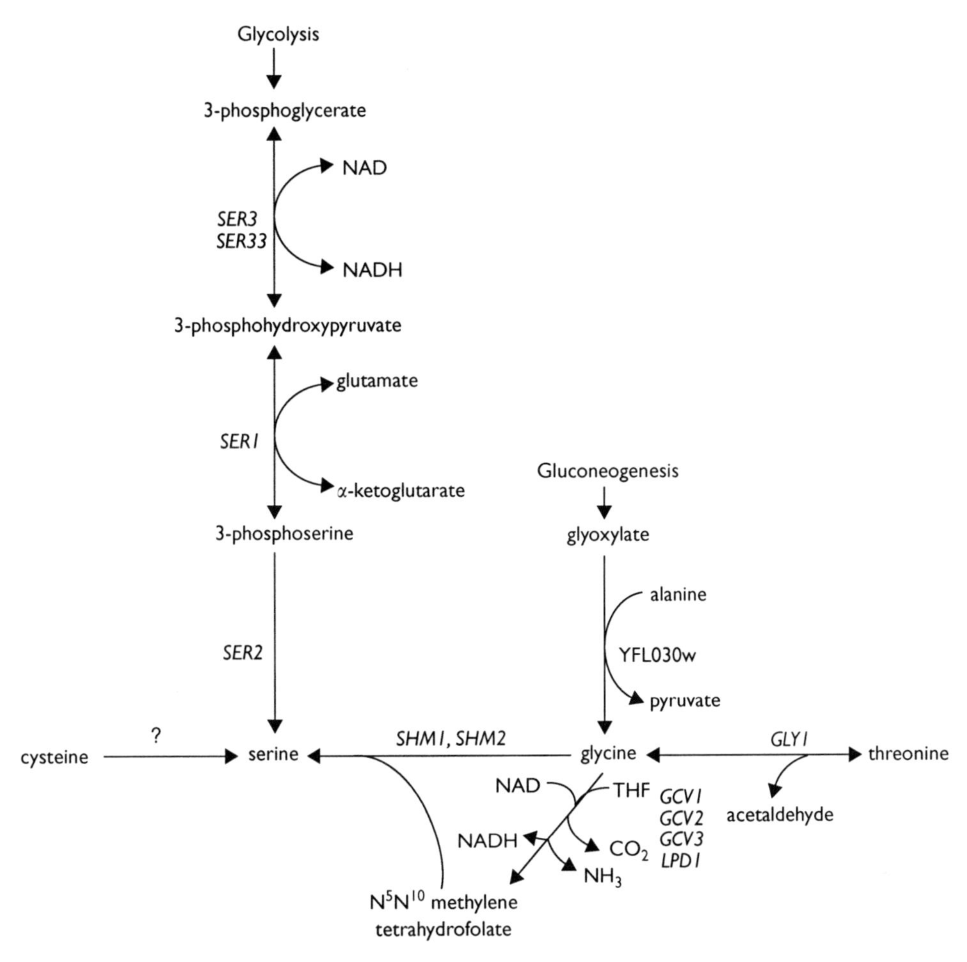

Figure 4.2 The pathways of serine biosynthesis in *S. cerevisiae*. Double-headed arrows indicate reversible reactions; single-headed arrows irreversible reactions. The italicized three-letter acronyms are the structural genes for the enzymes involved. *GCV1*, *GCV2*, *GCV3* and *LPD1* together form the glycine decarboxylase complex; *GLY1*: threonine aldolase; *SER1*: 3-phosphoserine aminotransferase; *SER2*: 3-phosphoserine phosphatase; *SER3*: phosphoglycerate dehydrogenase (minor isoenzyme); *SER33*: phosphoglycerate dehydrogenase (major isoenzyme); *SHM1*, *SHM2* serine hydroxymethyltransferases. The ORF YFL030w encodes alanine glyoxylate aminotransferase – apparently no gene name has yet been ascribed. The question mark against serine sulphydrase is to show that there is currently no known ORF or gene that encodes this activity.

mRNA level is very sensitive to additions of serine, glycine and threonine (Albers *et al.*, 2003). *ser3Δ ser33Δ* double mutants are reported as being auxotrophic for serine or glycine on glucose media (Albers *et al.*, 2003). This suggests that the supposed route of serine biosynthesis from cysteine via cysteine sulphydrase (Melcher and Entian, 1992) was not operative in the strains used by Albers *et al.* Such a conclusion would explain the lack of an identified gene encoding this activity and the absence from the yeast genome of any obvious ORF that could potentially encode it; although it should be stressed that Melcher and

Entian (1992) envisaged cysteine sulphydrase was a minor pathway. Alternatively, strains with different genetic backgrounds may have more of this activity. The auxotrophy of *ser3Δ ser33Δ* double mutants for glycine when on glucose is also at odds with the earlier assertion (Monschau *et al.*, 1997) that threonine aldolase is the main source of glycine when glucose is the carbon source.

Epiarginase regulation

As is the case for all amino acids, the anabolism and catabolism of arginine are tightly regulated at many levels. The so-called 'epiarginase control' operates at the level of enzyme activity by inhibiting the activity of the anabolic ornithine carbamoyltransferase by the catabolic enzyme arginase. Ornithine carbamoyltransferase catalyses the conversion of ornithine and carbamyl phosphate to citrulline. Arginase catalyses the cleavage of arginine to ornithine and urea. In the presence of both ornithine and arginine (the substrates of the two enzymes) ornithine carbamoyltransferase and arginase form a complex in a 1 : 1 ratio. In this complex ornithine carbamoyltransferase is inhibited but arginase remains active (Messenguy *et al.*, 1971). This exquisite fine-tuning of metabolism prevents the operation of a futile cycle in which ornithine produced by arginase is recycled by ornithine carbamoyltransferase when the cells are growing on arginine as nitrogen source (Messenguy and Wiame, 1969). Some workers believed that the inhibition of ornithine carbamoyltransferase was effected by the binding of ornithine at an allosteric site which is distinct from the catalytic site (Messenguy *et al.*, 1971). This was disputed by others because their results gave no evidence of such an allosteric site (Hensley, 1988). Instead, it was proposed that the binding of ornithine to the active site causes a conformational change required for formation of the complex with arginase. Recently, El Alami *et al.* (2003) confirmed this view by analysing mutations in both enzymes to identify residues that are required for their association and for catalysis. In arginase, two cysteines at the C-terminus are crucial for its epiarginase function but not for catalytic activity. In ornithine carbamoyltransferase mutations in D182, N184, N185, C289 and E256 greatly reduced the affinity for ornithine and impaired the interaction with arginase. In addition, four lysines at positions 260, 263, 265 and 268 in the so-called SMG loop, are very important in making ornithine carbamoyltransferase sensitive to arginase and ornithine. These lysines are most certainly involved in closing of the catalytic domain when arginine is present (El Alami *et al.*, 2003).

Amino acid catabolism

The catabolism of amino acids necessarily requires their initial uptake. A complete list of all known and putative transporters was published by Paulsen *et al.* (1998). The vast majority of amino acid transporters had been discovered previously by the combination of biochemistry and conventional yeast genetics. There are only three putative transporters of amino acids (into the cell) inferred from the genome sequence whose functions are unknown: Ydr160w, Yll061w and Ypl274w. Some transporters are general (Gap1, Agp2, Agp3) and some are class-specific e.g. Bap2 (branched-chain amino acids), Vap1 (also called Tat1 and Tap1, transports branched-chain amino acids, tyrosine and tryptophan), Apl1 (also known as Alp1, transports basic amino acids), Dip5 (dicarboxylic amino acids), Can1 (arginine, lysine and ornithine), Pap1 (isoleucine and valine) and Ycc5 (asparagine and glutamine). The remainder are specific for a single amino acid: Gnp1 (glutamine), Mup1 (methionine with high affinity), Mup3 (methionine with low affinity), Hip1 (histidine), Lyp1 (lysine), Tat2 (also known as Scm2, Tat2, Tap2 and Ltg3, tryptophan) and Put4 (proline). It should be noted that with the exception of Mup3, each specific amino acid carrier has a higher affinity for its cognate amino acid than the general carriers.

A prominent example of transporters whose function was not known when the yeast genome had just been sequenced were the family of seven membrane-spanning proteins now designated Avt1–7 (Russnak *et al.*, 2001). Four of these, Avt1 (Yjr001w), Avt3 (Ykl146w), Avt4 (Ynl101w) and Avt6 (Yer119c) have been demonstrated to mediate amino acid transport in vacuoles. Avt1, Avt3 and Avt4 transport glutamine, asparagine, isoleucine, leucine and tyrosine, Avt6 transports aspartate and glutamate. Avt1 carries the amino acids into the vacuole; Avt3, Avt4 and Avt6 carry amino acids out of the vacuole (Russnak *et al.*, 2001). The substrates for Avt2 (Yel064c), Avt5 (Ybl089w) and Avt7 (Yil088c) are unknown but it should be noted that the Avt proteins are apparently not involved in vacuolar transport of lysine, arginine and proline, the most abundant amino acids in the vacuole.

Many amino acids are catabolized to either ammonia or glutamate. Where glutamate is formed it can be used directly for the subsequent conversion into other amino acids. In the cases where ammonia is formed, the anabolic $NADP^+$-dependent glutamate dehydrogenases Gdh1 and Gdh3 are used to form glutamate (see Section 4.2). Arginine catabolism yields both glutamate and ammonia. These pathways have been studied intensively since the 1960s and 1970s and have been reviewed in great detail (Magasanik, 1992; Messenguy and Dubois, 2000; ter Schure *et al.*, 2000). However, at least five other amino acids (the branched-chain amino acids leucine, isoleucine and valine; and the aromatic amino acids phenylalanine and tryptophan) are catabolized quite differently. The carbon skeletons are not recycled, only ammonia is retrieved and the final product is a so-called 'fusel' alcohol. Consequently, these amino acids will not serve as a sole source of nitrogen and carbon; they will only serve as sole source of nitrogen. The pathway used is essentially the 'Ehrlich pathway'. The pathway is named after the originator of the idea (Ehrlich, 1907) although, in actual fact, the concept of the 'Ehrlich pathway' has for many years involved modifications suggested later by Neubauer and Fromherz (1911). The amino acids are first transaminated to the corresponding α-ketoacids: α-ketoglutarate accepts the ammonia group, hence forming glutamate. The second step is a decarboxylation reaction which produces an aldehyde. The final reaction requires an alcohol dehydrogenase.

For the decarboxylation stage there are five closely related decarboxylases (Pdc1, Pdc5, Pdc6, Ydl080c and Ydr380w). The spectrum of decarboxylase(s) used in the catabolism of each amino acid is subtly different in each case. In leucine catabolism Ydl080c is the major decarboxylase accounting for approximately 94% of the decarboxylation of α-ketoisocaproate; the minor decarboxylase is Ydr380w (Dickinson *et al.*, 1997, 2003). In valine catabolism, any one of the three isozymes of pyruvate decarboxylase will decarboxylate α-ketoisovalerate (Dickinson *et al.*, 1998) and in isoleucine catabolism any one of the five decarboxylases is sufficient for the conversion of isoleucine to the end product active amyl alcohol (Dickinson *et al.*, 2000). For the aromatic amino acids phenylalanine and tryptophan, Ydl080c has no role, the decarboxylation of 3-phenylpyruvate and 3-indolepyruvate (respectively), can be accomplished by any one of the remaining four decarboxylases (Dickinson *et al.*, 2003). Based on piecemeal evidence derived from studies in different yeast species, it had often previously been asserted (e.g. Large, 1986) that yeasts catabolize phenylalanine via *trans*-cinnamate to benzoate and protocatechuate and that the benzene ring is subsequently opened to give maleylacetate which is then supposedly reduced to 3-ketoadipate.

Tryptophan degradation was seen as involving ring opening by a dioxygenase to yield *N*-formyl-L-kynurenine. De-formylation was said to yield L-kynurenine which then has an alanine moiety removed to produce anthranilate. The pathway subsequently envisages 2,3-dihydroxybenzoate, catechol, *cis,cis*-muconate and finally 3-ketoadipate. The NMR study of Dickinson *et al.* (2003) showed no evidence of any of the intermediates or end products of either of these pathways. The only conclusion possible is that in *S. cerevisiae* phenylalanine and tryptophan, like leucine, valine and isoleucine, are catabolized according to the Ehrlich pathway. The final stage, culminating in fusel alcohol formation had always been assumed to be an alcohol dehydrogenase step (Webb and Ingraham, 1963) but this had never been rigorously examined with the benefit of a full knowledge of all of the potential alcohol dehydrogenases possessed by *S. cerevisiae*. The 20 alcohol dehydrogenases present in yeast are described in Section 3.3 of Chapter 3 and listed in Table 3.2. Using this up-to-date information, Dickinson *et al.* (2003) used strains containing all possible combinations of mutations affecting the seven *AAD* genes (putative aryl alcohol dehydrogenases, Delneri *et al.*, 1999a,b) five *ADH* genes and *SFA1*, and showed that the conversion of an aldehyde to a long chain or complex alcohol can be accomplished by any one of the ethanol dehydrogenases (Adh1, Adh2, Adh3, Adh4, Adh5) or by Sfa1 (formaldehyde dehydrogenase).

Thus, in *S. cerevisiae*, catabolism of the three branched-chain amino acids (leucine, valine and isoleucine) and the two aromatic amino acids (phenylalanine and tryptophan) proceeds essentially as Ehrlich proposed nearly 100 years ago (Ehrlich, 1907). However, it is much more complex than Ehrlich ever imagined because yeast uses at least three aminotransferases, five decarboxylases and six alcohol dehydrogenases. The precise combination of enzymes used at a particular time depends upon the amino acid, the carbon source and the stage of growth of the culture. However, the major specificity for this pathway clearly resides with the decarboxylases.

Salvage pathways

Recycling of methionine

A special importance of methionine is its role in one-carbon metabolism as the activated derivative *S*-adenosylmethionine. *S*-adenosylmethionine is the methyl donor in hundreds of transmethylation rections of nucleic acids, proteins and lipids. *S*-adenosylmethionine also serves as a precursor for the biosynthesis of polyamines and is a substrate in a wide range of other reactions. It is said that *S*-adenosylmethionine is second only to ATP in the number of reactions in which it participates. Prior to a recent study by Thomas *et al.* (2000), it had been believed that yeast was similar to other eukaryotes in that the majority of *S*-adenosylmethionine was recycled into methionine using the methyl and 5-methylthioadenosine cycles. In the methyl cycle, *S*-adenosylhomoscysteine, which is the product of transmethylation, is converted into homocysteine which is then directly methylated by methionine synthase to yield methionine. In the 5-methylthioadenosine cycle, methionine is directly synthesized from 5-methyladenosine. Surprisingly, *S*-adenosylmethionine recycling has been found to occur mainly through the *S*-adenosylmethionine-dependent remethylation of homocysteine (Thomas *et al.*, 2000). The *S*-adenosylmethionine–homocysteine methyltransferase is encoded by *SAM4* (YPL273w) and *S*-methylmethionine–homocysteine methyltransferase is encoded by *MHT1* (YLL062c). Homologues of these genes exist in other eukaryotes, so the newly discovered activities in yeast are likely to be present in those other organisms too. It seems likely that different organisms will use the alternative pathways to differing extents reflecting their overall metabolic and nutrient status. Yeast does use the 5-methylthioadenosine cycle but its activity is insufficient for the recycling of *S*-adenosylmethionine into methionine in *met6* mutants (which lack methionine synthase) (Thomas *et al.*, 2000).

4.4 Concluding remarks

Nitrogen metabolism in yeast is very precisely controlled at all levels to achieve highly efficient metabolic pathways and integration with all other aspects of metabolism and cellular physiology. This chapter has concentrated almost exclusively on amino acid anabolism and catabolism but this efficiency extends to the whole of nucleotide metabolism too. An example of this was shown by Shaw *et al.* (2001) who found that yeast gauges nucleotide pools and cell growth rate and responds by unique gene expression.

References

Albers, E., Laizé, V., Blomberg, A., Hohmann, S. and Gustafsson, L. (2003) Ser3p (YER081wp) and Ser33p (YIL074cp) are phosphoglycerate dehydrogenases in *Saccharomyces cerevisiae*. *Journal of Biological Chemistry* **278**, 10264–10272.

Beck, T. and Hall, M. N. (1999) The TOR signalling pathway controls nuclear localization of nutrient-regulated transcription factors. *Nature* **402**, 689–692.

Bertram, P. G., Choi, J. H., Carvalho, J., Ai, W. D., Zeng, C. B., Chan, T.-F. and Zheng, X. F. S. (2000) Tripartite regulation of Gln3 by TOR, Ure2, and phosphatases. *Journal of Biological Chemistry* **275**, 35727–35733.

Bertram, P. G., Choi, J. H., Carvalho, J., Ai, W. D., Zeng, C. B., Chan, T.-F. and Zheng, X. F. S. (2002) Convergence of TOR-nitrogen and Snf1-glucose signalling pathways onto Gln3. *Molecular and Cellular Biology* **22**, 1246–1252.

Cardenas, M. E., Cutler, N. S., Lorenz, M C., Di Como, C. J. and Heitman, J. (1999) The TOR signalling cascade regulates gene expression in response to nutrients. *Genes and Development* **13**, 3271–3279.

Carvalho, J., Bertram, P. G., Wente, S. R., Chan, T.-F., Ai, W. D. and Zheng, X. F. S. (2001) Phosphorylation regulates the interaction between Gln3p and the nuclear import factor Srp1p. *Journal of Biological Chemistry* **276**, 25359–25365.

Coffman, J. A., Rai, R., Cunningham, T., Svetlov, V. and Cooper, T. G. (1996) Gat1p, a GATA family protein whose production is sensitive to nitrogen catabolite repression, participates in transcriptional activation of nitrogen-catabolic genes in *Saccharomyces cerevisiae*. *Molecular and Cellular Biology* **16**, 847–858.

Coornaert, D., Vissers, S. and André, B. (1991) The pleiotropic *UGA35(DURL)* regulatory gene of *Saccharomyces cerevisiae*: cloning, sequence and identity with the *DAL81* gene. *Gene* **97**, 163–171.

Courchesne, W. E. and Magasanik, B. (1988) Regulation of nitrogen assimilation in *Saccharomyces cerevisiae*: roles of the *URE2* and *GLN3* genes. *Journal of Bacteriology* **170**, 708–713.

Cox, K. H., Tate, J. J. and Cooper, T. G. (2002) Cytoplasmic compartmentation of Gln3 during nitrogen catabolite repression and the mechanism of its nuclear localization during carbon starvation in *S. cerevisiae*. *Journal of Biological Chemistry* **277**, 37559–37566.

Crespo, J. L., Daicho, K., Ushimaru, T. and Hall, M. N. (2001) The GATA transcription factors GLN3 and GAT1 link TOR to salt stress in *Saccharomyces cerevisiae*. *Journal of Biological Chemistry* **276**, 34441–34444.

Delneri, D., Gardner, D. C. J. and Oliver, S. G. (1999a) Analysis of the seven-member *AAD* gene set demonstrates that genetic redundancy in yeast may be more apparent than real. *Genetics* **153**, 1591–1600.

Delneri, D., Gardner, D. C. J., Bruschi, C. V. and Oliver, S. G. (1999b) Disruption of seven hypothetical aryl alcohol dehydrogenase genes from *Saccharomyces cerevisiae* and construction of a multiple knock-out strain. *Yeast* **15**, 1681–1689.

DeLuna, A., Avendaño, A., Riego, L. and Gonzalez, A. (2001). NADP-glutamate dehydrogenase isoenzymes of *Saccharomyces cerevisiae. Journal of Biological Chemistry* **276**, 43775–43783.

Dickinson, J. R., Lanterman, M. M., Danner, D. J., Pearson, B. M., Sanz, P., Harrison, S. J. and Hewlins, M. J. E. (1997) A ^{13}C nuclear magnetic resonance investigation of the metabolism of leucine to isoamyl alcohol in *Saccharomyces cerevisiae. Journal of Biological Chemistry* **272**, 26871–26878.

Dickinson, J. R., Harrison, S. J. and Hewlins, M. J. E. (1998) An investigation of the metabolism of valine to isobutyl alcohol in *Saccharomyces cerevisiae. Journal of Biological Chemistry* **273**, 25751–25756.

Dickinson, J. R., Harrison, S. J., Dickinson, J. A. and Hewlins, M. J. E. (2000) An investigation of the metabolism of isoleucine to active amyl alcohol in *Saccharomyces cerevisiae. Journal of Biological Chemistry* **275**, 10937–10942.

Dickinson, J. R., Salgado, L. E. J. and Hewlins, M. J. E. (2003) The catabolism of amino acids to long chain and complex alcohols in *Saccharomyces cerevisiae. Journal of Biological Chemistry* **278**, 8028–8034.

Ehrlich, F. (1907) Über die bedingungen der fuselölbildung und über ihren zusammenhang mit dem eiweissaubau der hefe. *Berichte der Deutschen Chemischen Gesellschaft* **40**, 1027–1047.

El Alami, M., Dubois, E., Oudjama, Y., Tricot, C., Wouters J., Stalon, V. and Messenguy, F. (2003) Yeast epiarginase regulation, an enzyme–enzyme activity control: identification of residues of ornithine carbamoyltransferase (OTCase) and arginase responsible for their catalytic and regulatory activities. *Journal of Biological Chemistry* **278**, 21550–21558.

Goossens, A., Dever, T. E., Pascual-Ahuir, A. and Serrano, R. (2001) The protein kinase Gcn2 mediates sodium toxicity in yeast. *Journal of Biological Chemistry* **276**, 30753–30760.

Hensley, P. (1988) Ligand-binding and multienzyme complex formation between ornithine carbamoyltransferase and arginase from *Saccharomyces cerevisiae. Current Topics in Cellular Regulation* **29**, 35–75.

Hinnebusch, A. G. (1992) General and pathway-specific regulatory mechanisms controlling the synthesis of amino acid biosynthetic enzymes in *Saccharomyces cerevisiae.* In: E. W. Jones, J. R. Pringle and J. R. Broach (eds) *The Molecular Biology of the Yeast Saccharomyces: Vol. 2. Gene Expression*, pp. 319–414, Cold Spring Harbor: Cold Spring Harbor Laboratory.

Hinnebusch, A. G. (1997) Translational regulation of yeast *GCN4* – a window on factors that control initiator-t-RNA binding to the ribosome. *Journal of Biological Chemistry* **272**, 21661–21664.

Holmes, A. R., Collings, A., Farnden, K. J. F. and Shepherd, M. D. (1989) Ammonium assimilation by *Candida albicans* and other yeasts: evidence for activity of glutamate synthase. *Journal of General Microbiology* **135**,1423–1430.

Holmes, A. R., McNaughton, G. S., More, R. D. and Shepherd, M. D. (1991) Ammonium assimilation by *Candida albicans* and other yeasts: a ^{15}N isotope study. *Canadian Journal of Microbiology* **37**, 226–232.

Iraqui, I., Vissers, S., Bernard, F., Craene, J. O. D., Boles, E., Urrestarazu, A. and André, B. (1999) Amino acid signalling in *Saccharomyces cerevisiae:* a permease-like sensor of external amino acids and F-box protein Grr1p are required for transcriptional induction of the AGP1 gene, which encodes a broad-specificity amino acid permease. *Molecular and Cellular Biology* **19**, 989–1001.

Jones, E. W. and Fink, G. R. (1982) Regulation of amino acid and nucleotide biosynthesis in yeast. In: J. N. Strathern, E. W. Jones and J. R. Broach (eds) *The Molecular Biology of the Yeast Saccharomyces: Vol. 2. Metabolism and Gene expression*, pp. 181–299. Cold Spring Harbor: Cold Spring Harbor Laboratory.

KEGG (2003) The Kyoto Encyclopedia of Genes and Genomes, http://www.genome.ad.jp/kegg

Klasson, H., Fink, G. R. and Ljungdahl, P. O. (1999) Ssy1p and Ptr3p are plasma membrane components of a yeast system that senses extracellular amino acids. *Molecular and Cellular Biology* **19**, 5404–5416.

Kubota, H., Ota, K., Sakaki, Y. and Ito, T. (2001) Budding yeast GCN1 binds the GI domain to activate the eIF2α kinase GCN2. *Journal of Biological Chemistry* **276**, 17591–17596.

Kulkarni, A. A., Abdul-Hamd, A. T., Rai, R., El Berry, H. and Cooper, T. C. (2001) Gln3 nuclear localization and interaction with Ure2 in *Saccharomyces cerevisiae. Journal of Biological Chemistry* **276**, 32136–32144.

Kunz, J., Henriquez, R., Schneider, U., Deuter-Reinhard, M., Movva, N. R. and Hall. M. N. (1993) Target of rapamycin in yeast, TOR2, is an essential phosphatidylinositol kinase homolog required for G1 progression. *Cell* **73**, 585–596.

Large, P. J. (1986) Degradation of organic nitrogen compounds by yeasts. *Yeast* **2**, 1–34.

Lorenz, M.C. and Heitman, J. (1998) The MEP2 ammonium permease regulates pseudohyphal differentiation in *Saccharomyces cerevisiae. EMBO Journal* **17**, 1236–1247.

Magasanik, B. (1992) Regulation of nitrogen utilization. In: E. W. Jones, J. R. Pringle and J. R. Broach (eds) *The Molecular Biology of the Yeast Saccharomyces: Vol. 2. Gene Expression*, pp. 283–317. Cold Spring Harbor: Cold Spring Harbor Laboratory.

Marbach, I., Licht, R., Frohnmeyer, H. and Engelberg, D. (2001) Gcn2 mediates Gcn4 activation in response to glucose stimulation, or UV radiation, not via *GCN4* translation. *Journal of Biological Chemistry* **276**, 16944–16951.

Marton, M. J., Crouch, D. and Hinnebusch, A. G. (1993) GCN1, a translational activator of GCN4 in *Saccharomyces cerevisiae*, is required for phosphorylation of eukaryotic translation initiation factor 2 by protein kinase GCN2. *Molecular and Cellular Biology* **13**, 3541–3556.

Marton, M. J., Vazques de Aldana, C. R., Qui, H., Chakraburtty, K. and Hinnebusch, A. G. (1997) Evidence that GCN1 and GCN20, translational regulators of GCN4, function on elongating ribosomes in activation of eIF-2α kinase GCN2. *Molecular and Cellular Biology* **17**, 4474–4489.

Mazón, M. J. and Hemmings, B.A. (1979) Regulation of *Saccharomyces cerevisiae* nicotinamide adenine dinucleotide phosphate-dependent glutamate dehydrogenase by proteolysis during carbon starvation. *Jounal of Bacteriology* **139**, 686–689.

Melcher, K. and Entian, K.-D. (1992) Genetic analysis of serine biosynthesis and glucose repression in yeast. *Current Genetics* **21**, 295–300.

Melcher, K., Rose, M., Kunzler, M., Braus, G. H. and Entian, K.-D. (1995) Molecular analysis of the yeast *SER1* gene encoding 3-phosphoserine aminotransferase: regulation by general control and serine repression. *Current Genetics* **27**, 501–508.

Messenguy, F. and Wiame, J. M. (1969) The control of ornithine transcarbamylase activity by arginase in *Saccharomyces cerevisiae. FEBS Letters* **3**, 47–49.

Messenguy, F., Penninckx, M. and Wiame, J. M. (1971) Interaction between arginase and ornithine carbamoyltransferase in *Saccharomyces cerevisiae. European Journal of Biochemistry* **22**, 277–286.

Messenguy, F. and Dubois, E. (2000) Regulation of arginine metabolism in *Saccharomyces cerevisiae*: a network of specific and pleiotropic proteins in response to multiple environmental signals. *Food Technology and Biotechnology* **4**, 277–285.

Monschau, N., Stahmann, K.-P., Sahm, H., McNeil, J. B. and Bognar, A. L. (1997) Identification of *Saccharomyces cerevisiae* GLY1 as a threonine aldolase: a key enzyme in glycine biosynthesis. *FEMS Microbiology Letters* **150**, 55–60.

Natarajan, J., Meyer, M. R., Jackson, B. M., Slade, D., Roberts, C., Hinnebusch, A. G. and Marton, M. J. (2001) Transcriptional profiling shows that Gcn4 is a master regulator of gene expression during amino acid starvation. *Molecular and Cellular Biology* **21**, 4347–4368.

Neubauer, O. and Fromherz, K. (1911) Über den abbau der aminosären bei der hefegärung. *Hoppe-Seyler's Zeitschrift fur Physiologische Chemie* **70**, 326–350.

Paulsen, I. T., Sliwinski, M. K., Nelissen, B., Goffeau, A. and Saier, M. H. Jnr. (1998) Unified inventory of established and putative transporters encoded within the complete genome of *Saccharomyces cerevisiae. FEBS Letters* **430**, 116–125.

Piper, M. D., Hong, S.-P., Ball, G. E. and Dawes, I. W. (2000) Regulation of the balance of one-carbon metabolism in *Saccharomyces cerevisiae. Journal of Biological Chemistry* **275**, 30987–30995.

Rai, R., Tate, J. T. and Cooper, T. C. (2003) Ure2, a prion precursor with homology to glutathione *S*-transferase, protects *Saccharomyces cerevisiae* cells from heavy metal ion and oxidant toxicity. *Journal of Biological Chemistry* **278**, 12826–12833.

Rohde, J., Heitman, J. and Cardenas, M. (2001) The TOR kinases link nutrient sensing to cell growth. *Journal of Biological Chemistry* **276**, 9583–9586.

Russnak, R., Konczal, D. and McIntire, S. L. (2001) A family of yeast proteins mediating bidirectional vacuolar amino acid transport. *Journal of Biological Chemistry* **276**, 23849–23857.

Samuel, C. E. (1993) The eIF2α protein kinases, regulators of translation in eukaryotes from yeasts to humans. *Journal of Biological Chemistry* **268**, 7603–7608.

Shaw, R. J., Wilson, J. L., Smith, K. T. and Reines D. (2001) Regulation of an IMP dehydrogenase gene and its overexpression in drug-sensitive transcription elongation mutants of yeast. *Journal of Biological Chemistry* **276**, 32905–32916.

Sinclair, D. A. and Dawes, I. W. (1995) Genetics of the synthesis of serine from glycine and the utilization of glycine as sole nitrogen source by *Saccharomyces cerevisiae*. *Genetics* **140**, 1213–1222.

Stanbrough, M., Rowen, D. W. and Magasanik, B. (1995) Role of the GATA factors Gln3p and Nil1p of *Saccharomyces cerevisiae* in the expression of nitrogen-regulated genes. *Proceedings of the National Academy of Sciences, USA* **92**, 9450–9454.

Takada, Y. and Noguchi, T. (1985) Characteristics of alanine glyoxylate aminotransferase from *Saccharomyces cerevisiae*, a regulatory enzyme in the glyoxylate pathway of glycine and serine biosynthesis from tricarboxylic acid cycle intermediates. *Biochemical Journal* **231**, 157–163.

Taylor, K. L., Cheng, N., Williams, R. W., Steven, A. C. and Wickner, R. B. (1999) Prion domain initiation of amyloid formation *in vitro* from native Ure2p. *Science* **283**, 1339–1343.

ter Schure, E. G., van Riel, N. A. W. and Verrips, C. T. (2000) The role of ammonia in nitrogen catabolite repression in *Saccharomyces cerevisiae*. *FEMS Microbiology Reviews* **24**, 67–83.

Thomas, D., Becker, A. and Surdin-Kerjan, Y. (2000) Reverse methionine biosynthesis from *S*-adenosylmethionine in eukaryotic cells. *Journal of Biological Chemistry* **275**, 40718–40724.

Torres, J., Di Como, C. J., Herrero, E. and De La Torre-Ruiz, M. A. (2002) Regulation of the cell integrity pathway by rapamycin-sensitive TOR function in budding yeast. *Journal of Biological Chemistry* **277**, 43495–43504.

Ulane, R. and Ogur, M. (1972) Genetic and physiological control of serine and glycine biosynthesis in *Saccharomyces cerevisiae*. *Journal of Bacteriology* **109**, 34–43.

van der Merwe, G. K., Cooper, T. G. and van Vuuren, H. J. J. (2001) Ammonia regulates *VID30* expression and Vid30p function shifts nitrogen metabolism towards glutamate formation especially when *Saccharomyces cerevisiae* is grown in low concentrations of ammonia. *Journal of Biological Chemistry* **276**, 28659–28666.

Vazques de Aldana, C. R., Marton, M. J. and Hinnebusch, A. G. (1995) GCN20, a novel ATP binding cassette protein, and GCN1 reside in a complex that mediates activation of the eIF-2α kinase GCN2 in amino acid starved cells. *EMBO Journal* **14**, 3184–3199.

Webb, A. D. and Ingraham, J. L. (1963). Fusel oil. *Advances in Applied Microbiology* **5**, 317–353.

Wek, R. C., Jackson, B. M. and Hinnebusch, A. G. (1989) Juxtaposition of domains homologous to protein kinases and histidyl transfer RNA synthetases in GCN2 protein suggests a mechanism for coupling GCN4 expression to amino acid availability. *Proceedings of the National Academy of Sciences USA* **86**, 4579–4583.

Yang, R., Wek, S. A. and Wek, R. C. (2000) Glucose limitation induces *GCN4* translation by activation of Gcn2 protein kinase. *Molecular and Cellular Biology* **20**, 2706–2717.

Zheng, X. F. S., Florentino, D., Chen, J., Crabtree, G. R. and Schreiber, S. L. (1995) TOR kinase domains are required for two distinct functions, only one of which is inhibited by rapamycin. *Cell* **82**, 121–130.

Chapter 5

Molecular organization and biogenesis of the cell wall

Frans M. Klis, Piet de Groot, Stanley Brul and Klaas Hellingwerf

5.1 Introduction

The cell wall of yeasts accounts for about 25–30% of the cell dry weight and thus represents a considerable investment of the cell in terms of metabolic energy (Fleet, 1991). It has several functions. Because of its mechanical strength it protects the cell against physical damage and plays a major role in morphogenesis. Other functions are related to the surface layer of glycoproteins, which is generally found in fungal walls (De Nobel *et al.*, 2001b). The proteins of this layer are implanted in the underlying skeletal layer. They may be involved in sexual and nonsexual cell–cell recognition, in adhesion to host cells or to inert surfaces, they may confer hydrophobic properties to the wall, and they may limit the permeability of the cell wall to life-threatening molecules such as cell wall-degrading enzymes secreted by other organisms as defense proteins. They may also be involved in the retention of water and offer protection against dehydration. In the past few years, a more and more detailed molecular description of the cell wall of baker's yeast has gradually emerged (Lipke and Ovalle, 1998; Kapteyn *et al.*, 1999a; Cabib *et al.*, 2001; Klis *et al.*, 2002). This chapter presents its current status, discusses the biosynthesis and assembly of cell wall polymers, and extends this model to other ascomycetous fungi. Particular emphasis is given to the notion that cell wall formation is an intrinsic part of cell metabolism. Indeed, cell wall construction is tightly controlled both in space and in time, thus affecting cell shape, and the composition and molecular organization of the wall strongly depend on environmental conditions. For an earlier discussion of the yeast cell wall and its biosynthesis, the extensive review by Orlean (1997) is strongly recommended.

5.2 General organization and composition of the cell wall

Electron microscopic studies of the cell wall of various yeast species show a layered structure consisting of an apparently homogenous inner layer and a fibrillar outer layer, implanted in the inner layer and emanating from the cell surface (Tokunaga *et al.*, 1986; Baba and Osumi, 1987; Osumi, 1998). We now know that the inner layer represents the skeletal layer and that the outer layer consists of proteins covalently linked to the skeletal layer (see later). Table 5.1 shows the cell wall composition of *Saccharomyces cerevisiae*. There are only four classes of macromolecules present in the wall, which are presented in the order in which they occur, going from the outside to the inside. Forty percent of the wall (by dry weight) is accounted for by highly glycosylated glycoproteins, which are referred to as mannoproteins because their carbohydrate side chains are rich in mannose. Furthermore,

Table 5.1 The cell wall macromolecules of *S. cerevisiae*. The data are from Fleet (1991) and Dallies *et al.* (1998). Adapted from Klis *et al.* (2002)

Macromolecule	*Wall dry weight (%)*
Mannoproteins	40–50
1,6-β-Glucan	5
1,3-β-Glucan	50–60
Chitin	1–3

there are two classes of β-glucans: the highly branched and water-soluble 1,6-β-glucan interconnects most of the mannoproteins to the water-insoluble 1,3-β-glucan network. The latter, together with chitin, is largely responsible for the mechanical strength of the wall. Jointly, these carbohydrates make up for about 60% of the cell wall. When compared in terms of dry weight to 1,3-β-glucan, 1,6-β-glucan may seem to be a minor component of the cell wall of *S. cerevisiae* (about one-tenth of the mass of 1,3-β-glucan) but this is not the case in terms of number of molecules. Because 1,3-β-glucan and 1,6-β-glucan are estimated to consist of on average 1500 and 140 glucose residues, respectively (Manners *et al.*, 1973a; Fleet and Manners, 1976), the number of 1,6-β-glucan molecules is comparable to that of 1,3-β-glucan. Assuming the total wall dry weight per cell is about $5 \cdot 10^{-12}$ g (Kathoda *et al.*, 1976), this would correspond to about 7 million copies/cell of both components. It is more difficult to estimate the number of mannoproteins per cell in the cell wall, because their average molecular mass is not known. Using the rather arbitrary estimate of 200 kDa, produces a figure of about 6 million molecules per cell. In contrast to mycelial fungi, yeasts have only minor amounts of chitin in their walls. One should realize that the cell wall composition may vary considerably as shown for example by analyzing chemostat-grown cultures under various growth limitations (McMurrough and Rose, 1967). Furthermore, in the case of cell wall stress the chitin content increases considerably, sometimes by as much as tenfold (Kapteyn *et al.*, 1999b; Dallies *et al.*, 1998; Magnelli *et al.*, 2002; Ram *et al.*, 1998). The cell wall composition of the pathogenic yeast *Candida albicans* growing in the yeast form is comparable to that of *S. cerevisiae* (Klis *et al.*, 2001).

Before cell wall-degrading enzymes with known specificity became available, it was customary to fractionate cell walls using alkali (Fleet, 1991). Thus, in the older literature one often encounters alkali-soluble and alkali-insoluble β-glucan. It has been shown for *S. cerevisiae* that the alkali-soluble β-glucan is converted gradually in time into alkali-insoluble β-glucan and that the covalent coupling of chitin molecules to β-glucan is responsible for this (Hartland *et al.*, 1994). This is consistent with earlier studies by Wessels and co-workers in other fungi, indicating that the insolubility of cell wall β-glucan in alkali is due to its linkage to chitin (Mol and Wessels, 1987; Sietsma and Wessels, 1981). Using such a classical fractionation scheme based on the use of cold 1 M potassium hydroxide, Nguyen and co-workers have shown that several ascomycetous budding yeasts including *S. cerevisiae* have a similar cell wall composition, indicating that the cell wall organization of *S. cerevisiae* is representative of most ascomycetous budding yeasts (Nguyen *et al.*, 1998). Fission yeast, however, is different. For example, like many mycelial ascomycetous species it contains 1,3-α-glucan in its wall (Bush *et al.*, 1974; Manners and Meyer, 1977; Hochstenbach *et al.*, 1998).

5.3 Methodology for cell wall analysis

Changes in cell wall composition and structure, and cell wall defects can be detected by assaying cell wall porosity (De Nobel *et al.*, 1990a), sensitivity to cell wall-degrading enzymes (De Nobel *et al.*, 1990b; Ovalle *et al.*, 1998), sensitivity to Calcofluor and SDS (Ram *et al.*, 1994; De Groot *et al.*, 2001), or to sonication (Ruiz *et al.*, 1999; De Groot *et al.*, 2001). Both the sugar composition (Dallies *et al.*, 1998) and the polymer composition of the wall can be determined (Magnelli *et al.*, 2002). Cell surface proteins including wall proteins can be specifically labeled with a sulphonated biotinylation reagent, which due to its negative electrical charge cannot pass the plasma membrane (Casonova *et al.*, 1992; Mrsa *et al.*, 1997). Cell wall proteins (CWPs) can be liberated with aqueous hydrofluoric acid for GPI-CWPs (glycosylphosphatidylinositol anchor-dependent cell wall proteins) (Kapteyn *et al.*, 1996) or mild alkali for Pir-CWPs (protein with internal repeats cell wall proteins) (Mrsa *et al.*, 1997). Protein–cell wall polysaccharide complexes obtained by enzymatic digestion of cell wall glycans, or isolated from the culture medium of cell wall mutants, can be analyzed by immunoblotting with antisera directed against specific CWPs, against 1,3-β-glucan or 1,6-β-glucan, or by lectin blotting using lectins such as Concanavalin A to detect mannoproteins and by using wheat germ agglutinin to detect the presence of chitin (Klis *et al.*, 1998; Kapteyn *et al.*, 1997, 1999b; De Groot *et al.*, 2001). Where the complete genome sequence is known, rapid identification of CWPs becomes feasible by liquid chromatography of the liberated polypeptides or polypeptide fragments, followed by mass spectrometry (De Groot *et al.*, unpublished data). Finally, fluorophore-assisted carbohydrate electrophoresis (FACE) can be used as a simple tool to analyze *O*-glycosylation of yeast proteins and the fine structure of *N*-linked carbohydrate side-chains (Goins and Cutler, 2000).

A promising but underutilized tool is dimethyl sulfoxide (DMSO). DMSO is known to solubilize alkali-soluble β-glucan and α-glucan, thus allowing organic phase size exclusion chromatography (Fleet and Manners, 1976; Iacomini *et al.*, 1988; Sietsma and Wessels, 1981; Williams *et al.*, 1994; Grün *et al.*, unpublished data). When chitin in the wall is first removed by enzymatic digestion, the use of DMSO should allow complete solubilization of the remainder of the cell wall and subsequent separation of free polymers (if present) and covalently coupled macromolecular complexes.

5.4 The 1,3-β-glucan network

The molecules of 1,3-β-glucan are branched at the 6-position with 3–4% branch points, for which the 6-position of the 1,3-β-linked glucose residues is used (Fleet, 1991) and forms a three-dimensional network that is kept together by hydrogen bonding and maintains the shape of the cell (Kreger and Kopecka, 1975; Kopecka *et al.*, 1974). The molecules of 1,3-β-glucan in stationary phase cells comprise about 1500 sugar residues (Fleet, 1991). However, it is unknown if their length in other growth phases is similar. How the cell determines the chain length of 1,3-β-glucan molecules is also unknown. The 1,3-β-glucan meshwork is highly elastic. When cells are transferred to hypertonic solutions, they shrink rapidly (Morris *et al.*, 1986; Martinez de Maranon *et al.*, 1996) and cell wall porosity decreases (De Nobel *et al.*, 1990a). When returned to water, they rapidly swell again. 1,3-β-glucan is only slightly crystalline in lateral walls (Kreger and Kopecka, 1975). This is consistent with the observation that 1,3-β-glucan molecules in the cell wall of *S. cerevisiae* may be visualized with aniline blue, which is believed to bind to the single-helix form of 1,3-β-glucan molecules only, but not to the triple-helix conformer found in the crystalline form of 1,3-β-glucan (Young and Jacobs, 1998). This also explains the observation that aniline blue only stains the apical regions of growing hyphae of *Aspergillus fumigatus* (Beauvais *et al.*, 2001).

Saccharomyces cerevisiae possesses two plasma membrane-located 1,3-β-glucan synthase complexes, which contain either Fks1 or the related Fks2/Gsc2, depending on environmental conditions, and the regulatory G protein Rho1 (Inoue *et al.*, 1995; Douglas, 2001; Cabib *et al.*, 2001; Dijkgraaf *et al.*, 2002). Disruption of both *FKS1* and *FKS2* is lethal, but the single deletants are viable, indicating that these genes can take over each other's function (Inoue *et al.*, 1995). Homologues of *FKS1* and *GSC2* have been found in several other fungi including the basidiomycetous fungus *Cryptococcus neoformans* (De Nobel *et al.*, 2001a; Douglas, 2001). Both Fks1 and Fks2 have 16 predicted transmembrane segments and a large central hydrophilic domain predicted to be located in the cytosol. Interestingly, this central domain is strongly conserved among fungal *FKS* genes (Douglas, 2001). It is generally assumed that *FKS* proteins are responsible for the catalytic activity of the 1,3-β-glucan synthase complexes and form a channel guiding the newly formed 1,3-β-glucan chains through the plasma membrane (catalytic activity + transport function), but this has not yet been validated experimentally. They do not possess a known UDP-glucose binding site, neither the [R/K]XGG motif found in glycogen synthases (Furukawa *et al.*, 1993) and in the putative 1,3-α-glucan synthase in *Schizosaccharomyces pombe* (Hochstenbach *et al.*, 1998) nor the QXXRW motif found in many processive β-glycosyltransferases (Saxena *et al.*, 2001). Alternatively, the FKS proteins may, in addition to providing a 1,3-β-glucan channel, function as a scaffold protein for other proteins of the complex, such as the regulatory protein Rho1 and a putative 1,3-β-glucan synthase, and position them correctly to obtain a fully functional complex (scaffold function + transport function). The highly conserved central domain would be an obvious candidate for such a scaffolding function.

5.5 Biogenesis of 1,6-β-glucan

Mature 1,6-β-glucan, which has a degree of polymerization of about 140 and is branched at the three-positions, has 15% branch points and is thus much more highly branched than 1,3-β-glucan (Manners *et al.*, 1973b; Magnelli *et al.*, 2002). The branches may function to increase the number of attachment sites for chitin chains (Kollar *et al.*, 1997). Kollar and co-workers (Kollar *et al.*, 1997; Fujii *et al.*, 1999) have shown that GPI-CWPs are linked through a GPI remnant to a nonreducing end of 1,6-β-glucan, which in turn is linked through its reducing end to a nonreducing end of 1,3-β-glucan (Figure 5.1). The high degree of branching of 1,6-β-glucan explains why it is soluble in water, indicating that it may function as a flexible tether to connect the CWPs to the water-insoluble 1,3-β-glucan meshwork.

A molecular description of 1,6-β-glucan synthesis is still lacking. Basic questions such as how biosynthesis is initiated, what type of nucleotide is involved in chain elongation, and how chain termination occurs, are still unanswered. Whereas *in vitro* assays for measuring 1,3-β-glucan synthase and chitin synthase activities have been available for a long time, no such assay has yet been developed for 1,6-β-glucan synthesis. This considerably hampers establishing where synthesis takes place and identifying the proteins involved in 1,6-β-glucan synthesis. It is also not known whether 1,6-β-glucan is produced as a linear polymer and then remodeled by branching enzymes or synthesized as a branched molecule. This is particularly relevant for the detection of 1,6-β-glucan formed in an *in vitro* assay. For *in vitro* assays of 1,3-β-glucan and chitin synthase activities, trichloroacetic acid is routinely used to precipitate and partially purify the reaction products. If 1,6-β-glucan is formed as a branched and thus water-soluble molecule, such an approach would necessarily fail. An alternative might be to use 80% ethanol as precipitating agent, as this is known to precipitate water-soluble polysaccharides efficiently.

Genetic screens have identified several proteins, located in the endoplasmic reticulum (ER), the Golgi, and at the cell surface that are required for normal levels of 1,6-β-glucan in the cell wall. Such screens make use of the positive correlation between the level of 1,6-β-glucan in the cell wall and the sensitivity to K1 killer toxin, a protein toxin that binds to 1,6-β-glucan and is secreted by killer strains infected with the M1 RNA virus (Orlean, 1997; Shahinian and Bussey, 2000; De Groot *et al.*, 2001). In other words, 1,6-β-glucan-deficient mutant cells become more resistant to K1 killer toxin (*kre* or killer-resistant mutants). This has led to the hypothesis that the synthesis of 1,6-β-glucan is a stepwise process beginning in the ER, continued in the Golgi and completed at the cell surface (Klis, 1994; Orlean, 1997; Simons *et al.*, 1998; Shahinian and Bussey, 2000; Azuma *et al.*, 2002). However, in an immunogold labeling study using a 1,6-β-glucan antiserum, no internal signal could be found, not even in a late secretory mutant kept at the restrictive temperature to allow accumulation of secretory vesicles (Montijn *et al.*, 1999). Montijn and co-workers have proposed that 1,6-β-glucan synthesis requires a primer, the formation of which would depend on ER and Golgi proteins. The need of a primer for the synthesis of cell wall polymers is not unprecedented. For example, it has been shown that the synthesis of cellulose in plants requires sitosterolglucoside as a primer (Peng *et al.*, 2002). Alternatively, mutations in genes that encode proteins residing in the ER or Golgi might have an indirect effect on 1,6-β-glucan synthesis. It may be expected that the activity of any 1,6-β-glucan synthetic or processing enzyme at the cell surface will be sensitive to defects in the secretory pathway. Indeed, in a partial genomic screen 29 deletant strains out of 620 deletants tested were K1 killer-resistant (DeGroot *et al.*, 2001), which would correspond to almost 300 genes in the entire genome affecting 1,6-β-glucan levels. It thus seems likely that most K1 killer-resistant strains have an indirect effect on 1,6-β-glucan synthesis. Nevertheless, Humbel and co-workers, using the same immunogold labeling technique as in *S. cerevisiae* but now in *S. pombe*, have found evidence for the presence of 1,6-β-glucan in the Golgi (Humbel *et al.*, 2001). It is clear that to resolve the biogenesis of 1,6-β-glucan more biochemical work is urgently needed, including the development of an *in vitro* assay of 1,6-β-glucan synthase activity to resolve the details of the biogenesis pathway of 1,6-β-glucan.

Information about how the levels of 1,6-β-glucan in the cell wall are controlled is scarce. Expression levels of *PBS2*, which encodes the MAP kinase kinase of the HOG (high osmolarity glycerol) pathway, that is activated in response to increases in molarity of the medium, are inversely correlated with the cell wall level of 1,6-β-glucan (Jiang *et al.*, 1995). In contrast, expression levels of *KIC1*, which encodes a PAK kinase involved in morphogenesis and cell integrity (Sullivan *et al.*, 1998), and of *RHO3*, which encodes a monomeric G protein and is a multicopy suppressor of *KIC1*-deficient cells, are positively correlated with 1,6-β-glucan levels (Vink *et al.*, 2002). As the effects observed are modest, these effects may be indirect.

5.6 The contribution of chitin to the mechanical strength of the lateral walls

Saccharomyces cerevisiae contains three different chitin synthases encoded by *CHS1*, *CHS2*, and *CHS3*, each with a different function (Orlean, 1997). Chs3 is responsible for synthesis of the chitin ring at the incipient bud site, for deposition of chitin in the lateral walls of young cells after cytokinesis has taken place and at the base of the mating projection, and is required for the deposition of chitosan in the ascospore walls (see Chapter 1) (Shaw *et al.*, 1991). Chs3 also synthesizes stress chitin, which is deposited in large amounts in the lateral walls in response to cell wall damage (reviewed by Popolo *et al.*, 2001). Chs2 synthesizes the primary septum dividing the daughter cell from the mother cell, whereas Chs1 seems to have a repair function during cell separation when the primary septum is dissolved. Yeast chitin synthases are multipass transmembrane proteins that utilize the nucleotide UDP-GlcNAc as sugar donor. Chain elongation takes place at the cytosolic side of the plasma membrane and the growing chitin chain is extruded through the plasma membrane. Chitin synthase activities in cell-free lysates are normally determined using radioactively labeled UDP-*N*-acetylglucosamine as substrate. Recently, a simple, nonradioactive microtiter plate assay for all three chitin synthases has become available (Lucero *et al.*, 2002). Chitin is produced as linear chains with an average length *in vivo* of 100–120 GlcNAc residues (Kang *et al.*, 1984). How the cell controls the average chain length of chitin is a mystery. The sequences of yeast chitin synthases are characterized by a common motif, QRRRW, which is probably involved in binding UDP-GlcNAc (Saxena *et al.*, 2001). Fungal chitin synthases have been grouped in five major classes. For their classification and a discussion of their phylogenetic relationships, the reader may consult the reviews by Ruiz-Herrera *et al.* (2002) and Roncero (2002).

Although synthesis of chitin (1,4-*β*-linked *N*-acetyl-D-glucosamine) is essential for cell viability, the incorporation of chitin into the lateral walls does not normally seem to be required for viability. For example, the wall of buds that are still in direct connection with the cytosol of the mother cell does not contain chitin, because chitin deposition in the bud wall begins only after cytokinesis (Shaw *et al.*, 1991). The chitin content of cell walls is normally about 1–2% of wall dry weight, but most of it is found in the chitin ring in the neck and in the region of the bud scars (see Chapter 1) and only a small fraction (less than 10% of total chitin) is found dispersed in the lateral walls where it is covalently linked to 1,3-*β*-glucan and to 1,6-*β*-glucan (Molano *et al.*, 1980; Kollar *et al.*, 1995, 1997; Kapteyn *et al.*, 1997). When the wall of a cell gets damaged, the cell wall integrity pathway (which is fully described in Section 8.6 of Chapter 8) is activated, resulting in massive deposition of chitin in the lateral walls (De Nobel *et al.*, 2001a; Popolo *et al.*, 2001; Carotti *et al.*, 2002). Chemical assays reveal chitin contents as high as 20% of the wall dry weight (Dallies *et al.*, 1998; Kapteyn *et al.*, 1999b; Magnelli *et al.*, 2002). This is accompanied by transcriptional activation of *GFA1*, which encodes a glutamine-6-phosphate aminotransferase and is the first committed enzyme in *N*-acetylglucosamine biosynthesis (Lagorce *et al.*, 2002). Hyperaccumulation of chitin in cells subject to cell wall stress results in a strongly increased sensitivity to inhibitors of chitin synthesis, indicating that in such cells the contribution of chitin to the strength of the wall is indispensable (Popolo *et al.*, 1997).

Several proteins strongly affect the activity of the major chitin synthase in *S. cerevisiae*, Chs3. Chs7 is an integral membrane protein located in the ER and is required for the exit of Chs3p from the ER (Trilla *et al.*, 1999). Chs5 and Chs6 are two Golgi proteins that seem to be required for the transport of Chs3-carrying vesicles from the trans-Golgi network or early endosome to the plasma membrane (Santos *et al.*, 1997; Valdivia *et al.*, 2002). Chs4 is a plasma membrane-associated protein that probably acts as a direct activator of the Chs3-catalyzed chitin synthase activity (Trilla *et al.*, 1997; Ono *et al.*, 2000). In addition, Chs4 anchors Chs3 at the site of septum formation by interconnecting it with the Bni4–septin complex (DeMarini *et al.*, 1997). For a detailed discussion of how the synthesis of chitin is regulated including the role of noncatalytic *CHS* genes the reader is referred to the recent review by Roncero (2002).

5.7 Cell wall proteins

The outer protein layer of the wall determines the surface properties of the cell such as immunogenicity, electrical charge, hydrophobicity, flocculence, sexual agglutinability, and porosity. The CWPs of *S. cerevisiae* and other yeasts are generally heavily glycosylated and may carry both extended *N*-linked carbohydrate side chains and much shorter, but also much more numerous, *O*-linked side chains (Gemmill and Trimble, 1999; Cutler, 2001). Mannose (and a minor amount of *N*-acetyl-glucosamine in the core structure of *N*-chains) are the only sugars found in glycoproteins of *S. cerevisiae*, which explains their designation as mannoproteins, but other yeasts may use a wider variety of sugars to construct their side chains (Gemmill and Trimble, 1999). Importantly, both *N*- and *O*-linked side chains may contain phosphodiester-linked mannose residues (Jigami and Odani, 1999). As a result, the external protein layer of the cell wall contains numerous negative charges at physiological pH values. One could imagine that the phosphorylated carbohydrate side chains may play an important role in water retention and protection against dehydration. The negatively charged phosphodiester linkages can be visualized with the positively charged dye Alcian blue (Ballou, 1990; Jigami and Odani, 1999). Interestingly, the presence of phosphodiester groups makes *S. cerevisiae* more sensitive to osmotin, an antimicrobial plant protein that is positively charged at physiological pH values (Ibeas *et al.*, 2000).

In contrast to mammalian cells, *O*-glycosylation in *S. cerevisiae* begins in the ER with the transfer of mannose from dolichol-phosphomannose to serine or threonine residues and is completed in the Golgi (Strahl-Bolsinger *et al.*, 1999). *O*-chains are often clustered in specific regions of CWPs conferring to these regions a relatively rigid structure and spacer-like properties (Jentoft, 1990; Jue and Lipke, 2002). In *S. cerevisiae* the *O*-chains are relatively short: they consist of up to five mannose residues (Strahl-Bolsinger *et al.*, 1999).

The initial biosynthetic and processing steps of *N*-chains of yeasts in the ER seem largely similar to those in other eukaryotic cells. However, in the Golgi a number of *N*-chains – but not necessarily all – are extended to far greater lengths than the *N*-chains in, for example, mammalian cells (Gemmill and Trimble, 1999). In the older literature these elongated *N*-chains have led to designations such as mannan, phosphomannan, or galactomannan. These extended *N*-chains are largely responsible for the limited porosity of the cell wall of *S. cerevisiae* and may thus offer protection against foreign proteins such as cell wall-degrading enzymes from the fungal environment (De Nobel *et al.*, 1990b). The best-studied CWP of *S. cerevisiae* is probably the sexual adhesion protein Sag1p in mating type α cells (Wojciechowicz *et al.*, 1993; Cappellaro *et al.*, 1994; Lu *et al.*, 1994, 1995; Chen *et al.*, 1995; De Nobel *et al.*, 1996). It is a GPI-dependent cell wall protein (see later) with multiple *N*-chains and is heavily *O*-glycosylated in its C-terminal stalk region (Lu *et al.*, 1994; Chen *et al.*, 1995). Interestingly, the N-terminal domain of this protein, which is involved in interaction with the complementary recognition protein of mating type **a** cells, belongs to the immunoglobulin superfamily and has three immunoglobulin-like regions (Chen *et al.*, 1995; De Nobel *et al.*, 1996). Similar GPI-dependent adhesion proteins with a predicted N-terminal recognition domain and a C-terminal rodlike domain have been detected in various *Candida* species (Cormack *et al.*, 1999; Hoyer, 2001; Hoyer and Hecht, 2001; Hoyer *et al.*, 2001; Frieman *et al.*, 2002).

When isolated cell walls are extracted with SDS at elevated temperatures, they release a large number (>60) of low molecular weight proteins most of which are either adsorbed to the wall during homogenization or are associated with plasma membrane fragments that remain attached to the wall. Only a few of these are authentic secretory proteins (Valentin *et al.*, 1984). To identify secretory proteins and to avoid the adsorption of intracellular proteins to the cell wall during cell breakage, Cappellaro *et al.* (1998) used a nonpermeant biotinylation reagent to label cell surface proteins first. They then extracted intact cells under reducing conditions and purified the released proteins by Concanavalin A chromatography. In this way they identified seven major bands, six of which corresponded to secretory proteins known or predicted to be carbohydrate-processing enzymes. The remaining one was the secretory protein Cis3/Pir4, a cell wall protein whose function is still unknown (Moukadiri *et al.*, 1999). The proteins that resist extraction with SDS under reducing conditions are assumed to be covalently linked to other cell wall polymers (see later). They can be divided into two main classes: (i) GPI-dependent cell wall proteins (GPI-CWPs), and (ii) Pir-CWPs. Neither in *S. cerevisiae, C. albicans*, nor in *S. pombe*, is there evidence for the presence of hydrophobins, a class of secretory proteins often found in the form of a hydrophobic layer covering aerial hyphae and conidia of mycelial fungi (P. De Groot, unpublished data; J. G. H. Sietsma, unpublished data).

GPI-dependent cell wall proteins form the most abundant class of CWPs in *S. cerevisiae* and *C. albicans*. Recently, an algorithm to predict all fungal GPI proteins has been developed (De Groot *et al.*, 2003), which predicts the presence of 69 GPI proteins in *S. cerevisiae*, 106 in *C. albicans*, 33 in *S. pombe*, and 97 in *Neurospora crassa*. These GPI proteins often contain repeats and serine- and threonine-rich regions. In *S. cerevisiae* the majority of the GPI proteins (~40) are destined for the plasma membrane (Caro *et al.*, 1997; Hamada *et al.*, 1998a,b, 1999). The remaining GPI proteins are destined for the cell wall, but many of them are expressed only under specific growth conditions. The biosynthesis of GPI proteins includes an early step in the ER, in which the hydrophobic C-terminal region of the protein is replaced by a GPI anchor. The new C-terminal amino acid that is linked to the GPI anchor is called the omega amino acid. In *S. cerevisiae* the region preceding the omega amino acid seems to affect the final destination of GPI proteins. In particular, a dibasic motif just before the omega amino acid seems to favour the plasma membrane as the final destination, whereas other specific amino acids at the ω-2 and ω-5 positions are important for targeting GPI proteins to the cell wall (Caro *et al.*, 1997; Hamada *et al.*, 1998a,b, 1999). It is not yet clear if similar rules apply to other fungi. One of the complications is that frequently multiple ω-sites seem possible (De Groot *et al.*, 2003). In an attempt to extrapolate these rules to *C. albicans*, Sundström (2002) tentatively identified 54 GPI-CWPs.

Most of the GPI-CWPs are indirectly linked to 1,3-β-glucan through a connecting 1,6-β-glucan moiety (Figure 5.1) (Kapteyn *et al.*, 1996; Kollar *et al.*, 1997). GPI-CWPs undergo a postsecretory processing step, resulting in cleavage of their GPI anchor and coupling to 1,6-β-glucan. As a result, mature GPI-CWPs have only a remnant [core structure: – ethanolamine – PO4 – (Man)4 –] of the original GPI anchor remaining that links them to 1,6-β-glucan (Kollar *et al.*, 1997; Fujii *et al.*, 1999). The core structure is presumably substituted with ethanolamine phosphate groups (Flury *et al.*, 2000; Taron *et al.*, 2000; Grimme *et al.*, 2001; Richard *et al.*, 2002).

Pir cell wall proteins are directly linked to 1,3-β-glucan through an uncharacterized alkali-sensitive linkage (Mrsa *et al.*, 1997; Kapteyn *et al.*, 1999b). In *S. cerevisiae* a family of four proteins has been identified (Pir1, Pir2/Hsp150, Pir3, and Pir4/Cis3). They are all similarly organized, consisting of an N-terminal signal peptide, a Kex2 site, up to 11 repeats, and a highly conserved C-terminal region with four cysteine residues in a conserved spacing pattern: C-X_{66}-C-X_{16}-C-X_{12}-C. They have been localized in the cell wall immunologically (Yun *et al.*, 1997; Moukadiri *et al.*, 1999; Kapteyn *et al.*, 2000). Expression levels of *PIR1*, *PIR2*, and *PIR3* are strongly cell cycle-regulated and peak at the M/G1 transition. As the gene products have to follow the secretory pathway, it seems likely that they become incorporated into the wall during the G1 phase; in other words, during a period of isotropic growth of the cell wall (which is described in Chapter 1). It is tempting to predict that the gene that codes for the assembly enzyme that couples these proteins to the 1,3-β-glucan network is also upregulated during M/G1. A possible candidate for this assembly step is the GPI-protein Pst1, which is a putative plasma membrane protein that is also strongly upregulated during M/G1 (Spellman *et al.*, 1998). Once CWPs have been incorporated into the cell wall, they show limited turnover. Kratky *et al.* (1975) found that about 10% of the cell wall mannoproteins were slowly released in the course of several hours into the medium as nondialyzable material. This is consistent with our own observations (P. Mol *et al.*, unpublished data).

5.8 Surface display of heterologous proteins

The C-terminal regions of GPI-CWPs have been extensively used to target heterologous proteins to the cell wall of *S. cerevisiae* (Schreuder *et al.*, 1996; Murai *et al.*, 1997a,b, 1998, 1999; Van Der Vaart *et al.*, 1997). Recently, this approach has been successfully extended to the industrial yeast *Hansenula polymorpha* (Kim *et al.*, 2002). For example, when glucose oxidase, a secretory glycoprotein from *Aspergillus niger*, was extended with sequences that contained a GPI addition signal from three different *H. polymorpha* GPI proteins, about 99% of the glucose oxidase activity was found associated with intact cells and the remainder was found in the extracellular medium. FACS analysis, using glucose oxidase antibodies, confirmed that the enzyme was associated with the cell surface. Pir4/Cis3, a Pir-CWP, and Aga2, the recognition unit of the Aga1 protein, which are both extractable by reducing agents because they are disulfide-linked to other proteins, have also been successfully used to incorporate heterologous proteins into the cell wall (Boder and Wittrup, 1997; Moukadiri *et al.*, 1999).

5.9 Nonconventional cell surface proteins

There is growing evidence for the presence of proteins at the cell surface of *S. cerevisiae* and *C. albicans* and other yeasts that are nominally considered to be cytosolic proteins. The evidence is based on immunoelectron microscopy, the use of a nonpermeant biotinylation reagent to specifically label cell surface proteins and their subsequent identification by mass spectrometry, and the measurement of specific enzymatic activities associated with intact cells (Alloush *et al.*, 1997; reviewed in Chaffin *et al.*, 1998). In other words, their presence at the cell surface seems beyond doubt. Frequently, they can be released from intact cells by using reducing agents (Lopez-Ribot *et al.*, 1996). They are generally abundant proteins and include mainly glycolytic enzymes and heat shock proteins (Motshwene *et al.*, 2002). They lack an N-terminal signal peptide or an internal transmembrane domain to target them to the ER membrane and to initiate translocation. There is also no evidence that they become glycosylated, which makes it even more unlikely that they enter the ER. This raises the question how these proteins reach the cell surface. One obvious possibility is that limited cell lysis has resulted in the release of these proteins (Eroles *et al.*, 1997; Klis *et al.*, 2002). As the cell wall acts as a cation exchanger due to its multiple negative charges originating from phosphodiester bridges in protein-bound carbohydrate side chains, released proteins may adsorb to the wall. This would indicate the use of *mnn4* or *mnn6* strains, which are unable to form such phosphodiester bridges (Jigami and Odani, 1999). Another frequently presented explanation is that yeast cells may use a so-called 'nonclassical' protein export mechanism. An alternative explanation might be that these abundant proteins enter the secretory pathway after the endoplasmatic reticulum, thus avoiding glycosylation. Conceivably, the process of forming transport vesicles by the ER or the Golgi or the fusion of these vesicles with their target membrane is not fully efficient and allows abundant cytosolic proteins to join the secretory proteins.

5.10 Molecular organization of the cell wall

Of all the fungi the cell wall of *S. cerevisiae* has been studied in most detail. Our current knowledge about the molecular organization of this wall is summarized in Figure 5.1. Figure 5.1(a) shows a slightly simplified picture of how the various components of the cell wall are interconnected. The external protein layer consists largely of GPI-CWPs, which represent the most abundant class of CWP (Caro *et al.*, 1997; Hamada *et al.*, 1999). They are covalently linked through a remnant of their GPI anchor to a highly branched 1,6-β-glucan, which in turn is covalently linked to the 1,3-β-glucan meshwork. A minor class of CWPs, Pir-CWPs, is directly linked through an alkali-sensitive linkage to the 1,3-β-glucan meshwork and is found dispersed over the inner skeletal layer (Kapteyn *et al.*, 2001). The 1,3-β-glucan meshwork surrounds the entire cell and is kept together by hydrogen bonding. 1,3-β-Glucan molecules are occasionally branched through 1,6-β-linkages, resulting in multiple nonreducing ends per molecule, which may function as covalent attachment site for either 1,6-β-glucan or chitin molecules. As discussed previously, the 1,3-β-glucan meshwork is highly flexible with relatively wide pores that allow the passage of large molecules. The external protein layer, on the other hand, forms a permeability barrier and restricts penetration of large molecules into the cell wall (Zlotnik *et al.*, 1984; De Nobel *et al.*, 1990b). The available evidence strongly indicates that the wall of *C. albicans* is similarly organized (Kapteyn *et al.*, 2001; Klis *et al.*, 2001; Richard *et al.*, 2002). Figure 5.1(b) shows the various CWP–polysaccharide complexes identified in the cell wall. Panel 1 of Figure 5.1(b) shows the most abundant CWP–polysaccharide complex in vegetatively growing cells, consisting of a GPI-CWP linked through its GPI-glycan to a nonreducing end of a 1,6-β-glucan molecule, which is linked through its reducing end to one of the nonreducing ends of a 1,3-β-glucan molecule (see also Table 5.2). Another common CWP–polysaccharide complex is depicted in panel 2 and consists of a Pir-CWP linked through an alkali-sensitive linkage of unknown nature to a 1,3-β-glucan molecule. Interestingly, some GPI-CWPs are also linked to a 1,3-β-glucan molecule in a Pir-CWP-like fashion, either without a GPI-dependent linkage or in combination with it (Figure 5.1(b), panels 3 and 4). Finally, chitin may also be linked to 1,6-β-glucan, resulting in the complex presented in panel 5. This complex is normally rare (Table 5.2), but becomes much more abundant in cells showing hyperaccumulation of chitin in the lateral walls in response to cell wall stress (Kapteyn *et al.*, 1997).

The molecular organization of the cell wall of *C. albicans* in the hyphal form is probably similar to that of the cell wall of the yeast form (Klis *et al.*, 2001). Importantly, in contrast to

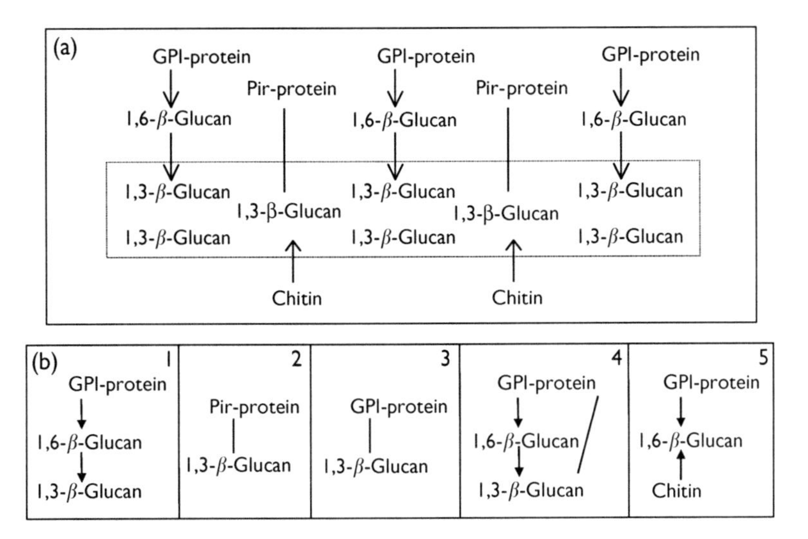

Figure 5.1 Molecular organization of the cell wall of *S. cerevisiae* and *C. albicans*. (a) Molecular model of the wall. GPI, glycosylphosphatidylinositol. CWP, cell wall protein. GPI-CWP, GPI-dependent cell wall protein. Pir proteins form a specific class of cell wall proteins with a conserved spacing pattern of four cysteine residues (see text). For clarity the model is slightly simplified (see Figure 5.1(b)). Note that in mature GPI-CWPs the GPI-anchor is processed such that the first mannose residue in the anchor is directly linked to 1,6-β-glucan instead of the glucosamine residue in the original GPI anchor (–AA_ω-EtN-$Man_{4/5}$→1,6-β-glucan, in which AA_ω represents the C-terminal amino acid of the mature protein and EtN represents ethanolamine). The arrows in the scheme indicate how the macromolecules are oriented with respect to their reducing end: the reducing end of polymer A→ a nonreducing end of polymer B. (b) CWP–polysaccharide complexes found in the cell wall. 1,6-β-Glc represents a 1,6-β-glucan molecule. 1,3-β-Glucan molecules (1,3-β-Glc) form a three-dimensional network kept together by hydrogen bonding. Panel 1 shows the most abundant CWP–polysaccharide complex. Pir-CWPs are directly linked to 1,3-β-glucan through an alkali-sensitive linkage of unknown nature (Panel 2). This complex seems to be favored during the G1 phase when the cells are growing isotropically. A limited number of GPI-CWPs are linked to 1,3-β-glucan in a similar way as Pir-CWPs (Panel 3). Such GPI-CWPs may also simultaneously be linked to 1,6-β-glucan (Panel 4). Panel 5 shows a complex that is rare under normal conditions, but becomes abundant when the cell wall has been damaged. This diagram is adapted from Klis *et al.* (2002).

S. cerevisiae, in which chitin is deposited in the lateral walls only after cytokinesis, chitin incorporation in hyphae of *C. albicans* takes place predominantly in the growing tips (Braun and Calderone, 1978). It is not known if chitin in growing hyphae of *C. albicans* becomes preferentially linked to 1,3-β-glucan (Figure 5.1(a)) or to 1,6-β-glucan (Figure 5.1(b), panel 5), nor is it known whether chitin chains are immediately linked to other cell wall polymers (co-synthetically) or later (post-synthetically).

As mentioned earlier, the available evidence strongly indicates that the walls of *S. cerevisiae* and *C. albicans* have a similar molecular organization. This suggests that the cell wall structure of *S. cerevisiae* may be found in many other fungi as well. In agreement with this, a cell wall adhesion protein has been described in *C. glabrata* that depends on its GPI anchor for cell wall incorporation and is linked through a 1,6-β-glucan moiety to 1,3-β-glucan

Table 5.2 Distribution of CWP–polysaccharide complexes in exponentially growing cells of *S. cerevisiae*. Cells were radioactively labeled with a protein hydrolysate and the isolated walls were extracted with enzymes of known specificity or using mild alkali extraction (Kapteyn *et al.*, 1997, 2001). 1,3-β-Glc: 1,3-β-glucan; 1,6-β-Glc: 1,3-β-glucan. The arrows in the complexes indicate how the macromolecules are oriented with respect to their reducing end: the reducing end of polymer A→ a nonreducing end of polymer B

Extraction	*Radioactivity released (%)*	*Corresponding CWP–polysaccharide complex*
1,6-β-glucanase	80	GPI-CWP→1,6-β-Glc
30 mM NaOH	15	Pir-CWP[a]–1,3-β-Glc
1,3-β-glucanase	98	GPI-CWP→1,6-β-Glc Pir-CWP[a]–1,3-β-Glc
1,3-β-glucanase + chitinase	100	GPI-CWP→1,6-β-Glc Pir-CWP[a]–1,3-β-Glc GPI-CWP→1,6-β-Glc ←chitin

Note
a The nature of the linkage between Pir-CWPs and 1,3-β-glucan is unknown.

(Cormack *et al.*, 1999; Frieman *et al.*, 2002). In addition, when Cwp2, a GPI-CWP from *S. cerevisiae*, is expressed in *C. glabrata*, it is incorporated in the cell wall and becomes covalently linked to a 1,6-β-glucan–1,3-β-glucan complex (Frieman *et al.*, 2002). These results are consistent with the notion that the wall of *C glabrata* is similarly organized as the wall of *S. cerevisiae*. 1,6-β-Glucosylated CWPs have also been identified in the pathogenic black yeast *Exophiala* (*Wangiella*) *dermatitidis* (Montijn *et al.*, 1997).

The model presented here describes the molecular organization of the walls of vegetatively growing cells. However, these walls are not entirely homogeneous in the lateral dimension. For example, the bud scar region of the mother cell is enriched in chitin (Molano *et al.*, 1980), and the birth scar region stains less well with the chitin-staining dye Calcofluor white than the lateral wall of the mother cell in general (G. Smits, personal communication). In the walls of the growing bud no chitin is incorporated under normal conditions, but chitin deposition is delayed until after cytokinesis. Conceivably, chitin in the bud scar is partially deposited as a free polymer that may form a crystalline structure, whereas at the outside it could be linked to 1,3-β-glucan molecules to integrate the scar in the 1,3-β-glucan. Also, the protein composition of the wall varies depending on the phase of the cell cycle (Shaw *et al.*, 1991; Caro *et al.*, 1998; Moukadiri *et al.*, 1999; Rodriguez-Pena *et al.*, 2000). In cells treated with mating pheromones, increased chitin deposition takes place at the base of the emerging mating projection, possibly, in the form of chitin→1,6-β-glucan←CWP complexes as depicted in Figure 5.1(b), panel 5 (Schekman and Brawley, 1979). These data indicate that the molecular organization of the cell wall may differ locally and may also depend on the phase of the cell cycle.

The model presented also allows us to formulate new questions. What is the difference between apical growth, as seen in small buds, and isotropic growth, as observed later in the cell cycle (see Chapter 1)? Is the cell wall locally weakened to allow insertion of new polymers or are newly made cell wall polymers deposited at the inside of the wall and do they move gradually to the outside of the cell wall slipping along each other to allow for cell wall extension? The latter does not seem consistent with the preferential location of most CWPs in an external protein layer. A similar question may be asked with respect to the widening of the cell neck that occurs in cells that form a mating projection in response to mating pheromone (see Chapter 1, Section 1.4). Another question is how are the rates of synthesis of the various cell wall components coordinated? Finally, it is of interest to know which enzymes are responsible for interconnecting the various types of cell wall macromolecules and in which order the assembly steps take place.

5.11 Cell wall assembly

The synthesis of chitin and 1,3-β-glucan takes place at the plasma membrane and is a vectorial process. The growing chains are extended at the cytosolic side of the plasma membrane and are believed to be guided through a protein pore to the noncytosolic side of the plasma membrane. Montijn *et al.* (1999) have presented indirect evidence that most or all 1,6-β-glucan in *S. cerevisiae* is synthesized at the plasma membrane as well. Finally, GPI-CWPs follow the secretory pathway and no evidence has ever been obtained for the intracellular attachment of 1,6-β-glucan to CWPs. Pir-CWPs also have the hallmarks of secretory proteins. Taken together, these data strongly indicate that the assembly of the cell wall takes place entirely at the cell surface. The molecular model of the cell wall presented in Figure 5.1 predicts the existence of at least five assembly steps and as many assembly enzymes (Table 5.3). None of them has yet been identified. In addition, as 1,3-β-glucan chains are synthesized as linear chains, but the mature chains are branched, β-glucan remodeling enzymes should be active at the cell surface. Several cell surface proteins are known to be able to remodel 1,3-β-glucan such as Bgl2, Gas1, Crh1, Crh2 and Crr1 (Rodriguez-Pena *et al.*, 2000; Popolo *et al.*, 2001). However, it remains to be determined for each of these enzymes if they are involved in the maturation of 1,3-β-glucan or in later remodeling steps to allow insertion of new cell wall material, for example, during isotropic growth.

Several dyes are known to bind preferentially to separate cell wall components. Congo red and Aniline blue seem to bind preferentially to 1,3-β-glucan molecules in the noncrystalline form, whereas Calcofluor white selectively binds to chitin (Pringle, 1991; Kopecka and Gabriel, 1992). Conceivably, these dyes may in addition to inhibiting crystallization of 1,3-β-glucan and chitin also inhibit specific assembly enzymes, because they interfere with binding of the assembly enzyme to their substrate. This might for example explain why Calcofluor white affects cell wall construction (Frost *et al.*, 1995; De Nobel *et al.*, 2001a). Cell

Table 5.3 Postulated cell wall assembly enzymes in *S. cerevisiae*. CWP: cell wall protein; GPI-CWP: GPI-anchor-dependent protein. The arrows in the complexes point from the reducing end of the polysaccharide to a nonreducing end of the other polysaccharide

Assembly step	*Proposed name*
GPI-CWP → (1,6)-β-glucan	CWP-(1,6)-glucan assembly enzyme
CWP[a]–(1,3)-β-glucan	CWP- (1,3)-glucan assembly enzyme
(1,6)-β-glucan → (1,3)-β-glucan	(1,6)-(1,3)-glucan assembly enzyme
Chitin → (1,3)-β-glucan	Chitin → (1,3)-glucan assembly enzyme
Chitin → (1,6)-β-glucan	Chitin → (1,6)-glucan assembly enzyme

Note

a Formation of an alkali-sensitive linkage between Pir-CWPs (and a limited number of GPI-CWPs) and (1,3)-β-glucan. The nature of the linkage is not yet known.

wall weakening is counteracted by hyperaccumulation of chitin in the lateral walls and strongly increased formation of chitin→1,6-*β*-glucan←CWP complexes as depicted in Figure 5.1(b), panel 5 (Kapteyn *et al.*, 1997) and possibly also the chitin→1,3-*β*-glucan complex (Figure 5.1(a)). Calcofluor might inhibit the formation of both complexes, explaining why so many cell wall mutants are hypersensitive to it (Ram *et al.*, 1994; Lussier *et al.*, 1997; De Groot *et al.*, 2001). One would predict that addition of Congo Red or Aniline Blue, and also of Calcofluor white in the case of chitin-overproducing cells, should result in the increased release of CWPs into the medium. In addition, one would expect that oligomers of 1,3-*β*-glucan, 1,6-*β*-glucan or chitin would also interfere with cell wall assembly. Although such experiments are complicated by the sensitivity of the sugar oligomers to enzymatic degradation by secretory proteins of *S. cerevisiae*, Bom *et al.* (2001) obtained evidence for such an inhibitory effect by 1,6-*β*-glucan oligosaccharides. They showed that cells grown in the presence of 0.2 M gentiobiose (Glc-1,6-*β*-Glc) release more of a GPI-dependent cell wall reporter protein into the medium than control cells, indicating that the incorporation of GPI proteins in the cell wall is affected under those conditions. Such cells also show a shorter lag phase with respect to cell lysis when treated with Zymolyase, an enzyme mixture consisting of a protease and an endo-1,3-*β*-glucanase. As the external protein layer of the cell wall determines the accessibility of the internal skeletal layer to the endo-1,3-*β*-glucanase in Zymolyase (De Nobel *et al.*, 1990a; Ovalle *et al.*, 1998), this is consistent with a thinner external mannoprotein layer. Although the concentration of gentiobiose used was rather high, it seems nevertheless likely that it had a direct effect on cell wall assembly, because in the presence of a membrane-active antimicrobial peptide a much lower gentiobiose concentration (25 mM) was sufficient to obtain strong growth inhibition. In the presence of gentiotriose, which probably has more affinity for the assembly enzyme and was thus a more effective inhibitor, only 2 mM was sufficient.

The precise order of events in cell wall assembly is still unknown, but a preliminary picture is beginning to emerge. One of the questions that have been studied in detail is when 1,6-*β*-glucan becomes attached to Sag1, the sexual adhesion protein of mating type *α* cells in *S. cerevisiae* (Lu *et al.*, 1994, 1995). The following observations are relevant. First, the intracellular precursor forms of Sag1 are not attached to 1,6-*β*-glucan. Second, three cell surface forms of Sag1 have been identified: a GPI-anchored form bound to the plasma membrane, a soluble periplasmic form with a truncated GPI anchor, and the mature form that is covalently bound to the cell wall. Third, only the wall-bound form reacts positively with 1,6-*β*-glucan antibodies. These observations are consistent with the notion that soluble, periplasmic forms of GPI-CWPs are linked to a preexisting 1,6-*β*-glucan→1,3-*β*-glucan network. In addition, Roh and co-workers have shown that an inhibitor of GPI-anchor synthesis prevents the incorporation of CWPs, but does not significantly affect the levels of 1,6-*β*-glucan→1,3-*β*-glucan (Roh *et al.*, 2002). Taken together, these results suggest that in a first assembly step 1,6-*β*-glucan is linked to the 1,3-*β*-glucan meshwork and that subsequently soluble, periplasmic forms of GPI-CWPs are connected to 1,6-*β*-glucan. In view of this, it is difficult to explain that the medium of the deletion mutant *ecm33Δ* contains considerable amounts of 1,6-*β*-glucosylated proteins (De Groot *et al.*, 2001), unless one assumes that Ecm33 is not involved in cell wall assembly itself, but acts as a transglucosylase that has a remodeling function that accompanies cell wall extension. Deposition of chitin in the lateral walls normally occurs after cytokinesis (Shaw *et al.*, 1991), indicating that linkage of chitin to other polymers in the cell wall network of the lateral walls represents the last step in cell wall assembly. This may not necessarily be the case during cell wall stress when the deposition of chitin in the lateral wall is not delayed until after cytokinesis and is strongly increased in conjunction with the formation of the chitin→1,6-β-glucan complex.

The existence of a soluble periplasmic form of Sag1 with an incomplete GPI anchor implies the action of GPI-anchor processing enzymes at the cell surface of *S. cerevisiae*. Kitagaki *et al.* (2002) have identified two related, presumably plasma membrane-bound GPI proteins Dcw1 and Dfg5 with an essential function in cell wall construction and belonging to the glycoside hydrolase family 76 (Coutinho and Henrissat, 1999). This family of proteins includes an α-1,6-mannanase from *Bacillus circulans* and also several proteins from the fission yeast *S. pombe* and the filamentous fungus *N. crassa*. It is tempting to predict that Dcw1 and Dfg5 are responsible for releasing plasma membrane-bound precursor forms of GPI-CWP like Sag1 from the plasma membrane.

5.12 Perspectives

Fungi may cause enormous economic damage in terms of harvest losses and food spoilage and often form a serious health threat. As the cell wall is essential for fungal viability and as its construction requires multiple steps that seem to be fungal-specific, the biogenesis of fungal cell wall components and the assembly steps at the cell surface seem promising targets for the design of antifungal compounds. The model fungus *S. cerevisiae* has offered, and will continue to offer, crucial insights that are valid for other fungi too.

Now that it is possible to describe the organization of the cell wall in the ascomycotinous yeasts *S. cerevisiae, C. albicans*, and other related yeasts in molecular terms, one of the next tasks will be to identify the enzymes that are responsible for the assembly of the cell wall network. The development of *in vitro* assays to measure and analyze the activities of these enzymes is urgently needed. *In vitro* assays for measuring 1,6-β-glucan synthase and 1,3-α-glucan synthase activities are also urgently needed. Another task will be to describe cell wall extension in molecular terms, both during isotropic growth and during apical growth. Additional questions include how the cell locally dissolves its wall and how this process is controlled, for example, during bud formation or when two mating partners fuse to form a zygote. Furthermore, the specifics of cell wall organization that correspond with particular morphogenetic processes, such as pseudohyphal growth or shmooing, are still unknown. Many related questions are still unresolved. Our knowledge of the molecular organization of the cell wall of fission yeast and of the mycelial species of the Ascomycotina is still far from complete. The cell wall ultrastructure of the basidiomycotinous yeasts and mycelial fungi is even less understood. Now that the genomes of several fungi are, or will soon be, known and better and more sensitive glycobiological techniques become available, the time is ripe for a concerted genomic and glycobiological approach to unravel the molecular architecture of the fungal cell wall, its dynamics, and the regulation of its formation. In addition, mass spectrometric analysis methods are expected to become the method of choice to determine the cell wall proteome and to monitor its dynamics both qualitatively and quantitatively in response to various environmental conditions. May *S. cerevisiae* continue to illuminate the way in these endeavors!

Acknowledgements

Our work was financially supported by EC contract QLK2-CT2000–00795 and by NWO (The Netherlands) and STW (The Netherlands).

References

Alloush, H. M., Lopez-Ribot, J. L., Masten, B. J., and Chaffin, W. L. (1997) 3-phosphoglycerate kinase: a glycolytic enzyme protein present in the cell wall of *Candida albicans*. *Microbiology* **143**, 321–330.

Azuma, M., Levinson, J. N., Page, N., and Bussey, H. (2002) *Saccharomyces cerevisiae* Big1p, a putative endoplasmic reticulum membrane protein required for normal levels of cell wall β-1,6-glucan. *Yeast* **19**, 783–793.

Baba, M. and Osumi, M. (1987) Transmission and scanning electron microscopic examination of intracellular organelles in freeze-substituted *Kloeckera* and *Saccharomyces cerevisiae* yeast cells. *Journal of Electron Microscopy Techniques* **5**, 249–261.

Ballou, C. E. (1990) Isolation, characterization, and properties of *Saccharomyces cerevisiae mnn* mutants with nonconditional protein glycosylation defects. *Methods of Enzymology* **185**, 440–470.

Beauvais, A., Bruneau, J. M., Mol, P. C., Buitrago, M. J., Legrand, R., and Latge, J. P. (2001) Glucan synthase complex of *Aspergillus fumigatus*. *Journal of Bacteriology* **183**, 2273–2279.

Boder, E. T., and Wittrup, K. D. (1997) Yeast surface display for screening combinatorial polypeptide libraries. *Nature and Biotechnology* **15**, 553–557.

Bom, I. J., Klis, F. M., de Nobel, H., and Brul, S. (2001) A new strategy for inhibition of the spoilage yeasts *Saccharomyces cerevisiae* and *Zygosaccharomyces bailii* based on combination of a membrane-active peptide with an oligosaccharide that leads to an impaired glycosylphosphatidylinositol (GPI)-dependent yeast wall protein layer. *FEMS Yeast Research* **1**, 187–194.

Braun, P. C. and Calderone, R. A. (1978) Chitin synthesis in *Candida albicans*: comparison of yeast and hyphal forms. *Journal of Bacteriology* **133**, 1472–1477.

Bush, D. A., Horisberger, M., Horman, I., and Wursch, P. (1974) The wall structure of *Schizosaccharomyces pombe*. *Journal of General Microbiology* **81**, 199–206.

Cabib, E., Roh, D. H., Schmidt, M., Crotti, L. B., and Varma, A. (2001) The yeast cell wall and septum as paradigms of cell growth and morphogenesis. *Journal of Biological Chemistry* **276**, 19679–19682.

Cappellaro, C., Baldermann, C., Rachel, R., and Tanner, W. (1994) Mating type-specific cell–cell recognition of *Saccharomyces cerevisiae*: cell wall attachment and active sites of **a**- and α-agglutinin. *EMBO Journal* **13**, 4737–4744.

Cappellaro, C., Mrsa, V., and Tanner, W. (1998) New potential cell wall glucanases of *Saccharomyces cerevisiae* and their involvement in mating. *Journal of Bacteriology* **180**, 5030–5037.

Casanova, M., Lopez-Ribot, J. L., Martinez, J. P., and Sentandreu, R. (1992) Characterization of cell wall proteins from yeast and mycelial cells of *Candida albicans* by labelling with biotin: comparison with other techniques. *Infection and Immunity* **60**, 4898–4906.

Caro, L. H., Tettelin, H., Vossen, J. H., Ram, A. F., van den Ende, H., and Klis, F. M. (1997) *In silicio* identification of glycosyl-phosphatidylinositol-anchored plasma-membrane and cell wall proteins of *Saccharomyces cerevisiae*. *Yeast* **13**, 1477–1489.

Caro, L. H., Smits, G. J., van Egmond, P., Chapman, J. W., and Klis, F. M. (1998) Transcription of multiple cell wall protein-encoding genes in *Saccharomyces cerevisiae* is differentially regulated during the cell cycle. *FEMS Microbiology Letters* **161**, 345–349.

Carotti, C., Ferrario, L., Roncero, C., Valdivieso, M. H., Duran, A., and Popolo, L. (2002) Maintenance of cell integrity in the *gas1* mutant of *Saccharomyces cerevisiae* requires the Chs3p-targeting and activation pathway and involves an unusual Chs3p localization. *Yeast* **19**, 1113–1124.

Chaffin, W. L., Lopez-Ribot, J. L., Casanova, M., Gozalbo, D., and Martinez, J. P. (1998) Cell wall and secreted proteins of *Candida albicans*: identification, function, and expression. *Microbiological and Molecular Biological Reviews* **62**, 130–180.

Chen, M. H., Shen, Z. M., Bobin, S., Kahn, P. C., and Lipke, P. N. (1995) Structure of *Saccharomyces cerevisiae* α-agglutinin. Evidence for a yeast cell wall protein with multiple immunoglobulin-like domains with atypical disulfides. *Journal of Biological Chemistry* **270**, 26168–26177.

Cormack, B. P., Ghori, N., and Falkow, S. (1999) An adhesin of the yeast pathogen *Candida glabrata* mediating adherence to human epithelial cells. *Science* **285**, 578–582.
Coutinho, P. M. and Henrissat, B. (1999) Carbohydrate-active enzymes: an integrated database approach. In: H.J. Gilbert, G. Davies, B. Henrissat, and B. Svensson (eds) *Recent Advances in Carbohydrate Bioengineering*, vols 3–12, Cambridge: The Royal Society of Chemistry.
Cutler, J. E. (2001) *N*-glycosylation of yeast, with emphasis on *Candida albicans*. *Medical Mycology* **39**(1), 75–86.
Dallies, N., Francois, J., and Paquet, V. (1998) A new method for quantitative determination of polysaccharides in the yeast cell wall. Application to the cell wall defective mutants of *Saccharomyces cerevisiae*. *Yeast* **14**, 1297–1306.
De Groot, P. W. J., Ruiz, C., Vázquez de Aldana, C. R., Dueas, E., Cid, V. J., Del Rey, F., Rodríquez-Peña, J. M., Pérez, P., Andel, A., Caubín, J., Arroyo, J., García, J. C., Gil, C., Molina, M., García, L. J., Nombela, C., and Klis, F. M. (2001) A genomic approach for the identification and classification of genes involved in cell wall formation and its regulation in *Saccharomyces cerevisiae*. *Comparative and Functional Genomics* **2**, 124–142.
De Groot, P. W. J., Hellingwerf, K. J., and Klis, F. M. (2003) Genome-wide identification of fungal GPI-proteins. *Yeast* **20**, 781–796.
DeMarini, D. J., Adams, A. E., Fares, H., De Virgilio, C., Valle, G., Chuang, J. S., and Pringle, J. R. (1997) A septin-based hierarchy of proteins required for localized deposition of chitin in the *Saccharomyces cerevisiae* cell wall. *Journal of Cellular Biology* **139**, 75–93.
De Nobel, J. G., Klis, F. M., Munnik, T., Priem, J., and van den Ende, H. (1990a) An assay of relative cell wall porosity in *Saccharomyces cerevisiae*, *Kluyveromyces lactis* and *Schizosaccharomyces pombe*. *Yeast* **6**, 483–490.
De Nobel, J. G., Klis, F. M., Priem, J., Munnik, T., and van den Ende, H. (1990b) The glucanase-soluble mannoproteins limit cell wall porosity *in Saccharomyces cerevisiae*. *Yeast* **6**, 491–499.
De Nobel, H., Lipke, P. N., and Kurjan, J. (1996) Identification of a ligand-binding site in an immunoglobulin fold domain of the *Saccharomyces cerevisiae* adhesion protein α-agglutinin. *Molecular Biology of the Cell* **7**, 143–153.
De Nobel, H., Ruiz, C., Martin, H., Morris, W., Brul, S., Molina, M., and Klis, F. M. (2001a) Cell wall perturbation in yeast results in dual phosphorylation of the Slt2/Mpk1 MAP kinase and in an Slt2-mediated increase in *FKS2*-lacZ expression, glucanase resistance and thermotolerance. *Microbiology* **146**, 2121–2132.
De Nobel, H., Sietsma, J. H., van den Ende, H., and Klis, F. M. (2001b) Molecular organization and construction of the fungal cell wall. In: R.J. Howard, and N.A.R. Gow (eds) *The Mycota VIII. Biology of the Fungal Cell*, pp. 181–200. Berlin: Springer Verlag.
Dijkgraaf, G. J., Li, H., and Bussey, H. (2002) Cell wall β-glucans of *Saccharomyces cerevisiae*. In: A. Steinbuchel (ed). *Biopolymers*, vol. 6, pp. 121–139, Chichester: Wiley-VCH.
Douglas, C. M. (2001) Fungal β(1,3)-glucan synthesis. *Medical Mycology* **39**(1), 55–66.
Eroles, P., Sentandreu, M., Elorza, M. V., and Sentandreu, R. (1997) The highly immunogenic enolase and Hsp70p are adventitious *Candida albicans* cell wall proteins. *Microbiology* **143**, 313–320.
Fleet, G. H. (1991) Cell walls. In: Rose, A.H. and J.S. Harrison (eds) *The Yeasts*, 2nd edn, vol. 4, *Yeast Organelles*, pp. 199–277. London: Academic Press.
Fleet, G. H. and Manners, D. J. (1976) Isolation and composition of an alkali-soluble glucan from the cell walls of *Saccharomyces cerevisiae*. *Journal of General Microbiology* **94**, 180–192.
Flury, I., Benachour, A., and Conzelmann, A. (2000) YLL031c belongs to a novel family of membrane proteins involved in the transfer of ethanolaminephosphate onto the core structure of glycosylphosphatidylinositol anchors in yeast. *Journal of Biological Chemistry* **275**, 24458–24465.
Frieman, M. B., McCaffery, J. M., and Cormack, B. P. (2002) Modular domain structure in the *Candida glabrata* adhesin Epa1p, a β1,6 glucan-cross-linked cell wall protein. *Molecular Microbiology* **46**, 479–492.
Frost, D. J., Brandt, K. D., Cugier, D., and Goldman, R. (1995) A whole-cell *Candida albicans* assay for the detection of inhibitors towards fungal cell wall synthesis and assembly. *Journal of Antibiotics* **48**, 306–310.

Fujii, T., Shimoi, H., and Iimura, Y. (1999) Structure of the glucan-binding sugar chain of Tip1p, a cell wall protein of *Saccharomyces cerevisiae*. *Biochimica and Biophysica Acta* **1427**, 133–144.

Furukawa, K., Tagaya, M., Tanizawa, K., and Fukui, T. (1993) Role of the conserved Lys-X-Gly-Gly sequence at the ADP-glucose-binding site in *Escherichia coli* glycogen synthase *Journal of Biological Chemistry* **268**, 23837–23842.

Gemmill, T. R. and Trimble, R. B. (1999) Overview of *N*- and *O*-linked oligosaccharide structures found in various yeast species. *Biochimica and Biophysica Acta* **1426**, 227–237.

Goins, T. L. and Cutler, J. E. (2000) Relative abundance of oligosaccharides in *Candida* species as determined by fluorophore-assisted carbohydrate electrophoresis. *Journal of Clinical Microbiology* **38**, 2862–2869.

Grimme, S. J., Westfall, B. A., Wiedman, J. M., Taron, C. H., and Orlean, P. (2001) The essential Smp3 protein is required for addition of the side-branching fourth mannose during assembly of yeast glycosylphosphatidylinositols. *Journal of Biological Chemistry* **276**, 27731–27739.

Hamada, K., Fukuchi, S., Arisawa, M., Baba, M., and Kitada, K. (1998a) Screening for glycosylphosphatidylinositol (GPI)-dependent cell wall proteins in *Saccharomyces cerevisiae*. *Molecular and General Genetics* **258**, 53–59.

Hamada, K., Terashima, H., Arisawa, M., and Kitada, K. (1998b) Amino acid sequence requirement for efficient incorporation of glycosylphosphatidylinositol-associated proteins into the cell wall of *Saccharomyces cerevisiae*. *Journal of Biological Chemistry* **273**, 26946–26953.

Hamada, K., Terashima, H., Arisawa, M., Yabuki, N., and Kitada, K. (1999) Amino acid residues in the omega-minus region participate in cellular localization of yeast glycosylphosphatidylinositol-attached proteins. *Journal of Bacteriology* **181**, 3886–3889.

Hartland, R. P., Vermeulen, C. A., Klis, F. M., Sietsma, J. H., and Wessels, J. G. (1994) The linkage of (1–3)-β-glucan to chitin during cell wall assembly in *Saccharomyces cerevisiae*. *Yeast* **10**, 1591–1599.

Hochstenbach, F., Klis, F. M., van den Ende, H., van Donselaar, E., Peters, P. J., and Klausner, R. D. (1998) Identification of a putative α-glucan synthase essential for cell wall construction and morphogenesis in fission yeast. *Proceedings of the National Acadamy of Sciences USA* **95**, 9161–9166.

Hoyer, L. L. (2001) The ALS gene family of *Candida albicans*. *Trends in Microbiology* **9**, 176–180.

Hoyer, L. L. and Hecht, J. E. (2001) The *ALS5* gene of *Candida albicans* and analysis of the Als5p N-terminal domain. *Yeast* **18**, 49–60.

Hoyer, L. L., Fundyga, R., Hecht, J. E., Kapteyn, J. C., Klis, F. M., and Arnold, J. (2001) Characterization of agglutinin-like sequence genes from non-albicans *Candida* and phylogenetic analysis of the *ALS* family. *Genetics* **157**, 1555–1567.

Humbel, B. M., Konomi, M., Takagi, T., Kamasawa, N., Ishijima, S. A., and Osumi, M. (2001) *In situ* localization of β-glucans in the cell wall of *Schizosaccharomyces pombe*. *Yeast* **18**, 433–444.

Iacomini, M., Gorin, P. A. J., Baron, M., Tulloch, A. P., and Mazurek, M. (1988) Novel D-glucans obtained by dimethyl sulfoxide extraction of the lichens *Letharia vulpina*, *Actinogyra muehlenbergii*, and an *Usnea* sp. *Carbohydrate Research* **176**, 117–126.

Ibeas, J. I., Lee, H., Damsz, B., Prasad, D. T., Pardo, J. M., Hasegawa, P. M., Bressan, R. A., and Narasimhan, M. L. (2000) Fungal cell wall phosphomannans facilitate the toxic activity of a plant PR-5 protein. *Plant Journal* **23**, 375–383.

Inoue, S. B., Takewaki, N., Takasuka, T., Mio, T., Adachi, M., Fujii, Y., Miyamoto, C., Arisawa, M., Furuichi, Y., and Watanabe, T. (1995) Characterization and gene cloning of 1,3-β-D-glucan synthase from *Saccharomyces cerevisiae*. *European Journal of Biochemistry* **231**, 845–854.

Jentoft, N. (1990) Why are proteins *O*-glycosylated? *Trends in Biochemical Sciences* **15**, 291–294.

Jiang, B., Ram, A. F., Sheraton, J., Klis, F. M., and Bussey, H. (1995) Regulation of cell wall β-glucan assembly: *PTC1* negatively affects *PBS2* action in a pathway that includes modulation of *EXG1* transcription. *Molecular and General Genetics* **248**, 260–269.

Jigami, Y. and Odani, T. (1999) Mannosylphosphate transfer to yeast mannan. *Biochimica and Biophysica Acta* **1426**, 335–345.

Jue, C. K. and Lipke, P. N. (2002) Role of Fig2p in agglutination in *Saccharomyces cerevisiae*. *Eukaryotic Cell* **1**, 843–845.

Kang, M. S., Elango, N., Mattia, E., Au-Young, J., Robbins, P. W., and Cabib, E. (1984) Isolation of chitin synthetase from *Saccharomyces cerevisiae*. Purification of an enzyme by entrapment in the product. *Journal of Biological Chemistry* **259**, 14966–14972.

Kapteyn, J. C., Montijn, R. C., Vink, E., de la Cruz, J., Llobell, A., Douwes, J. E., Shimoi. H., Lipke, P. N., and Klis, F. M. (1996) Retention of *Saccharomyces cerevisiae* cell wall proteins through a phosphodiester-linked β-1,3-/β-1,6-glucan heteropolymer. *Glycobiology* **6**, 337–345.

Kapteyn, J. C., Ram, A. F., Groos, E. M., Kollar, R., Montijn, R. C., van den Ende, H., Llobell, A., Cabib, E., and Klis, F. M. (1997) Altered extent of cross-linking of β1,6-glucosylated mannoproteins to chitin in *Saccharomyces cerevisiae* mutants with reduced cell wall β1,3-glucan content. *Journal of Bacteriology* **179**, 6279–6284.

Kapteyn, J. C., van den Ende, H., and Klis, F. M. (1999a) The contribution of cell wall proteins to the organization of the yeast cell wall. *Biochimica and Biophysica Acta* **1426**, 373–383.

Kapteyn, J. C., van Egmond, P., Sievi, E., van den Ende, H., Makarow, M., and Klis, F. M. (1999b) The contribution of the *O*-glycosylated protein Pir2p/Hsp150 to the construction of the yeast cell wall in wild-type cells and β 1,6-glucan-deficient mutants. *Molecular Microbiology* **31**, 1835–1844.

Kapteyn, J. C., Hoyer, L. L., Hecht, J. E., Muller, W. H., Andel, A., Verkleij, A. J., Makarow, M., van den Ende, H., and Klis, F. M. (2000) The cell wall architecture of *Candida albicans* wild-type cells and cell wall-defective mutants. *Molecular Microbiology* **35**, 601–611.

Kapteyn, J. C., ter Riet, B., Vink, E., Blad, S., de Nobel, H., van den Ende, H., and Klis, F. M. (2001) Low external pH induces *HOG1*-dependent changes in the organization of the *Saccharomyces cerevisiae* cell wall. *Molecular Microbiology* **39**, 469–479.

Kathoda, S., Abe, N., Matsui, M., and Hayashibe, M. (1976) Polysaccharide composition of the cell wall of baker's yeast with special reference to cell walls obtained from large- and small-sized cells. *Plant and Cell Physiology* **17**, 909–919.

Kim, S. Y., Sohn, J. H., Pyun, Y. R., and Choi, E. S. (2002) A cell surface display system using novel GPI-anchored proteins in *Hansenula polymorpha*. *Yeast* **19**, 1153–1163.

Kitagaki, H., Wu, H., Shimoi, H., and Ito, K. (2002) Two homologous genes, *DCW1* (*YKL046c*) and *DFG5*, are essential for cell growth and encode glycosylphosphatidylinositol (GPI)-anchored membrane proteins required for cell wall biogenesis in *Saccharomyces cerevisiae*. *Molecular Microbiology* **46**, 1011–1022.

Klis, F. M. (1994) Review: cell wall assembly in yeast. *Yeast* **10**, 851–869.

Klis, F. M., Ram, A. F. J., Montijn, R. C., Kapteyn, J. C., Caro, L. H. P., Vossen, J. H., Van Berkel, M. A. A., Brekelmans, S. S. C., and Van Den Ende, H. (1998) Posttranslational modifications of secretory proteins in: A. J. P. Brown and M. F. Tuite (eds) *Methods in Microbiology*, vol. 26, pp. 223–238, San Diego: Academic Press.

Klis, F. M., de Groot, P., and Hellingwerf, K. (2001) Molecular organization of the cell wall of *Candida albicans*. *Medical Mycology* **39** (supplement 1), 1–8.

Klis, F. M., Mol, P., Hellingwerf, K., and Brul, S. (2002) Dynamics of cell wall structure in *Saccharomyces cerevisiae*. *FEMS Microbiology Reviews* **26**, 239–256.

Kollar, R., Petrakova, E., Ashwell, G., Robbins, P. W., and Cabib, E. (1995) Architecture of the yeast cell wall. The linkage between chitin and β (1→3)-glucan. *Journal of Biological Chemistry* **270**, 1170–1178.

Kollar, R., Reinhold, B. B., Petrakova, E., Yeh, H. J., Ashwell, G., Drgonova, J., Kapteyn, J. C., Klis, F. M., and Cabib, E. (1997) Architecture of the yeast cell wall. β (1→6)-glucan interconnects mannoprotein, β (1→)3-glucan, and chitin. *Journal of Biological Chemistry* **272**, 17762–17775.

Kopecka, M. and Gabriel, M. (1992) The influence of congo red on the cell wall and (1–3)-β-D-glucan microfibril biogenesis in *Saccharomyces cerevisiae*. *Archives of Microbiology* **158**, 115–126.

Kopecka, M., Phaff, H. J., and Fleet, G. H. (1974) Demonstration of a fibrillar component in the cell wall of the yeast *Saccharomyces cerevisiae* and its chemical nature. *Journal of Cellular Biology* **62**, 66–76.

Kratky, Z., Biely, P., and Bauer, S. (1975) Wall mannan of *Saccharomyces cerevisiae*. Metabolic stability and release into growth medium. *Biochimica and Biophysica Acta* **404**, 1–6.

Kreger, D. R. and Kopecka, M. (1975) On the nature and formation of the fibrillar nets produced by protoplasts of *Saccharomyces cerevisiae* in liquid media: an electron microscopic, X-ray diffraction and chemical study. *Journal of General Microbiology* **92**, 207–220.

Lagorce, A., Le Berre-Anton, V., Aguilar-Uscanga, B., Martin-Yken, H., Dagkessamanskaia, A., and Francois, J. (2002) Involvement of *GFA1*, which encodes glutamine–fructose-6-phosphate amidotransferase, in the activation of the chitin synthesis pathway in response to cell-wall defects in *Saccharomyces cerevisiae. European Journal of Biochemistry* **269**, 1697–1707.

Lipke, P. N. and Ovalle, R. (1998) Cell wall architecture in yeast: new structure and new challenges. *Journal of Bacteriology* **180**, 3735–3740.

Lopez-Ribot, J. L., Alloush, H. M., Masten, B. J., and Chaffin, W. L. (1996) Evidence for presence in the cell wall of *Candida albicans* of a protein related to the hsp70 family. *Infection and Immununity* **64**, 3333–3340.

Lu, C. F., Kurjan, J., and Lipke, P. N. (1994) A pathway for cell wall anchorage of *Saccharomyces cerevisiae* α-agglutinin. *Molecular and Cellular Biology* **14**, 4825–4833.

Lu, C. F., Montijn, R. C., Brown, J. L., Klis, F., Kurjan, J., Bussey, H., and Lipke, P. N. (1995) Glycosyl phosphatidylinositol-dependent cross-linking of α-agglutinin and β 1,6-glucan in the *Saccharomyces cerevisiae* cell wall. *Journal of Cellular Biology* **128**, 333–340.

Lucero, H. A., Kuranda, M. J., and Bulik, D. A. (2002) A nonradioactive, high throughput assay for chitin synthase activity. *Analytical Biochemistry* **305**, 97–105.

Lussier, M., White, A. M., Sheraton, J., di Paolo, T., Treadwell, J., Southard, S. B., Horenstein, C. I., Chen-Weiner, J., Ram, A. F., Kapteyn, J. C., Roemer, T. W., Vo, D. H., Bondoc, D. C., Hall, J., Zhong, W. W., Sdicu, A. M., Davies, J., Klis, F. M., Robbins, P. W., and Bussey, H. (1997) Large scale identification of genes involved in cell surface biosynthesis and architecture in *Saccharomyces cerevisiae. Genetics* **147**, 435–450.

Magnelli, P., Cipollo, J. F., and Abeijon, C. (2002) A refined method for the determination of *Saccharomyces cerevisiae* cell wall composition and β-1,6-glucan fine structure. *Analytical Biochemistry* **301**, 136–150.

Manners, D. J. and Meyer, M. T. (1977) The molecular structures of some glucans from the cell walls of *Schizosaccharomyces pombe. Carbohydrate Research* **57**, 189–203.

Manners, D. J., Masson, A. J., and Patterson, J. C. (1973a) The structure of a β -(1→3)-glucan from yeast cell walls. *Biochemical Journal* **135**, 19–30.

Manners, D. J., Masson, A. J., Patterson, J. C., Björndal, H., and Lindberg, B. (1973b) The structure of a β -(1 →6)-glucan from yeast cell walls. *Biochemical Journal* **135**, 31–36.

Martinez de Maranon, I., Marechal, P. A., and Gervais, P. (1996) Passive response of *Saccharomyces cerevisiae* to osmotic shifts: cell volume variations depending on the physiological state. *Biochemical Biophysical Research Communications* **227**, 519–523.

McMurrough, I. and Rose, A. H. (1967) Effect of growth rate and substrate limitation on the composition and structure of the cell wall of *Saccharomyces cerevisiae. Biochemical Journal* **105**, 189–203.

Mol, P. C. and Wessels, J. G. H. (1987) Linkages between glucosaminoglycan and glucan determine alkali-insolubility of the glucan in walls of *S. cerevisiae. FEMS Microbiology Letters* **41**, 95–99.

Molano, J., Bowers, B., and Cabib, E. (1980) Distribution of chitin in the yeast cell wall. An ultrastructural and chemical study. *Journal of Cell Biology* **85**, 199–212.

Montijn, R. C., van Wolven, P., De Hoog, S., and Klis, F. M. (1997) β-Glucosylated proteins in the cell wall of the black yeast *Exophiala* (*Wangiella*) *dermatitidis. Microbiology* **143**, 1673–1680.

Montijn, R. C., Vink, E., Muller, W. H., Verkleij, A. J., Van Den Ende, H., Henrissat, B., and Klis, F. M. (1999) Localization of synthesis of β 1,6-glucan in *Saccharomyces cerevisiae. Journal of Bacteriology* **181**, 7414–7420.

Morris, G. J., Winters, L., Coulson, G. E., and Clarke, K. J. (1986) Effect of osmotic stress on the ultrastructure and viability of the yeast *Saccharomyces cerevisiae. Journal of General Microbiology* **132**, 2023–2034.

Motshwene, P., Brandt, W., and Lindsey, G. (2002) Significant quantities of the glycolytic enzyme phosphoglycerate mutase are present in the cell wall of yeast *Saccharomyces cerevisiae*. *Biochemical Journal* **369**, 357–362.

Moukadiri, I., Jaafar, L., and Zueco, J. (1999) Identification of two mannoproteins released from cell walls of a *Saccharomyces cerevisiae mnn1 mnn9* double mutant by reducing agents. *Journal of Bacteriology* **181**, 4741–4745.

Mrsa, V., Seidl, T., Gentzsch, M., and Tanner, W. (1997) Specific labelling of cell wall proteins by biotinylation. Identification of four covalently linked O-mannosylated proteins of *Saccharomyces cerevisiae*. *Yeast* **13**, 1145–1154.

Murai, T., Ueda, M., Atomi, H., Shibasaki, Y., Kamasawa, N., Osumi, M., Kawaguchi, T., Arai, M., and Tanaka, A. (1997a) Genetic immobilization of cellulase on the cell surface of *Saccharomyces cerevisiae*. *Applied Microbiology and Biotechnology* **48**, 499–503.

Murai, T., Ueda, M., Yamamura, M., Atomi, H., Shibasaki, Y., Kamasawa, N., Osumi, M., Amachi, T., and Tanaka, A. (1997b) Construction of a starch-utilizing yeast by cell surface engineering. *Applied and Environmental Microbiology* **63**, 1362–1366.

Murai, T., Ueda, M., Kawaguchi, T., Arai, M., and Tanaka, A. (1998) Assimilation of cell oligosaccharides by a cell surface-engineered yeast expressing β-glucosidase and carboxymethylcellulase from *Aspergillus aculeatus*. *Applied and Environmental Microbiology* **64**, 4857–4861.

Murai, T., Ueda, M., Shibasaki, Y., Kamasawa, N., Osumi, M., Imanaka, T., and Tanaka, A. (1999) Development of an arming yeast strain for efficient utilization of starch by co-display of sequential amylolytic enzymes on the cell surface. *Applied Microbiology and Biotechnology* **51**, 65–70.

Nguyen, T. H., Fleet, G. H., and Rogers, P. L. (1998) Composition of the cell walls of several yeast species. *Applied Microbiology and Biotechnology* **50**, 206–212.

Ono, N., Yabe, T., Sudoh, M., Nakajima, T., Yamada-Okabe, T., Arisawa, M., and Yamada-Okabe, H. (2000) The yeast Chs4 protein stimulates the trypsin-sensitive activity of chitin synthase 3 through an apparent protein–protein interaction. *Microbiology* **146**, 385–391.

Orlean, P. (1997) Biogenesis of yeast wall and surface components. In: J. R. Pringle J. R. Broach and E. W. Jones (eds) *The Molecular and Cellular Biology of the Yeast Saccharomyces. Cell Cycle and Biology*, pp. 229–362. Cold Spring Harbor, NY: Cold Spring Harbor Laboratory Press.

Osumi, M. (1998) The ultrastructure of yeast: cell wall structure and formation. *Micron* **29**, 207–233.

Ovalle, R., Lim, S. T., Holder, B., Jue, C. K., Moore, C. W., and Lipke, P. N. (1998) A spheroplast rate assay for determination of cell wall integrity in yeast. *Yeast* **14**, 1159–1166.

Peng, L., Kawagoe, Y., Hogan, P., and Delmer, D. (2002) Sitosterol-β-glucoside as primer for cellulose synthesis in plants. *Science* **295**, 147–150.

Popolo, L., Gilardelli, D., Bonfante, P., and Vai, M. (1997) Increase in chitin as an essential response to defects in assembly of cell wall polymers in the *ggp1Δ* mutant of *Saccharomyces cerevisiae*. *Journal of Bacteriology* **179**, 463–469.

Popolo, L., Gualtieri, T., and Ragni, E. (2001) The yeast cell-wall salvage pathway. *Medical Mycology* **39** (1), 111–121.

Pringle, J. R. (1991) Staining of bud scars and other cell wall chitin with calcofluor. *Methods of Enzymology* **194**, 732–735.

Ram, A. F., Wolters, A., ten Hoopen, R., and Klis, F. M. (1994) A new approach for isolating cell wall mutants in *Saccharomyces cerevisiae* by screening for hypersensitivity to calcofluor white. *Yeast* **10**, 1019–1030.

Ram, A. F., Kapteyn, J. C., Montijn, R. C., Caro, L. H., Douwes, J. E., Baginsky, W., Mazur, P., van den Ende, H., and Klis, F. M. (1998) Loss of the plasma membrane-bound protein Gas1p in *Saccharomyces cerevisiae* results in the release of β-1,3-glucan into the medium and induces a compensation mechanism to ensure cell wall integrity. *Journal of Bacteriology* **180**, 1418–1424.

Richard, M., de Groot, P., Courtin, O., Poulain, D., Klis, F., and Gaillardin, C. (2002) *GPI7* affects cell-wall protein anchorage in *Saccharomyces cerevisiae* and *Candida albicans*. *Microbiology* **148**, 2125–2133.

Rodriguez-Pena, J. M., Cid, V. J., Arroyo, J., and Nombela, C. (2000) A novel family of cell wall-related proteins regulated differently during the yeast life cycle. *Molecular and Cellular Biology* **20**, 3245–3255.

Roncero, C. (2002) The genetic complexity of chitin synthesis in fungi. *Current Genetics* **41**, 367–378.

Roh, D. H., Bowers, B., Riezman, H., and Cabib, E. (2002) Rho1p mutations specific for regulation of $\beta(1\rightarrow3)$glucan synthesis and the order of assembly of the yeast cell wall. *Molecular Microbiology* **44**, 1167–1183.

Roncero, C. (2002) The genetic complexity of chitin synthesis in fungi. *Current Genetics* **41**, 367–378.

Ruiz, C., Cid, V. J., Lussier, M., Molina, M., and Nombela, C. (1999) A large-scale sonication assay for cell wall mutant analysis in yeast. *Yeast* **15**, 1001–1008.

Ruiz-Herrera, J., Gonzalez-Prieto, J. M., and Ruiz-Medrano, R. (2002) Evolution and phylogenetic relationships of chitin synthases from yeasts and fungi. *FEMS Yeast Research* **1**, 247–256.

Santos, B., Duran, A., and Valdivieso, M. H. (1997) *CHS5*, a gene involved in chitin synthesis and mating in *Saccharomyces cerevisiae*. *Molecular and Cellular Biology* **17**, 2485–2496.

Saxena, I. M., Brown, R. M. Jr, and Dandekar, T. (2001) Structure–function characterization of cellulose synthase: relationship to other glycosyltransferases. *Phytochemistry* **57**, 1135–1148.

Schekman, R. and Brawley, V. (1979) Localized deposition of chitin on the yeast cell surface in response to mating pheromone. *Proceedings of the National Acadamy of Sciences USA* **76**, 645–649.

Schreuder, M. P., Mooren, A. T., Toschka, H. Y., Verrips, C. T., and Klis, F. M. (1996) Immobilizing proteins on the surface of yeast cells. *Trends in Biotechnology* **14**, 115–120.

Shahinian, S. and Bussey, H. (2000) β-1,6-Glucan synthesis in *Saccharomyces cerevisiae*. *Molecular Microbiology* **35**, 477–489.

Shaw, J. A., Mol, P. C., Bowers, B., Silverman, S. J., Valdivieso, M. H., Duran, A., and Cabib, E. (1991) The function of chitin synthases 2 and 3 in the *Saccharomyces cerevisiae* cell cycle. *Journal of Cell Biology* **114**, 111–123.

Sietsma, J. H. and Wessels, J. G. (1981) Solubility of $(1\rightarrow 3)$-β-D/$(1 \rightarrow 6)$-β-D-glucan in fungal walls: importance of presumed linkage between glucan and chitin. *Journal of General Microbiology* **125**, 209–212.

Simons, J. F., Ebersold, M., and Helenius, A. (1998) Cell wall 1,6-β-glucan synthesis in *Saccharomyces cerevisiae* depends on ER glucosidases I and II, and the molecular chaperone BiP/Kar2p. *EMBO Journal* **17**, 396–405.

Spellman, P. T., Sherlock, G., Zhang, M. Q., Iyer, V. R., Anders, K., Eisen, M. B., Brown, P. O., Botstein, D., and Futcher, B. (1998) Comprehensive identification of cell cycle-regulated genes of the yeast *Saccharomyces cerevisiae* by microarray hybridization. *Molecular Biology of the Cell* **9**, 3273–3297.

Strahl-Bolsinger, S., Gentzsch, M., and Tanner, W. (1999) Protein *O*-mannosylation. *Biochimica and Biophysica Acta* **1426**, 297–307.

Sullivan, D. S., Biggins, S., and Rose, M. D. (1998) The yeast centrin, Cdc31p, and the interacting protein kinase, Kic1p, are required for cell integrity. *Journal of Cell Biology* **143**, 751–765.

Sundström, P. (2002) Adhesion in *Candida* spp. *Cellular Microbiology* **4**, 461–469.

Taron, C. H., Wiedman, J. M., Grimme, S. J., and Orlean, P. (2000) Glycosylphosphatidylinositol biosynthesis defects in Gpi11p- and Gpi13p-deficient yeast suggest a branched pathway and implicate Gpi13p in phosphoethanolamine transfer to the third mannose. *Molecular Biology of the Cell* **11**, 1611–1630.

Tokunaga, M., Kusamichi, M., and Koike, H. (1986) Ultrastructure of outermost layer of cell wall in *Candida albicans* observed by rapid-freezing technique. *Journal of Electron Microscopy*, **35**, 237–246.

Trilla, J. A., Cos, T., Duran, A., and Roncero, C. (1997) Characterization of *CHS4* (*CAL2*), a gene of *Saccharomyces cerevisiae* involved in chitin biosynthesis and allelic to *SKT5* and *CSD4*. *Yeast* **13**, 795–807.

Trilla, J. A., Duran, A., and Roncero, C. (1999) Chs7p, a new protein involved in the control of protein export from the endoplasmic reticulum that is specifically engaged in the regulation of chitin synthesis in *Saccharomyces cerevisiae*. *Journal of Cell Biology* **145**, 1153–1163.

Valdivia, R. H., Baggott, D., Chuang, J. S., and Schekman, R. W. (2002) The yeast clathrin adaptor protein complex 1 is required for the efficient retention of a subset of late Golgi membrane proteins. *Developmental Cell* **2**, 283–294.

Valentin, E., Herrero, W., Pastor, J. F. I., and Sentandreu, R. (1984) Solubilization and analysis of mannoprotein molecules from the cell wall of *Saccharomyces cerevisiae*. *Journal of General Microbiology* **130**, 1419–1428.

Van der Vaart, J. M., te Biesebeke, R., Chapman, J. W., Toschka, H. Y., Klis, F. M., and Verrips, C. T. (1997) Comparison of cell wall proteins of *Saccharomyces cerevisiae* as anchors for cell surface expression of heterologous proteins. *Applied and Environmental Microbiology* **63**, 615–620.

Vink, E., Vossen, J. H., Ram, A. F., Van Den Ende, H., Brekelmans, S., de Nobel, H., and Klis, F. M. (2002) The protein kinase Kic1 affects 1,6-β-glucan levels in the cell wall of *Saccharomyces cerevisiae*. *Microbiology* **148**, 4035–4048.

Williams, D. L., Pretus, H. A., Ensley, H. E., and Browder, I. W. (1994) Molecular weight analysis of a water-insoluble, yeast-derived (1→3)-β-D-glucan by organic-phase size-exclusion chromatography. *Carbohydrate Research* **253**, 293–298.

Wojciechowicz, D., Lu, C. F., Kurjan, J., and Lipke, P. N. (1993) Cell surface anchorage and ligand-binding domains of the *Saccharomyces cerevisiae* cell adhesion protein α-agglutinin, a member of the immunoglobulin superfamily. *Molecular and Cellular Biology* **13**, 2554–2563.

Young, S. H. and Jacobs, R. R. (1998) Sodium hydroxide-induced conformational change in schizophyllan detected by the fluorescence dye, aniline blue. *Carbohydrate Research* **310**, 91–99.

Yun, D. J., Zhao, Y., Pardo, J. M., Narasimhan, M. L., Damsz, B., Lee, H., Abad, L. R., D'Urzo, M. P., Hasegawa, P. M., and Bressan, R. A. (1997) Stress proteins on the yeast cell surface determine resistance to osmotin, a plant antifungal protein. *Proceedings of the National Acadamy of Sciences USA* **94**, 7082–7087.

Zlotnik, H., Fernandez, M. P., Bowers, B., and Cabib, E. (1984) *Saccharomyces cerevisiae* mannoproteins form an external cell wall layer that determines wall porosity. *Journal of Bacteriology* **159**, 1018–1026.

Chapter 6

Lipids and membranes*

Michael Schweizer

6.1 Introduction

Fatty acyl CoAs represent bioactive compounds that are involved in intracellular transport, cellular signalling, endocytosis and transcriptional control, as well as serving as substrates for β-oxidation and biosynthesis of phospholipids, sphingolipids and sterols. Metabolic engineering of *Saccharomyces cerevisiae* has been successful for the production of metabolites and therapeutic proteins but has not yet been used for the production of lipid-derived compounds. Extensive incorporation of triglyceride fatty acids into yeast can be achieved by growing the cells in the presence of triacylglycerols and exogenously supplied lipase, thus opening up an avenue for the bioconversion of low-cost oils into value-added lipid products (Dyer *et al.*, 2002). Recently, a remarkable feat of metabolic manipulation in yeast (Szczebara *et al.*, 2003) has been achieved: researchers took advantage of the fact that yeast cannot synthesise complex steroids and introduced into the yeast cell an altered sterol pathway whose end product was hydrocortisone, an anti-inflammatory steroid hormone for treatment of arthritis or progesterone for the production of vitamin D_2. This 'tour de force' of metabolic engineering was successful because all the genes of the sterol biosynthetic pathway are known in yeast and genetic manipulation allowed removal of undesirable gene products and exploited the flexibility of localisation and trafficking of sterol intermediates. In this chapter, the central role and impact of lipid metabolism on yeast physiology will be presented.

6.2 Uptake, activation and intracellular transport of fatty acids

Yeast cells take up fatty acids from the growth media by a process which can be saturated, implying the involvement of a protein (DiRusso and Black, 1999). On the basis of their chain lengths fatty acids are classified as short chain fatty acids (SCFA, i.e. C8–C12), medium chain fatty acids (MCFA, i.e. C14–C16), long chain fatty acids (LCFA, i.e. C16–C18) and very long chain fatty acids (VLCFA, i.e. ≥C20), which in some instances may be mono-unsaturated due to the action of the $\Delta 9$-desaturase. It is also known that uptake and activation of fatty acyl CoA thioesters are two separate functions; therefore, one can postulate the existence of a membrane-bound fatty acid transporter protein (FATP) in the plasma membrane. In fact an open reading frame (ORF) has been identified in *S. cerevisiae* which, based on sequence similarity, albeit only 33% identity and 54% similarity to FATP from murine 3T3-Ll adipocytes, could theoretically encode a fatty acid transporter

* This chapter is dedicated to my scientific mentor, Professor Dr Eckhart Schweizer, Lehrstuhl für Biochemie, Friedrich-Alexander-Universität Erlangen-Nürnberg in recognition of his distinguished contribution to the field of yeast fatty acid biosynthesis on the occasion of his forthcoming retirement.

(Faergeman *et al.*, 1997). Disruption of this ORF, designated *FAT1*, caused severely impaired growth on rich medium containing myristate (C14:0), palmitate (C16:0) or oleate (C18:1) and cerulenin, an inhibitor of fatty acid synthase (Fas), that is, under conditions of no fatty acid synthesis and in the presence of oleate, suppression of fatty acid desaturation. The most severe growth effect was observed in the presence of cerulenin and oleate suggesting that a *fat1Δ* strain is apparently deficient in fatty acid transport, activation by acyl CoA synthetase or synthesis of lipids. By confocal laser scanning microscopy it was shown that the uptake of a fluorophore-labelled LCFA analogue was impaired in the *fat1Δ* strain. It was demonstrated that the phospholipid pool in the *fat1Δ* strain was unchanged and therefore, there must be a defect in the uptake of oleate prior to its metabolic utilisation. Furthermore, the decrease in uptake of oleate was not due to a decrease in long chain acyl CoA synthetase (LCS) activity, despite some sequence similarity between Fat1 and acyl CoA synthetases since both are members of the family of AMP-binding proteins (DiRusso and Black, 1999; DiRusso *et al.*, 2000). It could be that the function of Fatl comes into play when cells are grown under anaerobic conditions and have a requirement for unsaturated MCFAs. Strains deleted for *FAT1* grown on dextrose accumulate C22:0, C24:0 and C26:0 at increased levels compared to the wild type (Watkins *et al.*, 1998). Following growth on oleate there is a smaller but significant increase in the same compounds. However, the levels of MCFAs and LCFAs are similar in both wild type and *fat1Δ*. It has been shown that disruption of *FAT1* reduces very long chain acyl CoA synthetase (VLCS) activity by 70% whereas LCS activity was unaffected (Choi and Martin, 1999). Fractionation studies have indicated that VLCS activity is associated mainly with the peroxisomal fraction with very little in the mitochondria. This suggests that the primary role of Fat1 is activation of VLCFAs, a step essential for their incorporation into complex lipids or for *β*-oxidation (Watkins *et al.*, 1998). Data from experiments described in Choi and Martin (1999) suggest that Fat1 is responsible for mobilising VLCFA to sites of *β*-oxidation. If one considers that the likelihood of VLCFAs having to be activated is small, since they constitute only 1% of the cellular fatty acid pool (Welch and Burlingame, 1973) and the majority are endogenously produced rather than coming from exogenous sources, then it is more likely that the cellular role of Fat1 is to maintain VLCFAs at a level which is non-toxic to the cell by facilitating their *β*-oxidation. More recent studies based on site-directed mutagenesis of *FAT1* have shown that the uptake of a fluorescent LCFA analogue or labelled oleate and VLCFA activation can be separated suggesting that Fat1 is involved in both the transmembrane trafficking and in the maintenance and turnover of VLCFAs (Zou *et al.*, 2002).

Four fatty acyl CoA synthetase genes *FAA* (fatty acid activation) have been identified and characterised from strains unable to grow at 37 °C in the presence of cerulenin and palmitate but able to grow on palmitate alone. These mutants were defective in acyl CoA synthetase activity and unable to repress acetyl CoA carboxylase (Acc). It was shown that *FAA1* is an essential gene when Fas is intact. Further phenotypic analysis of *faa1Δ* strains suggested that *S. cerevisiae* contains other LCS activities which are defective at 24 °C and 30°C since they can be rescued at these temperatures by myristate and palmitate in the presence of cerulenin. Strains lacking Faa1 have a 2–5-fold reduction in the incorporation of labelled acyl chains into all phospholipid species at 36 °C. It has been shown *in vitro* that Faal prefers C12:0–Cl6:0, Faa2, a peroxisome-associated enzyme (Hettema *et al.*, 1996) can accommodate chain lengths between C7:0 and Cl7:0, with C9:0–C13:0 being its preferred substrates, whereas Faa3 has a preference for unsaturated Cl6:1, Cl8:1 fatty acids and can weakly activate VLCFAs (Watkins *et al.*, 1998). Furthermore, it has been shown that all three Faas have myristoyl CoA synthetase activities *in vitro* and the enzymes have different temperature optima, 30 °C for Faal and 25 °C for Faa2 and Faa3 (Knoll *et al.*, 1994).

Fortuitously, a mutation in the fourth *FAA* gene was obtained during the creation of triple *FAA* deletants; one isolate, unlike the other triple deletants, was unable to grow in the presence of cerulenin and myristate. This strain was used to clone *FAA4* by complementation. Faa1 and Faa4 are functionally interchangeable in the cell and provide an example for which two members of a gene family are encoded by duplicated regions of the *S. cerevisiae* genome since their flanking ORFs are also duplicated (Pearson *et al.*, 1996). *FAA1* and *FAA4* are the most important members of the Faa family since a drastic reduction in the incorporation of ^{3}H-labelled fatty acids into cellular phospholipids and a marked increase in free fatty acids are to be found in *faa1Δ* and *faa4Δ* strains (Johnson *et al.*, 1994) leading to diminished cellular utilisation of nutrient fatty acids. Faa4 is not the last member of the Faa family since the presence of Faa5 can be implied by the fact that a *nmt1 faa1Δ faa2Δ faa3Δ faa4Δ* mutant can grow at 30 °C on myristate, palmitate or oleate as the sole carbon source. The most likely candidate for *FAA5* is *FAT1* which is capable of activating VLCFAs (Choi and Martin, 1999; Zou *et al.*, 2002). The evidence to date is consistent with fatty acid import being a complex multicomponent process involving at least the proteins Fat1 and Faa1 and Faa4 (Choi and Martin, 1999; DiRusso *et al.*, 2000; Faergeman *et al.*, 2001). A physical interaction between Fat1 and Faa1 or Faa4 has been demonstrated by negative dominance, two-hybrid analysis and co-immunoprecipitation (Zou *et al.*, 2003). The current working hypothesis for Fat1-mediated transport is that Fat1 increases fatty acid binding to the membrane and this has a positive effect on diffusion across the membrane. The interaction between Fat1 and Faa1/Faa4 implies acylation of the imported fatty acid. This coupling of import and activation of fatty acids is reminiscent of vectorial acylation encountered in bacteria (Klein *et al.*, 1971). These results are in good agreement with the finding that Fat1, and Faa1 and Faa4 are found in the storage granule (lipid particles) which may serve as a repository of proteins involved in lipid and sterol metabolism, emphasises the relatedness of their functions (Athenstaedt *et al.*, 1999b). In humans loss of control of VLCFA metabolism and accumulation of VLCFAs is associated with X-linked adrenoleukodystrophy (ALD) (Choi and Martin, 1999).

With the one exception noted previously, these tests have been carried out in a wild-type strain with an active myristoyl-CoA:protein N-myristoyltransferase encoded by *NMT1* (Farazi *et al.*, 2001). Nmt1 is essential during vegetative growth (Ashrafi *et al.*, 1998) and deletion of the corresponding gene in a *faa1Δ faa2Δ faa3Δ* background renders the strain inviable on complete medium even upon supplementation with myristate and in the presence of active Fas. Strains carrying *nmt1 faa1* mutations are slow growers in the presence of myristate but deletion of *FAA2, FAA3* or *FAA4* in the same background has no effect. *S. cerevisiae* cannot survive loss of *FAA1 FAA4* in the *nmt1* background. In fact it has been shown that Faa4 is responsible for regulating protein N-myristoylation during the transition from logarithmic to stationary phase. Furthermore, N-myristoylation of proteins including the ADP-ribosylation factors, Arf1 and Arf2 contribute to stationary phase survival (Ashrafi *et al.*, 1998). Loss of *ARF1* or *ARF2* in combination with a *nmt1* mutant compromised for myristoyl CoA binding results in a dramatic loss of colony-forming units after time spent in the stationary phase suggesting there are certain myristoylated proteins, for example Arf1, which are important for stationary phase survival. Overexpression of Faa2 can rescue *nmt1* mutants, probably due to the activation of the endogenous C14:0 pool. This pool is partly derived from membrane phospholipids since overexpression of phospholipase B (PLB) encoded by *PLB1* also suppresses the growth defect and C14:0 auxotrophy in a *nmt1* mutant and PLB hydrolyses phosphatidylcholine (PtdCho) and phosphatidylethanolamine (PtdEtn) into glycerophosphocholine (Gro*P*Cho), glycerophosphoethanolamine (Gro*P*Etn), respectively and free fatty acids which are then metabolised in the cell to myristoyl CoA (Johnson *et al.*, 1994). The structure of Nmt1 has recently been determined and the Nmt fold resembles a similar structure found in the Gcn5-related N-acetyltransferase superfamily (Bhatnagar *et al.*, 1999). Nmt proteins may turn out to be useful therapeutic targets since they are essential for the survival of human pathogens, such as *Candida albicans* (Lodge *et al.*, 1997).

There is a requirement for fatty acyl CoAs to be transported within the cell. Overexpression of the bovine acyl CoA binding protein in yeast has shown that it can function as an intracellular acyl CoA pool former. A threefold increase in $\Delta 9$-desaturase mRNA levels was observed in a yeast *acb1Δ* strain, suggesting that the repression of the *OLE1* expression is caused by acyl CoA esters and not by fatty acids (Knudsen *et al.*, 2000).

ACB1 was originally recognised as the yeast homologue of the diazepam binding inhibitor (Rose *et al.*, 1992). Disruption of this ORF causes no obvious phenotype, although *acb1Δ* strains have a selective disadvantage against wild-type cells when the two are grown together in a mixed culture. One would therefore expect that *acb1Δ* strains would have reduced acyl CoA levels. However, the level of acyl CoA in the *acb1Δ* strain was 2.5-fold higher than in the wild type which may explain the high rate of adaptation (1 in 10^5) of the initial slow growing phenotype to a faster growing phenotype (Knudsen *et al.*, 2000; Gaigg *et al.*, 2001). Use of a conditional *ACB1* deletant strain under the control of a galactose promoter permitted a more accurate investigation of the role of Acb1. Since the general phospholipid pattern, the rate of phospholipid synthesis and the turnover of individual phospholipid classes were not affected, it would appear that Acb1 does not play a role in general glycerophospholipid synthesis. However, sphingolipid synthesis was drastically reduced as was the content of C26:0 in total fatty acids pointing to a more specialised role for this protein in VLCFA synthesis. ^{3}H-inositol labelling of inositolphosphorylceramide (IPC) and mannosyl-inositolphosphorylceramide (MIPC) but not of mannosyl-di-inositolphosphorylceramide ($M(IP)_2C$) is reduced in the conditional deletant strain suggesting an imbalance consistent with blocking of sphingolipid transport between ER and Golgi. Depletion of Acb1 leads to vacuoles with severely altered morphology and accumulation of autophagosomes or autophagocytotic/endosome-like bodies. Results obtained by lipid mass spectrometry suggest, however, that subtle changes in the level of PtdCho, phosphatidylserine (PtdSer) and PtdEtn exist in the *acb1Δ* strain in comparison to the wild type (Faergeman *et al.*, 2002). It can be concluded that Acb1 is involved in supplying acyl moieties for lipid modification of proteins and protein remodelling essential for the transport of distinct proteins to, and assembly of, the vacuole (Gaigg *et al.*, 2001).

6.3 Synthesis, elongation and desaturation of fatty acids

In common with fatty acids, acetate and malonate, the substrates of the multienzyme complexes Acc and Fas, respectively, have to be activated in order to carry out their cellular functions. The concentration of intracellular acetyl CoA, the initial substrate, is regulated by its rates of synthesis, utilisation and degradation. Acetyl CoA is primarily synthesised from glucose via pyruvate, from β-oxidation of fatty acids, by the breakdown of lysine and, to a lesser extent, by the activation of acetate by acetyl CoA synthetase (Acs). Acetate, the substrate for Acs, is generated from pyruvate which is produced by pyruvate decarboxylase and acetaldehyde dehydrogenase and by the anaplerotic carboxylation of pyruvate to oxaloacetate, catalysed by pyruvate carboxylase (Pronk *et al.*, 1996).

Acs is encoded by two genes, *ACS1* and *ACS2*, whose products are enzymes of different catalytic properties (van den Berg *et al.*, 1996). When S. *cerevisiae* is grown on acetate or ethanol, ATP-dependent activation of acetate to acetyl CoA is catalysed by the products of both genes. The K_M for acetate of Acsl is 30-fold lower than that of Acs2 and unlike Acs2, Acsl can use propionate as a substrate. Under aerobic, glucose-limited conditions Acs1 is repressed. Addition of glucose to respiring *S. cerevisiae* cells results in the accumulation of acetate which is not counteracted by overexpression of either Acs1 or Acs2 implying that another approach has to be taken to reduce acetate production which has a negative effect on biomass-directed applications of *S. cerevisiae* (de Jong-Gubbels *et al.*, 1998). The carbon-source regulation of *ACS1* is mediated by the positive regulators Cat8 and Adr1. In addition, glucose repression of *ACS1* is controlled by interplay of the negative regulator Ume6 and the ubiquitous transcription factor Abfl (Kratzer and Schüller, 1997).

A further enzyme, acetyl CoA hydrolase (Ach), which catalyses the hydrolysis of acetyl CoA, probably has the biological function of maintaining the cytosolic acetyl CoA concentration and CoASH pool for both synthesis and oxidation of fatty acids and sterol synthesis (Lee *et al.*, 1990). By examining the effects of various carbon sources it would appear that Ach expression is repressed by glucose. *ACH1* mRNA and enzyme activity is increased in stationary phase but decreased in early and late log phases. The regulated expression of Ach is consistent with the reaction it carries out – the scission of acetyl CoA's high energy thioester bond – which has apparently no metabolic advantage and is therefore kept to a minimum. Interestingly, in a proteomic analysis Ach was one of ten upregulated enzymes following exposure of *S. cerevisiae* to 0.9 mM sorbic acid at pH 4 (de Nobel *et al.*, 2001).

The *de novo* synthesis of saturated fatty acyl CoAs from acetyl CoA and malonyl CoA requires the enzymes Acc and Fas. Fatty acyl CoAs synthesised by Fas in seven sequential additions of two carbon atoms from malonyl CoA produced by Acc to the growing acyl chain have, as a rule, a chain length of C16:0 and to a lesser extent, C18:0. Acc is an α_4 homomer and Fas is an $\alpha_6\beta_6$ heteromer (Schweizer, 1986). The trifunctional monomer of Acc is encoded by *ACC1*, the trifunctional α polypeptide of Fas by the *FAS2* gene and the pentafunctional β polypeptide of Fas by *FAS1*. Each enzyme complex must be converted from the *apo-* to the *holo*-form. In the case of Acc this is performed by a biotin–apoprotein ligase, the product of the *ACC2* gene, which attaches the essential biotin prosthetic group to the biotin carboxyl-carrier domain of Acc. Not surprisingly, deletion of this gene is a lethal event (Hoja *et al.*, 1998). The conversion of *apo*-FAS to *holo*-FAS requires a phosphopantetheinyl transferase activity, which attaches the 4′-phosphopantetheine arm to a serine residue in the acyl carrier protein (ACP) domain located at the NH_2-terminus of Fas2. To date, it has not been shown that there is a gene product capable of carrying out this transferase activity. However, the product of the ORF YPL148c is responsible for pantetheinylation of the ACP function of mitochondrial Fas. Disruption of this ORF has no effect whatsoever on the pantetheinylation of the cytoplasmic Type I Fas. (Stuible *et al.*, 1998). Therefore, the genetic information for phosphopantetheinyl transferase activity of cytoplasmic Fas may lie virtually *incognito* in the C-terminal region of Fas2 as suggested by sequence similarity to equivalent enzymes of Fas/Pks (polyketide synthase) systems (Lambalot *et al.*, 1996) and is indeed the case (Fichtlscherer *et al.*, 2000).

The fact that strains lacking *ACC1* and *ACC2* genes cannot grow even after supplementation with fatty acids suggests that Acc activity is required for a process other than fatty acid synthesis *de novo* and indeed studies with two conditional alleles, *mtr7* (*acc1*^{ts}) and *acc1*^{cs}, confirm this suspicion (Schneiter *et al.*, 1996, 1999b). Strains containing either of these mutant alleles are defective in the nuclear export of poly(A) RNA. Interestingly, *acc1*^{cs} was isolated in a screen for mutants which have a synthetic lethal interaction with a hyperrecombination mutant *hpr1*Δ. Further investigation showed that *mtr7* and *hpr1*Δ are also synthetically lethal. The basis for the synthetic lethal interaction could be due to a defect in export of nuclear poly(A) RNA combined with the altered lipid composition of the nuclear envelope in cells containing a compromised *ACC1* gene. Immunofluorescence microscopy indicated that Acc1 is associated with the cytoplasmic surface of the ER membrane but is independent of the nuclear pore complex (Ivessa *et al.*, 1997). In addition, conditional *acc1* mutants display altered vacuolar membrane morphology and inheritance which may be caused by faulty acylation of Vac8 (Pan *et al.*, 2000; Schneiter *et al.*, 2000a). Strains carrying the *mtr7* allele are defective in malonyl CoA-dependent elongation of VLCFA (C26:0), found in ceramides, the fatty acyl derivatives of sphingolipids and the lipid moiety of glycosyl phosphatidylinositol (GPI)-proteins. The alterations in the nuclear envelope and vacuolar morphology as well as the requirement for C26:0 suggests there may be a function for VLCFAs, which account for 1% of the cellular fatty acid content, in stabilising or lengthening curved membrane structures (Schneiter and Kohlwein, 1997).

A further example of the importance of fatty acyl species in organelle structure and function has been documented in *Schizosaccharomyces pombe*. In this yeast two mutants, *cut6* and *lsd1*, are impaired in non-chromosomal nuclear components. The two essential genes, *cut6*$^{+}$ and *lsd1*$^{+}$, encode Acc and the α subunit of Fas, respectively (Saitoh *et al.*, 1996). Conditional *lsd1/fas2* mutants are characterised by large and small daughter nuclei and a thick septum and accumulate abnormal phospholipids which disappear when the cells are cultured in the presence of C16:0 whereas *cut6*, like *mtr7*, is not supplementable by this fatty acid (Yokoyama *et al.*, 2001).

Fas also appears to play a role in the virulence of *Candida albicans* since mice infected with wild-type *C. albicans* (2n) died within three days. However, when the diploid contained one mutated allele of *FAS2*, death was delayed by 24 h. Application of a strain in which both *FAS2* alleles were disrupted, allowed the mice to survive 22 days, suggesting the strain was avirulent. This has implications for Fas as an appropriate target for antifungal drugs (Zhao *et al.*, 1997).

The fatty acyl composition of *S. cerevisiae* varies in response to environmental stress, such as temperature and ethanol concentration. Low temperature reduces the fluidity of the *S. cerevisiae* plasma membrane and it has been suggested that as an adaptation to low temperatures yeast increases the amount of unsaturated fatty acyl chains in the membrane lipid to compensate for reduced fluidity. The yeast genome encodes only one fatty acid desaturase, $\Delta 9$-desaturase, the product of the *OLE1* gene, which can introduce a double bond into fatty acyl chains under aerobic conditions (Zhang *et al.*, 1999). The major monounsaturated fatty acids (MUFAs) in *S. cerevisiae*, palmitoleic (C16:1,$\Delta 9$) and oleic (C18:1,$\Delta 9$) acids, exist in roughly equal proportions and together make up at least 70% of lipids with strain- and carbon source-dependent variation in the relative composition (Tuller *et al.*, 1999). Ole1 is a modular protein with a cytochrome b_5-like domain at its C-terminus. It has been demonstrated that neither the cytochrome b_5-dependent rat liver desaturase nor the yeast cytochrome P450 sterol biosynthetic enzymes have such a domain. The disruption of the cytochrome b_5-like domain of Olel resulted in a requirement for oleic acid, suggesting that the dedicated domain plays an essential role in the desaturase reaction (Mitchell and Martin, 1995). As is the case with Acc it would appear that the *OLE1* gene product is responsible for a structural function in the yeast cell; the mutation *mdm2* preventing segregation of mitochondria from the mother to the daughter cell, has been shown to be allelic to *OLE1*. A role for acyl chain desaturation in localisation of Ole1 has been demonstrated using a *ts* mutant (Stewart and Yaffe, 1991). Under non-permissive conditions Ole1, but no other soluble membrane-bound ER resident proteins alter their subcellular distribution from the ER membrane to the plasma membrane. This protein-specific relocalisation of MUFAs in lipids was shown to be temperature- and energy-independent but lipid-dependent. The reduction of MUFA synthesis results in local alteration of membrane fluidity supporting the idea that membrane lipid composition can affect protein sorting (Tatzer *et al.*, 2002).

Olel is responsive to nutritional and physiological conditions; expression of *OLE1* mRNA level is sharply reduced and remains so until most of the added MUFAs are incorporated into cellular lipids. The half life of *OLE1* mRNA is reduced from 10 to 2.5 min following the addition of MUFAs to the growth medium whereas saturated fatty acids (SFAs) have no influence. The MUFA-triggered reduction in the half life of *OLE1* mRNA requires protein synthesis since the addition of cycloheximide abrogated the effect (Gonzalez and Martin, 1996). Distinct, but interacting promoter elements, the fatty acid response (FAR) element and a downstream low-oxygen response element (LORE) regulate the transcription of *OLE1* under hypoxic growth conditions or in response to cobalt stimulation (Martin *et al.*, 2001, 2002) *via* the interaction of 90 kDa N-terminal fragments of the ER membrane-bound proteins, Mga2 and Spt23 (Zhang *et al.*, 1999; Hoppe *et al.*, 2000; Rape *et al.*, 2001). The ubiquitin-dependent proteolysis of Spt23 but not that of Mga2 is repressed by polyunsaturated fatty acids (PUFAs) (Martin *et al.*, 2002). This proteolysis-dependent release of a peptide fragment which turns out to be a transcription factor is reminiscent of the proteolytic process of sterol regulatory element binding protein-1 in lipid and cholesterol metabolism in mammals (Brown and Goldstein, 1999). Simultaneous deletion of *MGA2* and *SPT23* causes fatty acid auxotrophy which is corrected by overexpression of Ole1 (Zhang *et al.*, 1999). However, deletion of *MGA2* alone decreases the half life of *OLE1* mRNA from 10 to 7 min and it has been shown that the elements responsible for *OLE1* mRNA stability reside in the 5′- and 3′-untranslated regions as well as in the transmembrane region and the C-terminus of Ole1 (Martin *et al.*, 2002).

It has been reported that the extraplastidial oleate desaturase gene *(FAD2)* of *Arabidopsis thaliana* has been expressed in *S. cerevisiae*, resulting in the identification of linoleic acid (C18:2,$\Delta 9$,$\Delta 12$) and hexadecadienoic acid (C16:2,$\Delta 9$,$\Delta 12$) by GC-MS (Covello and Reed, 1996). The expression of the yeast $\Delta 9$-desaturase gene in tomato has resulted in an alteration of the fatty acid profile leading to changes in flavour compounds (Wang *et al.*, 2001). More recently, it was demonstrated that co-expression of plant fatty acid desaturases FAD2 and FAD3 under the control of a fatty acid-inducible peroxisomal gene promoter, coupled the processes of fatty acid uptake with the induction of a new metabolic pathway leading from oleic acid (18:1) to linolenic acid (18:3,$\Delta 9$,$\Delta 12$,$\Delta 15$) (Dyer *et al.*, 2002). A further example of altering the fatty acid profile in yeast is provided by the heterologous expression of the $\Delta 12$-acetylenase gene from *Crepis alpina* in *S. cerevisiae.* Yeast containing this gene under the control of a yeast promoter were capable of introducing a triple bond into linoleic acid (C18:2,$\Delta 9$,$\Delta 12$) taken up from the growth medium. Crepenynic acid (9-octadecen-12-ynoic acid) was identified by GC-MS. Furthermore, this strain grown in the absence of linoleic acid was capable of desaturating endogenously produced oleic acid by the introduction of a double bond at C12. However, the amount of endogenous linoleic acid was not sufficient to support the production of crepenynic acid (Lee *et al.*, 1998).

In addition to Olel there are other lipid-biosynthetic enzymes which modify fatty acids prior to integration into membrane bilayer lipids and these include the fatty acyl elongation systems. Since the fatty acyl elongation process relies on the addition of two carbon atoms to a growing fatty acyl chain, it is likely that the essential function associated with Acc which cannot be corrected by the addition of MCFAs to the growth medium, is the provision of malonate to produce VLCFAs. Strains defective in Fas can grow on C14:0 due to the existence of a membrane-bound elongation system and it has been shown that the product of the *ELO1* gene is responsible for the Fas-independent elongation of C14:0 to C16:0 (Toke and Martin, 1996; Dittrich *et al.*, 1998). The protein encoded by *ELO1* has at least five potential membrane-spanning domains and a putative NADPH-binding site. The gene is regulated by added fatty acids; it is repressed when the medium is supplemented with C16:0 and C18:0 and is induced 3–5-fold when C14:0 is added transiently over a period of four hours (Toke and Martin, 1996). *ELO1* also functions as an elongase for MUFAs (C14:1,$\Delta 9$ to C16:1, $\Delta 11$) explaining why the acyl chain supply influences the lipid composition of the cell since in the presence of exogenous MUFAs the phospholipid composition of the cell is altered, in particular for PtdEtn and PtdCho (Schneiter *et al.*, 2000b).

Cell fractionation and *in vitro* studies indicated the possible existence of multiple fatty acyl elongation systems associated with the ER and the mitochondrial membrane. Indeed, two homologues of *ELO1* were found in the *S. cerevisiae* genome and it has been shown that the products of these genes do indeed function as fatty acyl elongases and that they are required for the formation of sphingolipids. Elo2 has the highest affinity for LCFA whereas Elo3 with a broader substrate specificity is necessary for the conversion of C24:0 acyl chains to C26:0 acyl chains. Strains lacking *ELO2* accumulate hydroxylated species of C16:0 and C18:0 fatty acids (HO-C16:0 and HO-C18:0) which are not detectable in the wild type. In an *elo2Δ* background overexpression of Elo3 counteracts the accumulation of hydroxylated fatty acids; there is a 50-fold decrease in the amount of HO-C16:0 and HO-C16:1, whereas HO-C18:0, HO-C18:1 species are decreased to a lesser extent. C20 species of fatty acids are not detected in the *elo2Δ* strains but the total C22–C26 species found are equivalent to 30% of wild-type levels. Disruption of *ELO3* abolishes C26:0 and HO-C26:0 (2-hydroxyhexacosanoic acid) species. From these observations it is not surprising that overexpression of Elo3 can compensate for the loss of *ELO2* and *vice versa*. Loss of either Elo2 or Elo3 reduces the cellular phospholipid levels and causes accumulation of phytosphingosine, an early intermediate in the ceramide biosynthetic pathway. The most abundant species of VLCFA in wild-type *S. cerevisiae* are C26:0 and α HO-C26:0, synthesised by the action of Scs7/Fah1 (Mitchell and Martin, 1997), which make up approximately 2% and 1.5%, respectively, of the total fatty acid mass in mid-log phase cells grown in glucose-containing medium. HO-fatty acids and 22- and 24-carbon SFA/MUFA species represent approximately 0.6% of the total fatty acids and are apparently derived from metabolic intermediates of the fatty acyl elongation pathway (Oh *et al.*, 1997).

Strains deleted for *ELO2* and *ELO3* are inviable showing that the two proteins perform overlapping functions in sphingolipid and GPI synthesis. *ELO2* was originally cloned as *GNS1* which corrected a defect in β-glucan synthase and also as *FEN1*, mutants of which exhibited bud localisation defects and resistance to SR31747A, an inhibitor of sterol isomerase. *ELO3* was also previously identified in three independent screens: (1) as *APA1* which corrected a decrease in the level of the plasma membrane H^+-ATPase; (2) as *SUR4* which when mutated suppresses reduced viability of the starvation mutant phenotype *(rvs161)* (Desfarges *et al.*, 1993) and (3) as *SRE1* whose mutants suppress the inhibition of SR31747A (Oh *et al.*, 1997). Further evidence for the major role of these two proteins in sphingolipid metabolism is the identification of mutant *elo2* and *elo3* alleles which allowed yeast to grow normally and secrete in the absence of synaptobrevin/VAMP (vesicle-associated membrane protein) v-SNAREs (vesicle SNAP (soluble NSF-(NEM-sensitive fusion protein) attachment protein) receptor) thus implicating VLCFAs in membrane trafficking (David *et al.*, 1998).

The activities encoded by *ELO2* and *ELO3* are paralogous to the condensing step of Fas. However, addition of a two-carbon unit to a growing fatty acyl chain requires three other activities, namely, ketoreductase, dehydratase and enoyl reductase. To elongate a C16:0 to C26:0 fatty acyl chain a total of five reaction cycles is required. The gene encoding the enoyl reductase, *TSC13* was identified in a screen for suppressors of the Ca^{2+}-sensitivity associated with *csg2Δ* strains which are defective in sphingolipid synthesis, a selection strategy which also led to the identification of *ELO2* and *ELO3* (Kohlwein *et al.*, 2001). Strains mutated in *TSC13* accumulate long chain sphingoid bases (LCBs), for example, phytosphingosine and ceramides containing fatty acids with chain lengths shorter than C26:0; deletion of either *ELO2* or *ELO3* in *tsc13* mutants intensifies the phenotype and in fact Tsc13 co-precipitates with Elo2 and Elo3 suggesting the existence of a complex of proteins involved in the elongation process. The synthetic lethality of a *tsc13* mutant and *mtr7* is consistent with a role of Tcs13 in fatty acid elongation. GFP-tagged Tsc13 showed, in common with Elo2 and Elo3, nuclear rim-ER staining and in addition was enriched in structures corresponding to nuclear–vacuolar junctions, contact sites between the nucleus and the vacuole, which are dependent on the presence of Vac8 (Pan *et al.*, 2000; Schneiter *et al.*, 2000a).

6.4 Fatty acid metabolism in mitochondria and peroxisomes

It is well established that lipids are an important factor in mitochondrial biogenesis but it is a generally held view that lipids are synthesised in the cytoplasm and then transported into the mitochondria. However, evidence is accumulating which suggests that mitochondria may in fact possess a type II Fas, in which the constituent functions are not part of a multifunctional polypeptide chain, consistent with the prokaryotic origin of the mitochondrion (Schneider *et al.*, 1997b). Several genes encoding component proteins of mitochondrial Fas have been identified: *ACP1* and *YPL148c* (see before), *CEM1* (3-ketoacyl ACP synthase), *OAR1* (3-ketoacyl ACP reductase). *MCT1* (malonyl CoA ACP transferase) and *ETR1* (2-enoyl thioester reductase) (Harington *et al.*, 1994; Schneider *et al.*, 1995, 1997a; Stuible *et al.*, 1998; Torkko *et al.*, 2001). The *S. cerevisiae acp1Δ*, *ypl148cΔ* and *cem1Δ* strains all showed a respiratory-deficient rho-zero phenotype. The *cem1Δ* strain could be complemented by Faa2 redirected to mitochondria by a mutation that provided the enzyme with a mitochondrial import sequence (Harington *et al.*, 1994). Strains lacking *ETR1* have rudimentary mitochondria with decreased contents of cytochromes and a respiration-deficient phenotype. Immunolocalisation and *in vivo* targeting experiments showed these proteins to be predominantly mitochondrial. Mitochondrial targeting was essential for complementation of the mutant phenotype. The phenotype of strains carrying a mutated *ETR1* allele was complemented by a mitochondrially targeted FabI protein from *E. coli.* FabI represents a nonhomologous 2-enoyl-ACP reductase which participates in the last step of the type II fatty acid synthesis (Torkko *et al.*, 2001).

Further information on the fatty acid synthesising machinery in mitochondria is provided by the Hfa1 protein encoded by *YMR207c* which has 56% identity and 73% similarity to Acc1 and therefore may be regarded as providing malonyl CoA required for mitochondrial fatty acid biosynthesis (Marthol *et al.*, 2002). Disruption of the mitochondrial *ACP1* gene has little or no effect on *de novo* overall fatty acid synthesis in yeast. The picture was different when SCFAs were examined in the *acp1Δ* strain. Octanoic acid (C8:0) has been shown to be one of the major products associated with mitochondrial ACP in *Neurospora crassa*; in *E. coli* lipoic acid biosynthesis occurs by way of ACP-bound intermediates (Morris *et al.*, 1995). Upon examination of the *acp1Δ* yeast strain it was found that it contained 10–20-fold lower amounts of lipoic acid than the wild-type cells whereas *petite* mutants have a normal lipoic acid content (Brody *et al.*, 1997). Therefore, the rho-negative phenotype is obviously not responsible for lipoic acid deficiency. Isotopic labelling experiments in *E. coli* have shown that lipoic acid is synthesised from C8:0. Lipoic acid is a coenzyme covalently bound to proteins which are involved in the oxidative decarboxylation of α-ketoacids. The cofactor function of lipoic acid in the pyruvate dehydrogenase and α-ketoglutarate dehydrogenase reactions could well explain the *petite* phenotype. A *lip5Δ* strain (Sulo and Martin, 1993) is incapable of synthesising lipoic acid but is still capable of attaching lipoic acid to protein. Interestingly, *lip5Δ* and *acp1Δ* cannot be supplemented by the addition of lipoic acid to the medium. Recently, a lipoic acid ligase, the product of *YLR239c/LIP2*, specific for glycine decarboxylase and pyruvate dehydrogenase has been characterised (Marvin *et al.*, 2001). This protein has similarity to the lipoyl ligase B of *E.coli* and *Kluyveromyces lactis*. There is another ORF, *YJL046w*, homologous to the lipoyl ligase A of *E. coli*. It has been postulated that the product of *YJL046w* is responsible for the transfer of lipoic acid imported into the mitochondria whereas the product of *YLR239c/LIP2* transfers endogenously produced lipoic acid (Hofmann *et al.*, 2002).

Fatty acids are degraded in *S. cerevisiae* only in specialised organelles, the peroxisomes, and not as in animals in both mitochondria and peroxisomes (Lazarow and Kunau, 1997; Trotter, 2001). Growth, albeit poor, on oleate induces a complete set of peroxisomal β-oxidation enzymes in *S. cerevisiae*. Mutants defective in fatty acid oxidation were used to clone the genes encoding the β-oxidation enzymes. *POX1/FOX1* encodes an acyl CoA oxidase, the enzyme catalysing the first and rate-limiting reaction of the β-oxidation cycle converting fatty acyl CoA chains to *trans*-2-enoyl CoA and H_2O_2 at the expense of molecular oxygen. This reaction involves participation of FAD as a cofactor but differs from its mitochondrial counterpart in animals in that the electrons are directly transferred to molecular oxygen rather than travelling down the electron transport chain with its concomitant oxidative phosphorylation.

The next two catalytic steps are carried out by the bifunctional protein (Fox2/Pox2) which has *trans*-2-enoyl CoA hydratase-2 and the D-specific 3-hydroxyacyl CoA dehydrogenase activities. *Trans*-2-enoyl CoA esters are converted into 3-ketoacyl CoA by way of D- instead of L-3-hydroxyacyl CoA esters. The L-isomers are found in mammalian, plant and bacterial systems. The D-specific 3-hydroxyacyl CoA dehydrogenase is located in the N-terminal part of Fox2/Pox2 whereas the *trans*-2-enoyl CoA hydratase-2 activity is located in its C-terminal portion (Lazarow and Kunau, 1997). The 3-ketoacyl CoA thiolase (Pot1/Fox3) catalyses the fourth and final step of fatty acid β-oxidation by shortening the acyl CoA chain thiolytically by a C2-unit to release acetyl CoA. The acetyl CoA formed by β-oxidation can either be further oxidised by the mitochondrial tricarboxylic acid cycle or converted to succinate *via* the glyoxylate cycle, thus allowing yeast to grow on fatty acids as a sole carbon source (Geisbrecht *et al.*, 1998).

Three types of peroxisomal targeting sequence (PTS) have been defined (Lazarow and Kunau, 1997); PTS1 a C-terminal tripeptide SKL, PTS2 a conserved N-terminal nonapeptide and PTS3 an internal sequence. Pox1/Fox1 probably has an internal targeting sequence whereas Fox2/Pox2 has a PTS1 sequence and Pot1/Fox3 has the N-terminal PTS2 recognition motif (Lazarow and Kunau, 1997). It would appear that Fox2 is transported into the peroxisome by the intra-peroxisomal protein Pex7 (Marzioch *et al.*, 1994; Distel *et al.*, 1996). Cells mutated for *PEX7* have shown that the thiolase is mislocated to the cytoplasm (Rehling *et al.*, 1996). Pex19, essential for peroxisome biogenesis, appears to be required for importing both PTS1 and PTS2 matrix proteins (Gotte *et al.*, 1998; Hettema *et al.*, 1999, 2000). More recent studies have shown that *PEX3* and *PEX19* appear to be required for the assembly of peroxisomal membrane proteins (PMPs) (Hettema *et al.*, 2000). This implies that other *PEX* genes are responsible for matrix protein localisation. The human orthologue of the farnesylated yeast protein, Pex19 (Gotte *et al.*, 1998), has been shown to be required for peroxisomal membrane import. Strains lacking *PEX3* and *PEX19* are likely to be useful for investigating the role of chaperones in targeting and insertion of PMP.

Pex19 is an oleate-inducible protein and shares this property with all three β-oxidation enzymes. Oleate-responsive elements (ORE) are found in their promoters and those of many other peroxisomal genes. The ORE consists of a $CGGN_{17-18}CCG$ motif (Karpichev and Small, 1998). The transcriptional regulation is brought about by binding of the heterodimer consisting of Oaf1 and Pip2 to ORE. Oaf1, like Gal4 contains a Zn[2]-Cys[6] cluster motif and receives the oleate signal. Under non-inducing conditions Oaf1 is inactive and inhibits Oaf1/Pip2 activity. In the presence of oleate Oaf1 activates the heterodimer and as a result of this release of inhibition peroxisomal protein genes are expressed (Baumgartner *et al.*, 1999). The Oaf1/Pip2 regulatory mechanism requires at least one additional transcription factor and Adr1 has been identified as important for yeast growth on fatty acids (Gurvitz *et al.*, 2000). Cells lacking Adr1 do not grow on fatty acids and they have a deficiency in transcribing genes encoding peroxisomal proteins.

In addition, two glucose-responsive elements in the promoter of *POX1/FOX1* are involved in glucose repression of this gene. In the *POT1/FOX3* promoter the glucose-responsive element consists of overlapping Abf1 and replication protein A (RP-A) binding sites which binds the *UME6* gene product (Einerhand *et al.*, 1995). It would appear that the coordinated regulation of the expression of genes involved in peroxisomal function in response to glucose is mediated by proteins associated with the RP-A site. Three proteins, Abfl, RP-A and Ume6, are necessary for glucose repression. Adrl, Snfl and Snf4 function as global regulators for the derepression of glucose-repressible genes and act as positive regulators of *FOX2/POX2*, *POT1/FOX3* – albeit each to a different extent – and catalase A encoded by *CTA1*. Cta1 is responsible for coping with H_2O_2, released by the first reaction catalysed by Poxl/Fox1 (Trotter, 2001).

An interesting link between mitochondria and peroxisomes is provided by the RTG transcription factors (Komeili *et al.*, 2000; Epstein *et al.*, 2001). The three genes *RTG1, RTG2* and *RTG3* were identified as being required for the expression of Cit2, a peroxisomal isoform of citrate synthase, under conditions of mitochondrial respiratory insufficiency. The RTG transcription factors have also been shown to be involved in the expression of genes encoding peroxisomal proteins and glyoxylate cycle enzymes. It would appear that RTG-dependent gene expression links mitochondrial and peroxisomal functions (Komeili *et al.*, 2000). Rtg1 and Rtg3 bind as a heterodimer to the upstream region of the *CIT2* gene causing a dramatic increase in *CIT2* expression in rho-zero cells (Sekito *et al.*, 2000). Rtg2 seems to be required for the transfer of Rtg1 and Rtg3 from the cytoplasma to the nucleus.

Saccharomyces cerevisiae as mentioned above, is capable of degrading saturated fatty acids in a multistep process catalysed by Pox1/Fox1, Fox2/Pox2 and Pot1/Fox3. For unsaturated intermediates, however, the classical β-oxidation pathway can catabolise only MUFAs containing a double bond in the *trans*-configuration at the C2-position. Therefore, the participation of auxiliary enzymes are required for the removal of *cis*-double bonds. The degradation of MUFAs requires two auxiliary enzymes since neither *trans*-2,4-dienoyl CoA nor *cis*-2,4-dienoyl CoA is a substrate for *trans*-2-enoyl CoA hydratase-2 (Fox2/Pox2): 2,4-dienoyl CoA reductase (Sps19) or 3,2-enoyl CoA isomerase (Eci1) (Gurvitz *et al.*, 2001). For the degradation of fatty acids containing even-numbered *cis*-double bonds, for example petroselinoyl CoA (*cis*-C18:1,$\Delta 6$) the problem is solved by using 2,4-dienoyl CoA reductase to yield 3-*trans*-enoyl CoA which is then converted by the 3,2-enoyl CoA isomerase into 2-*trans*-enoyl CoA (Figure 6.1). The former enzyme is encoded by the oleate-inducible gene *SPS19* which was cloned in a screen for genes expressed during sporulation (Gurvitz *et al.*, 1997). The deduced amino acid sequence of *SPS19* contains a PTS1 sequence and the promoter of the gene contains potential OREs as well as a canonical Adr1-binding site (Gurvitz *et al.*, 1999a; 2000). These sequence signatures underline the fact that β-oxidation enzymes in *S. cerevisiae* are found exclusively in peroxisomes. The peroxisomal $NADP^+$-dependent isocitrate dehydrogenase, Idp3 (Figure 6.1), is involved in regeneration of NADPH consumed by Sps19 (Geisbrecht *et al.*, 1998; Henke *et al.*, 1998). Interestingly, Sps19 is not essential for ascosporogenesis on acetate and oleate media but is essential for this process to occur on medium containing petroselinate either because the strain mutated for *SPS19* cannot obtain energy for completion of sporulation or because of the potential toxicity of the substrate (Gurvitz *et al.*, 1997).

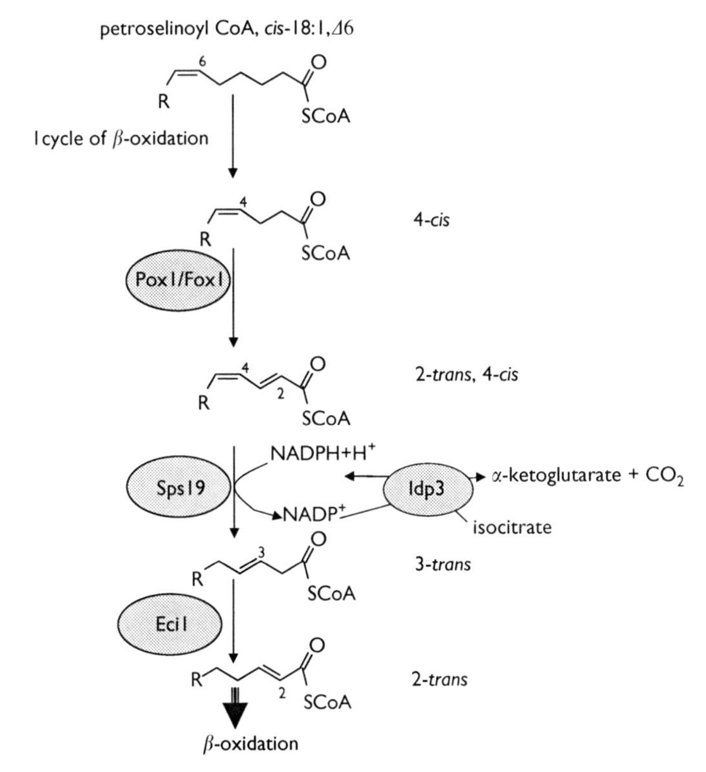

Figure 6.1 β-oxidation of petroselinoyl CoA.

The breakdown of MUFAs with a *cis*-double bond at an odd-numbered position (oleic acid: *cis*-C18:1,$\Delta 9$) can succeed in either of two ways (Figure 6.2): (i) with Eci1 alone or (ii) by the NADPH-dependent reductive pathway using Sps19, Eci1 and Dci1 (3,5-2,4-dienoyl CoA isomerase) in concert with the core set of β-oxidation enzymes, Pox1/Fox1, Fox2/Pox2 and Pot1/Fox3. It has been shown that *DCI1* is dispensable for the utilisation of oleoyl CoA since a strain lacking *DCI1* can still grow on oleate and a strain deleted for *ECI1* and *DCI1* grows as poorly as a strain carrying only a mutated version of *ECI1* (Gurvitz *et al.*, 1999b; 2001). The observation that *sps19Δ* strains are capable of growth on oleate is consistent with the fact that 2,4-dienoyl CoA reductase is not required for the breakdown of oleic acid. 3,2-enoyl CoA isomerase encoded by *ECI1* can cope with both 3-*cis* and 3-*trans*-enoyl CoA derivatives converting them to 2-*trans*-enoyl CoA which can then be degraded *via* Fox2/Pox2 and Pot1/Fox3 (Geisbrecht *et al.*, 1998; Gurvitz *et al.*, 1998; Henke *et al.*, 1998). Interestingly, Eci1 contains both PTS1 and PTS2 sequences and is targeted to the peroxisome by its own PTS1 or by dimerising with Dci1 and making use of that protein's PTS1 (Yang *et al.*, 2001). The PTS2 sequence of Eci1 is apparently non-functional (Gurvitz *et al.*, 2001).

An understanding of the metabolism of *trans*-MUFAs is important because it has been shown that diets high in *trans*-unsaturated fatty acids may increase the risk of coronary heart disease (Mensink *et al.*, 1992). *S. cerevisiae* can utilise *trans*-double bonds at even-numbered positions in the absence of known auxiliary enzymes (Gurvitz *et al.*, 2001).

The question remains as to how fatty acids, both saturated and unsaturated, are transported into the peroxisome. Entry into the peroxisomes for LCFAs is dependent on the peroxisomal

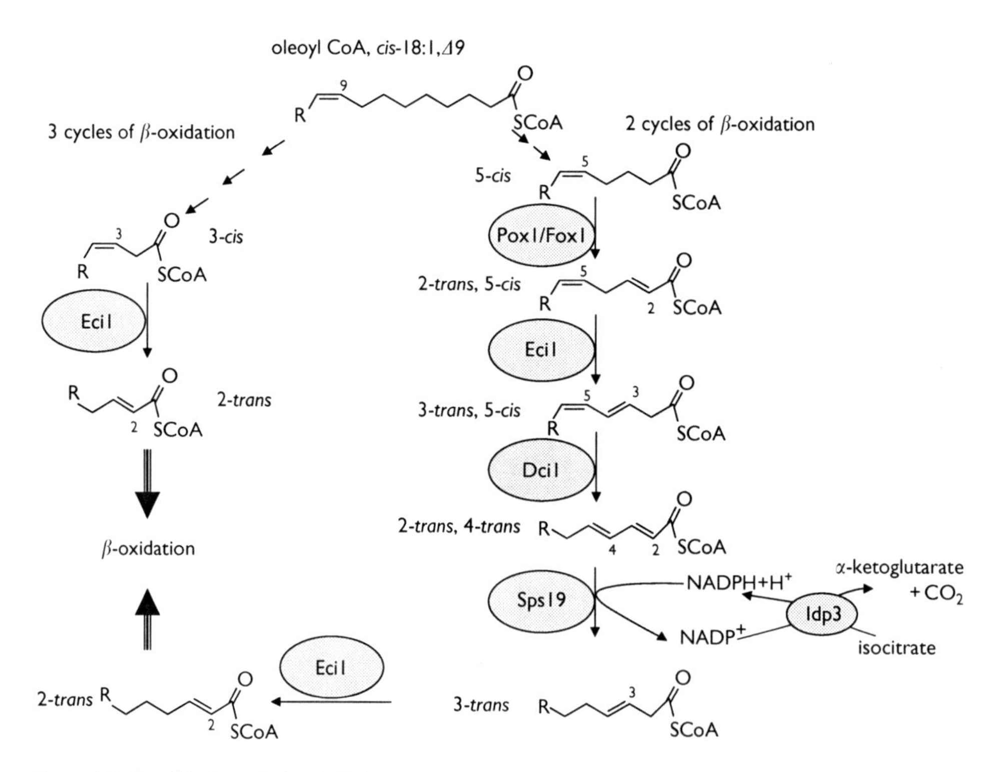

Figure 6.2 β-oxidation of oleoyl CoA.

transporters Pxa2/Patl and Pxa1/Pat2 (Tabak *et al.*, 1995). Disruption of either *PXA2/PAT1* or *PXA1/PAT2* or both genes leads to latency of LCFA β-oxidation. The transport systems for MCFA and LCFA overlap partially since Faa2, although its preferred substrate is laureate (C12:0), does have a broad substrate specificity (Knoll *et al.*, 1994) and is able to activate oleate because of a limited oleoyl CoA synthetase activity as illustrated by the fact that *pxa2/pat1Δ faa2Δ* or *pxa1/pat2Δ faa2Δ* mutants, in contrast to *pxa2/pat1Δ* or *pxa1/pat2Δ* strains, have no capacity for the β-oxidation of oleate (Hettema *et al.*, 1996; Hettema and Tabak, 2000). From this it can be concluded that fatty acids can cross the peroxisomal membrane by two parallel pathways, for example, as MCFA or LCFA in an *FAA2*-dependent system which may well require the involvement of Fat2. It has been suggested that Fat2/Psc60 which belongs to the same family of the AMP-binding proteins as Fatl (Schjerling *et al.*, 1996), may have an accessory function necessary for the intraperoxisomal fatty acyl activation by Faa2 (Johnson *et al.*, 1994). The sequence of Fat2/Pcs60 contains a PTS1 peroxisomal targeting signal and the protein is localised both in the peroxisomal membrane and matrix. The other pathway for LCFA is taken only by activated fatty acids and is dependent on *PXA2/PAT1* and *PXA1/PAT2*. Pxa1/Pat2 is the yeast orthologue of the human ALD protein which is also an ABC (ATP-binding cassette) transporter and when defective as in X chromosome-linked ALD causes a neuro-degenerative disorder with impaired peroxisomal oxidation of LCFAs (Foury, 1997).

The fact that the peroxisomal membrane in *S. cerevisiae* is impermeable to both NAD(H) and acetyl CoA raises the question of how do peroxisomes oxidise NADH to NAD^+ during β-oxidation of fatty acids and when oxidation is complete how does acetyl CoA leave the peroxisome? There are three malate dehydrogenases in *S. cerevisiae*, one of which, Mdh3 has a PTS1 targeting sequence and it is the activity of this enzyme which converts oxaloacetate to malate by oxidation of NADH to NAD^+ (McAlister-Henn *et al.*, 1995; van Roermund *et al.*, 1995). The reaction carried out by Mdh3 is the reverse of that carried out in the glyoxylate cycle (Lazarow and Kunau, 1997). Deletion of *MDH3* prevented growth on oleate and resulted in impaired β-oxidation capacity in intact cells but β-oxidation was normal in cell lysates. This is not the only 'reverse' reaction found in peroxisomes in *S. cerevisiae* – α-ketoglutarate and aspartate produce glutamate and oxaloacetate rather than glutamate and oxaloacetate being used to produce α-ketoglutarate and aspartate (van Roermund *et al.*, 1995; van der Klei and Veenhuis, 1997). Therefore, a shuttle system has to be postulated that requires α-ketoglutarate/malate and glutamate/aspartate carriers. *S. cerevisiae* is different from non-conventional yeasts in that its complete glyoxylate cycle is not peroxisomal bound. However, it has been shown that the *AAT2* gene which codes for cytosolic and peroxisomal aspartate aminotransferase is not essential for growth on oleate or for β-oxidation suggesting that Aat2 is not necessary for the peroxisomal NAD(H) redox shuttle (Verleur *et al.*, 1997).

The transport of acetyl CoA from the peroxisome is dependent on two pathways, one in which it is transferred to the carrier molecule, carnitine by the *CAT2*-encoded carnitine-O-acetyltransferase active in peroxisomes and mitochondria. This protein is interesting in that it carries a mitochondrial-targeting sequence at its N-terminus and a slightly altered PTS1 signal at its C-terminus. Furthermore, it has an internal peroxisomal targeting sequence which is revealed on deletion of both the N- and C-terminal motifs. In cells grown on oleate, *CAT2* transcripts initiate at the second AUG, bypassing the mitochondrial targeting signal (Elgersma *et al.*, 1995).

The second pathway depends on peroxisomal Cit2 activity which converts acetyl CoA and oxaloacetate into citrate (van Roermund *et al.*, 1995). Disruption of either *CIT2* or

CAT2 does not impair the peroxisomal β-oxidation in *S. cerevisiae*. However, the double null mutant cannot grow on oleate because acetyl CoA accumulates in the peroxisome and brings β-oxidation to a standstill.

The sequestering of the β-oxidation enzymes in peroxisomes increases the effectivity of this metabolic pathway and by taking advantage of the culturing techniques that are typically employed for studies of peroxisomal biogenesis, that is growth in media containing fatty acids as a sole carbon source it has been shown that there is extensive uptake of these fatty acids from the media and a subsequent increase in total cellular lipid content from 2% to 15% dry cell weight (Dyer *et al.*, 2002).

6.5 The roles of phospholipids in yeast physiology and membrane homeostasis

Phospholipid synthesis in *S. cerevisiae* is dependent on water-soluble precursors, nucleotides, in particular CTP, lipids and growth phase (Carman and Henry, 1999) for regulating both the mRNA and protein levels and activity of the biosynthetic enzymes. The biosynthetic routes of the various phospholipid species differ, for example, PtdEtn and PtdCho are synthesised by the coupling of diacylglycerol (DAG) with CDP-ethanolamine and CDP-choline, respectively in the so-called Kennedy or *salvage* pathway (Figure 6.3) (Henry and Patton-Vogt, 1998). On the other hand, phosphatidyl serine (PtdSer) and phosphatidylinositol (PtdIns) are synthesised *de novo* from CDP-DAG: PtdSer is produced by the exchange of CMP from CDP-DAG for L-serine and PtdIns is the product when inositol replaces CMP in CDP-DAG. These two branches of the phospholipid biosynthetic pathway (Zinser *et al.*, 1991) are linked by the fact that PtdSer can be decarboxylated to PtdEtn which in turn can be methylated to produce PtdCho. Inspection of the phospholipid biosynthetic pathways implies that a coordinate regulation of the participating enzymes has to exist and this has

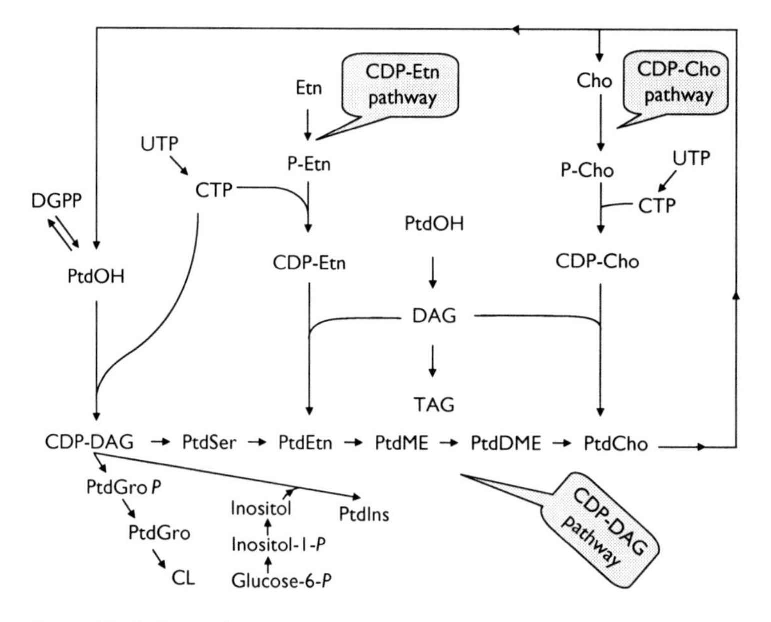

Figure 6.3 Pathways for phospholipid synthesis (adapted from Carman and Henry, 1999).

been shown for phosphatidyl serine (PS) synthase and phosphatidic acid (PA) phosphatase which converts phosphatidic acid (PtdOH) to DAG. PS synthase encoded by *CHO1/PSS1* catalyses the committed step for the synthesis of PtdSer from CDP-DAG (Carman and Henry, 1999). PtdOH activates PS synthase whereas CDP-DAG, PtdIns and DGPP (diacyl-glycerol pyrophosphate) activate PA phosphatase (Bae-Lee and Carman, 1990). In addition, DAG and cardiolipin (CL) inhibit PS synthase (Carman and Henry, 1999). Therefore, the differential regulation of PS synthase, PA phosphatase and CDP-DAG synthase from PtdOH and CTP plays a central role in partitioning of PtdOH to the different branches of phospholipid and neutral lipid biosynthetic pathways.

The membrane phospholipids of *S. cerevisiae*, in descending order of abundance, are PtdCho, PtdEtn, PtdIns and PtdSer (Voelker, 2000). These phospholipids play important roles in various cellular processes including signal transduction in addition to being structural components of cellular membranes (Carman and Henry, 1999; Howe and McMaster, 2001). Mitochondrial membranes contain, in addition to the above-mentioned phospholipids and CL, phosphatidylglycerol (PtdGro) (Daum and Vance, 1997). Phospholipid composition can vary dramatically according to culture conditions. For instance, it was found that during the transition to stationary phase the PtdCho level increases at the expense of PtdEtn. This increase in PtdCho is most pronounced in the presence of lactate. Lactate-grown cells also contain higher levels of PtdIns and CL (Tuller *et al.*, 1999). One explanation for the increase of PtdCho at the expense of PtdEtn could be that prior to stationary phase the bulk of cellular methionine is used for protein synthesis, however, once the cells are in stationary phase methionine becomes available for methylation. An alternative explanation is that during exponential growth PtdEtn may be preferred for cellular membranes whereas in the stationary phase it is PtdCho. The increased level of PtdIns may be caused by the reduction in PS synthase activity which accompanies the transition to stationary phase allowing CDP-DAG to be used for the synthesis of PtdIns (Janssen *et al.*, 2000). In a high throughput assay (DeRisi *et al.*, 1997) it has been shown that in glucose-depleted medium the expression of *INO1* which encodes inositol-1-phosphate synthase increases at least 2-fold which could also explain the increase in PtdIns content.

High levels of PtdEtn in mitochondria correlate with the existence of a mitochondrial localisation of PS decarboxylase (Psd1) (Figure 6.4). In *psd1Δ* strains the growth rate on non-fermentable carbon sources correlates with the mitochondrial content of PtdEtn suggesting that the import of PtdEtn into the mitochondria is growth limiting but this does not cause obvious morphological and biochemical defects. However, GPI-anchor synthesis must be compromised since Gas1 (GPI-anchored surface glycoprotein) maturation was delayed although neither carboxypeptidase Y nor invertase processing was affected. This implies that high levels of non-bilayer forming lipids, for example, PtdEtn, are not essential for membrane vesicle fusion processes *in vivo*, possibly as yet unknown compensatory effects may be responsible for the maintenance of the biophysical properties of the membranes (Birner *et al.*, 2001). This is in contrast to the situation in bacteria where PtdEtn is involved in vesicle-mediated protein trafficking (de Kruijff, 1997).

In the absence of exogenous ethanolamine and choline aminoglycerophospholipids, PtdSer, PtdEtn and PtdCho are synthesised *de novo* from L-serine and CDP-DAG in the ER or mitochondria-associated membrane (MAM) and PtdSer is then subsequently transported to the mitochondria or to the Golgi/vacuole for conversion to PtdEtn by Psd1 and Psd2, respectively (Figure 6.4). A *psd1Δ* strain which has only 10% of PS decarboxylase activity is auxotrophic for ethanolamine when grown at 37 °C or on lactate. PtdEtn, unlike PtdCho

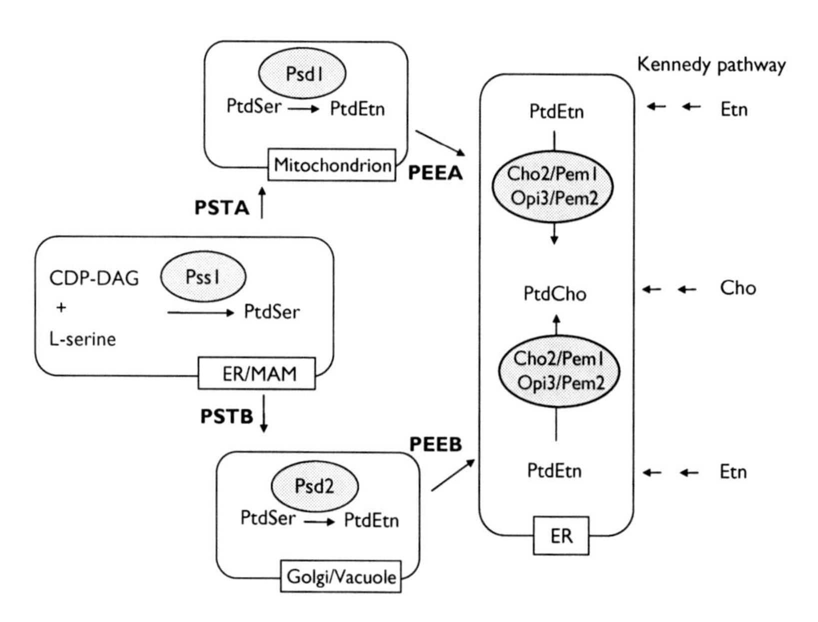

Figure 6.4 Aminoglycerophospholipid transport (adapted from Voelker, 2003).

and PtdSer which are bilayer lipids, has an essential function which is independent of its ability as a non-bilayer lipid to form hexagonal phase structures. This conditional ethanolamine auxotrophy is not due to a decrease in the Psd2 activity located in the Golgi/vacuole compartment. Measurement of phospholipid pools showed that the PtdEtn content of a *psd1Δ* strain was drastically reduced in the absence of ethanolamine; the addition of ethanolamine increased the PtdEtn pool suggesting that PtdEtn produced by the Kennedy pathway cannot fully compensate for the lack of Psd1 activity. The PtdCho pools in the *psd1Δ* strain were unaffected and therefore did not contribute to the ethanolamine auxotrophy. A strain deleted for *PSD1* is an ethanolamine auxotroph at 35 °C and PtdEtn content was reduced by 50% compared with wild-type cells at 30 °C (Storey *et al.*, 2001a).

A *psd1Δ psd2Δ dpl1Δ* triple mutant is an ethanolamine auxotroph which cannot be rescued by choline, clearly demonstrating that either PtdEtn or one of its metabolites is essential. Dihydrosphingosine-1-phosphate lyase (Dpl1) permits low levels of PtdEtn synthesis without ethanolamine supplementation. As stated above PtdEtn is required for many membrane-associated functions, for example, GPI anchor addition. It would appear that PtdEtn carries out an essential cellular function which is independent of its physical properties (Storey *et al.*, 2001a). A similar genetic screen but using a *psd1*^{ts} allele revealed mutations in *PHB1* and *PHB2* encoding the two subunits of the prohibitin complex. This complex is required for stability of mitochondrially encoded proteins and is located in the inner mitochondrial membrane. Strains lacking *PHB1* and *PHB2* contain increased amounts of mitochondrial PtdEtn. Deletants of *PSD1* in combination with deletion of either *PHB1* or *PHB2* when grown on glucose can only survive for a few generations when the medium is supplemented with ethanolamine suggesting that the level of PtdEtn is important for prohibitin mutants. Both *phb* mutants and *psd1Δ* strains exhibit destabilisation of mitochondrially encoded polypeptides; in *phb1Δ phb2Δ psd1*^{ts} strains the destabilisation phenomenon is so drastic that the mitochondrial genome is lost and assembly of nuclear-encoded proteins of the mitochondrial inner membrane (MIM) is impaired. Therefore low levels of mitochondrial PtdEtn are likely to be responsible for the synthetic lethality of *psd1Δ* and prohibitin mutants (Birner *et al.*, 2003).

All the genes encoding the proteins necessary for aminoglycerophospholipid synthesis are known. It is postulated that specific gene products are involved in the interorganelle transport of specific lipids; mutants thereof are designated *pst* (phosphatidylserine transport) and *pee* (phosphatidylethanolamine export) (Figure 6.4) (Voelker, 2000). Potential lipid transport mechanisms are: (i) monomer diffusion, (ii) lipid transfer protein, (iii) vesicle traffic, (iv) membrane apposition and (v) fusion. To isolate a mitochondrial phospholipid transport mutant a *psd2Δ* strain was used to select for ethanolamine auxotrophy on a non-fermentable carbon source following mutagenesis. Among the mutants isolated *pstA1-1* has a defect in transport between MAM and the mitochondrial outer membrane (MOM). Areas of close physical association exist between MAM and the mitochondria as shown by morphological and biochemical experiments (Schumacher *et al.*, 2002). The *pstA1-1* mutant is complemented by *MET30* which encodes a protein belonging to the SCF (suppressor of kinetochore protein 1, cullin, F-box) family of ubiquitinated ligases. Therefore, ubiquitination is closely allied to the regulation of PtdSer transport to the mitochondria. These data identify a novel mechanism involving ubiquitination in the regulation of interorganelle phospholipid transport. Reconstitution experiments with MAM and mitochondria from wild type and *pstA1-1* strains reveal that the mutation affects both MAM and mitochondria raising the intriguing possibility that targets in both organelles are ubiquitinated and form the basis of areas of apposition between them (Voelker, 2003).

Using the same genetic screen for ethanolamine auxotrophy with a *psd1Δ* strain a novel mutation in the morphogenesis checkpoint-dependent gene *MCD4* was identified (Storey *et al.*, 2001b). The ethanolamine auxotrophy is not due to a defect in GPI anchor synthesis but reveals important interactions between Psd1 and Mcd4 and suggests that the latter may have a role in aminoglycerophospholipid metabolism.

Genes of the PSTB pathway (Figure 6.4) were found in a similar genetic screen using *psd1Δ* strains. The ethanolamine auxotrophy of a *pstB1* mutant was complemented by *STT4* which encodes 1-phosphatidylinositol (PI) 4-kinase activity implying that PtdIns(4)P or PtdIns(4,5)P_2 are involved in the regulation of PtdSer transport to the Golgi/vacuole compartment (Trotter *et al.*, 1998). A further mutation, *pstB2* was complemented by the Sec14 homologue *PDR17* (pleiotropic drug resistance 17) implying lipid transfer exchange in the delivery of PtdSer to the Golgi/vacuole. Pstb2 is found in the soluble fraction and tightly associated with membranes (Wu *et al.*, 2000). Reconstitution experiments showed that Pstb2 must be present on the acceptor membrane in order for PtdSer transport to occur from the donor. Further experiments revealed that the C2 domain of Psd2 which is known to bind *inter alia* PtdIns(4,5)P_2 (Chung *et al.*, 1998) but is not required for Psd2 activity, is essential for the transfer of nascent PtdSer from the donor to the acceptor membrane (Mehrotra *et al.*, 2000; Voelker, 2003).

Auxotrophy for the other water-soluble phospholipid precursor, choline does not appear to involve interorganelle transport. Choline auxotrophy arises when the methylation pathway is disrupted but not when the genes in the Kennedy pathway are mutated (McMaster and Bell, 1994). Choline kinase is a cytosolic enzyme that catalyses the committed step in the Kennedy pathway for the synthesis of PtdCho by the CDP-choline pathway (Figure 6.3). Loss of regulation of this enzyme is thought to play a role in the generation of human tumours by Ras oncogenes (Yu *et al.*, 2002). Choline kinase activity could therefore be a marker for cancer and a target for anticancer drugs. Furthermore, protein kinase A phosphorylation of choline kinase at Ser(30) and Ser(85) increases the activity of this enzyme and this is confirmed by mutations of Ser(30) and Ser(85) resulting in a decrease of phosphorylation of the choline kinase protein *in vivo*. These findings indicate that protein kinase A phosphorylation of choline kinase regulates PtdCho synthesis in the Kennedy pathway. Interestingly, mutation of these phosphorylation sites does not suppress the *sec14*^{ts} phenotype or cause choline excretion; possibly because PtdCho synthesis is not stopped and the overall PtdCho content was not significantly altered. This is consistent with results showing that the overall PtdCho content rather than PtdCho synthesised by either the Kennedy or the methylation pathway is important for Golgi-secretory function (Xie *et al.*, 2001). These phosphorylation site mutants of choline kinase will be a useful tool for studying the regulation of phospholipid synthesis and the mechanism by which the RAS-cAMP pathway mediates cell growth.

There are two ORFs *GAT1* (YKR067w) and *GAT2* (YBL011w), previously known as *SCT1* (choline transporter suppressor) (Zheng and Zou, 2001). These gene products acylate both glycerol-3-phosphate (Gro3*P*) and dihydroxyacetone phosphate (DHAP) in a stereospecific manner at the *sn-1* position. Surprisingly, although simultaneous deletion of *GAT1* and *GAT2* is a lethal event, deletion of either gene causes no growth defect although the PtdOH pool is decreased and the PtdSer/PtdIns ratio is increased. Gat1 utilises a broad range of fatty acyl CoAs as acyl donors whereas Gat2 has a preference for fatty acyl CoAs with a chain length of 16-carbon fatty acids. Gat1 is found in both lipid particles and the ER whereas Gat2 is associated mainly with the ER. The alternative pathway for the production of lyso-PtdOH (1-acyl-Gro3*P*) is the conversion of DHAP to lyso-PtdOH by the enzymes DHAP acyltransferase (DHAPAT) and 1-acyl-DHAP reductase (ADR) (Athenstaedt and Daum, 1999, 2000; Athenstaedt *et al.*, 1999a). In animal cells the DHAP pathway is obligatory for ether lipid biosynthesis (Athenstaedt *et al.*, 1999a). It was shown by lipid mass spectrometry that there are trace amounts of alkylether lipids present in yeast (Schneiter *et al.*, 1999a) and there is experimental evidence that a mitochondrial DHAPAT activity may be responsible for the shift to the DHAP pathway in the absence of Gat1. In mitochondria DHAP rather than Gro3*P* is acylated and this may explain why there is a greater contribution of the DHAP pathway to the synthesis of CL in this organelle. Since 1-acyl-DHAP has to be reduced by ADR (Ayr1), an activity not present in the mitochondria one has to invoke transport of 1-acyl-DHAP from the mitochondria to a site of ADR activity, most likely the so-called MAM fraction (Gaigg *et al.*, 1995; Athenstaedt *et al.*, 1999a). Ayr1 activity is also found in lipid particles. Deletion of *AYR1* affects neither lipid composition nor growth in a haploid strain but its overexpression results in growth arrest (Athenstaedt and Daum, 2000). However, the enzyme does appear to have an essential role in the germination of spores. It would appear that, in addition to Ayr1, there is another ADR isoenzyme in the microsomal fraction.

Lyso-PtdOH synthesised by either of these pathways is then acylated a second time by Slc1 (Nagiec *et al.*, 1993; Athenstaedt *et al.*, 1999a) to produce PtdOH. *SLC1* was originally isolated as a suppressor of an essential step in sphingolipid biosynthesis in which a C16/C18 fatty acyl chain is introduced at the *sn-2* position of a glycerophospholipid (Athenstaedt and Daum, 1997).

This biosynthetic pathway for PtdOH provides a good example of interorganelle transport and possible membrane contact being an integral part of lipid metabolism. In a strain containing deletions of both *GAT1* and *GAT2* and a plasmid-borne copy of *GAT1* under the control of the galactose promoter Gro3*P* acyltransferase levels dropped by 98% whereas lyso-PtdOH acyltransferase, encoded by *SLC1*, remained unchanged. Following transfer to glucose-containing medium the *de novo* fatty acid pool in such a strain is used for the synthesis of steroyl esters rather than glycerophospholipids suggesting that there is a residual GAT activity. There are several ORFs whose products contain an acyltransferase motif and which could step in and take over the acylation of Gro3*P* under the extenuating circumstances of a double *GAT1/GAT2* deletion (Zaremberg and McMaster, 2002).

TAG synthesis was reduced by 50% in a *gat2Δ* strain with a concomitant decrease in PtdCho deacylation. This implies that PtdCho produced by the Kennedy pathway is preferentially deacylated when it is produced from Gat2-synthesised DAG whereas Gat1-synthesised DAG is less susceptible. The production of glycerophosphocholine (Gro*P*Cho) in *gat1Δ* and *gat2Δ* strains occurs *via* the action of PLB encoded by *PLB1-3* (Fyrst *et al.*, 1999; Merkel *et al.*, 1999). PLB proteins are located in the plasma membrane or periplasmatic space and their activity releases Gro*P*Cho into the medium. However, Gro*P*Cho from *gat1Δ* and *gat2Δ* strains was intracellular. Intracellular Gro*P*Cho is produced independently of PLB-encoded lipases, implying the existence of a hitherto unidentified lipase capable of PtdCho deacylation in *gat1Δ* and *gat2Δ* strains (Zaremberg and McMaster, 2002).

In addition to deacylation by PLB PtdCho can be metabolised in two other ways: (i) head group removal *via* phospholipase D to give choline and PtdOH which is dephosphorylated to DAG (Figure 6.3) (Howe and McMaster, 2001) or (ii) *via* the action of Lro1 (PC:DAG transferase) to produce TAG from DAG and PtdCho (Dahlqvist *et al.*, 2000; Oelkers *et al.*, 2000; Sandager *et al.*, 2002). In fact there are at least four enzymes capable of DAG acylation (Figure 6.5). TAG can be synthesised through the action of Dga1 which acylates DAG using acyl CoA as a donor. The *S. cerevisiae* gene encoding acyl CoA:DAG acyltransferase (DAGAT) was found on the basis of sequence similarity to a DAGAT recently detected in *Mortierella ramanniana* (Lardizabal *et al.*, 2001). This reaction produces about 30% of the cellular content of TAG whereas Lro1 contributes 40%. The remaining DAG acylation

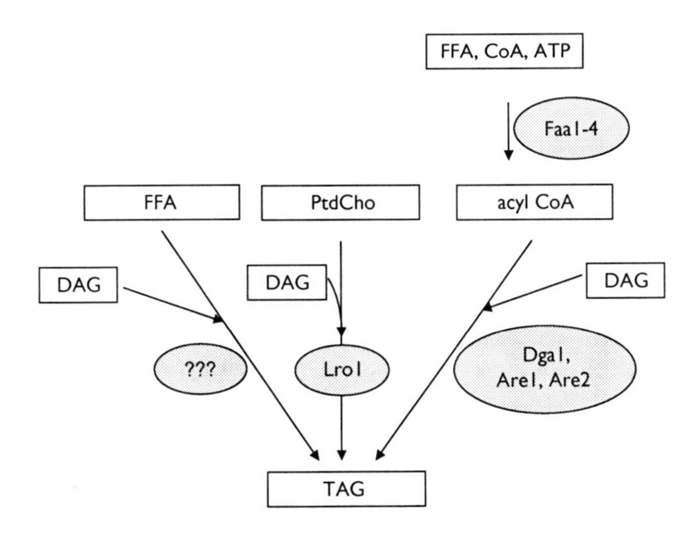

Figure 6.5 Pathways of TAG synthesis (redrawn from Sorger and Daum, 2002).

activity appears to be provided by the products of the *ARE1* and *ARE2* genes which encode steroyl esterases (ACAT, acyl CoA:cholesterol-*O*-acyl transferase-related enzymes) since a *dga1 lro1 are1 are2* strain is virtually devoid of TAG. This suggests that lipid storage and lipid bodies are dispensable in yeast (Sandager *et al.*, 2002). An acyl CoA-independent TAG synthesis pathway in lipid particles and microsomes which is independent of Dga1 and Lro1 utilises free fatty acids for TAG formation without the requirement of ATP (Figure 6.5) (Sorger and Daum, 2002).

The conversion of PtdOH to DAG is a dephosphorylation step carried out by two Mg^{2+}-dependent phosphatase species, the 104 and 45 kDa enzymes, however, the cognate genes have not yet been identified. It is postulated that these activities will be responsible for phospholipid and TAG synthesis (Figure 6.3) whereas the Mg^{2+}-independent phosphatases are likely to play a role in lipid signalling (Carman, 1997; Furneisen and Carman, 2000). Two Mg^{2+}-independent DGPP phosphatases which convert DGPP to PtdOH have been isolated from *S. cerevisiae* and are encoded by *DPP1* and *LPP1*, both located on chromosome IV (Carman and Henry, 1999). PA and the novel phospholipid DGPP are interconvertible due to the reactions of PA kinase and DGPP phosphatase. DGPP phosphatase encoded by *DPP1* is a promiscuous enzyme accepting lyso-PtdOH, ceramide-1-phosphate, PtdGro*P* and isoprenoid phosphates as substrates (Faulkner *et al.*, 1999). Purified DGPP phosphatase catalyses the removal of the β-phosphate of DGPP to yield PtdOH followed by removal of the phosphate from PtdOH to yield DAG. The second Mg^{2+}-independent PA phosphatase, Lpp1 prefers PtdOH as a substrate but will also utilise lyso-PtdOH and DGPP. Neither of these genes is essential and the double deletant has no obvious growth defects. However, in *lpp1* and *dpp1* mutants the cellular levels of PtdIns and PtdOH are altered (Toke *et al.*, 1998).

Inositol and growth phase, both of which are known to affect the expression of several phospholipid biosynthetic enzymes were tested for their influence on Dpp1 and were found, in contrast to their negative influence on most phospholipid-biosynthetic enzymes, to exert a positive influence on Dpp1 activity (Oshiro *et al.*, 2000). These regulatory effects are UAS_{INO} (inositol-sensitive upstream activation sequence)-independent since no such *cis*-element is found in the promoter of *DPP1*. This is not the only case in which inositol or growth phase regulates enzymes in a fashion opposite to the UAS_{INO}-mediated pathway; other examples are inositol-1-phosphatase (Murray and Greenberg, 1997) and the 45 kDa Mg^{2+}-dependent PA phosphatase (Morlock *et al.*, 1988). In spite of the absence of a UAS_{INO}-element the gene products Opi1, Ino2 and Ino4 are required for the inositol-regulated expression of Dpp1 but not for the growth phase-dependent regulation (Oshiro *et al.*, 2000).

PtdOH is also the substrate for CDP-DAG synthase. The gene, *CDS1*, encoding this enzyme is essential for vegetative growth and spore germination (Shen *et al.*, 1996), not surprisingly since the product of the reaction, the liponucleotide intermediate CDP-DAG, is the starting material for the synthesis of PtdIns and PtdCho *via* PtdSer and PtdEtn, not forgetting CL (Figure 6.3). In accordance with its central biosynthetic role Cds1 has a multi-organelle distribution and is found in the ER, the cytoplasmic faces of MOM and MIM and it is also enriched in post-Golgi secretory vesicles and the plasma membrane. At high levels of CDP-DAG synthase activity the synthesis of PtdIns is favoured over that of PtdSer whereas the reverse is true at low levels of activity. Reduced CDP-DAG synthase levels also result in inositol secretion (Shen and Dowhan, 1996), one of the phenotypes observed in the original *cds1* mutant. This mutation also exhibited a reduction both in the amount and activity of CDP-DAG synthase, an elevated level of PtdOH, constitutive expression of inositol-l-phosphate and an increase in PS synthase protein and activity. The intracellular concentration of CDP-DAG influences not only the level of gene expression but also enzyme stability. The steady-state levels of PS synthase and *INO1* mRNA are inversely proportional to the activity of CDP-DAG synthase levels but PI synthase activity falls off when CDP-DAG synthase is reduced (Carman and Zeimetz, 1996; Shen and Dowhan, 1997).

PtdIns is the starting point for the pathways of phosphoinositide metabolism, products of which are important for signal transduction and play a role in the anchoring of proteins in the plasma membrane. PtdIns is synthesised from CDP-DAG and *myo*-inositol by PI synthase (Figure 6.3) which is localised in MOM and microsomes and is encoded by *PIS1*, an essential gene (Nikawa and Yamashita, 1997). It is one of the three enzymes which uses CDP-DAG as a substrate for phospholipid synthesis. Furthermore, derivatives of PtdIns, for example, sphingolipids and polyphosphoinositides, are essential metabolites. The regulation of this enzyme is an exception to the canonical inositol/choline regulation of the phospholipid-synthesising enzymes since its expression is constitutive (Henry and Patton-Vogt, 1998). The activity of PI synthase but not its mRNA level is regulated by CDP-DAG synthase. The transcription of PI synthase is repressed in the presence of glycerol as the sole carbon source and induced by galactose, confirming the functionality of at least one of the Mcm1-binding sites in its promoter. The galactose-mediated induction requires the product of the *SLN1* gene which is a two-component regulator (Posas *et al.*, 1996; Nikawa and Yamashita, 1997).

Inositol which in addition to CDP-DAG is required for the synthesis of PtdIns is produced from inositol-1-phosphate, the product of the *INO1* gene in the first committed step of the biosynthesis of inositol-containing compounds linking carbon and phospholipid metabolism since it converts glucose-6-phosphate to inositol-1-phosphate (Figure 6.3). The *INO1* gene is tightly regulated at the level of its transcription and activity (Majumder *et al.*, 1997).

PS synthase carries out the synthesis of PtdSer from CDP-DAG and L-serine. This pathway exists in prokaryotes and yeast but not in animals or plants where PtdSer is synthesised by a base-exchange reaction between PtdEtn or PtdCho and L-serine (Carman and Henry, 1999). The subcellular localisation of the protein encoded by *CHO1/PSS1* is MOM and microsomes. This is in agreement with the localisation of the following enzyme PS decarboxylase (Psd1) found at the contact sites between MOM and MIM. Mutants of *CHO1/PSS1* are choline or ethanolamine auxotrophs and they have both mitochondrial and vacuolar defects (Henry and Patton-Vogt, 1998). The presence of inositol inhibits PS synthase activity allowing more CDP-DAG to be used for PtdIns biosynthesis. It has been shown that PS synthase can be phosphorylated by cAMP-dependent protein kinase A *in vitro* and *in vivo* resulting in a decrease in its activity. It is not known whether phosphorylation is responsible for the decrease in PS synthase activity observed in stationary phase which increases the ratio of PtdIns to PtdSer. The regulation of PS synthase by sphingoid bases, for example, phytosphingosine, occurs at K_a' and K_i' which lie within the range of phytosphingosine concentration in the cell indicating a physiological relevance for this type of regulation. *CHO1/PSS1* is negatively regulated by Sin3 but positively by Ume6. These two transcription factors together with Rpd3 constitute the Ume6-Sin3-Rpd3 complex which is active in the repression of gene expression. It has been shown recently that Rpd3 and Sin3 regulate phospholipid biosynthesis *via* the UAS_{INO} element in an indirect fashion since there is no evidence to support direct binding of either Rpd3 or Sin3 to UAS_{INO}. A striking phenotype is that in a *sin3* mutant the membranes of the cell are completely lacking in PtdEtn, possibly because the strain overexpresses *CHO1/PSS1*, *CHO2/PEM1* and *OPI3/PEM2* genes resulting in increased PtdCho/PtdEtn ratio. In the absence of experimental evidence it cannot be said whether the lack of PtdEtn is the result of repression of *PSD1* and/or *PSD2* (Elkhaimi *et al.*, 2000).

The conversion of PtdEtn to PtdCho requires three methylation steps catalysed by two enzymes. PE methyltransferase (PEMT) converts PtdEtn to phosphatidylmonomethylethanolamine (PtdME); the second and third methylation steps are carried out by the product of the *OPI3/PEM2* gene (Henry and Patton-Vogt, 1998). *CHO2/PEM1* and *OPI3/PEM2* are regulated at the level of transcription in response to the water-soluble precursors inositol and choline. Both enzymes are localised in the ER (Janssen *et al.*, 2002). However, it remained unclear why experiments in which membrane extracts prepared from *opi3/pem2* and *cho2/pem1* strains to which the phospholipid substrates were added as lipid suspensions exhibited the expected saturation kinetics; either the substrates are transported to the ER membranes or there is vesicle fusion. Mixing experiments using membranes from *opi3/pem2* and *cho2/pem1 opi3/pem2* strains showed that more methylated PtdEtn derivatives were formed than expected from the activity of each enzyme in the separate fractions, supporting the hypothesis that the methyltransferases, although localised in different microsomal membranes, are capable of cooperating. Futhermore using ESI-MS/MS (Electrospray Ionisation Tandem Mass Spectrometry) it has been shown that PtdCho synthesised by the methylation pathway has more molecular diversity than that produced by the Kennedy pathway. Yeast can remodel PtdCho by exchange of acyl chains varying in length and saturation and this also contributes to steady state distribution of PtdCho species (Boumann *et al.*, 2002).

It is postulated that phospholipid synthesis in *S. cerevisiae* exists in either the 'on' or 'off' state as defined by maximal expression or repression, respectively of UAS_{INO}-regulated genes (Carman and Henry, 1999). In the 'on' state Dpp1 is 'off' and *vice versa.* It is hypothesised that PtdOH or a metabolite thereof (Henry and Patton-Vogt, 1998; Carman and Henry, 1999) generates the signal causing derepression of UAS_{INO} genes. Since Dpp1 is involved in the metabolism of PtdOH it could be that DGPP plays a role in the regulation of UAS_{INO}-containing genes. DGPP stimulates PS synthase, but not PI synthase and it has been shown that regulation of PS synthase by DGPP would favour the synthesis of PtdSer at the expense of PtdIns. Therefore, the partitioning of CDP-DAG between PtdSer and PtdIns may be somehow linked to the expression of DGPP phosphatase in that reduced levels of DGPP would favour synthesis of PtdSer. In fact *dpp1Δ* mutants have a reduced PtdIns content in comparison to wild type and therefore an altered PtdIns/PtdCho ratio, an index of phopholipid synthesis regulation (Oshiro *et al.*, 2000). Interestingly *DPP1* is regulated by zinc and Dpp1 is localised to the vacuolar membrane (Han *et al.*, 2001).

Although the reactions described earlier represent the main biosynthetic pathway in yeast for PtdCho it can also be synthesised from choline and ethanolamine in the so-called Kennedy or *salvage* pathway (Figure 6.3). Ethanolamine and choline kinases, encoded by the non-essential *EKI1* and *CKI1* genes respectively, have overlapping functions but Eki1 is primarily responsible for PtdEtn synthesis and Cki1 for PtdCho synthesis (Carman and Henry, 1999; Kim and Carman, 1999). The next step is the conversion of the phosphorylated compounds to cytidylated compounds; specifically phosphocholine (P-Cho) is converted to CDP-choline and phosphoethanolamine (P-Etn) to CDP-ethalonamine. These two reactions involving CTP are carried out by the products of the P-Cho cytidyltransferase-encoding gene *PCT1* and P-Etn cytidyltransferase-encoding gene *MUQ1*, respectively. The final step of the Kennedy pathway for PtdCho synthesis can be carried out by either cholinephosphotransferase (Cpt1) or by the dual specific choline/ethanolaminephosphotransferase (Ept1) (Howe and McMaster, 2001). The latter enzyme is responsible for synthesising 5–10% of PtdCho and all of the PtdEtn produced by the Kennedy pathway (Henneberry *et al.*, 2001). Inactivation of *CPT1* results in bypass of Sec14 but inactivation of *EPT1* does not, confirming the importance of PtdCho synthesised by the Kennedy pathway for Sec14's role as a phospholipid transfer protein. Furthermore, the synthesis of Gro*P*Cho by deacylation of PtdCho requires the substrate to have been synthesised *via* the CDP-choline pathway since this reaction still takes place in the absence of the methylation pathway (Dowd *et al.*, 2001). Interestingly, the latter pathway does not accelerate the synthesis of PtdCho in response to increased temperature, but the CDP-choline pathway does.

There are two genes capable of encoding CTP synthetase, *URA7* and *URA8*. Ura7 is more abundant than Ura8 and is therefore thought to be responsible for the majority of CTP synthesised *in vivo*. A mutation (E161K) of the *URA7*-encoded CTP synthetase renders CTP synthetase resistant to product inhibition. This lack of product inhibition results in an increased utilisation of the CDP-choline pathway for PtdCho synthesis as well as an increase in PtdOH synthesis with a decrease in PtdSer synthesis (Figure 6.3) (Carman and Henry, 1999). The E161K mutant also increases the cellular CTP levels at least 20-fold to well above the K_M-value for P-Cho cytidyltransferase, Pct1 and this results in an increase of Pct1 activity and the cellular concentration of CDP-choline (Ostrander *et al.*, 1998). The decrease in PS synthesis is attributable to a direct inhibition of the enzyme by CTP. Cells containing the E161K mutation excrete inositol which reflects altered regulation of PtdCho metabolism and the total lipid content is increased at the expense of phospholipids, the latter is a scenario encountered when wild-type cells enter stationary phase. These observations are an excellent example of the complex relationships involved in the regulation of lipid metabolism and emphasise that PtdCho is an important metabolite for yeast cell physiology. Like choline kinase, CTP synthetase activity is regulated *via* phosphorylation by the RAS-cAMP pathway (Ostrander *et al.*, 1998; Park *et al.*, 1999).

The enzymes of phospholipid synthesis are membrane bound and are located in the ER with their active sites orientated towards the cytosol which contains the substrates, for example, Gro3*P*, inositol, L-serine and CDP-choline. For instance, the majority of PtdEtn is synthesised in mitochondria by the product of the *PSD1* gene (Trotter and Voelker, 1995; Nicolson and Mayinger, 2000; Voelker, 2003). The substrate for this enzyme PtdSer has to be transferred from the ER to the MIM. PtdEtn then has to be returned to the ER *via* MOM. The transbilayer movement, although energy-independent, seems to be protein-mediated. Drs2 thought to function as an aminophospholipid translocase was originally postulated to be located at the plasma membrane, but is in fact found in the late Golgi membranes (Chen *et al.*, 1999). Recently the substance Ro09-0198, a tetracyclic polypeptide which forms a tight equimolar complex with PtdEtn has become a useful tool in measuring the transmembrane movement of this phospholipid (Kato *et al.*, 2002). Studies with this polypeptide have led to the identification of Ros3, which plays an important role in regulating phospholipid translocation across bilayer membranes and is found in areas of sphingolipid/ergosterol-rich membranes known as lipid rafts.

PtdIns and PtdCho can also be synthesised from exogenous inositol and choline, respectively. The genes of the two inositol-specific transporters *(ITR1* and *ITR2)* have been cloned (Robinson *et al.*, 1996). Strains lacking both *ITR1* and *ITR2* cannot repress *INO1* (Greenberg and Lopes, 1996) suggesting that the rate-limiting step of the inositol response is the uptake of inositol. The *ITR1* gene is the more highly expressed of the two and contains a potential UAS_{INO} element in its promoter. In addition to the transcriptional regulation by Ino2-Ino4-Opi1, Itr1 can be inactivated irreversibly by degradation through the endocytic pathway providing evidence of enzyme inactivation and protein degradation as mechanisms for regulation of phospholipid biosynthesis.

Glycerophosphoinositol (Gro*P*Ins) produced from PtdIns by the action of PLB is released into the medium and can be re-used by inositol-starved but not inositol-supplemented cells. Labelling data have shown that the inositol moiety but not the glycerol moiety of Gro*P*Ins is incorporated into lipids. There are two mechanisms by which Gro*P*Ins can enter the cell: either inositol-transporter dependent or inositol-transporter independent. An inositol-independent transporter, Git1 (Patton-Vogt and Henry, 1998), has been postulated for Gro*P*Ins. It belongs to the sugar permease family, a subfamily of the major facilitator superfamily (Goffeau *et al.*, 1997).

HNM1 encodes a choline transporter and deletion analysis of its promoter defined an UAS_{INO} which is required for the inositol response. Furthermore, it was shown that accumulation of choline in yeast cells leads to decreased uptake which implies that choline inhibits transport through either allosteric or covalent binding to its transporter (Li and Brendel, 1993).

Cardiolipin (CL), first isolated from heart tissue, is found primarily in MIM and MOM, and particularly in the mitochondrial membrane contact sites (Zinser *et al.*, 1991). CL is a structurally unique phospholipid carrying four acyl residues and two negative charges lending it both acidic and hydrophobic properties. It is required for cytochrome oxidase activity and may be involved in mitochondrial protein import. The synthesis of CL in yeast involves three sequential reactions (Figure 6.6); the committed step is catalysed by the enzyme phosphatidyl glycerophosphate (PGP) synthase from the liponucleotide CDP-DAG and Gro3*P* to yield phosphatidylglycerophosphate (PtdGro*P*). The next step is the dephosphorylation of PtdGro*P* to PtdGro, which in a final step is converted to CL by the CL synthase (Chang *et al.*, 1998b).

The gene encoding PGP synthase in *S. cerevisiae*, *PGS1*, was originally proposed to encode a second PS synthase based on sequence similarity. However, overexpression of this gene in a *CHO1/PSS1* deletant strain exhibited no PS synthase activity. More recent studies have shown that a *pgs1* null mutant contains neither CL nor PtdGro, although CL synthase activity was still present. The viability of *pgs1* mutants in spite of their seriously compromised mitochondrial function implies that neither PtdGro nor CL is absolutely essential for cell viability but may be required for critical function of the mitochondria (Chang *et al.*, 1998a).

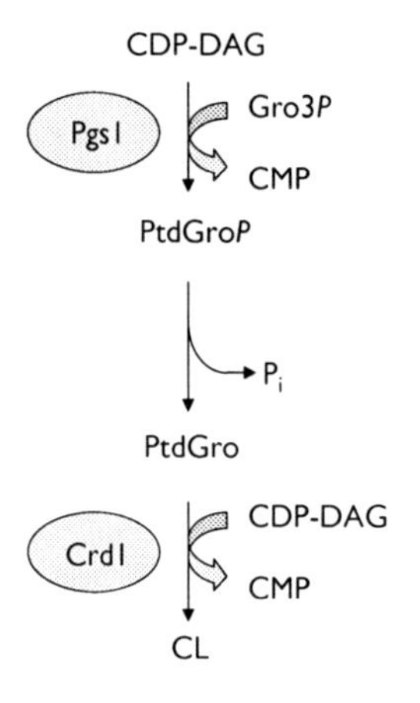

Figure 6.6 Cardiolipin synthesis (adapted from Carman and Henry, 1999).

The role of CL synthesis in mitochondrial function has been investigated by characterising a *crd1* null mutant. Such a mutant is *ts*-sensitive and has pleiotropic effects which are less apparent when the mitochondrial membranes contain PtdGro. It would appear that CL is required for essential cellular functions and for maintenance of mitochondrial DNA, for example, ATP/ADP carrier activity may be CL-dependent at elevated temperature (Jiang *et al.*, 2000).

PGP synthase is regulated by cross pathway control and by factors affecting the mitochondrial development (Minskoff and Greenberg, 1997). In contrast to enzymes for general phospholipid biosynthesis PGP synthase is not under the control of the *INO2, INO4* and *OPI1* regulatory genes. However, expression of PGP synthase is decreased by the addition of inositol to the growth medium only when cells are capable of synthesising PtdCho, invoking flux in the PtdCho biosynthetic pathway as part of the regulatory machinery. Evidence supporting this is provided by the fact that no inositol-mediated decrease in PGP synthase activity is observed in *cho2/pem1* and *opi3/pem2* strains which are mutated in the methylation pathway for the production of PtdCho. This post-translational control is likely to involve inactivation or degradation of the enzyme. It can be said that PGP synthase is coordinately regulated with inositol-1-phosphate synthase, PS synthase, PS decarboxylase and the phospholipid N-methyl transferases since they all have in common the fact that they are repressed in inositol-grown cells. Neither PGP phosphatase nor CL synthase is subject to inositol-mediated repression. *PGS1* mRNA is fully derepressed in the presence of inositol and choline and in *ino2Δ* and *opi1Δ* strains. However, in an *ino4Δ* background there is partial derepression suggesting that Ino4 may have a negative role in the transcriptional regulation of *PGS1* (Carman and Henry, 1999). Expression of *CRD1* is regulated by mitochondrial developmental factors since mRNA levels were 7–10-fold higher in stationary than in the logarithmic phase and 3-fold higher in wild type than in rho-zero cells. However, Crd1 expression is not affected by inositol and choline or in *ino2Δ*, *ino4Δ* and *opi1Δ* backgrounds (Jiang *et al.*, 1999).

6.6 UAS_{INO} – the common denominator of phospholipid regulation

Genetic studies have shown that the mitochondrial PS decarboxylase encoded by *PSD1* is responsible for correct regulation of *INO1* in response to inositol in the absence of ethanolamine, however disruption of *PSD2* has no effect (Griac, 1997). This confirms the cell's ability to monitor flux through the phospholipid biosynthetic pathway by adjusting the transcription of the genes therein. Mutations in every step of the synthesis from PtdSer to PtdCho lead to an Opi^- (overproduction of inositol) phenotype. It has been shown that the regulatory response is independent of the synthesis route of PtdCho (Griac *et al.*, 1996), implying that the transcription response to inositol is generated by the formation of PtdCho and not by the availability of choline or any other intermediate in PtdCho biosynthesis. Therefore, the signal must lie between PtdOH and PtdSer. The original *cds1* mutant produces a reduced amount of CDP-DAG and has an Opi^- phenotype which was not reversed by the addition of choline to the medium. Deletion of the *CDS1* gene is lethal (Shen and Dowhan, 1996, 1997) but by using a conditional copy of the *CDS1* gene in a *cds1Δ* strain it has been shown that *INO1, CHO1/PSS1, CHO2/PEM1* and *OPI3/PEM2* genes are all derepressed at reduced levels of the *CDS1* gene product, supporting the theory that the regulatory signal is to be found at the start of the synthetic pathway leading from PtdOH to

PtdCho. Taken together it can be said that the misregulation of *INO1* is caused by two metabolic conditions, each resulting in an Opi$^-$ phenotype; one involves increased PtdCho turnover (*sec14^{ts} cki1*) and the other involves reduced PtdCho synthesis as in *cds1, psd1, cho1/pss1, cho2/pem1* and *opi3/pem2*.

A model has been postulated explaining the coordinated regulation of genes containing UAS_{INO} (Henry and Patton-Vogt, 1998; Carman and Henry, 1999). The signal produced by PtdOH or a precursor/metabolite thereof is transmitted *via* the *OPI1* gene product, the negative regulator of *INO2* and *INO4* (White *et al.*, 1991). The positive regulators Ino2 and Ino4 form a heterodimer (Ambroziak and Henry, 1994; Carman and Henry, 1999) and have been shown to bind to the E-box-containing UAS_{INO} motif. The transcription of genes with a UAS_{INO} is increased in the absence of inositol and choline and causes the pool of PtdOH to be reduced because derepression of *INO1* leads to inositol production for use in PtdIns synthesis (Figure 6.3). Addition of inositol to the medium causes repression of the system because of the increased production of PtdIns. The combined presence of inositol and choline leads to a further repression since inactivation of the pathway from PtdEtn to PtdCho results in the production of PtdCho from choline and DAG, produced by dephosphorylation of PtdOH, and this again reduces the PtdOH pool (Carman and Henry, 1999).

The importance of inositol for the cell is shown by the fact that genes whose products have apparently nothing to do with phospholipid biosynthesis, for example, *snf* and *swi* mutants, defective in invertase expression and mating type switching, respectively, show inositol auxotrophy (Carman and Henry, 1999). A connection between fatty acid synthesis and the regulation of *INO1* transcription relies on Snf1, the AMP-activated protein kinase which is activated under conditions of reduced cellular energy and increased AMP-to-ATP ratio. A *snf1Δ* strain is auxotrophic for inositol and displays a 3-fold increase in Acc1 activity. The inositol auxotrophy can be suppressed by addition of exogenous fatty acids and a reduction in Acc1 activity suggesting that metabolic flux through the *de novo* fatty acid biosynthetic pathway can exert a regulatory function on gene expression (Shirra *et al.*, 2001).

Repression of the *INO1* gene expression occurs in the absence of *OPI1, INO2, INO4* or *SIN3*. Specifically, lack of Opi1 renders the *INO1* promoter more accessible to the basal transcription machinery. It could be speculated that Opi1 creates a cellular environment conducive to selective binding of Ino2/Ino4 at UAS_{INO} but neither Opi1 nor Ino2/Ino4 are responsible for transmission of the signal resulting in inositol-induced repression of genes carrying a UAS_{INO} in their promoters (Graves and Henry, 2000). Opi1 can be phosphorylated by protein kinase C and this post-translational modification inactivates Opi1 and attenuates its activity (Sreenivas *et al.*, 2001).

It has been demonstrated that both lithium and the branched fatty acid valproate influence molecular targets in the phosphoinositide metabolism increasing the expression of *INO1* and *INO2* (Vaden *et al.*, 2001). Bipolar disorder which is characterised by recurring bouts of mania and depression can be treated by lithium and valproate, however neither compound provides a completely effective treatment. Understanding the mechanisms of action of lithium and valproate may uncover the link between inositol metabolism and manic depression. An *opi1* mutant which constitutively expresses *INO1* is more resistant to lithium-, but not valproate-induced growth inhibition suggesting that valproate may inhibit synthesis of inositol-1-phosphate. Both drugs decrease the rate of membrane PtdIns synthesis and increase PtdCho synthesis and valproate but not lithium increases the expression of *CHO1/PSS1* and *OPI3/PEM2*. The drug-induced reduction of the PtdIns to PtdCho ratio may affect secretory vesicle formation which may underlie the therapeutic mechanisms of these drugs (Ding and Greenberg, 2003). Interestingly, *epi*-inositol, one of the nine possible inositol isomers, unlike the major naturally occurring isomer *myo*-inositol, is not a substrate for PtdIns synthesis. Both *epi*- and *myo*-inositol exert similar behavioural effects suggesting that these may be due to the effects on gene expression. It was shown that *epi*-inositol can reverse the lithium- or valproate-induced *INO1* expression (Shaldubina *et al.*, 2002).

The current model for regulation of phospholipid biosynthesis involves the products of the *INO2, INO4, OPI1* and *SIN3* genes. Opi1 and Sin3 interact *via* one of the four paired amphiphatic helix (PAH) motifs, present in the latter protein. In addition, Opi1 and Ino2 interact *via* the central domain of Ino2. The ability of Opi1 to bind the general repressor protein Sin3 as well as the specific activator Ino2 defines the transcription complex binding to the promoters of genes regulated by UAS_{INO} (Wagner *et al.*, 2001). However, evidence that other factors may be involved in specific instances is provided by the observation that expression of *FAS1* is Ino2-dependent at the permissive temperature in a strain with a *ts* defect in protein myristoylation, but at the restrictive temperature *FAS1* expression is Ino2 independent (Cok *et al.*, 1998; Graves and Henry, 2000). Opi1's negative regulation of phospholipid biosynthetic genes is dictated by its phosphorylation state. Phosphorylation at Ser(26) inactivates Opi1 and attenuates its activity as shown by the 50% reduction in the activity of an *INO1-lacZ* reporter gene expressed in a strain in which the *OPI1* gene contains a S26A mutant (Sreenivas *et al.*, 2001).

The transcriptional activation activity of Ino2/Ino4 resides solely in the N-terminus of Ino2 and it has been shown that Ino4 is essential for DNA binding (Schwank *et al.*, 1995). The *INO2* and *INO4* genes themselves contain UAS_{INO} in their promoter sequences (Ashburner and Lopes, 1995). Using reporter constructs it has been shown that *INO2* expression is dependent on the presence of *INO2* and *INO4* genes and is constitutively overexpressed in an *opi*$^-$ background. Therefore, the pattern of *INO2* expression reflects that of *INO1* and *INO2* may be said to be autoregulated (Ashburner and Lopes, 1995). On the other hand, an *INO4* reporter construct is constitutively expressed. Ino2-independent regulation described for *INO4* is also found in the presence of inositol for the gene encoding the sphingolipid biosynthetic enzyme, IPC synthase which has a UAS_{INO} in its promoter (Ko *et al.*, 1994). With the exception of *CTR1* (Li and Brendel, 1993) and *INO4* (Ashburner and Lopes, 1995) all phospholipid biosynthetic genes are regulated by *INO2* and *INO4;* however, the amount of derepression is highest for *INO1* (30-fold) (Lopes *et al.*, 1993; Howe and McMaster, 2001) but is only 2–3-fold for *CHO2/PEM1* and *OPI3/PEM2*. A similar modest effect is also found for *ACC1* (3-fold) (Hasslacher *et al.*, 1993), and *FAS1* and *FAS2* (2-fold) (Schüller *et al.*, 1992). The *INO2–INO4–OPI1* regulatory circuit is an excellent example of cell-wide communication between nutrient availability, general transcription and membrane biogenesis.

6.7 Phospholipid turnover, remodelling and signalling

Membrane phospholipid is an important reservoir of lipid products involved in signalling the induction of functional responses. The lipid products are released by action of phospholipases. Phospholipase D (PLD) catalyses the hydrolysis of PtdCho to PtdOH and choline (Figure 6.3) (Ella *et al.*, 1996). PtdOH is the starting material for many lipid second messengers, for example, DAG and PA phosphatases. PLD has been detected in yeast mitochondria and shown to be activated in response to glucose repression and is encoded by *PLD1*. *PLD1* was found to be allelic to *SPO14*, a gene whose product is required for meiosis and spore formation (Waksman *et al.*, 1996). A defect in phospholipase activity which affects sporulation demonstrates the far-reaching effects of phospholipid turnover, for example, response to nutritional signals which initiate sporulation. Deletion of *PLD1* is not a lethal event and mutational studies have shown that it is possible to separate the PLD functions related to sporulation from those involved in Sec14-independent growth indicating that PLD plays distinct roles in sporulation and secretion (Li *et al.*, 2000a; Rudge *et al.*, 2001).

There is a second gene, *PLD2*, encoding PLD activity (Waksman *et al.*, 1997; Tang *et al.*, 2002). However, this time the preferred substrate is PtdEtn rather than PtdCho and furthermore this gene product is not essential for sporulation. *PLD2* encodes a Ca^{2+}-dependent enzyme and in contrast to Pld1 is not activated by PtdIns(4,5)P_2 (Sciorra *et al.*, 1999).

In mammalian systems hydrolysis of polyphosphoinositides by phospholipase C (PLC) is a major signal transduction pathway responsive to hormones, growth factors and other regulatory signals (Flick and Thorner, 1998). Mammals have four PLC isoforms (Rebecchi and Pentyala, 2000) but to date only one PLC species, the mammalian δ-homologue, has been identified in yeast and this is encoded by the *PLC1* gene. This enzyme is Ca^{2+}-dependent and cells deficient in Plcl display several phenotypes from lethality in certain genetic backgrounds to slow growth at 30°C, sensitivity to hyperosmotic conditions and missegregation of chromosomes during mitosis (Flick and Thorner, 1993). To identify the roles that Plcl plays in cellular physiology, suppressors have been isolated; one of these is *PHO81*, an important regulator of the *PHO* regulon and *SPL2* which encodes a novel protein that is also regulated in part by the *PHO* regulon. A synthetic lethal phenotype of *plc1Δ* in combination with either *pho81Δ* or *spl2Δ* has been found and confirms the interrelationship of these two pathways, emphasising that hydrolysis of PtdIns(4,5)P_2, is required for a number of nutritional and stress-related responses (Flick and Thorner, 1998). It was also shown that there is a physical interaction between Gpr1, PLC and Gpa in mitogen-controlled signalling pathways leading to pseudohyphal differentiation (Ansari *et al.*, 1999). Further discussion of PLC is found in Rebecchi and Pentyala (2000).

The third phospholipase, phospholipase B (PLB), presents a complicated picture since three isoforms have been detected, two from the plasma membrane and a third from the culture supernatant and the periplasmic space (Lee *et al.*, 1994). PLB hydrolyses fatty acids both from the *sn-1* and *sn-2* positions of glycerophospholipids and also possesses an acyltransferase activity catalysing the synthesis of glycerophospholipids from lyso-glycerophospholipids. The phospholipase B/lysophospholipase activity of Plb1 greatly exceeds the activity of the first step of hydrolysis, therefore lyso-phospholipids are not found as intermediate products of Plb1 activity (Merkel *et al.*, 1999). The same enzyme can synthesise PtdCho from two molecules of lyso-PtdCho by its transacylation activity. It is suggested that Plb1's main function is the deacylation of PtdCho and PtdEtn but not of PtdIns implying the existence of other enzyme(s) to carry out this reaction. Similarity searches have revealed the existence of two candidate genes *PLB2* and *PLB3* which are capable of transacylation of PtdIns. Deletion of *PLB1* resulted in a significant reduction in Gro*P*Cho release from the cells whereas the production of Gro*P*Ins was hardly affected. Deletion of *PLB3* reduces Gro*P*Ins production by 50% but Gro*P*Cho release is increased by over 50%. Deletion of *PLB2* affects the production of neither Gro*P*Cho nor Gro*P*Ins. Mutants lacking *PLB1, PLB2* and *PLB3* release Gro*P*Cho and Gro*P*Ins at 10% of the wild-type rate suggesting the presence of alternative pathways for the release of Gro*P*Cho and Gro*P*Ins in such a strain. The product of the *SPO1* gene is a possible candidate for this alternative pathway, however it is expressed only during sporulation and this may explain why overexpression of *SPO1* in a *plb1Δ plb2Δ plb3Δ* triple mutant did not increase PLB activity *in vitro*. Nevertheless given Spo1's similarity to

PLB, its unique role in spindle pole body (SPB) duplication, and pleiotropic effects on meiosis II, late gene expression and spore formation, it is suggested that Spo1 participates in a novel meiotic pathway that functions through the SPB to coordinate nuclear division with spore development (Tevzadze *et al.*, 2000). Although, unlike Plb1 and Plb3, Plb2 has no *in vitro* activity for the production of Gro*P*Ins or Gro*P*Cho, cell free extracts of a strain overproducing Plb2 does have plasma membrane-associated PLB activity. However, the enzyme must be active *in vivo* since it has a protective effect against toxicity of extracellular lyso-PtdCho (Merkel *et al.*, 1999). In another report it was found that Plb2 has no transacylase activity but did have lyso-phospholipase activity (Fyrst *et al.*, 1999). These differences may be associated with strain background.

Phospholipase A_1 and A_2 (Pla_1, Pla_2) hydrolyse fatty acids from the *sn-1* and *sn-2* positions of phospholipids, respectively, to produce lyso-phospholipid (Divecha and Irvine, 1995; Leslie, 1997; Gijon and Leslie, 1997). Pla_2 activity encoded by the *SPO22* gene was shown by transcript profiling to be expressed early in meiosis (Primig *et al.*, 2000) and its product is found to be associated with mitochondrial membranes (Wagner and Paltauf, 1994). The enzyme deacetylates phospholipids in the order of PtdCho = PtdEtn > PtdSer ≫ PtdIns. *In vivo* studies on the incorporation of exogenous radiolabelled fatty acids, for example, C18:0 and C18:1 into cellular phospholipids suggest a remodelling of the lyso-phospholipids – deacetylation followed by reacetylation. The deacetylation of PtdCho and PtdEtn could be carried out by either Pla_2 or one of the PLBs which also deacetylates at both the *sn-1* and the *sn-2* positions (Wagner and Paltauf, 1994).

The lipid molecules, DAG and DGPP involved in cell signalling, in addition to being synthesised from PtdOH can also be mobilised from the storage lipid TAG by the action of the lipases encoded by *TGL1* and *TGL2*. Evidence suggests that Tgl2 is less highly expressed than Tgl1 which is localised to lipid particles (Athenstaedt *et al.*, 1999b). Further lipases may exist since unknown ORFs have been found with similarity to *TGL1* and *TGL2* (Van Heusden *et al.*, 1998).

6.8 Phosphoinositides – key players in signalling and regulation of membrane trafficking

Phosphoinositides, the collective term for PtdIns and its phosphorylated derivatives, play a decisive role in cell signalling and growth as well as vesicular trafficking, transcription and the maintenance of the actin skeleton (Stefan *et al.*, 2002). The diversity of the phenotypes resulting from the mutation of genes involved in the metabolism of phosphoinositides is a paradigm for the ramifications encountered in the regulation of yeast physiology.

The balance of lipid kinases and phosphatases ensures that phosphoinositides are subjected to a turnover allowing them to regulate the various cellular processes in a temporal manner. Furthermore, the existence of different pools for some phosphoinositides adds a spatial dimension to their regulatory properties. Yeast lacks the phosphoinositides PtdIns(3,4)P_2 and PtdIns(3,4,5)P_3 found in mammals (Wera *et al.*, 2001).

There are two PI 4-kinases encoded by the essential genes *PIK1* and *STT4* and a recently discovered third form, encoded by *LSB6*, which is not an essential gene (Han *et al.*, 2002b) (Figure 6.7). The involvement of these phosphoinositides in intracellular signalling is shown by the fact that the *PIK1* mRNA level increases threefold in 30 min as a response to α-factor (Flanagan *et al.*, 1993). Pik1 exists as a heterodimer with the yeast frequenin homologue Frq1. Frequenins together with recoverins and neurocalcins are Ca^{2+}-binding calmodulin-related

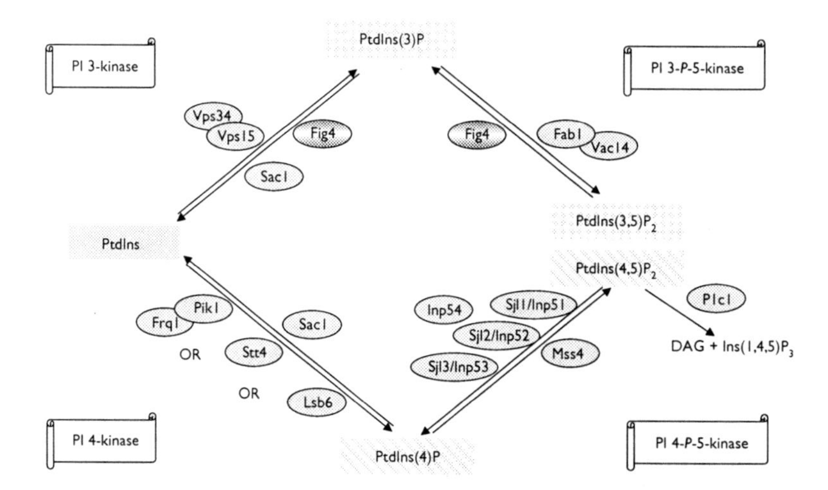

Figure 6.7 Synthesis and turnover of phosphoinositides.

proteins found in excitable cells of metazoans but their physiological roles have not been characterised. It has now been shown that the myristoylated Frq1 which is an essential protein, is required to stimulate the activity of Pik1 (Hendricks *et al.*, 1999; Ames *et al.*, 2000; Huttner *et al.*, 2003).

Evidence of involvement of Stt4 in membrane transport is that a mutation therein can suppress the phenotype of a *psd2Δ* strain which lacks the vacuolar form of PS decarboxylase (Trotter *et al.*, 1998). *STT4* was originally discovered in a screen for mutants (Yoshida *et al.*, 1994; Hama *et al.*, 1999) defective in a pathway whose target is the protein kinase Pkc1. Stt4 produces PtdIns(4)P which is then phosphorylated by Mss4 to PtdIns(4,5)P_2. Stt4 has been shown to be localised at the plasma membrane and to function in the Pkc1-mediated MAP kinase cascade. It is proposed that the PtdIns(4,5)P_2 pool at the plasma membrane is required to recruit and/or activate effector proteins, such as Rom2 (Audhya and Emr, 2002).

Mss4 and the yeast synaptojanins control the level of PtdIns(4,5)P_2 which governs the actin skeleton by recruiting actin-binding proteins (Stefan *et al.*, 2002). Synaptojanin is found in mammalian nerve cells and is an inositol polyphosphate 5-phosphatase and is thought to play a role in synaptic-vesicle endocytosis and recycling (Odorizzi *et al.*, 2000). PtdIns(4,5)P_2 binds to several actin-regulatory proteins including profilin (Stefan *et al.*, 2002). It would appear that the synaptojanins do not regulate the amount of actin but are involved primarily in controlling the organisation of actin filaments (Karpova *et al.*, 1998).

Further evidence of the involvement of phosphorylated derivatives of PtdIns in the maintenance of the actin skeleton is provided by the physical interaction of the Lsb6 protein with Las17/Bee1 which plays a role in actin patch assembly and polymerisation (Munn, 2001; Han *et al.*, 2002b). Lsb6 has been identified as the third PI 4-kinase localised to the plasma and vacuolar membranes. In contrast to *STT4, LSB6* does not encode an essential protein and can partially suppress an *STT4* deletion but not the lethal phenotype of *pik1Δ* mutants thus providing evidence that Lsb6 functions as a PI 4-kinase *in vivo*. Since deletion of *LSB6* caused neither an obvious phenotype nor did it appear to affect the level of total phosphoinositides it can be concluded at least for the moment that Stt4 and Pik1 are the major

PI 4-kinases in *S. cerevisiae*. However, it is interesting to speculate that *LSB6* may encode the 55-kDa PI 4-kinase which has been isolated from Triton X-100-treated membranes of *S. cerevisiae*. Supporting this speculation is the observation that overexpression of Lsb6 localises it to the plasma membrane as does Stt4. Pik1 is on the other hand a soluble 120-kDa enzyme which has been shown by indirect immunofluorescence to co-localise to the Golgi (Odorizzi *et al.*, 2000). These observations suggest the existence of two pools of PtdIns(4)P, one generated by Pik1 is involved in protein secretion and the other generated by Stt4, possibly with some contribution from Lsb6, is involved in the maintenance of vacuolar morphology and organisation of the actin cytoskeleton (Foti *et al.*, 2001).

Studies involving the yeast synaptojanin-like proteins provide evidence for the compartmentalisation of different phosphoinositide species. There are three yeast paralogues of the mammalian synaptojanin, *SJL1/INP51, SJL2/INP52* and *SJL3/INP53*. Each protein carries the signature motif of inositol polyphosphate 5-phosphatase in the C-terminal half of the polypeptide (Stolz *et al.*, 1998; Guo *et al.*, 1999) whereas the N-terminal region contains a Sac1 domain corresponding to polyphosphoinositide (PPI) phosphatase. This combination of a PtdIns(4,5)P_2 5-phosphatase activity and a PPI phosphatase may facilitate localised phosphoinositide turnover which accepts PtdIns(3,5)P_2, PtdIns(3)P and PtdIns(4)P (Fruman *et al.*, 1998; Simonsen *et al.*, 2001; Wiradjaja *et al.*, 2001). Deletion of *SJL1/INP51* whose Sac1-like domain has no intrinsic PPI phosphatase activity causes an increase in PtdIns(4,5)P_2 levels (Stefan *et al.*, 2002). On the other hand strains lacking *SJL2/INP52* and *SJL3/INP53* accumulate PtdIns(3,5)P_2 (Guo *et al.*, 1999). A triple deletion of *SJL1/INP51, SJL2/INP52* and *SJL3/INP53* is lethal suggesting that these proteins have essential overlapping functions.

In addition to the three synaptojanin-like paralogues there is one other polyphosphate 5-phosphatase in yeast encoded by *INP54* and which does not contain a Sac1-like domain (Wiradjaja *et al.*, 2001). All four inositol polyphosphate 5-phosphatases hydrolyse PtdIns(4,5)P_2 to generate PtdIns(4)P. The need for four versions of this type of enzyme for the regulation of the cellular concentration of PtdIns(4,5)P_2 may reflect a different localisation for each of them. It is known that Sjl2/Inp52 and Sjl3/Inp53 are found everywhere in the cell except in the nucleus; under hyperosmotic stress they translocate to cortical actin patches. The localisation of Inp54 to the ER is the first evidence for targeting of a yeast inositol polyphosphate 5-phosphatase to a specific subcellular compartment. It is postulated that Inp54 plays a regulatory role in secretion by modulating PtdIns(4,5)P_2 on the cytosolic surface of the ER membrane. This would place the enzyme in an optimal position to regulate PtdIns(4,5)P_2 levels on vesicle budding from the ER (Wiradjaja *et al.*, 2001). Sjl1/Inp51 has not yet been localised since the recombinant protein is very susceptible to proteolysis. Sjl3/Inp53 appears to play a role in the regulation of the rate of trafficking from the *trans*-Golgi network to the prevacuolar compartment but not in the opposite direction. Furthermore this effect is cargo-specific since only the transport of alkaline phosphatase and Kex2, but not that of Vps10, is compromised in an *SJL3/INP53* mutant (Ha *et al.*, 2001).

PtdIns(4,5)P_2 binds to the pleckstrin homology (PH) domain of PLC (Plc1) (Stefan *et al.*, 2002) and is converted into DAG and Ins(1,4,5)P_3 (Figure 6.7). In humans Ins(1,4,5)P_3 mediates Ca^{2+} release from internal stores and DAG is involved in the activation of protein kinase Pkc1 (Wera *et al.*, 2001). In yeast *PLC1* has pleiotropic effects since *PLC1* deletants are extremely sensitive to high osmolarity and grow slowly, particularly on carbon sources other than glucose (Flick and Thorner, 1993). Furthermore, Plc1 has a role in nuclear processes since the kinase, Ipk2/Arg82/ArgRIII, which converts Ins(1,4,5)P_3 into Ins(1,4,5,6)P_4 and then into $Ins(1,3,4,5,6)P_5$ and the kinase Ipk1 which converts $Ins(1,3,4,5,6)P_5$ into $Ins(1,2,3,4,5,6)P_6$ are located in the nucleus (York *et al.*, 1999). Surprisingly, the PLC-dependent inositol polyphosphate kinase pathway is required for efficient mRNA export from the nucleus. The IP_5 2-kinase (Ipk1) and Plc1 were identified in a genetic screen with a mutant deficient in the factor Gle1 which is associated with the nuclear pore complex and which is required for mRNA export from the nucleus. The suggestion that the transcriptional regulator Ipk2/Arg82/ArgRIII (Zhang *et al.*, 2001) is the ancestor of an $InsP_3$ kinase and diphosphoinositol polyphosphate ($[PP]_2$-$InsP_4$) synthase activities provides an interesting link between transcription, mRNA export and arginine metabolism. Yor163 has been identified as containing phosphohydrolase activity towards PP-$InsP_5$ and $[PP]_2$-$InsP_4$ (Safrany *et al.*, 1999a,b). The implication that Plc1 has a nuclear location ties in well with the discovery that it is associated with the kinetochore complex CBF3. Chromatin immunoprecipitation (CHIP) analysis has shown that Plc1 is localised to centromeric loci *in vivo* (Lin *et al.*, 2000).

A further vindication of the involvement of phospholipid metabolism in protein trafficking was the finding that one of the large set of *vps* mutants encodes a PI 3-kinase (Odorizzi *et al.*, 2000) (Figure. 6.7). The main function of this enzyme is to create localised areas of the *trans*-Golgi network which are enriched for PtdIns(3)P. Vps34 is activated and recruited to the membrane by Vps15 which is a serine/threonine protein kinase. Vps34 PI 3-kinase shows a striking similarity to the catalytic subunit of p85/p110 mammalian PI 3-kinase (Sotsios and Ward, 2000). The lipid kinase Vps34 activity is fundamental since point mutations in the kinase domain are sufficient to cause a vacuolar sorting defect. The localised production of PtdIns(3)P by Vps34 may serve to attract and/or activate effector molecules such as vesicle coat proteins which catalyse the transport reaction. The Vps15–Vps34 complex functions at the Golgi to endosome stage (Odorizzi *et al.*, 2000) of vacuolar protein transport. PtdIns(3)P is responsible for recruiting and possibly activating the proteins Vac1 and Vps27. Like the human counterpart EEA1 (human early endosome autoantigen 1), Vac1 contains a FYVE (Fab1, Ygl023, Vps27 and EEA1) domain which binds PtdIns(3)P with high specificity (Vanhaesebroeck *et al.*, 2001; Stefan *et al.*, 2002) and interacts with other proteins, for example, Rab5-GTPase homologues, Vps21 and Pep12 which is an endosomal t-SNARE protein. PtdIns(3)P is required for Golgi to vacuole protein sorting *via* its interactions with proteins containing FYVE and PX (phagocyte oxidase) homology domains (Ago *et al.*, 2001). Fifteen yeast proteins including Spo14/Pld1 contain PX domains (Yu and Lemmon, 2001). Vps34 has been found to co-immunoprecipitate with Vps30/Apg6 and Apg14, proteins involved in autophagy, the transport pathway which delivers cytosolic proteins and organelles to the vacuole for degradation by hydrolysis (Simonsen *et al.*, 2001). Therefore there is a likelihood that PtdIns(3)P plays a role in autophagy (Kihara *et al.*, 2001).

Accumulation of PtdIns(3,5)P_2 is also encountered when yeast cells are exposed to osmotic stress. The osmoregulation of PtdIns(3,5)P_2, which is independent of the *HOG1* pathway, may be associated with vacuolar volume. Fab1, a PI 3-*P*-5-kinase whose upregulation rather than an alteration in the turnover of PtdIns(3,5)P_2 is responsible for the increase in PtdIns(3,5)P_2 (Bonangelino *et al.*, 2002). Increased levels of PtdIns(3,5)P_2 may protect cells from osmotic stress by modulating water efflux and ion influx across the vacuolar membrane, thus regulating vacuole membrane fusion and retrograde traffic. Two additional gene products, those of the *VAC14* and *VAC7* genes, are required for the increase of PtdIns(3,5)P_2. Vac7 has an unknown function whereas Vac14 is a positive regulator of Fab1 activity. The localisation of these three proteins suggests that PtdIns(3,5)P_2 is produced and functions in the late endosome and vacuole.

Clearly there are several turnover pathways for PtdIns(4,5)P_2 and PtdIns(3,5)P_2. PtdIns(4,5)P_2 can be dephosphorylated by Sjl1/Inp51, Sjl2/Inp52, Sjl3/Inp53 and Inp54 because each contains an inositol polyphosphate 5-phosphatase domain at its C-terminus (Stolz *et al.*, 1998; Guo *et al.*, 1999). Sjl2/Inp52 and Sjl3/Inp53 carry a Sac1-like domain at their N-termini which allows them to dephosphorylate PtdIns(3)P, PtdIns(4)P and PtdIns(3,5)P_2. A fourth protein Fig4 which contains a Sac1-domain at its N-terminus is also capable of dephosphorylating PtdIns(3)P and PtdIns(3,5)P_2. Fig4 is a peripheral membrane protein and it may act on a PtdIns(3,5)P_2 pool which is inaccessible to the ER-localised Sac1 (Hughes *et al.*, 2000b; Gary *et al.*, 2002).

6.9 Sec14 – the connection between phospholipids and the Golgi secretory pathway

The role of Sec14 in yeast physiology is to maintain a balance in the levels of PtdIns and PtdCho in the Golgi membrane. PtdCho-loaded Sec14 inhibits P-Cho cytidyltransferase (Pct1), the rate-limiting enzyme of the CDP-choline pathway but PtdIns-loaded Sec14 does not (Figure 6.8). This provides experimental evidence for Sec14 acting as a sensor for maintenance of the correct PtdIns/PtdCho ratio which is critical for the secretory competence of the Golgi membranes. Since Sec14 binds PtdIns 16-fold more strongly than PtdCho, PtdCho is only loaded when the PtdCho/PtdIns ratio in the membrane is high (Sha and Luo, 1999). A three-dimensional structure for Sec14 at a 2.5 Å resolution has been obtained and the data provide insight into the architecture of Sec14. The protein contains a hydrophobic pocket which is thought to be critical for Sec14 to carry out phospholipid exchange by making contact with the lipid bilayer (Sha *et al.*, 1998).

Deletion of *SEC14* is lethal (Bankaitis *et al.*, 1989). Suppressors of *sec14*^{ts} mutants at the restrictive temperature of 37°C have been selected and it has been found that mutations in a number of genes are capable of rendering the cell independent of the Sec14 requirement. These bypass (suppressor) mutations are found in genes of the CDP-choline pathway as well as in *SAC1* and *KES1/OSH4*. Sac1 is a PPI phosphatase which converts PtdIns(4)P to PtdIns

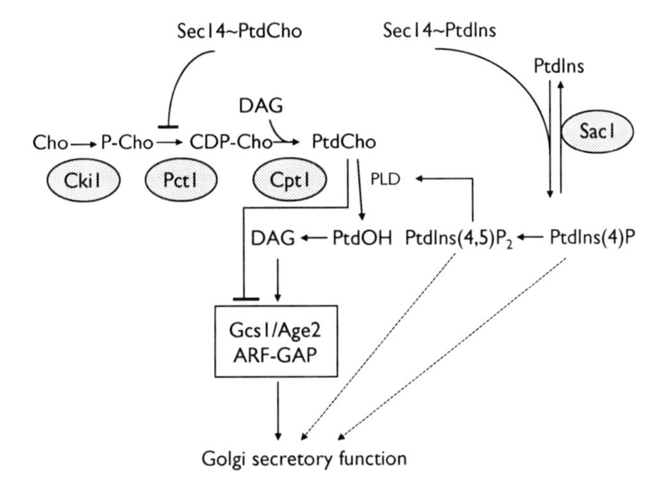

Figure 6.8 The role of Sec14 in Golgi secretory function (adapted from Li *et al.*, 2002).

(Odorizzi *et al.*, 2000) and regulates metabolism of phosphoinositides in the ER and the Golgi (Konrad *et al.*, 2002). The bypass of Sec14 activity of *SAC1* mutants is due to the accumulation of PtdIns(4)P which can then be converted to PtdIns(4,5)P_2 required for PLD activity (Sciorra *et al.*, 1999; Yanagisawa *et al.*, 2002). Kes1 is unique in its action of Sec14 bypass because it is the only gene product whose overproduction abrogates its bypass function, particularly in *sac1* mutants which overproduce PtdIns(4)P (Guo *et al.*, 1999). Kes1 has been classified as an oxysterol binding protein (Osh4), preferentially binding PtdIns(4,5)P_2 *in vitro*, however *in vivo* it prefers to bind PtdIns(4)P produced by the action of Pik1 (Li *et al.*, 2002). A new facet has been brought to the Sec14 story by the investigation of the function of Kes1/Osh4. There may be some correlation between PtdIns(4)P levels and Kes1/Osh4 overproduction and it was shown that the amount of Kes1/Osh4 overproduction required to bypass Sec14 is proportional to the cellular level of PtdIns(4)P. Furthermore, Kes1/Osh4 is mislocalised in *sac1* mutants. Kes1/Osh4 function is linked to the yeast ADP-ribosylation factor (ARF) cycle since *arf1Δ* and *gcs1Δ* alleles counteract *kes1/osh4*-mediated Sec14 bypass. This fits better into the scheme of Sec14 bypass than the original mis-identification that the substrate of Sac1 was IPC (Guo *et al.*, 1999; Stock *et al.*, 1999).

Inactivation of the *SEC14* gene product increases PtdCho turnover (Patton-Vogt *et al.*, 1997). Opi^- and Opc^- (overproduction of choline) phenotypes were observed in double *sec14*^{ts} and *cki1* mutants. The elevated choline excretion permits the secretory pathway to function when Sec14 is inactivated. The free choline in the growth medium of *sec14*^{ts} *cki1* strains is produced by the action of PLD on PtdCho, not by the action of PLB or PLC since neither of these produces free choline. This hypothesis is supported by the observation that total ^{32}P-orthophosphate is lost only gradually from the chloroform-soluble pool which contains the PLD-mediated choline and PtdOH pools. The PtdOH pool can be reused directly leading to the formation of PtdCho *via* the PtdEtn methylation pathway and for the production of sphingolipids. The reason that *sec14*^{ts} *cki1* double mutants survive at the restrictive temperature for Sec14 is because resynthesis of PtdCho is prevented following accelerated turnover. Furthermore, in a *sec14*^{ts} *cki1* strain at the restrictive temperature the *INO1* gene is derepressed even in the presence of inositol and this probably correlates to, and is caused by, the increase in the rate of PtdCho turnover resulting from inactivation of Sec14. The association between inositol metabolism and PtdCho is well established; mutations in structural genes for every step in the PtdOH to PtdCho pathway exhibit an Opi^- phenotype and misregulate *INO1* and other co-regulated genes. The *INO1* derepression occurs in response to a metabolic signal generated in the course of the overall alteration in phospholipid metabolism resulting in increased PtdCho turnover and the bypass of the Sec14 requirement. Increased PtdCho turnover in *sac1Δ* strains is thought to be the cause of the observed inositol auxotrophy in these strains. Mutations in the CDP-choline pathway can relieve the inositol auxotrophy whereas constitutive transcription of *INO1* does not (Rivas *et al.*, 1999).

In spite of *SEC14* encoding an essential function, five *SEC14* homologues have been identified in the genome of *S. cerevisiae* with sequence identity varying from 21% to 64%. Interestingly, the homologue Sfh1/Ykl091 with the highest identity to Sec14 is not a phospholipid transfer protein but the others, Sfh2–Sfh5, are apparently novel phosphatidylinositol binding proteins (Li *et al.*, 2000b). Two of the homologues, *SFH3/PDR16* and *SFH4/PDR17* encode products which when lacking cause multiple drug sensitivity (van den Hazel *et al.*, 1999; Wu *et al.*, 2000). Furthermore Sfh4 has been isolated in a screen for genes involved in the metabolic steps between PtdSer formation and methylation of PtdEtn; these steps include vectorial lipid transport of PtdSer between the ER and the Golgi/vacuole and in the reverse direction for PtdEtn. Among the ethanolamine auxotrophs one was complemented by *PSTB2* which encodes a protein capable of PtdIns transfer activity and turned out to be allelic to *SFH4* (van den Hazel *et al.*, 1999). Overexpression of Pstb2/Sfh4 suppresses *sec14*^{ts} mutants. However, overexpression of Sec14 does not suppress the defect in *pstb2/sfh4* conditional mutants. Strains mutated in *PSTB2* accumulate PtdSer in the Golgi thus preventing interaction between PtdSer and Psd2 located in MAM (Wu *et al.*, 2000).

A further interesting development in the Sec14 story involves the identification of synthetic lethality between *sec14-1*^{ts} and a *gcs1* mutation (Yanagisawa *et al.*, 2002) (Figure 6.8). Gcs1 and Age2 are ARF GTPase-activating proteins and it has been demonstrated that they act downstream of Sec14 and PLD in both Sec14-dependent and -independent pathways for Golgi function. The mechanism by which Sec14 regulates membrane trafficking interfaces with the activity of proteins involved in the control of ADP ribosylation factor cycling remains to be elucidated. In spite of the identification of downstream targets of Sec14, it still remains to be determined whether PtdCho, PtdOH or DAG (Howe and McMaster, 2001) are the regulators for vesicle transport. Since DAG is a non-bilayer forming lipid, dramatic changes in the physical properties of the bilayer, for example, membrane curvature, could initiate the process of membrane budding (Ktistakis *et al.*, 1996). An *E. coli* DAG kinase, converting DAG to PtdOH, when expressed in a yeast strain with compromised Sec14 activity reduces viability and vesicle trafficking arguing that increased PtdOH has a negative effect on Sec14-mediated transport. Further evidence that PtdOH may be involved in this process, is the identification of a human homologue, ORP1, of Kes1/Osh4 (McMaster, 2001) which preferentially binds PtdOH, produced when PLD acts on PtdCho. PLD-independent turnover – by Plb1 which metabolises PtdCho to generate Gro*P*Cho and free fatty acids – can partially compensate for loss of functional Sec14 suggesting that it is PtdCho which is toxic (Xie *et al.*, 2001). This argument, however, is flawed: Plb1 is normally resident in the plasma membrane thus Plb1 turnover and Sec14-mediated vesicle transport are physically separated within the cell. One possible explanation is that overproduction of Plb1 may result in it being mislocated to the Golgi. The essential requirement for an intact Spo14/Pld1 in effecting bypass of *sec14*^{ts} demonstrates that elevated production of PtdOH by Spo14/Pld1 is necessary but not sufficient for bypassing Sec14 (Liscovitch *et al.*, 2000). A localised concentration of PtdOH in the vicinity of PtdIns(4,5)P_2, both of which are negatively charged, may allow the coatomer to bind to its receptor in the Golgi membrane and transform the planar bilayer into a curved bud (Martin, 1997). The role of PLD in *cki1*-mediated bypass of the *sec14*^{ts} defect has been confirmed by constructing a *sec14*^{ts} *cki1 pld1* triple mutant. This mutant is unable to grow at 37°C and has neither an Opi^- nor Opc^- phenotype (Henry and Patton-Vogt, 1998).

Pct1 is a target for Sec14~PtdCho (McMaster, 2001). However, recent findings suggest that the interaction between these two proteins is likely to be indirect since Pct1 has been shown by immunofluorescence to be located in the nucleus/nuclear membrane at all stages of the cell cycle whereas Sec14 has been shown to have a cytosolic/Golgi membrane location. The same study in which both pathways for the production of PtdCho were inactivated, showed that cell growth rate slowed down when PtdCho synthesis *via* the Kennedy pathway was prevented by a *pct*^{ts} allele in conjunction with an inactive PtdEtn methylation pathway. This could imply that cells at 37°C are unable to sense the switching-off of PtdCho synthesis, possibly due to the existence of a still unknown heat-activated phospholipase which causes the increase in the amount of DAG associated with cell death (Howe *et al.*, 2002).

6.10 Sterols – metabolism, uptake, trafficking and regulation

The primary sterol of yeast is ergosterol which shares many structural similarities with cholesterol found in animal plasma membranes and which to a lesser extent is also found in their organelle membranes (Parks and Casey, 1995). There is no obvious reason why yeast uses ergosterol in preference to other sterols which would pack in certain phospholipid bilayers just as effectively. It is probably a case of ergosterol being best able to satisfy a variety of functions more effectively than other sterols. One of the structural differences between ergosterol (sterol C-28) and cholesterol (sterol C-27) is the existence of a C7–C8 double bond which creates a conjugated diene system in the B-ring (Figure 6.9). Furthermore, ergosterol has an unsaturated side chain which carries a methyl group on C24. Ergosterol and its biosynthetic steps are the major targets for antifungal compounds. There are a number of drugs, for example, polyene antibiotics, that specifically block ergosterol biosynthesis but which have minor effects on cholesterol synthesis of the host organism.

One of the main functions of ergosterol is to maintain a membrane fluidity in the face of environmental changes, for example, medium, temperature. Since sterols are metabolically expensive cellular components it may be that they have other critical roles in the cell. The biosynthetic pathway leading to ergosterol can be divided into two sections: the first is the synthesis of mevalonate from acetyl CoA and the second is the conversion of mevalonate to ergosterol (Figure 6.10). Ergosterol synthesis is an aerobic process requiring molecular oxygen and haem (Parks and Casey, 1995; Lees *et al.*, 1999). Sterol auxotrophic mutants take up sterols from the medium and can grow anaerobically in the presence of added sterols.

The pathway of mevalonate synthesis involves the condensation of two acetyl CoAs by a thiolase, working in the reverse of the final step of β-oxidation. The enzyme is encoded by *ERG10*, the transcription of which is regulated by cellular sterol levels since genetically or pharmacologically-induced reduction of flux through the mevalonate pathway results in an induction of *ERG10* mRNA (Dimster-Denk and Rine, 1996). *ERG10* was one of the genes whose expression was upregulated in freeze-stressed cells and a strain in which *ERG10* was overexpressed has increased freezing tolerance (Rodriguez-Vargas *et al.*, 2002). It could be that the up-regulation of *ERG10* at low temperature satisfies a requirement of higher sterol levels under these conditions. (See Chapter 9 for further information on cold and freeze stress.)

The second addition of acetyl CoA to acetoacetyl CoA catalysed by Erg13 produces 3-hydroxy-3-methylglutaryl CoA (HMG CoA). HMG CoA is then reduced to mevalonate in a reaction catalysed by HMG CoA reductase, an integral ER membrane protein. *S. cerevisiae* possesses two isozymes of HMG CoA reductase encoded by the genes *HMG1* and *HMG2*. Disruption of both *HMG1* and *HMG2* results in inviability and the cells can be rescued by expression of the mammalian HMG CoA reductase. Haem activates HMG CoA reductase and its effect is mediated through a complex with Hap1 that enhances transcription of *HMG1*.

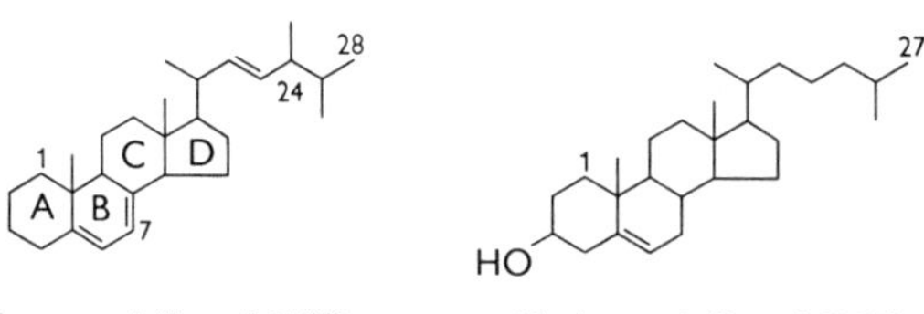

Figure 6.9 Comparison of ergosterol and cholesterol structures.

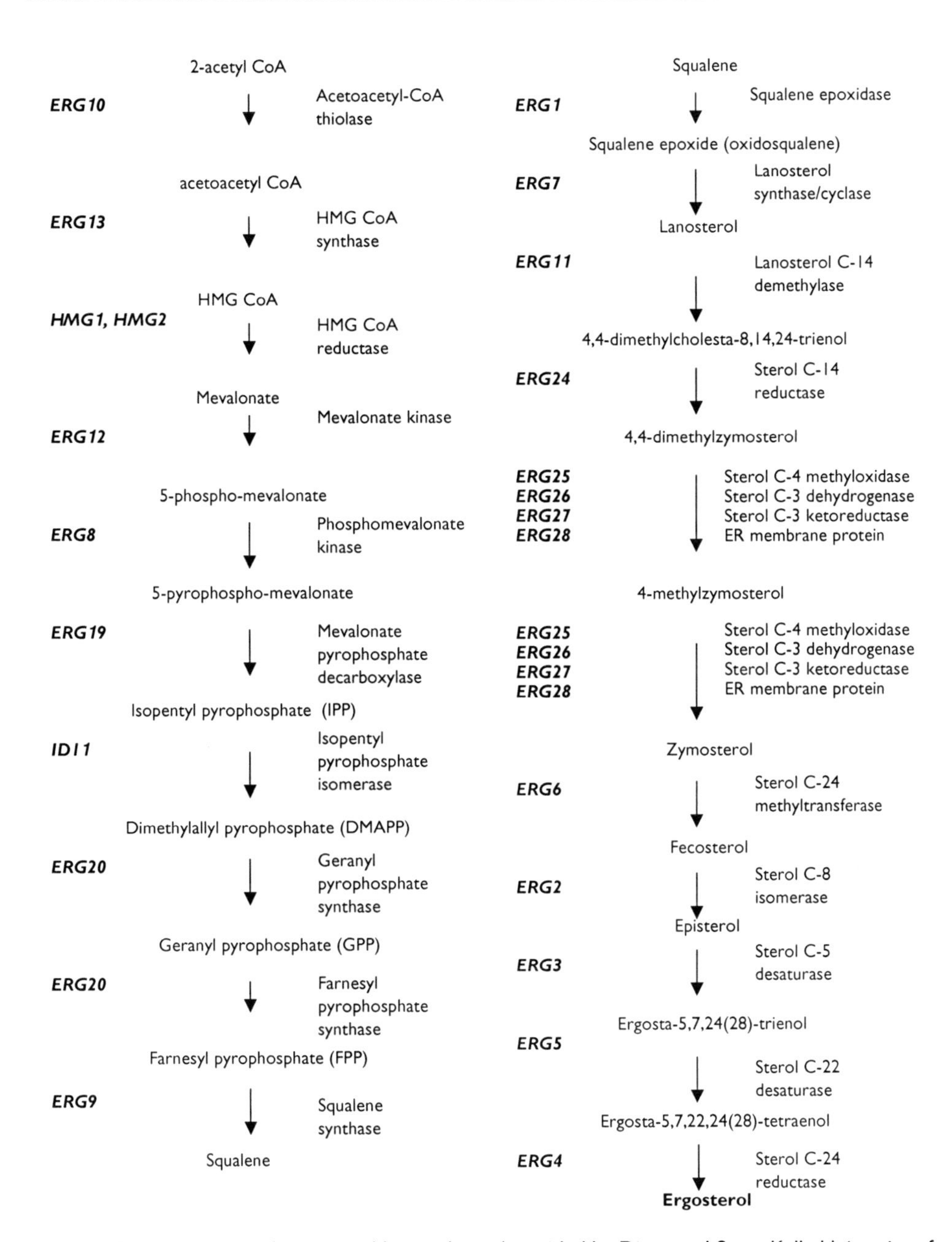

Figure 6.10 Mevalonate and ergosterol biosynthesis (provided by Diane and Steve Kelly, University of Wales, Aberystwyth).

Hmg1 is expressed during logarithmic aerobic growth whereas Hmg2 is induced as the cells enter stationary phase and under anaerobic conditions. This regulation takes place at the translational level and mevalonate is thought to be the signal (Dimster-Denk *et al.*, 1994). Hmg1, like Hmg2, contains eight predicted transmembrane helices and when Hmg2 is over-expressed peripheral ER membrane arrays and short nuclear-associated membrane sterols accumulate. On the other hand, overexpression of Hmg1 induces paired nuclear-associated membranes, known as karmellae and it has been shown that the signal for karmellae induction lies in the eighth loop, Loop G, of the protein (Profant *et al.*, 2000). Other gene products involved in the karmellae biogenesis are those encoded by *VPS* genes (Koning *et al.*, 2002). The reason for this may be that inefficient flux through the secretory and vacuole biogenesis pathways gives rise to altered ER composition and/or function which at restricted temperatures prevents the assembly of karmellae and this is consistent with the fact that many vacuole biogenesis mutants are sensitive to substances, such as tunicamycin and dithiothreitol (DTT) which inhibit protein processing in the ER. A correct environment of lipid and/or protein in the ER appears to be essential for karmellae assembly.

Hmg1 is fairly stable whereas Hmg2 is rapidly degraded through the ubiquitination-mediated degradation pathway in the ER, a process also required for ER quality control. The *UBC7* gene plays a central role in ubiquitination and feedback-regulated degradation of Hmg2. Inhibition of squalene synthase for which farnesyl pyrophosphate (FPP) is the substrate also increased ubiquitination of Hmg2 and its degradation rate, suggesting that FPP is the signal controlling *UBC7*-dependent Hmg2-ubiquitination and degradation. Taking as an example the involvement of HRD (HMG CoA reductase degradation) which is responsible for ERAD (ER-associated degradation) it has been shown that the HRD ubiquitination ligase complex associates with both ERAD substrates and correctly folded stable proteins; but ubiquitination only occurs in the presence of ERAD substrates (Bays *et al.*, 2001; Gardner *et al.*, 2001b). The HRD complex is apparently capable of distinguishing between misfolded, that is, degradation-competent and the correctly folded states.

FPP has been defined as the primary positive signal for Hmg2 degradation. A secondary positive signal is derived from the alternative oxysterol pathway which branches off from the sterol biosynthetic pathway at the level of the *ERG1* gene and is induced by inhibition of squalene epoxidase and increased expression of lanosterol synthase/oxidosqualene lanosterol cyclase (Gardner *et al.*, 2001a). The oxysterol-derived signal in yeast may alter the structure of the Hmg2 transmembrane domain to allow degradation at lower levels of the FPP-derived signal. The other possibility is that oxysterols affect a global biophysical property of the ER membrane which causes instability of HMG CoA reductase. These results indicate that squalene epoxidase acts upon 2,(3*S*)-oxidosqualene (squalene epoxide) and converts it to 2,3(*S*),22(*S*),23-dioxidosqualene which is also a substrate for lanosterol synthase/oxidosqualene lanosterol cyclase. This alternative pathway leads in yeast to the production of the oxysterol, 24(*S*),25-epoxycholesta-5,7,22-triene-3-ol. The oxysterol pathway may be induced by partially inhibiting lanosterol synthase/oxidosqualene lanosterol cyclase which has a higher affinity for dioxidosqualene than oxidosqualene. Therefore, oxysterols may well serve as a source for HMG CoA reductase degradation in both yeast and mammals.

After this short detour into the oxysterol pathway we return to the ergosterol biosynthetic pathway at the step catalysed by the product of *ERG12*. In this step mevalonate is phosphorylated by mevalonate kinase. The mevalonate pathway produces, in addition to sterols involved in membrane structure, several types of isoprenoids which are essential for many cellular functions, for example, the side chains of ubiquinone and haem A involved in electron transport, dolichol for glycoprotein synthesis and farnesyl/geranylgeranyl groups for protein prenylation required, *inter alia* for the control of cell proliferation and mating (Omer and Gibbs, 1994). Erg12 is a cytosolic enzyme which is inhibited by FPP and geranyl pyrophosphate (GPP), intermediates downstream of the reaction. The enzyme is not rate-limiting because its overexpression does not significantly increase the amount of ergosterol produced. The *ERG12* gene has been found to be identical to the *RAR1* gene which was cloned by complementation of mutations which increase the mitotic stability of plasmids carrying a weak ARS as origin of replication (Oulmouden and Karst, 1991). Therefore, it would appear that *erg12* mutants affect DNA replication implying that an intermediate of the sterol pathway or ergosterol itself may be involved in DNA synthesis. It has been shown that mevalonate kinase plays an essential role in DNA replication in animal cells. The 'sparking' function of ergosterol has been described frequently and supposes that infinitely small amounts of ergosterol are required for the cell cycle. Recent studies have shown that this function of sterols can be replaced by haem (Smith and Parks, 1997).

The phosphate group of 5-phospho-mevalonate is converted to a pyrophosphate by phosphomevalonate kinase encoded by *ERG8*. It has been shown recently that there are two classes of phosphomevalonate kinases; the fungal class, also found in eubacteria and plants and the human class also functioning in animals making this enzyme a very attractive target for the development of antibacterial and/or antifungal drugs (Houten and Waterham, 2001).

5-pyrophospho-mevalonate is converted to isopentenyl pyrophosphate (IPP), a 5-carbon isoprene molecule, by mevalonate pyrophosphate decarboxylase encoded by *ERG19*. *ERG19* is an essential gene and overexpression of the enzyme appears to impair growth slightly and results in a low sterol content suggesting that Erg19 is a rate-limiting enzyme whose overproduction causes accumulation of IPP, GPP and/or FPP which may lead to feedback inhibition of Erg12 or HMG CoA reductase (Berges *et al.*, 1997).

IPP isomerase encoded by *IDI1* catalyses the interconversion of the primary 5-carbon homoallylic- and allylic-diphosphate building blocks, IPP and dimethylallyl pyrophosphate (DMAPP). DMAPP is the ultimate source of allylic substrates required to prime the prenyl transferases for chain elongation by condensation with IPP. Thus, the relative demand for DMAPP is higher during periods of maximal sterol biosynthesis which utilises FPP *per se* than during biosynthesis of prenyl pyrophosphates needed for the production of dolichols and ubiquinones where addition of IPP is required. If differential needs exist for isoprenoids of different chain lengths control of the activity of IPP isomerase would provide a mechanism for maintaining a balance of the levels of IPP and DMAPP consistent with metabolic requirements. The biochemical reaction catalysed by IPP isomerase has created substrates for the production of GPP and FPP carried out by the product of the *ERG20* gene.

FPP lies at the meeting point of two important biosynthetic pathways: one for sterol and the other for dolichols, ubiquinones as well as isoprenoids and haem a (Rip *et al.*, 1985; Grabowska *et al.*, 1998). A head-to-head condensation of two FPP molecules and the cleavage of both pyrophosphate moieties in the presence of NADPH is catalysed by the product of the *ERG9* gene to produce squalene. It is the first committed step in ergosterol biosynthesis and regulates the flux of isoprene intermediates through the sterol pathway. Inhibition of squalene synthase by zaragozic acid did not induce Hmg1 expression in contrast to lovastatin treatment which does (Dimster-Denk and Rine, 1996). There is a strong PEST sequence found in the deduced amino acid sequence of Erg9; such sequences are found in proteins which experience rapid intracellular turnover and have sites for serine/threonine phosphorylation which may be used to regulate specific degradation by proteinases. It has been shown that manipulation of squalene synthase does have an effect on polyprenol synthesis suggesting that squalene synthase activity may be the key step in directing FPP to either the sterol or polyprenol pathways rather than the different affinities of the branch point enzymes for FPP (Grabowska *et al.*, 1998).

A mutation in the *ERG20* gene causes the yeast strain to synthesise dolichols containing 14–31 isoprene units rather than the 14–19 isoprene units characteristic for the wild type.

The same strain excretes increased amounts of geraniol, dimethylallyl and isopentenyl alcohols but not farnesol. It has been suggested that the longer dolichols produced *in vivo* is a result of an alteration in the IPP/FPP ratio which affects the activity of the *cis*-prenyltransferase, the first committed step in dolichol biosynthesis. Increased IPP causes *cis*-prenyltransferase to synthesise longer dolichols, a situation observed only to date *in vitro* but now shown *in vivo* (Plochocka *et al.*, 2000).

Longer chain length dolichols are also encountered upon oleate induction (Szkopinska *et al.*, 2001). Deletion of the *PEX1* gene which results in deficiency of peroxisome biogenesis abolishes this effect. It was also shown that in a *pex1Δ* strain there is a reduction in the amount of FPP synthase and HMG CoA reductase. These observations confirm the hypothesis that it is the ratio of IPP/FPP which influences the chain length of dolichols and extends it to suggest that dolichol synthesis occurs in peroxisomes rather than the ER. Supporting evidence is supplied by the finding that FPP synthase and *cis*-prenyltransferase have been detected in mammalian peroxisomes. Although there is evidence that the two genes, *RER2* and *SRT1*, capable of encoding *cis*-prenyltransferase are localised to a subcompartment of the ER in the former case and, in the latter case to lipid particles (Sato *et al.*, 2001). Furthermore, the *RER2* gene is expressed in the early logarithmic phase whereas *SRT1* expression is induced in stationary phase. Further studies show that FPP synthase, independently of HMG CoA reductase and, to a lesser extent, of squalene synthase is the step in ergosterol synthesis which shows the greatest response to changes in the internal and external environmental conditions contributing to the adaptation of the yeast cell (Szkopinska *et al.*, 2000).

The substrate for *ERG20* is also the substrate of the Mod5 protein which is responsible for the isopentenylation of Adenosine (A) to 6-isopentyladenosine (i^6A) on tRNAs in the nucleus (Mod5-II), cytosol and mitochondria (Mod5-I). In a search for genes which when overexpressed affect a nonsense suppression as a result of reduced Mod5-I, the isoform found in both the mitochondria and cytoplasm, *ERG20* was one of the genes recovered (Benko *et al.*, 2000). The reason for this was probably the fact that there is a limited pool of DMAPP and when more is required for sterol biosynthesis less is available for tRNA modification causing a reduction in the amount of nonsense suppression.

Squalene synthase, the product of *ERG9* is regulated at the level of transcription by *HAP1,2,3* and *HAP4, ROX1, YAP1* and *INO2/INO4* emphasising its position at the junction of the O_2-independent/dependent sections of the ergosterol biosynthetic pathway, directly following the branch point to polyprenol synthesis (Kennedy *et al.*, 1999). *INO2/INO4* regulation of this gene implies a link between sterol and phospholipid/fatty acid biosynthesis. Experiments using *ERG9-lacZ* reporter constructs have identified three genes, *TPO1, SLK19* and YER064c whose products specifically regulate squalene synthesis. *TPO1*, which positively regulates *ERG9* expression, is a polyamine transporter and a member of the drug–H^+ antiporter DHA12 family of multidrug efflux proteins in the major facilitator superfamily (Sa-Correia and Tenreiro, 2002). Yer064 and Slk19 decrease *ERG9* expression: Yer064 is predicted to be a membrane-associated small molecule transporter whereas Slk19 is involved in karyogamy and it contains several putative leucine zipper domains suggesting it may have a regulatory role in more than one pathway. An altered sterol profile was observed in strains carrying single deletions of these genes. Furthermore, strains deleted for these regulatory factors are altered in their responses to various antifungal reagents (Kennedy and Bard, 2001).

Investigation of the *MOT3* gene which encodes a nuclear protein capable of repression and activation of transcription revealed that it represses transcription of *ERG9*, as well as that of *ERG2* and *ERG6* during aerobic and hypoxic growth. Ergosterol is known to be required for endocytosis and homotypic vacuole fusion and this may well explain why *mot3Δ* strains are synthetically lethal with *PAN1*, a gene required for endocytosis and *VPS41* whose product is involved in vacuole fusion and protein targeting. Furthermore, *mot3Δ* mutants have increased levels of sterol. These data provide evidence that transcriptional regulation of genes required for ergosterol biosynthesis significantly affects vacuolar function. Overexpression of *UPC2* and *ECM22* which results in an increase in *ERG2* expression gives rise to abnormal vacuolar morphology, probably as a result of the elevated *ERG9* and *ERG2* mRNA levels, providing further evidence that the increase in expression of *ERG9, ERG6* and *ERG2* observed in *mot3Δ* mutants contributes to the *mot3Δ* vacuolar effect (Hongay *et al.*, 2002).

Squalene, an open-chain C_{30} hydrocarbon, is cyclised in two steps to form the tetracyclic sterol skeleton. Squalene epoxidase encoded by *ERG1* catalyses the oxidation of squalene to form squalene epoxide (2,3-oxidosqualene) which is then oxidised by the product of the *ERG7* gene to yield lanosterol. Post-squalene biosynthetic steps are, in contrast to those prior to squalene, dependent on oxygen.

ERG1 is an essential gene and its product is the target of the antifungal compound terbinafine, a non-competive inhibitor of squalene epoxidase. When cells are treated with this compound depletion of ergosterol and accumulation of squalene occur. Furthermore, in the presence of terbinafine yeast cells overexpress Erg1. Erg1 is distributed between the ER and lipid particles and the amount of squalene epoxidase was induced several fold in both compartments following treatment with terbinafine. However, squalene epoxide in lipid particles is apparently enzymatically inactive. It is suggested that the ER contains factors required for enzyme activity and in fact close contact between lipid particles and ER may be necessary in sterol biosynthesis providing a means of trafficking and regulating sterol synthesis in growing cells (Leber *et al.*, 1998). It could be that there is a coordination of sterol biosynthesis with sterol transport from the ER to the plasma membrane and this is effected by lipid particles. In the absence of oxygen yeast becomes auxotrophic for ergosterol and MUFAs. It can be shown that the sterol biosynthetic pathway appears to contribute significantly to the oxygen consumption under anaerobic conditions, possibly as a means for the cell to detoxify potentially harmful intermediates of ergosterol biosynthesis. There is some evidence that there is an alternative oxygen consumption pathway unlinked to the respiratory chain, haem, sterol or MUFA synthesis which has implications for oenological fermentation (Rosenfeld *et al.*, 2003).

EMSA analysis and promoter deletion studies have revealed a unique sequence within the *ERG1* promoter which consists of two 6 bp direct repeats separated by 4 bp (AGCTCGGC-CGAGCTCG) (Leber *et al.*, 2001). In addition to this DNA region there is at least one other sequence of the *ERG1* promoter which specifically binds proteins from induced cells suggesting that the sterol-mediated feedback regulation of *ERG1* expression involves at least two regulatory regions.

ERG7 was not cloned by complementation of an *erg7* mutant but was amplified using degenerate PCR primers based on the known sequence of *ERG7* from *C. albicans* and the gene encoding the lanosterol synthase/oxidosqualene lanosterol cyclase from *A. thaliana* (Shi *et al.*, 1994). The substrate of Erg7, squalene epoxide, which acculumates in an *erg7* deletion strain, allowed the demonstration of the presence of Erg7 in lipid particles (Athenstaedt *et al.*, 1999b; Milla *et al.*, 2002), again indicating that lipid particles have a role in the regulation and trafficking of sterol compounds as well as acting as a reservoir of potentially toxic intermediates which may compromise membrane integrity.

The next step of the ergosterol biosynthesis is catalysed by the cytochrome P450 lanosterol C-14 demethylase (Lees *et al.*, 1999); the gene encoding it, *ERG11*, was isolated by creating a gene library from a strain resistant to ketoconazole; a similar approach used for cloning *ERG1* in which the gene library was prepared from a strain resistant to allylamine (Jandrositz *et al.*, 1991). It has been noted recently that Ca^{2+}-signalling modulates the action of terbinafine and ketoconazole which target Erg1 and Erg11, respectively (Edlind *et al.*, 2002). It may be that Ca^{2+}-signalling plays a regulatory role in biogenesis and maintenance of membranes.

The isolation of *ERG24* involved a strain carrying two recessive mutations *fen1* and *fen2*, which allow survival of cells producing 4,4-dimethylcholesta-8,14,24-trienol whose accumulation renders the cells nystatin-resistant. Transformation of the strain and selection for ergosterol production and nystatin sensitivity yielded a plasmid containing the *ERG24* ORF (Parks and Casey, 1995). Complementation of an *erg24* mutant by the human lamin B receptor confirms that the latter is in fact a sterol C-14 reductase and expression of the human gene increases the cell's susceptibility to fenpropimorph (Silve *et al.*, 1998). The *fen1* and *fen2* mutants confer resistance to fenpropimorph, an inhibitor of Erg24 and Erg2, and were used in order to identify genes involved in the regulation of ergosterol biosynthesis. *FEN1/ELO2* confers general resistance to sterol biosynthesis inhibitors and *fen2*, in addition to its resistance phenotype, results in a reduced sterol content in comparison to the wild type. Deletion of *FEN1* eliminates the ergosterol 'sparking' effect whereas overexpression of *FEN1* results in an increased sensitivity to inhibitors of sterol biosynthesis. *FEN2* has now been identified as the plasma membrane H^+–pantothenate symporter (Stolz and Sauer, 1999).

A mutation in *erg11* in combination with *hem2* or *hem4* can suppress an *erg25* mutation (Gachotte *et al.*, 1997). In the triple mutant (*erg11 erg25 hem2* or *erg11 erg25 hem4*) which accumulates lanosterol, growth is slow but its survival indicates that endogenously synthesised lanosterol can act as a substitute, albeit poor, for ergosterol or the haem mutation permits the uptake of sufficient sterol for the 'sparking' function.

The next two steps in ergosterol biosynthesis involve consecutive removal of two methyl groups from the sterol C-4 position by the concerted action of three enzymes. The *ERG25* gene encodes a sterol C-4 methyloxidase which oxidises one of the methyl groups of 4,4-dimethylzymosterol to a carboxyl group. The carboxylic group is then removed by a second enzyme, sterol C-3 dehydrogenase which is encoded by *ERG26*. The third enzyme required to complete the demethylation is a sterol C-3 ketoreductase encoded by *ERG27* (Figure 6.11). This sequence of reactions catalysed by Erg25, Erg26 and Erg27 has to be repeated to remove the second methyl group thus creating zymosterol (Gachotte *et al.*, 1998, 1999) (Figure 6.10). The *ERG26* gene was identified on account of 30% sequence similarity between the uncharacterised ORF YGL001c and the deduced amino acid sequence of a cholesterol dehydrogenase isolated from *Nocardia* which oxidises the 3-hydroxyl group to a 3-ketone, a reaction deemed to be exactly that capable of removing a C-4 carboxylic acid. Genes responsible for the conversion of lanosterol to zymosterol are essential and their deletion results in ergosterol auxotrophy. However, *ERG26* disruptions are viable in a *hem1Δ* or *hem3Δ* strain background implying that the accumulation of *ERG26* intermediates is toxic to growing haem-competent yeast cells. It can be hypothesised that the haem mutation is required to facilitate sterol uptake and to limit the production of toxic substances. Epitope tagging showed that there is a multienzyme complex containing Erg25, Erg26, Erg27 and a fourth protein dubbed Erg28 (Mo *et al.*, 2002). It is thought that Erg28 acts as a membrane anchor to tether Erg27 and possibly Erg25 and Erg26, thus creating a demethylation complex in the ER

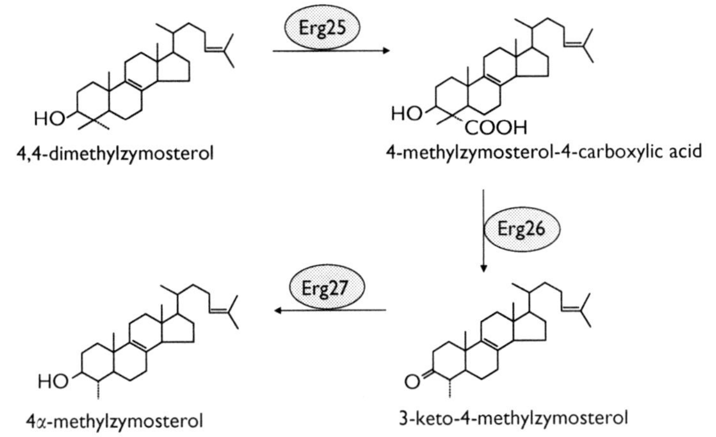

Erg25

HO 4,4-dimethylzymosterol

HO COOH 4-methylzymosterol-4-carboxylic acid

Erg26

Erg27

HO 4α-methylzymosterol

O 3-keto-4-methylzymosterol

Figure 6.11 Steps in sterol C-4 α demethylation (adapted from Mo *et al.*, 2002).

(Figure 6.10). Among mutants with *ts* defects in the incorporation of ^{3}H-inositol into the cell membrane, a sphingolipid biosynthetic mutant was isolated whose phenotype was suppressed by *ERG26* carried on a low copy plasmid. The mutant strain, designated *erg26-1*, has increased rates of biosynthesis of mono- and diacylglycerides (MAG, DAG) but a reduction in the rate of TAG synthesis at the permissive temperature. It was shown that the accumulation of toxic sterol intermediates is responsible for the *ts* loss of the growth phenotype of *erg26-1*. The rate of biosynthesis of PtdOH was almost doubled in *erg26-1* cells at the restrictive temperature. This may be due to a temperature-dependent inhibition of PA phosphatase activities responsible for the conversion of PtdOH to DAG. PtdIns levels were decreased about 2-fold in the temperature-shifted *erg26-1* cells (Baudry *et al.*, 2001). It could be that under such conditions the inositol uptake, mediated by Itr1, may be compromised by the accumulation of zymosterol intermediates. Further experiments are necessary to confirm or refute this.

It has been shown recently that *erg26-1* cells have a defect in sphingolipid metabolism and blocking sterol synthesis results in altered sphingolipid metabolism. Specifically the biosynthetic rate and steady state levels of IPCs were decreased in comparison to that of the wild type. Cells lacking *ERG26* appear to be affected in ceramide-B rather than ceramide-C production. This could be regarded as mimicking the effect of fumonisin (Swain *et al.*, 2002a). Genetic studies have also shown that lowering the level of LCB and increased phosphorylated LCB (LCBP) improved the growth of *erg26-1* cells. Yeast cells lacking the *ARV1* gene are also defective in sphingolipid metabolism. The *ARV1* gene was isolated from a strain in which steroyl esters cannot be synthesised (*are1Δ are2Δ*) (Swain *et al.*, 2002b). It was demonstrated that Arv1 is required for sterol uptake and trafficking in yeast. In addition the rate of synthesis and steady state levels of PtdSer and PtdEtn are reduced, whereas the levels of PtdCho and PtdGro are increased and the levels of sterol esters were increased in *arv1Δ* cells. Such cells also had decreased levels of C18:1. It can be said that Arv1 which localises to the ER and Golgi, plays a role in regulating sterol and sphingolipid homeostasis and it may also be involved in trafficking sphingolipids and sterols between the ER, Golgi and the plasma membrane.

ERG6 is the structural gene for *S*-adenosyl-L-methionine (SAM):C-24 sterol methyltransferase which catalyses the formation of fecosterol from zymosterol at the expense of SAM. Erg6 is not required for normal growth, meiosis and sporulation indicating that the sterol C-24 methyltransferase is not required for the 'sparking' function of sterols in the yeast cell cycle. An *erg6* mutation displays a variety of phenotypes associated with membrane permeability and fluidity and not surprisingly, the *ERG6* gene has been isolated several times in various guises, for example, as a protein counteracting hypersensitivity to Li^+ and Na^+ and as a suppressor of *erd2Δ* which is lacking the receptor responsible for the retrieval of ER proteins from the Golgi (Parks *et al.*, 1999). Interestingly, Erd2-depleted cells accumulate lipid droplets and sterol C-24 methyltransferase is most enriched in the cellular fraction known as lipid microdroplets.

It has long been observed that strains defective in Erg6 are hypersensitive to many drugs, for example small lipophilic compounds. The reason for this is that Erg6 limits the rates of passive drug diffusion across the membrane. Drug uptake is increased directly in an *ERG6* deletant and not via Pdr5, which is a plasma membrane ATP-binding transporter actively involved in exporting drugs (Emter *et al.*, 2002). This property of yeast strains lacking Erg6 has been exploited recently for the construction of a strain devoid of ergosterol biosynthesis. Such a strain will facilitate studies for instance with small molecule inhibitors of protein tyrosine kinases using a tri-hybrid system (Clark and Peterson, 2003).

The involvement of Erg6 in membrane integrity is consistent with the finding that *erg6* mutants are sensitive to brefeldin A, which causes the collapse of Golgi structures and inhibits protein trafficking to the plasma membrane (Parks *et al.*, 1999). The essentiality of *ERG6* has been assessed by competition experiments in which congenic strains differing only in the presence or absence of *ERG* genes involved in the late steps of the ergosterol pathway; with the exception of *ERG5* all strains carrying the wild-type alleles of *ERG24*, *ERG6*, *ERG2* and *ERG3* had a measurable advantage over the strains deleted for the respective gene, underlining the importance of ergosterol biosynthesis to the physiology of *S. cerevisiae* (Palermo *et al.*, 1997).

The next step in ergosterol biosynthesis involves the sterol C-8 isomerase encoded by *ERG2*. Disruption of *ERG2* is lethal in cells grown in the absence of exogenous ergosterol except in mutants resistant to the immunosuppressant SR31747A because they lack either *ELO3/SUR4* or the *ELO2/FEN1* gene product. Microarray analysis has shown that the *ERG2* gene product is the target for SR31747A since there was good agreement between the expression profile of the wild type treated with SR31747A and that of the *ERG2* deletion mutant. Furthermore treatment with SR31747A of an *ERG2* deletion mutant revealed no significant change in the expression profile. The expression of 121 genes was modulated by SR31747A (Cinato *et al.*, 2002). Using a compendium of expression profiles it has been shown that Erg2 is the target of the topical anaesthetic dycylonine. Interestingly the human gene with the greatest similarity to *ERG2* is the so-called σ-receptor which is a neuro-steroid interacting protein which positively regulates K^+-conductance. This suggests that a potential mechanism for the anaesthetic property of dycylonine is by binding to the σ-receptor it inhibits nerve conductance by reducing K^+ current (Hughes *et al.*, 2000a).

Endogenously synthesised ergosterol represses *ERG2* transcription whereas ergosterol exogenously fed to the cell has little or no effect, possibly suggesting that ergosterol or an intermediate metabolite has to be present in a particular cell compartment for the correct regulation of ergosterol biosynthetic genes (Soustre *et al.*, 2000). Recently, it was shown that ergosterol levels are critical for endocytosis in yeast (D'Hondt *et al.*, 2000). Furthermore sterol-compromised cells have altered uptake of various cations and macromolecules. The discovery that *ERG2* and *END11* are identical underlines the connection between sterol metabolism and endocytosis (Munn, 2001). This is confirmed by the report that *ERG3, ERG4, ERG5* and *ERG6* gene products are involved in homotypic vacuole fusion (Munn *et al.*, 1999; Heese-Peck *et al.*, 2002).

ERG3 is the structural gene for sterol C-5 desaturase, probably a mixed-function oxidase which converts episterol to ergosta-5,7,24(28)-trienol. There is some indication that Δ5-sterols may be important for respiration because *erg3Δ* mutants are unable to grow on defined medium containing glycerol or ethanol. Three hundred mutants resistant to the polyene antibiotics, nystatin and filipin, all retained $\Delta 5,7$-sterols whereas an equivalent number of mutants under fermentative conditions revealed no predominance of mutants retaining $\Delta 5$, 7-sterols (Parks and Casey, 1995). It was shown that *erg3*, like *erg6* mutants, are supersensitive to nalidixic acid. Nalidixic acid is an inhibitor of START of the cell cycle. Therefore, it is likely that it is the membrane sterol composition rather than the enzyme activities themselves, which mediates the adaptation to nalidixic acid which leads to restoration of nuclear activity and cell proliferation. Nalidixic acid may also modify membrane lipid composition (Prendergast *et al.*, 1995). Furthermore, it was demonstrated using an *ERG3-lacZ* reporter gene that the absence of ergosterol caused a 35-fold increase in the expression of *ERG3* as measured by β-galactosidase activity. Antifungal agents in mutants and wild-type strains cause up to a 9-fold increase in *ERG3* transcription. These results show that ergosterol exerts a regulatory effect on gene transcription in *S. cerevisiae*. *ERG3* is shown to be allelic to *SYR1* which was cloned by complementation of mutants resistant to the phytotoxin syringomycin E and which are sensitive to growth in the presence of a high concentration of Ca^{2+} (Wangspa and Takemoto, 1998). Erg3 also seems to play a role in resistance to fluconazole, possibly by decreasing efficiency of the Pdr5 efflux pump (Kontoyiannis, 2000a). Mitochondria could function as a physiological partner for Erg3 since mitochondria are postulated to mediate the conversion of non-toxic 14α-methylfecosterol into toxic 14α-methylsterols in the presence of fluconazole (Kontoyiannis, 2000b).

The *ERG5* gene encoding a sterol C-22 desaturase was cloned using a negative selection protocol which involves selection of transformants unable to grow on nystatin-containing medium at 37 °C (Skaggs *et al.*, 1996). The cytochrome P450 superfamily are involved in a variety of monooxygenase reactions including lanosterol demethylase. This cytochrome P450 enzyme, encoded by *ERG5*, is not essential for cell viability whereas *ERG11* is an essential gene since mutants therein result in the requirement of sterol supplementation for growth. It was shown that in addition to its housekeeping role, Erg5 is responsible for benzopyrene promutagen activation, possibly due to an as yet undiscovered activity of P450 enzymes in xenobiotic metabolism (Kelly *et al.*, 1997b). Erg5 is also a target for antifungals (Kelly *et al.*, 1997a).

Deletion strains of *ERG2, ERG5* or *ERG6* were subjected to microarray analysis and the profiles obtained were found to be similar to those obtained by microarray analysis of strains exposed to antifungal agents which target the ergosterol pathway. In addition to genes involved in ergosterol biosynthesis genes required for mitochondrial functions were also identified. For instance, the transcript levels of *COX4, COX5A, COX8* and *COX13* encoding the cytochrome oxidase complex were increased. A similar situation was found for the cytochrome c reductase complex and ATP synthase. Decreased levels were observed for some members of the hypoxic gene family (*CYC7, HEM13*). Three genes involved in the oxidative stress response (*SOD2, GRE2,* YDR453c) also had increased levels of transcripts.

Uncharacterised ORFs were found which responded to ergosterol perturbations and are being investigated further (Bammert and Fostel, 2000). The responsive genes related to lipid and fatty acid syntheses, for example, *FAS1* deletants, but not *FAS2* deletants, have reduced levels of ergosterol esters and sphingolipids. Another study in which isoprenoid biosynthesis was blocked by *inter alia* antifungal azoles (Dimster-Denk *et al.*, 1999) came to similar conclusions, although the direction of change of the transcript levels may have been opposite (Bammert and Fostel, 2000).

The *ERG4* gene product carries out the final step in ergosterol biosynthesis. The sterols accumulated in *erg4* mutants are similar in structure to ergosterol and the effects of sterol substitution are insufficient to distinguish mutant and wild type in regard to differences in permeability or antibiotic sensitivity. Had it not been for the systematic sequencing of the yeast genome and the identification of sequence similarity between *ERG24* and YGL022w coupled with the observation that deletant strains of YGL022w accumulate sterol C-24(28), it would have been a difficult procedure to identify the *ERG4* gene. Activity measurements in subcellular fractions *in vitro* and microscopic analysis of an Erg4-EGFP (enhanced GFP) fusion *in vivo* have shown that Erg4 is localised to the ER as is to be expected since it contains eight transmembrane regions (Zweytick *et al.*, 2000a). From the ER, ergosterol is transported to the plasma membrane, very likely *via* the secretory pathway. The presence of Erg4 in the ER facilitates the further metabolism of ergosterol to ergosterol esters by Are1 and Are2, both of which are localised in the ER. However, the ER does not contain a high concentration of steroyl esters because they are rapidly and efficiently transported to the lipid particles. Ergosterol precursors seem frequently to be esterified prior to their storage in lipid particles emphasising the depot function of this cellular compartment. The lack of ergosterol in the membrane in *erg4* cells may allow the import of certain drugs, for example, terbinafine or nystatin or the cells may become more sensitive to drugs because membrane-bound pumps of the multidrug resistance family may, due to a lack of ergosterol, be less efficient. In a screen for the identification of yeast mutants which have an altered sensitivity to 10-undecanoic acid (C11:1,*Δ10*), it was found that *erg4* mutants are more growth-sensitive to MCFAs than wild-type cells. The association of ergosterol with the yeast plasma membrane H^+-ATPase, Pma1, in lipid rafts suggest an indirect effect (McDonough *et al.*, 2002).

Although the majority of enzymes involved in ergosterol biosynthesis are located in the ER some are found in lipid particles and some in both compartments. This implies that the intermediates in ergosterol biosynthesis have to migrate between the compartments and necessitates a structural and functional relationship between the ER and lipid particles to maintain sterol homeostasis in the yeast cell (Milla *et al.*, 2002).

Much has been written about sterol/cholesterol regulation in mammals, however, until recently it has not yet been shown that yeast possesses a comparable regulatory mechanism for sterol biosynthesis. There are two transcription factors, Upc2 and Ecm22, which bind to the yeast equivalent of a sterol response element (SRE) located in the promoters of many *ERG* genes (Vik and Rine, 2001). Both DNA-binding proteins are members of the Zn(2)-Cys(6) binuclear cluster family of fungal transcription factors. Upc2 and Ecm22 activate *ERG2* and *ERG3* expression by binding directly to the SRE in the relevant promoters. *UPC2* was identified as a mutant which allows sterol uptake under aerobic conditions (Crowley *et al.*, 1998). *ECM22* was discovered because its mutation causes sensitivity to calcofluor white, a substance known to cause cell wall perturbance. Upc2 and Ecm22 have also been implicated in the regulation of the anaerobic DAN/TIR mannoproteins. A strain deleted for *ERG2* is synthetically lethal with an *ecm22Δ upc2Δ* double deletant suggesting that it is a sterol-specific block which generates the signal activating Ecm22 and Upc2. This concept is supported by the induced *ERG2* expression in an *ERG24* deletion strain (Soustre *et al.*, 2000). Deletion of *ERG2* increases expression of an *ERG3-lacZ* reporter construct. These observations support the hypothesis that it is an intermediate acting late in the ergosterol biosynthetic pathway which supplies the signal to Upc2 and Ecm22. Since the SRE (TCGTATA) motif has been found upstream of *ERG1, ERG2, ERG3, ERG6, ERG8, ERG11, ERG12, ERG13* and *ERG25* it is likely that Upc2 and Ecm22 are the major transcriptional regulators of sterol-responsive genes. Interestingly this 7 bp motif is also found in the promoter of *LCB1/TSC2* suggesting that sphingolipid and sterol biosynthesis may be regulated coordinately. In common with the mammalian SREBP, Upc2 and Ecm22 are integral membrane proteins thus making release from the membrane a requirement for their processing (Shianna *et al.*, 2001). Upc2 and Ecm22 may also be transcriptional repressors since *CYC7, BAP2* and *DPP1* are repressed by the level of sterols.

In contrast to the plasma membrane, ER and mitochondrial membranes are poor in sterols (Zinser and Daum, 1995). Yeast cells must therefore carry out intracellular trafficking to direct the sterols into specific compartments. If one assumes that the lack of sterol uptake under aerobic conditions is due to transcriptional inhibition of the genes involved in the uptake, overexpression of such genes on high copy vectors might allow 'illegal' sterol uptake, that is under aerobic conditions. Mutant strains permeable to exogenous sterols in aerobiosis were transformed with a high copy gene library and selection was made for growth on fenpropimorph and ergosterol. Using this protocol the *SUT1* gene was isolated. *SUT1* and the structurally related *SUT2* (YPR009) are two non-essential genes suspected to encode regulatory proteins implicated in the regulation of sterol uptake, biosynthesis and trafficking. It seems that *SUT1* promotes an increase in the flux of the sterol pathway and may increase intracellular sterol trafficking. Sut1which accumulates in the nucleus contains a Zn(2)-Cys(6) binuclear cluster DNA binding domain, common to many transcription factors (Ness *et al.*, 2001). Furthermore *SUT1* is a hypoxic gene regulated by Rox1 whereas this is not the case for *SUT2*. The target genes of *SUT1* include *DAN1* and *ANB1* (Abramova *et al.*, 2001). *DAN1* is a hypoxic gene which encodes a putative GPI-anchored membrane protein whose transcription is not repressed by Rox1 whereas the transcription of *ANB1* is negatively regulated by this transcription factor. Sut1 interacts with Cyc8/Ssn6 thus converting it to a transcriptional activation complex capable of exerting regulatory effects on the expression of *DAN1* (Regnacq *et al.*, 2001). It could be that *SUT1* expression facilitates cell adaptation to anaerobiosis.

The gene products of *Aus1, Pdr11* or *Dan1* are required for anaerobic growth and sterol uptake. Aus1 and Pdr11 are members of the ABC superfamily of membrane transporters. Other genes involved include *HES1/OSH5*, a member of the oxysterol-binding protein family (Wilcox *et al.*, 2002). Four mannoproteins, Dan1, Dan3, Dan4, Tir4, members of the PAU (seripauperin: so-named on account of their low serine content) family were tested for their expression in *upc2* mutants. *DAN1* was highly upregulated in the absence of *UPC2*, however, its deletion significantly reduced sterol uptake. It has been speculated that these mannoproteins contribute to the cholesterol binding proteins of the cell wall and provide a bridge rendering sterol accessible for import processes.

Sterols *per se* can be toxic to the cell, and for storage purposes as well as detoxification they are esterified preferentially with C18:1 fatty acyl chains thus demonstrating substrate specificity in the esterification process in yeast. Two gene products, *ARE1/SAT2* and *ARE2/SAT1* are known to be responsible for 25% and 75% of sterol ester function, respectively. Are1/Sat2 esterifies mainly sterol intermediates whereas Are2/Sat1 seems to prefer ergosterol. Physiologically, Are1 is more important during anaerobiosis and correspondingly Are1 is upregulated under haem-deficient growth conditions whereas Are2 is downregulated (Zweytick *et al.*, 2000b; Jensen-Pergakes *et al.*, 2001). Neither of the genes, *ARE1/SAT2* and *ARE2/SAT1* encoding these enzymes is essential and deletion of one or both genes does not compromise mitotic cell growth or spore germination. However, measurement of ^{14}C-acetate incorporation into saponifiable lipids indicates downregulation of sterol biosynthesis in double deletants. The leucine zipper motifs identified in Are1/Sat2 and Are2/Sat1 could mediate protein–protein interaction, not unexpected since esterification could well be carried out by a multienzyme complex (Sturley, 2000; Guo *et al.*, 2001). The analysis of esterification reactions in yeast is likely to help the understanding of sterol homeostasis and atherosclerosis in humans (Sturley, 1998).

6.11 Oxysterols – their role in lipid metabolism, vesicle transport and cell signalling

Oxysterols are 27-carbon products of cholesterol oxidation which play a role in apoptosis, cellular ageing and sphingolipid metabolism (Schroepfer, 2000). Human oxysterol binding protein (OSBP) was proposed to mediate feedback control of the mevalonate pathway (Lehto and Olkkonen, 2003), however, this is performed by the SRE binding protein (SREBP) and the SREBP cleavage-activating protein (SCAP) (Brown and Goldstein, 1999).

Homologues of OSBP are present in many eukaryotes and the yeast genome contains seven OSBP homologues (Oshs). The three largest Osh proteins in yeast contain a PH domain at their NH_2-termini and a ligand binding (LB) domain at their COOH-ends. Ankyrin and GOLD (Golgi dynamics) domains mediating protein–protein interactions are also found at the NH_2-termini of Osh1/Swh1, Osh2 and Osh3. The four remaining Osh proteins (Osh4–Osh7) consist only of a LB domain. None of the *OSH* genes is essential. However, strains containing disruptions of these homologues have phenotypes similar to viable mutants defective in sterol biosynthesis, for example, cold sensitivity and nystatin resistance. Since elimination of all *OSH* genes results in lethality it may be concluded that all seven yeast *OSHs* perform at least one essential function in common. Viable combinations of *OSH* deletion alleles exhibit specific sterol-related defects (Beh *et al.*, 2001). Expression profiles of individual *OSH* deletion mutants were tested for their influence using promoter fusion reporter plasmids. Loss of *KES1/OSH4* had the greatest effect on the expression of 96 selected genes associated with lipid and sterol metabolism whereas deletion of *OSH2* had the least effect and in general the expression profiles demonstrated clear differences between the various deletion mutants and between deletion mutants and the wild type. In summary it may be said that Osh1/Swh1, Osh5/Hes1 and Osh6 are required for proper regulation of sterol biosynthesis and Osh2 and Osh4/Kes1 may facilitate transfer of ergosterol to the cell membrane. Interestingly disruption of *KES1/OSH4* allows bypass of Sec14. It could be predicted that cells lacking Osh proteins would accumulate aberrant organelles. Although it is not known how *osh4/kes1* mutants mediate the bypass of Sec14, it has been suggested that other *sec14* bypass suppressors re-establish Golgi membrane budding by creating a favourable lipid composition. If, however, Osh4/Kes1 changes Golgi lipid composition this is not brought about by altering the cellular level of ergosterol or PtdCho. Therefore *OSH4* function is independent of PtdCho synthesis. Some of the *OSH* deletants were sensitive to NaCl, a further indication that the cell membrane is affected (Beh *et al.*, 2001).

It was shown that expression of *OSH3* was induced by α-factor and this was dependent on Ste12. Interestingly, *OSH3* was found in a screen for unfolded protein response (UPR) target genes thus implicating it in the control of membrane growth and/or membrane remodelling (Kaufman, 1999). *OSH1* and *OSH5* are negatively controlled by the transcription factors Rox1 and Upc2. *OSH5* expression was elevated in a calcineurin-dependent way after exposure of cells to Ca^{2+} suggesting that Osh5 participates in Ca^{2+}-dependent signalling (Yoshimoto *et al.*, 2002). Osh1 and other PH-containing Osh proteins are responsible for the targeting of proteins to the plasma membrane. However, it was shown that the PH domain of Osh1 targets to the late Golgi and to the nuclear–vacuolar junction whereas Osh2 and Osh3, despite each having a PH domain organisation similar to Osh1, are localised differently (Levine and Munro, 2001, 2002). This difference in intracellular targeting suggests that Osh proteins may perform similar tasks at various cellular sites. Phosphorylation at the 4-position of inositol in PtdIns seems to contribute to recruitment of PH domains to the Golgi apparatus although the PH domain also recognises a second determinant that is ARF dependent (Levine and Munro, 2002).

6.12 Sphingolipids – important for cellular structure and signalling

There are four types of lipids in *S. cerevisiae* membranes: phospholipids, neutral lipids (TAG), sphingolipids and ergosterol. Sphingolipids also function as lipid anchors for numerous cell surface proteins accounting for 30% of the plasma membrane and make up 7–8% of the total membrane mass (Patton and Lester, 1991). The biosynthetic pathways for sphingolipids and glycosphingolipids are detailed in Figures 6.12(a) and (b).

Sphingolipids are defined by long chain bases (LCBs) of which the two in yeast are dihydrosphingosine (DHS; sphinganine) and phytosphingosine (PHS) which differ in their degree of hydroxylation. The LCBs are linked *via* their 2-amino group to a VLCFA to yield ceramides, the precursors of the complex sphingolipids (Dickson and Lester, 2002). Before going on to describe sphingolipid synthesis a short appraisal of the synthesis of VLCFA will be given. The synthesis of VLCFAs requires four reactions; elongation of the fatty acyl chain by two carbon atoms in each cycle carried Elo2/Fen1 and Elo3/Sur4 (see earlier text) and in this respect it follows the same pattern as the reaction cycle of Fas, that is a condensation reaction, a ketoreduction, dehydration and an enoyl reduction. In contrast to Fas, the enzymes catalysing these reactions are not contained on a multifunctional polypeptide located in the cytoplasma but are to be found as individual entities in the microsomal fraction. The first reduction step which uses 3-ketostearoyl CoA as the substrate is carried out by the product of the ORF YBR159w (Han *et al.*, 2002a). In the absence of this protein, yeast strains are compromised in growth, have reduced VLCFAs and an elevated amount of sphingoid bases and medium-chain ceramides. In addition, there is a relief of the Ca^{2+}-sensitivity associated with the *csg2Δ* mutant. Furthermore, the ybr159wΔ strain has reduced dehydratase activity, that is, the next step in the elongation cycle, it could be that Ybr159 is required for the stability or function of the dehydratase activity of the elongation cycle. It has been shown that the ketoreductase co-localises and co-immunoprecipitates with Elo2/Fen1 and Tsc13 which carries out the second reduction step of the cycle. Strains lacking YBR159w display a slightly leaky phenotype; they grow extremely slowly and synthesise some VLCFAs suggesting the existence of a functional homologue. Deletion of the postulated orthologue of YBR159w, *AYR1* which is responsible for the reduction of 1-acyl-DHAP to lyso-PtdOH (Athenstaedt and Daum, 2000), in a strain already carrying a deletion of YBR159w results in inviability suggesting that Ayr1 may be responsible for the residual-3-ketoreductase activity. YBR159w deletants are also synthetically lethal with *fas*-mutants.

The gene encoding the enoyl reductase of the elongation cycle is *TCS13* and it was also identified in the screen for suppressors of Ca^{2+}-sensitivity of *csg2Δ* which are defective in sphingolipid synthesis. The *tsc13* mutant accumulates high levels of LCBs and ceramides containing fatty acids less than C26. Inactivation of Acc1 in a *tsc13* mutant is lethal. Tsc13 is localised to the ER and is highly enriched at nuclear–vacuolar junctions (Kohlwein *et al.*, 2001).

The synthesis of sphingolipids commences in the ER with the condensation of palmitoyl CoA and L-serine to yield 3-ketodihydrosphingosine (3-ketosphinganine) (Figure 6.12(a)).

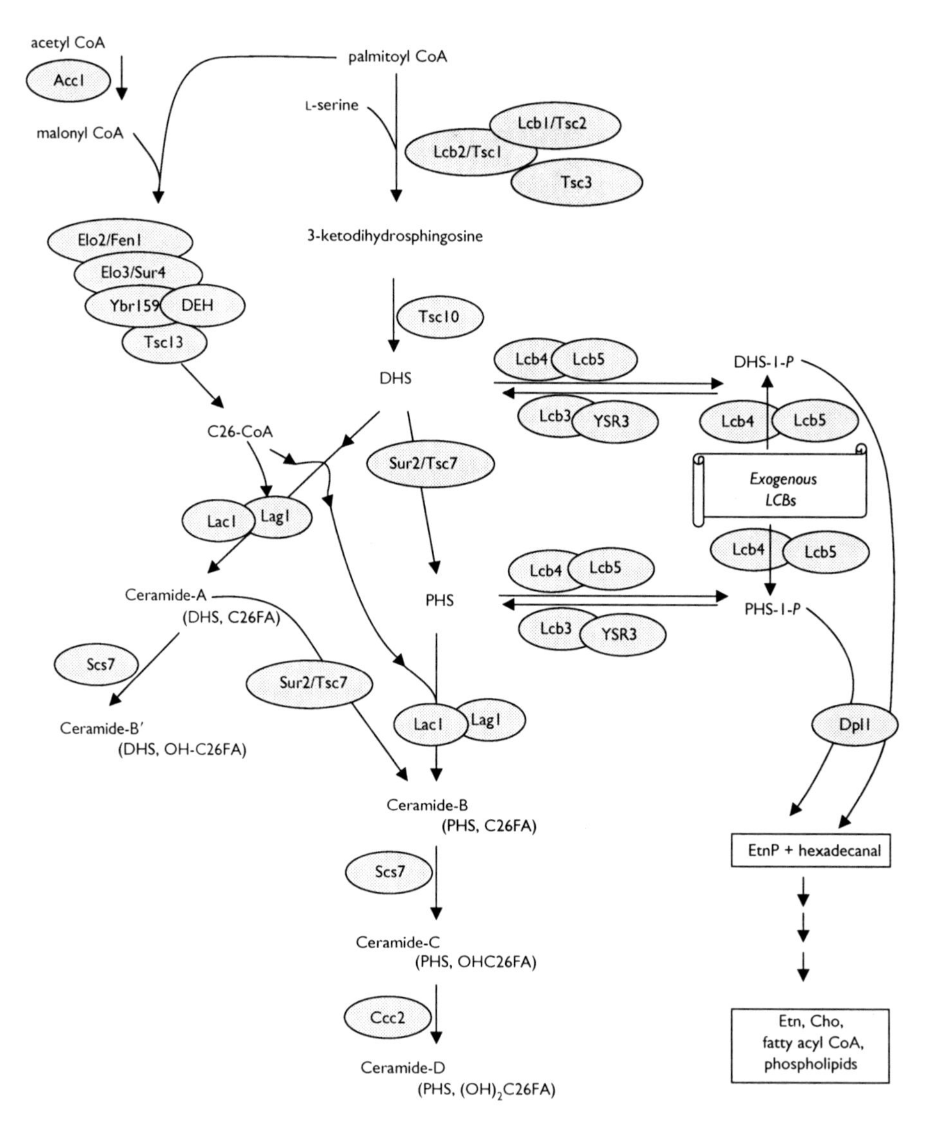

Figure 6.12(a) Synthesis of VLCFs, LCBs and sphingolipids (adapted from Dickson and Lester, 2002).

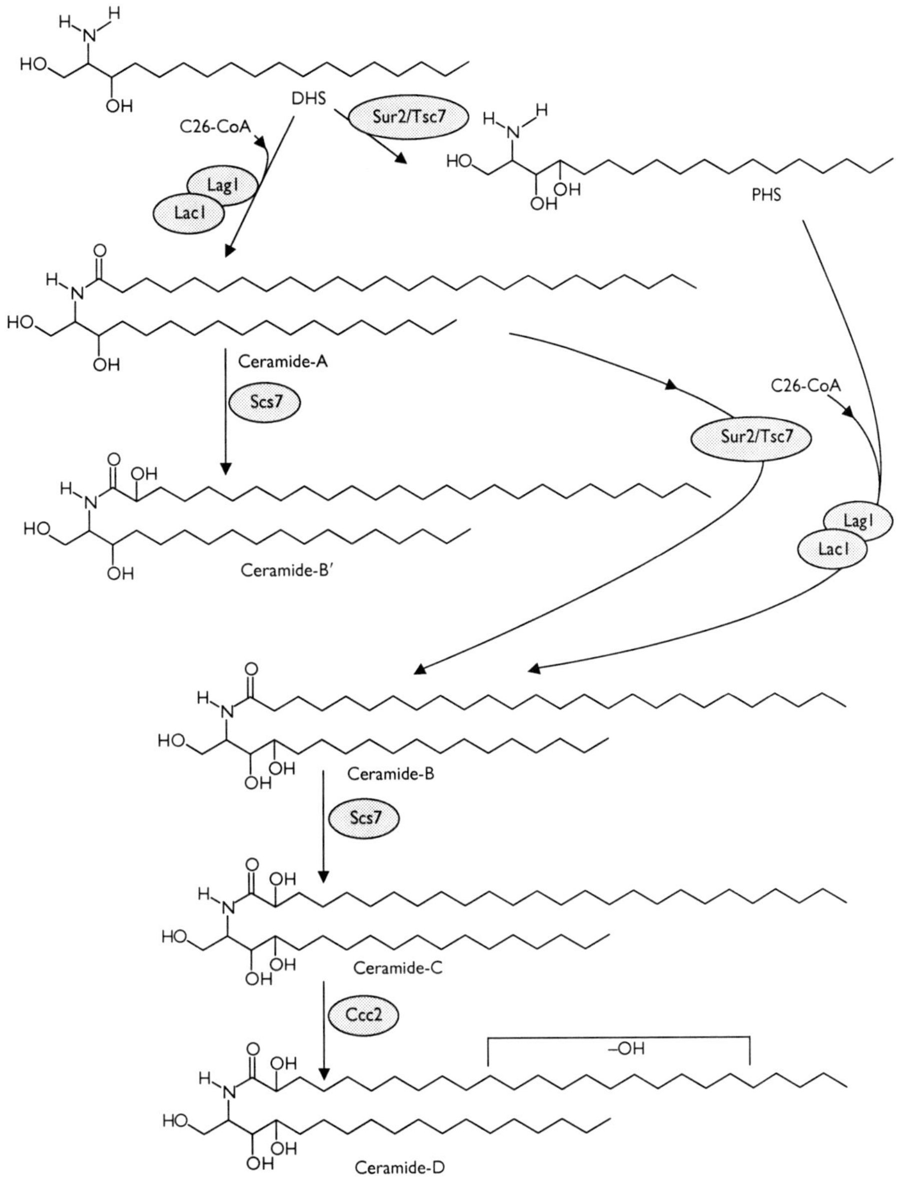

Figure 6.12(b) Ceramide hydroxylation.

The enzyme carrying out the reaction is a membrane-bound L-serine palmitoyl transferase which represents the first committed step in sphingolipid synthesis. There are two essential genes *LCB1/TSC1* and *LCB2/TSC2* which are likely to be the regulatory and catalytic components, respectively, as judged by amino acid similarity to other enzymes catalysing a similar reaction (Dickson and Lester, 2002). In addition Tsc3 is required at elevated temperatures (Gable *et al.*, 2002). The next step in the reaction pathway is the reduction of 3-ketodihydrosphingosine to DHS. The discovery of the gene *TSC10* encoding this enzyme was another success of the screen for *ts* mutants which suppress the Ca^{2+}-sensitivity of *csg2Δ* mutants. The *TSC* supressors uncovered in this screen fall into 15 complementation groups which include *FAS1/TSC4*, *FAS2/TSC5* and *TSC13* (Beeler *et al.*, 1998). His-tagged Tsc10 was expressed in *E. coli* and the recombinant protein was capable of catalysing NADPH-dependent reduction of 3-ketodihydrosphingosine to DHS. Furthermore, *tsc10* mutants accumulate 3-ketodihydrosphingosine and microsomal membranes isolated from such mutants have low 3-ketodihydrosphingosine reductase activity (Beeler *et al.*, 1998). DHS is then subsequently hydroxylated at C4 to yield PHS by the product of *SUR2/TSC7* (Figures 6.12(a) and (b)). PHS and the endproduct of VLCFA synthesis, that is C26-fatty acyl CoA, combine under the action of ceramide synthase encoded by *LAG1* and *LAC1* to yield ceramide-B (Figures 6.12(a) and (b)). Interestingly deletion of *LAG1* which is a longevity-assurance gene, extends life span but it is not known what effect its product has on levels of ceramide and phytosphingosine. Lag1 is likely to influence growth, proliferation and apoptosis in human cells since it may have a role in ceramide signalling (Jazwinski and Conzelmann, 2002). The ceramide level of the cell is regulated not only by the synthesis but also by hydrolysis of ceramide. This latter reaction is catalysed by the products of two genes *YPC1* and *YDC1*. In addition to their ceramidase activity both of these enzymes can generate ceramide by reversing this reaction (Mao *et al.*, 2000a,b). The reverse reaction occurs more readily with Ypc1 than Ydc1. In the ceramidase reaction Ypc1 prefers phytoceramide (ceramide-B: PHS, C26FA) whereas Ydc1 prefers dihydroceramide (ceramide-A: DHS, C26FA) as the substrate. Although neither gene is essential, deletion of *YPC1* renders the strain sensitive to heat stress indicating that it is dihydroceramide but not phytoceramide which plays a role in the heat stress response (which is fully described in Chapter 9) (Jenkins *et al.*, 1997; Mao *et al.*, 2000b). Overexpression of either of these two genes can, on account of their acyl CoA-independent ceramide synthesis, partially correct the sphingolipid synthesis defect in *lag1Δ lac1Δ* mutants. However, *ypc1Δ ydc1Δ lag1Δ lac1Δ* quadruple mutants, in spite of not producing any sphingolipids are viable, possibly because of the novel lipids they produce. The *lag1Δ lac1Δ* double deletant strain is resistant to aureobasidin A, an inhibitor of IPC synthase and displays a 50% reduction in the rate of transport of GPI-anchored proteins to the Golgi (Schorling *et al.*, 2001). In addition to producing ceramide-B this pathway can also produce ceramide-A (DHS, C26FA) by coupling C26-CoA to DHS. Ceramide-A can then be converted to ceramide-B by hydroxylation at C4 of DHS (Swain *et al.*, 2002a) (Figures 6.12(a) and (b)). The hydroxylation of DHS by *SUR2/TSC7* either before or after the addition of C26-CoA gives rise to ceramide-B or ceramide-A, respectively (Figure 6.12(a)).

Saccharomyces cerevisiae synthesises only three complex sphingolipids, but four different ceramides (Dickson and Lester, 2002). In addition to ceramide-A and ceramide-B there is also ceramide-C (PHS OHC26-CoA) and ceramide-D (PHS, $(OH)_2$C26-CoA), derivatives of ceramide-B hydroxylated at the C4 position of the C26 fatty acyl chain and again at an unidentified position more distal in the fatty acyl chain (Figures 6.12(a) and (b)).

There are four hydroxylation steps involved in sphingolipid synthesis: *SCS7/FAH1* introduces the hydroxyl group at the C4 position of the C26-fatty acyl chain of ceramide-A to give ceramide-B′ and at the equivalent position of ceramide-B to give ceramide-C (Figures 6.12(a) and (b)). *SUR2/TSC7* introduces the hydroxyl group into LCB of DHS to give PHS or carries out the same reaction on ceramide-A which is produced by the addition of the C26-fatty acyl CoA to DHS; *CCC2* hydroxylates ceramide-C at varying positions of its C26-fatty acyl chain. *SUR7* was identified as a suppressor of *rvs161* and/or *rvs167* (reduced viability of starvation). Overexpression of *SUR7* suppressed defects in actin polarisation, bud selection and growth in these two mutants. In addition to Sur7 there are two other ORFs YDL222 and YNL194, members of what has become known as the *SUR7* family (Young *et al.*, 2002). Mutants of these integral membrane proteins have altered sphingolipid composition: changes in the length of LCB and the degree of hydroxylation (Swain *et al.*, 2002a; Young *et al.*, 2002). The *SUR7* family defines novel domains in the yeast plasma membrane which do not correspond to actin patches, Num1 patches, Std1 patches or chitin domains induced by osmotic shock; they are stable and have a polarised distribution, being present throughout the cortex of the mother and the older portion of the bud but are not found in small buds. In addition, mutants of the *SUR7* family are compromised in sporulation. Therefore this gene family may be involved in a novel ceramide-based signalling pathway which affects sporulation calling to mind the involvement of Spo14/Pld1 in the phospholipid pathway (Rudge *et al.*, 1998). It can be concluded that *SUR7* on account of its localisation at the plasma membrane is not involved in the major sphingolipid pathway located in the ER and the Golgi.

Each of the ceramides is modified by the addition of *myo*-inositol-1-phosphate from PtdIns with the concomitant release of DAG to form IPC (Figure 6.13). Mutations in the IPC synthase gene, *AUR1*, generate resistance to the antifungal drug aureobasidin A suggesting that this drug inhibits IPC synthase. This enzyme activity is lacking in humans and could provide an ideal target for antifungal drugs (Nagiec *et al.*, 1997). Aureobasidin A is an antifungal antibiotic produced by *Aureobasidium pullulans* R106 and is active against growing cells and might affect microtubule organisation in the cytoskeleton (Hashida-Okado *et al.*, 1996). IPC is in turn mannosylated to give MIPC, a step catalysed by Sur1/Csg1 and Csg2 (Beeler *et al.*, 1994, 1997). The final step in the synthesis of sphingolipids is a repeat of the first step and is carried out by the product of *IPT1* (Dickson *et al.*, 1997b; Leber *et al.*, 1997). Not surprisingly, the gene encoding this enzyme shows 27% identity to Aur1. However, Iptl

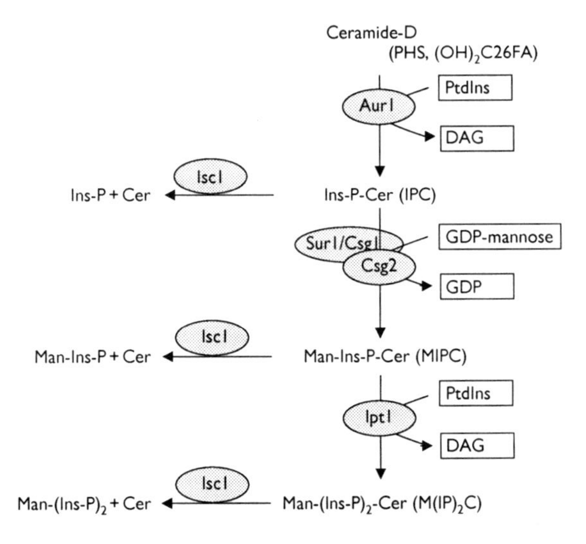

Figure 6.13 Ceramide mannosylation.

is specifically the $M(IP)_2C$ synthase since a strain with a deletion of *IPT1* accumulates MIPC but not $M(IP)_2C$. Interestingly, this enzyme is 1000-fold more resistant than IPC synthase to aureobasidin A (Beeler *et al.*, 1997; Dickson *et al.*, 1997b). Yeast cells unable to carry out this final step in sphingolipid synthesis are resistant to the lipodepsinonapeptide syringomycin E (Grilley *et al.*, 1998) and to a plant defensin from *Dahlia merckii* (Thevissen *et al.*, 2000). It has been shown that the *IPT1* gene is responsible for membrane binding, membrane permeabilisation and growth inhibition caused by the plant defensin suggesting that membrane patches containing $M(IP)_2C$ may constitute binding sites for the defensin itself or function as anchor sites for membrane or cell wall-associated proteins with the defensin.

The intermediates of sphingolipid metabolism including sphingoid bases and ceramides impinge on several cellular signalling pathways involved in endocytosis, heat shock response, growth inhibition, cell integrity and Ca^{2+}-sensitivity (Dickson and Lester, 2002; Young *et al.*, 2002). A specific allele of *LCB1*, *lcb1-100*, originally isolated as the *end8-1*^{ts} is defective in endocytosis (Munn and Riezman, 1994). The *lcb1-100* allele encodes a very labile L-serine palmitoyl transferase activity and sphingolipid synthesis is greatly reduced when the cells are shifted to 37°C (Zanolari *et al.*, 2000). Endocytosis is the process commonly used for recycling of components of the secretory pathway, uptake of micronutrients (Friant *et al.*, 2000) and downregulation of cell surface receptors. In *lcb1-100* strains the levels of the five species of the sphingoid LCB intermediates are reduced in cells grown at the permissive temperature and the level of complex sphingolipids is reduced by 50% (Hearn *et al.*, 2003). Furthermore, the phosphorylated sphingoid bases (DHS-1-*P*, PHS-1-*P*) (Figure 6.12(a)) cannot be detected. LCBs may play a role in endocytosis because they control protein phosphorylation since in *lcb1-100* strains either overexpression of *YCK2* or *PKC1* or inactivation of PP2A protein phosphatase restores endocytosis but does not correct the *ts* phenotype (Shaw *et al.*, 2001). *In vitro* it has been shown that DHS or PHS interacts directly with Pkh1 and Pkh2 which are thought to be involved in the regulation of endocytosis and organisation of the actin cytoskeleton (Dickson and Lester, 2002). Ypk1 and Ykr2 are essential for polarisation of the actin skeleton and they may be regulated by Pkh1 and Pkh2 in an LCB-dependent manner (Schmelzle *et al.*, 2002).

Sphingolipids are also believed to be signalling molecules specific for heat stress. Shifting cells from 25°C to 37°C causes a slight transient increase in C18-DHS and C18-PHS and a dramatic increase in C20-DHS and C20-PHS (Dickson *et al.*, 1997a; Dickson, 1998; Jenkins *et al.*, 1997). Sphingolipid compensatory (Slc) cells fail to accumulate the thermoprotectant trehalose when transferred from 25°C to 37°C in the absence of sphingolipid synthesis implying that sphingolipids play a role in heat-induced trehalose induction (Dickson, 1998). Treatment of wild-type cells with DHS mimics heat stress and results in trehalose accumulation. It has been demonstrated that the phosphorylated species of LCB increase during heat stress suggesting that they and their unphosphorylated derivatives are candidates for the signalling molecules. Strains lacking LCB kinase activity (*lcb4Δ lcb5Δ*) (Figure 6.12(a)) have the same level of resistance to heat stress as the wild type but reduced thermotolerance arguing that DHS-1-*P* and PHS-1-*P* are not required for heat stress resistance while they may play a role during the induction of thermotolerance (Dickson, 1998; Dickson and Lester, 2002).

DHS and PHS can be phosphorylated by the major lipid kinase Lcb4 and the minor one, Lcb5, respectively (Figure 6.12(a)) (Nagiec *et al.*, 1998). Lcb4 oscillates between the *trans*-Golgi network and late endosomes (Hait *et al.*, 2002) and regulates synthesis of ceramide from exogenously added LCB (Funato *et al.*, 2003). DHS-1-*P* and PHS-1-*P* can then be converted into EtnP and fatty aldehydes (hexadecanal) thus providing a link to the phospholipid pathway and fatty acyl lipids. A deletion of the gene *DPL1* encoding the DHS-1-*P* and PHS-1-*P* lyase has several phenotypes including accumulation of phosphorylated sphingoid bases, slow growth in the exponential phase and increased growth in the stationary phase suggesting an involvement of DHS-1-*P* and PHS-1-*P* in growth regulation (Dickson, 1998; Gottlieb *et al.*, 1999). Phosphorylated DHS and PHS can also be dephosphorylated by either of two phosphatases Lcb3 or Ysr3 (Mao *et al.*, 1997; Qie *et al.*, 1997; Mandala *et al.*, 1998). Deletion of *DPL1* and *LCB3* is lethal in vegetative cells suggesting there is normally a flux of these phosphorylated sphingolipids.

Screening for a multicopy suppressor of the LCB-sensitive phenotype caused by an LCBP lyase mutant (*dpl1Δ*) identified a novel gene *RSB1* (resistance to sphingolipid LCB). Rsb1 is postulated to be a transporter or flippase that translocates LCBs from the cytosolic side outward towards the plasma membrane in an ATP-dependent manner. Microarray analysis showed that *RSB1* is one of the genes regulated by Pdr1 and Pdr3 (DeRisi *et al.*, 2000). Indirect immunofluorescence microscopy has demonstrated that Rsb1 is located at the plasma membrane, the ER and other organelles (Kihara and Igarashi, 2002).

An inositol phosphosphingolipid phospholipase C activity releases ceramide from IPC, MIPC and $M(IP)_2C$ (Figure 6.13) and deletion of the *ISC1* gene encoding this activity decreased cellular Na^+ and Li^+ tolerance but K^+ tolerance and general osmostress regulation remained unaffected (Sawai *et al.*, 2000). The increased salt sensitivity of the *isc1Δ* mutant is related to a significant reduction of Na^+/Li^+-stimulated *ENA1* expression. It is suggested that Isc1-dependent hydrolysis of an unidentified yeast inositol phosphosphingolipid represents an early event in one of the salt-induced signalling pathways of *ENA1* transcriptional activation (Betz *et al.*, 2002). There is an Isc1 homologue in *S. pombe* which also releases ceramide from IPC and MIPC and is an essential gene, apparently involved in cell wall formation and division (Feoktistova *et al.*, 2001).

The observations described here make it clear that sphingolipids are very prominent in yeast cell physiology acting as signalling molecules and structural components and which apparently also undergo remodelling.

6.13 GPI anchors – their rise to the surface

Glycosyl-phosphatidylinositol (GPI) membrane anchors also known as lipid tails are evolutionarily well conserved from yeast to mammals (Kinoshita and Inoue, 2000). GPI anchors were first observed in *Trypanosoma* spp. where they contribute to the variant surface glycoprotein (VSG) on the outer leaflet of the plasma membrane (McConville and Menon, 2000). In yeast GPI anchor mutants were originally isolated as being unable to incorporate tritiated inositol into protein (Leidich *et al.*, 1994). GPIs have the conserved core structure: EtnP-6Manα1,2Manα1, 6-Manα1,4-GlcNα1,6Ins-P-lipid. Following their stepwise pre-assembly in the ER, GPIs are transferred to proteins bearing a C-terminal GPI attachment signal sequence. The GPI-anchored proteins are then incorporated into vesicles and transported to the external face of the plasma membrane. Evidence suggests that following attachment to the protein the GPI anchor undergoes remodelling in which DAG is replaced by ceramide (Stevens, 1995). Furthermore, as described in Chapter 5, the GPI on some of the surface glycoproteins becomes attached *via trans*-glycosylation to cell wall β-glucan (Orlean, 1997). Mutants blocking GPI assembly are lethal. The GPI biosynthetic pathway in yeast (Figure 6.14) differs from that in many other organisms in that a fourth mannose is added prior to the addition of

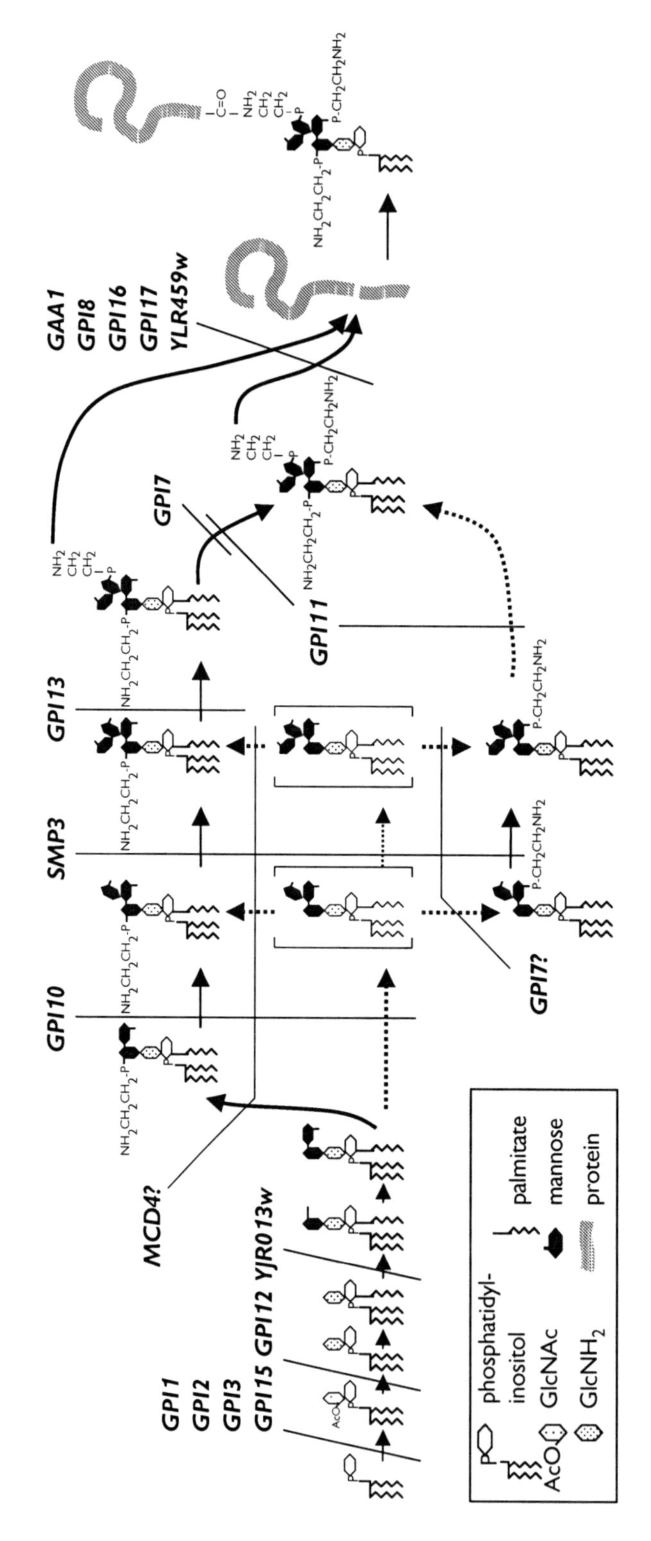

Figure 6.14 Pathway of GPI anchor synthesis (provided by Peter Orlean, University of Illinois).

EtnP to the third mannose. The activity responsible for the addition of the fourth mannose is encoded by *SMP3* (Taron *et al.*, 2000).

The transfer of *N*-acetylglucosamine (GlcNAc) from UDP-GlcNAc to the acceptor PtdIns takes place on the cytoplasmic face of the ER. This transfer is carried out by a multiprotein complex consisting of four subunits: Gpi1, Gpi2, Gpi3 and Gpi15/Ynl038. The last-named ORF encodes an essential gene product and was isolated on the basis of its sequence similarity to the human PIG-H. Gpi3 which belongs to the family of glycosyltransferases, binds UDP-GlcNAc and is probably the catalytic subunit. Membranes from a strain carrying *GPI15*/YNL038w under the expression of a regulatable promoter displayed no GlcNAc-PtdIns transferase activity when the ORF is not expressed. Depletion of Gpi15/Ynl038 in an *smp3* mutant background prevented the formation of tri-mannosylated GPI intermediates normally found in *smp3* strains; thus demonstrating the requirement of Gpi15/Ynl038 for GPI synthesis *in vivo* (Grimme *et al.*, 2001).

The second step in GPI-anchor synthesis is the removal of the *N*-acetyl group from GlcNAc-PtdIns. Human and *S. cerevisiae* homologues are known and the yeast gene *GPI12* is essential. The yeast de-N-acetylase can rescue a mammalian de-N-acetylase mutant (Watanabe *et al.*, 1999).

The next step is carried out by the product of YJR013w, the yeast homologue of PIG-M (Maeda *et al.*, 2001). This gene, sometimes referred to as *GPI14* encodes α-1,4 mannosyltransferase which transfers mannose from dolichol-P-mannose (Dol-P-Man) to the fourth position of $GlcNH_2$. This glycosyltransferase contains a DXD motif found in a lumenal domain suggesting that the transfer of the first mannose occurs on the lumenal side of the ER rather than on the cytoplasmic face where the previous reactions have taken place. This implies the existence of a flippase (Kinoshita and Inoue, 2000) which to my knowledge has not yet been found. In *dpm1* mutants Dol-P-Man synthase is defect and therefore GPI precursor assembly is blocked (Orlean *et al.*, 1988).

Following the second mannosylation EtnP is added to the first mannose by the action of Mcd4. The mammalian orthologue of Mcd4 is PIG-N (Hong *et al.*, 1999). The *MCD4* gene is essential and a mutant of it has cell wall defects, probably due to the lack of cell wall GPI-anchored mannoproteins, for example, Gas1 (Kalebina *et al.*, 2002). Mcd4 consists of a hydrophilic N-terminal lumenal domain and several transmembrane domains (Gaynor *et al.*, 1999). In addition to Mcd4 there are two other homologous proteins, Gpi13 and Gpi7 with a possibility of a third, Gpi11, which are likely to be involved in the transfer of EtnP to Man1, Man2 and Man3 (Hong *et al.*, 1999). Biochemical analysis of Gpi7, located mainly at the plasma membrane, but also in minor amounts in the ER, has shown that this protein is responsible for the addition of EtnP to the second mannose of the GPI core (Benachour *et al.*, 1999). Following the action of Gpi10 which is responsible for attaching the third mannose in α-1,2-linkage to the GPI core structure intermediate (Sutterlin *et al.*, 1998), Gpi13 was shown to be involved in transferring EtnP to the third mannose of the GPI core (Flury *et al.*, 2000).

There is evidence that the GPI assembly pathway in yeast is branched since it would appear that the GPIs accumulating in a *GPI11*-defective strain are likely to have been generated independently of each other (Figure 6.14). It is feasible that Gpi11 functions in concert with each family member, that is Mcd4, Gpi7 and Gpi13, of the homologous EtnP transferases. Interestingly, the postulated branching of the GPI-biosynthetic pathway may well be correlated with a physical separation raising the possibility that the GPIs synthesised by the different branches of the pathway are destined for different functions in the yeast cell (Taron *et al.*, 2000; Richard *et al.*, 2002).

Deletion of any one of the three mutants encoding EtnP transferases (Mcd4, Gpi7 and Gpi13) results in constitutive activation of the Hog1 MAP kinase cascade (see Section 8.6 of Chapter 8, Chapter 9 and Toh-e and Oguchi, 2001). Interestingly, *GPI7* is not an essential gene whereas *MCD4* and *GPI13* are, suggesting that EtnP on Man2 may not be essential for its function or that Man2 can receive its EtnP by the action of Gpi11 or Mcd4.

The GPI transamidase consists of four proteins: Gaa1, Gpi8, Gpi16 and Gpi17 (Fraering *et al.*, 2001; Ohishi *et al.*, 2001). Gpi8 is likely to be the catalytic component because of homology to members of a family of cysteine proteases which have transamidase activity. The stable expression of Gaa1 and Gpi8 is dependent on Gpi17 and Gpi16. The finding that a stable complex lacking PIG-S, the homologue of Gpi17, could be found in human cells is in agreement with the fact that Fraering *et al.* (2001) isolated a GPI transamidase complex from yeast which consisted of Gaa1, Gpi8 and Gpi16. Peter Orlean (personal communication) and Erfei Bi have found that there may be a further activity involved: a *ts* allele of the ORF YLR459w was discovered to have a GPI transamidase phenotype when tested for the accumulation of ^{3}H-inositol-labelled GPIs and also has an unusual actin organisation phenotype suggesting that YLR459w functions both in GPI synthesis and actin organisation (P. Orlean, unpublished results). (Note added in proof: Hong *et al.*, 2003 have suggested that Ylr459/Cdc91 may recognise the GPI attachment signal or the lipid moiety of GPI.)

The addition of the fourth mannose is catalysed in yeast by the product of the essential gene *SMP3*. Strains carrying an *smp3*^{ts} mutation accumulate Man3-GPIs and prevent the production of substrates for GPI transamidase. The presence of the fourth mannose in the yeast GPI core structure represents a potential drug target, even though there is some suggestion of a fourth mannose having been detected in the murine Thy-1 glycoprotein (Homans *et al.*, 1988) and the human renal membrane dipeptidase (Brewis *et al.*, 1995). Paradoxically the presence of the fourth mannose is required for the action of Gpi13 which transfers EtnP to Man3. The fourth mannose may be necessary for the incorporation of certain mannoproteins into the cell wall *via* crosslinking between one of the mannoses in the GPI and β-1,6-glucan. A partial defect in the processing of Gas1 in the presence of an *smp3* allele is consistent with the fourth mannose being required for the export of GPI-anchored proteins from the ER.

In yeast it has been reported that the acylation of inositol at position 2 of the GPI core structure is required for subsequent mannosylation of the GlcNAc residue. An acyl CoA-dependent activity in yeast preparations has been described (Costello and Orlean, 1992; Doerrler and Lehrman, 2000). However, since so far no ORF encoding deacetylase has been described the acyl chain attached to the inositol ring is not indicated in Figure 6.14. (Note added in proof: Two recent papers (Tsukahara *et al.*, 2003; Umemura *et al.*, 2003) describe the identification of a gene, *GWT1*, whose product is responsible for the acylation of inositol in GPI anchors.)

After the GPI moiety has been synthesised and attached to the relevant protein *via* the bridging EtnP on Man3 the GPI protein is exported from the ER *via* the Golgi to the plasma membrane in the form of vesicles. Efficient packaging of GPI protein into such vesicles requires Emp24 and Erv25 as cargo receptors (Muniz *et al.*, 2000). These vesicles make up a different population from that used by the secretory proteins, pro-α-factor and Gap1 (Watanabe *et al.*, 2002). This transport is dependent on ongoing sphingolipid synthesis, in particular ceramide and/or inositol sphingolipid are indispensable for the transport of GPI-anchored proteins (Guillas *et al.*, 2001). GPI anchor proteins could function either as membrane components or as substrates for remodelling of GPI-lipid moieties. It could be that sphingolipid/sterol-enriched detergent-insoluble microdomains, also known as lipid rafts, may play a role in protein sorting and lipid traffic (Brown and London, 1998).

Ceramide has been shown to promote membrane microdomain formation *in vitro* (Xu *et al.*, 2001) or it may be that the hydrophobic interaction between ceramide and the GPI-lipid moiety is required for membrane association. It should, however, be noted that *erg3/6* mutants are not affected in the Gas1 transport from the ER to the Golgi. It could be, however, that ceramides play a more active role by acting as a substrate for remodelling of the GPI-lipid moiety which occurs in both the ER and Golgi structures (Watanabe *et al.*, 2002).

From the Golgi, the vesicles fuse with the plasma membrane and the GPI-anchored proteins are despatched to their functional compartments, either the plasma membrane, or as in the case for **a**-agglutinin, transferred from the GPI anchor to cell wall glycan thus exposing its binding domain for **a**-agglutinin on the surface of the MATα cell (Lipke and Kurjan, 1992). This implies the important role of lipids in transmembrane signalling pathways where they act as a sorting signal in protein targeting. The remodelling of the GPI anchor may contribute to the lateral mobility and conformation of a protein in the membrane. GPI proteins are not only attached to the plasma membrane but also are found as an intrinsic part of the cell wall. GPI-proteins in the cell wall lack at least the phospholipid part of the GPI-molecule (Caro *et al.*, 1997).

In connection with this discussion on the synthesis of GPI anchors and their transfer to protein it is appropriate to mention an idea proposed in the late 1980s that the transport and sorting of GPI-anchored proteins depends on interaction with glycosphingolipids (Futerman, 1995). Support for this hypothesis is provided by the fact that inhibition of sphingolipid synthesis reduces the rate of transport of GPI-anchored proteins in yeast. In the ER GPI-anchored proteins are associated with IPC and after further modification in the Golgi apparatus to form M(IP)$_2$C, the microdomains, that is, the physical association of IPC and GPI-anchored proteins, are transported as a vesicular entity to the surface of the cell (Futerman, 1995; Hooper, 1998). By blocking the transport of the GPI-anchored proteins out of the ER, the synthesis of IPC and M(IP)$_2$C is reduced as was L-serine palmitoyltransferase. This can be explained by postulating that inhibition of the movement of GPI-anchor protein out of the ER leads to a block in the glycosphingolipid transport which then results in an accumulation of IPC in the ER, either by direct inhibition of enzyme activity or by downregulation of L-serine palmitoyltransferase.

The physiological significance of GPI membrane anchorage is not immediately obvious. Possibly, the spatial requirements, clustering and diffusion properties in the outer leaflet of the plasma membrane determine how a protein is fixed in the membrane. A regulatory role is possible since the concentration of certain proteins at the cell surface could be altered by the action of dedicated phospholipases cleaving the GPI anchor. Furthermore, GPI anchors may confer properties on the respective proteins not provided by anchorage *via* transmembrane domains.

6.14 Lipid rafts – the way in and out

Lipid rafts are clusters of membrane lipids which are found in ergosterol/cholesterol and sphingolipid-rich membranes, such as the plasma membrane (Harder and Simons, 1997; Simons and Ikonen, 1997; Xu *et al.*, 2001). The tight packaging of lipids is due to the fact that sphingolipids have highly saturated acyl chains. In the presence of ergosterol these sphingolipids can exist in the liquid-ordered (L_0) phase as opposed to the liquid crystalline (L_α) phase in which unsaturated phospholipid-rich domains are found (London and Brown, 2000). This tight packaging of lipids allows the isolation of lipid rafts as detergent-resistant membranes.

Sphingolipid and ergosterol biosynthetic mutants are deficient in mating and fail to polarise proteins to the tip of the *shmoo* (Bagnat and Simons, 2002). The structure of sterols and sphingolipids confers a species specificity on lipid raft formation (Xu *et al.*, 2001). It is known that Gas1 and Pma1, a plasma membrane H^+-ATPase are associated with lipid rafts (Bagnat *et al.*, 2000, 2001; Gong and Chang, 2001; Lee *et al.*, 2002). In the case of Pma1 oligomerisation is linked to lipid rafts since in cells depleted of ceramide Pma1 exists in the monomeric state. This association of Pma1 with lipid rafts has been confirmed by using a conditional *elo3Δ erg6*^{ts} mutant since in this genetic background lipid rafts associated with pre-existing Pma1 were compromised. Furthermore at 37°C in an *elo3Δ* strain newly synthesised Pma1 did not enter raft domains at the beginning of the biosynthetic pathway and when it arrived at the plasma membrane it was sent to the vacuole for degradation indicating that the C26 fatty acid substitution is important for raft association of Pma1. Furthermore in the *elo3Δ erg6*^{ts} mutant a selective enrichment of ergosterol was discovered in the lipid rafts indicating that both sterols and sphingolipids play a role in maintaining raft domains. An *erg6Δ* mutant apparently compensates the lack of ergosterol by increasing both sterol and sphingolipid levels. A modest increase in sterol levels also appears to compensate for the altered sphingolipid composition of *elo3Δ* mutants. In an *elo3Δ erg6Δ* strain supplemented with ergosterol, raft association of Gas1 is not affected but that of Pma1 is, suggesting that GPI-anchored proteins associate with domains with a different lipid composition from that of the typical raft (Eisenkolb *et al.*, 2002). It was also found in *lcb1–100* cells that the uracil transporter Fur4 was drastically reduced when the cells are heat-shocked: uracil transport is not restored as in the wild type implying that a functional secretory pathway and synthesis of complex lipids was involved and Fur4 is associated with lipid rafts (Hearn *et al.*, 2003).

6.15 Conclusions

The wealth of information gained in the past 4–5 years on the biosynthesis and regulation of lipids and membrane homeostasis goes to prove that high throughput analysis is not the only way of gaining insight into the workings of a cell. While there is a reliance on modern technology, for example, lipid mass spectrometry and confocal microscopy, many of the results and conclusions have been obtained by 'old-fashioned' suppressor selection and measurement of enzyme activity and the ability to 'ask yeast the right questions' with the result that we can now begin to appreciate the full impact of lipids on cell signalling, trafficking, membrane physiology and cellular structures. Anyone who believed that lipids are just 'fat' is in for a surprise!

Acknowledgements

I am especially grateful to Professor Peter Orlean, University of Illinois for discussion, sharing unpublished results on GPI anchor synthesis and providing Figure 6.14. I thank Professor Steve Kelly and Dr Diane Kelly, University of Wales, Aberystwyth for providing Figure 6.10, Dr Regina Leber, Universitaet Graz for clarifying the terbinafine story and Professor Vytas Bankaitis, University of Alabama for discussions concerning Sec14. I also greatly appreciated the timeliness of the 43rd International Conference on the Bioscience of Lipids, Graz, Austria, 2002 organised by Professor Guenther Daum, TU Graz. Among the interesting discussions I had there I would like to single out the one with Professor Dennis Voelker, National Jewish Medical and Research Centre, Denver.

References

Abramova, N. E., Cohen, B. D., Sertil, O., Kapoor, R., Davies, K. J. *et al.* (2001) Regulatory mechanisms controlling expression of the DAN/TIR mannoprotein genes during anaerobic remodeling of the cell wall in *Saccharomyces cerevisiae. Genetics* **157**, 1169–1177.

Ago, T., Takeya, R., Hiroaki, H., Kuribayashi, F., Ito, T. *et al.* (2001) The PX domain as a novel phosphoinositide-binding module. *Biochemical and Biophysical Research Communications* **287**, 733–738.

Ambroziak, J. and Henry, S. A. (1994) *INO2* and *INO4* gene products, positive regulators of phospholipid biosynthesis in *Saccharomyces cerevisiae*, form a complex that binds to the *INO1* promoter. *Journal of Biological Chemistry* **269**, 15344–15349.

Ames, J. B., Hendricks, K. B., Strahl, T., Hüttner, I. G., Hamasaki, N. *et al.* (2000) Structure and calcium-binding properties of Frq1, a novel calcium sensor in the yeast *Saccharomyces cerevisiae. Biochemistry* **39**, 12149–12161.

Ansari, K., Martin, S., Farkasovsky, M., Ehbrecht, I. M. and Küntzel, H. (1999) Phospholipase C binds to the receptor-like GPR1 protein and controls pseudohyphal differentiation in *Saccharomyces cerevisiae. Journal of Biological Chemistry* **274**, 30052–30058.

Ashburner, B. P. and Lopes, J. M. (1995) Regulation of yeast phospholipid biosynthetic gene expression in response to inositol involves two superimposed mechanisms. *Proceedings of the National Academy of Sciences USA* **92**, 9722–9726.

Ashrafi, K., Farazi, T. A. and Gordon, J. I. (1998) A role for *Saccharomyces cerevisiae* fatty acid activation protein 4 in regulating protein N-myristoylation during entry into stationary phase. *Journal of Biological Chemistry* **273**, 25864–25874.

Athenstaedt, K. and Daum, G. (1997) Biosynthesis of phosphatidic acid in lipid particles and endoplasmic reticulum of *Saccharomyces cerevisiae. Journal of Bacteriology* **179**, 7611–7616.

Athenstaedt, K. and Daum, G. (1999) Phosphatidic acid, a key intermediate in lipid metabolism. *European Journal of Biochemistry* **266**, 1–16.

Athenstaedt, K. and Daum, G. (2000) 1-Acyldihydroxyacetone-phosphate reductase (Ayr1p) of the yeast *Saccharomyces cerevisiae* encoded by the open reading frame YIL124w is a major component of lipid particles. *Journal of Biological Chemistry* **275**, 235–240.

Athenstaedt, K., Weys, S., Paltauf, F. and Daum, G. (1999a) Redundant systems of phosphatidic acid biosynthesis via acylation of glycerol-3-phosphate or dihydroxyacetone phosphate in the yeast *Saccharomyces cerevisiae. Journal of Bacteriology* **181**, 1458–1463.

Athenstaedt, K., Zweytick, D., Jandrositz, A., Kohlwein, S. D. and Daum, G. (1999b) Identification and characterization of major lipid particle proteins of the yeast *Saccharomyces cerevisiae. Journal of Bacteriology* **181**, 6441–6448.

Audhya, A. and Emr, S. D. (2002) Stt4 PI 4-kinase localizes to the plasma membrane and functions in the Pkc1-mediated MAP kinase cascade. *Developmental Cell* **2**, 593–605.

Bae-Lee, M. and Carman, G. M. (1990) Regulation of yeast phosphatidylserine synthase and phosphatidylinositol synthase activities by phospholipids in Triton X-100/phospholipid mixed micelles. *Journal of Biological Chemistry* **265**, 7221–7226.

Bagnat, M. and Simons, K. (2002) Cell surface polarization during yeast mating. *Proceedings of the National Academy of Sciences USA* **99**, 14183–14188.

Bagnat, M., Keranen, S., Shevchenko, A. and Simons, K. (2000) Lipid rafts function in biosynthetic delivery of proteins to the cell surface in yeast. *Proceedings of the National Academy of Sciences USA* **97**, 3254–3259.

Bagnat, M., Chang, A. and Simons, K. (2001) Plasma membrane proton ATPase Pma1p requires raft association for surface delivery in yeast. *Molecular Biology of the Cell* **12**, 4129–4138.

Bammert, G. F. and Fostel, J. M. (2000) Genome-wide expression patterns in *Saccharomyces cerevisiae*, comparison of drug treatments and genetic alterations affecting biosynthesis of ergosterol. *Antimicrobial Agents and Chemotherapy* **44**, 1255–1265.

Bankaitis, V. A., Malehorn, D. E., Emr, S. D. and Greene, R. (1989) The *Saccharomyces cerevisiae SEC14* gene encodes a cytosolic factor that is required for transport of secretory proteins from the yeast Golgi complex. *Journal of Cell Biology* **108**, 1271–1281.

Baudry, K., Swain, E., Rahier, A., Germann, M., Batta, A. *et al.* (2001) The effect of the *erg26-1* mutation on the regulation of lipid metabolism in *Saccharomyces cerevisiae*. *Journal of Biological Chemistry* **276**, 12702–12711.

Baumgartner, U., Hamilton, B., Piskacek, M., Ruis, H. and Rottensteiner, H. (1999) Functional analysis of the Zn(2)Cys(6) transcription factors Oaf1p and Pip2p. Different roles in fatty acid induction of beta-oxidation in *Saccharomyces cerevisiae*. *Journal of Biological Chemistry* **274**, 22208–22216.

Bays, N. W., Wilhovsky, S. K., Goradia, A., Hodgkiss-Harlow, K. and Hampton, R. Y. (2001) *HRD4/NPL4* is required for the proteasomal processing of ubiquitinated ER proteins. *Molecular Biology of the Cell* **12**, 4114–4128.

Beeler, T., Gable, K., Zhao, C. and Dunn, T. (1994) A novel protein, CSG2p, is required for Ca2+ regulation in *Saccharomyces cerevisiae*. *Journal of Biological Chemistry* **269**, 7279–7284.

Beeler, T. J., Fu, D., Rivera, J., Monaghan, E., Gable, K. *et al.* (1997) *SUR1* (*CSG1/BCL21*), a gene necessary for growth of *Saccharomyces cerevisiae* in the presence of high Ca2+ concentrations at 37 °C, is required for mannosylation of inositolphosphorylceramide. *Molecular and General Genetics* **255**, 570–579.

Beeler, T., Bacikova, D., Gable, K., Hopkins, L., Johnson, C. *et al.* (1998) The *Saccharomyces cerevisiae* TSC10/YBR265w gene encoding 3-ketosphinganine reductase is identified in a screen for temperature-sensitive suppressors of the Ca2+-sensitive *csg2Δ* mutant. *Journal of Biological Chemistry* **273**, 30688–30694.

Beh, C. T., Cool, L., Phillips, J. and Rine, J. (2001) Overlapping functions of the yeast oxysterol-binding protein homologues. *Genetics* **157**, 1117–1140.

Benachour, A., Sipos, G., Flury, I., Reggiori, F., Canivenc-Gansel, E. *et al.* (1999) Deletion of GPI7, a yeast gene required for addition of a side chain to the glycosylphosphatidylinositol (GPI) core structure, affects GPI protein transport, remodeling, and cell wall integrity. *Journal of Biological Chemistry* **274**, 15251–15261.

Benko, A. L., Vaduva, G., Martin, N. C. and Hopper, A. K. (2000) Competition between a sterol biosynthetic enzyme and tRNA modification in addition to changes in the protein synthesis machinery causes altered nonsense suppression. *Proceedings of the National Academy of Sciences USA* **97**, 61–66.

Berges, T., Guyonnet, D. and Karst, F. (1997) The *Saccharomyces cerevisiae* mevalonate diphosphate decarboxylase is essential for viability, and a single Leu-to-Pro mutation in a conserved sequence leads to thermosensitivity. *Journal of Bacteriology* **179**, 4664–4670.

Betz, C., Zajonc, D., Moll, M. and Schweizer, E. (2002) ISC1-encoded inositol phosphosphingolipid phospholipase C is involved in Na^+/Li^+ halotolerance of *Saccharomyces cerevisiae*. *European Journal of Biochemistry* **269**, 4033–4039.

Bhatnagar, R. S., Futterer, K., Waksman, G. and Gordon, J. I. (1999) The structure of myristoyl-CoA:protein N-myristoyltransferase. *Biochimica et Biophysica Acta* **1441**, 162–172.

Birner, R., Burgermeister, M., Schneiter, R. and Daum, G. (2001) Roles of phosphatidylethanolamine and of its several biosynthetic pathways in *Saccharomyces cerevisiae*. *Molecular Biology of the Cell* **12**, 997–1007.

Birner, R., Nebauer, R., Schneiter, R. and Daum, G. (2003) Synthetic lethal interaction of the mitochondrial phosphatidylethanolamine biosynthetic machinery with the prohibitin complex of *Saccharomyces cerevisiae*. *Molecular Biology of the Cell* **14**, 370–383.

Bonangelino, C. J., Nau, J. J., Duex, J. E., Brinkman, M., Wurmser, A. E. *et al.* (2002) Osmotic stress-induced increase of phosphatidylinositol 3,5-bisphosphate requires Vac14p, an activator of the lipid kinase Fab1p. *Journal of Cell Biology* **156**, 1015–1028.

Boumann, H. A., Heck, A. J. R., de Kruijff, B. and De Kroon, A. I. P. M. (2002) Generation of the steady state molecular species profile of PC in *S. cerevisiae*. In: R. M. Epand and P. K. Kinnuen (eds) *43rd International Conference on the Bioscience of Lipids*, p. 21. Graz, Austria: Elsevier.

Brewis, I. A., Ferguson, M. A., Mehlert, A., Turner, A. J. and Hooper, N. M. (1995) Structures of the glycosyl-phosphatidylinositol anchors of porcine and human renal membrane dipeptidase. Comprehensive structural studies on the porcine anchor and interspecies comparison of the glycan core structures. *Journal of Biological Chemistry* **270**, 22946–22956.

Brody, S., Oh, C., Hoja, U. and Schweizer, E. (1997) Mitochondrial acyl carrier protein is involved in lipoic acid synthesis in *Saccharomyces cerevisiae*. *FEBS Letters* **408**, 217–220.

Brown, D. A. and London, E.(1998) Functions of lipid rafts in biological membranes. *Annual Reviews of Cell and Developmental Biology* **14**, 111–136.

Brown, M. S. and Goldstein, J. L. (1999) A proteolytic pathway that controls the cholesterol content of membranes, cells, and blood. *Proceedings of the National Academy of Sciences USA* **96**, 11041–11048.

Carman, G. M. (1997) Phosphatidate phosphatases and diacylglycerol pyrophosphate phosphatases in *Saccharomyces cerevisiae* and *Escherichia coli*. *Biochimica et Biophysica Acta* **1348**, 45–55.

Carman, G. M. and Henry, S. A. (1999) Phospholipid biosynthesis in the yeast *Saccharomyces cerevisiae* and interrelationship with other metabolic processes. *Progress in Lipid Research* **38**, 361–399.

Carman, G. M. and Zeimetz, G. M. (1996) Regulation of phospholipid biosynthesis in the yeast *Saccharomyces cerevisiae*. *Journal of Biological Chemistry* **271**, 13293–13296.

Caro, L. H., Tettelin, H., Vossen, J. H., Ram, A. F., van den Ende, H. *et al.* (1997) *In silicio* identification of glycosyl-phosphatidylinositol-anchored plasma-membrane and cell wall proteins of *Saccharomyces cerevisiae*. *Yeast* **13**, 1477–1489.

Chang, S. C., Heacock, P. N., Clancey, C. J. and Dowhan, W. (1998a) The *PEL1* gene (renamed *PGS1*) encodes the phosphatidylglycerophosphate synthase of *Saccharomyces cerevisiae*. *Journal of Biological Chemistry* **273**, 9829–9836.

Chang, S. C., Heacock, P. N., Mileykovskaya, E., Voelker, D. R. and Dowhan, W. (1998b) Isolation and characterization of the gene (*CLS1*) encoding cardiolipin synthase in *Saccharomyces cerevisiae*. *Journal of Biological Chemistry* **273**, 14933–14941.

Chen, C. Y., Ingram, M. F., Rosal, P. H. and Graham, T. R. (1999) Role for Drs2p, a P-type ATPase and potential aminophospholipid translocase, in yeast late Golgi function. *Journal of Cell Biology* **147**, 1223–1236.

Choi, J. Y. and Martin, C. E. (1999) The *Saccharomyces cerevisiae FAT1* gene encodes an acyl-CoA synthetase that is required for maintenance of very long chain fatty acid levels. *Journal of Biological Chemistry* **274**, 4671–4683.

Chung, S. H., Song, W. J., Kim, K., Bednarski, J. J., Chen, J. *et al.* (1998) The C2 domains of Rabphilin3A specifically bind phosphatidylinositol 4,5-bisphosphate containing vesicles in a Ca2+-dependent manner. *In vitro* characteristics and possible significance. *Journal of Biological Chemistry* **273**, 10240–10248.

Cinato, E., Peleraux, A., Silve, S., Galiegue, S., Dhers, C. *et al.* (2002) A DNA microarray-based approach to elucidate the effects of the immunosuppressant SR31747A on gene expression in *Saccharomyces cerevisiae*. *Gene Expression* **10**, 213–230.

Clark, D. D. and Peterson, B. R. (2003) Analysis of protein tyrosine kinase inhibitors in recombinant yeast lacking the *ERG6* gene. *Chembiochem* **4**, 101–107.

Cok, S. J., Martin, C. G. and Gordon, J. I. (1998) Transcription of *INO2* and *INO4* is regulated by the state of protein N-myristoylation in *Saccharomyces cerevisiae*. *Nucleic Acids Research* **26**, 2865–2872.

Costello, L. C. and Orlean, P. (1992) Inositol acylation of a potential glycosyl phosphoinositol anchor precursor from yeast requires acyl coenzyme A. *Journal of Biological Chemistry* **267**, 8599–8603.

Covello, P. S. and Reed, D. W. (1996) Functional expression of the extraplastidial *Arabidopsis thaliana* oleate desaturase gene (*FAD2*) in *Saccharomyces cerevisiae*. *Plant Physiology* **111**, 223–226.

Crowley, J. H., Leak, F. W., Jr., Shianna, K. V., Tove, S. and Parks, L. W. (1998) A mutation in a purported regulatory gene affects control of sterol uptake in *Saccharomyces cerevisiae*. *Journal of Bacteriology* **180**, 4177–4183.

Dahlqvist, A., Stahl, U., Lenman, M., Banas, A., Lee, M. *et al.* (2000) Phospholipid:diacylglycerol acyltransferase: an enzyme that catalyzes the acyl-CoA-independent formation of triacylglycerol in yeast and plants. *Proceedings of the National Academy of Sciences USA* **97**, 6487–6492.

Daum, G. and Vance, J. E. (1997) Import of lipids into mitochondria. *Progress in Lipid Research* **36**, 103–130.

David, D., Sundarababu, S. and Gerst, J. E. (1998) Involvement of long chain fatty acid elongation in the trafficking of secretory vesicles in yeast. *Journal of Cell Biology* **143**, 1167–1182.

de Jong-Gubbels, P., van den Berg, M. A., Luttik, M. A., Steensma, H. Y., van Dijken, J. P. *et al.* (1998) Overproduction of acetyl-coenzyme A synthetase isoenzymes in respiring *Saccharomyces cerevisiae* cells does not reduce acetate production after exposure to glucose excess. *FEMS Microbiology Letters* **165**, 15–20.

de Kruijff, B. (1997) Lipid polymorphism and biomembrane function. *Current Opinion in Chemical Biology* **1**, 564–569.

de Nobel, H., Lawrie, L., Brul, S., Klis, F., Davis, M. *et al.* (2001) Parallel and comparative analysis of the proteome and transcriptome of sorbic acid-stressed *Saccharomyces cerevisiae*. *Yeast* **18**, 1413–1428.

DeRisi, J. L., Iyer, V. R. and Brown, P. O. (1997) Exploring the metabolic and genetic control of gene expression on a genomic scale. *Science* **278**, 680–686.

DeRisi, J., van den Hazel, B., Marc, P., Balzi, E., Brown, P. *et al.* (2000) Genome microarray analysis of transcriptional activation in multidrug resistance yeast mutants. *FEBS Letters* **470**, 156–160.

Desfarges, L., Durrens, P., Juguelin, H., Cassagne, C., Bonneu, M. *et al.* (1993) Yeast mutants affected in viability upon starvation have a modified phospholipid composition. *Yeast* **9**, 267–277.

D'Hondt, K., Heese-Peck, A. and Riezman, H. (2000) Protein and lipid requirements for endocytosis. *Annual Reviews of Genetics* **34**, 255–295.

Dickson, R. C. (1998) Sphingolipid functions in *Saccharomyces cerevisiae*: comparison to mammals. *Annual Reviews of Biochemistry* **67**, 27–48.

Dickson, R. C. and Lester, R. L. (2002) Sphingolipid functions in *Saccharomyces cerevisiae*. *Biochimica et Biophysica Acta* **1583**, 13–25.

Dickson, R. C., Nagiec, E. E., Skrzypek, M., Tillman, P., Wells, G. B. *et al.* (1997a) Sphingolipids are potential heat stress signals in *Saccharomyces*. *Journal of Biological Chemistry* **272**, 30196–30200.

Dickson, R. C., Nagiec, E. E., Wells, G. B., Nagiec, M. M. and Lester, R. L. (1997b) Synthesis of mannose-(inositol-P)2-ceramide, the major sphingolipid in *Saccharomyces cerevisiae*, requires the *IPT1* (YDR072c) gene. *Journal of Biological Chemistry* **272**, 29620–29625.

Dimster-Denk, D. and Rine, J. (1996) Transcriptional regulation of a sterol-biosynthetic enzyme by sterol levels in *Saccharomyces cerevisiae*. *Molecular and Cellular Biology* **16**, 3981–3989.

Dimster-Denk, D., Thorsness, M. K. and Rine, J. (1994) Feedback regulation of 3-hydroxy-3-methylglutaryl coenzyme A reductase in *Saccharomyces cerevisiae*. *Molecular Biology of the Cell* **5**, 655–665.

Dimster-Denk, D., Rine, J., Phillips, J., Scherer, S., Cundiff, P. *et al.* (1999) Comprehensive evaluation of isoprenoid biosynthesis regulation in *Saccharomyces cerevisiae* utilizing the Genome Reporter Matrix. *Journal of Lipid Research* **40**, 850–860.

Ding, D. and Greenberg, M. L. (2003) Lithium and valproate decrease the membrane phosphatidylinositol/phosphatidylcholine ratio. *Molecular Microbiology* **47**, 373–381.

DiRusso, C. C. and Black, P. N. (1999) Long-chain fatty acid transport in bacteria and yeast. Paradigms for defining the mechanism underlying this protein-mediated process. *Molecular and Cellular Biochemistry* **192**, 41–52.

DiRusso, C. C., Connell, E. J., Faergeman, N. J., Knudsen, J., Hansen, J. K. *et al.* (2000) Murine FATP alleviates growth and biochemical deficiencies of yeast fat1Delta strains. *European Journal of Biochemistry* **267**, 4422–4433.

Distel, B., Erdmann, R., Gould, S. J., Blobel, G., Crane, D. I. *et al.* (1996) A unified nomenclature for peroxisome biogenesis factors. *Journal of Cell Biology* **135**, 1–3.

Dittrich, F., Zajonc, D., Hühne, K., Hoja, U., Ekici, A. *et al.* (1998) Fatty acid elongation in yeast – biochemical characteristics of the enzyme system and isolation of elongation-defective mutants. *European Journal of Biochemistry* **252**, 477–485.

Divecha, N. and Irvine, R. F. (1995) Phospholipid signaling. *Cell* **80**, 269–278.

Doerrler, W. T. and Lehrman, M. A. (2000) A water-soluble analogue of glucosaminylphosphatidylinositol distinguishes two activities that palmitoylate inositol on GPI anchors. *Biochemical and Biophysical Research Communications* **267**, 296–299.

Dowd, S. R., Bier, M. E. and Patton-Vogt, J. L. (2001) Turnover of phosphatidylcholine in *Saccharomyces cerevisiae*. The role of the CDP-choline pathway. *Journal of Biological Chemistry* **276**, 3756–3763.

Dyer, J. M., Chapital, D. C., Kuan, J. W., Mullen, R. T. and Pepperman, A. B. (2002) Metabolic engineering of *Saccharomyces cerevisiae* for production of novel lipid compounds. *Applied Microbiology and Biotechnology* **59**, 224–230.

Edlind, T., Smith, L., Henry, K., Katiyar, S. and Nickels, J. (2002) Antifungal activity in *Saccharomyces cerevisiae* is modulated by calcium signalling. *Molecular Microbiology* **46**, 257–268.

Einerhand, A. W., Kos, W., Smart, W. C., Kal, A. J., Tabak, H. F. *et al.* (1995) The upstream region of the *FOX3* gene encoding peroxisomal 3-oxoacyl-coenzyme A thiolase in *Saccharomyces cerevisiae* contains ABF1- and replication protein A-binding sites that participate in its regulation by glucose repression. *Molecular and Cellular Biology* **15**, 3405–3414.

Eisenkolb, M., Zenzmaier, C., Leitner, E. and Schneiter, R. (2002) A specific structural requirement for ergosterol in long-chain fatty acid synthesis mutants important for maintaining raft domains in yeast. *Molecular Biology of the Cell* **13**, 4414–4428.

Elgersma, Y., van Roermund, C. W., Wanders, R. J. and Tabak, H. F. (1995) Peroxisomal and mitochondrial carnitine acetyltransferases of *Saccharomyces cerevisiae* are encoded by a single gene. *EMBO Journal* **14**, 3472–3479.

Elkhaimi, M., Kaadige, M. R., Kamath, D., Jackson, J. C., Biliran, H., Jr. *et al.* (2000) Combinatorial regulation of phospholipid biosynthetic gene expression by the *UME6*, *SIN3* and *RPD3* genes. *Nucleic Acids Research* **28**, 3160–3167.

Ella, K. M., Dolan, J. W., Qi, C. and Meier, K. E. (1996) Characterization of *Saccharomyces cerevisiae* deficient in expression of phospholipase D. *Biochemical Journal* **314**, 15–19.

Emter, R., Heese-Peck, A. and Kralli, A. (2002) *ERG6* and *PDR5* regulate small lipophilic drug accumulation in yeast cells via distinct mechanisms. *FEBS Letters* **521**, 57–61.

Epstein, C. B., Waddle, J. A., Hale, W. T., Dave, V., Thornton, J. *et al.* (2001) Genome-wide responses to mitochondrial dysfunction. *Molecular Biology of the Cell* **12**, 297–308.

Faergeman, N. J., DiRusso, C. C., Elberger, A., Knudsen, J. and Black, P. N. (1997) Disruption of the *Saccharomyces cerevisiae* homologue to the murine fatty acid transport protein impairs uptake and growth on long-chain fatty acids. *Journal of Biological Chemistry* **272**, 8531–8538.

Faergeman, N. J., Black, P. N., Zhao, X. D., Knudsen, J. and DiRusso, C. C. (2001) The Acyl-CoA synthetases encoded within *FAA1* and *FAA4* in *Saccharomyces cerevisiae* function as components of the fatty acid transport system linking import, activation, and intracellular utilization. *Journal of Biological Chemistry* **276**, 37051–37059.

Faergeman, N. J., Mutenda, K., Roepstorff, P. and Knudsen, J. (2002) Mass spectrometric lipid analysis suggests a role of acyl CoA binding protein in phospholipid remodeling in *Saccharomyces cerevisiae*. In: R. M. Epand and P. K. Kinnuen (eds) *43rd International Conference on the Bioscience of Lipids*, p. 24. Graz, Austria: Elsevier.

Farazi, T. A., Waksman, G. and Gordon, J. I. (2001) The biology and enzymology of protein N-myristoylation. *Journal of Biological Chemistry* **276**, 39501–39504.

Faulkner, A., Chen, X., Rush, J., Horazdovsky, B., Waechter, C. J. *et al.* (1999) The *LPP1* and *DPP1* gene products account for most of the isoprenoid phosphate phosphatase activities in *Saccharomyces cerevisiae*. *Journal of Biological Chemistry* **274**, 14831–14837.

Feoktistova, A., Magnelli, P., Abeijon, C., Perez, P., Lester, R. L. *et al.* (2001) Coordination between fission yeast glucan formation and growth requires a sphingolipase activity. *Genetics* **158**, 1397–1411.

Fichtlscherer, F., Wellein, C., Mittag, M. and Schweizer, E. (2000) A novel function of yeast fatty acid synthase. Subunit α is capable of self-pantetheinylation. *European Journal of Biochemistry* **267**, 2666–2671.

Flanagan, C. A., Schnieders, E. A., Emerick, A. W., Kunisawa, R., Admon, A. *et al.* (1993) Phosphatidylinositol 4-kinase: gene structure and requirement for yeast cell viability. *Science* **262**, 1444–1448.

Flick, J. S. and Thorner, J. (1993) Genetic and biochemical characterization of a phosphatidylinositol-specific phospholipase C in *Saccharomyces cerevisiae*. *Molecular and Cellular Biology* **13**, 5861–5876.

Flick, J. S. and Thorner, J. (1998) An essential function of a phosphoinositide-specific phospholipase C is relieved by inhibition of a cyclin-dependent protein kinase in the yeast *Saccharomyces cerevisiae*. *Genetics* **148**, 33–47.

Flury, I., Benachour, A. and Conzelmann, A. (2000) YLL031c belongs to a novel family of membrane proteins involved in the transfer of ethanolaminephosphate onto the core structure of glycosylphosphatidylinositol anchors in yeast. *Journal of Biological Chemistry* **275**, 24458–24465.

Foti, M., Audhya, A. and Emr, S. D. (2001) Sac1 lipid phosphatase and Stt4 phosphatidylinositol 4-kinase regulate a pool of phosphatidylinositol 4-phosphate that functions in the control of the actin cytoskeleton and vacuole morphology. *Molecular Biology of the Cell* **12**, 2396–2411.

Foury, F. (1997) Human genetic diseases: a cross-talk between man and yeast. *Gene* **195**, 1–10.

Fraering, P., Imhof, I., Meyer, U., Strub, J. M., van Dorsselaer, A. *et al.* (2001) The GPI transamidase complex of *Saccharomyces cerevisiae* contains Gaa1p, Gpi8p, and Gpi16p. *Molecular Biology of the Cell* **12**, 3295–3306.

Friant, S., Zanolari, B. and Riezman, H. (2000) Increased protein kinase or decreased PP2A activity bypasses sphingoid base requirement in endocytosis. *EMBO Journal* **19**, 2834–2844.

Fruman, D. A., Meyers, R. E. and Cantley, L. C. (1998) Phosphoinositide kinases. *Annual Reviews of Biochemistry* **67**, 481–507.

Funato, K., Lombardi, R., Vallee, B. and Riezman, H. (2003) Lcb4p is a key regulator of ceramide synthesis from exogenous long chain sphingoid base in *Saccharomyces cerevisiae. Journal of Biological Chemistry* **278**, 7325–7334.

Furneisen, J. M. and Carman, G. M. (2000) Enzymological properties of the LPP1-encoded lipid phosphatase from *Saccharomyces cerevisiae. Biochimica et Biophysica Acta* **1484**, 71–82.

Futerman, A. H. (1995) Inhibition of sphingolipid synthesis: effects on glycosphingolipid-GPI-anchored protein microdomains. *Trends in Cell Biology* **5**, 377–379.

Fyrst, H., Oskouian, B., Kuypers, F. A. and Saba, J. D. (1999) The *PLB2* gene of *Saccharomyces cerevisiae* confers resistance to lysophosphatidylcholine and encodes a phospholipase B/lysophospholipase. *Biochemistry* **38**, 5864–5871.

Gable, K., Han, G., Monaghan, E., Bacikova, D., Natarajan, M. *et al.* (2002) Mutations in the yeast *LCB1* and *LCB2* genes, including those corresponding to the hereditary sensory neuropathy type I mutations, dominantly inactivate serine palmitoyltransferase. *Journal of Biological Chemistry* **277**, 10194–10200.

Gachotte, D., Pierson, C. A., Lees, N. D., Barbuch, R., Koegel, C. *et al.* (1997) A yeast sterol auxotroph (*erg25*) is rescued by addition of azole antifungals and reduced levels of heme. *Proceedings of the National Academy of Sciences USA* **94**, 11173–11178.

Gachotte, D., Barbuch, R., Gaylor, J., Nickel, E. and Bard, M. (1998) Characterization of the *Saccharomyces cerevisiae ERG26* gene encoding the C-3 sterol dehydrogenase (C-4 decarboxylase) involved in sterol biosynthesis. *Proceedings of the National Academy of Sciences USA* **95**, 13794–13799.

Gachotte, D., Sen, S. E., Eckstein, J., Barbuch, R., Krieger, M. *et al.* (1999) Characterization of the *Saccharomyces cerevisiae* ERG27 gene encoding the 3-keto reductase involved in C-4 sterol demethylation. *Proceedings of the National Academy of Sciences USA* **96**, 12655–12660.

Gaigg, B., Simbeni, R., Hrastnik, C., Paltauf, F. and Daum, G. (1995) Characterization of a microsomal subfraction associated with mitochondria of the yeast, *Saccharomyces cerevisiae.* Involvement in synthesis and import of phospholipids into mitochondria. *Biochimica et Biophysica Acta* **1234**, 214–220.

Gaigg, B., Neergaard, T. B., Schneiter, R., Hansen, J. K., Faergeman, N. J. *et al.* (2001) Depletion of acyl-coenzyme A-binding protein affects sphingolipid synthesis and causes vesicle accumulation and membrane defects in *Saccharomyces cerevisiae. Molecular Biology of the Cell* **12**, 1147–1160.

Gardner, R. G., Shan, H., Matsuda, S. P. and Hampton, R. Y. (2001a) An oxysterol-derived positive signal for 3-hydroxy-3-methylglutaryl-CoA reductase degradation in yeast. *Journal of Biological Chemistry* **276**, 8681–8694.

Gardner, R. G., Shearer, A. G. and Hampton, R. Y. (2001b) *In vivo* action of the HRD ubiquitin ligase complex: mechanisms of endoplasmic reticulum quality control and sterol regulation. *Molecular and Cellular Biology* **21**, 4276–4291.

Gary, J. D., Sato, T. K., Stefan, C. J., Bonangelino, C. J., Weisman, L. S. *et al.* (2002) Regulation of Fab1 phosphatidylinositol 3-phosphate 5-kinase pathway by Vac7 protein and Fig4, a polyphosphoinositide phosphatase family member. *Molecular Biology of the Cell* **13**, 1238–1251.

Gaynor, E. C., Mondesert, G., Grimme, S. J., Reed, S. I., Orlean, P. *et al.* (1999) MCD4 encodes a conserved endoplasmic reticulum membrane protein essential for glycosylphosphatidylinositol anchor synthesis in yeast. *Molecular Biology of the Cell* **10**, 627–648.

Geisbrecht, B. V., Zhu, D., Schulz, K., Nau, K., Morrell, J. C. *et al.* (1998) Molecular characterization of *Saccharomyces cerevisiae* Δ3, Δ2-enoyl-CoA isomerase. *Journal of Biological Chemistry* **273**, 33184–33191.

Gijon, M. A. and Leslie, C. C. (1997) Phospholipases A2. *Seminars in Cell and Developmental Biology* **8**, 297–303.

Goffeau, A., Park, J., Paulsen, I. T., Jonniaux, J. L., Dinh, T. *et al.* (1997) Multidrug-resistant transport proteins in yeast: complete inventory and phylogenetic characterization of yeast open reading frames with the major facilitator superfamily. *Yeast* **13**, 43–54.

Gong, X. and Chang, A. (2001) A mutant plasma membrane ATPase, Pma1-10, is defective in stability at the yeast cell surface. *Proceedings of the National Academy of Sciences USA* **98**, 9104–9109.

Gonzalez, C. I. and Martin, C. E. (1996) Fatty acid-responsive control of mRNA stability. Unsaturated fatty acid-induced degradation of the *Saccharomyces OLE1* transcript. *Journal of Biological Chemistry* **271**, 25801–25809.

Gotte, K., Girzalsky, W., Linkert, M., Baumgart, E., Kammerer, S. *et al.* (1998) Pex19p, a farnesylated protein essential for peroxisome biogenesis. *Molecular and Cellular Biology* **18**, 616–628.

Gottlieb, D., Heideman, W. and Saba, J. D. (1999) The *DPL1* gene is involved in mediating the response to nutrient deprivation in *Saccharomyces cerevisiae*. *Molecular and Cellular Biology Research Communications* **1**, 66–71.

Grabowska, D., Karst, F. and Szkopinska, A. (1998) Effect of squalene synthase gene disruption on synthesis of polyprenols in *Saccharomyces cerevisiae*. *FEBS Letters* **434**, 406–408.

Graves, J. A. and Henry, S. A. (2000) Regulation of the yeast *INO1* gene. The products of the *INO2*, *INO4* and *OPI1* regulatory genes are not required for repression in response to inositol. *Genetics* **154**, 1485–1495.

Greenberg, M. L. and Lopes, J. M. (1996) Genetic regulation of phospholipid biosynthesis in *Saccharomyces cerevisiae*. *Microbiological Reviews* **60**, 1–20.

Griac, P. (1997) Regulation of yeast phospholipid biosynthetic genes in phosphatidylserine decarboxylase mutants. *Journal of Bacteriology* **179**, 5843–5848.

Griac, P., Swede, M. J. and Henry, S. A. (1996) The role of phosphatidylcholine biosynthesis in the regulation of the *INO1* gene of yeast. *Journal of Biological Chemistry* **271**, 25692–25698.

Grilley, M. M., Stock, S. D., Dickson, R. C., Lester, R. L. and Takemoto, J. Y. (1998) Syringomycin action gene *SYR2* is essential for sphingolipid 4-hydroxylation in *Saccharomyces cerevisiae*. *Journal of Biological Chemistry* **273**, 11062–11068.

Grimme, S. J., Westfall, B. A., Wiedman, J. M., Taron, C. H. and Orlean, P. (2001) The essential Smp3 protein is required for addition of the side-branching fourth mannose during assembly of yeast glycosylphosphatidylinositols. *Journal of Biological Chemistry* **276**, 27731–27739.

Guillas, I., Kirchman, P. A., Chuard, R., Pfefferli, M., Jiang, J. C. *et al.* (2001) C26-CoA-dependent ceramide synthesis of *Saccharomyces cerevisiae* is operated by Lag1p and Lac1p. *EMBO Journal* **20**, 2655–2665.

Guo, S., Stolz, L. E., Lemrow, S. M. and York, J. D. (1999) SAC1-like domains of yeast *SAC1*, *INP52*, and *INP53* and of human synaptojanin encode polyphosphoinositide phosphatases. *Journal of Biological Chemistry* **274**, 12990–12995.

Guo, Z., Cromley, D., Billheimer, J. T. and Sturley, S. L. (2001) Identification of potential substrate-binding sites in yeast and human acyl-CoA sterol acyltransferases by mutagenesis of conserved sequences. *Journal of Lipid Research* **42**, 1282–1291.

Gurvitz, A., Rottensteiner, H., Hiltunen, J. K., Binder, M., Dawes, I. W. *et al.* (1997) Regulation of the yeast *SPS19* gene encoding peroxisomal 2,4-dienoyl-CoA reductase by the transcription factors Pip2p and Oaf1p: β-oxidation is dispensable for *Saccharomyces cerevisiae* sporulation in acetate medium. *Molecular Microbiology* **26**, 675–685.

Gurvitz, A., Mursula, A. M., Firzinger, A., Hamilton, B., Kilpelainen, S. H. *et al.* (1998) Peroxisomal Δ3-cis-Δ2-trans-enoyl-CoA isomerase encoded by *ECI1* is required for growth of the yeast *Saccharomyces cerevisiae* on unsaturated fatty acids. *Journal of Biological Chemistry* **273**, 31366–31374.

Gurvitz, A., Hamilton, B., Hartig, A., Ruis, H., Dawes, I. W. *et al.* (1999a) A novel element in the promoter of the *Saccharomyces cerevisiae* gene *SPS19* enhances ORE-dependent up-regulation in oleic acid and is essential for derepression. *Molecular and General Genetics* **262**, 481–492.

Gurvitz, A., Mursula, A. M., Yagi, A. I., Hartig, A., Ruis, H. *et al.* (1999b) Alternatives to the isomerase-dependent pathway for the beta-oxidation of oleic acid are dispensable in *Saccharomyces cerevisiae.* Identification of YOR180c/*DCI1* encoding peroxisomal Δ(3,5)- Δ(2,4)-dienoyl-CoA isomerase. *Journal of Biological Chemistry* **274**, 24514–24521.

Gurvitz, A., Wabnegger, L., Rottensteiner, H., Dawes, I. W., Hartig, A. *et al.*, (2000) Adr1p-dependent regulation of the oleic acid-inducible yeast gene *SPS19* encoding the peroxisomal β-oxidation auxiliary enzyme 2,4-dienoyl-CoA reductase. *Molecular and Cellular Biology Research Communications* **4**, 81–89.

Gurvitz, A., Hamilton, B., Ruis, H. and Hartig, A. (2001) Peroxisomal degradation of trans-unsaturated fatty acids in the yeast *Saccharomyces cerevisiae. Journal of Biological Chemistry* **276**, 895–903.

Ha, S. A., Bunch, J. T., Hama, H., DeWald, D. B. and Nothwehr, S. F. (2001) A novel mechanism for localizing membrane proteins to yeast trans-Golgi network requires function of synaptojanin-like protein. *Molecular Biology of the Cell* **12**, 3175–3190.

Hait, N. C., Fujita, K., Lester, R. L. and Dickson, R. C. (2002) Lcb4p sphingoid base kinase localizes to the Golgi and late endosomes. *FEBS Lettersers* **532**, 97–102.

Hama, H., Schnieders, E. A., Thorner, J., Takemoto, J. Y. and DeWald, D. B. (1999) Direct involvement of phosphatidylinositol 4-phosphate in secretion in the yeast *Saccharomyces cerevisiae. Journal of Biological Chemistry* **274**, 34294–34300.

Han, G. S., Johnston, C. N., Chen, X., Athenstaedt, K., Daum, G. *et al.* (2001) Regulation of the *Saccharomyces cerevisiae DPP1*-encoded diacylglycerol pyrophosphate phosphatase by zinc. *Journal of Biological Chemistry* **276**, 10126–10133.

Han, G., Gable, K., Kohlwein, S. D., Beaudoin, F., Napier, J. A. *et al.* (2002a) The *Saccharomyces cerevisiae* YBR159w gene encodes the 3-ketoreductase of the microsomal fatty acid elongase. *Journal of Biological Chemistry* **277**, 35440–35449.

Han, G. S., Audhya, A., Markley, D. J., Emr, S. D. and Carman, G. M. (2002b) The *Saccharomyces cerevisiae LSB6* gene encodes phosphatidylinositol 4-kinase activity. *Journal of Biological Chemistry* **277**, 47709–47718.

Harder, T. and Simons, K. (1997) Caveolae, DIGs, and the dynamics of sphingolipid-cholesterol microdomains. *Current Opinion in Cell Biology* **9**, 534–542.

Harington, A., Schwarz, E., Slonimski, P. P. and Herbert, C. J. (1994) Subcellular relocalization of a long-chain fatty acid CoA ligase by a suppressor mutation alleviates a respiration deficiency in *Saccharomyces cerevisiae. EMBO Journal* **13**, 5531–5538.

Hashida-Okado, T., Ogawa, A., Endo, M., Yasumoto, R., Takesako, K. *et al.* (1996) *AUR1*, a novel gene conferring aureobasidin resistance on *Saccharomyces cerevisiae*: a study of defective morphologies in Aur1p-depleted cells. *Molecular and General Genetics* **251**, 236–244.

Hasslacher, M., Ivessa, A. S., Paltauf, F. and Kohlwein, S. D. (1993) Acetyl-CoA carboxylase from yeast is an essential enzyme and is regulated by factors that control phospholipid metabolism. *Journal of Biological Chemistry* **268**, 10946–10952.

Hearn, J. D., Lester, R. L. and Dickson, R. C. (2003) The uracil transporter Fur4p associates with lipid rafts. *Journal of Biological Chemistry* **278**, 3679–3686.

Heese-Peck, A., Pichler, H., Zanolari, B., Watanabe, R., Daum, G. *et al.* (2002) Multiple functions of sterols in yeast endocytosis. *Molecular Biology of the Cell* **13**, 2664–2680.

Hendricks, K. B., Wang, B. Q., Schnieders, E. A. and Thorner, J. (1999) Yeast homologue of neuronal frequenin is a regulator of phosphatidylinositol-4-OH kinase. *Nature Cell Biology* **1**, 234–241.

Henke, B., Girzalsky, W., Berteaux-Lecellier, V. and Erdmann, R. (1998) *IDP3* encodes a peroxisomal NADP-dependent isocitrate dehydrogenase required for the beta-oxidation of unsaturated fatty acids. *Journal of Biological Chemistry* **273**, 3702–3711.

Henneberry, A. L., Lagace, T. A., Ridgway, N. D. and McMaster, C. R. (2001) Phosphatidylcholine synthesis influences the diacylglycerol homeostasis required for SEC14p-dependent Golgi function and cell growth. *Molecular Biology of the Cell* **12**, 511–520.

Henry, S. A. and Patton-Vogt, J. L. (1998) Genetic regulation of phospholipid metabolism: yeast as a model eukaryote. *Progress in Nucleic Acid Research and Molecular Biology* **61**, 133–179.

Hettema, E. H., Distel, B. and Tabak, H. F. (1999) Import of proteins into peroxisomes. *Biochimica et Biophysica Acta* **1451**, 17–34.

Hettema, E. H. and Tabak, H. F. (2000) Transport of fatty acids and metabolites across the peroxisomal membrane. *Biochimica et Biophysica Acta* **1486**, 18–27.

Hettema, E. H., van Roermund, C. W., Distel, B., van den Berg, M., Vilela, C. *et al.* (1996) The ABC transporter proteins Pat1 and Pat2 are required for import of long-chain fatty acids into peroxisomes of *Saccharomyces cerevisiae. EMBO Journal* **15**, 3813–3822.

Hettema, E. H., Girzalsky, W., van Den Berg, M., Erdmann, R. and Distel, B. (2000) *Saccharomyces cerevisiae* Pex3p and Pex19p are required for proper localization and stability of peroxisomal membrane proteins. *EMBO Journal* **19**, 223–233.

Hofmann, J., Friedrich, R., Kurz, N. and Schweizer, E. (2002) Isolation and characterization of mitochondrial lipoate ligases from *Saccharomyces cerevisiae.* In: R. M. Epand and P. K. Kinnuen (eds) *43rd International Conference on the Bioscience of Lipids*, p. 26. Graz, Austria: Elsevier.

Hoja, U., Wellein, C., Greiner, E. and Schweizer, E. (1998) Pleiotropic phenotype of acetyl-CoA-carboxylase-defective yeast cells: viability of a *bpl1-amber* mutation depending on its readthrough by normal tRNA(Gln)(CAG). *European Journal of Biochemistry* **254**, 520–526.

Homans, S. W., Ferguson, M. A., Dwek, R. A., Rademacher, T. W., Anand, R. *et al.* (1988) Complete structure of the glycosyl phosphatidylinositol membrane anchor of rat brain Thy-1 glycoprotein. *Nature* **333**, 269–272.

Hong, Y., Maeda, Y., Watanabe, R., Ohishi, K., Mishkind, M. *et al.* (1999) Pig-N, a mammalian homologue of yeast Mcd4p, is involved in transferring phosphoethanolamine to the first mannose of the glycosylphosphatidylinositol. *Journal of Biological Chemistry* **274**, 35099–35106.

Hong, Y., Ohishi, K., Kang, J. Y., Tanaka, S., Inoue, N., Nishimura, J., Maeda, Y., Kinoshita, T. (2003) Human PIG-U and yeast Cdc91p are the fifth subunit of GPI transamidase that attaches GPI-anchors to proteins. *Molecular Biology of the Cell* **14**, 1780–1789.

Hongay, C., Jia, N., Bard, M. and Winston, F. (2002) Mot3 is a transcriptional repressor of ergosterol biosynthetic genes and is required for normal vacuolar function in *Saccharomyces cerevisiae. EMBO Journal* **21**, 4114–4124.

Hooper, N. M. (1998) Membrane biology: do glycolipid microdomains really exist? *Current Biology* **8**, R114–R116.

Hoppe, T., Matuschewski, K., Rape, M., Schlenker, S., Ulrich, H. D. *et al.* (2000) Activation of a membrane-bound transcription factor by regulated ubiquitin/proteasome-dependent processing. *Cell* **102**, 577–586.

Houten, S. M. and Waterham, H. R. (2001) Nonorthologous gene displacement of phosphomevalonate kinase. *Molecular Genetics of Metabolism* **72**, 273–276.

Howe, A. G. and McMaster, C. R. (2001) Regulation of vesicle trafficking, transcription, and meiosis: lessons learned from yeast regarding the disparate biologies of phosphatidylcholine. *Biochimica et Biophysica Acta* **1534**, 65–77.

Howe, A. G., Zaremberg, V. and McMaster, C. R. (2002) Cessation of growth to prevent cell death due to inhibition of phosphatidylcholine synthesis is impaired at 37 degrees C in *Saccharomyces cerevisiae. Journal of Biological Chemistry* **277**, 44100–44107.

Hughes, T. R., Marton, M. J., Jones, A. R., Roberts, C. J., Stoughton, R. *et al.* (2000a) Functional discovery via a compendium of expression profiles. *Cell* **102**, 109–126.

Hughes, W. E., Cooke, F. T. and Parker, P. J. (2000b) Sac phosphatase domain proteins. *Biochemical Journal* **350**, 337–352.

Hüttner, I. G., Strahl, T., Osawa, M., King, D. S., Ames, J. B. *et al.* (2003) Molecular interactions of yeast frequenin (Frq1) with the phosphatidylinositol 4-kinase isoform, Pik1. *Journal of Biological Chemistry* **278**, 4862–4874.

Ivessa, A. S., Schneiter, R. and Kohlwein, S. D. (1997) Yeast acetyl-CoA carboxylase is associated with the cytoplasmic surface of the endoplasmic reticulum. *European Journal of Cell Biology* **74**, 399–406.

Jandrositz, A., Turnowsky, F. and Hogenauer, G. (1991) The gene encoding squalene epoxidase from *Saccharomyces cerevisiae*: cloning and characterization. *Gene* **107**, 155–160.

Janssen, M. J., Koorengevel, M. C., de Kruijff, B. and de Kroon, A. I. (2000) The phosphatidylcholine to phosphatidylethanolamine ratio of *Saccharomyces cerevisiae* varies with the growth phase. *Yeast* **16**, 641–650.

Janssen, M. J., de Jong, H. M., de Kruijff, B. and de Kroon, A. I. (2002) Cooperative activity of phospholipid-N-methyltransferases localized in different membranes. *FEBS Letters* **513**, 197–202.

Jazwinski, S. M. and Conzelmann, A. (2002) *LAG1* puts the focus on ceramide signaling. *International Journal of Biochemistry and Cell Biology* **34**, 1491–1495.

Jenkins, G. M., Richards, A., Wahl, T., Mao, C., Obeid, L. *et al.* (1997) Involvement of yeast sphingolipids in the heat stress response of *Saccharomyces cerevisiae*. *Journal of Biological Chemistry* **272**, 32566–32572.

Jensen-Pergakes, K., Guo, Z., Giattina, M., Sturley, S. L. and Bard, M. (2001) Transcriptional regulation of the two sterol esterification genes in the yeast *Saccharomyces cerevisiae*. *Journal of Bacteriology* **183**, 4950–4957.

Jiang, F., Gu, Z., Granger, J. M. and Greenberg, M. L. (1999) Cardiolipin synthase expression is essential for growth at elevated temperature and is regulated by factors affecting mitochondrial development. *Molecular Microbiology* **31**, 373–379.

Jiang, F., Ryan, M. T., Schlame, M., Zhao, M., Gu, Z. *et al.* (2000) Absence of cardiolipin in the crd1 null mutant results in decreased mitochondrial membrane potential and reduced mitochondrial function. *Journal of Biological Chemistry* **275**, 22387–22394.

Johnson, D. R., Knoll, L. J., Levin, D. E. and Gordon, J. I. (1994) *Saccharomyces cerevisiae* contains four fatty acid activation (*FAA*) genes: an assessment of their role in regulating protein N-myristoylation and cellular lipid metabolism. *Journal of Cell Biology* **127**, 751–762.

Kalebina, T. S., Laurinavichiute, D. K., Packeiser, A. N., Morenkov, O. S., Ter-Avanesyan, M. D. *et al.* (2002) Correct GPI-anchor synthesis is required for the incorporation of endoglucanase/glucanosyltransferase Bgl2p into the *Saccharomyces cerevisiae* cell wall. *FEMS Microbiology Letters* **210**, 81–85.

Karpichev, I. V. and Small, G. M. (1998) Global regulatory functions of Oaf1p and Pip2p (Oaf2p), transcription factors that regulate genes encoding peroxisomal proteins in *Saccharomyces cerevisiae*. *Molecular and Cellular Biology* **18**, 6560–6570.

Karpova, T. S., McNally, J. G., Moltz, S. L. and Cooper, J. A. (1998) Assembly and function of the actin cytoskeleton of yeast: relationships between cables and patches. *Journal of Cell Biology* **142**, 1501–1517.

Kato, U., Emoto, K., Fredriksson, C., Nakamura, H., Ohta, A. *et al.* (2002) A novel membrane protein, Ros3p, is required for phospholipid translocation across the plasma membrane in *Saccharomyces cerevisiae*. *Journal of Biological Chemistry* **277**, 37855–37862.

Kaufman, R. J. (1999) Stress signaling from the lumen of the endoplasmic reticulum: coordination of gene transcriptional and translational controls. *Genes and Development* **13**, 1211–1233.

Kelly, S. L., Lamb, D. C., Baldwin, B. C., Corran, A. J. and Kelly, D. E. (1997a) Characterization of *Saccharomyces cerevisiae* CYP61, sterol Δ22-desaturase, and inhibition by azole antifungal agents. *Journal of Biological Chemistry* **272**, 9986–9988.

Kelly, S. L., Lamb, D. C. and Kelly, D. E. (1997b) Sterol 22-desaturase, cytochrome P45061, possesses activity in xenobiotic metabolism. *FEBS Letters* **412**, 233–235.

Kennedy, M. A. and Bard, M. (2001) Positive and negative regulation of squalene synthase (*ERG9*), an ergosterol biosynthetic gene, in *Saccharomyces cerevisiae*. *Biochimica et Biophysica Acta* **1517**, 177–189.

Kennedy, M. A., Barbuch, R. and Bard, M. (1999) Transcriptional regulation of the squalene synthase gene (*ERG9*) in the yeast *Saccharomyces cerevisiae*. *Biochimica et Biophysica Acta* **1445**, 110–122.

Kihara, A. and Igarashi, Y. (2002) Identification and characterization of a *Saccharomyces cerevisiae* gene, *RSB1*, involved in sphingoid long-chain base release. *Journal of Biological Chemistry* **277**, 30048–30054.

Kihara, A., Noda, T., Ishihara, N. and Ohsumi, Y. (2001) Two distinct Vps34 phosphatidylinositol 3-kinase complexes function in autophagy and carboxypeptidase Y sorting in *Saccharomyces cerevisiae*. *Journal of Cell Biology* **152**, 519–530.

Kim, K. H. and Carman, G. M. (1999) Phosphorylation and regulation of choline kinase from *Saccharomyces cerevisiae* by protein kinase A. *Journal of Biological Chemistry* **274**, 9531–9538.
Kinoshita, T. and Inoue, N. (2000) Dissecting and manipulating the pathway for glycosylphosphatidylinositol-anchor biosynthesis. *Current Opinion in Chemical Biology* **4**, 632–638.
Klein, K., Steinberg, R., Fiethen, B. and Overath, P. (1971) Fatty acid degradation in *Escherichia coli*. An inducible system for the uptake of fatty acids and further characterization of old mutants. *European Journal of Biochemistry* **19**, 442–450.
Knoll, L. J., Johnson, D. R. and Gordon, J. I. (1994) Biochemical studies of three *Saccharomyces cerevisiae* acyl-CoA synthetases, Faa1p, Faa2p, and Faa3p. *Journal of Biological Chemistry* **269**, 16348–16356.
Knudsen, J., Neergaard, T. B., Gaigg, B., Jensen, M. V. and Hansen, J. K. (2000) Role of acyl-CoA binding protein in acyl-CoA metabolism and acyl-CoA-mediated cell signaling. *Journal of Nutrition* **130**, 294S–298S.
Ko, J., Cheah, S. and Fischl, A. S. (1994) Regulation of phosphatidylinositol:ceramide phosphoinositol transferase in *Saccharomyces cerevisiae*. *Journal of Bacteriology* **176**, 5181–5183.
Kohlwein, S. D., Eder, S., Oh, C.-S., Martin, C. E., Gable, K. *et al.* (2001) Tsc13p is required for fatty acid elongation and localizes to a novel structure at the nuclear–vacuolar interface in *Saccharomyces cerevisiae*. *Molecular and Cellular Biology* **21**, 109–125.
Komeili, A., Wedaman, K. P., O'Shea, E. K. and Powers, T. (2000) Mechanism of metabolic control. Target of rapamycin signaling links nitrogen quality to the activity of the Rtg1 and Rtg3 transcription factors. *Journal of Cell Biology* **151**, 863–878.
Koning, A. J., Larson, L. L., Cadera, E. J., Parrish, M. L. and Wright, R. L. (2002) Mutations that affect vacuole biogenesis inhibit proliferation of the endoplasmic reticulum in *Saccharomyces cerevisiae*. *Genetics* **160**, 1335–1352.
Konrad, G., Schlecker, T., Faulhammer, F. and Mayinger, P. (2002) Retention of the yeast Sac1p phosphatase in the endoplasmic reticulum causes distinct changes in cellular phosphoinositide levels and stimulates microsomal ATP transport. *Journal of Biological Chemistry* **277**, 10547–10554.
Kontoyiannis, D. P. (2000a) Efflux-mediated resistance to fluconazole could be modulated by sterol homeostasis in *Saccharomyces cerevisiae*. *Journal of Antimicrobial Chemotherapy* **46**, 199–203.
Kontoyiannis, D. P. (2000b) Modulation of fluconazole sensitivity by the interaction of mitochondria and Erg3p in *Saccharomyces cerevisiae*. *Journal of Antimicrobial Chemotherapy* **46**, 191–197.
Kratzer, S. and Schüller, H. J. (1997) Transcriptional control of the yeast acetyl-CoA synthetase gene, *ACS1*, by the positive regulators *CAT8* and *ADR1* and the pleiotropic repressor *UME6*. *Molecular Microbiology* **26**, 631–641.
Ktistakis, N. T., Brown, H. A., Waters, M. G., Sternweis, P. C. and Roth, M. G. (1996) Evidence that phospholipase D mediates ADP ribosylation factor-dependent formation of Golgi coated vesicles. *Journal of Cell Biology* **134**, 295–306.
Lambalot, R. H., Gehring, A. M., Flugel, R. S., Zuber, P., LaCelle, M. *et al.* (1996) A new enzyme superfamily – the phosphopantetheinyl transferases. *Chemical Biology* **3**, 923–936.
Lardizabal, K. D., Mai, J. T., Wagner, N. W., Wyrick, A., Voelker, T. *et al.* (2001) *DGAT2* is a new diacylglycerol acyltransferase gene family: purification, cloning, and expression in insect cells of two polypeptides from *Mortierella ramanniana* with diacylglycerol acyltransferase activity. *Journal of Biological Chemistry* **276**, 38862–38869.
Lazarow, P. B. and Kunau, W. H. (1997) Peroxisomes. In: J. R. Pringle, J. R. Broach and E. W. Jones (eds) *The Molecular and Cellular Biology of the Yeast Saccharomyces*, pp. 547–605. New York: Cold Spring Harbor Laboratory Press.
Leber, A., Fischer, P., Schneiter, R., Kohlwein, S. D. and Daum, G. (1997) The yeast *mic2* mutant is defective in the formation of mannosyl-diinositolphosphorylceramide. *FEBS Letters* **411**, 211–214.
Leber, R., Landl, K., Zinser, E., Ahorn, H., Spok, A. *et al.* (1998) Dual localization of squalene epoxidase, Erg1p, in yeast reflects a relationship between the endoplasmic reticulum and lipid particles. *Molecular Biology of the Cell* **9**, 375–386.
Leber, R., Zenz, R., Schrottner, K., Fuchsbichler, S., Puhringer, B. *et al.* (2001) A novel sequence element is involved in the transcriptional regulation of expression of the *ERG1* (squalene epoxidase) gene in *Saccharomyces cerevisiae*. *European Journal of Biochemistry* **268**, 914–924.

Lee, F. J., Lin, L. W. and Smith, J. A. (1990) A glucose-repressible gene encodes acetyl-CoA hydrolase from *Saccharomyces cerevisiae*. *Journal of Biological Chemistry* **265**, 7413–7418.

Lee, K. S., Patton, J. L., Fido, M., Hines, L. K., Kohlwein, S. D. *et al.* (1994) The *Saccharomyces cerevisiae PLB1* gene encodes a protein required for lysophospholipase and phospholipase B activity. *Journal of Biological Chemistry* **269**, 19725–19730.

Lee, M., Lenman, M., Banas, A., Bafor, M., Singh, S. *et al.* (1998) Identification of non-heme diiron proteins that catalyze triple bond and epoxy group formation. *Science* **280**, 915–918.

Lee, M. C., Hamamoto, S. and Schekman, R. (2002) Ceramide biosynthesis is required for the formation of the oligomeric H+-ATPase Pma1p in the yeast endoplasmic reticulum. *Journal of Biological Chemistry* **277**, 22395–22401.

Lees, N. D., Bard, M. and Kirsch, D. R. (1999) Biochemistry and molecular biology of sterol synthesis in *Saccharomyces cerevisiae*. *Critical Reviews in Biochemistry and Molecular Biology* **34**, 33–47.

Lehto, M. and Olkkonen, V. M. (2003) The OSBP-related proteins: a novel protein family involved in vesicle transport, cellular lipid metabolism, and cell signalling. *Biochimica et Biophysica Acta* **1631**, 1–11.

Leidich, S. D., Drapp, D. A. and Orlean, P. (1994) A conditionally lethal yeast mutant blocked at the first step in glycosyl phosphatidylinositol anchor synthesis. *Journal of Biological Chemistry* **269**, 10193–10196.

Leslie, C. C. (1997) Properties and regulation of cytosolic phospholipase A2. *Journal of Biological Chemistry* **272**, 16709–16712.

Levine, T. P. and Munro, S. (2001) Dual targeting of Osh1p, a yeast homologue of oxysterol-binding protein, to both the Golgi and the nucleus–vacuole junction. *Molecular Biology of the Cell* **12**, 1633–1644.

Levine, T. P. and Munro, S. (2002) Targeting of Golgi-specific pleckstrin homology domains involves both PtdIns 4-kinase-dependent and -independent components. *Current Biology* **12**, 695–704.

Li, X., Routt, S. M., Xie, Z., Cui, X., Fang, M. *et al.* (2000a) Identification of a novel family of nonclassic yeast phosphatidylinositol transfer proteins whose function modulates phospholipase D activity and Sec14p-independent cell growth. *Molecular Biology of the Cell* **11**, 1989–2005.

Li, X., Xie, Z. and Bankaitis, V. A. (2000b) Phosphatidylinositol/phosphatidylcholine transfer proteins in yeast. *Biochimica et Biophysica Acta* **1486**, 55–71.

Li, X., Rivas, M. P., Fang, M., Marchena, J., Mehrotra, B. *et al.* (2002) Analysis of oxysterol binding protein homologue Kes1p function in regulation of Sec14p-dependent protein transport from the yeast Golgi complex. *Journal of Cell Biology* **157**, 63–77.

Li, Z. and Brendel, M. (1993) Co-regulation with genes of phospholipid biosynthesis of the *CTR/HNM1*-encoded choline/nitrogen mustard permease in *Saccharomyces cerevisiae*. *Molecular and General Genetics* **241**, 680–684.

Lin, H., Choi, J. H., Hasek, J., DeLillo, N., Lou, W. *et al.* (2000) Phospholipase C is involved in kinetochore function in *Saccharomyces cerevisiae*. *Molecular and Cellular Biology* **20**, 3597–3607.

Lipke, P. N. and Kurjan, J. (1992) Sexual agglutination in budding yeasts: structure, function, and regulation of adhesion glycoproteins. *Microbiological Reviews* **56**, 180–194.

Liscovitch, M., Czarny, M., Fiucci, G. and Tang, X. (2000) Phospholipase D: molecular and cell biology of a novel gene family. *Biochemical Journal* **345**, 401–415.

Lodge, J. K., Jackson-Machelski, E., Devadas, B., Zupec, M. E., Getman, D. P. *et al.* (1997) N-myristoylation of Arf proteins in *Candida albicans*: an *in vivo* assay for evaluating antifungal inhibitors of myristoyl-CoA: protein N-myristoyltransferase. *Microbiology* **143**, 357–366.

London, E. and Brown, D. A. (2000) Insolubility of lipids in Triton X-100: physical origin and relationship to sphingolipid/cholesterol membrane domains (rafts). *Biochimica et Biophysica Acta* **1508**, 182–195.

Lopes, J. M., Schulze, K. L., Yates, J. W., Hirsch, J. P. and Henry, S. A. (1993) The *INO1* promoter of *Saccharomyces cerevisiae* includes an upstream repressor sequence (*URS1*) common to a diverse set of yeast genes. *Journal of Bacteriology* **175**, 4235–4238.

Maeda, Y., Watanabe, R., Harris, C. L., Hong, Y., Ohishi, K. *et al.* (2001) PIG-M transfers the first mannose to glycosylphosphatidylinositol on the lumenal side of the ER. *EMBO Journal* **20**, 250–261.

Majumder, A. L., Johnson, M. D. and Henry, S. A. (1997) 1L-*myo*-inositol-1-phosphate synthase. *Biochimica et Biophysica Acta* **1348**, 245–256.

Mandala, S. M., Thornton, R., Tu, Z., Kurtz, M. B., Nickels, J. *et al.* (1998) Sphingoid base 1-phosphate phosphatase: a key regulator of sphingolipid metabolism and stress response. *Proceedings of the National Academy of Sciences USA* **95**, 150–155.

Mao, C., Wadleigh, M., Jenkins, G. M., Hannun, Y. A. and Obeid, L. M. (1997) Identification and characterization of *Saccharomyces cerevisiae* dihydrosphingosine-1-phosphate phosphatase. *Journal of Biological Chemistry* **272**, 28690–28694.

Mao, C., Xu, R., Bielawska, A. and Obeid, L. M. (2000a) Cloning of an alkaline ceramidase from *Saccharomyces cerevisiae.* An enzyme with reverse (CoA-independent) ceramide synthase activity. *Journal of Biological Chemistry* **275**, 6876–6884.

Mao, C., Xu, R., Bielawska, A., Szulc, Z. M. and Obeid, L. M. (2000b) Cloning and characterization of a *Saccharomyces cerevisiae* alkaline ceramidase with specificity for dihydroceramide. *Journal of Biological Chemistry* **275**, 31369–31378.

Marthol, S., Hoffmann, J., Stegner, S., Hoja, U., Schulz, R. *et al.* (2002) An organelle-specific acetyl CoA carboxylase controls mitochondrial fatty acid synthesis in *Saccharomyces cerevisie.*, In: R. M. Epand, and P. K. Kinnuen *43rd International Conference on the Bioscience of Lipids*, p. 30. Graz, Austria: Elsevier.

Martin, C. E., Oh, C.-S., Kandasamy, P., Chellapa, R. and Vemula, M. (2001) Yeast desaturases. *Biochemical Society Transactions* **30**, 1080–1082.

Martin, C. E., Vemula, M., Oh, C.-S., Kandasamy, P. and Chelappa, A. (2002) *Saccharomyces cerevisiae OLE1* gene expression is regulated by N-terminal proteolytic fragments of two membrane proteins that are functionally linked to fatty acid and oxygen-mediated regulation of transcription and mRNA stability. In: R. M. Epand and P. K. Kinnuen (eds) *43rd International Conference on the Bioscience of Lipids*, pp. 13–14. Graz, Austria: Elsevier.

Martin, T. F. (1997) Protein transport. Greasing the Golgi budding machine. *Nature* **387**, 21–22.

Marvin, M. E., Williams, P. H. and Cashmore, A. M. (2001) The isolation and characterisation of a *Saccharomyces cerevisiae* gene (*LIP2*) involved in the attachment of lipoic acid groups to mitochondrial enzymes. *FEMS Microbiology Letters* **199**, 131–136.

Marzioch, M., Erdmann, R., Veenhuis, M. and Kunau, W. H. (1994) *PAS7* encodes a novel yeast member of the WD-40 protein family essential for import of 3-oxoacyl-CoA thiolase, a PTS2-containing protein, into peroxisomes. *EMBO Journal* **13**, 4908–4918.

McAlister-Henn, L., Steffan, J. S., Minard, K. I. and Anderson, S. L. (1995) Expression and function of a mislocalized form of peroxisomal malate dehydrogenase (*MDH3*) in yeast. *Journal of Biological Chemistry* **270**, 21220–21225.

McConville, M. J. and Menon, A. K. (2000) Recent developments in the cell *bi*ology and biochemistry of glycosylphosphatidylinositol lipids (review). *Molecular Membrane Biology* **17**, 1–16.

McDonough, V., Stukey, J. and Cavanagh, T. (2002) Mutations in erg4 affect the sensitivity of *Saccharomyces cerevisiae* to medium-chain fatty acids. *Biochimica et Biophysica Acta* **1581**, 109–118.

McMaster, C. R. (2001) Lipid metabolism and vesicle trafficking: more than just greasing the transport machinery. *Biochemical Cell Biology* **79**, 681–692.

McMaster, C. R. and Bell, R. M. (1994) Phosphatidylcholine biosynthesis in *Saccharomyces cerevisiae.* Regulatory insights from studies employing null and chimeric *sn-1,2*-diacylglycerol choline- and ethanolaminephosphotransferases. *Journal of Biological Chemistry* **269**, 28010–28016.

Mehrotra, B., Myszka, D. G. and Prestwich, G. D. (2000) Binding kinetics and ligand specificity for the interactions of the C2B domain of synaptogamin II with inositol polyphosphates and phosphoinositides. *Biochemistry* **39**, 9679–9686.

Mensink, R. P., Zock, P. L., Katan, M. B. and Hornstra, G. (1992) Effect of dietary *cis* and *trans* fatty acids on serum lipoprotein[a] levels in humans. *Journal of Lipid Research* **33**, 1493–1501.

Merkel, O., Fido, M., Mayr, J. A., Pruger, H., Raab, F. *et al.* (1999) Characterization and function *in vivo* of two novel phospholipases B/lysophospholipases from *Saccharomyces cerevisiae. Journal of Biological Chemistry* **274**, 28121–28127.

Milla, P., Athenstaedt, K., Viola, F., Oliaro-Bosso, S., Kohlwein, S. D. *et al.* (2002) Yeast oxidosqualene cyclase (Erg7p) is a major component of lipid particles. *Journal of Biological Chemistry* **277**, 2406–2412.

Minskoff, S. A. and Greenberg, M. L. (1997) Phosphatidylglycerophosphate synthase from yeast. *Biochimica et Biophysica Acta* **1348**, 187–191.

Mitchell, A. G. and Martin, C. E. (1995) A novel cytochrome b5-like domain is linked to the carboxyl terminus of the *Saccharomyces cerevisiae* Δ-9 fatty acid desaturase. *Journal of Biological Chemistry* **270**, 29766–29772.

Mitchell, A. G. and Martin, C. E. (1997) Fah1p, a *Saccharomyces cerevisiae* cytochrome b5 fusion protein, and its *Arabidopsis thaliana* homolog that lacks the cytochrome b5 domain both function in the α-hydroxylation of sphingolipid-associated very long chain fatty acids. *Journal of Biological Chemistry* **272**, 28281–28288.

Mo, C., Valachovic, M., Randall, S. K., Nickels, J. T. and Bard, M. (2002) Protein-protein interactions among C-4 demethylation enzymes involved in yeast sterol biosynthesis. *Proceedings of the National Academy of Sciences USA* **99**, 9739–9744.

Morlock, K. R., Lin, Y. P. and Carman, G. M. (1988) Regulation of phosphatidate phosphatase activity by inositol in *Saccharomyces cerevisiae*. *Journal of Bacteriology* **170**, 3561–3566.

Morris, T. W., Reed, K. E. and Cronan, J. E. Jr (1995) Lipoic acid metabolism in *Escherichia coli*: the *lplA* and *lipB* genes define redundant pathways for ligation of lipoyl groups to apoprotein. *Journal of Bacteriology* **177**, 1–10.

Muniz, M., Nuoffer, C., Hauri, H. P. and Riezman, H. (2000) The Emp24 complex recruits a specific cargo molecule into endoplasmic reticulum-derived vesicles. *Journal of Cell Biology* **148**, 925–930.

Munn, A. L. (2001) Molecular requirements for the internalisation step of endocytosis: insights from yeast. *Biochimica et Biophysica Acta* **1535**, 236–257.

Munn, A. L. and Riezman, H. (1994) Endocytosis is required for the growth of vacuolar H(+)-ATPase- defective yeast: identification of six new *END* genes. *Journal of Cell Biology* **127**, 373–386.

Munn, A. L., Heese-Peck, A., Stevenson, B. J., Pichler, H. and Riezman, H. (1999) Specific sterols required for the internalization step of endocytosis in yeast. *Molecular Biology of the Cell* **10**, 3943–3957.

Murray, M. and Greenberg, M. L. (1997) Regulation of inositol monophosphatase in *Saccharomyces cerevisiae*. *Molecular Microbiology* **25**, 541–546.

Nagiec, M. M., Wells, G. B., Lester, R. L. and Dickson, R. C. (1993) A suppressor gene that enables *Saccharomyces cerevisiae* to grow without making sphingolipids encodes a protein that resembles an *Escherichia coli* fatty acyltransferase. *Journal of Biological Chemistry* **268**, 22156–22163.

Nagiec, M. M., Nagiec, E. E., Baltisberger, J. A., Wells, G. B., Lester, R. L. *et al.* (1997) Sphingolipid synthesis as a target for antifungal drugs. Complementation of the inositol phosphorylceramide synthase defect in a mutant strain of *Saccharomyces cerevisiae* by the *AUR1* gene. *Journal of Biological Chemistry* **272**, 9809–9817.

Nagiec, M. M., Skrzypek, M., Nagiec, E. E., Lester, R. L. and Dickson, R. C. (1998) The *LCB4* (YOR171c) and *LCB5* (YLR260w) genes of *Saccharomyces* encode sphingoid long chain base kinases. *Journal of Biological Chemistry* **273**, 19437–19442.

Ness, F., Bourot, S., Regnacq, M., Spagnoli, R., Berges, T. *et al.* (2001) *SUT1* is a putative Zn[II]2Cys6-transcription factor whose upregulation enhances both sterol uptake and synthesis in aerobically growing *Saccharomyces cerevisiae* cells. *European Journal of Biochemistry* **268**, 1585–1595.

Nicolson, T. and Mayinger, P. (2000) Reconstitution of yeast microsomal lipid flip-flop using endogenous aminophospholipids. *FEBS Letters* **476**, 277–281.

Nikawa, J. and Yamashita, S. (1997) Phosphatidylinositol synthase from yeast. *Biochimica et Biophysica Acta* **1348**, 173–178.

Odorizzi, G., Babst, M. and Emr, S. D. (2000) Phosphoinositide signaling and the regulation of membrane trafficking in yeast. *Trends in Biochemical Sciences* **25**, 229–235.

Oelkers, P., Tinkelenberg, A., Erdeniz, N., Cromley, D., Billheimer, J. T. *et al.* (2000) A lecithin cholesterol acyltransferase-like gene mediates diacylglycerol esterification in yeast. *Journal of Biological Chemistry* **275**, 15609–15612.

Oh, C.-S., Toke, D. A., Mandala, S. and Martin, C. E. (1997) *ELO2* and *ELO3*, homologues of the *Saccharomyces cerevisiae ELO1* gene, function in fatty acid elongation and are required for sphingolipid formation. *Journal of Biological Chemistry* **272**, 17376–17384.

Ohishi, K., Inoue, N. and Kinoshita, T. (2001) PIG-S and PIG-T, essential for GPI anchor attachment to proteins, form a complex with GAA1 and GPI8. *EMBO Journal* **20**, 4088–4098.

Omer, C. A. and Gibbs, J. B. (1994) Protein prenylation in eukaryotic microorganisms: genetics, biology and biochemistry. *Molecular Microbiology* **11**, 219–225.

Orlean, P. (1997) Biogenesis of yeast wall and surface components, In: J. R. Pringle, J. R. Broach and E. W. Jones (eds) *The Molecular and Cellular Biology of the Yeast Saccharomyces: Cell Cycle and Cell Biology*, pp. 229–362. New York: Cold Spring Harbor Laboratory Press.

Orlean, P., Albright, C. and Robbins, P. W. (1988) Cloning and sequencing of the yeast gene for dolichol phosphate mannose synthase, an essential protein. *Journal of Biological Chemistry* **263**, 17499–17507.

Oshiro, J., Rangaswamy, S., Chen, X., Han, G. S., Quinn, J. E. *et al.* (2000) Regulation of the DPP1-encoded diacylglycerol pyrophosphate (DGPP) phosphatase by inositol and growth phase. Inhibition of DGPP phosphatase activity by CDP-diacylglycerol and activation of phosphatidylserine synthase activity by DGPP. *Journal of Biological Chemistry* **275**, 40887–40896.

Ostrander, D. B., O'Brien, D. J., Gorman, J. A. and Carman, G. M. (1998) Effect of CTP synthetase regulation by CTP on phospholipid synthesis in *Saccharomyces cerevisiae*. *Journal of Biological Chemistry* **273**, 18992–19001.

Oulmouden, A. and Karst, F. (1991) Nucleotide sequence of the *ERG12* gene of *Saccharomyces cerevisiae* encoding mevalonate kinase. *Current Genetics* **19**, 9–14.

Palermo, L. M., Leak, F. W., Tove, S. and Parks, L. W. (1997) Assessment of the essentiality of *ERG* genes late in ergosterol biosynthesis in *Saccharomyces cerevisiae*. *Current Genetics* **32**, 93–99.

Pan, X., Roberts, P., Chen, Y., Kvam, E., Shulga, N. *et al.* (2000) Nucleus-vacuole junctions in *Saccharomyces cerevisiae* are formed through the direct interaction of Vac8p with Nvj1p. *Molecular Biology of the Cell* **11**, 2445–2457.

Park, T. S., Ostrander, D. B., Pappas, A. and Carman, G. M. (1999) Identification of Ser424 as the protein kinase A phosphorylation site in CTP synthetase from *Saccharomyces cerevisiae*. *Biochemistry* **38**, 8839–8848.

Parks, L. W. and Casey, W. M. (1995) Physiological implications of sterol biosynthesis in yeast. *Annual Reviews of Microbiology* **49**, 95–116.

Parks, L. W., Crowley, J. H., Leak, F. W., Smith, S. J. and Tomeo, M. E. (1999) Use of sterol mutants as probes for sterol functions in the yeast, *Saccharomyces cerevisiae*. *Critical Reviews in Biochemistry and Molecular Biology* **34**, 399–404.

Patton, J. L. and Lester, R. L. (1991) The phosphoinositol sphingolipids of *Saccharomyces cerevisiae* are highly localized in the plasma membrane. *Journal of Bacteriology* **173**, 3101–3108.

Patton-Vogt, J. L. and Henry, S. A. (1998) *GIT1*, a gene encoding a novel transporter for glycerophosphoinositol in *Saccharomyces cerevisiae*. *Genetics* **149**, 1707–1715.

Patton-Vogt, J. L., Griac, P., Sreenivas, A., Bruno, V., Dowd, S. *et al.* (1997) Role of the yeast phosphatidylinositol/phosphatidylcholine transfer protein (Sec14p) in phosphatidylcholine turnover and *INO1* regulation. *Journal of Biological Chemistry* **272**, 20873–20883.

Pearson, B. M., Hernando, Y., Payne, J., Wolf, S. S., Kalogeropoulos, A. *et al.* (1996) Sequencing of a 35.71 kb DNA segment on the right arm of yeast chromosome *XV* reveals regions of similarity to chromosomes *I* and *XIII*. *Yeast* **12**, 1021–1031.

Plochocka, D., Karst, F., Swiezewska, E. and Szkopinska, A. (2000) The role of *ERG20* gene (encoding yeast farnesyl diphosphate synthase) mutation in long dolichol formation. Molecular modeling of FPP synthase. *Biochimie* **82**, 733–738.

Posas, F., Wurgler-Murphy, S. M., Maeda, T., Witten, E. A., Thai, T. C. *et al.* (1996) Yeast *HOG1* MAP kinase cascade is regulated by a multistep phosphorelay mechanism in the SLN1-YPD1-SSK1 "two-component" osmosensor. *Cell* **86**, 865–875.

Prendergast, J. A., Singer, R. A., Rowley, N., Rowley, A., Johnston, G. C. *et al.* (1995) Mutations sensitizing yeast cells to the start inhibitor nalidixic acid. *Yeast* **11**, 537–547.

Primig, M., Williams, R. M., Winzeler, E. A., Tevzadze, G. G., Conway, A. R. *et al.* (2000) The core meiotic transcriptome in budding yeasts. *Nature Genetics* **26**, 415–423.

Profant, D. A., Roberts, C. J. and Wright, R. L. (2000) Mutational analysis of the karmellae-inducing signal in Hmg1p, a yeast HMG-CoA reductase isozyme. *Yeast* **16**, 811–827.

Pronk, J. T., Steensma, H. Y. and Van Dijken, J. P. (1996) Pyruvate metabolism in *Saccharomyces cerevisiae. Yeast* **12**, 1607–1633.

Qie, L., Nagiec, M. M., Baltisberger, J. A., Lester, R. L. and Dickson, R. C. (1997) Identification of a *Saccharomyces gene, LCB3*, necessary for incorporation of exogenous long chain bases into sphingolipids. *Journal of Biological Chemistry* **272**, 16110–16117.

Rape, M., Hoppe, T., Gorr, I., Kalocay, M., Richly, H. *et al.* (2001) Mobilization of processed, membrane-tethered SPT23 transcription factor by CDC48(UFD1/NPL4), a ubiquitin-selective chaperone. *Cell* **107**, 667–677.

Rebecchi, M. J. and Pentyala, S. N. (2000) Structure, function, and control of phosphoinositide-specific phospholipase C. *Physiological Reviews* **80**, 1291–1335.

Regnacq, M., Alimardani, P., El Moudni, B. and Berges, T. (2001) SUT1p interaction with Cyc8p(Ssn6p) relieves hypoxic genes from Cyc8p-Tup1p repression in *Saccharomyces cerevisiae. Molecular Microbiology* **40**, 1085–1096.

Rehling, P., Marzioch, M., Niesen, F., Wittke, E., Veenhuis, M. *et al.* (1996) The import receptor for the peroxisomal targeting signal 2 (PTS2) in *Saccharomyces cerevisiae* is encoded by the *PAS7* gene. *EMBO Journal* **15**, 2901–2913.

Richard, M., De Groot, P., Courtin, O., Poulain, D., Klis, F. *et al.* (2002) GPI7 affects cell-wall protein anchorage in *Saccharomyces cerevisiae* and *Candida albicans. Microbiology* **148**, 2125–2133.

Rip, J. W., Rupar, C. A., Ravi, K. and Carroll, K. K. (1985) Distribution, metabolism and function of dolichol and polyprenols. *Progress in Lipid Research* **24**, 269–309.

Rivas, M. P., Kearns, B. G., Xie, Z., Guo, S., Sekar, M. C. *et al.* (1999) Pleiotropic alterations in lipid metabolism in yeast *sac1* mutants: relationship to "bypass Sec14p" and inositol auxotrophy. *Molecular Biology of the Cell* **10**, 2235–2250.

Robinson, K. S., Lai, K., Cannon, T. A. and McGraw, P. (1996) Inositol transport in *Saccharomyces cerevisiae* is regulated by transcriptional and degradative endocytic mechanisms during the growth cycle that are distinct from inositol-induced regulation. *Molecular Biology of the Cell* **7**, 81–89.

Rodriguez-Vargas, S., Estruch, F. and Randez-Gil, F. (2002) Gene expression analysis of cold and freeze stress in Baker's yeast. *Applied and Environmental Microbiology* **68**, 3024–3030.

Rose, T. M., Schultz, E. R. and Todaro, G. J. (1992) Molecular cloning of the gene for the yeast homolog (ACB) of diazepam binding inhibitor/endozepine/acyl-CoA-binding protein. *Proceedings of the National Academy of Sciences USA* **89**, 11287–11291.

Rosenfeld, E., Beauvoit, B., Blondin, B. and Salmon, J. M. (2003) Oxygen consumption by anaerobic *Saccharomyces cerevisiae* under enological conditions: effect on fermentation kinetics. *Applied and Environmental Microbiology* **69**, 113–121.

Rudge, S. A., Morris, A. J. and Engebrecht, J. (1998) Relocalization of phospholipase D activity mediates membrane formation during meiosis. *Journal of Cell Biology* **140**, 81–90.

Rudge, S. A., Pettitt, T. R., Zhou, C., Wakelam, M. J. and Engebrecht, J. A. (2001) *Spo14* separation-of-function mutations define unique roles for phospholipase D in secretion and cellular differentiation in *Saccharomyces cerevisiae. Genetics* **158**, 1431–1444.

Sa-Correia, I. and Tenreiro, S. (2002) The multidrug resistance transporters of the major facilitator superfamily, 6 years after disclosure of *Saccharomyces cerevisiae* genome sequence. *Journal of Biotechnology* **98**, 215–226.

Safrany, S. T., Caffrey, J. J., Yang, X. and Shears, S. B. (1999a) Diphosphoinositol polyphosphates: the final frontier for inositide research? *Biological Chemistry* **380**, 945–951.

Safrany, S. T., Ingram, S. W., Cartwright, J. L., Falck, J. R., McLennan, A. G. *et al.* (1999b) The diadenosine hexaphosphate hydrolases from *Schizosaccharomyces pombe* and *Saccharomyces cerevisiae* are homologues of the human diphosphoinositol polyphosphate phosphohydrolase. Overlapping substrate specificities in a MutT-type protein. *Journal of Biological Chemistry* **274**, 21735–21740.

Saitoh, S., Takahashi, K., Nabeshima, K., Yamashita, Y., Nakaseko, Y. *et al.* (1996) Aberrant mitosis in fission yeast mutants defective in fatty acid synthetase and acetyl CoA carboxylase. *Journal of Cell Biology* **134**, 949–961.

Sandager, L., Gustavsson, M. H., Stahl, U., Dahlqvist, A., Wiberg, E. *et al.* (2002) Storage lipid synthesis is non-essential in yeast. *Journal of Biological Chemistry* **277**, 6478–6482.

Sato, M., Fujisaki, S., Sato, K., Nishimura, Y. and Nakano, A. (2001) Yeast *Saccharomyces cerevisiae* has two cis-prenyltransferases with different properties and localizations. Implication for their distinct physiological roles in dolichol synthesis. *Genes to Cells* **6**, 495–506.

Sawai, H., Okamoto, Y., Luberto, C., Mao, C., Bielawska, A. *et al.* (2000) Identification of *ISC1* (YER019w) as inositol phosphosphingolipid phospholipase C in *Saccharomyces cerevisiae. Journal of Biological Chemistry* **275**, 39793–39798.

Schjerling, C. K., Hummel, R., Hansen, J. K., Borsting, C., Mikkelsen, J. M. *et al.* (1996) Disruption of the gene encoding the acyl-CoA-binding protein (ACB1) perturbs acyl-CoA metabolism in *Saccharomyces cerevisiae. Journal of Biological Chemistry* **271**, 22514–22521.

Schmelzle, T., Helliwell, S. B. and Hall, M. N. (2002) Yeast protein kinases and the RHO1 exchange factor TUS1 are novel components of the cell integrity pathway in yeast. *Molecular and Cellular Biology* **22**, 1329–1339.

Schneider, R., Massow, M., Lisowsky, T. and Weiss, H. (1995) Different respiratory-defective phenotypes of *Neurospora crassa* and *Saccharomyces cerevisiae* after inactivation of the gene encoding the mitochondrial acyl carrier protein. *Current Genetics* **29**, 10–17.

Schneider, R., Brors, B., Burger, F., Camrath, S. and Weiss, H. (1997a) Two genes of the putative mitochondrial fatty acid synthase in the genome of *Saccharomyces cerevisiae. Current Genetics* **32**, 384–388.

Schneider, R., Brors, B., Massow, M. and Weiss, H. (1997b) Mitochondrial fatty acid synthesis: a relic of endosymbiontic origin and a specialized means for respiration. *FEBS Letters* **407**, 249–252.

Schneiter, R. and Kohlwein, S. D. (1997) Organelle structure, function, and inheritance in yeast: a role for fatty acid synthesis? *Cell* **88**, 431–434.

Schneiter, R., Hitomi, M., Ivessa, A. S., Fasch, E. V., Kohlwein, S. D. *et al.* (1996) A yeast acetyl coenzyme A carboxylase mutant links very-long-chain fatty acid synthesis to the structure and function of the nuclear membrane-pore complex. *Molecular and Cellular Biology* **16**, 7161–7172.

Schneiter, R., Brugger, B., Sandhoff, R., Zellnig, G., Leber, A. *et al.* (1999a) Electrospray ionization tandem mass spectrometry (ESI-MS/MS) analysis of the lipid molecular species composition of yeast subcellular membranes reveals acyl chain-based sorting/remodeling of distinct molecular species en route to the plasma membrane. *Journal of Cell Biology* **146**, 741–754.

Schneiter, R., Guerra, C. E., Lampl, M., Gogg, G., Kohlwein, S. D. *et al.* (1999b) The *Saccharomyces cerevisiae* hyperrecombination mutant hpr1Delta is synthetically lethal with two conditional alleles of the acetyl coenzyme A carboxylase gene and causes a defect in nuclear export of polyadenylated RNA. *Molecular and Cellular Biology* **19**, 3415–3422.

Schneiter, R., Guerra, C. E., Lampl, M., Tatzer, V., Zellnig, G. *et al.* (2000a) A novel cold-sensitive allele of the rate-limiting enzyme of fatty acid synthesis, acetyl coenzyme A carboxylase, affects the morphology of the yeast vacuole through acylation of Vac8p. *Molecular and Cellular Biology* **20**, 2984–2995.

Schneiter, R., Tatzer, V., Gogg, G., Leitner, E. and Kohlwein, S. D. (2000b) Elo1p-dependent carboxy-terminal elongation of C14:1Δ(9) to C16:1Delta(11) fatty acids in *Saccharomyces cerevisiae. Journal of Bacteriology* **182**, 3655–3660.

Schorling, S., Vallee, B., Barz, W. P., Riezman, H. and Oesterhelt, D. (2001) Lag1p and Lac1p are essential for the acyl-CoA-dependent ceramide synthase reaction in *Saccharomyces cerevisae. Molecular Biology of the Cell* **12**, 3417–3427.

Schroepfer, G. J., Jr. (2000) Oxysterols: modulators of cholesterol metabolism and other processes. *Physiological Reviews* **80**, 361–554.

Schüller, H. J., Hahn, A., Troster, F., Schütz, A. and Schweizer, E. (1992) Coordinate genetic control of yeast fatty acid synthase genes *FAS1* and *FAS2* by an upstream activation site common to genes involved in membrane lipid biosynthesis. *EMBO Journal* **11**, 107–114.

Schumacher, M. M., Choi, J. Y. and Voelker, D. R. (2002) Phosphatidylserine transport to the mitochondria is regulated by ubiquitination. *Journal of Biological Chemistry* **277**, 51033–51042.

Schwank, S., Ebbert, R., Rautenstrauss, K., Schweizer, E. and Schüller, H. J. (1995) Yeast transcriptional activator *INO2* interacts as an Ino2p/Ino4p basic helix-loop-helix heteromeric complex with the inositol/choline- responsive element necessary for expression of phospholipid biosynthetic genes in *Saccharomyces cerevisiae. Nucleic Acids Research* **23**, 230–237.

Schweizer, M. (1986) Yeast fatty acid synthase genes. In: Hardie, D. G. and Coggins, J. R. (eds) *Multidomain Proteins – Structure and Evolution*, pp. 195–227, Amsterdam, New York, Oxford: Elsevier.

Sciorra, V. A., Rudge, S. A., Prestwich, G. D., Frohman, M. A., Engebrecht, J. *et al.* (1999) Identification of a phosphoinositide binding motif that mediates activation of mammalian and yeast phospholipase D isoenzymes. *EMBO Journal* **18**, 5911–5921.

Sekito, T., Thornton, J. and Butow, R. A. (2000) Mitochondria-to-nuclear signaling is regulated by the subcellular localization of the transcription factors Rtg1p and Rtg3p. *Molecular Biology of the Cell* **11**, 2103–2115.

Sha, B. and Luo, M. (1999) PI transfer protein: the specific recognition of phospholipids and its functions. *Biochimica et Biophysica Acta* **1441**, 268–277.

Sha, B., Phillips, S. E., Bankaitis, V. A. and Luo, M. (1998) Crystal structure of the *Saccharomyces cerevisiae* phosphatidylinositol- transfer protein. *Nature* **391**, 506–510.

Shaldubina, A., Ju, S., Vaden, D. L., Ding, D., Belmaker, R. H. *et al.* (2002) *Epi*-inositol regulates expression of the yeast *INO1* gene encoding inositol-1-P synthase. *Molecular Psychiatry* **7**, 174–180.

Shaw, J. D., Cummings, K. B., Huyer, G., Michaelis, S. and Wendland, B. (2001) Yeast as a model system for studying endocytosis. *Experimental Cell Research* **271**, 1–9.

Shen, H. and Dowhan, W. (1996) Reduction of CDP-diacylglycerol synthase activity results in the excretion of inositol by *Saccharomyces cerevisiae. Journal of Biological Chemistry* **271**, 29043–29048.

Shen, H. and Dowhan, W. (1997) Regulation of phospholipid biosynthetic enzymes by the level of CDP-diacylglycerol synthase activity. *Journal of Biological Chemistry* **272**, 11215–11220.

Shen, H., Heacock, P. N., Clancey, C. J. and Dowhan, W. (1996) The *CDS1* gene encoding CDP-diacylglycerol synthase in *Saccharomyces cerevisiae* is essential for cell growth. *Journal of Biological Chemistry* **271**, 789–795.

Shi, Z., Buntel, C. J. and Griffin, J. H. (1994) Isolation and characterization of the gene encoding 2, 3-oxidosqualene- lanosterol cyclase from *Saccharomyces cerevisiae. Proceedings of the National Academy of Sciences USA* **91**, 7370–7374.

Shianna, K. V., Dotson, W. D., Tove, S. and Parks, L. W. (2001) Identification of a *UPC2* homolog in *Saccharomyces cerevisiae* and its involvement in aerobic sterol uptake. *Journal of Bacteriology* **183**, 830–834.

Shirra, M. K., Patton-Vogt, J., Ulrich, A., Liuta-Tehlivets, O., Kohlwein, S. D. *et al.* (2001) Inhibition of acetyl coenzyme A carboxylase activity restores expression of the *INO1* gene in a *snf1* mutant strain of *Saccharomyces cerevisiae. Molecular and Cellular Biology* **21**, 5710–5722.

Silve, S., Dupuy, P. H., Ferrara, P. and Loison, G. (1998) Human lamin B receptor exhibits sterol C14-reductase activity in *Saccharomyces cerevisiae. Biochimica et Biophysica Acta* **1392**, 233–244.

Simons, K. and Ikonen, E. (1997) Functional rafts in cell membranes. *Nature* **387**, 569–572.

Simonsen, A., Wurmser, A. E., Emr, S. D. and Stenmark, H. (2001) The role of phosphoinositides in membrane transport. *Current Opinion in Cell Biology* **13**, 485–492.

Skaggs, B. A., Alexander, J. F., Pierson, C. A., Schweitzer, K. S., Chun, K. T., Koegel, C., Barbuch, R. and Bard, M. (1996) Cloning and characterization of the *Saccharomyces cerevisiae* C-22 sterol desaturase gene, encoding a second cytochrome P-450 involved in ergosterol biosynthesis. *Gene* **169**, 105–109.

Smith, S. J. and Parks, L. W. (1997) Requirement of heme to replace the sparking sterol function in the yeast *Saccharomyces cerevisiae. Biochimica et Biophysica Acta* **1345**, 71–76.

Sorger, D. and Daum, G. (2002) Triacylgycerol synthesis in lipid particles of the yeast. In: Epand, R. M. and Kinnuen, P. K. (eds) *43rd International Conference on the Bioscience of Lipids*, p. 36. Graz, Austria: Elsevier.

Sotsios, Y. and Ward, S. G. (2000) Phosphoinositide 3-kinase: a key biochemical signal for cell migration in response to chemokines. *Immunological Reviews* **177**, 217–235.

Soustre, I., Dupuy, P. H., Silve, S., Karst, F. and Loison, G. (2000) Sterol metabolism and *ERG2* gene regulation in the yeast *Saccharomyces cerevisiae. FEBS Letters* **470**, 102–106.

Sreenivas, A., Villa-Garcia, M. J., Henry, S. A. and Carman, G. M. (2001) Phosphorylation of the yeast phospholipid synthesis regulatory protein Opi1p by protein kinase C. *Journal of Biological Chemistry* **276**, 29915–29923.

Stefan, C. J., Audhya, A. and Emr, S. D. (2002) The yeast synaptojanin-like proteins control the cellular distribution of phosphatidylinositol (4,5)-bisphosphate. *Molecular Biology of the Cell* **13**, 542–557.

Stevens, V. L. (1995) Biosynthesis of glycosylphosphatidylinositol membrane anchors. *Biochemical Journal* **310**, 361–370.

Stewart, L. C. and Yaffe, M. P. (1991) A role for unsaturated fatty acids in mitochondrial movement and inheritance. *Journal of Cell Biology* **115**, 1249–1257.

Stock, S. D., Hama, H., DeWald, D. B. and Takemoto, J. Y. (1999) Sec14-dependent secretion in *Saccharomyces cerevisiae.* Nondependence on sphingolipid synthesis-coupled diacylglycerol production. *Journal of Biological Chemistry* **274**, 12979–12983.

Stolz, J. and Sauer, N. (1999) The fenpropimorph resistance gene *FEN2* from *Saccharomyces cerevisiae* encodes a plasma membrane H^+-pantothenate symporter. *Journal of Biological Chemistry* **274**, 18747–18752.

Stolz, L. E., Huynh, C. V., Thorner, J. and York, J. D. (1998) Identification and characterization of an essential family of inositol polyphosphate 5-phosphatases (*INP51*, *INP52* and *INP53* gene products) in the yeast *Saccharomyces cerevisiae. Genetics* **148**, 1715–1729.

Storey, M. K., Clay, K. L., Kutateladze, T., Murphy, R. C., Overduin, M. *et al.* (2001a) Phosphatidylethanolamine has an essential role in *Saccharomyces cerevisiae* that is independent of its ability to form hexagonal phase structures. *Journal of Biological Chemistry* **276**, 48539–48548.

Storey, M. K., Wu, W. I. and Voelker, D. R. (2001b) A genetic screen for ethanolamine auxotrophs in *Saccharomyces cerevisiae* identifies a novel mutation in Mcd4p, a protein implicated in glycosylphosphatidylinositol anchor synthesis. *Biochimica et Biophysica Acta* **1532**, 234–247.

Stuible, H. P., Meier, S., Wagner, C., Hannappel, E. and Schweizer, E. (1998) A novel phosphopantetheine: protein transferase activating yeast mitochondrial acyl carrier protein. *Journal of Biological Chemistry* **273**, 22334–22339.

Sturley, S. L. (1998) A molecular approach to understanding human sterol metabolism using yeast genetics. *Current Opinion in Lipidology* **9**, 85–91.

Sturley, S. L. (2000) Conservation of eukaryotic sterol homeostasis: new insights from studies in budding yeast. *Biochimica et Biophysica Acta* **1529**, 155–163.

Sulo, P. and Martin, N. C. (1993) Isolation and characterization of *LIP5*. A lipoate biosynthetic locus of *Saccharomyces cerevisiae. Journal of Biological Chemistry* **268**, 17634–17639.

Sutterlin, C., Escribano, M. V., Gerold, P., Maeda, Y., Mazon, M. J. *et al.* (1998) *Saccharomyces cerevisiae* GPI10, the functional homologue of human PIG-B, is required for glycosylphosphatidylinositol-anchor synthesis. *Biochemical Journal* **332**, 153–159.

Swain, E., Baudry, K., Stukey, J., McDonough, V., Germann, M. *et al.* (2002a) Sterol-dependent regulation of sphingolipid metabolism in *Saccharomyces cerevisiae. Journal of Biological Chemistry* **277**, 26177–26184.

Swain, E., Stukey, J., McDonough, V., Germann, M., Liu, Y. *et al.* (2002b) Yeast cells lacking the *ARV1* gene harbor defects in sphingolipid metabolism. Complementation by human *ARV1*. *Journal of Biological Chemistry* **277**, 36152–36160.

Szczebara, F. M., Chandelier, C., Villeret, C., Masurel, A., Bourot, S. *et al.* (2003) Total biosynthesis of hydrocortisone from a simple carbon source in yeast. *Nature Biotechnology* **21**, 143–149.

Szkopinska, A., Swiezewska, E. and Karst, F. (2000) The regulation of activity of main mevalonic acid pathway enzymes: farnesyl diphosphate synthase, 3-hydroxy-3-methylglutaryl-CoA reductase, and squalene synthase in yeast *Saccharomyces cerevisiae. Biochemical and Biophysical Research Communications* **267**, 473–477.

Szkopinska, A., Swiezewska, E. and Skoneczny, M. (2001) A novel family of longer chain length dolichols present in oleate-induced yeast *Saccharomyces cerevisiae. Biochimie* **83**, 427–432.

Tabak, H. F., Elgersma, Y., Hettema, E., Franse, M. M., Voorn-Brouwer, T. *et al.* (1995) Transport of proteins and metabolites across the impermeable membrane of peroxisomes. *Cold Spring Harbor Symposia on Quanitative Biology* **60**, 649–655.

Tang, X., Waksman, M., Ely, Y. and Liscovitch, M. (2002) Characterization and regulation of yeast $Ca2^{+}$-dependent phosphatidylethanolamine-phospholipase D activity. *European Journal of Biochemistry* **269**, 3821–3830.

Taron, C. H., Wiedman, J. M., Grimme, S. J. and Orlean, P. (2000) Glycosylphosphatidylinositol biosynthesis defects in Gpi11p- and Gpi13p-deficient yeast suggest a branched pathway and implicate Gpi13p in phosphoethanolamine transfer to the third mannose. *Molecular Biology of the Cell* **11**, 1611–1630.

Tatzer, V., Zellnig, G., Kohlwein, S. D. and Schneiter, R. (2002) Lipid-dependent subcellular relocalization of the acyl chain desaturase in yeast. *Molecular Biology of the Cell* **13**, 4429–4442.

Tevzadze, G. G., Swift, H. and Esposito, R. E. (2000) Spo1, a phospholipase B homolog, is required for spindle pole body duplication during meiosis in *Saccharomyces cerevisiae. Chromosoma* **109**, 72–85.

Thevissen, K., Cammue, B. P., Lemaire, K., Winderickx, J., Dickson, R. C. *et al.* (2000) A gene encoding a sphingolipid biosynthesis enzyme determines the sensitivity of *Saccharomyces cerevisiae* to an antifungal plant defensin from dahlia (*Dahlia merckii*). *Proceedings of the National Academy of Sciences USA* **97**, 9531–9536.

Toh-e, A. and Oguchi, T. (2001) Defects in glycosylphosphatidylinositol (GPI) anchor synthesis activate Hog1 kinase and confer copper-resistance in *Saccharomyces cerevisisae. Genes and Genetic Systems* **76**, 393–410.

Toke, D. A. and Martin, C. E. (1996) Isolation and characterization of a gene affecting fatty acid elongation in *Saccharomyces cerevisiae. Journal of Biological Chemistry* **271**, 18413–18422.

Toke, D. A., Bennett, W. L., Oshiro, J., Wu, W. I., Voelker, D. R. *et al.* (1998) Isolation and characterization of the *Saccharomyces cerevisiae LPP1* gene encoding a Mg2+-independent phosphatidate phosphatase. *Journal of Biological Chemistry* **273**, 14331–14338.

Torkko, J. M., Koivuranta, K. T., Miinalainen, I. J., Yagi, A. I., Schmitz, W. *et al.* (2001) *Candida tropicalis* Etr1p and *Saccharomyces cerevisiae* Ybr026p (Mrf1p), 2-enoyl thioester reductases essential for mitochondrial respiratory competence. *Molecular and Cellular Biology* **21**, 6243–6253.

Trotter, P. J. (2001) The genetics of fatty acid metabolism in *Saccharomyces cerevisiae. Annual Reviews of Nutrition* **21**, 97–119.

Trotter, P. J. and Voelker, D. R. (1995) Identification of a non-mitochondrial phosphatidylserine decarboxylase activity (*PSD2*) in the yeast *Saccharomyces cerevisiae. Journal of Biological Chemistry* **270**, 6062–6070.

Trotter, P. J., Wu, W. I., Pedretti, J., Yates, R. and Voelker, D. R. (1998) A genetic screen for aminophospholipid transport mutants identifies the phosphatidylinositol 4-kinase, STT4p, as an essential component in phosphatidylserine metabolism. *Journal of Biological Chemistry* **273**, 13189–13196.

Tsukahara, K., Hata, K., Nakamoto, K., Sagane, K., Watanabe, N.A. *et al.* (2003) Medicinal genetics approach towards identifying the molecular target of a novel inhibitor of fungal cell wall assembly. *Molecular Microbiology* **48**, 1029–1042.

Tuller, G., Nemec, T., Hrastnik, C. and Daum, G. (1999) Lipid composition of subcellular membranes of an FY1679-derived haploid yeast wild-type strain grown on different carbon sources. *Yeast* **15**, 1555–1564.

Umemura, M., Okamoto, M., Nakayama, K., Sagane, K., Tsukahara, K. *et al.* (2003) *GWT1* gene is required for inositol acylation of glycosylphosphatidylinositol anchors in yeast. *Journal of Biological Chemistry* **278**, 23639–23647.

Vaden, D. L., Ding, D., Peterson, B. and Greenberg, M. L. (2001) Lithium and valproate decrease inositol mass and increase expression of the yeast *INO1* and *INO2* genes for inositol biosynthesis. *Journal of Biological Chemistry* **276**, 15466–15471.

van den Berg, M. A., de Jong-Gubbels, P., Kortland, C. J., van Dijken, J. P., Pronk, J. T. *et al.* (1996) The two acetyl-coenzyme A synthetases of *Saccharomyces cerevisiae* differ with respect to kinetic properties and transcriptional regulation. *Journal of Biological Chemistry* **271**, 28953–28959.

van den Hazel, H. B., Pichler, H., do Valle Matta, M. A., Leitner, E., Goffeau, A. *et al.* (1999) *PDR16* and *PDR17*, two homologous genes of *Saccharomyces cerevisiae*, affect lipid biosynthesis and resistance to multiple drugs. *Journal of Biological Chemistry* **274**, 1934–1941.

van der Klei, I. J. and Veenhuis, M. (1997) Yeast peroxisomes: function and biogenesis of a versatile cell organelle. *Trends in Microbiology* **5**, 502–509.

Van Heusden, G. P., Nebohacova, M., Overbeeke, T. L. and Steensma, H. Y. (1998) The *Saccharomyces cerevisiae TGL2* gene encodes a protein with lipolytic activity and can complement an *Escherichia coli* diacylglycerol kinase disruptant. *Yeast* **14**, 225–232.

van Roermund, C. W., Elgersma, Y., Singh, N., Wanders, R. J. and Tabak, H. F. (1995) The membrane of peroxisomes in *Saccharomyces cerevisiae* is impermeable to NAD(H) and acetyl-CoA under in vivo conditions. *EMBO Journal* **14**, 3480–3486.

Vanhaesebroeck, B., Leevers, S. J., Ahmadi, K., Timms, J., Katso, R. *et al.* (2001) Synthesis and function of 3-phosphorylated inositol lipids. *Annual Reviews of Biochemistry* **70**, 535–602.

Verleur, N., Elgersma, Y., van Roermund, C. W., Tabak, H. F. and Wanders, R. J. (1997) Cytosolic aspartate aminotransferase encoded by the *AAT2* gene is targeted to the peroxisomes in oleate-grown *Saccharomyces cerevisiae. European Journal of Biochemistry* **247**, 972–980.

Vik, A. and Rine, J. (2001) Upc2p and Ecm22p, dual regulators of sterol biosynthesis in *Saccharomyces cerevisiae. Molecular and Cellular Biology* **21**, 6395–6405.

Voelker, D. R. (2000) Interorganelle transport of aminoglycerophospholipids. *Biochimica et Biophysica Acta* **1486**, 97–107.

Voelker, D. R. (2003) New perspectives on the regulation of intermembrane glycerophospholipid traffic. *Journal of Lipid Research* **44**, 441–449.

Wagner, C., Dietz, M., Wittmann, J., Albrecht, A. and Schüller, H. J. (2001) The negative regulator Opi1 of phospholipid biosynthesis in yeast contacts the pleiotropic repressor Sin3 and the transcriptional activator Ino2. *Molecular Microbiology* **41**, 155–166.

Wagner, S. and Paltauf, F. (1994) Generation of glycerophospholipid molecular species in the yeast *Saccharomyces cerevisiae*. Fatty acid pattern of phospholipid classes and selective acyl turnover at *sn*-1 and *sn*-2 positions. *Yeast* **10**, 1429–1437.

Waksman, M., Eli, Y., Liscovitch, M. and Gerst, J. E. (1996) Identification and characterization of a gene encoding phospholipase D activity in yeast. *Journal of Biological Chemistry* **271**, 2361–2364.

Waksman, M., Tang, X., Eli, Y., Gerst, J. E. and Liscovitch, M. (1997) Identification of a novel Ca2+-dependent, phosphatidylethanolamine-hydrolyzing phospholipase D in yeast bearing a disruption in *PLD1*. *Journal of Biological Chemistry* **272**, 36–39.

Wang, C., Xing, J., Chin, C. K., Ho, C. T. and Martin, C. E. (2001) Modification of fatty acids changes the flavor volatiles in tomato leaves. *Phytochemistry* **58**, 227–232.

Wangspa, R. and Takemoto, J. Y. (1998) Role of ergosterol in growth inhibition of *Saccharomyces cerevisiae* by syringomycin E. *FEMS Microbiology Letters* **167**, 215–220.

Watanabe, R., Ohishi, K., Maeda, Y., Nakamura, N. and Kinoshita, T. (1999) Mammalian PIG-L and its yeast homologue Gpi12p are N- acetylglucosaminylphosphatidylinositol de-N-acetylases essential in glycosylphosphatidylinositol biosynthesis. *Biochemical Journal* **339**, 185–192.

Watanabe, R., Funato, K., Venkataraman, K., Futerman, A. H. and Riezman, H. (2002) Sphingolipids are required for the stable membrane association of glycosylphosphatidylinositol-anchored proteins in yeast. *Journal of Biological Chemistry* **277**, 49538–49544.

Watkins, P. A., Lu, J. F., Steinberg, S. J., Gould, S. J., Smith, K. D. *et al.* (1998) Disruption of the *Saccharomyces cerevisiae FAT1* gene decreases very long-chain fatty acyl-CoA synthetase activity and elevates intracellular very long-chain fatty acid concentrations. *Journal of Biological Chemistry* **273**, 18210–18219.

Welch, J. W. and Burlingame, A. L. (1973) Very long-chain fatty acids in yeast. *Journal of Bacteriology* **115**, 464–466.

Wera, S., Bergsma, J. C. T. and Thevelein, J. M. (2001) Phosphoinositides in yeast: genetically tractable signalling. *FEMS Yeast Research* **1**, 9–13.

White, M. J., Hirsch, J. P. and Henry, S. A. (1991) The *OPI1* gene of *Saccharomyces cerevisiae*, a negative regulator of phospholipid biosynthesis, encodes a protein containing polyglutamine tracts and a leucine zipper. *Journal of Biological Chemistry* **266**, 863–872.

Wilcox, L. J., Balderes, D. A., Wharton, B., Tinkelenberg, A. H., Rao, G. *et al.* (2002) Transcriptional profiling identifies two members of the ATP-binding cassette transporter superfamily required for sterol uptake in yeast. *Journal of Biological Chemistry* **277**, 32466–32472.

Wiradjaja, F., Ooms, L. M., Whisstock, J. C., McColl, B., Helfenbaum, L. *et al.* (2001) The yeast inositol polyphosphate 5-phosphatase Inp54p localizes to the endoplasmic reticulum via a C-terminal hydrophobic anchoring tail: regulation of secretion from the endoplasmic reticulum. *Journal of Biological Chemistry* **276**, 7643–7653.

Wu, W. I., Routt, S., Bankaitis, V. A. and Voelker, D. R. (2000) A new gene involved in the transport-dependent metabolism of phosphatidylserine, *PSTB2/PDR17*, shares sequence similarity with the gene encoding the phosphatidylinositol/phosphatidylcholine transfer protein, Sec14. *Journal of Biological Chemistry* **275**, 14446–14456.

Xie, Z., Fang, M. and Bankaitis, V. A. (2001) Evidence for an intrinsic toxicity of phosphatidylcholine to Sec14p- dependent protein transport from the yeast Golgi complex. *Molecular Biology of the Cell* **12**, 1117–1129.

Xu, X., Bittman, R., Duportail, G., Heissler, D., Vilcheze, C. *et al.* (2001) Effect of the structure of natural sterols and sphingolipids on the formation of ordered sphingolipid/sterol domains (rafts). Comparison of cholesterol to plant, fungal, and disease-associated sterols and comparison of sphingomyelin, cerebrosides, and ceramide. *Journal of Biological Chemistry* **276**, 33540–33546.

Yanagisawa, L. L., Marchena, J., Xie, Z., Li, X., Poon, P. P. *et al.* (2002) Activity of specific lipid-regulated ADP ribosylation factor-GTPase-activating proteins is required for Sec14p-dependent Golgi secretory function in yeast. *Molecular Biology of the Cell* **13**, 2193–2206.

Yang, X., Purdue, P. E. and Lazarow, P. B. (2001) Eci1p uses a PTS1 to enter peroxisomes: either its own or that of a partner, Dci1p. *European Journal of Cell Biology* **80**, 126–138.

Yokoyama, K., Saitoh, S., Ishida, M., Yamakawa, Y., Nakamura, K. *et al.* (2001) Very-long-chain fatty acid-containing phospholipids accumulate in fatty acid synthase temperature-sensitive mutant strains of the fission yeast *Schizosaccharomyces pombe* fas2/lsd1. *Biochimica et Biophysica Acta* **1532**, 223–233.

York, J. D., Odom, A. R., Murphy, R., Ives, E. B. and Wente, S. R. (1999) A phospholipase C-dependent inositol polyphosphate kinase pathway required for efficient messenger RNA export. *Science* **285**, 96–100.

Yoshida, S., Ohya, Y., Goebl, M., Nakano, A. and Anraku, Y. (1994) A novel gene, *STT4*, encodes a phosphatidylinositol 4-kinase in the *PKC1* protein kinase pathway of *Saccharomyces cerevisiae*. *Journal of Biological Chemistry* **269**, 1166–1172.

Yoshimoto, H., Saltsman, K., Gasch, A. P., Li, H. X., Ogawa, N. *et al.* (2002) Genome-wide analysis of gene expression regulated by the calcineurin/Crz1p signaling pathway in *Saccharomyces cerevisiae*. *Journal of Biological Chemistry* **277**, 31079–31088.

Young, M. E., Karpova, T. S., Brugger, B., Moschenross, D. M., Wang, G. K. *et al.* (2002) The Sur7p family defines novel cortical domains in *Saccharomyces cerevisiae*, affects sphingolipid metabolism, and is involved in sporulation. *Molecular and Cellular Biology* **22**, 927–934.

Yu, J. W. and Lemmon, M. A. (2001) All phox homology (PX) domains from *Saccharomyces cerevisiae* specifically recognize phosphatidylinositol 3-phosphate. *Journal of Biological Chemistry* **276**, 44179–44184.

Yu, Y., Sreenivas, A., Ostrander, D. B. and Carman, G. M. (2002) Phosphorylation of *Saccharomyces cerevisiae* choline kinase on Ser30 and Ser85 by protein kinase A regulates phosphatidylcholine synthesis by the CDP-choline pathway. *Journal of Biological Chemistry* **277**, 34978–34986.

Zanolari, B., Friant, S., Funato, K., Sutterlin, C., Stevenson, B. J. *et al.* (2000) Sphingoid base synthesis requirement for endocytosis in *Saccharomyces cerevisiae. EMBO Journal* **19**, 2824–2833.

Zaremberg, V. and McMaster, C. R. (2002) Differential partitioning of lipids metabolized by separate yeast glycerol-3-phosphate acyltransferases reveals that phospholipase D generation of phosphatidic acid mediates sensitivity to choline – containing lysolipids and drugs. *Journal of Biological Chemistry* **277**, 39035–39044.

Zhang, S., Skalsky, Y. and Garfinkel, D. J. (1999) *MGA2* or *SPT23* is required for transcription of the Δ9 fatty acid desaturase gene, *OLE1*, and nuclear membrane integrity in *Saccharomyces cerevisiae. Genetics* **151**, 473–483.

Zhang, T., Caffrey, J. J. and Shears, S. B. (2001) The transcriptional regulator, Arg82, is a hybrid kinase with both monophosphoinositol and diphosphoinositol polyphosphate synthase activity. *FEBS Letters* **494**, 208–212.

Zhao, X. J., McElhaney-Feser, G. E., Sheridan, M. J., Broedel, S. E. Jr. and Cihlar, R. L. (1997) Avirulence of *Candida albicans FAS2* mutants in a mouse model of systemic candidiasis. *Infection and Immunity* **65**, 829–832.

Zheng, Z. and Zou, J. (2001) The initial step of the glycerolipid pathway: identification of glycerol 3-phosphate/dihydroxyacetone phosphate dual substrate acyltransferases in *Saccharomyces cerevisiae. Journal of Biological Chemistry* **276**, 41710–41716.

Zinser, E. and Daum, G. (1995) Isolation and biochemical characterization of organelles from the yeast, *Saccharomyces cerevisiae. Yeast* **11**, 493–536.

Zinser, E., Sperka-Gottlieb, C. D., Fasch, E. V., Kohlwein, S. D., Paltauf, F. *et al.* (1991) Phospholipid synthesis and lipid composition of subcellular membranes in the unicellular eukaryote *Saccharomyces cerevisiae. Journal of Bacteriology* **173**, 2026–2034.

Zou, Z., DiRusso, C. C., Ctrnacta, V. and Black, P. N. (2002) Fatty acid transport in *Saccharomyces cerevisiae.* Directed mutagenesis of *FAT1* distinguishes the biochemical activities associated with Fat1p. *Journal of Biological Chemistry* **277**, 31062–31071.

Zou, Z., Tong, F., Faergeman, N. J., Borsting, C., Black, P. N. *et al.* (2003) Vectorial acylation in *Saccharomyces cerevisiae*: Fat1p and fatty acyl-CoA synthetase are interacting components of a fatty acid import complex. *Journal of Biological Chemistry* **278**, 16414–16422.

Zweytick, D., Hrastnik, C., Kohlwein, S. D. and Daum, G. (2000a) Biochemical characterization and subcellular localization of the sterol C-24(28) reductase, Erg4p, from the yeast *Saccharomyces cerevisiae. FEBS Letters* **470**, 83–87.

Zweytick, D., Leitner, E., Kohlwein, S. D., Yu, C., Rothblatt, J. *et al.* (2000b) Contribution of Are1p and Are2p to steryl ester synthesis in the yeast *Saccharomyces cerevisiae. European Journal of Biochemistry* **267**, 1075–1082.

Chapter 7

Protein trafficking

Jeremy D. Brown

7.1 Introduction

Eukaryotic cells rely for their viability on the partitioning of many basic cellular processes into membrane-bounded organelles. These are the nucleus, endoplasmic reticulum (ER), Golgi apparatus, endosomes, vacuolar compartment, mitochondria and peroxisomes (Figure 7.1). Most molecules destined for the vacuole, cell surface and outside the cell are routed through the ER and Golgi, which, together with the vesicular intermediates between them, comprise the secretory pathway (Palade, 1975). The targeting and trafficking problems posed by this complex organization have been the subject of intense study in yeast and other systems. Proteins must be recognized as destined for partitioning outside the cytosol, targeted to and then translocated across the correct membrane. In the early secretory pathway compartments – the ER and Golgi – proteins are sorted, modified and often assembled into complexes *en route* to their final destination. Incorrectly assembled proteins are retained in the ER until they fold correctly or are targeted for degradation.

In yeast nearly 2000 of the approximately 6000 open reading frames in the genome encode membrane proteins. Continued sequencing of genomes of diverse organisms has revealed that this is typical – one-third of all protein species synthesized by cells are membrane proteins. Additional proteins are translocated into and function within the lumenal spaces of organelles or are secreted. Thus a significant proportion of the proteins synthesized by each cell require targeting to membranes either for insertion into or transport across them. A major purpose of this effort is growth, increasing the surface area and hence the volume of the cell. The whole secretory pathway is organized towards the growing bud tip and this organization is lost when the actin cytoskeleton is perturbed indicating a close link between the cytoskeleton, polarized growth and secretion (Novick and Botstein, 1985; Ayscough *et al.*, 1997). The secretory pathway is also closely linked to general metabolism, such that many secretory pathway mutants perturb ribosome and tRNA biogenesis, thereby reducing the cell's protein synthetic capacity (Mizuta and Warner, 1994; Li and Warner, 1996; Li *et al.*, 2000). The pathway through which this effect is mediated has been partly characterized. Transcriptional repression requires components of a pathway that monitors cell integrity – protein kinase C and cell membrane sensor proteins. Downstream components are, however, only partly understood though repression of ribosomal protein gene transcription is known to require the dual-function Rap1p transcriptional activator/silencer protein (Mizuta *et al.*, 1998; Li *et al.*, 2000).

Screens, both simple and elaborate, as well as biochemical and sequence gazing approaches have contributed to the identification and analysis of factors required for targeting, translocation and sorting. More recently transcription profiling and proteomic

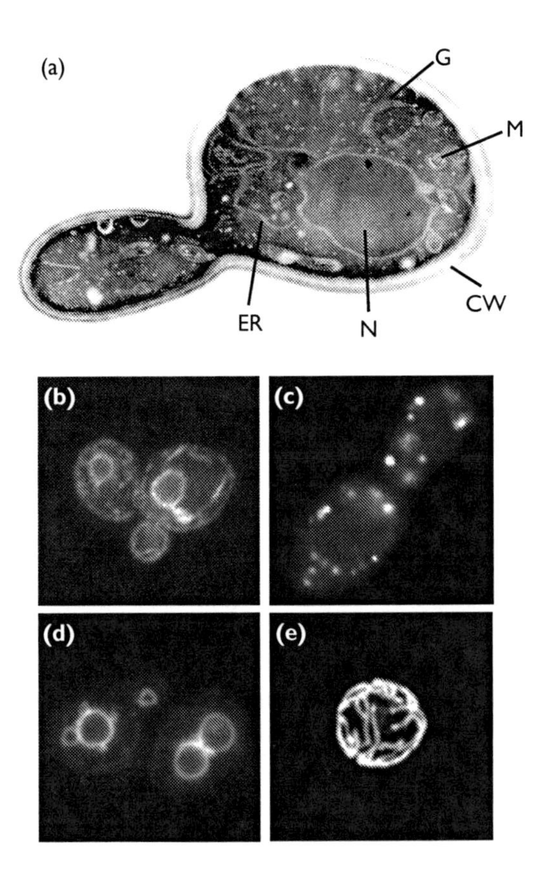

Figure 7.1 Organelles of *S. cerevisiae*. (a) Electron micrograph of a budding yeast cell stained with tungstosilicic acid hydrate and polyvinylalcohol: G – Golgi; M – mitochondrion; N – nucleus; ER – endoplasmic reticulum; CW – cell wall. (b–e) Live yeast cells labelled to show different organelles. (b) endoplasmic reticulum, (c) late Golgi, (d) vacuolar membrane and (e) mitochondria. Except for vacuolar membranes, which were visualized by labelling with the lipophilic dye FM4-64 (Molecular Probes), organelles were visualized by fusion of green fluorescent protein to sequences that targeted it to the organelle. Thanks to Frank Verner for the image in (a), James Whyte and Sean Munro for the Sec7p–GFP fusion used to obtain (c) and Jodi Nunnari for the image in (e) which is reproduced from *Molecular Biology of the Cell* (1997, **8**, 1233–1242), with permission of the American Society for Cell Biology.

approaches have added a more global perspective of the effects of insult and injury to trafficking pathways and the interactions between components of these pathways. In this review, the emphasis is on recent findings and I have attempted to show how both specific and more often general mechanisms are used to carry out the processes of targeting, translocation and sorting.

7.2 Protein targeting and translocation

The first step in sorting is recognition of *cis*-acting targeting or signal sequences that organelle-targeted proteins contain. This is carried out by cytosolic targeting factors and/or receptors on the membrane to which the protein is targeted. In some cases the primary sequences are extremely degenerate, with only the overall character being conserved (hydrophobicity for an ER signal sequence; helical amphiphilicity for mitochondrial targeting sequence (Kaiser *et al.*, 1987; Lemire *et al.*, 1989)). Therefore, how signal sequences are efficiently recognized, despite this variability, is a particularly intriguing question not yet fully resolved. Structural studies are providing insight into signal sequence binding pockets/surfaces on targeting factors and this information is invaluable in understanding how they recognize their substrates. Following targeting, proteins are either inserted into or transported across the membrane (translocated) through a proteinaceous translocation apparatus. For both the ER and mitochondria, components of the translocation apparatus are, in the main, defined and research is focussed on how these interact with each other, the targeting apparatus and the substrate, which in both cases imports in extended conformation. It is clear from these studies that the ER and mitochondrial translocons are modular and play an active role in sorting substrates to different destinations. These translocons include or recruit motors to drive the translocation process in the correct direction (discussed by Schatz and Dobberstein, 1996). Peroxisomes are able to import folded and oligomerized proteins and though proteins required for translocation have been found, no specific 'translocon' complex has been identified. It is also clear that membrane and lumenal proteins are imported into peroxisomes by different pathways. In addition the motor (if there is one) is still unknown, although the process is ATP- and temperature-dependent. Several alternative mechanisms have been proposed for how proteins are transported into this organelle. Though far from demonstrated, perhaps the most attractive model is that the translocation apparatus assembles at the site of import in response to docked substrate(s) and disassembles immediately following translocation.

The endoplasmic reticulum

Targeting

Two targeting routes to the ER operate side-by-side in yeast (Feldheim and Schekman, 1994; Ng *et al.*, 1996) (Figure 7.2). These are the co-translational signal recognition particle (SRP)-dependent and post-translational or SRP-independent routes. The post-translational route was the first found. Schekman and colleagues noted that genetic depletion of the cytosolic Hsp70 Ssa1/2p resulted in defects in the import of carboxypeptidase Y (CPY) and α-Factor (α-F) precursors into the ER (Deshaies *et al.*, 1988), while several laboratories reported that yeast ER membranes could import proteins post-translationally (Hansen *et al.*, 1986; Rothblatt and Meyer, 1986). The DnaJ homologue, Ydj1p is also required for efficient translocation of certain proteins into the ER (Caplan *et al.*, 1992; Becker *et al.*, 1996). Like other DnaJ proteins, Ydj1p is a co-chaperone that modifies both the peptide binding and ATPase activities of its partner Hsp70 proteins, in this case Ssa1p (Cyr and Douglas, 1994). The requirement for Ssa1p and Ydj1p in targeting indicates that the substrate proteins are exposed in the cytosol and the most likely requirement is in maintaining these polypeptides in an unfolded conformation compatible with translocation across the ER membrane.

SRP is a cytosolic ribonucleoprotein, first isolated from dog pancreas extracts, which captures ribosomes synthesizing presecretory proteins and targets them to the translocation apparatus in the ER membrane, through interaction with the two-subunit SRP receptor (SR) in a GTP-dependent fashion (see Keenan *et al.*, 2001 for a detailed description of the activities of

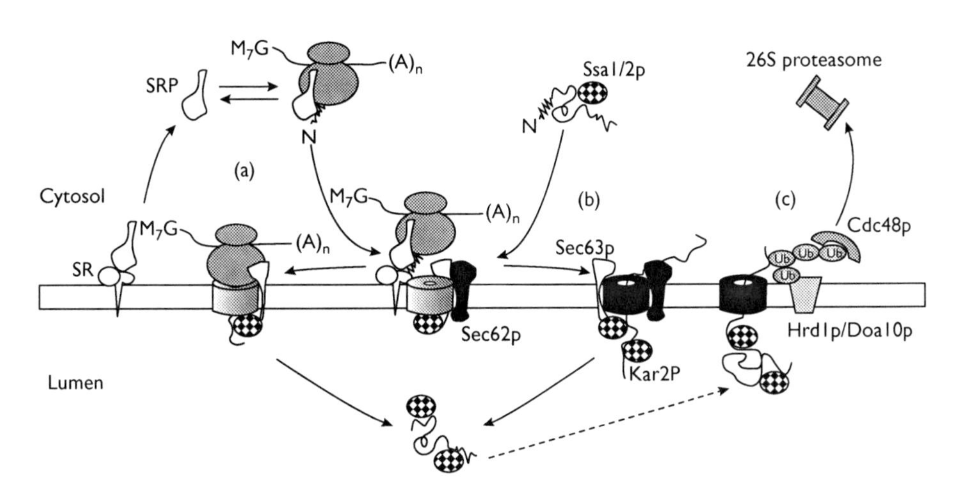

Figure 7.2 The translocon is the conduit for all protein transport across the ER membrane. Three pathways are shown, two biosynthetic and one degradative, by which proteins cross the ER membrane via the translocon complex. (a) SRP dependent targeting and co-translational transport across or insertion into the ER membrane. SRP examines nascent chains as they emerge from ribosomes for the presence of hydrophobic signal sequences. If a signal sequence is present SRP targets the ribosome to the translocon via interaction with the SRP receptor (SR). Translocons containing either Sec61p or Ssh1p act as acceptors for proteins targeted by this route. (b) SRP-independent targeting. In this post-translational pathway the signal sequence of the protein is recognized by the Sec62p complex and translocation is through the Sec61p transolcon only. Depending on the substrate, translocation may require cytosolic chaperones such as Ssa1/2p. Translocation in both pathways requires Sec63p and the ER-lumenal Kar2p protein. (c) Dislocation of proteins from the ER for destruction by the ERAD pathway (including membrane proteins, not shown here). This also requires Sec61p and is depicted as being triggered by prolonged interaction of a luminal substrate with Kar2p. Ubiquitin ligase complexes 'tag' substrate proteins, which are then extracted from the ER by the action of the Cdc48p/Ufd1p/Npl4p complex before degradation by the proteasome. In this and Figure 7.3 chaperones are indicated with cross-hatching.

SRP). Yeast SRP consists of the scR1 RNA and six proteins, five of which are homologues of mammalian SRP subunits (Hann and Walter, 1991; Hann *et al.*, 1992; Brown *et al.*, 1994). Signal sequence binding is a function of the Srp54p component of the particle, homologues of which are found in all cellular organisms and chloroplasts. In yeast lack of this protein, while not disruptive to the assembly of the remainder of the particle, leads, not surprisingly to complete loss of SRP function (Hann and Walter, 1991; Brown *et al.*, 1994; Ciufo and Brown, 2000). Srp54p contains a methionine-rich domain which has been cross-linked to signal sequences, and crystal structures of both mammalian and bacterial homologues of Srp54p reveal that this domain of the protein contains a hydrophobic groove lined with methionine residues – the signal sequence binding pocket (discussed further in Wild *et al.*, 2002). Binding of SRP to the ribosome–nascent chain complex results in a slowing of translation – a phenomenon termed elongation arrest. This function is conserved to the yeast particle and is crucial for full SRP activity (Mason *et al.*, 2000). As in mammalian SRP, elongation arrest requires the Srp14p subunit.

An intriguing study suggests that yeast SRP may interact with ribosomes at or just prior to the translocation step in the translation elongation cycle where the ribosome moves along the mRNA (Ogg and Walter, 1995). This inference was made through the observation that defects in ER targeting seen on lowering levels of SRP (by genetic depletion or through the *sec65-1* temperature-sensitive SRP mutant) can be suppressed by cycloheximide, a competitive inhibitor of the translocation step in polypeptide elongation, but not by antibiotics that inhibit other steps in this cycle. The effect of cycloheximide in this model is to increase the window of opportunity that SRP has to interact with the ribosome, thus allowing a reduced amount of SRP to carry out its function. Intriguingly other antibiotics that block the translocation step in elongation (sordarins) also suppress the temperature sensitivity of *sec65-1* cells (JDB, unpublished observation). The target of sordarins is well characterized: they bind specifically to eEF-2, the GTPase that catalyses translocation, when it is bound to the ribosome (Justice *et al.*, 1998; Gomez-Lorenzo *et al.*, 2000). Thus SRP may bind the ribosome–EF-2 complex to effect elongation arrest. The translocation defects associated with conditional mutations in either subunit of the SRP receptor and deletion of the SRP pathway-specific translocon component Ssh1p (see later) are also suppressed by sub-lethal doses of cycloheximide (Prinz *et al.*, 2000; Wilkinson *et al.*, 2001). Both these scenarios are expected to reduce the capacity of the SRP pathway and these results agree with the model put forward by (Ogg and Walter, 1995).

It is the signal sequence of a protein that determines whether a protein will translocate co- or post-translationally. Yeast SRP recognizes a subset of signal sequences that are more hydrophobic (as does the *E. coli* SRP homologue (Ulbrandt *et al.*, 1997)) whereas the post-translational pathway shows a preference for those that are less hydrophobic (Feldheim and Schekman, 1994; Ng *et al.*, 1996). Signal sequence recognition in post-translational targeting requires a complex of the membrane proteins Sec62p, Sec71p and Sec72p (the Sec62p-complex) (Musch *et al.*, 1992; Feldheim and Schekman, 1994; Lyman and Schekman, 1997). This complex, along with the translocon component Sec61p, interacts with the signal sequence (see later). Only some proteins that access the ER through this pathway require Ssa1/2p and Ydj1p for translocation (Becker *et al.*, 1996). These proteins may fold particularly quickly or be prone to aggregation. Alternatively, Ssa1/2p and Ydj1p could represent a targeting mechanism for these proteins. Ydj1p is farnesylated, a modification that targets it to membranes, and yeast strains with only non-farnesylated Ydj1p are defective in ER targeting at high temperatures (Caplan *et al.*, 1992). Membrane localized Ydj1p may stimulate release of substrate from Ssa1p to the translocation machinery (Cyr and Douglas, 1994; Ziegelhoffer *et al.*, 1995).

Neither removal of SRP activity nor substantial (95%) decrease in activity of the post-translational targeting pathway is lethal, although loss of SRP results in very slow growth, suggesting that co-translational targeting is extremely important for some proteins (Hann and Walter, 1991; Brown *et al.*, 1994; Ng *et al.*, 1996). It is unclear whether there is any communication between the two targeting pathways – although some proteins (e.g. pre-Kar2p) can use either. Loss of SRP does, however, lead to increased expression of cytosolic and ER stress proteins including Ssa1/2p and Ydj1p (Arnold and Wittrup, 1994). This forms at least part of an adaptation mechanism that reduces the amount of untranslocated precursors at long times after SRP depletion, and which also includes a drastic reduction in ribosome synthesis and hence cellular protein synthetic capacity (Mutka and Walter, 2001). As well as its role as a chaperone Ydj1p has been implicated in aiding degradation of both abnormal and short-lived proteins so the decrease in precursors could be due to increased proteolysis

(Lee *et al.*, 1996; Yaglom *et al.*, 1996). Whether or not proteins normally targeted by SRP are translocated by the post-translational pathway when SRP is missing is unclear – though this is the most obvious possibility and one perhaps supported by the observation that combinations of SRP gene deletions with mutations in the post-translational pathway are synthetic negative for growth and lethal for spore germination (Ng *et al.*, 1996). Arguing against this is the observation that SRP-dependent signal sequences are not recognized by the post-translational pathway (Lyman and Schekman, 1997; Dunnwald *et al.*, 1999). An alternative is that translocation may still be co-translational but not catalysed by SRP, taking place much less efficiently and relying on the inherent ability of the translocation apparatus to both bind ribosomes and discriminate signal sequences. Whatever the mechanism employed in the absence of SRP, cells trade off growth rate for accurate partitioning of proteins into the ER.

Translocation

Translocation into the ER requires the Sec61p-complex consisting of the polytopic membrane protein Sec61p and two smaller membrane proteins Sss1p and Sbh1p. Sec61p is conserved (homologues include the bacterial SecY and mammalian SEC61α proteins) and contains 10 transmembrane domains (Wilkinson *et al.*, 1996). Sss1p and Sbh1p are small proteins with single transmembrane domains. Like Sec61p, Sss1p is conserved (SecE in bacteria). However, Sbh1p homologues are not found in eubacteria where it is 'replaced' by SecG in the SecYEG translocon. Three distinct pools of the Sec61p-complex are present in ER membranes; free heterotrimers and larger complexes which can be purified separately and are responsible for co- and post-translational translocation (Panzner *et al.*, 1995). The co-translational complex contains the Sec61p-complex attached to ribosomes, whereas the post-translational complex consists of the Sec61p-complex along with the Sec62p-complex and Sec63p. Sec63p is a membrane protein which, in its ER-lumenal portion, contains homology to DnaJ proteins (Deshaies *et al.*, 1991; Feldheim *et al.*, 1992). This 'DnaJ-loop' provides a binding site for the ER-lumenal Hsp70-like protein Kar2p (Corsi and Schekman, 1997), which in post-translational translocation provides a driving force for translocation. A second ER-localized translocon complex has been identified consisting of Ssh1p and Sbh2p (homologues of Sec61p and Sbh1p, respectively) and Sss1p (Finke *et al.*, 1996). This complex is dedicated to co-translational translocation and is only found as part of co-translational complexes with ribosomes (Finke *et al.*, 1996). Deletion of *ssh1* is synthetic lethal with the *sec65-1* SRP mutation but not a *sec62* mutation (Wilkinson *et al.*, 2001). In addition, in certain genetic backgrounds *ssh1Δ* cells have similar translocation defects and slow growth phenotypes as SRP-deficient cells (Wilkinson *et al.*, 2001). Extending these observations Ssh1p was recently shown to interact only with translocation substrates that are targeted by SRP (Wittke *et al.*, 2002).

Electron microscopic observations have shown that the purified Sec61p-complex forms oligomeric rings in membranes – but only in the presence of either ribosomes or the Sec62p/Sec63p complex (Hanein *et al.*, 1996). In a recent study, ribosomes exposing short nascent chains containing a signal sequence were assembled with purified Sec61p to form 'active' ribosome–translocon complexes (Beckmann *et al.*, 2001). Reconstructions at ~15 Å resolution were possible, these revealing the translocon as a compact disk of sufficient size to comprise three Sec61p heterotrimers positioned under the nascent chain exit site of the ribosome. Surprisingly, a gap of at least 15 Å was seen between ribosome and translocon. This observation and similar data from examination of mammalian ribosome–translocon complexes (e.g. Menetret *et al.*, 2000) does not sit well with data from other studies that suggest that there is a tight seal between ribosome and translocon during co-translational translocation (Crowley *et al.*, 1994) and it may be that these complexes lack some regulatory or other factors present in native complexes.

It is unclear from present data whether Sec61p oligomers seen attached to either ribosomes or Sec62p complexes are normally stable in the ER membrane or whether they assemble and dissociate as required. Perhaps significantly the Sec61p-complex released from ribosomes will form rings in the presence of Sec62p/63p, indicating that there is no intrinsic difference between the two Sec61p populations (Hanein *et al.*, 1996). An attractive hypothesis is that Sec61p-complexes are recruited either to a ribosome bound to SRP and its receptor or the Sec62p/Sec63p complex when it has a signal sequence docked to it. This would provide an exquisitely sensitive mechanism to ensure that pores (holes) are only present in the membrane when there is a molecule to transport. A piece of the puzzle that has been missing is identification of interaction between the Sec61p-complex and either SRP or the SRP receptor. Such an interaction might be expected since these components provide the initial binding site for the ribosome on the membrane before it is 'passed' to the translocon. Recently, it was demonstrated that both Ssh1p and Sec61p are in close proximity to the SRP receptor in the ER membrane whereas Sec62p is not (Wittke *et al.*, 2002). This is substantiated by the finding that Sec61p interacts directly with the β-subunit of the SRP receptor in two-hybrid experiments (JDB and A. Robb unpublished observation). This interaction is likely to be transient and an attractive model, supported in part by experiments on the mammalian SRP receptor (Fulga *et al.*, 2001), is that the β-subunit of the SRP receptor transduces information on the readiness of the ribosome–SRP–SRP receptor–translocon complex for transfer of the nascent chain from SRP to the translocon, acting as a crucial molecular switch in the targeting process.

Signal sequence recognition in the post-translational pathway involves both components of the Sec62p complex and Sec61p itself (Musch *et al.*, 1992; Lyman and Schekman, 1997; Plath *et al.*, 1998). Detailed analysis of these interactions using a series of substrate proteins containing only one cross-linkable residue in the signal sequence of each revealed a pattern of cross-links to both Sec61p and Sec62p-complex proteins consistent with the signal peptide forming an alpha helix oriented in a specific way at the Sec61p–Sec62p interface (Plath *et al.*, 1998). The interactions of the Sec62p complex with substrate are only seen in the absence of ATP. Once ATP is added interactions are only to Sec61p as the nascent chain moves through the translocon into the ER lumen (Musch *et al.*, 1992; Lyman and Schekman, 1997). This relies on the acivities of Sec63p and Kar2p, and Kar2p is the ATP-binding component. In the absence of Kar2p, ATP does not promote release of substrate from the Sec62p-complex, and when the Kar2p–Sec63p interaction is impaired, for example, in the *sec63-1* mutant which is in the Sec63p DnaJ-loop, substrate is released from Sec62p much less efficiently (Lyman and Schekman, 1997). In addition the non-hydrolysable ATP analogue AMPPNP does not promote release of substrate to Sec61p (Matlack *et al.*, 1997) and these data suggest a model in which Kar2p binding to Sec63p and hydrolysing ATP provides a signal indicating that the lumenal proteins necessary for translocation are in place ready to receive the polypeptide chain.

In addition to its role in initiating translocation, Kar2p is also crucial for further events during post-translational translocation. Membranes isolated from *sec63-1* or specific *kar2* mutants accumulate stalled precursors that do not complete transit into the ER lumen (Lyman and Schekman, 1995). This again appears to be due to defective interaction between Kar2p and

Sec63p, in this case resulting in failure of Kar2p to be properly positioned to bind the incoming polypeptide as it emerges from the translocon. A major question is whether Kar2p is a motor that pulls the substrate through the translocon, or whether it simply binds segments of polypeptide as they move randomly in and out of the pore (the Brownian ratchet model) conferring unidirectionality to an otherwise random process. Examination of translocation in a solubilized system revealed that the Brownian ratchet may be sufficient (Matlack *et al.*, 1999). Thus when substrate was bound to Sec61p, Sec62p and Sec63p and then Kar2p, the added substrate protein was translocated through the Sec complexes with multiple Kar2p molecules associated with it. Significantly, antibodies against the substrate protein could also affect translocation in this way though less efficiently.

Reconstitution of co-translational translocation with purified mammalian Sec61p complex and SRP receptor led to general acceptance that these were the only ER-derived components required (Görlich and Rapoport, 1993). However, both genetic and biochemical data indicate that, at least in yeast Sec63p and Kar2p are also necessary for co-translational translocation (Brodsky *et al.*, 1995; Young *et al.*, 2001). Most strikingly depletion of Sec62p *in vivo* through use of a *GAL* promoter-driven *SEC62* leads to only post-translational translocation defects, whereas depletion of Sec63p results in a block to all translocation. Sec63p and Kar2p are unlikely to be needed in the same way as during post-translational translocation as the ribosome may provide the force to move the nascent chain through the translocon. Perhaps Sec63p and particularly Kar2p are required to prevent the translocon from becoming blocked by aggregated nascent chains. Alternatively, and consistent with experiments carried out using mammalian components, Kar2p may regulate translocon opening and closing (Hamman *et al.*, 1998; Haigh and Johnson, 2002).

A second Hsp70-like protein is involved in ER protein import. This component has been variously termed Lhs1p, Ssi1p or Cer1p and is most homologous to mammalian GRP170. Cells lacking Lhs1p reveal partial translocation defects for a number of proteins, particularly those that use the post-translation pathway to the ER (Hamilton and Flynn, 1996). In addition *lhs1Δ* cells are defective in repair of mis-folded proteins in the ER (Saris *et al.*, 1997). Genetic analysis has led to the isolation of the ER-localized Kar2p-interacting Sil1p as a multi-copy suppressor of the translocation defects of *lhs1Δ* cells (Tyson and Stirling, 2000). While *SIL1*, like *LHS1*, is not an essential gene, lack of both is a lethal event that correlates with a block to all ER import. Thus Sil1p and Lhs1p provide an important part of the ER import apparatus. Given that Sil1p and Lhs1p are conserved proteins this function is likely to be important in other organisms. In this context it is worth noting that the *Yarrowia lipolytica* Sil1p homologue, Sls1p, was identified as a Kar2p-interacting and translocon-associated protein in this yeast, mutation of which was synthetic lethal with an SRP mutation (Boisrame *et al.*, 1996, 1998).

Events that occur at the completion of translocation have not been examined. Questions that require answering include how the channel is prevented from staying open, how the ribosome disengages, and how the translocon is assembled and disassembled if required.

Mitochondria

The majority, though not all, of mitochondrial proteins are encoded by nuclear genes. Additionally, mitochondria are bounded by two – the outer and inner – membranes. This arrangement necessitates both significant protein import into these organelles and sorting to the four destinations possible, as well as insertion of proteins from the mitochondrial matrix

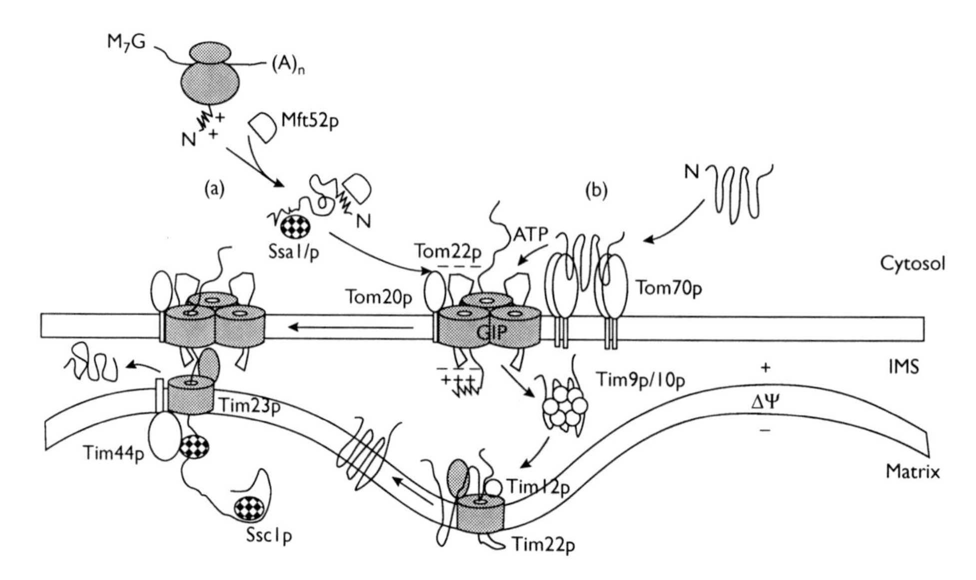

Figure 7.3 Targeting to and translocation across the mitochondrial membranes follows several routes. mRNAs encoding mitochondrial proteins may be targeted to mitochondria by sequences in their 3′ untranslated regions. The mechanism for this is currently unknown. (a) Targeting of proteins that contain cleavable presequences. Cytosolic Mft52p may bind to presequences as, or soon after, they emerge from the ribosome. It is unknown how Mft52p interacts with other targeting factors and chaperones but, together with ribosome associated factors such as NAC and RAC promotes targeting of proteins and ribosomes synthesizing them to mitochondria. As with ER-targeting cytosolic chaperones Ssa1p/2p and Ydj1p are important for the translocation of some substrates. Presequences are recognized sequentially by Tom20p, Tom22p and then the Tom40p containing general insertion pore (GIP) through which the protein translocates in extended conformation to bind the intermembrane space (IMS) domain of Tom22p. The inner and outer membrane translocation machineries become attached and the translocating protein is passed across the inner membrane Tim23p complex in a process that requires Tim50p, Tim44p and its interaction with Ssc1p, ATP and $\Delta\Psi$, or released into the IMS. Note that the portion of Tim23p that spans the IMS and inserts into the outer membrane is not shown. (b) Proteins with internal targeting sequences such as the inner membrane carrier proteins are bound to multiple Tom70p receptors. The proteins are released from Tom70p to Tom22p and the GIP in a reaction that requires ATP and are inserted through the outer membrane in loops. In the IMS these proteins are bound first by the Tim9p/10p protein, which passes them to the inner membrane associated Tim12p before integration into the membrane by the Tim22p complex.

into the inner membrane (Figure 7.3). At least four translocation and/or insertion machineries are required for all these events. As with ER translocation recent insight has stemmed in part from purification and study of translocon components and complexes, and this has yielded important biochemical and structural information.

Targeting

Research in the field of mitochondrial protein import has focused on membrane-associated factors and events, largely ignoring targeting. This is despite the early observation that mRNAs encoding certain mitochondrial proteins are enriched in mitochondrial fractions and that several proteins import more efficiently co-translationally than post-translationally (Ades and Butow, 1980; Suissa and Schatz, 1982; Verner, 1993). This area has recently been revisited and extended to whole organelle level through the use of microarray technology (Marc *et al.*, 2002). In this study about half the nuclear transcribed mRNAs known to encode mitochondrial proteins were significantly enriched in the pool of mRNAs extracted from isolated mitochondria, suggesting coupling of translation and translocation for the proteins encoded by them. However, a crucial difference in targeting to mitochondria over targeting to the ER membrane is that no factors are known that explicitly couple translation and translocation (as seen with SRP-dependent targeting). Indeed at least in some and perhaps in many cases the mRNA itself is targeted to the vicinity of mitochondria. Thus 3′ untranslated regions of several mRNAs encoding mitochondrial proteins are sufficient to target heterologous mRNAs to mitochondria (Corral-Debrinski *et al.*, 2000; Marc *et al.*, 2002). The basis of this is unknown, but, as discussed by Lithgow mutations in several genes encoding proteins known to interact with the cytoskeleton, RNA, or both, affect mitochondrial protein targeting (Lithgow *et al.*, 1997; Lithgow, 2000). One example is Npl3p, an RNA binding protein that shuttles between nucleus and cytoplasm. A mutation in the RNA-binding domain of Npl3p results in its prolonged residence in the cytoplasm, increasing both stability of a variety of mRNAs and mitochondrial import of a protein that imports poorly post-translationally (Gratzer *et al.*, 2000). Similarly, mutations in nuclear transport factors that increase the cytoplasmic pool of Npl3p suppress mitochondrial protein import defects (Corral-Debrinski *et al.*, 1999). How this is brought about is unclear, but stabilizing transcripts (and hence the time available to target them to mitochondria) may be sufficient to increase the co-translational import of the proteins that they encode.

Other proteins and protein complexes have been implicated more directly in protein targeting to mitochondria. These are Mft52p, and the ribosome-associated nascent polypeptide associated complex (NAC) and ribosome associated complex (RAC). Mft52p is a cytosolic mitochondrial targeting sequence binding factor. It was identified by the *mft1* mutant (mitochondrial fusion targeting), isolated in a screen for strains that could not efficiently target a protein consisting of F_1-β ATPase targeting information fused to β-galactosidase to the mitochondrion (Garrett *et al.*, 1991). Purified Mft52p binds mitochondrial targeting sequences (Cartwright *et al.*, 1997). NAC is an abundant heterodimer originally postulated to be crucial for ER targeting (Wiedmann *et al.*, 1994). However, yeast mutants lacking this complex have negligible ER import defects, but do show defects in mitochondrial targeting, which are particularly pronounced for proteins such as fumarase that require co-translational targeting (George *et al.*, 1998; Knox *et al.*, 1998). Combining deletions in genes encoding NAC and Mft52p results in severe mitochondrial targeting defects (George *et al.*, 1998) and these phenotypes led Lithgow and colleagues to propose that NAC and Mft52p play a combined role in committing precursors to mitochondrial targeting. Consistent with this *in vitro* NAC stimulates the attachment of ribosomes synthesizing mitochondrial precursors to mitochondria (Funfschilling and Rospert, 1999) and *in vivo* NAC mutants have decreased mitochondrially bound mRNA (George *et al.*, 2002). RAC is a stable complex of the proteins Ssz1p and zuotin, which are DnaK and DnaJ homologues respectively. RAC has a similar ability as that of NAC, to stimulate translocation of a ribosome-bound mitochondrial precursor (Gautschi *et al.*, 2001). However, at present there is no further evidence linking RAC to mitochondrial import and deletion mutants have no detected mitochondrial import defects.

Both mRNA localization and co-translational recognition of signal sequences has the advantage that, even if targeting and translocation are not strictly coupled, ribosomes will be concentrated around the organelle to which the proteins they are synthesizing are targeted, improving the rate of transport into the organelle. It is also likely that short nascent chains emanating from these translating ribosomes will encounter the translocation apparatus leading to docking of the ribosomes and the establishment of co-translational translocation (Lithgow, 2000). Another cytosolic factor that recognizes mitochondrially targeted proteins – presumably post-translationally – is MSF (mitochondrial import-stimulating factor), a heterodimeric protein identified in mammalian cell extracts that targets precursors to receptors on the mitochondrial membrane *in vitro* (Hachiya *et al.*, 1993). Although no homologues of MSF subunits have been identified in the yeast genome, an activity similar to MSF has been identified biochemically (Komiya *et al.*, 1997). As with ER targeting lack of Ssa1/2p or Ydj1p results in accumulation of some mitochondrial precursors, e.g. F_1-β ATPase, *in vivo* (Caplan *et al.*, 1992; Becker *et al.*, 1996). ATP is required for F_1-β ATPase import and, intriguingly, removal of a segment of F_1-β ATPase required for its tetramerization eliminates the requirement for ATP in F_1-β ATPase import (Chen and Douglas, 1988), consistent with the idea that chaperones act to maintain this substrate in an unfolded, monomeric and transport-competent state. It is unclear how general this model is, but it confirms the important point that, as with ER import, it is crucial that proteins do not fold extensively otherwise the ability of the import machinery to translocate them is decreased.

The outer membrane of mitochondria contains several receptors more or less closely associated with the translocation machinery. These are Tom20p, Tom22p, Tom37p and Tom70p (translocase of the outer mitochondrial membrane). These receptors between them bind the various substrates that import into mitochondria. These fall broadly into two classes. The first is proteins that contain a 'typical' cleavable mitochondrial targeting presequence. These have the capacity to form amphipathic helices positively charged on one face and hydrophobic on the other. These are bound by Tom20p and Tom22p through interaction with both the hydrophobic and basic faces of the presequence. Both Tom20p and Tom22p contain clusters of negatively charged residues (acid-bristles) that were proposed to bind to the positive charges in presequences (Bolliger *et al.*, 1995; Haucke *et al.*, 1995). However, analysis of a rat Tom20p-presequence complex by NMR revealed that the presequence was bound into a hydrophobic groove on the protein (Abe *et al.*, 2000). As Tom20p is only peripherally associated with the translocation machinery whereas Tom22p is an integral part of it (Dekker *et al.*, 1998) the current model is of a two-step recognition of presequences, first by Tom20p which then passes them to Tom22p.

The second class of proteins imported into mitochondria lack cleavable presequences. These proteins typically have internal targeting information and a systematic search of the binding specificity of receptor proteins revealed that both Tom22p and Tom70p bind peptides derived from one such protein, the phosphate carrier (Brix *et al.*, 1999). The basis of this binding is unclear as both hydrophobic and hydrophilic peptides were bound – though the results appear relevant as peptides from the Tom22p-targeted cytochrome oxidase subunit IV did not bind Tom70p. Tom70p forms dimers in the outer membrane (Millar and Shore, 1994) and several Tom70p bind each precursor protein (Wiedemann *et al.*, 2001). This concerted binding could have a chaperone-like effect, preventing aggregation or folding of the substrate. Release of precursors from Tom70p requires ATP (Komiya *et al.*, 1997) and once released they are passed to Tom22p and thence to the translocation machinery.

Translocation

The translocation machinery of the mitochondrion is doubly complex in that there are two membranes through which many substrates pass, and four possible final locations for them (outer and inner membranes, and intermembrane and matrix spaces). As described earlier, substrates bind to the receptor protein and all are likely to be transferred to Tom22p for import through the outer membrane. Tom22p is also an important component of the translocon. In addition to the acid-bristle in its cytosolic domain Tom22p also contains an acid-bristle protruding into the intermembrane space. This intermembrane space domain of the protein is important for protein transport, and binds to presequences, possibly anchoring them in the intermembrane space at early times in the translocation process, before recruitment of the inner membrane translocon (Bolliger *et al.*, 1995; Moczko *et al.*, 1997) (see later).

The Tom machinery contains several components in addition to the receptor subunits (reviewed in Pfanner and Meijer, 1997). These components, which form a general insertion pore (GIP), are a large subunit Tom40p, and several small proteins Tom5p, Tom6p and Tom7p. Tom6p and Tom7p modulate the interactions between the receptor subcomplexes and Tom40p. In the absence of Tom6p receptors are less well associated with the pore (Alconada *et al.*, 1995). Deletion of *TOM7* reveals an opposite phenotype – an increased association of Tom20p and Tom22p with Tom40p (Honlinger *et al.*, 1996). Cells lacking Tom7p are defective in the insertion of the outer membrane protein porin at a step after binding to the receptors. Thus dynamic interactions between receptor and pore components are presumably required for effective lateral release of outer membrane proteins into the membrane. Tom5p is closely associated with Tom40p (Dekker *et al.*, 1996), and like the Tom20p/22p receptor subunits has a negatively charged cytoplasmic domain. Although loss of Tom5p does not appear to affect interactions between other Tom components, *tom5Δ* cells are defective in the transport of a wide variety of proteins across the outer membrane (Dietmeier *et al.*, 1997). These results led to the proposal that Tom5p was part of the chain of negatively charged surfaces that bind mitochondrial targeting sequences and direct them to and through the Tom machinery.

Purification and study of Tom protein complexes by electron microscopy revealed that the whole complex (GIP plus Tom20p and Tom22p) consists of three pore-like structures of sufficient diameter to allow an extended polypeptide to traverse the complex (Kunkele *et al.*, 1998; Model *et al.*, 2002). Removal of Tom20p yielded structures comprising two pores and after removal of Tom22p the smaller complexes that remained consisted mainly of single pores (Model *et al.*, 2002). Thus the receptors may act to anchor individual GIP complexes together. An additional role for Tom22p is revealed by the fact that the pores that remain when Tom22p is absent are no longer correctly gated (van Wilpe *et al.*, 1999). This function may be related to the role of Tom22p in recognizing and anchoring targeting sequences on the intermembrane space side of the GIP.

Passing a protein across or integrating it into the inner membrane of the mitochondrion is accomplished by two distinct sets of translocation components that comprise the Tim (translocase of the inner mitochondrial membrane) machinery. One of these consists of Tim17p, Tim23p, Tim50p and Tim44p (Tim23p complex) all of which are essential proteins. The Tim23p complex is responsible for the recognition of proteins that contain cleavable presequences – proteins targeted to the matrix and the intermembrane space. A further protein, Tim11p, was proposed to be a component of this complex (Tokatlidis *et al.*, 1996). However, this has subsequently been identified as the e subunit of the F1F0 ATPase

(Arnold *et al.*, 1997) and does not purify with the Tim23p complex (Geissler *et al.*, 2002). Full import of proteins into the matrix of the mitochondrion requires both ATP and a potential across the inner membrane ($\Delta\psi$), which drives basic matrix-targeting presequences through the membrane (Ungermann *et al.*, 1996).

Immunoprecipitation studies showed that Tim23p becomes attached to the outer membrane TOM complex when a translocating substrate is present (Horst *et al.*, 1995). A signal must therefore be generated by the precursor that recruits the Tim23p complex. This signal is not known, but could be a signal sequence bound to Tom22p in the intermembrane space. Intriguingly the N-terminus of Tim23p is exposed on the cytosolic face of the outer membrane, indicating that Tim23p spans both membranes (Donzeau *et al.*, 2000). This region of Tim23p is only essential at high temperatures indicating that it may be important for stabilizing the interaction between TOM and TIM complexes. Tim50p contains a large intermembrane space domain that contacts Tim23p (Geissler *et al.*, 2002; Yamamoto *et al.*, 2002) and thus Tim50p has been proposed to guide proteins to the Tim23p channel (see later). In addition Tim50p can be cross-linked to translocating substrates and depletion of Tim50p leads to import defects for matrix targeted proteins but not intermembrane space proteins indicating a specific role for this protein in matrix protein import (Geissler *et al.*, 2002).

The Tim23p complex is extremely dynamic. Tim23p reversibly dimerizes, an interaction stimulated by $\Delta\psi$, and broken by import substrate (Bauer *et al.*, 1996). Dimerization of Tim23p is through Leucine-zipper motifs in the intermembrane space domain of Tim23p and dimers can be formed *in vitro* using fusion proteins containing this domain. This dimerization domain can also be cross-linked to a peptide corresponding to a matrix targeting sequence. These data suggested that Tim23p may form part, or all, of the pore in the inner membrane which is opened by matrix-targeting sequences. In agreement with this purified Tim23p reconstituted into liposomes forms presequence sensitive channels (Truscott *et al.*, 2001).

Tim23p also provides a link, through the associated Tim44p to Ssc1p the motor that drives import into mitochondria. Ssc1p is an Hsp70 that functions in conjunction with the nucleotide exchange factor Mge1p, homologous to the *E. coli* GrpE protein (Westermann *et al.*, 1995; Horst *et al.*, 1997). Like Kar2p in the ER, Ssc1p has been proposed to function either passively, as a molecular ratchet, or actively to pull proteins across the membrane. Discussion of this has led to a number of papers and reviews (e.g. Matouschek *et al.*, 2000; Pfanner and Wiedemann, 2002) and despite this some of the important contacts are still unclear. Thus Ssc1p has been reported to bind Tim44p through both the peptide binding domain (that also interacts with translocation substrates) and the ATPase domain (Moro *et al.*, 2002; Strub *et al.*, 2002). The latter of these would be more consistent with Ssc1p acting as a motor as it would provide a platform for the generation of a power stroke while the peptide binding domain interacts with substrate. It is clear that Ssc1p *can* exert pulling power to unfold protein domains on the outside of mitochondria (Huang *et al.*, 1999; Voisine *et al.*, 1999; Geissler *et al.*, 2001; Lim *et al.*, 2001). However, it is important to question, particularly in the light of the number of chaperone and other activities that aid in targeting nascent proteins to mitochondria *in vivo* (discussed earlier) whether this 'unfoldase' activity is strictly necessary. Thus, any pulling force exerted by Ssc1p may be used to improve speed of import (pulling proteins through the membranes) rather than unfolding substrates.

Not all protein import requires ATP hydrolysis. In particular the insertion of inner membrane proteins such as ADP-ATP carrier proteins, synthesized with no cleavable signal sequence, requires $\Delta\psi$ only and does not require either Ssc1p or Tim44p (Haucke and Schatz, 1997). Insertion of these inner membrane proteins requires the second Tim complex containing Tim22p, Tim18p and Tim54p (Sirrenberg *et al.*, 1996; Kerscher *et al.*, 1997, 2000). Of these only Tim22p has a clearly defined role. It is homologous to Tim23p and, like Tim23p, purified Tim22p forms gated channels when reconstituted into liposomes (Kovermann *et al.*, 2002). As with Tim23p opening of these channels is favoured by $\Delta\psi$, but instead of matrix targeting presequences they respond to the presence of carrier protein targeting signals. Carrier proteins insert into the outer membrane Tom complex via internal loops rather than in extended conformation (Wiedemann *et al.*, 2001) and these loops may insert directly into the Tim22p complex and thence into the membrane through lateral movement. Other components of this pathway are several small intermembrane space proteins Tim9p, Tim10p and Tim12p (Koehler *et al.*, 1998a,b; Sirrenberg *et al.*, 1998; Adam *et al.*, 1999). These are small, cysteine-rich proteins that appear to act sequentially – a soluble complex of Tim9p and Tim10p aids transfer of the substrate across the outer membrane and passes it on the inner membrane associated Tim12p before insertion via the Tim22p complex. The Tim9p/Tim10p complex preferentially binds hydrophobic segments of carrier proteins (Curran *et al.*, 2002) and thus may (like Tom70p on the outer surface of the mitochondrion) act as a chaperone, preventing aggregation of the substrate proteins as they pass through the intermembrane space.

Last but not least the insertion of proteins synthesized in the mitochondria into the inner membrane requires the activity of Oxa1p and Mba1p (Hell *et al.*, 1998, 2001; Preuss *et al.*, 2001). Insertion of substrates by this system is dependent on $\Delta\psi$, but substrates interact with the insertion machinery in its absence (Hell *et al.*, 2001). Intriguingly, interaction of mitochondrially encoded membrane proteins with Oxa1p is tightly linked to their translation on mitochondrial ribosomes, suggesting a co-translational mechanism for their insertion (Hell *et al.*, 2001). Oxa1p activity is also important for the biogenesis of some nuclear encoded inner membrane proteins (Hell *et al.*, 1998). These are imported fully into the mitochondrial matrix via the GIP and Tim23p complexes and are then inserted into the membrane. This pathway is termed 'conservative sorting' as it seems to have evolved from the bacterial progenitors of mitochondria, and consistent with this Oxa1p is homologous to the YidC proteins of bacteria that are required for insertion of proteins into the bacterial plasma membrane (Samuelson *et al.*, 2000; Scotti *et al.*, 2000).

Peroxisomes

Peroxisomal biogenesis and protein targeting has been studied in both *S. cerevisiae* and several other yeast species – *Pichia pastoris, Yarrowia lipolytica* and *Hansenula polymorpha.* This is because specific metabolic conditions, for example growth on oleic acid as sole carbon source, results in extensive proliferation of peroxisomes (as the enzymes responsible for β-oxidation of fatty acids are peroxisomal) and in these yeast species they can make up a large proportion (>80%) of the cytoplasmic volume, making them amenable to morphological and biochemical studies. Thus this discussion draws on literature from these other yeast species as well as *S. cerevisiae.* The vast changes in gene expression on induction of peroxisomes have also been used as a means of heterologous expression, and the methylotrophic *P. pastoris* in particular has become a commonly used system for expression of foreign proteins in the laboratory (Cregg *et al.*, 2000). Most known genes involved in peroxisomal protein targeting (currently 24, though not all of these are conserved in every system – Purdue and Lazarow, 2001a; Tam and Rachubinski, 2002) have been identified from collections of mutants that cannot grow under conditions requiring functional peroxisomes.

Targeting

Nearly all peroxisomal matrix proteins carry a peroxisomal targeting sequence (PTS) that can be recognized as falling into one of two classes, PTS1 (the consensus tripeptide – (S/A/C)(K/R/H)L at the C-terminus of the protein) and PTS2 (the consensus nonapeptide RLX_5H/QL most often located near the N-terminus of the protein). Mutations in two complementation groups were identified as failing to import specifically either PTS1- (*pex5* peroxin) or PTS2-targeted (*pex7*) proteins, and further work showed that Pex5p and Pex7p are indeed targeting factors for these two classes of import substrate (McCollum *et al.*, 1993; Rehling *et al.*, 1996; Zhang and Lazarow, 1996). Both Pex5p and Pex7p have dual localizations – cytoplasmic and intra-peroxisomal (Elgersma *et al.*, 1996, 1998; Zhang and Lazarow, 1996; Purdue *et al.*, 1998). This, combined with other experimental evidence has led to the view now becoming more and more accepted that they are shuttling proteins rather than purely cytosolic targeting factors, importing into the peroxisomal matrix along with substrate proteins before being recycled to the cytoplasm (discussed in Purdue and Lazarow, 2001a; Smith and Schnell, 2001). Pex5p contains a tetratricopeptide repeat (TPR) domain and both mutagenesis (in yeast) and structural studies (of human Pex5p) have revealed that it is this region of the protein that binds PTS1 motifs (Gatto *et al.*, 2000; Klein *et al.*, 2001). Intriguingly, import of acyl-CoA oxidase, a matrix protein without a PTS1 or PTS2 sequence, is also reliant on Pex5p. In this case binding to the targeting factor is in a region outside the TPR repeats (Klein *et al.*, 2002). Thus Pex5p is a multifunctional PTS binding factor and it will be of interest to determine whether other non-PTS1/2 proteins are targeted by this (PTS3) branch of peroxisomal import. Interaction between Pex7p and PTS2 sequences is not well characterized. In addition to Pex5p and Pex7p the cytosolic DnaJ homologue, Djp1p, is required for efficient PTS1- and PTS2-dependent import. In the absence of Djp1p *S. cerevisiae* is unable to grow under conditions that require functional peroxisomes (Hettema *et al.*, 1998). This requirement for chaperone activity in peroxisomal protein import is not well understood, but may point to regulation of protein–protein interactions in import complexes.

Two further proteins are specifically required for PTS2-dependent targeting. These are Pex18p and Pex21p, redundant proteins that form ternary complexes with Pex7p and substrates. In the absence of both Pex18p and Pex21p, PTS2-dependent import is abrogated and Pex7p becomes exclusively cytoplasmic (Purdue *et al.*, 1998; Stein *et al.*, 2002). Remarkably, turnover of Pex18p and Pex21p is linked to their function in peroxisomal targeting (Purdue and Lazarow, 2001b). Thus, if matrix protein import is impaired then Pex18p is stabilized. Intriguingly a proportion of Pex18p is ubiquitinated and turnover of Pex18p requires both Doa4p (proteasome) function and another peroxin, Pex4p, a ubiquitin-conjugating enzyme (Wiebel and Kunau, 1992) anchored on the peroxisomal membrane by a further protein, Pex22p (Koller *et al.*, 1999). These data suggest that Pex18p/Pex21p may be substrates of Pex4p and that somehow their degradation is related to the cycling of Pex7p between cytoplasmic and peroxisome-associated pools. Pex4p is required for both PTS1- and PTS2-dependent import and intriguingly in *P. pastoris* Pex4p is required to maintain levels of the PTS1 binding factor Pex5p (Koller *et al.*, 1999). Thus ubiquitination and regulated proteolysis may play a role in both PTS1 and PTS2 targeting pathways.

Pex5p and Pex7p have membrane associated receptors. These are Pex13p and Pex14p (Erdmann and Blobel, 1996; Albertini *et al.*, 1997). Pex5p binds to Pex13p and Pex14p, while Pex7p only binds Pex14p. Pex14p and Pex13p themselves interact in the two-hybrid assays, and also interact with a third peroxin, Pex17p (Albertini *et al.*, 1997; Huhse *et al.*, 1998). Deletion of the gene encoding any of these three proteins results in loss of both PTS1 and PTS2 import. This suggest the presence of an oligomeric receptor that acts, as in mitochondria, as a gathering point for proteins targeted through different pathways. However, whether they act in a concerted or consecutive manner is unclear.

Translocation

Peroxisomes, unlike the ER and mitochondria, import folded and oligomerized protein complexes. Physiological substrates as large as ~450 kDa are imported into yeast peroxisomes (Titorenko *et al.*, 2002) and mammalian peroxisomes will even import PTS1-coated gold beads of 4–9 nm diameter (Walton *et al.*, 1995). Thus a very different type of translocation apparatus is required to those found in ER or mitochondrial membranes. Epistasis analysis in *P. pastoris* revealed an order of action of various peroxins required for matrix protein import (Collins *et al.*, 2000) and this combined with other data support a model of import in which the zinc RING-domain containing proteins Pex10p, Pex12p and Pex2p interact to form an 'import complex' that incorporates or contacts the receptor protein complex of Pex13p, Pex14p and Pex17p. These proteins can all be co-immunoprecipitated along with the PTS1 binding factor Pex5p (Albertini *et al.*, 2001; Hazra *et al.*, 2002). The zinc RING-domains of Pex10p, Pex12p and Pex2p are cytoplasmic and those of Pex10p and Pex12p interact in two-hybrid experiments (Albertini *et al.*, 2001). RING-domains have had a number of activities ascribed to them, including both homo- and hetero-oligomerization and conjugation of ubiquitin to substrate proteins (Joazeiro and Weissman, 2000; Kentsis and Borden, 2000; Kentsis *et al.*, 2002). A plausible suggestion would be that the RING-domains provide (at least part of) a means whereby large and transient assemblies of import factors assemble to accommodate folded and/or oligomerized proteins during import. As discussed above, Pex4p is a ubiquitin ligase and the RING-domains of one or more of the import proteins could also provide a link between the ligase and its substrates. A fourth protein classified as a translocation protein is Pex8p. Unlike Pex2p, Pex10 and Pex12p, Pex8p is a peripheral membrane protein. It localizes to the matrix side of the membrane and provides a binding site for Pex5p (Rehling *et al.*, 2000). Given this, Pex8p may be a site of disassembly for imported substrate–Pex5p complexes.

The discussion made here is, unfortunately, largely speculative. There is currently no morphological evidence for any translocation complex or pore in the peroxisomal membrane. Another part of peroxisomal import that is not understood is the requirement for ATP. However, it is tempting to attribute this at least in part to Pex1p and Pex6p. These interacting AAA ATPase, the only known ATPases involved in peroxisomal import, act late in the import process (Collins *et al.*, 2000). AAA ATPases have very diverse cellular functions, though a commonality is that they often occur where protein unfolding or dissociation occurs (Vale, 2000). Pex1p and Pex6p are localized to the cytoplasmic surface of the organelle and mutants lacking these proteins share some phenotypes with *pex4* mutants. Thus they have reduced levels of Pex5p and in *pex1* and *pex6* cells peroxisomes are present, but with drastically reduced levels of matrix protein import. Like Pex4p, then, these proteins have been proposed to be important in recycling at least Pex5p.

The targeting sequences of peroxisomal membrane proteins are poorly characterized. However, consensus sequences are emerging and a number of peroxisomal membrane proteins contain targeting sequences that consist a hydrophilic peptide containing a patch of basic amino acids next to a hydrophobic stretch or transmembrane domain (see Purdue and Lazarow, 2001a). Two proteins, Pex3p and Pex19p, are required for membrane protein insertion and a third, Pex16p, is important in other yeasts but not conserved to *S. cerevisiae*. Confusion arose over the fact that *pex3* and *pex19* strains appeared not to contain any remnants of peroxisomes, leading to the suggestion that these organelles could be derived *de novo* from other sources such as the ER. Indeed, deficit in membrane insertion leads, in turn, to deficit in matrix protein import and thus the organelle might be predicted to be 'non-existent' in mutants that fail to insert peroxisomal membrane proteins. In addition over-expression of either Pex3p (Baerends *et al.*, 1996) or another membrane protein Pex15p (Elgersma *et al.*, 1997) results in proliferation of the ER and localization of these proteins to the ER. However, more recent studies have identified peroxisomal remnants in cells lacking either Pex3p or Pex19p (Snyder *et al.*, 1999; Hazra *et al.*, 2002) and indicated that the over-expression studies resulted in aberrant, non-physiological localizations of Pex3p and Pex15p (Hettema *et al.*, 2000). In addition experiments in *S. cerevisiae* in which either protein import into the ER or trafficking from the ER are blocked have shown that neither of these impact on peroxisome biogenesis (South *et al.*, 2000, 2001). Thus it is unlikely that the ER plays a role in this process. An exception, or perhaps a warning that there are significant differences in peroxisome biogenesis in different systems, is *Y. lipolytica* where there is strong evidence for both N-glycosylation of peroxisomal membrane proteins and effects of mutations in the ER targeting pathway on peroxisomal biogenesis (Titorenko and Rachubinski, 1998a,b).

As can be seen from the above discussion we are still a considerable way from understanding peroxisomal biogenesis. Given the lack of explicit detail of the function of many of the components of the import machinery it is quite possible that some data will be re-interpreted in the light of future results and that the functions of some factors are currently incorrectly assigned. Last, it should be noted that the peroxisome is not the only system that translocates folded and/or oligomerized proteins. The twin-arginine transport pathway of bacteria (and the homologous ΔpH pathway of chloroplast thylakoids) both translocate proteins with co-factors bound to them. Thus maintaining membrane integrity while translocating wide or bulky substrates is not a problem solved only once in evolution.

7.3 The general secretory pathway

Protein modification, folding and quality control in the ER

The import of a protein into the ER is coupled to several important events in its biogenesis. Depending on the protein these may include signal sequence cleavage, N-glycosylation, folding, and addition of disulphide bonds and/or glycosyl phosphatidlyinositol (GPI) anchors. Significant discussion of these events is not within the scope of this review. However, it is important to note that defects in these processes result in general in delay or failure in further sorting and transport events. For example, if cleavable signal sequences are not removed from proteins this can lead to their accumulation in the ER (Schauer *et al.*, 1985; Kaiser *et al.*, 1987). Similarly GPI anchors are essential for further biogenesis of proteins that receive them, for example, Gas1p transport to the Golgi is defective if GPI attachment is prevented (Hamburger *et al.*, 1995; Doering and Schekman, 1996).

A number of ER resident proteins are known or proposed to aid in protein folding. These include the Hsp70 homologues Kar2p (Simons *et al.*, 1995) and Lhs1p (Saris *et al.*, 1997), the DnaJ proteins Scj1p and Jem1p (Silberstein *et al.*, 1998), protein disulphide isomerase (PDI) and Ero1p (ER oxidation) both important in introduction and shuffling of disulphide bonds (Frand *et al.*, 2000) and the calreticulin/calnexin Cne1p (Parlati *et al.*, 1995). If the demands for protein folding in the ER are overwhelmed and/or proteins fail to fold correctly and have to be disposed of, two important mechanisms are called on. These are the unfolded protein response (UPR) and ER-associated protein degradation (ERAD).

The UPR (reviewed in Patil and Walter, 2001; Spear and Ng, 2001) is a stress response mechanism whereby components of the ER involved in protein folding such as those listed above are upregulated to deal with accumulating unfolded proteins within the lumen of the organelle. It is mediated by the transmembrane serine/threonine kinase Ire1p which initiates an atypical mRNA splicing event on the message encoding the transcription factor Hac1p (Sidrauski and Walter, 1997). This results in translation of an active Hac1p that increases transcription of UPR-responsive genes by binding UPR-specific upstream activating sequences (Cox and Walter, 1996). The signal that generates the UPR is proposed to be titration of available Kar2p onto unfolded protein substrates resulting in decreased levels of Kar2p free to bind the lumenal domain of Ire1p. Ire1p not bound to Kar2p would then be active. The UPR is a conserved stress pathway and this (Kar2p binding by Ire1p) has been seen in both the yeast and mammalian system (Bertolotti *et al.*, 2000; Okamura *et al.*, 2000). Global analysis of genes responsive to the UPR has led to the realization that it is a major response – transcription of genes encoding proteins that function throughout the secretory pathway as well as in lipid metabolism and numerous non-secretory pathway factors are induced (Travers *et al.*, 2000). An integrated view of this has not yet been assimilated, but a crucial point gained from this and other studies is that the UPR is, perhaps not surprising as this is how the cell rids itself of protein from the ER, linked to ERAD.

Analysis of ERAD (or ER quality control) rapidly led to the conclusion that degradation takes place in the cytosol, and that Sec61p, in addition to its role in import, is crucial for retro-translocation of proteins to the cytosol (Pilon *et al.*, 1997; Plemper *et al.*, 1997) (Figure 7.2). Sec61p thus forms a bidirectional channel in the membrane and the two functions of Sec61p can be dissected genetically indicating that the interactions required for each are not equivalent (Zhou and Schekman, 1999). Glycopeptides are also exported across the yeast ER membrane to the cytosol and this export is also mediated by Sec61p (Gillece *et al.*, 2000). Thus all known proteinaceous transport substrates that cross the ER membrane do so through Sec61p.

Proteins implicated in ERAD include many of the ER chaperones involved in folding (Plemper *et al.*, 1997; Brodsky *et al.*, 1999; Nishikawa *et al.*, 2001) as well as cytosolic chaperones (particularly for destruction of membrane proteins) (Zhang *et al.*, 2001). Thus cells deficient in chaperone activities have specific or general defects in ERAD. The role of these proteins may be to prevent aggregation of (mis-folded) ERAD substrates, which would preclude retro-translocation, rather than directly targeting them for destruction (Nishikawa *et al.*, 2001). One of the most interesting aspects of ERAD is how the signal is generated that labels proteins for destruction. Perhaps the most likely possibility is that it is a stochastic process – an unfolded protein has a certain likelihood of being degraded and thus the longer it remains unfolded the higher the chance that it will be degraded. Another possibility is that the proteins responsible for ERAD directly recognize substrates – and some evidence for this exists (see later) – though how this might be accomplished is a mystery.

Despite the problem of defining how a protein is designated as an ERAD substrate great progress has been made in understanding the proteins that carry out the process. The final destination of exported proteins is the proteasome and mutations that affect its function result in failure to degrade exported proteins (Hiller *et al.*, 1996). Proteins are targeted to the proteasome by ubiquitination and, not surprisingly, an ER-associated ubiquitin ligase complex was identified (the Hrd1p complex) through screens for mutants that affect degradation of either constitutive or regulated ERAD substrates (reviewed in Hampton, 2002). A second ER-localized ubiquitin ligase complex containing Doa10p is induced by the UPR and plays a role in destruction of proteins that are not substrates of the Hrd1p complex (Swanson *et al.*, 2001). Both complexes contain integral membrane proteins containing RING-domains that have ubiquitin ligase activity (Hrd1p/Der3p and Doa10p) as well as ubiquitin conjugating enzymes (Ubc7p and either Ubc1p or Ubc6p) and the membrane protein Cue1p (coupling ubiquitination to ER degradation) which anchors Ubc7p to the membrane (Biederer *et al.*, 1996, 1997; Bordallo *et al.*, 1998; Bordallo and Wolf, 1999; Swanson *et al.*, 2001; Bays *et al.*, 2001a). Hrd1p will ubiquitinate unfolded protein over folded substrates *in vitro* (Bays *et al.*, 2001a), supporting the proposal that ERAD proteins scan the ER for unfolded proteins making them into substrates for ERAD by adding ubiquitin to them.

Another issue under intense investigation is since proteins are degraded in the cytosol how are they extracted from the lumen or membrane of the ER. A number of studies have identified the AAA-ATPase Cdc48p–Ufd1p–Npl4p complex as required after ubiquitination for degradation of ERAD substrates (Bays *et al.*, 2001b; Ye *et al.*, 2001; Braun *et al.*, 2002). Though this complex is cytosolic, in its absence abstraction of ubiquitinated membrane and lumenal ERAD substrates from the ER is defective (Jarosch *et al.*, 2002). Thus the ATPase activity of the Cdc48p complex may provide the force to actively remove substrates from the ER through the Sec61p channel. As the ubiquitin ligase activity of the Hrd1p (and Doa10p) complexes are on the cytosolic side of the membrane the question remains, particularly for lumenal proteins, of how the initial part of the protein that becomes ubiquitinated becomes exposed in the cytosol.

There are yet more twists to ER quality control and ERAD. First, ERAD substrates continue to be degraded in the absence of, for example, Hrd1p. Thus alternative pathways for their destruction exist. A third ubiquitin ligase Rps5p has recently been demonstrated to be required for ER quality control and this may provide an answer to this particular puzzle. In cells deficient for both Hrd1p and Rps5p activity ERAD substrates (both lumenal and membrane) are stabilized significantly over what is seen when either one alone is missing (Haynes *et al.*, 2002). The Rps5p-dependent pathway, termed HIP (Hrd1p independent proteolysis) requires trafficking between the ER and Golgi to function (Haynes *et al.*, 2002). Other studies have also identified ER–Golgi transport as important for ERAD, though whether all substrate proteins must transit between these compartments prior to their destruction or whether defects in transport result in impaired ER (and hence ERAD) function is unclear (Caldwell *et al.*, 2001; Vashist *et al.*, 2001; Taxis *et al.*, 2002). Further investigation is required to determine why ER–Golgi transport is important for ERAD.

Vesicle mediated trafficking

Once a protein has been correctly folded and modified in the ER it must be sorted and transported to its correct destination. This means retention of proteins resident in the ER or Golgi plus retrieval if they escape to the wrong compartment, and routing to the vacuole or cell surface for other proteins. These events are mediated by active sorting and packaging of proteins into vesicles. Vesicular trafficking occurs in both directions between compartments of the secretory pathway, forward (anterograde) as well as backward (retrograde). Retrograde transport, from the Golgi to the ER, is crucial for several reasons: first to prevent loss of proteins which have escaped from the ER; second to replenish the ER membrane with proteins essential for anterograde transport; third to replace the lipids lost as vesicles are budded from the ER. Structurally different vesicles mediate different trafficking steps.

Early stages in the secretory pathway require the function of two types of vesicle. These are the COP I and COP II (Coat Protein) coated vesicles, which mediate transport between the ER and Golgi and intra-Golgi traffic. Anterograde transport is mediated by COP II vesicles, whereas retrograde transport requires COP I (Pelham, 2001). Other coats include that formed by clathrin, important for both endocytosis and trafficking of proteins from the Golgi apparatus to the cell surface and vacuole via endosomal compartments, and the retromer coat required for retrieval of proteins from late endosomes to the Golgi. Clathrin is not essential in yeast indicating that alternative pathways exist for events that are normally clathrin-mediated or that these processes are not essential in yeast.

COP II – anterograde transport and mechanisms of cargo selection

The development and subsequent dissection of an *in vitro* assay for transport from the ER to the Golgi in the Schekman laboratory allowed identification of the vesicles responsible for anterograde protein movement between these organelles. The factors described from these studies, and others, as required for formation of these COP II coated vesicles, and a model for how these assemble is presented in Figure 7.4.

Initiation of vesicle formation is through recruitment of the soluble GTPase Sar1p by the ER membrane protein Sec12p that also acts as a guanine nucleotide exchange factor (GEF) for this protein. Sar1p-GTP bound to the membrane recruits further components, Sec23p/24p and then Sec13p/31p, which form the vesicle coat (Salama *et al.*, 1993; Barlowe *et al.*, 1994). Sec23p is a GTPase activating protein (GAP) for Sar1p (Yoshihisa *et al.*, 1993), and, as the GDP-bound form of Sar1p does not interact stably with Sec23/24p, Sar1p is not a component of the final vesicle coat. Vesicles do form in the presence of non-hydrolysable GTP analogues, thus GTP hydrolysis by Sar1p and its dissociation from the coat proteins is not required for vesicle formation. GTP hydrolysis is required, however, for vesicle uncoating as stably coated, fusion-incompetent vesicles are produced in the presence of GTPγS (Barlowe *et al.*, 1994; Oka and Nakano, 1994). Both Sec23p/24p and Sec13p/31p interact with Sec16p, a protein tightly associated with the ER membrane that is also a component of COP II vesicles (Espenshade *et al.*, 1995; Gimeno *et al.*, 1996). Sec16p has been proposed to act as a scaffold that organizes COP II coat subunits, and *in vitro* it stimulates budding reactions, though only in the presence of GTP and not GTPγS (Supek *et al.*, 2002). Sec16p itself binds to the ER membrane through interaction with Sed4p, a Sec12p homologue required for efficient ER to Golgi transport (Gimeno *et al.*, 1995). Genetic data indicate that Sed4p may delay or counteract the GAP action of Sec23p on Sar1p (Saito-Nakano and Nakano, 2000) and thus Sec16p may play a pivotal role in nucleating coat assembly, both bringing together components of the coat and stabilizing it, preventing premature disassembly.

Initial analysis suggested that Sec23/24p and Sec13/31p were both large oligomeric complexes (Hicke *et al.*, 1992; Salama *et al.*, 1993). However, more recent results have shown that while Sec13/31p is a heterotetramer containing two copies of each subunit, Sec23/24p is in fact a heterodimer (Lederkremer *et al.*, 2001). Thus individual Sec23/24p heterodimers may interact with Sar1p (and cargo molecules; see later) before coming together to form the platform on which Sec13/31p assembles. Both Sec23p and Sec24p interact with Sec31p

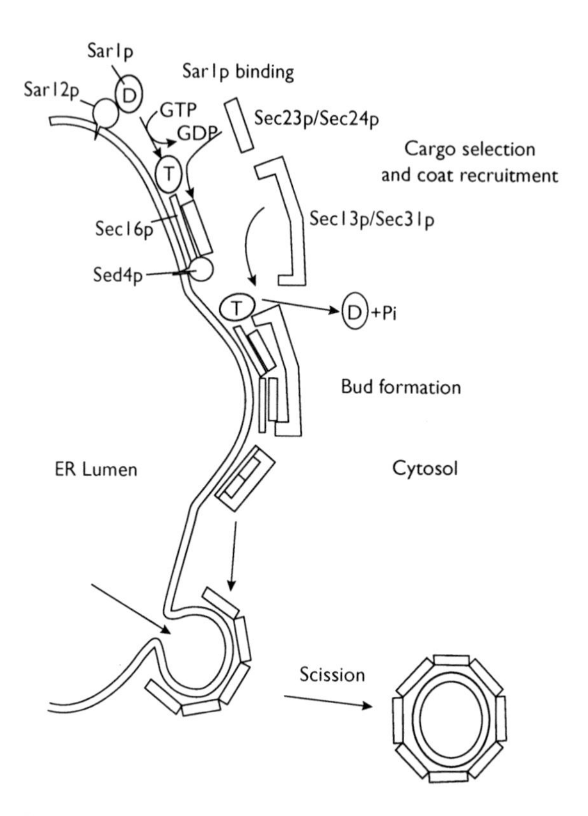

Figure 7.4 Formation of COP II coated vesicles. Vesicle formation is initiated by recruitment of Sar1p by the GTP exchange factor Sec12p. GTP-Sar1p binds to the membrane and recruits the Sec23/24p complex. Sec23/24p also contacts Sec16p, which is itself bound to the Sec12p homologue Sed4p which, by inhibiting premature activity of the Sar1p GAP activity of Sec23p, aids in the formation of a stable coat. Sec13/31p binds to the membrane through interactions with Sec23/24p and Sec16p. Further binding of Sec23/24p and particularly Sec13/31p result in deformation of the membrane and eventual formation of a vesicle which is pinched off from the ER membrane. Sec12p and Sed4p are excluded from the vesicle. Cargo and cargo-adapters are not shown in this model, but see text for a discussion of cargo packaging.

(Shaywitz *et al.*, 1997) and this, together with the fact that there are 2 Sec31p molecules in each Sec13/31p complex provides ample scope for assembly of a coat with multiple contacts between the components. The Sec13/31p complex is both the final component of the coat to assemble, and the one that drives membrane curvature and hence budding (Antonny *et al.*, 2001). Moreover, in staged budding reactions addition of Sec13/31p stimulates GTP hydrolysis by Sar1p (Antonny *et al.*, 2001) suggesting that it may rearrange the Sar1p–Sec23/24p interaction thereby allowing full access of the GAP activity of Sec23p to Sar1p. Structural information, in the shape of electron microscopy, is contributing to understanding of the COP II coat. Of particular interest the coat itself is seen to be a polygonal mesh, into which individual Sec13/31p complexes can be modelled (Matsuoka *et al.*, 2001; see discussion in Barlowe, 2002). Consistent with the proposal that the Sec13/31p complex indeed forms the outer layer of the coat, attempts to cross-link COP II components to phospholipids yielded efficient cross-linking to both Sar1p and Sec23/24p, but not Sec13/31p suggesting that it is indeed further from the membrane.

Despite the great amount of detail that is known, several important questions remain. Thus how the site of vesicle formation is selected, and how coat polymerization is regulated are not well defined. *S. cerevisiae* is peculiar in this respect in that unlike other organisms (including other budding yeasts such as *P. pastoris*) COP II vesicles do not bud from specialized ER exit sites. The assembly of vesicle coats from large complexes is a recurring theme. This is likely to speed up the process, appears to aid in membrane deformation as large, relatively inflexible protein complexes bind and form curved lattices that force the membrane to bend, and may also aid in pinching-off the vesicle from the ER. So far no 'pinchase' or severing protein has been identified as required to complete this final step in COP II vesicle formation.

An aspect of vesicular trafficking that has received much attention in the last few years is sorting and loading proteins into vesicles, and much of this has focused on COP II vesicle formation at the ER. COP II vesicles are enriched for proteins that follow the secretory pathway, but lack ER-resident proteins such as Kar2p and Sec61p (Barlowe *et al.*, 1994; Bednarek *et al.*, 1995). Therefore, a distinction is made between resident proteins and those to be transported prior to or during packaging. Recent data indicate that COP II proteins themselves act as sorting factors – both directly and through interaction with cargo-bound transmembrane receptor or adapter proteins. Active sorting requires the presence of *cis*-acting structural, or sequence motifs within transported proteins, which allow their selection by COP II components and/or adaptors. Such motifs have indeed been noted and selection, combined with other 'sieving' or 'gating' mechanisms, ensure that cargo molecules are greatly enriched over ER resident proteins in vesicles.

Incorporation of some cargo into vesicles is initiated by Sar1p. Thus interactions have been identified between Sar1p and the SNARE proteins required at the terminal docking and fusion stages of vesicle transport (see later) (Springer and Schekman, 1998). Complexes of Sar1p and SNARE proteins can be assembled *in vitro* which then recruit Sec23/24p and Sec13/31p (neither of which bind the SNARE proteins on their own) in a GTP-dependent fashion, suggesting that they are physiologically relevant. SNARE proteins are present in the final vesicles whereas Sar1p is not (see earlier text) and Springer and Schekman (1998) found that different regions of the cytosolic tail of the Bet1p SNARE were required for Sar1p and Sec23/24p binding. Thus the model of recruitment of these SNAREs into vesicles is by binding to Sar1p, which then passes them to Sec23/24p. In broader experiments Kuehn *et al.* (1998) isolated complexes that formed on ER-derived microsomes in the presence of Sar1p and Sec23/24p (but not Sec13/31p) and found that they contained cargo such as the SNARE proteins, glycosylated pro-α-Factor and amino acid permeases but not ER resident proteins. These experiments demonstrate the important point that much cargo is collected into the pre-budding complex before completion of the coat.

Sec24p plays a crucial role in this accumulation of vesicle contents. Yeast has three Sec24p homologues, and the contents of vesicles depend on which Sec24p homologues are present (Kurihara *et al.*, 2000; Miller *et al.*, 2002). Thus, for example, in the absence of the Sec24p homologue Lst1p the plasma membrane ATPase Pma1p is not packaged into vesicles, resulting in its accumulation in the ER and pH sensitive growth (Roberg *et al.*, 1999; Shimoni *et al.*, 2000). Similarly while Lst1p is found (like Sec24p) as part of a heterodimer with Sec23p and can function alone in vesicle budding reactions (Shimoni *et al.*, 2000), vesicles produced with only Lst1p do not contain the SNARE proteins vital for vesicle docking (Miller *et al.*,

2002). In addition, unlike Sec24p, Lst1p does not bind Bet1p *in vitro*, and the message from these studies is that the correct mix of Sec24p homologues is crucial for packaging the full complement of cargo into functional vesicles.

Several proteins are either known or proposed to be packaging factors or adapter proteins that incorporate cargo into vesicles. These include the p24 family, Erv14p, Erv29p, Shr3p and Chs7p (Schimmoller *et al.*, 1995; Trilla *et al.*, 1999; Belden and Barlowe, 2001b; Powers and Barlowe, 2002). The eight-member p24 family, typified by Emp24p, are membrane protein components of COP II vesicles, and cycle between the ER and the Golgi (Schimmoller *et al.*, 1995; Marzioch *et al.*, 1999). Deletion of *EMP24* results in slowing of transport of certain cargo proteins (invertase and Gas1p) from the ER to the Golgi. As yet there has been no observed interaction between p24 proteins and cargo, but some studies are suggestive of a direct role in cargo selection (Schimmoller *et al.*, 1995; Muniz *et al.*, 2000). In addition to defects in cargo recruitment strains lacking p24 proteins reveal reduced fidelity of sorting resulting in transport of ER residents such as Kar2p and PDI to the Golgi (Elrod-Erickson and Kaiser, 1996). Thus these proteins may play both positive and negative roles in cargo selection and represent a gating function in vesicle formation. Perhaps surprisingly a viable strain lacking all members of this family has been constructed indicating that the role of the p24 family is non-essential or redundant with other sorting events (Springer *et al.*, 2000). Because of these somewhat confusing results the p24 proteins have been suggested to be, in addition to cargo receptors, structural components of vesicles and negative regulators of vesicle budding. The mechanism by which the p24 proteins Emp24p and Erv25p themselves are recruited into COP II vesicles was addressed by Belden and Barlowe (2001a). Both these proteins bound Sec23/24p and particularly Sec13/31p through their cytoplasmic tails. Binding to Sec13/31p required a di-aromatic motif in the cytoplasmic tails of Emp24p/Erv25p and, unlike binding of SNARE proteins to Sec23/24p, is independent of Sar1p. Thus, as well as being a structural component of the coat, Sec13/31p aids cargo assembly into the vesicle.

Erv14p and Erv29p also cycle between the ER and Golgi (Belden and Barlowe, 2001b; Powers and Barlowe, 2002). Erv14p, like Emp24p and Erv25p, contains hydrophobic residues in a cytoplasmic segment that are crucial for ER export, and this protein is an adapter for transport of the membrane protein Axl2p (Powers and Barlowe, 2002). Similarly Erv29p binds to glycosylated pro-α-Factor and is required for its incorporation into COP II vesicles (Belden and Barlowe, 2001b). In the absence of Erv29p several soluble secretory proteins accumulate in the ER (Caldwell *et al.*, 2001). Thus both Evr14p and Erv29p are *bona fide* adapter proteins.

Shr3p is required for the functional expression of amino acid permeases and, more specifically, its role is in directing these polytopic membrane proteins into COP II vesicles (Ljungdahl *et al.*, 1992; Kuehn *et al.*, 1996, 1998). In contrast to the previously discussed proteins Shr3p does not enter vesicles, but remains in the ER, even though it interacts with both coat components as well as permeases such as Gap1p (Gilstring *et al.*, 1999). Shr3p does not aid permease folding as high level expression of permeases in the absence of Shr3p (resulting in their accumulation in the ER) does not lead to activation of the UPR (Gilstring *et al.*, 1999). Thus Shr3p is proposed to act as a packaging chaperone for the mature permeases, ensuring their incorporation into COP II vesicles by transient interaction with coat components at the site of vesicle formation.

The above discussion centres mostly on adapter and packaging proteins for which specific functions have been demonstrated. A further gating function into vesicles may be provided by ER resident chaperones, which release and rebind proteins as they fold, only permanently releasing them once they are fully folded. This event makes proteins available for incorporation into vesicles, but does not explain the discrimination against resident proteins in vesicle formation. Given the diversity of cargo proteins incorporated into vesicles it would seem that a large array of adapters and packaging proteins have to exist if all cargo proteins are actively sorted. This is unlikely to be the case (where would the cargo fit if there are so many adapters?) and the puzzle remains as to how general or specific these events and mechanisms are. Perhaps Shr3p and similar proteins that direct vesicle incorporation of several substrates with similar structures or motifs will turn out to be a general rule.

COP I – retrograde transport through recognition of KKXX, HDEL and transmembrane domains

The COP I coated vesicle was first identified and analysed in mammalian *in vitro* systems by Rothman and colleagues using an intra-Golgi transport assay (Orci *et al.*, 1993; Ostermann *et al.*, 1993). COP I coats consist of seven proteins; α-, β-, β′-, γ-, δ-, ε- and ζ-COP the 'coatomer complex' that bind *en bloc* to the membrane (Hara-Kuge *et al.*, 1994). Homologues of all the coatomer components have been identified in yeast – Ret1p, Sec26p, Sec27p, Sec21p, Ret2p, YIL076w and Ret3p (α- to ζ-COP, respectively) (Stenbeck *et al.*, 1992; Duden *et al.*, 1994; Letourneur *et al.*, 1994; Gerich *et al.*, 1995; Cosson *et al.*, 1996). As with COP II vesicles an activated GTP-binding protein, in this case ARF (ADP-ribosylation factor), is required to initiate vesicle formation. However, unlike Sar1p, ARF remains on vesicles after their formation, and hydrolysis of GTP is required for vesicle uncoating. Yeast has three ARF proteins, Arf1-3p (Stearns *et al.*, 1990; Lee *et al.*, 1994) along with four ARF GEFs (Gea1p/2p, Sec7p, and Syt1p) and six proteins predicted or demonstrated to have ARF GAP activity (Peyroche *et al.*, 1996; Poon *et al.*, 1996, 1999; Zhang *et al.*, 1998; Jones *et al.*, 1999). In addition to formation of COP I coats ARF is important for other vesicle trafficking events such as those mediated by clathrin, and mutations that impair ARF activity result in multiple defects in post-ER trafficking (Gaynor *et al.*, 1998; Deitz *et al.*, 2000; Peyroche *et al.*, 2001; Spang *et al.*, 2001; Yahara *et al.*, 2001). Roles for the different ARF GAPs are being dissected indicating that the regulators of ARF specify how this general factor functions in specific trafficking events. Thus Glo3p appears specific for retrograde transport from the Golgi to the ER (Dogic *et al.*, 1999; Poon *et al.*, 1999) and Age1p is specific for sorting from the *trans*-Golgi (Poon *et al.*, 2001).

COP I mediates retrograde transport from the Golgi to ER in yeast. The first evidence for this came from genetic screens in which mutants (*ret*) defective in retrieval of proteins from the ER to Golgi were isolated. Several of the mutations isolated mapped either to genes encoding COP I components (Letourneur *et al.*, 1994; Cosson *et al.*, 1996) or regulators of ARF (Glo3p) (Dogic *et al.*, 1999). COP I mutants fail to retrieve type-I membrane proteins with the di-lysine motif KKXX at their cytoplasmic C-terminal tail and this motif is bound by components of the COP I coat itself (Cosson and Letourneur, 1994). Other retrieval motifs have also been identified that bind to COP I. These are two separated or neighbouring aromatic residues (Cosson *et al.*, 1998; Belden and Barlowe, 2001a). Peptides containing these motifs can be used to isolate coatomer from cell lysates.

Several other mechanisms exist for returning ER resident proteins to the ER. The best characterized of these is the HDEL-mediated pathway. The tetrapeptide HDEL is found at the C-terminus of eleven yeast proteins, all of which reside in the ER. The HDEL motif is in the ER lumen as these are either lumenal (e.g. Kar2p, PDI) or type II integral membrane proteins (Sec20p, Sed4p). The importance of the HDEL sequence is clear as its removal from luminal proteins results in their secretion from the cell (Hardwick *et al.*, 1990) or transport to the vacuole for the membrane protein Sec20p (Sweet and Pelham, 1992). HDEL-containing proteins are normally transported to the Golgi to some extent as they obtain Golgi modifications to their glycosyl side chains (Pelham *et al.*, 1988). The retrieval receptor that directs their return to the ER is Erd2p (Semenza *et al.*, 1990), a Golgi resident whose mammalian homologue binds directly to peptides containing the retrieval motif (Wilson *et al.*, 1993). Mammalian Erd2p recruits ARF GAP to membranes thereby presumably recruiting COP I components (Aoe *et al.*, 1997) suggesting a direct route for these proteins into retrograde transport vesicles; it remains to be seen whether this mechanism is conserved to yeast.

Another mechanism of protein retrieval from the Golgi recognizes transmembrane regions of proteins (e.g. Sec12p, Sec63p, Sec71p) and requires the function of the *RER1* gene product (Boehm *et al.*, 1997; Sato *et al.*, 1997; Letourneur and Cosson, 1998). Rer1p is a Golgi-localized membrane protein that cycles to and from the ER, promotes packaging of proteins such as Sec12p into COP I vesicles and interacts with the α- and γ-subunits of coatomer (Letourneur and Cosson, 1998; Sato *et al.*, 2001a). Thus, in addition to actively excluding ER residents from anterograde vesicles, yeast has several systems to ensure their recovery from later compartments if they do escape, all of which rely on COP I coated vesicles. Retrieval systems also function later in the secretory pathway – for example from late to early Golgi (Harris and Waters, 1996), and from the endosome to the late Golgi (Bryant and Stevens, 1997). Thus protein retrieval after escape from the compartment in which they function is a general phenomenon, crucial for maintenance of compartment-specific protein composition.

Vesicle docking and fusion

The later stages in trafficking between compartments of the secretory pathway, docking and fusion, follow a generic mechanism, regardless of the source or target membrane or the vesicle coat, which is shed before fusion. The central tenet of this model is the SNARE hypothesis (Sollner *et al.*, 1993). This states that vesicles contain proteins on their surface (v-SNAREs) specifying the target membrane to which they bind by complementary interaction with t-SNAREs. While formation of complexes between SNAREs on different membranes can be sufficient to allow mixing of the lipid bilayers *in vitro* (Weber *et al.*, 1998) other factors contribute to and control the process *in vivo*. Thus the interactions of SNAREs, and the fusion of the vesicle with the membrane are governed by other, general fusion proteins; *N*-ethylmaleimide-sensitive fusion protein (NSF) and soluble NSF attachment protein (SNAP). Yeast NSF and SNAP are the products of *SEC18* and *SEC17* respectively and, as far as is known, Sec18p and Sec17p are required for all vesicle trafficking steps. Other proteins and protein complexes, including effectors (TRAPP, Sec34/35p, VFT, HOPS and exocyst complexes – Lowe, 2000; Short and Barr, 2002), GTPases of the Rab family, Sec1p/Munc18 (SM) homologues (Jahn, 2000) and LMA1 are also required. The attractive hypothesis that individual Rab, SM and SNARE proteins specify particular targeting events has been partly confirmed (e.g. Pelham, 1999; McNew *et al.*, 2000) though a number of the proteins in these families have been implicated in more than one event. Thus specificity of vesicle targeting is a cumulative effect of the actions of several components, and layers of regulation mediated by the action of general and unique factors required for each docking and fusion event.

A model for the involvement of general players in vesicle targeting and docking is presented in Figure 7.5, with emphasis on mechanism. Specific components are indicated for docking and fusion of COP II vesicles as a wealth of genetic and biochemical data are available for this example. For further discussion of other examples and particularly models of membrane fusion see Mayer (2002). As described before, during vesicle formation

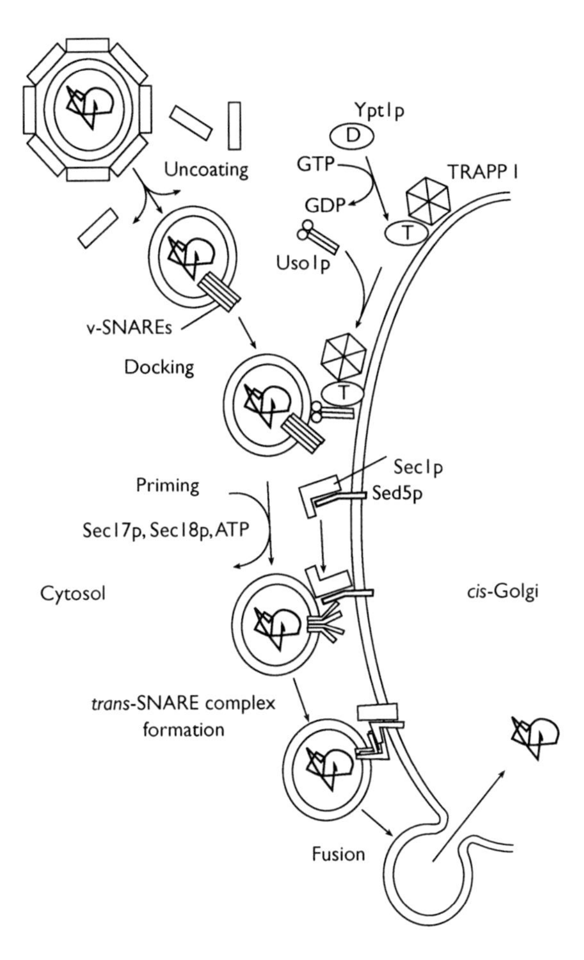

Figure 7.5 Docking and fusion of COP II coated vesicles. Uncoating of the COP II vesicle reveals the v-SNARE molecules on the surface of the vesicle, and allows docking and tethering of the vesicle to the membrane through the activities of Uso1p and the TRAPP I complex. The Rab GTPase Ypt1p is recruited to the membrane by the GEF function of the TRAPP I complex and is required to release the t-SNARE Sed5p from a complex with Sec1p. This activates Sec5p for formation of *trans*-SNARE complexes with v-SNAREs (Sec22p, Bet1p and Bos1p). Before these *trans*-SNARE complexes can be formed the v-SNAREs must be primed by the action of the general fusion priming factors Sec18p (NSF) and Sec17p (SNAP), which disrupt the *cis*-interactions between them. Formation of the *trans*-SNARE complexes is aided by the action of LMA1 and following this the fusion reaction, which also requires calcium, calmodulin and protein phosphatase I, can proceed.

v-SNAREs (Sec22p, Bet1p, Bos1p) are loaded into the nascent vesicle. SNAREs have intrinsic ability to form complexes in which Leucine-zipper motifs in these proteins associate in a four-helix bundle (Sutton *et al.*, 1998) and at this point these factors are bound together in inactive *cis*-SNARE complexes. The first stage in the fusion reaction is docking of the vesicle onto the target membrane. Biochemically docking requires both the coiled-coil protein Uso1p and the Rab Ypt1p (Cao *et al.*, 1998), though genetically Uso1p has been defined as acting prior to Ypt1p (Sapperstein *et al.*, 1996). Also acting at, or even before this stage is the multi-subunit TRAPP (transport protein particle) complex, which includes a GEF function for Ypt1p (Jones *et al.*, 2000; Wang *et al.*, 2000; Sacher *et al.*, 2001). The TRAPP complex is of two types, TRAPP I and TRAPP II, both of which stably localize to the early Golgi (Barrowman *et al.*, 2000; Sacher *et al.*, 2001). TRAPP I is required for ER–Golgi transport whereas TRAPP II is proposed to be important in later transport steps (Sacher *et al.*, 2001). Membrane binding of Ypt1p is linked to nucleotide exchange of GDP for GTP, and thus it can be suggested that a signal is passed from the docked vesicle through TRAPP I to mark the docking site through Ypt1p membrane binding. Hydrolysis of GTP leads to Ypt1p-GDP dissociating from the membrane, a process aided by Rab-GDP dissociation inhibitor (GDI, or Sec19p – Garrett *et al.*, 1994) and this is important for recycling of Ypt1p and Rab proteins in general.

Lupashin and Waters (1997) showed that Ypt1p interacts with the t-SNARE Sed5p, releasing it from a complex with the SM protein Sly1p, which allows it to interact with v-SNARES. These and other data suggested that Sec1p family members may associate with t-SNAREs to prevent inappropriate or irreversible interactions. However, more recent examination has suggested that Sly1p may in fact contribute directly to the specificity of interactions of Sed5p, preventing it from making non-physiological SNARE complexes and can be found within SNARE complexes containing Sed5p (Peng and Gallwitz, 2002). The binding site for Sly1p on Sed5p maps to the N-terminus of the protein (Bracher and Weissenhorn, 2002; Dulubova *et al.*, 2002; Yamaguchi *et al.*, 2002). A similar interaction has been proposed both for the interaction of Sly1p with its other t-SNARE partner Ufe1p on the ER membrane (Yamaguchi *et al.*, 2002) and Vps45p (the SM protein required for endosomal fusion) with the t-SNARE Tlg2p (Dulubova *et al.*, 2002). Similar to the Sly1p–Sed5p interaction the Vps45p–Tlg2p interaction is required to promote entry of Tlg2p into SNARE complexes (Bryant and James, 2001). Incorporating this model into the existing data on Sed5p it can be proposed that Ypt1p-GTP activated by TRAPP I passes a signal to Sly1p that then activates Sed5p for interaction with the incoming v-SNAREs.

SNARE complexes accumulate in *sec18-1* mutant cells incubated at the non-permissive temperature (Wilson *et al.*, 1989; Sogaard *et al.*, 1994) and Sec18p functions along with Sec17p (SNAP) to disrupt *cis*-SNARE complexes on the docked vesicles, which is a prerequisite for membrane fusion. Unpaired, activated v-SNAREs interact with the t-SNAREs on the target membrane and this formation of *trans*-SNARE complexes is proposed to be a crucial step in the fusion reaction (Weber *et al.*, 1998). The activated state of SNAREs is unstable and the action of LMA1 (a heterodimer of thioredoxin and Pbi2p), which acts after Sec18p, is thought to be in stabilizing this (Barlowe, 1997; Xu *et al.*, 1997). The final steps towards membrane fusion are still poorly understood though in many cases calcium, calmodulin and protein phosphatase are required (Peters and Mayer, 1998; Peters *et al.*, 1999; Muller *et al.*, 2002).

Intriguingly, though the above model places components of the system in particular places, very few of them are restricted in space within the cell. Indeed the tight localization of the

TRAPP complex to the Golgi is almost unique amongst ER–Golgi transport factors, other components required cycling between either the ER and Golgi membranes (e.g. v- and t-SNAREs). However, these factors are somehow only active on the correct membrane. This was clearly demonstrated through use of an *in vivo* system in which donor and target membranes can be derived from different cells. Cao and Barlowe (2000) were able to show requirement for the Bet1p and Bos1p v-SNAREs solely on the vesicle, whereas the t-SNARE Sed5p and Ypt1p were only required on the target membrane despite all these components being found on both membranes.

Differences between the scheme outlined here for ER–Golgi transport and other transport steps include the function of SM proteins (discussed in Mayer, 2002) and the roles played by effector complexes. In some cases the SM protein is part of the effector complex (Vps33p in the HOPS complex required in vacuolar sorting) and in most cases where an effector complex is known a separate coiled-coil tethering factor like Uso1p has not been described. All effector complexes do, however, have the common function of defining the target membrane and, in particular, specify where vesicles dock on the target membrane. This is most clearly seen in targeting to the cell membrane where the exocyst is localized and targets secretory vesicles to the growing bud tip despite the t-SNARE Sso1/2p being localized throughout the cell membrane (Finger *et al.*, 1998).

Transport through the Golgi apparatus and sorting at the trans-Golgi

The Golgi apparatus in yeast is morphologically less well defined than the mammalian Golgi complex, and stacked cisternae are only occasionally seen. However, there are definite and separable *cis*-, *medial*- and *trans*-Golgi compartments that can be identified on the basis of particular marker proteins and operationally by glycosyl side chain modifications and proteolytic removal of pro-sequences (Franzusoff *et al.*, 1991; Morin-Ganet *et al.*, 2000 and see Lupashin *et al.*, 1996 for purification of *cis*-Golgi). The early Golgi compartment is also defined by the presence of the essential t-SNARE Sed5p (Hardwick and Pelham, 1992; Banfield *et al.*, 1994) which is required for both anterograde and retrograde transport to this compartment. In addition to those in the ER-derived vesicles that fuse with the Sed5p-compartment four additional SNARE molecules interact with Sed5p and are implicated in intra-Golgi transport. These are Sft1p, Vti1p and Gos1p, Ykt6p (Banfield *et al.*, 1995; McNew *et al.*, 1997, 1998; von Mollard *et al.*, 1997; Holthuis *et al.*, 1998b). Vti1p is important for other trafficking events (von Mollard *et al.*, 1997; Holthuis *et al.*, 1998) but Gos1p, Sft1p and Ykt6p are only known to interact with Sed5p indicating that they are specifically required for retrograde traffic to the *cis*-Golgi. Temperature sensitive mutations within Sft1p or Sed5p result in failure to add late, but not early, Golgi modifications to glycoproteins and dispersal of late Golgi markers into vesicles (Banfield *et al.*, 1995; Wooding and Pelham, 1998). An attractive model for Golgi function, that would accommodate retrograde transport only, and which is supported by the above data is the maturation of individual Golgi elements by transport of processing enzymes from more mature to less mature compartments (the 'cisternal maturation' model). In this model, a net forward flux of membrane and secretory cargo is achieved by a greater number of COP II vesicles fusing to form *cis*-Golgi elements than COP I vesicles budding from it to return to the ER (a likely scenario since these only have to carry escaped ER residents and cycling proteins) combined with breakdown of the most mature compartments into post-Golgi vesicles. Resident Golgi proteins are retained by retrograde transport and/or exclusion from post-Golgi vesicles.

The cisternal maturation model has gained considerable support over the last few years and is accepted by most researchers as how most if not all protein transport through the Golgi occurs. A final and necessary aspect of Golgi function is that Golgi resident proteins, all of which are membrane proteins, must be retained and/or retrieved. The *trans*-Golgi contains two t-SNAREs, Tlg1p and Tlg2p, also found on endosomal compartments and which are required for correct localization of *trans*-Golgi proteins such as Kex2p (Abeliovich *et al.*, 1998; Holthuis *et al.*, 1998a). Analysis of mutants that failed to retain late-Golgi membrane proteins (Nothwehr *et al.*, 1996; Redding *et al.*, 1996) has led to the view that two process contribute to their maintenance in the Golgi – slow delivery to endosomal compartments and retrieval via the same route as sorting receptors for soluble vacuolar proteases such as CPY (see later) (Nothwehr *et al.*, 2000; Ha *et al.*, 2001). Retrieval is dependent on aromatic amino acid-based sorting signals in the cytosolic domains of these proteins (Nothwehr *et al.*, 1993; Cooper and Stevens, 1996; Bryant and Stevens, 1997) and, at least in some cases, is through their binding to the Vps35p component of the 'retromer' coat that forms on the endosome–Golgi vesicles (Nothwehr *et al.*, 2000 and see later).

The *trans*-Golgi is the major post-ER sorting point in the secretory pathway and at least four different anterograde transport routes from the *trans*-Golgi lead to endosomal and vacuolar compartments and the plasma membrane (Figure 7.6). Not surprisingly, a whole slew of factors and families of proteins are required for post-Golgi trafficking. An important feature of the membranes of the Golgi and endosomal system that has been appreciated in the last few years is that they are defined by the differently phosphorylated forms of phosphatidyl inositol (PI) that they contain. Thus the only PI3-kinase in yeast, Vps34p, crucial for sorting to the vacuole via endosomes (Stack *et al.*, 1995a) 'marks' endosomal membranes with PI3-phosophate (Gillooly *et al.*, 2000). Similarly PI4-phosphate provides a marker for Golgi membranes and is required for Golgi function. Thus the Pik1p PI4-kinase is required for sorting from the Golgi to the plasma membrane (Audhya *et al.*, 2000; Schorr *et al.*, 2001; Levine and Munro, 2002). Last, the PI5-kinase Fab1p which produces PI3,5-phosphate from PI3 is important for the specific delivery of at least some proteins to the lumen of the vacuole (Odorizzi *et al.*, 1998). Several protein domains bind to phosphorylated PI and these are found in numerous proteins involved in protein trafficking. 'Sorting nexins', many of which are involved in retrieval of proteins from endosomal compartments, contain the PI3-binding PX domain (Nothwehr and Hindes, 1997; Voos and Stevens, 1998; Sato *et al.*, 2001b; Seaman and Williams, 2002). Five yeast proteins contain the FYVE domain which inserts into membranes in the presence of PI3 (Misra and Hurley, 1999; Stahelin *et al.*, 2002). Of these proteins, which include Fab1p, at least four are required for vacuolar sorting (reviewed in Stenmark *et al.*, 2002).

Two transport routes to the vacuole

There are three distinguishable post-Golgi compartments in yeast. These are the early endosome or post-Golgi endosomes (PGE) into which much endocytosed cargo is first deposited and through which Golgi membrane proteins traffic before recycling, late endosomes or prevacuolar compartments (PVC) and the degradative vacuole itself. Two signal-dependent routes have been identified by which proteins traffic from the Golgi to the vacuole (Figure 7.6). One appears to be direct from the Golgi, while the other two-step pathway is via the PVC (Vida *et al.*, 1993; Piper *et al.*, 1995). There is no evidence for proteins normally trafficking to the yeast vacuole via the cell surface as is the case for proteins such as acid phosphatase

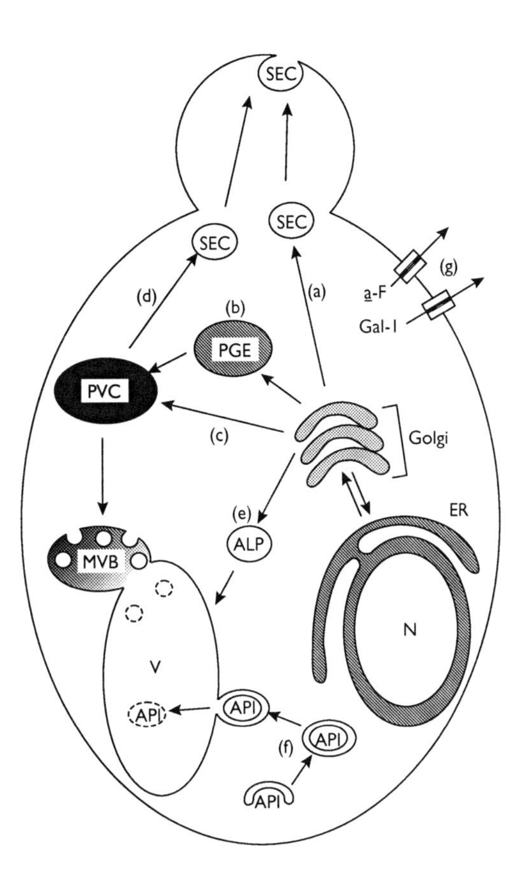

Figure 7.6 Post-Golgi transport routes and non-classical transport pathways. A number of pathways are indicated along with, in some cases, specific substrates (see text for details). All vesicle mediated events are thought to require the SNARE machinery, and all vesicles are likely to be produced with a coat that is shed before fusion with the target membrane. The Golgi is indicated as a stacked organelle here for convenience – it is not usually seen in this form in yeast *in vivo*. Recycling events are not indicated, but have been demonstrated for many of the inter-organelle transport events. (a) Secretory cargo (SEC), particularly membrane proteins, is packaged into vesicles at the *trans*-Golgi which are targeted to the growing bud tip. (b) A default, non-signal mediated, route for membrane proteins to the vacuole is via the post-Golgi endosome (PGE) and two signal mediated pathways from the *trans*-Golgi to the vacuole (V) have been dissected. (c) The two-step pathway followed by CPY is via the pre-vacuolar compartment (PVC) and this may also be taken by a separate class of (soluble) secretory cargo (d) that are routed from the PVC to the cell surface. Maturation of the PVC into a multi-vesicular body (MVB) sorts proteins into those destined for the lumen of the vacuole and those that will reside in its limit membrane as the MVB fuses with the vacuole. (e) ALP containing vesicles are targeted directly from the Golgi to the vacuolar membrane. (f) A third pathway to the vacuole, followed by API, is via double-membraned vesicles that form around API in the cytoplasm. This, the cytoplasm to vacuole targeting pathway is modified to form the autophagic degradative pathway under conditions of poor nutrient availability. (g) Direct secretion from the cytoplasm occurs through several routes, two of which are indicated. **a**-factor is secreted through the Ste6p ABC transporter, whereas other non-classical export substrates follow a route identified through use of the Galectin-1 protein and requires the membrane protein Nce2p which may form an export channel.

in mammalian cells (Braun *et al.*, 1989; Gottschalk *et al.*, 1989), although if mis-sorted to the cell surface they can obtain a vacuolar localization through endocytosis.

The two-step transport pathway is taken by soluble hydrolases such as CPY. The sorting signals for CPY and other proteins that follow this route are localized in the propeptide near their N-terminus and a family of transmembrane receptors, Vps10p/Vth1p/Vth2p with overlapping substrate specificities recognize these signals (Marcusson *et al.*, 1994; Cooper and Stevens, 1996; Westphal *et al.*, 1996). This contrasts with sorting based on recognition of mannose-6-phosphate on vacuolar hydrolases in the mammalian *trans*-Golgi. Vps10p is a *bona fide* receptor for CPY since its lumenal domain binds CPY in a 1:1 ratio (Cooper and Stevens, 1996). A twist to this story has emerged recently in that yeast does have a protein, Mrl1p, distantly related to mannose-6-phosphate receptors. This protein is localized to the Golgi and endosomal compartments and though it has no effect on the sorting of CPY, in its absence (and particularly in combination with a *vps10* deletion) sorting of other abundant proteinases that follow the 'CPY' pathway is drastically reduced (Whyte and Munro, 2001). It remains to be seen whether Mrl1p recognizes glycan- or peptide-based structures or whether it is indeed a *bona fide* transport receptor but, as pointed out by Whyte and Munro (2001) the existence of Mrl1p points to a more ancient origin of mannose-6-phosphate based sorting than previously thought.

In addition to sorting receptors, other factors have been identified that are important for delivery of CPY and other soluble hydrolases to the PGE. In particular, vesicle budding from the Golgi requires the dynamin Vps1p, serine/threonine kinase Vps1p and PI3-kinase Vps34p. Roles of these proteins are extensively discussed in (Stack *et al.*, 1995b). A temperature sensitive allele of the clathrin heavy chain gene (*chc1ts*) secretes CPY at the non-permissive temperature (Seeger and Payne, 1992). However, this defect is corrected after several hours indicating that an adaptive response compensates for loss of clathrin based sorting. Immuno-isolation of clathrin-coated vesicles from yeast lysates resulted in the co-isolation of Vps10p significantly strengthening the argument for CPY sorting in clathrin-coated vesicles (Deloche *et al.*, 2001). While AP-1, the only clathrin-binding adapter protein complex in yeast is not important in Vps10p-dependent sorting of CPY, the clathrin-interacting GGA proteins Gga1/2p are crucial for this (Dell'Angelica *et al.*, 2000; Hirst *et al.*, 2000; Costaguta *et al.*, 2001). Thus these are likely to be the packaging/recruitment factors that ensure that these cargo (and others (Black and Pelham, 2000)) are incorporated into vesicles destined for the PGE.

After carrying CPY to the PVC, Vps10p is recycled to the *trans*-Golgi in a manner dependent on a tyrosine-based signal in its cytosolic domain (Cereghino *et al.*, 1995; Cooper and Stevens, 1996). Recycling also depends on the products of several other *VPS* genes and a complex termed the 'retromer', comprising Vps5p, Vps17p, Vps26p, Vps29p and Vps35p, was identified that forms a coat on the retrograde vesicles from the PGE to the Golgi (Seaman *et al.*, 1997, 1998). Vps35p provides binding sites in the retromer, for cargo such as Vps10p and Golgi enzymes that are retrieved through this route (Nothwehr *et al.*, 1999, 2000). Recent studies indicate that the Vps26p component of the retromer may aid or regulate the membrane association and cargo binding activities of Vps35p (Reddy and Seaman, 2001) providing a potential mechanism whereby Vps35p may be cycled on and off cargo proteins in cycles of vesicles formation and uncoating.

The second step along the CPY pathway is from the PVC to the vacuole. Along with recycling of proteins to the Golgi, a second and fundamental sorting step takes place at the PVC which determines whether or not a protein ends up on the limit membrane of the vacuole or is internalized into it. This is mediated by the generation of vesicles inside the PVC by internalization (hence its other name, the multivesicular body or MVB). A variety of factors required for this are defined by the class E *vps* mutants that accumulate aberrant multilamellar late endosomes (Raymond *et al.*, 1992; Piper *et al.*, 1995; Babst *et al.*, 1997). Most class E mutants are specific to anterograde transport to the vacuole but cells lacking Vps27p function are defective in both anterograde and retrograde transport from the late endosome (Piper *et al.*, 1995).

A crucial step in sorting of many, though not all, proteins into internal membranes or the PVC and hence to the lumen of the vacuole is the addition of single ubiquitin molecules (Katzmann *et al.*, 2001; Reggiori and Pelham, 2001; Urbanowski and Piper, 2001). If ubiquitination is prevented then these proteins, such as Cps1p and Phm5p, relocalize to the outer membrane of the vacuole. That ubiquitin is indeed a sorting signal is clearly demonstrated by the fact that addition of the ubiquitin coding sequence (or sites for the addition of ubiquitin) to the open reading frames of proteins, such as DPAP-A and Vps10p that normally localize to the Golgi or cycle between the Golgi and PVC, results in their redistribution to the vacuolar lumen (Katzmann *et al.*, 2001; Urbanowski and Piper, 2001). Several questions immediately come to mind – What is/are the ubiquitin ligases responsible for this, and how does it recognize its target proteins? How is ubiquitin itself recognized as a targeting signal? Is ubiquitin recycled as is seen with proteasomal degradation?

The answer to the last of these questions is yes – in the absence of the de-ubiquitinating enzyme Doa4p, which itself partially localizes to endosomes (Amerik *et al.*, 2000), stably ubiquitinated substrates accumulate (Katzmann *et al.*, 2001). These proteins are still internalized and so removal of ubiquitin is not a prerequisite for completion of this targeting pathway. So far two ubiquitin ligases have been implicated in addition of the single ubiquitin moieties. These are the RING-domain multispanning membrane protein Tul1p and Rps5p (Helliwell *et al.*, 2001; Reggiori and Pelham, 2002). Of these Tul1p, which is localized to the Golgi and functions in concert with the ubiquitin conjugating enzyme Ubc4p, recognizes proteins, at least in part, through their transmembrane domains (Reggiori and Pelham, 2002). In the absence of Tul1p proteins such as Cps1p are not ubiquitinated and fail to be sorted. Interestingly, insertion of single polar residues into transmembrane domains of reporter proteins is sufficient to induce Tul1p-dependent ubiquitination and internalization in the PVC suggesting that Tul1p may also form another layer of quality control beyond the ER that targets aberrant membrane proteins for vacuolar degradation (Reggiori *et al.*, 2000; Reggiori and Pelham, 2002). Rsp5p is required for the regulated destruction of the amino acid permease Gap1p and in this case polyubiquitin, rather than monoubiquitin is required for vacuolar targeting (Helliwell *et al.*, 2001). Intriguingly, regulation of cell surface expression of another permease Tat2p is also at the level of sorting to the vacuole from the Golgi (Beck *et al.*, 1999). Gap1p and Tat2p are expressed at the cell surface under diametrically opposite growth conditions (poor and good nitrogen sources, see Chapter 4). Thus, the sorting apparatus in the Golgi may be able to receive input from several different signalling pathways and respond by appropriate ubiquitination of various substrate proteins. This would be a very rapid way of adapting to different growth conditions.

Recognition of ubiquitin in Golgi–vacuole sorting has been attributed to two factors. Vps23p, Vps28p and Vps37p form a large soluble complex termed the ESCRT-I (endosomal sorting complex required for transport I) that is required for internalization of proteins in the PVC, localizes to endosomes and can bind ubiquitinated cargo proteins in cell extracts (Katzmann *et al.*, 2001). Consistent with a role in ubiquitin recognition Vps23p contains a critical 'ubiquitin conjugating like' domain that lacks only one critical active site cysteine residue when compared to genuine ubiquitin conjugating proteins (Babst *et al.*, 2000; Katzmann *et al.*, 2001). The second factor that recognizes ubiquitin is a complex of Vps27p and Hse1p (Bilodeau *et al.*, 2002; Shih *et al.*, 2002). Vps27p and Hse1p both contains the recently identified ubiquitin interaction motif (UIM) and in the absence of these internalization of ubiquitinated cargo in the PVC is abrogated (Bilodeau *et al.*, 2002). Intriguingly, Vps27p is required for other events (recycling of Golgi proteins and the generation of the luminal membranes in PVCs themselves) and these are unaffected when the UIMs of Vps27p are removed confirming the multifunctional nature of this protein.

In addition to ESCRT-I, two further multi-subunit complexes termed ESCRT-II and -III function along with the AAA ATPase Vps4p to internalize proteins in the PVC (Babst *et al.*, 2002a,b). The three ESCRT complexes are proposed to act sequentially (I, II then III) to select and concentrate cargoes destined for internalization in the MVB pathway. Loss of Vps4p activity results in accumulation of ESCRT-III components on the membrane (Babst *et al.*, 1998) suggesting a role for this protein in disassembly of complexes that assemble on the PVC membrane during cargo sorting. Another activity important for internalization at the PVC is, as mentioned earlier, Fab1p and its activity in generating IP3,5-phosphate. In the absence of Fab1p, vacuoles are excessively large, leading to the suggestion that the defect is in membrane invagination (Odorizzi *et al.*, 1998). However, cargo that do not require ubiquitination, and stably ubiquitinated cargo are still targeted to the lumen in *fab1Δ* cells suggesting that the defect may be in inappropriate removal of ubiquitin from substrates prior to packaging into internalizing membranes (Reggiori and Pelham, 2001; Urbanowski and Piper, 2001). Mechanistically it can be speculated that in the absence of Fab1p the failure to convert PI3-phosphate to PI3,5-phosphate allows incorrect binding, regulation or segregation of deubiquitinating and/or packaging activities at the PVC membrane leading to the observed sorting deficiency.

Once the PVC has matured through the sorting of outer membrane from internalized cargo and recycling of factors out of the compartment, it fuses with the vacuole to deliver its contents to either the limit membrane or the interior of the organelle. In the absence of vacuolar hydrolase activity, accomplished by either increasing the pH of the organelle by disruption of the vacuolar ATPase (*vma4Δ*) or removal of the processing protease that activates luminal hydrolases (*pep4Δ*) vesicles can be seen in the lumen of the vacuole (Wurmser and Emr, 1998). Thus the final step of the PVC/MVB sorting pathway is the destruction of these vesicles by the degradative machinery of the vacuole.

The direct Golgi to vacuole transport route was identified through study of the membrane protein alkaline phosphatase (ALP). This route was initially identified through the observation that sorting of ALP is normal in many of the Golgi to vacuole transport mutants that affect CPY transport. The identification of mutations affecting ALP trafficking led to the realization that an adaptor protein complex (AP-3) homologous to those that direct assembly of clathrin-coated vesicles from the *trans*-Golgi and cytoplasmic membranes is required for this process (Cowles *et al.*, 1997a). However, AP-3 does not co-purify with clathrin (Vowels and Payne, 1998; Rehling *et al.*, 1999; Yeung *et al.*, 1999) and has thus been assumed to direct assembly of a different coat. Another factor required for the ALP sorting pathway is Vps41p (Radisky *et al.*, 1997; Rehling *et al.*, 1999). Deletion of *VPS41* also leads to defects in other sorting pathways to the vacuole, but this is a secondary defect as Vam3p, the t-SNARE which is required for all fusion events at the vacuole, achieves its vacuolar localization through the ALP pathway (Piper *et al.*, 1997) and defects in this route are thus predicted to result in eventual loss of all vacuolar targeting. Vps41p contains two functional domains (Darsow *et al.*, 2001). The N-terminus of the protein contains an AP-3 binding site, whereas the C-terminus contains a domain homlogous to the clathrin heavy chain repeat required for homo-oligomerization of clathrin heavy chain molecules. Vps41p itself homo-oligomerizes and this suggests that Vps41p may itself form all or part of the coat on the vesicles that transport ALP and other cargo that follow this route. The sorting information in ALP and Vam3p resides in their cytoplasmic tails and comprise a di-leucine motif that is bound by the AP-3 complex (Cowles *et al.*, 1997b; Darsow *et al.*, 1998; Vowels and Payne, 1998).

Transport to the cell surface

Mutants that impose a post-Golgi block on secretion (e.g. *sec4, sec6* and *sec8*) accumulate vesicles that can be fractionated by density into two populations (Harsay and Bretscher, 1995; David *et al.*, 1998). Coats have not been identified on these vesicles, but their different densities indicate that their composition is not the same. The two vesicle types carry different cargos; soluble invertase and acid phosphatase are found in higher density vesicles, whereas lower density vesicles contain membrane and cell wall proteins such as Pma1p and Bgl2p. The initial supposition that both vesicle types are derived directly from the Golgi turns out to be incorrect. Mutations that block the Golgi–endosome trafficking pathway followed by CPY and dependent on clathrin (Chs1p), dynamin (Vps1p) and the PVC-localized t-SNARE Pep12p result in loss of the higher density vesicles and re-routing of their cargo to the lower density vesicles (Gurunathan *et al.*, 2002; Harsay and Schekman, 2002). Sorting was not dependent on the CPY receptor indicating other requirements for sorting these cargo at the Golgi (Harsay and Schekman, 2002). Differences in the pathways taken by the invertase and Pma1p containing vesicles are also revealed by mutations that disrupt the actin cytoskeleton. Thus *act1–1* and *sla2/end4* mutants accumulate dense vesicles and not the Pma1p containing lighter vesicles (Harsay and Bretscher, 1995; Mulholland *et al.*, 1997). Interestingly the two vesicle classes have different Rab proteins associated with them – Ypt1p on dense vesicles, Sec4p on light vesicles (Mulholland *et al.*, 1997). These differences have not been explained at the level of the function of the vesicles, and why yeast routes a significant proportion of its (soluble) secretory cargo through the endosomal system is something of a mystery. Harsay and Schekman (2002) speculate that this may allow the cell the possibility of separately regulating requirements for cell surface expansion (membrane proteins) and secretion of soluble proteins.

7.4 Specialized transport routes

Although the major route that proteins follow is the secretory pathway, yeast possesses several other important 'non-classical' protein trafficking pathways. Here, I discuss both biogenic and degradative transport pathways from cytoplasm to vacuole and two pathways by which proteins are secreted direct from the cytoplasm.

Cytoplasm to vacuole protein targeting and autophagy

Analysis of the trafficking of vacuolar aminopeptidase I (API) has revealed a specialized and constitutive cytoplasm to vacuole targeting (CVT) pathway (Segui-Real *et al.*, 1995). This takes place through formation of double-membrane bound vesicles in the cytoplasm that are then targeted to the vacuole (Scott *et al.*, 1997). Comparison of mutants defective in the delivery of API to the vacuole (*cvt* mutants) with those defective in a mechanism for delivery of cytoplasmic proteins and organelles to the vacuole, the inducible autophagy pathway (*apg*, and *aut*) (Baba *et al.*, 1994), led to the conclusion that these processes use much of the same machinery (Harding *et al.*, 1996; Scott *et al.*, 1996; reviewed in Khalfan and Klionsky, 2002).

Autophagy may be considered as an inducible and non-selective version of the CVT pathway. A key player in the switch between the two pathways is Apg13p. Autophagy is regulated by the Tor serine/threonine kinase, which operates as a nutrient sensing switch (Noda and Ohsumi, 1998). Under high nutrient conditions Tor kinase is active, and Apg13p is phosphorylated. In low nutrient conditions Tor-kinase is inactive and Apg13p becomes dephosphorylated. This leads to it being able to form a complex with the serine/threonine kinase Apg1p. In this complex Apg1p displays high kinase activity and this is suggested to drive the initiation of autophagy at the expense of the CVT pathway (Kamada *et al.*, 2000; Scott *et al.*, 2000). There are several important differences between the vesicles formed in the CVT and autophagy pathways. Thus CVT vesicles are small and form specifically around CVT cargo, whereas autophagic vesicles are larger and non-selective engulfing apparently random cytoplasmic contents in addition to CVT substrates. A further difference may be the source of the membranes used to make the vesicles. Autophagy requires Sec23/24p and Sec12p, both important in COP II formation (Ishihara *et al.*, 2001), whereas CVT requires the cognate t-SNARE/SM pair Tlg2p and Vps45p (Abeliovich *et al.*, 1999). Thus the two pathways do not differ simply at the level of packaging. The outer membrane of the vesicles formed in both pathways fuses with the vacuole membrane in a reaction that requires the vacuolar t-SNARE Vam3p and other components of the general and vacuolar fusion machinery (Darsow *et al.*, 1997; Ishihara *et al.*, 2001) releasing subvacuolar vesicles that are then degraded.

Only one integral membrane protein has been identified as necessary for CVT and autophagy. This is Apg9p, which is found dispersed in multiple punctate structures throughout the cytosol that do not correspond to other membranes or organelles (Lang *et al.*, 2000; Noda *et al.*, 2000). While Apg9p does not accumulate in autophagosomes and its function in generating CVT and autophagic vesicles is currently mysterious it is required for the accumulation of other factors required for CVT vesicle/autophagosome formation to a prevacuolar compartment (Suzuki *et al.*, 2001). Another factor required for CVT and autophagy is the PI3-kinase Vps34p (Kihara *et al.*, 2001). The function of Vps34p in these processes is distinct from its function in vacuolar protein sorting and here it is found complexed with the CVT/autophagy-specific factor Apg14p. As Apg14p localizes to the prevacuolar compartment it may be that the activity of Vps34p is harnessed here to mark the membrane of this compartment with PI3-phosphate much in the same way as it does endosomal membranes. In support of this, several proteins required for CVT are PI3-phosphate-binding proteins. Cvt13p and Cvt20p contain PX domains and are required for only the CVT pathway (not autophagy) and associate with Apg17p, a component of the Apg1p-kinase complex (Nice *et al.*, 2002). Mutation of the PX domains of Cvt13p and Cvt20p disrupts CVT and in a *vps34Δ* strain these proteins, normally found in the prevacuolar compartment, redistribute to the cytoplasm (Nice *et al.*, 2002). A further PI3-phosphate binding protein that is required specifically for Cvt is Eft1p (Wurmser and Emr, 2002). Eft1p, Cvt13p and Cvt20p are three of the small class of proteins required only for CVT and perhaps there are other, as yet unidentified, PI3-phosphate binding proteins required for autophagy.

Targeting information in the prepropeptide of API is sufficient to direct sorting of proteins into CVT vesicles (Martinez *et al.*, 1997) and a specific receptor, Cvt19p, has been identified that recruits pre-API to both CVT vesicles and autophagosomes (Scott *et al.*, 2001). Cvt19p localizes to the prevacuolar compartment (Kim *et al.*, 2002) and on the basis of this localization it can be speculated that it functions along with other factors to organize the early events in CVT vesicle formation.

Another vesicle-mediated targeting pathway to the vacuole is regulated destruction of the cytosolic gluconeogenic enzyme fructose-1,6-bisphosphatase (FBPase). This enzyme is targeted to the vacuole when cells are supplemented with fresh glucose (Chiang and Schekman, 1991). The isolation of mutants defective in this process (*vid*) enabled (Huang and Chiang, 1997) to identify sequestered FBPase in vesicles that are an intermediate in its targeting to vacuoles. The *vid* mutants do not have defects in other vacuolar targeting events and so represent a novel, regulated route to the vacuole. Morphologically 'Vid' vesicles are smaller (30–40 nm) than those generated by the cytoplasm to vacuole pathway (100 nm). Cloning of several of the *VID* genes has revealed Vid24p as a peripheral membrane protein that is a component of the Vid vesicles. In the absence of Vid24p FBPase accumulates in Vid vesicles and is not imported into the vacuole suggesting a role for Vid24p in targeting (Chiang and Chiang, 1998). Vid22p is a glycosylated integral membrane protein localized to the plasma membrane. Its targeting to the membrane is independent of the secretory pathway and in the absence of this protein FBPase remains in the cytosol and susceptible to proteinase K following glucose shift, indicating that it does not enter vesicles (Brown *et al.*, 2002).

Import of FBPase into Vid vesicles also requires the cytosolic proteins cyclophilin A (Cpr1p), the ubiquitin conjugating enzyme Ubc1p and Ssa2p (but not other cytosolic Ssa (Hsp70p) proteins) (Brown *et al.*, 2000, 2001; Shieh *et al.*, 2001). How all these factors fit together is currently not known, but the involvement of ubiquitin conjugation confirms an active sorting and targeting mechanism, and the involvement of Hsp70p may account for the ATPase requirement observed for *in vitro* packaging of FBPase into Vid vesicles.

Protein export from the cytoplasm

The secretion of the **a**-factor mating pheromone is one of the better-understood non-classical transport pathways that is likely representative of a class of export routes. **a**-factor is synthesized as a precursor that undergoes extensive modification and proteolysis (Chen *et al.*, 1997). It is exported from the cell as a farnesylated, carboxymethylated peptide by the Ste6p ABC (ATP-binding cassette)-transporter (Kuchler *et al.*, 1989; McGrath and Varshavsky, 1989). A second peptide has been described also derived from the **a**-factor precursor (Chen *et al.*, 1997). This AFRP (**a**-factor related peptide) relies on the same processing enzymes as **a**-factor up to the late proteolytic steps, but is exported by a different, and as yet unidentified, transporter. No physiological role for AFRP has been identified and it is essentially inert in assays for mating responses and mating inhibition. Chen *et al.* (1997) speculated that, since no cellular mechanism is known in yeast for the metabolic turnover of prenyl groups, the proteases that process AFRP (which appear to be sequence non-specific) could, in conjunction with the AFRP transporter, be a way for cells to eliminate prenylated peptides that might otherwise accumulate and become toxic. However, why yeast should divert so much **a**-factor precursor into the AFRP pathway (AFRP is present at approximately the same level as **a**-factor) is a mystery. Perhaps the **a**-factor precursor is scavenged by the prenyl-group turnover machinery in a non-specific way. If AFRP does identify a general metabolic mechanism then the proteolytic and export machinery should be present in α-cells, which could be tested relatively easily. Twenty-nine potential ABC-transporter proteins are identified in the yeast protein database, of which eight are likely to be plasma membrane proteins. Several of these have unknown functions and could therefore be the AFRP exporter.

A second non-classical export pathway was uncovered by Cleves *et al.* (1996). Yeast exports a number of proteins independently of the secretory pathway (i.e. independent of Sec18p function). To analyse this, and to see if similar export machinery was used in yeast and higher eukaryotes where such pathways are also known Cleves *et al.* (1996) expressed a characterized mammalian non-classical export substrate (Galectin-1) in yeast. Galectin-1 was expressed and moreover exported. Using the fortunate observation that expression of an epitope-tagged Galectin-1 was detrimental to the growth of *ste6Δ* cells (although there was no evidence for Ste6p being the exporter) Cleves *et al.* (1996) were able, by screening for suppressors of this phenotype, to identify a potential multi-spanning membrane protein designated Nce102p (non-classical export) as responsible for the export of Galectin-1. Deletion of *NCE102* shows no adverse effects on cell growth, nor does it alter the spectrum of non-classically secreted proteins: however, export of Galectin-1 is reduced by about two/thirds in these cells compared to wild type (Cleves *et al.*, 1996). Thus there are probably redundant exporters for non-classical substrates in yeast, a supposition supported by the presence of a close homologue of *NCE102* in the yeast genome (YGR131w). Since Galectin-1 expression is detrimental to the growth of *ste6Δ* cells there may be some interplay between different export pathways: possibly the exporters are found together in the cell membrane or Nce102p exports a range of peptides including some also recognized by Ste6p and that are toxic. High level expression of Galectin-1 would lead to accumulation of such peptides in the cytoplasm, a situation aggravated by lack of Ste6p.

7.5 Summary and concluding remarks

A recurring theme in study of protein targeting and trafficking in yeast and other organisms is that of functional redundancy. This provides, at one level, a fail-safe mechanism, such that cells can adapt and survive if one pathway is debilitated. However, multiple pathways could also play an active role in organellar biogenesis and regulated transport allowing large demand on one pathway (through upregulation of, say, a set of secretory proteins) to not interfere with overall constitutive events. A clear example of a targeting route being used in this way is the biogenesis of peroxisomal membrane proteins, separate from other peroxisomal protein import pathways. Changing growth conditions may require rapid proliferation of peroxisomes. The induction of peroxisomal membrane proteins and their insertion into membranes can be accommodated without taxing the pre-existing matrix protein import machinery. Since the import machinery for matrix proteins consists mainly of membrane proteins, increase in membrane protein insertion also increases the import capacity of the peroxisomes.

The secretory pathway in and the mechanisms used by yeast for protein trafficking in general are very similar to those used by higher eukaryotes. For example, mammalian homologues of Sec62p, Sec63p and Sec72p are known and these interact with the mammalian SEC61 translocon (Meyer *et al.*, 2000). Thus the post-translational ER-targeting pathway may be of general importance. However, the marked contrast between protein targeting to the ER in yeast and higher eukaryotic systems where most proteins use SRP is a factor of concern for researchers expressing proteins heterologously in yeast. Co-translational targeting may be extremely important for many such proteins while their signal sequences may not be recognized by yeast SRP.

Several topics have not been discussed here. In particular, I have not covered the rapidly growing and large field of endocytosis. For a discussion of this readers are referred to Munn (2001) and Shaw *et al.* (2001). Other issues are poorly understood. One example is membrane insertion of tail-anchored membrane proteins (a class of proteins that includes SNAREs). *In vitro* studies using mammalian microsomes indicated that at least some of these proteins require a proteinaceous insertion machinery non-identical to that used by other proteins (Kutay *et al.*, 1995), and this has been confirmed in yeast (Steel *et al.*, 2002). A second example is sub-compartmentalization of organelles into specialized regions. In addition to the sub-compartments in the Golgi, there is evidence that the yeast ER may have functionally distinct regions. This has been suggested both by fractionation of ER membranes and by immuno-electron microscopy (Sanderson and Meyer, 1991; Bachem and Mendgen, 1995). Thus, while not readily seen at the light microscopy level, and as in other organisms including closely related yeast species (Bevis *et al.*, 2002) there may be parts of the ER more dedicated to or even specialized for protein import, folding and assembly, or vesicle formation.

This chapter has been modified extensively from what was written for the first edition of this book. A number of models have either been thrown out completely or drastically changed in the light of new data and whole areas of trafficking – particularly peroxisomal import and vacuolar targeting have blossomed in remarkable ways. The involvement of ubiquitin in extracting proteins from the ER, sorting at the Golgi and into the PVC highlights the fact that sorting signals in individual proteins are not always enough and that generic 'tags' have to be used in a number of settings to ensure correct sorting. The field of protein trafficking in yeast remains an exciting and active area of research. I expect that a number of the models presented here will be modified again in the (perhaps very) near future. I encourage the reader to explore further, and particularly by searching for new studies that will update and expand on the models presented here.

Acknowledgements

Many thanks to Christopher Blackwell and particularly Nia Bryant for comments on the manuscript and suggesting areas that needed improvement and clarification. Many primary references have not been cited here in an effort to limit the number of citations, and I apologize to anyone who feels slighted for their work not having been acknowledged. Work in the author's laboratory is supported by the Medical Research Council.

References

Abe, Y., Shodai, T., Muto, T., Mihara, K., Torii, H., Nishikawa, S., Endo, T. and Kohda, D. (2000) Structural basis of presequence recognition by the mitochondrial protein import receptor Tom20. *Cell* **100**, 551–560.

Abeliovich, H., Darsow, T. and Emr, S. D. (1999) Cytoplasm to vacuole trafficking of aminopeptidase I requires a t-SNARE-Sec1p complex composed of Tlg2p and Vps45p. *EMBO Journal* **18**, 6005–60016.

Abeliovich, H., Grote, E., Novick, P. and Ferro-Novick, S. (1998) Tlg2p, a yeast syntaxin homolog that resides on the Golgi and endocytic structures. *Journal of Biological Chemistry* **273**, 11719–11727.

Adam, A., Endres, M., Sirrenberg, C., Lottspeich, F., Neupert, W. and Brunner, M. (1999) Tim9, a new component of the TIM22.54 translocase in mitochondria. *EMBO Journal* **18**, 313–319.

Ades, I. Z. and Butow, R. A. (1980) The products of mitochondria-bound cytoplasmic polysomes in yeast. *Journal of Biological Chemistry* **255**, 9918–9924.

Albertini, M., Rehling, P., Erdmann, R., Girzalsky, W., Kiel, J. A., Veenhuis, M. and Kunau, W. H. (1997) Pex14p, a peroxisomal membrane protein binding both receptors of the two PTS-dependent import pathways. *Cell* **89**, 83–92.

Albertini, M., Girzalsky, W., Veenhuis, M. and Kunau, W. H. (2001) Pex12p of *Saccharomyces cerevisiae* is a component of a multi-protein complex essential for peroxisomal matrix protein import. *European Journal of Cell Biology* **80**, 257–270.

Alconada, A., Kubrich, M., Moczko, M., Honlinger, A. and Pfanner, N. (1995) The mitochondrial receptor complex: the small subunit Mom8b/Isp6 supports association of receptors with the general insertion pore and transfer of preproteins. *Molecular and Cellular Biology* **15**, 6196–6205.

Amerik, A. Y., Nowak, J., Swaminathan, S. and Hochstrasser, M. (2000) The Doa4 deubiquitinating enzyme is functionally linked to the vacuolar protein-sorting and endocytic pathways. *Molecular Biology of the Cell* **11**, 3365–3380.

Antonny, B., Madden, D., Hamamoto, S., Orci, L. and Schekman, R. (2001) Dynamics of the COPII coat with GTP and stable analogues. *Nature Cell Biology* **3**, 531–537.

Aoe, T., Cukierman, E., Lee, A., Cassel, D., Peters, P. J. and Hsu, V. W. (1997) The KDEL receptor, ERD2, regulates intracellular traffic by recruiting a GTPase-activating protein for ARF1. *EMBO Journal* **16**, 7305–7316.

Arnold, C. E. and Wittrup, K. D. (1994) The stress response to loss of signal recognition particle function in *Saccharomyces cerevisiae*. *Journal of Biological Chemistry* **269**, 30412–30418.

Arnold, I., Bauer, M. F., Brunner, M., Neupert, W. and Stuart, R. A. (1997) Yeast mitochondrial F1F0-ATPase: the novel subunit e is identical to Tim11. *FEBS Letters* **411**, 195–200.

Audhya, A., Foti, M. and Emr, S. D. (2000) Distinct roles for the yeast phosphatidylinositol 4-kinases, Stt4p and Pik1p, in secretion, cell growth, and organelle membrane dynamics. *Molecular Biology of the Cell* **11**, 2673–2689.

Ayscough, K. R., Stryker, J., Pokala, N., Sanders, M., Crews, P. and Drubin, D. G. (1997) High rates of actin filament turnover in budding yeast and roles for actin in establishment and maintenance of cell polarity revealed using the actin inhibitor latrunculin-A. *Journal of Cell Biology* **137**, 399–416.

Baba, M., Takeshige, K., Baba, N. and Ohsumi, Y. (1994) Ultrastructural analysis of the autophagic process in yeast: detection of autophagosomes and their characterization. *Journal of Cell Biology* **124**, 903–913.

Babst, M., Sato, T. K., Banta, L. M. and Emr, S. D. (1997) Endosomal transport function in yeast requires a novel AAA-type ATPase, Vps4p. *EMBO Journal* **16**, 1820–1831.

Babst, M., Wendland, B., Estepa, E. J. and Emr, S. D. (1998) The Vps4p AAA ATPase regulates membrane association of a Vps protein complex required for normal endosome function. *EMBO Journal* **17**, 2982–2993.

Babst, M., Odorizzi, G., Estepa, E. J. and Emr, S. D. (2000) Mammalian tumor susceptibility gene 101 (TSG101) and the yeast homologue, Vps23p, both function in late endosomal trafficking. *Traffic* **1**, 248–258.

Babst, M., Katzmann, D. J., Estepa-Sabal, E. J., Meerloo, T. and Emr, S. D. (2002a) Escrt-III: an endosome-associated heterooligomeric protein complex required for mvb sorting. *Developmental Cell* **3**, 271–282.

Babst, M., Katzmann, D. J., Snyder, W. B., Wendland, B. and Emr, S. D. (2002b) Endosome-associated complex, ESCRT-II, recruits transport machinery for protein sorting at the multivesicular body. *Developmental Cell* **3**, 283–289.

Bachem, U. and Mendgen, K. (1995) Endoplasmic reticulum subcompartments in a plant parasitic fungus and in baker's yeast: differential distribution of lumenal proteins. *Experimental Mycology* **19**, 137–152.

Baerends, R. J., Rasmussen, S. W., Hilbrands, R. E., van der Heide, M., Faber, K. N., Reuvekamp, P. T., Kiel, J. A., Cregg, J. M., van der Klei, I. J. and Veenhuis, M. (1996) The *Hansenula polymorpha PER9* gene encodes a peroxisomal membrane protein essential for peroxisome assembly and integrity. *Journal of Biological Chemistry* **271**, 8887–8894.

Banfield, D. K., Lewis, M. J. and Pelham, H. R. (1995) A SNARE-like protein required for traffic through the Golgi complex. *Nature* **375**, 806–809.

Banfield, D. K., Lewis, M. J., Rabouille, C., Warren, G. and Pelham, H. R. B. (1994) Localization of Sed5, a putative vesicle targeting molecule, to the cis-Golgi network involves both its transmembrane and cytoplasmic domains. *Journal of Cell Biology* **127**, 357–371.

Barlowe, C. (1997) Coupled ER to Golgi transport reconstituted with purified cytosolic proteins. *Journal of Cell Biology* **139**, 1097–1108.

Barlowe, C. (2002) COPII-dependent transport from the endoplasmic reticulum. *Current Opinion in Cell Biology* **14**, 417–422.

Barlowe, C., Orci, L., Yeung, T., Hosobuchi, M., Hamamoto, S., Salama, N., Rexach, M. F., Ravazzola, M., Amherdt, M. and Schekman, R. (1994) COPII: a membrane coat formed by Sec proteins that drive vesicle budding from the endoplasmic reticulum. *Cell* **77**, 895–907.

Barrowman, J., Sacher, M. and Ferro-Novick, S. (2000) TRAPP stably associates with the Golgi and is required for vesicle docking. *EMBO Journal* **19**, 862–869.

Bauer, M. F., Sirrenberg, C., Neupert, W. and Brunner, M. (1996) Role of Tim23 as voltage sensor and presequence receptor in protein import into mitochondria. *Cell* **87**, 33–41.

Bays, N. W., Gardner, R. G., Seelig, L. P., Joazeiro, C. A. and Hampton, R. Y. (2001a) Hrd1p/Der3p is a membrane-anchored ubiquitin ligase required for ER-associated degradation. *Nature Cell Biology* **3**, 24–29.

Bays, N. W., Wilhovsky, S. K., Goradia, A., Hodgkiss-Harlow, K. and Hampton, R. Y. (2001b) HRD4/NPL4 is required for the proteasomal processing of ubiquitinated ER proteins. *Molecular Biology of the Cell* **12**, 4114–4128.

Beck, T., Schmidt, A. and Hall, M. N. (1999) Starvation induces vacuolar targeting and degradation of the tryptophan permease in yeast. *Journal of Cell Biology* **146**, 1227–1238.

Becker, J., Walter, W., Yan, W. and Craig, E. A. (1996) Functional interaction of cytosolic hsp70 and a DnaJ-related protein, Ydj1p, in protein translocation in vivo. *Molecular and Cellular Biology* **16**, 4378–4386.

Beckmann, R., Spahn, C. M., Eswar, N., Helmers, J., Penczek, P. A., Sali, A., Frank, J. and Blobel, G. (2001) Architecture of the protein-conducting channel associated with the translating 80S ribosome. *Cell* **107**, 361–372.

Bednarek, S. Y., Ravazzola, M., Hosobuchi, M., Amherdt, M., Perrelet, A., Schekman, R., and Orci, L. (1995) COPI-coated and COPII-coated vesicles bud directly from the endoplasmic-reticulum in yeast. *Cell* **83**, 1183–1196.

Belden, W. J. and Barlowe, C. (2001a) Distinct roles for the cytoplasmic tail sequences of Emp24p and Erv25p in transport between the endoplasmic reticulum and Golgi complex. *Journal of Biological Chemistry* **276**, 43040–43048.

Belden, W. J. and Barlowe, C. (2001b) Role of Erv29p in collecting soluble secretory proteins into ER-derived transport vesicles. *Science* **294**, 1528–1531.

Bertolotti, A., Zhang, Y., Hendershot, L. M., Harding, H. P. and Ron, D. (2000) Dynamic interaction of BiP and ER stress transducers in the unfolded-protein response. *Nature Cell Biology* **2**, 326–332.

Bevis, B. J., Hammond, A. T., Reinke, C. A. and Glick, B. S. (2002) *De novo* formation of transitional ER sites and Golgi structures in *Pichia pastoris*. *Nature Cell Biology* **4**, 750–756.

Biederer, T., Volkwein, C. and Sommer, T. (1996) Degradation of subunits of the Sec61p complex, an integral component of the ER membrane, by the ubiquitin-proteasome pathway. *EMBO Journal* **15**, 2069–2076.

Biederer, T., Volkwein, C. and Sommer, T. (1997) Role of Cue1p in ubiquitination and degradation at the ER surface. *Science* **278**, 1806–1809.

Bilodeau, P. S., Urbanowski, J. L., Winistorfer, S. C. and Piper, R. C. (2002) The Vps27p Hse1p complex binds ubiquitin and mediates endosomal protein sorting. *Nature Cell Biology* **4**, 534–539.

Black, M. W. and Pelham, H. R. (2000) A selective transport route from Golgi to late endosomes that requires the yeast GGA proteins. *Journal of Cell Biology* **151**, 587–600.

Boehm, J., Letourneur, F., Ballensiefen, W., Ossipov, D., Demolliere, C. and Schmitt, H. D. (1997) Sec12p requires Rer1p for sorting to coatomer (COPI)-coated vesicles and retrieval to the ER. *Journal of Cell Science* **110**, 991–1003.

Boisrame, A., Beckerich, J. M. and Gaillardin, C. (1996) Sls1p, an endoplasmic-reticulum component, is involved in the protein translocation process in the yeast *Yarrowia lipolytica*. *Journal of Biological Chemistry* **271**, 11668–11675.

Boisrame, A., Kabani, M., Beckerich, J. M., Hartmann, E. and Gaillardin, C. (1998) Interaction of Kar2p and Sls1p is required for efficient co- translational translocation of secreted proteins in the yeast *Yarrowia lipolytica*. *Journal of Biological Chemistry* **273**, 30903–30908.

Bolliger, L., Junne, T., Schatz, G. and Lithgow, T. (1995) Acidic receptor domains on both sides of the outer membrane mediate translocation of precursor proteins into yeast mitochondria. *EMBO Journal* **14**, 6318–6326.

Bordallo, J. and Wolf, D. H. (1999) A RING-H2 finger motif is essential for the function of Der3/Hrd1 in endoplasmic reticulum associated protein degradation in the yeast *Saccharomyces cerevisiae*. *FEBS Letters* **448**, 244–248.

Bordallo, J., Plemper, R. K., Finger, A. and Wolf, D. H. (1998) Der3p/Hrd1p is required for endoplasmic reticulum-associated degradation of misfolded lumenal and integral membrane proteins. *Molecular Biology of the Cell* **9**, 209–222.

Bracher, A. and Weissenhorn, W. (2002) Structural basis for the Golgi membrane recruitment of Sly1p by Sed5p. *EMBO Journal* **21**, 6114–6124.

Braun, M., Waheed, A. and von Figura, K. (1989) Lysosomal acid phosphatase is transported to lysosomes via the cell surface. *EMBO Journal* **8**, 3633–3640.

Braun, S., Matuschewski, K., Rape, M., Thoms, S. and Jentsch, S. (2002) Role of the ubiquitin-selective CDC48(UFD1/NPL4) chaperone (segregase) in ERAD of OLE1 and other substrates. *EMBO Journal* **21**, 615–621.

Brix, J., Rudiger, S., Bukau, B., Schneider-Mergener, J. and Pfanner, N. (1999) Distribution of binding sequences for the mitochondrial import receptors Tom20, Tom22, and Tom70 in a presequence-carrying preprotein and a non-cleavable preprotein. *Journal of Biological Chemistry* **274**, 16522–16530.

Brodsky, J. L., Goeckeler, J. and Schekman, R. (1995) BiP and Sec63p are required for both co- and posttranslational protein translocation into the yeast endoplasmic reticulum. *Proceedings of the National Academy of Sciences USA* **92**, 9643–9646.

Brodsky, J. L., Werner, E. D., Dubas, M. E., Goeckeler, J. L., Kruse, K. B. and McCracken, A. A. (1999) The requirement for molecular chaperones during endoplasmic reticulum-associated protein degradation demonstrates that protein export and import are mechanistically distinct. *Journal of Biological Chemistry* **274**, 3453–3460.

Brown, C. R., McCann, J. A. and Chiang, H. L. (2000) The heat shock protein Ssa2p is required for import of fructose-1, 6-bisphosphatase into Vid vesicles. *Journal of Cell Biology* **150**, 65–76.

Brown, C. R., Cui, D. Y., Hung, G. G. and Chiang, H. L. (2001) Cyclophilin A mediates Vid22p function in the import of fructose-1,6-bisphosphatase into Vid vesicles. *Journal of Biological Chemistry* **276**, 48017–48026.

Brown, C. R., McCann, J. A., Hung, G. G., Elco, C. P. and Chiang, H. L. (2002) Vid22p, a novel plasma membrane protein, is required for the fructose-1,6-bisphosphatase degradation pathway. *Journal of Cell Science* **115**, 655–666.

Brown, J. D., Hann, B. C., Medzihradszky, K. F., Niwa, M., Burlingame, A. L. and Walter, P. (1994) Subunits of the *Saccharomyces cerevisiae* signal recognition particle required for its functional expression. *EMBO Journal* **13**, 4390–4400.

Bryant, N. J. and James, D. E. (2001) Vps45p stabilizes the syntaxin homologue Tlg2p and positively regulates SNARE complex formation. *EMBO Journal* **20**, 3380–3388.

Bryant, N. J. and Stevens, T. H. (1997) Two separate signals act independently to localize a yeast late Golgi membrane protein through a combination of retrieval and retention. *Journal of Cell Biology* **136**, 287–297.

Caldwell, S. R., Hill, K. J. and Cooper, A. A. (2001) Degradation of endoplasmic reticulum (ER) quality control substrates requires transport between the ER and Golgi. *Journal of Biological Chemistry* **276**, 23296–23303.

Cao, X. and Barlowe, C. (2000) Asymmetric requirements for a Rab GTPase and SNARE proteins in fusion of COPII vesicles with acceptor membranes. *Journal of Cell Biology* **149**, 55–66.

Cao, X., Ballew, N. and Barlowe, C. (1998) Initial docking of ER-derived vesicles requires Uso1p and Ypt1p but is independent of SNARE proteins. *EMBO Journal* **17**, 2156–2165.

Caplan, A. J., Cyr, D. M. and Douglas, M. G. (1992) YDJ1p facilitates polypeptide translocation across different intracellular membranes by a conserved mechanism. *Cell* **71**, 1143–1155.

Cartwright, P., Beilharz, T., Hansen, P., Garrett, J. and Lithgow, T. (1997) Mft52, an acid-bristle protein in the cytosol that delivers precursor proteins to yeast mitochondria. *Journal of Biological Chemistry* **272**, 5320–5325.

Cereghino, J. L., Marcusson, E. G. and Emr, S. D. (1995) The cytoplasmic tail domain of the vacuolar protein sorting receptor Vps10p and a subset of VPS gene products regulate receptor stability, function, and localization. *Molecular Biology of the Cell* **6**, 1089–1102.

Chen, P., Choi, J. D., Wang, R., Cotter, R. J. and Michaelis, S. (1997) A novel a-factor-related peptide of *Saccharomyces cerevisiae* that exits the cell by a Ste6p-independent mechanism. *Molecular Biology of the Cell* **8**, 1273–1291.

Chen, W. J. and Douglas, M. G. (1988) An F1-ATPase beta-subunit precursor lacking an internal tetramer-forming domain is imported into mitochondria in the absence of ATP. *Journal of Biological Chemistry* **263**, 4997–5000.

Chiang, H. L. and Schekman, R. (1991) Regulated import and degradation of a cytosolic protein in the yeast vacuole. *Nature* **350**, 313–318.

Chiang, M. C. and Chiang, H. L. (1998) Vid24p, a novel protein localized to the fructose-1, 6-bisphosphatase-containing vesicles, regulates targeting of fructose-1,6-bisphosphatase from the vesicles to the vacuole for degradation. *Journal of Cell Biology* **140**, 1347–1356.

Ciufo, L. F. and Brown, J. D. (2000) Nuclear export of yeast signal recognition particle lacking Srp54p via the Xpo1p/Crm1p, NES-dependent pathway. *Current Biology* **10**, 1256–1264.

Cleves, A. E., Cooper, D. N., Barondes, S. H. and Kelly, R. B. (1996) A new pathway for protein export in *Saccharomyces cerevisiae*. *Journal of Cell Biology* **133**, 1017–1026.

Collins, C. S., Kalish, J. E., Morrell, J. C., McCaffery, J. M. and Gould, S. J. (2000) The peroxisome biogenesis factors pex4p, pex22p, pex1p, and pex6p act in the terminal steps of peroxisomal matrix protein import. *Molecular and Cellular Biology* **20**, 7516–7526.

Cooper, A. A. and Stevens, T. H. (1996) Vps10p cycles between the late-Golgi and prevacuolar compartments in its function as the sorting receptor for multiple yeast vacuolar hydrolases. *Journal of Cell Biology* **133**, 529–541.

Corral-Debrinski, M., Belgareh, N., Blugeon, C., Claros, M. G., Doye, V. and Jacq, C. (1999) Overexpression of yeast karyopherin Pse1p/Kap121p stimulates the mitochondrial import of hydrophobic proteins in vivo. *Molecular Microbiology* **31**, 1499–1511.

Corral-Debrinski, M., Blugeon, C. and Jacq, C. (2000) In yeast, the 3' untranslated region or the presequence of ATM1 is required for the exclusive localization of its mRNA to the vicinity of mitochondria. *Molecular and Cellular Biology* **20**, 7881–7892.

Corsi, A. K. and Schekman, R. (1997) The lumenal domain of Sec63p stimulates the ATPase activity of BiP and mediates BiP recruitment to the translocon in *Saccharomyces cerevisiae*. *Journal of Cell Biology* **137**, 1483–1493.

Cosson, P. and Letourneur, F. (1994) Coatomer interaction with di-lysine endoplasmic reticulum retention motifs. *Science* **263**, 1629–1631.

Cosson, P., Demolliere, C., Hennecke, S., Duden, R. and Letourneur, F. (1996) Delta- and zeta-COP, two coatomer subunits homologous to clathrin-associated proteins, are involved in ER retrieval. *EMBO Journal* **15**, 1792–1798.

Cosson, P., Lefkir, Y., Demolliere, C. and Letourneur, F. (1998) New COP1-binding motifs involved in ER retrieval. *EMBO Journal* **17**, 6863–6870.

Costaguta, G., Stefan, C. J., Bensen, E. S., Emr, S. D. and Payne, G. S. (2001) Yeast Gga coat proteins function with clathrin in Golgi to endosome transport. *Molecular Biology of the Cell* **12**, 1885–1896.

Cowles, C. R., Odorizzi, G., Payne, G. S. and Emr, S. D. (1997a) The AP-3 adaptor complex is essential for cargo-selective transport to the yeast vacuole. *Cell* **91**, 109–118.

Cowles, C. R., Snyder, W. B., Burd, C. G. and Emr, S. D. (1997b) Novel Golgi to vacuole delivery pathway in yeast: identification of a sorting determinant and required transport component. *EMBO Journal* **16**, 2769–2782.

Cox, J. S. and Walter, P. (1996) A novel mechanism for regulating activity of a transcription factor that controls the unfolded protein response. *Cell* **87**, 391–404.

Cregg, J. M., Cereghino, J. L., Shi, J. and Higgins, D. R. (2000) Recombinant protein expression in *Pichia pastoris. Molecular Biotechnology* **1**, 23–52.

Crowley, K. S., Liao, S., Worrell, V. E., Reinhart, G. D. and Johnson, A. E. (1994) Secretory proteins move through the ER membrane via an aqueous, gated pore. *Cell* **7**, 461–471.

Curran, S. P., Leuenberger, D., Oppliger, W. and Koehler, C. M. (2002) The Tim9p-Tim10p complex binds to the transmembrane domains of the ADP/ATP carrier. *EMBO Journal* **21**, 942–953.

Cyr, D. M. and Douglas, M. G. (1994) Differential regulation of Hsp70 subfamilies by the eukaryotic DnaJ homologue YDJ1. *Journal of Biological Chemistry* **269**, 9798–9804.

Darsow, T., Rieder, S. E. and Emr, S. D. (1997) A multispecificity syntaxin homologue, Vam3p, essential for autophagic and biosynthetic protein transport to the vacuole. *Journal of Cell Biology* **138**, 517–529.

Darsow, T., Burd, C. G. and Emr, S. D. (1998) Acidic di-leucine motif essential for AP-3-dependent sorting and restriction of the functional specificity of the Vam3p vacuolar t-SNARE. *Journal of Cell Biology* **142**, 913–922.

Darsow, T., Katzmann, D. J., Cowles, C. R. and Emr, S. D. (2001) Vps41p function in the alkaline phosphatase pathway requires homo-oligomerization and interaction with AP-3 through two distinct domains. *Molecular Biology of the Cell* **12**, 37–51.

David, D., Sundarababu, S. and Gerst, J. E. (1998) Involvement of long chain fatty acid elongation in the trafficking of secretory vesicles in yeast. *Journal of Cell Biology* **143**, 1167–1182.

Deitz, S. B., Rambourg, A., Kepes, F. and Franzusoff, A. (2000) Sec7p directs the transitions required for yeast Golgi biogenesis. *Traffic* **1**, 172–83.

Dekker, P. J., Muller, H., Rassow, J. and Pfanner, N. (1996) Characterization of the preprotein translocase of the outer mitochondrial membrane by blue native electrophoresis. *Biological Chemistry* **377**, 535–538.

Dekker, P. J., Ryan, M. T., Brix, J., Muller, H., Honlinger, A. and Pfanner, N. (1998) Preprotein translocase of the outer mitochondrial membrane: molecular dissection and assembly of the general import pore complex. *Molecular and Cellular Biology* **18**, 6515–6524.

Dell'Angelica, E. C., Puertollano, R., Mullins, C., Aguilar, R. C., Vargas, J. D., Hartnell, L. M. and Bonifacino, J. S. (2000) GGAs: a family of ADP ribosylation factor-binding proteins related to adaptors and associated with the Golgi complex. *Journal of Cell Biology* **149**, 81–94.

Deloche, O., Yeung, B. G., Payne, G. S. and Schekman, R. (2001) Vps10p transport from the *trans*-Golgi network to the endosome is mediated by clathrin-coated vesicles. *Molecular Biology of the Cell* **12**, 475–485.

Deshaies, R. J., Sanders, S. L., Feldheim, D. A. and Schekman, R. (1991) Assembly of yeast Sec proteins involved in translocation into the endoplasmic reticulum into a membrane-bound multisubunit complex. *Nature* **349**, 806–808.

Deshaies, R. J., Koch, B. D., Werner, W. M., Craig, E. A. and Schekman, R. (1988) A subfamily of stress proteins facilitates translocation of secretory and mitochondrial precursor polypeptides. *Nature* **332**, 800–805.

Dietmeier, K., Honlinger, A., Bomer, U., Dekker, P. J., Eckerskorn, C., Lottspeich, F., Kubrich, M. and Pfanner, N. (1997) Tom5 functionally links mitochondrial preprotein receptors to the general import pore. *Nature* **388**, 195–200.

Doering, T. L. and Schekman, R. (1996) GPI anchor attachment is required for Gas1p transport from the endoplasmic reticulum in COP II vesicles. *EMBO Journal* **15**, 182–191.

Dogic, D., de Chassey, B., Pick, E., Cassel, D., Lefkir, Y., Hennecke, S., Cosson, P. and Letourneur, F. (1999) The ADP-ribosylation factor GTPase-activating protein Glo3p is involved in ER retrieval. *European Journal of Cell Biology* **78**, 305–310.

Donzeau, M., Kaldi, K., Adam, A., Paschen, S., Wanner, G., Guiard, B., Bauer, M. F., Neupert, W. and Brunner, M. (2000) Tim23 links the inner and outer mitochondrial membranes. *Cell* **101**, 401–412.

Duden, R., Hosobuchi, M., Hamamoto, S., Winey, M., Byers, B. and Schekman, R. (1994) Yeast beta- and beta'-coat proteins (COP). Two coatomer subunits essential for endoplasmic reticulum-to-Golgi protein traffic. *Journal of Biological Chemistry* **269**, 24486–24495.

Dulubova, I., Yamaguchi, T., Gao, Y., Min, S. W., Huryeva, I., Sudhof, T. C. and Rizo, J. (2002) How Tlg2p/syntaxin 16 'snares' Vps45. *EMBO Journal* **21**, 3620–3631.

Dunnwald, M., Varshavsky, A. and Johnsson, N. (1999) Detection of transient in vivo interactions between substrate and transporter during protein translocation into the endoplasmic reticulum. *Molecular Biology of the Cell* **10**, 329–344.

Elgersma, Y., Kwast, L., Klein, A., Voorn-Brouwer, T., van den Berg, M., Metzig, B., America, T., Tabak, H. F. and Distel, B. (1996) The SH3 domain of the *Saccharomyces cerevisiae* peroxisomal membrane protein Pex13p functions as a docking site for Pex5p, a mobile receptor for the import PTS1-containing proteins. *Journal of Cell Biology* **135**, 97–109.

Elgersma, Y., Kwast, L., van den Berg, M., Snyder, W. B., Distel, B., Subramani, S. and Tabak, H. F. (1997) Overexpression of Pex15p, a phosphorylated peroxisomal integral membrane protein required for peroxisome assembly in *S. cerevisiae*, causes proliferation of the endoplasmic reticulum membrane. *EMBO Journal* **16**, 7326–7341.

Elgersma, Y., Elgersma-Hooisma, M., Wenzel, T., McCaffery, J. M., Farquhar, M. G. and Subramani, S. (1998) A mobile PTS2 receptor for peroxisomal protein import in Pichia pastoris. *Journal of Cell Biology* **140**, 807–820.

Elrod-Erickson, M. J. and Kaiser, C. A. (1996) Genes that control the fidelity of endoplasmic reticulum to Golgi transport identified as suppressors of vesicle budding mutations. *Molecular Biology of the Cell* **7**, 1043–1058.

Erdmann, R. and Blobel, G. (1996) Identification of Pex13p a peroxisomal membrane receptor for the PTS1 recognition factor. *Journal of Cell Biology* **135**, 111–121.

Espenshade, P., Gimeno, R. E., Holzmacher, E., Teung, P. and Kaiser, C. A. (1995) Yeast SEC16 gene encodes a multidomain vesicle coat protein that interacts with Sec23p. *Journal of Cell Biology* **131**, 311–324.

Feldheim, D. and Schekman, R. (1994) Sec72p contributes to the selective recognition of signal peptides by the secretory polypeptide translocation complex. *Journal of Cell Biology* **126**, 935–943.

Feldheim, D., Rothblatt, J. and Schekman, R. (1992) Topology and functional domains of Sec63p, an endoplasmic reticulum membrane protein required for secretory protein translocation. *Molecular and Cellular Biology* **12**, 3288–3296.

Finger, F. P., Hughes, T. E. and Novick, P. (1998) Sec3p is a spatial landmark for polarized secretion in budding yeast. *Cell* **92**, 559–571.

Finke, K., Plath, K., Panzner, S., Prehn, S., Rapoport, T. A., Hartmann, E. and Sommer, T. (1996) A second trimeric complex containing homologs of the Sec61p complex functions in protein transport across the ER membrane of *S. cerevisiae*. *EMBO Journal* **15**, 1482–1494.

Frand, A. R., Cuozzo, J. W. and Kaiser, C. A. (2000) Pathways for protein disulphide bond formation. *Trends in Cellular Biology* **10**, 203–210.

Franzusoff, A., Redding, K., Crosby, J., Fuller, R. S. and Schekman, R. (1991) Localization of components involved in protein transport and processing through the yeast Golgi apparatus. *Journal of Cell Biology* **112**, 27–37.

Fulga, T. A., Sinning, I., Dobberstein, B. and Pool, M. R. (2001) SRbeta coordinates signal sequence release from SRP with ribosome binding to the translocon. *EMBO Journal* **20**, 2338–2347.

Funfschilling, U. and Rospert, S. (1999) Nascent polypeptide-associated complex stimulates protein import into yeast mitochondria. *Molecular Biology of the Cell* **10**, 3289–3299.

Garrett, J. M., Singh, K. K., von der Haar, R. A. and Emr, S. D. (1991) Mitochondrial protein import: isolation and characterization of the *Saccharomyces cerevisiae MFT1* gene. *Molecular and General Genetics* **225**, 483–491.

Garrett, M. D., Zahner, J. E., Cheney, C. M. and Novick, P. J. (1994) *GDI1* encodes a GDP dissociation inhibitor that plays an essential role in the yeast secretory pathway. *EMBO Journal* **13**, 1718–1728.

Gatto, G. J., Jr., Geisbrecht, B. V., Gould, S. J. and Berg, J. M. (2000) Peroxisomal targeting signal-1 recognition by the TPR domains of human PEX5. *Nature Structural Biology* **7**, 1091–1095.

Gautschi, M., Lilie, H., Funfschilling, U., Mun, A., Ross, S., Lithgow, T., Rucknagel, P. and Rospert, S. (2001) RAC, a stable ribosome-associated complex in yeast formed by the DnaK- DnaJ homologs Ssz1p and zuotin. *Proceedings of the National Academy of Sciences USA* **98**, 3762–3767.

Gaynor, E. C., Chen, C. Y., Emr, S. D. and Graham, T. R. (1998) ARF is required for maintenance of yeast Golgi and endosome structure and function. *Molecular Biology of the Cell* **9**, 653–670.

Geissler, A., Rassow, J., Pfanner, N. and Voos, W. (2001) Mitochondrial import driving forces, enhanced trapping by matrix Hsp70 stimulates translocation and reduces the membrane potential dependence of loosely folded preproteins. *Molecular and Cellular Biology* **21**, 7097–7104.

Geissler, A., Chacinska, A., Truscott, K. N., Wiedemann, N., Brandner, K., Sickmann, A., Meyer, H. E., Meisinger, C., Pfanner, N. and Rehling, P. (2002) The mitochondrial presequence translocase: an essential role of Tim50 in directing preproteins to the import channel. *Cell* **111**, 507–518.

George, R., Beddoe, T., Landl, K. and Lithgow, T. (1998) The yeast nascent polypeptide-associated complex initiates protein targeting to mitochondria in vivo. *Proceedings of the National Academy of Sciences USA* **95**, 2296–2301.

George, R., Walsh, P., Beddoe, T. and Lithgow, T. (2002) The nascent polypeptide-associated complex (NAC) promotes interaction of ribosomes with the mitochondrial surface in vivo. *FEBS Letters* **516**, 213–216.

Gerich, B., Orci, L., Tschochner, H., Lottspeich, F., Ravazzola, M., Amherdt, M., Wieland, F. and Harter, C. (1995) Non-clathrin-coat protein alpha is a conserved subunit of coatomer and in *Saccharomyces cerevisiae* is essential for growth. *Proceedings of the National Academy of Sciences USA* **92**, 3229–3233.

Gillece, P., Pilon, M. and Romisch, K. (2000) The protein translocation channel mediates glycopeptide export across the endoplasmic reticulum membrane. *Proceedings of the National Academy of Sciences USA* **97**, 4609–4614.

Gillooly, D. J., Morrow, I. C., Lindsay, M., Gould, R., Bryant, N. J., Gaullier, J. M., Parton, R. G. and Stenmark, H. (2000) Localization of phosphatidylinositol 3-phosphate in yeast and mammalian cells. *EMBO Journal* **19**, 4577–4588.

Gilstring, C. F., Melin-Larsson, M. and Ljungdahl, P. O. (1999) Shr3p mediates specific COPII coatomer-cargo interactions required for the packaging of amino acid permeases into ER-derived transport vesicles. *Molecular Biology of the Cell* **10**, 3549–3565.

Gimeno, R. E., Espenshade, P. and Kaiser, C. A. (1995) *SED4* encodes a yeast endoplasmic reticulum protein that binds Sec16p and participates in vesicle formation. *Journal of Cell Biology* **131**, 325–338.

Gimeno, R. E., Espenshade, P. and Kaiser, C. A. (1996) COPII coat subunit interactions: Sec24p and Sec23p bind to adjacent regions of Sec16p. *Molecular Biology of the Cell* **7**, 1815–1823.

Gomez-Lorenzo, M. G., Spahn, C. M., Agrawal, R. K., Grassucci, R. A., Penczek, P., Chakraburtty, K., Ballesta, J. P., Lavandera, J. L., Garcia-Bustos, J. F. and Frank, J. (2000) Three-dimensional cryo-electron microscopy localization of EF2 in the *Saccharomyces cerevisiae* 80S ribosome at 17.5 A resolution. *EMBO Journal* **19**, 2710–2718.

Görlich, D. and Rapoport, T. A. (1993) Protein translocation into proteoliposomes reconstituted from purified components of the endoplasmic reticulum membrane. *Cell* **75**, 615–630.

Gottschalk, S., Waheed, A., Schmidt, B., Laidler, P. and von Figura, K. (1989) Sequential processing of lysosomal acid phosphatase by a cytoplasmic thiol proteinase and a lysosomal aspartyl proteinase. *EMBO Journal* **8**, 3215–3219.

Gratzer, S., Beilharz, T., Beddoe, T., Henry, M. F. and Lithgow, T. (2000) The mitochondrial protein targeting suppressor (*mts1*) mutation maps to the mRNA-binding domain of Npl3p and affects translation on cytoplasmic polysomes. *Molecular Microbiology* **35**, 1277–1285.

Gurunathan, S., David, D. and Gerst, J. E. (2002) Dynamin and clathrin are required for the biogenesis of a distinct class of secretory vesicles in yeast. *EMBO Journal* **21**, 602–614.

Ha, S. A., Bunch, J. T., Hama, H., DeWald, D. B. and Nothwehr, S. F. (2001) A novel mechanism for localizing membrane proteins to yeast trans-Golgi network requires function of synaptojanin-like protein. *Molecular Biology of the Cell* **12**, 3175–3190.

Hachiya, N., Alam, R., Sakasegawa, Y., Sakaguchi, M., Mihara, K. and Omura, T. (1993) A mitochondrial import factor purified from rat liver cytosol is an ATP-dependent conformational modulator for precursor proteins. *EMBO Journal* **12**, 1579–1586.

Haigh, N. G. and Johnson, A. E. (2002) A new role for BiP: closing the aqueous translocon pore during protein integration into the ER membrane. *Journal of Cell Biology* **156**, 261–270.

Hamburger, D., Egerton, M. and Riezman, H. (1995) Yeast Gaa1p is required for attachment of a completed GPI anchor onto proteins. *Journal of Cell Biology* **129**, 629–639.

Hamilton, T. G. and Flynn, G. C. (1996) Cer1p, a novel Hsp70-related protein required for posttranslational endoplasmic-reticulum translocation in yeast. *Journal of Biological Chemistry* **271**, 30610–30613.

Hamman, B. D., Hendershot, L. M. and Johnson, A. E. (1998) BiP maintains the permeability barrier of the ER membrane by sealing the lumenal end of the translocon pore before and early in translocation. *Cell* **92**, 747–758.

Hampton, R. Y. (2002) ER-associated degradation in protein quality control and cellular regulation. *Current Opinion in Cell Biology* **14**, 476–482.

Hanein, D., Matlack, K. E. S., Jungnickel, B., Plath, K., Kalies, K. U., Miller, K. R., Rapoport, T. A. and Akey, C. W. (1996) Oligomeric rings of the Sec61p complex induced by ligands required for protein translocation. *Cell* **87**, 721–732.

Hann, B. C. and Walter, P. (1991) The signal recognition particle in *S. cerevisiae*. *Cell* **67**, 131–144.

Hann, B., C. Stirling, J. and Walter, P. (1992) *SEC65* gene product is a subunit of the yeast signal recognition particle required for its integrity. *Nature* **356**, 532–533.

Hansen, W., Garcia, P. D. and Walter, P. (1986) In vitro protein translocation across the yeast endoplasmic reticulum: ATP-dependent post-translational translocation of the prepro-a factor. *Cell* **45**, 397–406.

Hara-Kuge, S., Kuge, O., Orci, L., Amherdt, M., Ravazzola, M., Wieland, F. T. and Rothman, J. E. (1994) En bloc incorporation of coatomer subunits during the assembly of COP-coated vesicles. *Journal of Cell Biology* **124**, 883–892.

Harding, T. M., Hefner-Gravink, A., Thumm, M. and Klionsky, D. J. (1996) Genetic and phenotypic overlap between autophagy and the cytoplasm to vacuole protein targeting pathway. *Journal of Biological Chemistry* **271**, 17621–17624.

Hardwick, K. G. and Pelham, H. R. (1992) *SED5* encodes a 39-kD integral membrane protein required for vesicular transport between the ER and the Golgi complex. *Journal of Cell Biology* **119**, 513–521.

Hardwick, K. G., Lewis, M. J., Semenza, J., Dean, N. and Pelham, H. R. (1990) *ERD1*, a yeast gene required for the retention of luminal endoplasmic reticulum proteins, affects glycoprotein processing in the Golgi apparatus. *EMBO Journal* **9**, 623–630.

Harris, S. L. and Waters, M. G. (1996) Localization of a yeast early Golgi mannosyltransferase, Och1p, involves retrograde transport. *Journal of Cell Biology* **132**, 985–998.

Harsay, E. and Bretscher, A. (1995) Parallel secretory pathways to the cell surface in yeast. *Journal of Cell Biology* **131**, 297–310.

Harsay, E. and Schekman, R. (2002) A subset of yeast vacuolar protein sorting mutants is blocked in one branch of the exocytic pathway. *Journal of Cell Biology* **156**, 271–285.

Haucke, V. and Schatz, G. (1997) Reconstitution of the protein insertion machinery of the mitochondrial inner membrane. *EMBO Journal* **16**, 4560–4567.

Haucke, V., Lithgow, T., Rospert, S., Hahne, K. and Schatz, G. (1995) The yeast mitochondrial protein import receptor Mas20p binds precursor proteins through electrostatic interaction with the positively charged presequence. *Journal of Biological Chemistry* **270**, 5565–5570.

Haynes, C. M., Caldwell, S. and Cooper, A. A. (2002) An HRD/DER-independent ER quality control mechanism involves Rsp5p-dependent ubiquitination and ER-Golgi transport. *Journal of Cell Biology* **158**, 91–101.

Hazra, P. P., Suriapranata, I., Snyder, W. B. and Subramani, S. (2002) Peroxisome remnants in pex3delta cells and the requirement of pex3p for interactions between the peroxisomal docking and translocation subcomplexes. *Traffic* **3**, 560–574.

Hell, K., Herrmann, J. M., Pratje, E., Neupert, W. and Stuart, R. A. (1998) Oxa1p, an essential component of the N-tail protein export machinery in mitochondria. *Proceedings of the National Academy of Sciences USA* **95**, 2250–2255.

Hell, K., Neupert, W. and Stuart, R. A. (2001) Oxa1p acts as a general membrane insertion machinery for proteins encoded by mitochondrial DNA. *EMBO Journal* **20**, 1281–1288.

Helliwell, S. B., Losko, S. and Kaiser, C. A. (2001) Components of a ubiquitin ligase complex specify polyubiquitination and intracellular trafficking of the general amino acid permease. *Journal of Cell Biology* **153**, 649–662.

Hettema, E. H., Ruigrok, C. C., Koerkamp, M. G., van den Berg, M., Tabak, H. F., Distel, B. and Braakman, I. (1998) The cytosolic DnaJ-like protein djp1p is involved specifically in peroxisomal protein import. *Journal of Cell Biology* **142**, 421–434.

Hettema, E. H., Girzalsky, W., van Den Berg, M., Erdmann, R. and Distel, B. (2000) *Saccharomyces cerevisiae* pex3p and pex19p are required for proper localization and stability of peroxisomal membrane proteins. *EMBO Journal* **19**, 223–233.

Hicke, L., Yoshihisa, T. and Schekman, R. (1992) Sec23p and a novel 105-kDa protein function as a multimeric complex to promote vesicle budding and protein transport from the endoplasmic reticulum. *Molecular Biology of the Cell* **3**, 667–676.

Hiller, M. M., Finger, A., Schweiger, M. and Wolf, D. H. (1996) ER degradation of a misfolded luminal protein by the cytosolic ubiquitin-proteasome pathway. *Science* **273**, 1725–1728.

Hirst, J., Lui, W. W., Bright, N. A., Totty, N., Seaman, M. N. and Robinson, M. S. (2000) A family of proteins with gamma-adaptin and VHS domains that facilitate trafficking between the trans-Golgi network and the vacuole/lysosome. *Journal of Cell Biology* **149**, 67–80.

Holthuis, J. C., Nichols, B. J., Dhruvakumar, S. and Pelham, H. R. (1998a) Two syntaxin homologues in the TGN/endosomal system of yeast. *EMBO Journal* **17**, 113–126.

Holthuis, J. C., Nichols, B. J. and Pelham, H. R. (1998b) The syntaxin Tlg1p mediates trafficking of chitin synthase III to polarized growth sites in yeast. *Molecular Biology of the Cell* **9**, 3383–3397.

Honlinger, A., Bomer, U., Alconada, A., Eckerskorn, C., Lottspeich, F., Dietmeier, K. and Pfanner, N. (1996) Tom7 modulates the dynamics of the mitochondrial outer membrane translocase and plays a pathway-related role in protein import. *EMBO Journal* **15**, 2125–2137.

Horst, M., Hilfiker-Rothenfluh, S., Oppliger, W. and Schatz, G. (1995) Dynamic interaction of the protein translocation systems in the inner and outer membranes of yeast mitochondria. *EMBO Journal* **14**, 2293–2297.

Horst, M., Oppliger, W., Rospert, S., Schonfeld, H. J., Schatz, G. and Azem, A. (1997) Sequential action of two hsp70 complexes during protein import into mitochondria. *EMBO Journal* **16**, 1842–1849.

Huang, P. H. and Chiang, H. L. (1997) Identification of novel vesicles in the cytosol to vacuole protein degradation pathway. *Journal of Cell Biology* **136**, 803–810.

Huang, S., Ratliff, K. S., Schwartz, M. P., Spenner, J. M. and Matouschek, A. (1999) Mitochondria unfold precursor proteins by unraveling them from their N-termini. *Nature Structural Biology* **6**, 1132–1138.

Huhse, B., Rehling, P., Albertini, M., Blank, L., Meller, K. and Kunau, W. H. (1998) Pex17p of *Saccharomyces cerevisiae* is a novel peroxin and component of the peroxisomal protein translocation machinery. *Journal of Cell Biology* **140**, 49–60.

Ishihara, N., Hamasaki, M., Yokota, S., Suzuki, K., Kamada, Y., Kihara, A., Yoshimori, T., Noda, T. and Ohsumi, Y. (2001) Autophagosome requires specific early Sec proteins for its formation and NSF/SNARE for vacuolar fusion. *Molecular Biology of the Cell* **12**, 3690–3702.

Jahn, R. (2000) Sec1/Munc18 proteins: mediators of membrane fusion moving to center stage. *Neuron* **27**, 201–204.

Jarosch, E., Taxis, C., Volkwein, C., Bordallo, J., Finley, D., Wolf, D. H. and Sommer, T. (2002) Protein dislocation from the ER requires polyubiquitination and the AAA-ATPase Cdc48. *Nature Cell Biology* **4**, 134–139.

Joazeiro, C. A. and Weissman, A. M. (2000) RING finger proteins: mediators of ubiquitin ligase activity. *Cell* **102**, 549–552.

Jones, S., Jedd, G., Kahn, R. A., Franzusoff, A., Bartolini, F. and Segev, N. (1999) Genetic interactions in yeast between Ypt GTPases and Arf guanine nucleotide exchangers. *Genetics* **152**, 1543–1556.

Jones, S., Newman, C., Liu, F. and Segev, N. (2000) The TRAPP complex is a nucleotide exchanger for Ypt1 and Ypt31/32. *Molecular Biology of the Cell* **11**, 4403–4411.

Justice, M. C., Hsu, M. J., Tse, B., Ku, T., Balkovec, J., Schmatz, D. and Nielsen, J. (1998) Elongation factor 2 as a novel target for selective inhibition of fungal protein synthesis. *Journal of Biological Chemistry* **273**, 3148–3151.

Kaiser, C. A., Preuss, D., Grisafi, P. and Botstein, D. (1987) Many random sequences functionally replace the secretion signal sequence of yeast invertase. *Science* **235**, 312–317.

Kamada, Y., Funakoshi, T., Shintani, T., Nagano, K., Ohsumi, M. and Ohsumi, Y. (2000) Tor-mediated induction of autophagy via an Apg1 protein kinase complex. *Journal of Cell Biology* **150**, 1507–1513.

Katzmann, D. J., Babst, M. and Emr, S. D. (2001) Ubiquitin-dependent sorting into the multivesicular body pathway requires the function of a conserved endosomal protein sorting complex, ESCRT-I. *Cell* **106**, 145–155.

Keenan, R. J., Freymann, D. M., Stroud, R. M. and Walter, P. (2001) The signal recognition particle. *Annual Reviews of Biochemistry* **70**, 755–775.

Kentsis, A. and Borden, K. L. (2000) construction of macromolecular assemblages in eukaryotic processes and their role in human disease: linking RINGs together. *Current Protein and Peptide Science* **1**, 49–73.

Kentsis, A., Gordon, R. E. and Borden, K. L. (2002) Control of biochemical reactions through supramolecular RING domain self-assembly. *Proceedings of the National Academy of Sciences USA* **99**, 15404–15409.

Kerscher, O., Holder, J., Srinivasan, M., Leung, R. S. and Jensen, R. E. (1997) The Tim54p-Tim22p complex mediates insertion of proteins into the mitochondrial inner membrane. *Journal of Cell Biology* **139**, 1663–1675.

Kerscher, O., Sepuri, N. B. and Jensen, R. E. (2000) Tim18p is a new component of the Tim54p-Tim22p translocon in the mitochondrial inner membrane. *Molecular Biology of the Cell* **11**, 103–116.

Khalfan, W. A. and Klionsky, D. J. (2002) Molecular machinery required for autophagy and the cytoplasm to vacuole targeting (Cvt) pathway in *S. cerevisiae*. *Current Opinion in Cell Biology* **14**, 468–475.

Kihara, A., Noda, T., Ishihara, N. and Ohsumi Y. (2001) Two distinct Vps34 phosphatidylinositol 3-kinase complexes function in autophagy and carboxypeptidase Y sorting in *Saccharomyces cerevisiae*. *Journal of Cell Biology* **152**, 519–530.

Kim, J., Huang, W. P., Stromhaug, P. E. and Klionsky, D. J. (2002) Convergence of multiple autophagy and cytoplasm to vacuole targeting components to a prevacuolar membrane compartment prior to de novo vesicle formation. *Journal of Biological Chemistry* **277**, 763–773.

Klein, A. T., Barnett, P., Bottger, G., Konings, D., Tabak, H. F. and Distel, B. (2001) Recognition of peroxisomal targeting signal type 1 by the import receptor Pex5p. *Journal of Biological Chemistry* **276**, 15034–15041.

Klein, A. T., van den Berg, M., Bottger, G., Tabak, H. F. and Distel, B. (2002) *Saccharomyces cerevisiae* acyl-CoA oxidase follows a novel, non-PTS1, import pathway into peroxisomes that is dependent on pex5p. *Journal of Biological Chemistry* **277**, 25011–25019.

Knox, C., Sass, E., Neupert, W. and Pines, O. (1998) Import into mitochondria, folding and retrograde movement of fumarase in yeast. *Journal of Biological Chemistry* **273**, 25587–25593.

Koehler, C. M., Jarosch, E., Tokatlidis, K., Schmid, K., Schweyen, R. J. and Schatz, G. (1998a) Import of mitochondrial carriers mediated by essential proteins of the intermembrane space. *Science* **279**, 369–373.

Koehler, C. M., Merchant, S., Oppliger, W., Schmid, K., Jarosch, E., Dolfini, L., Junne, T., Schatz, G. and Tokatlidis, K. (1998b) Tim9p, an essential partner subunit of Tim10p for the import of mitochondrial carrier proteins. *EMBO Journal* **17**, 6477–6486.

Koller, A., Snyder, W. B., Faber, K. N., Wenzel, T. J., Rangell, L., Keller, G. A. and Subramani, S. (1999) Pex22p of *Pichia pastoris*, essential for peroxisomal matrix protein import, anchors the ubiquitin-conjugating enzyme, Pex4p, on the peroxisomal membrane. *Journal of Cell Biology* **146**, 99–112.

Komiya, T., Rospert, S., Schatz, G. and Mihara, K. (1997) Binding of mitochondrial precursor proteins to the cytoplasmic domains of the import receptors Tom70 and Tom20 is determined by cytoplasmic chaperones. *EMBO Journal* **16**, 4267–4275.

Kovermann, P., Truscott, K. N., Guiard, B., Rehling, P., Sepuri, N. B., Muller, H., Jensen, R. E., Wagner, R. and Pfanner, N. (2002) Tim22, the essential core of the mitochondrial protein insertion complex, forms a voltage-activated and signal-gated channel. *Molecules to Cells* **9**, 363–373.

Kuchler, K., Sterne, R. E. and Thorner, J. (1989) *Saccharomyces cerevisiae STE6* gene product: a novel pathway for protein export in eukaryotic cells. *EMBO Journal* **8**, 3973–3984.

Kuehn, M. J., Schekman, R. and Ljungdahl, P. O. (1996) Amino acid permeases require COPII components and the ER resident membrane protein Shr3p for packaging into transport vesicles in vitro. *Journal of Cell Biology* **135**, 585–595.

Kuehn, M. J., Herrmann, J. M. and Schekman, R. (1998) COPII-cargo interactions direct protein sorting into ER-derived transport vesicles. *Nature* **391**, 187–190.

Kunkele, K. P., Heins, S., Dembowski, M., Nargang, F. E., Benz, R., Thieffry, M., Walz, J., Lill, R., Nussberger, S. and Neupert, W. (1998) The preprotein translocation channel of the outer membrane of mitochondria. *Cell* **93**, 1009–1119.

Kurihara, T., Hamamoto, S., Gimeno, R. E., Kaiser, C. A., Schekman, R. and Yoshihisa, T. (2000) Sec24p and Iss1p function interchangeably in transport vesicle formation from the endoplasmic reticulum in *Saccharomyces cerevisiae*. *Molecular Biology of the Cell* **11**, 983–998.

Kutay, U., Ahnerthilger, G., Hartmann, E., Wiedenmann, B. and Rapoport, T. A. (1995) Transport route for synaptobrevin via a novel pathway of insertion into the endoplasmic-reticulum membrane. *EMBO Journal* **14**, 217–223.

Lang, T., Reiche, S., Straub, M., Bredschneider, M. and Thumm, M. (2000) Autophagy and the cvt pathway both depend on AUT9. *Journal of Bacteriology* **182**, 2125–2133.

Lederkremer, G. Z., Cheng, Y., Petre, B. M., Vogan, E., Springer, S., Schekman, R., Walz, T. and Kirchhausen, T. (2001) Structure of the Sec23p/24p and Sec13p/31p complexes of COPII. *Proceedings of the National Academy of Sciences USA* **98**, 10704–10709.

Lee, F. J., Stevens, L. A., Kao, Y. L., Moss, J. and Vaughan, M. (1994) Characterization of a glucose-repressible ADP-ribosylation factor 3 (ARF3) from *Saccharomyces cerevisiae*. *Journal of Biological Chemistry* **269**, 20931–20937.

Lee, D. H., Sherman, M. Y. and Goldberg, A. L. (1996) Involvement of the molecular chaperone Ydj1 in the ubiquitin-dependent degradation of short-lived and abnormal proteins in *Saccharomyces cerevisiae*. *Molecular and Cellular Biology* **16**, 4773–4781.

Lemire, B. D., Fankhauser, C., Baker, A. and Schatz, G. (1989) The mitochondrial targeting function of randomly generated peptide sequences correlates with predicted helical amphiphilicity. *Journal of Biological Chemistry* **264**, 20206–20215.

Letourneur, F. and Cosson, P. (1998) Targeting to the endoplasmic reticulum in yeast cells by determinants present in transmembrane domains. *Journal of Biological Chemistry* **273**, 33273–33278.

Letourneur, F., Gaynor, E. C., Hennecke, S., Demolliere, C., Duden, R., Emr, S. D., Riezman, H. and Cosson, P. (1994) Coatomer is essential for retrieval of dilysine-tagged proteins to the endoplasmic reticulum. *Cell* **79**, 1199–1207.

Levine, T. P. and Munro, S. (2002) Targeting of Golgi-specific pleckstrin homology domains involves both PtdIns 4-kinase-dependent and -independent components. *Current Biology* **12**, 695–704.

Li, B. and Warner, J. R. (1996) Mutation of the Rab6 homologue of *Saccharomyces cerevisiae*, YPT6, inhibits both early Golgi function and ribosome biosynthesis. *Journal of Biological Chemistry* **271**, 16813–16819.

Li, Y., Moir, R. D., Sethy-Coraci, I. K., Warner, J. R. and Willis, I. M. (2000) Repression of ribosome and tRNA synthesis in secretion-defective cells is signaled by a novel branch of the cell integrity pathway. *Molecular and Cellular Biology* **20**, 3843–3851.

Lim, J. H., Martin, F., Guiard, B., Pfanner, N. and Voos, W. (2001) The mitochondrial Hsp70-dependent import system actively unfolds preproteins and shortens the lag phase of translocation. *EMBO Journal* **20**, 941–950.

Lithgow, T. (2000) Targeting of proteins to mitochondria. *FEBS Letters* **476**, 22–26.

Lithgow, T., Cuezva, J. M. and Silver, P. A. (1997) Highways for protein delivery to the mitochondria. *Trends in Biochemical Sciences* **22**, 110–113.

Ljungdahl, P. O., Gimeno, C. J., Styles, C. A. and Fink, G. R. (1992) SHR3: a novel component of the secretory pathway specifically required for localization of amino acid permeases in yeast. *Cell* **71**, 463–478.

Lowe, M. (2000) Membrane transport: tethers and TRAPPs. *Current Biology* **10**, R407–409.

Lupashin, V. V., Hamamoto, S. and Schekman, R. W. (1996) Biochemical requirements for the targeting and fusion of ER-derived transport vesicles with purified yeast Golgi membranes. *Journal of Cell Biology* **132**, 277–289.

Lupashin, V. V. and Waters, M. G. (1997) t-SNARE activation through transient interaction with a rab-like guanosine triphosphatase. *Science* **276**, 1255–1258.

Lyman, S. K. and Schekman, R. (1995) Interaction between BiP and Sec63p is required for the completion of protein translocation into the ER of *Saccharomyces cerevisiae*. *Journal of Cell Biology* **131**, 1163–1171.

Lyman, S. K. and Schekman, R. (1997) Binding of secretory precursor polypeptides to a translocon subcomplex is regulated by BiP. *Cell* **88**, 85–96.

Marc, P., Margeot, A., Devaux, F., Blugeon, C., Corral-Debrinski M. and Jacq, C. (2002) Genome-wide analysis of mRNAs targeted to yeast mitochondria. *EMBO Reports* **3**, 159–164.

Marcusson, E. G., Horazdovsky, B. F., Cereghino, J. L., Gharakhanian, E. and Emr, S. D. (1994) The sorting receptor for yeast vacuolar carboxypeptidase Y is encoded by the VPS10 gene. *Cell* **77**, 579–586.

Martinez, E., Jimenez, M. A., Segui-Real, B., Vandekerckhova, J. and Sandoval, I.V. (1997) Folding of the presequence of yeast pAPI into an amphipathic helix determines transport of the protein from the cytosol to the vacuole. *Journal of Molecular Biology* **267**, 1124–1138.

Marzioch, M., Henthorn, D. C., Herrmann, J. M., Wilson, R., Thomas, D. Y., Bergeron, J. J., Solari, R. C. and Rowley, A. (1999) Erp1p and Erp2p, partners for Emp24p and Erv25p in a yeast p24 complex. *Molecular Biology of the Cell* **10**, 1923–1938.

Mason, N., Ciufo, L. F. and Brown, J. D. (2000) Elongation arrest is a physiologically important function of signal recognition particle. *EMBO Journal* **19**, 4164–4174.

Matlack, K. E. S., Plath, K., Misselwitz, B. and Rapoport, T. A. (1997) Protein transport by purified yeast sec complex and Kar2p without membranes. *Science* **277**, 938–941.

Matlack, K. E., Misselwitz, B., Plath, K. and Rapoport, T. A. (1999) BiP acts as a molecular ratchet during posttranslational transport of prepro-alpha factor across the ER membrane. *Cell* **97**, 553–564.

Matouschek, A., Pfanner, N. and Voos, W. (2000) Protein unfolding by mitochondria. The Hsp70 import motor. *EMBO Reports* **1**, 404–410.

Matsuoka, K., Schekman, R., Orci, L. and Heuser, J. E. (2001) Surface structure of the COPII-coated vesicle. *Proceedings of the National Academy of Sciences USA* **98**, 13705–13709.

Mayer, A. (2002) Membrane fusion in eukaryotic cells. *Annual Reviews in Cell and Developmental Biology* **18**, 289–314.

McCollum, D., Monosov, E. and Subramani, S. (1993) The *pas8* mutant of *Pichia pastoris* exhibits the peroxisomal protein import deficiencies of Zellweger syndrome cells-the PAS8 protein binds to the COOH-terminal tripeptide peroxisomal targeting signal, and is a member of the TPR protein family. *Journal of Cell Biology* **121**, 761–774.

McGrath, J. P. and Varshavsky, A. (1989) The yeast *STE6* gene encodes a homologue of the mammalian multidrug resistance P-glycoprotein. *Nature* **340**, 400–404.

McNew, J. A., Sogaard, M., Lampen, N. M., Machida, S., Ye, R. R., Lacomis, L., Tempst, P., Rothman, J. E. and Sollner, T. H. (1997) Ykt6p, a prenylated SNARE essential for endoplasmic reticulum golgi transport. *Journal of Biological Chemistry* **272**, 17776–17783.

McNew, J. A., Coe, J. G., Sogaard, M., Zemelman, B. V., Wimmer, C., Hong, W. and Sollner, T. H. (1998) Gos1p, a *Saccharomyces cerevisiae* SNARE protein involved in Golgi transport. *FEBS Letters* **435**, 89–95.

McNew, J. A., Parlati, F., Fukuda, R., Johnston, R. J., Paz, K., Paumet, F., Sollner, T. H. and Rothman, J. E. (2000) Compartmental specificity of cellular membrane fusion encoded in SNARE proteins. *Nature* **407**, 153–159.

Menetret, J., Neuhof, A., Morgan, D. G., Plath, K., Radermacher, M., Rapoport, T. A. and Akey, C. W. (2000) The structure of ribosome-channel complexes engaged in protein translocation. *Molecules to Cells* **6**, 1219–1232.

Meyer, H. A., Grau, H., Kraft, R., Kostka, S., Prehn, S., Kalies, K. U. and Hartmann, E. (2000) Mammalian Sec61 is associated with Sec62 and Sec63. *Journal of Biological Chemistry* **275**, 14550–14557.

Millar, D. G. and Shore, G. C. (1994) Mitochondrial Mas70p signal anchor sequence. Mutations in the transmembrane domain that disrupt dimerization but not targeting or membrane insertion. *Journal of Biological Chemistry* **269**, 12229–12232.

Miller, E., Antonny, B., Hamamoto, S. and Schekman, R. (2002) Cargo selection into COPII vesicles is driven by the Sec24p subunit. *EMBO Journal* **21**, 6105–6113.

Misra, S. and Hurley, J. H. (1999) Crystal structure of a phosphatidylinositol 3-phosphate-specific membrane-targeting motif, the FYVE domain of Vps27p. *Cell* **97**, 657–666.

Mizuta, K. and Warner, J. R. (1994) Continued functioning of the secretory pathway is essential for ribosome synthesis. *Molecular and Cellular Biology* **14**, 2493–2502.

Mizuta, K., Tsujii, R., Warner, J. R. and Nishiyama, M. (1998) The C-terminal silencing domain of Rap1p is essential for the repression of ribosomal protein genes in response to a defect in the secretory pathway. *Nucleic Acids Research* **26**, 1063–1069.

Moczko, M., Bomer, U., Kubrich, M., Zufall, N., Honlinger, A. and Pfanner, N. (1997) The intermembrane space domain of mitochondrial Tom22 functions as a trans binding site for preproteins with N-terminal targeting sequences. *Molecular and Cellular Biology* **17**, 6574–6584.

Model, K., Prinz, T., Ruiz, T., Radermacher, M., Krimmer, T., Kuhlbrandt, W., Pfanner, N. and Meisinger, C. (2002) Protein translocase of the outer mitochondrial membrane: role of import receptors in the structural organization of the TOM complex. *Journal of Molecular Biology* **316**, 657–666.

Morin-Ganet, M. N., Rambourg, A., Deitz, S. B., Franzusoff, A. and Kepes, F. (2000) Morphogenesis and dynamics of the yeast Golgi apparatus. *Traffic* **1**, 56–68.

Moro, F., Okamoto, K., Donzeau, M., Neupert, W. and Brunner, M. (2002) Mitochondrial protein import: molecular basis of the ATP-dependent interaction of MtHsp70 with Tim44. *Journal of Biological Chemistry* **277**, 6874–6880.

Mulholland, J., Wesp, A., Riezman, H. and Botstein, D. (1997) Yeast actin cytoskeleton mutants accumulate a new class of Golgi-derived secretary vesicle. *Molecular Biology of the Cell* **8**, 1481–1499.

Muller, O., Bayer, M. J., Peters, C., Andersen, J. S., Mann, M. and Mayer, A. (2002) The Vtc proteins in vacuole fusion: coupling NSF activity to V(0) trans-complex formation. *EMBO Journal* **21**, 259–269.

Muniz, M., Nuoffer, C., Hauri, H. P. and Riezman, H. (2000) The Emp24 complex recruits a specific cargo molecule into endoplasmic reticulum-derived vesicles. *Journal of Cell Biology* **148**, 925–930.

Munn, A. L. (2001) Molecular requirements for the internalisation step of endocytosis: insights from yeast. *Biochimica et Biophysica Acta* **1535**, 236–257.

Musch, A., Wiedmann, M. and Rapoport, T. (1992) Yeast sec proteins interact with polypeptides traversing the endoplasmic reticulum membrane. *Cell* **69**, 343–352.

Mutka, S. C. and Walter, P. (2001) Multifaceted physiological response allows yeast to adapt to the loss of the signal recognition particle-dependent protein-targeting pathway. *Molecular Biology of the Cell* **12**, 577–588.

Ng, D. T. W., Brown, J. D. and Walter, P. (1996) Signal sequences specify the targeting route to the endoplasmic- reticulum membrane. *Journal of Cell Biology* **134**, 269–278.

Nice, D. C., Sato, T. K., Stromhaug, P. E., Emr, S. D. and Klionsky, D. J. (2002) Cooperative binding of the cytoplasm to vacuole targeting pathway proteins, Cvt13 and Cvt20, to phosphatidylinositol 3-phosphate at the pre-autophagosomal structure is required for selective autophagy. *Journal of Biological Chemistry* **277**, 30198–30207.

Nishikawa, S. I., Fewell, S. W., Kato, Y., Brodsky, J. L. and Endo, T. (2001) Molecular chaperones in the yeast endoplasmic reticulum maintain the solubility of proteins for retrotranslocation and degradation. *Journal of Cell Biology* **153**, 1061–1070.

Noda, T. and Ohsumi, Y. (1998) Tor, a phosphatidylinositol kinase homologue, controls autophagy in yeast. *Journal of Biological Chemistry* **273**, 3963–3966.

Noda, T., Kim, J., Huang, W. P., Baba, M., Tokunaga, C., Ohsumi, Y. and Klionsky, D. J. (2000) Apg9p/Cvt7p is an integral membrane protein required for transport vesicle formation in the Cvt and autophagy pathways. *Journal of Cell Biology* **148**, 465–480.

Nothwehr, S. F., Bruinsma, P. and Strawn, L. A. (1999) Distinct domains within Vps35p mediate the retrieval of two different cargo proteins from the yeast prevacuolar/endosomal compartment. *Molecular Biology of the Cell* **10**, 875–890.

Nothwehr, S. F. and Hindes, A. E. (1997) The yeast *VPS5/GRD2* gene encodes a sorting nexin-1-like protein required for localizing membrane proteins to the late Golgi. *Journal of Cell Science* **110**, 1063–1072.

Nothwehr, S. F., Roberts, C. J. and Stevens, T. H. (1993) Membrane protein retention in the yeast Golgi apparatus: dipeptidyl aminopeptidase A is retained by a cytoplasmic signal containing aromatic residues. *Journal of Cell Biology* **121**, 1197–1209.

Nothwehr, S. F., Bryant, N. J. and Stevens, T. H. (1996) The newly identified yeast GRD genes are required for retention of late-Golgi membrane proteins. *Molecular and Cellular Biology* **16**, 2700–2707.

Nothwehr, S. F., Ha, S. A. and Bruinsma, P. (2000) Sorting of yeast membrane proteins into an endosome-to-Golgi pathway involves direct interaction of their cytosolic domains with Vps35p. *Journal of Cell Biology* **151**, 297–310.

Novick, P. and Botstein, D. (1985) Phenotypic analysis of temperature-sensitive yeast actin mutants. *Cell* **40**, 405–416.

Odorizzi, G., Babst, M. and Emr, S. D. (1998) Fab1p PtdIns(3)P 5-kinase function essential for protein sorting in the multivesicular body. *Cell* **95**, 847–858.

Ogg, S. C. and Walter, P. (1995) SRP samples nascent chains for the presence of signal sequences by interacting with ribosomes at a discrete step in translation elongation. *Cell* **81**, 1075–1084.

Oka, T. and Nakano, A. (1994) Inhibition of GTP hydrolysis by Sar1p causes accumulation of vesicles that are a functional intermediate of the ER-to-Golgi transport in yeast. *Journal of Cell Biology* **124**, 425–434.

Okamura, K., Kimata, Y., Higashio, H., Tsuru, A. and Kohno, K. (2000) Dissociation of Kar2p/BiP from an ER sensory module, Ire1p, triggers the unfolded protein response in yeast. *Biochemical and Biophysical Reearchs Communications* **279**, 445–450.

Orci, L., Palmer, D. J., Ravazzola, M., Perrelet, A., Amherdt, M. and Rothman, J. E. (1993) Budding from Golgi membranes requires the coatomer complex of non-clathrin coat proteins. *Nature* **362**, 648–652.

Ostermann, J., Orci, L., Tani, K., Amherdt, M., Ravazzola, M., Elazar, Z. and Rothman, J. E. (1993) Stepwise assembly of functionally active transport vesicles. *Cell* **75**, 1015–1025.

Palade, G. (1975) Intracellular aspects of the process of protein synthesis. *Science* **189**, 347–358.

Panzner, S., Dreier, L., Hartmann, E., Kostka, S. and Rapoport, T. A. (1995) Posttranslational protein transport in yeast reconstituted with a purified complex of sec proteins and Kar2p. *Cell* **81**, 561–570.

Parlati, F., Dominguez, M., Bergeron, J. J. and Thomas, D. Y. (1995) *Saccharomyces cerevisiae CNE1* encodes an endoplasmic reticulum (ER) membrane protein with sequence similarity to calnexin and calreticulin and functions as a constituent of the ER quality control apparatus. *Journal of Biological Chemistry* **270**, 244–253.

Patil, C. and Walter, P. (2001) Intracellular signaling from the endoplasmic reticulum to the nucleus: the unfolded protein response in yeast and mammals. *Current Opinion in Cell Biology* **13**, 349–355.

Pelham, H. R. (1999) SNAREs and the secretory pathway-lessons from yeast. *Experimental Cell Research* **247**, 1–8.

Pelham, H. R. (2001) Traffic through the Golgi apparatus. *Journal of Cell Biology* **155**, 1099–1101.

Pelham, H. R., Hardwick, K. G. and Lewis, M. J. (1988) Sorting of soluble ER proteins in yeast. *EMBO Journal* **7**, 1757–1762.

Peng, R. and Gallwitz, D. (2002) Sly1 protein bound to Golgi syntaxin Sed5p allows assembly and contributes to specificity of SNARE fusion complexes. *Journal of Cell Biology* **157**, 645–655.

Peters, C. and Mayer, A. (1998) Ca2+/calmodulin signals the completion of docking and triggers a late step of vacuole fusion. *Nature* **396**, 575–580.

Peters, C., Andrews, P. D., Stark, M. J., Cesaro-Tadic, S., Glatz, A., Podtelejnikov, A., Mann, M. and Mayer, A. (1999) Control of the terminal step of intracellular membrane fusion by protein phosphatase 1. *Science* **285**, 1084–1087.

Peyroche, A., Paris, S. and Jackson, C. L. (1996) Nucleotide exchange on ARF mediated by yeast Gea1 protein. *Nature* **384**, 479–481.

Peyroche, A., Courbeyrette, R., Rambourg, A. and Jackson, C. L. (2001) The ARF exchange factors Gea1p and Gea2p regulate Golgi structure and function in yeast. *Journal of Cell Science* **114**, 2241–2253.

Pfanner, N. and Meijer, M. (1997) Mitochondrial biogenesis: the Tom and Tim machine. *Current Biology* **7**, 100–107.

Pfanner, N. and Wiedemann, N. (2002) Mitochondrial protein import: two membranes, three translocases. *Current Opinion in Cell Biology* **14**, 400.

Pilon, M., Schekman, R. and Romisch, K. (1997) Sec61p mediates export of a misfolded secretory protein from the endoplasmic reticulum to the cytosol for degradation. *EMBO Journal* **16**, 4540–4548.

Piper, R. C., Cooper, A. A., Yang, H. and Stevens, T. H. (1995) VPS27 controls vacuolar and endocytic traffic through a prevacuolar compartment in *Saccharomyces cerevisiae*. *Journal of Cell Biology* **131**, 603–617.

Piper, R. C., Bryant, N. J. and Stevens, T. H. (1997) The membrane protein alkaline phosphatase is delivered to the vacuole by a route that is distinct from the VPS-dependent pathway. *Journal of Cell Biology* **138**, 531–545.

Plath, K., Mothes, W., Wilkinson, B. M., Stirling, C. J. and Rapoport, T. A. (1998) Signal sequence recognition in posttranslational protein transport across the yeast ER membrane. *Cell* **94**, 795–807.

Plemper, R. K., Bohmler, S., Bordallo, J., Sommer, T. and Wolf, D. H. (1997) Mutant analysis links the translocon and BiP to retrograde protein transport for ER degradation. *Nature* **388**, 891–895.

Poon, P. P., Wang, X., Rotman, M., Huber, I., Cukierman, E., Cassel, D., Singer, R. A. and Johnston, G. C. (1996) *Saccharomyces cerevisiae* Gcs1 is an ADP-ribosylation factor GTPase-activating protein. *Proceedings of the National Academy of Sciences USA* **93**, 10074–10077.

Poon, P. P., Cassel, D., Spang, A., Rotman, M., Pick, E., Singer, R. A. and Johnston, G. C. (1999) Retrograde transport from the yeast Golgi is mediated by two ARF GAP proteins with overlapping function. *EMBO Journal* **18**, 555–564.

Poon, P. P., Nothwehr, S. F., Singer, R. A. and Johnston, G. C. (2001) The Gcs1 and Age2 ArfGAP proteins provide overlapping essential function for transport from the yeast trans-Golgi network. *Journal of Cell Biology* **155**, 1239–1250.

Powers, J. and Barlowe, C. (2002) Erv14p directs a transmembrane secretory protein into COPII-coated transport vesicles. *Molecular Biology of the Cell* **13**, 880–891.

Preuss, M., Leonhard, K., Hell, K., Stuart, R. A., Neupert, W. and Herrmann, J. M. (2001) Mba1, a novel component of the mitochondrial protein export machinery of the yeast *Saccharomyces cerevisiae*. *Journal of Cell Biology* **153**, 1085–1096.

Prinz, W. A., Grzyb, L., Veenhuis, M., Kahana, J. A., Silver, P. A. and Rapoport, T. A. (2000) Mutants affecting the structure of the cortical endoplasmic reticulum in *Saccharomyces cerevisiae*. *Journal of Cell Biology* **150**, 461–474.

Purdue, P. E. and Lazarow, P. B. (2001a) Peroxisome biogenesis. *Annual Reviews in Cell and Developmental Biology* **17**, 701–752.

Purdue, P. E. and Lazarow, P. B. (2001b) Pex18p is constitutively degraded during peroxisome biogenesis. *Journal of Biological Chemistry* **276**, 47684–47689.

Purdue, P. E., Yang, X. and Lazarow, P. B. (1998) Pex18p and Pex21p, a novel pair of related peroxins essential for peroxisomal targeting by the PTS2 pathway. *Journal of Cell Biology* **143**, 1859–1869.

Radisky, D. C., Snyder, W. B., Emr, S. D. and Kaplan, J. (1997) Characterization of *VPS41*, a gene required for vacuolar trafficking and high-affinity iron transport in yeast. *Proceedings of the National Academy of Sciences USA* **94**, 5662–5666.

Raymond, C. K., Howald-Stevenson, I., Vater, C. A. and Stevens, T. H. (1992) Morphological classification of the yeast vacuolar protein sorting mutants: evidence for a prevacuolar compartment in class E vps mutants. *Molecular Biology of the Cell* **3**, 1389–1402.

Redding, K., Brickner, J. H., Marschall, L. G., Nichols, J. W. and Fuller, R. S. (1996) Allele-specific suppression of a defective trans-Golgi network (TGN) localization signal in Kex2p identifies three genes involved in localization of TGN transmembrane proteins. *Molecular and Cellular Biology* **16**, 6208–6217.

Reddy, J. V. and Seaman, M. N. (2001) Vps26p, a component of retromer, directs the interactions of Vps35p in endosome-to-Golgi retrieval. *Molecular Biology of the Cell* **12**, 3242–3256.

Reggiori, F. and Pelham, H. R. (2001) Sorting of proteins into multivesicular bodies: ubiquitin-dependent and -independent targeting. *EMBO Journal* **20**, 5176–5186.

Reggiori, F. and Pelham, H. R. (2002) A transmembrane ubiquitin ligase required to sort membrane proteins into multivesicular bodies. *Nature Cell Biology* **4**, 117–123.

Reggiori, F., Black, M. W. and Pelham, H. R. (2000) Polar transmembrane domains target proteins to the interior of the yeast vacuole. *Molecular Biology of the Cell* **11**, 3737–3749.

Rehling, P., Marzioch, M., Niesen, F., Wittke, E., Veenhuis, M. and Kunau, W. H. (1996) The import receptor for the peroxisomal targeting signal 2 (PTS2) in *Saccharomyces cerevisiae* is encoded by the *PAS7* gene. *EMBO Journal* **15**, 2901–2913.

Rehling, P., Darsow, T., Katzmann, D. J. and Emr, S. D. (1999) Formation of AP-3 transport intermediates requires Vps41 function. *Nature Cell Biology* **1**, 346–353.

Rehling, P., Skaletz-Rorowski, A., Girzalsky, W., Voorn-Brouwer, T., Franse, M. M., Distel, B., Veenhuis, M., Kunau, W. H. and Erdmann, R. (2000) Pex8p, an intraperoxisomal peroxin of *Saccharomyces cerevisiae* required for protein transport into peroxisomes binds the PTS1 receptor pex5p. *Journal of Biological Chemistry* **275**, 3593–3602.

Roberg, K. J., Crotwell, M., Espenshade, P., Gimeno, R. and Kaiser, C. A. (1999) LST1 is a SEC24 homologue used for selective export of the plasma membrane ATPase from the endoplasmic reticulum. *Journal of Cell Biology* **145**, 659–672.

Rothblatt, J. A. and Meyer, D. I. (1986) Secretion in yeast: translocation and glycosylation of prepro-a factor *in vitro* can occur via ATP-dependent post-translational mechanism. *EMBO Journal* **5**, 1031–1036.

Sacher, M., Barrowman, J., Wang, W., Horecka, J., Zhang, Y., Pypaert, M. and Ferro-Novick, S. (2001) TRAPP I implicated in the specificity of tethering in ER-to-Golgi transport. *Molecules to Cells* **7**, 433–442.

Saito-Nakano, Y. and Nakano, A. (2000) Sed4p functions as a positive regulator of Sar1p probably through inhibition of the GTPase activation by Sec23p. *Genes to Cells.* **5**, 1039–1048.

Salama, N. R., Yeung, T. and Schekman, R. W. (1993) The Sec13p complex and reconstitution of vesicle budding from the ER with purified cytosolic proteins. *EMBO Journal* **12**, 4073–4082.

Samuelson, J. C., Chen, M., Jiang, F., Moller, I., Wiedmann, M., Kuhn, A., Phillips, G. J. and Dalbey, R. E. (2000) YidC mediates membrane protein insertion in bacteria. *Nature* **40**, 637–641.

Sanderson, C. M. and Meyer, D. I. (1991) Purification and functional characterization of membranes derived from the rough endoplasmic reticulum of *Saccharomyces cerevisiae. Journal of Biological Chemistry* **266**, 13423–13430.

Sapperstein, S. K., Lupashin, V. V., Schmitt, H. D. and Waters, M. G. (1996) Assembly of the ER to Golgi SNARE complex requires Uso1p. *Journal of Cell Biology* **132**, 755–767.

Saris, N., Holkeri, H., Craven, R. A., Stirling, C. J. and Makarow, M. (1997) The Hsp70 homologue Lhs1p is involved in a novel function of the yeast endoplasmic reticulum, refolding and stabilization of heat-denatured protein aggregates. *Journal of Cell Biology* **137**, 813–824.

Sato, K., Sato, M. and Nakano, A. (1997) Rer1p as common machinery for the endoplasmic reticulum localization of membrane proteins. *Proceedings of the National Academy of Sciences USA* **94**, 9693–9698.

Sato, K., Sato, M. and Nakano, A. (2001a) Rer1p, a retrieval receptor for endoplasmic reticulum membrane proteins, is dynamically localized to the Golgi apparatus by coatomer. *Journal of Cell Biology* **152**, 935–944.

Sato, T. K., Overduin, M. and Emr, S. D. (2001b) Location, location, location: membrane targeting directed by PX domains. *Science* **294**, 1881–1885.

Schatz, G. and Dobberstein, B. (1996) Common principles of protein translocation across membranes. *Science* **271**, 1519–1526.

Schauer, I., Emr, S., Gross, C. and Schekman, R. (1985) Invertase signal and mature sequence substitutions that delay intercompartmental transport of active enzyme. *Journal of Cell Biology* **100**, 1664–1675.

Schimmoller, F., Singer-Kruger, B., Schroder, S., Kruger, U., Barlowe, C. and Riezman, H. (1995) The absence of Emp24p, a component of ER-derived COPII-coated vesicles, causes a defect in transport of selected proteins to the Golgi. *EMBO Journal* **14**, 1329–1339.

Schorr, M., Then, A., Tahirovic, S., Hug, N. and Mayinger, P. (2001) The phosphoinositide phosphatase Sac1p controls trafficking of the yeast Chs3p chitin synthase. *Current Biology* **11**, 1421–1426.

Scott, S. V., Hefner-Gravink, A., Morano, K. A., Noda, T., Ohsumi, Y. and Klionsky, D. J. (1996) Cytoplasm-to-vacuole targeting and autophagy employ the same machinery to deliver proteins to the yeast vacuole. *Proceedings of the National Academy of Sciences USA* **93**, 12304–12308.

Scott, S. V., Baba, M., Ohsumi, Y. and Klionsky, D. J. (1997) Aminopeptidase I is targeted to the vacuole by a nonclassical vesicular mechanism. *Journal of Cell Biology* **138**, 37–44.

Scott, S. V., Nice, D. C., III Nau, J. J., Weisman, L. S., Kamada, Y., Keizer-Gunnink, I., Funakoshi, T., Veenhuis, M., Ohsumi, Y. and Klionsky, D. J. (2000) Apg13p and Vac8p are part of a complex of phosphoproteins that are required for cytoplasm to vacuole targeting. *Journal of Biological Chemistry* **275**, 25840–25849.

Scott, S. V., Guan, J., Hutchins, M. U., Kim, J. and Klionsky, D. J. (2001) Cvt19 is a receptor for the cytoplasm-to-vacuole targeting pathway. *Molecules to Cells* **7**, 1131–1141.

Scotti, P. A., Urbanus, M. L., Brunner, J., deGier, J. W. L., vonHeijne, G., vanderDoes, C., Driessen, A. J. M., Oudega, B. and Luirink, J. (2000) YidC, the *Escherichia coli* homologue of mitochondrial Oxa1p, is a component of the Sec translocase. *EMBO Journal* **19**, 542–549.

Seaman, M. N., Marcusson, E. G., Cereghino, J. L. and Emr, S. D. (1997) Endosome to Golgi retrieval of the vacuolar protein sorting receptor, Vps10p, requires the function of the VPS29, VPS30, and VPS35 gene products. *Journal of Cell Biology* **137**, 79–92.

Seaman, M. N., McCaffery, J. M. and Emr, S. D. (1998) A membrane coat complex essential for endosome-to-Golgi retrograde transport in yeast. *Journal of Cell Biology* **142**, 665–681.

Seaman, M. N. and Williams, H. P. (2002) Identification of the functional domains of yeast sorting nexins Vps5p and Vps17p. *Molecular Biology of the Cell* **13**, 2826–2840.

Seeger, M. and Payne, G. S. (1992) A role for clathrin in the sorting of vacuolar proteins in the Golgi complex of yeast. *EMBO Journal* **11**, 2811–2818.

Segui-Real, B., Martinez, M. and Sandoval, I. V. (1995) Yeast aminopeptidase I is post-translationally sorted from the cytosol to the vacuole by a mechanism mediated by its bipartite N-terminal extension. *EMBO Journal* **14**, 5476–5484.

Semenza, J. C., Hardwick, K. G., Dean, N. and Pelham, H. R. (1990) *ERD2*, a yeast gene required for the receptor-mediated retrieval of luminal ER proteins from the secretory pathway. *Cell* **61**, 1349–1357.

Shaw, J. D., Cummings, K. B., Huyer, G., Michaelis, S. and Wendland, B. (2001) Yeast as a model system for studying endocytosis. *Experimental Cell Research* **271**, 1–9.

Shaywitz, D. A., Espenshade, P. J., Gimeno, R. E. and Kaiser, C. A. (1997) COPII subunit interactions in the assembly of the vesicle coat. *Journal of Biological Chemistry* **272**, 25413–25416.

Shieh, H. L., Chen, Y., Brown, C. R. and Chiang, H. L. (2001) Biochemical analysis of fructose-1,6-bisphosphatase import into vacuole import and degradation vesicles reveals a role for UBC1 in vesicle biogenesis. *Journal of Biological Chemistry* **276**, 10398–10406.

Shih, S. C., Katzmann, D. J., Schnell, J. D., Sutanto, M., Emr, S. D. and Hicke, L. (2002) Epsins and Vps27p/Hrs contain ubiquitin-binding domains that function in receptor endocytosis. *Nature Cell Biology* **4**, 389–393.

Shimoni, Y., Kurihara, T., Ravazzola, M., Amherdt, M., Orci, L. and Schekman, R. (2000) Lst1p and Sec24p cooperate in sorting of the plasma membrane ATPase into COPII vesicles in *Saccharomyces cerevisiae*. *Journal of Cell Biology* **151**, 973–984.

Short, B. and Barr, F. A. (2002) Membrane traffic: exocyst III—makes a family. *Current Biology* **12**, R18–R20.

Sidrauski, C. and Walter, P. (1997) The transmembrane kinase Ire1p is a site-specific endonuclease that initiates mRNA splicing in the unfolded protein response. *Cell* **90**, 1031–1039.

Silberstein, S., Schlenstedt, G., Silver, P. A. and Gilmore, R. (1998) A role for the DnaJ homologue Scj1p in protein folding in the yeast endoplasmic reticulum. *Journal of Cell Biology* **143**, 921–933.

Simons, J. F., Ferro-Novick, S., Rose, M. D. and Helenius, A. (1995) BiP/Kar2p serves as a molecular chaperone during carboxypeptidase Y folding in yeast. *Journal of Cell Biology* **130**, 41–49.

Sirrenberg, C., Bauer, M. F., Guiard, B., Neupert, W. and Brunner, M. (1996) Import of carrier proteins into the mitochondrial inner membrane mediated by Tim22. *Nature* **384**, 582–585.

Sirrenberg, C., Endres, M., Folsch, H., Stuart, R. A., Neupert, W. and Brunner, M. (1998) Carrier protein import into mitochondria mediated by the intermembrane proteins Tim10/Mrs11 and Tim12/Mrs5. *Nature* **391**, 912–915.

Smith, M. D. and Schnell, D. J. (2001) Peroxisomal protein import. the paradigm shifts. *Cell* **105**, 293–296.

Snyder, W. B., Faber, K. N., Wenzel, T. J., Koller, A., Luers, G. H., Rangell, L., Keller, G. A. and Subramani, S. (1999) Pex19p interacts with Pex3p and Pex10p and is essential for peroxisome biogenesis in *Pichia pastoris*. *Molecular Biology of the Cell* **10**, 1745–1761.

Sogaard, M., Tani, K., Ye, R. R., Geromanos, S., Tempst, P., Kirchhausen, T., Rothman, J. E. and Sollner, T. (1994) A rab protein is required for the assembly of SNARE complexes in the docking of transport vesicles. *Cell* **78**, 937–948.

Sollner, T., Bennett, M. K., Whiteheart, S. W., Scheller, R. H. and Rothman, J. E. (1993) A protein assembly-disassembly pathway in vitro that may correspond to sequential steps of synaptic vesicle docking, activation, and fusion. *Cell* **75**, 409–418.

South, S. T., Sacksteder, K. A., Li, X., Liu, Y. and Gould, S. J. (2000) Inhibitors of COPI and COPII do not block PEX3-mediated peroxisome synthesis. *Journal of Cell Biology* **149**, 1345–1360.

South, S. T., Baumgart, E. and Gould, S. J. (2001) Inactivation of the endoplasmic reticulum protein translocation factor, Sec61p, or its homolog, Ssh1p, does not affect peroxisome biogenesis. *Proceedings of the National Academy of Sciences USA* **98**, 12027–12031.

Spang, A., Herrmann, J. M., Hamamoto, S. and Schekman, R. (2001) The ADP ribosylation factor-nucleotide exchange factors Gea1p and Gea2p have overlapping, but not redundant functions in retrograde transport from the Golgi to the endoplasmic reticulum. *Molecular Biology of the Cell* **12**, 1035–1045.

Spear, E. and Ng, D. T. (2001) The unfolded protein response: no longer just a special teams player. *Traffic* **2**, 515–523.

Springer, S. and Schekman, R. (1998) Nucleation of COPII vesicular coat complex by endoplasmic reticulum to Golgi vesicle SNAREs. *Science* **281**, 698–700.

Springer, S., Chen, E., Duden, R., Marzioch, M., Rowley, A., Hamamoto, S., Merchant, S. and Schekman, R. (2000) The p24 proteins are not essential for vesicular transport in *Saccharomyces cerevisiae. Proceedings of the National Academy of Sciences USA* **97**, 4034–4039.

Stack, J. H., Herman, P. K., DeWald, D. B., Marcusson, E. G., Lin Cereghino, J., Horazdovsky, B. F. and Emr, S. D. (1995a) Novel protein kinase/phosphatidylinositol 3-kinase complex essential for receptor-mediated protein sorting to the vacuole in yeast. *Cold Spring Harbor Symposia on Quantitative Biol*ogy **60**, 157–170.

Stack, J. H., Horazdovsky, B. and Emr, S. D. (1995b) Receptor-mediated protein sorting to the vacuole in yeast: roles for a protein kinase, a lipid kinase and GTP-binding proteins. *Annual Reviews of Cell and Developmental Biol*ogy **11**, 1–33.

Stahelin, R. V., Long, F., Diraviyam, K., Bruzik, K. S., Murray, D. and Cho, W. (2002) Phosphatidylinositol 3-phosphate induces the membrane penetration of the FYVE domains of Vps27p and Hrs. *Journal of Biological Chemistry* **277**, 26379–26388.

Stearns, T., Kahn, R. A., Botstein, D. and Hoyt, M. A. (1990) ADP ribosylation factor is an essential ,protein in *Saccharomyces cerevisiae* and is encoded by two genes. *Molecular and Cellular Biology* **10**, 6690–6699.

Steel, G. J., Brownsword, J. and Stirling, C. J. (2002) Tail-anchored protein insertion into yeast ER requires a novel posttranslational mechanism which is independent of the SEC machinery. *Biochemistry* **41**, 11914–11920.

Stein, K., Schell-Steven, A., Erdmann, R. and Rottensteiner, H. (2002) Interactions of Pex7p and Pex18p/Pex21p with the peroxisomal docking machinery: implications for the first steps in PTS2 protein import. *Molecular and Cellular Biology* **22**, 6056–6069.

Stenbeck, G., Schreiner, R., Herrmann, D., Auerbach, S., Lottspeich, F., Rothman, J. E. and Wieland, F. T. (1992) Gamma-COP, a coat subunit of non-clathrin-coated vesicles with homology to Sec21p. *FEBS Letters* **314**, 195–198.

Stenmark, H., Aasland, R. and Driscoll, P. C. (2002) The phosphatidylinositol 3-phosphate-binding FYVE finger. *FEBS Letters* **513**, 77–84.

Strub, A., Rottgers, K. and Voos, W. (2002) The Hsp70 peptide-binding domain determines the interaction of the ATPase domain with Tim44 in mitochondria. *EMBO Journal* **21**, 2626–2635.

Suissa, M. and Schatz, G. (1982) Import of proteins into mitochondria. Translatable mRNAs for imported mitochondrial proteins are present in free as well as mitochondria-bound cytoplasmic polysomes. *Journal of Biological Chemistry* **257**, 13048–13055.

Supek, F., Madden, D. T., Hamamoto, S., Orci, L. and Schekman, R. (2002) Sec16p potentiates the action of COPII proteins to bud transport vesicles. *Journal of Cell Biology* **158**, 1029–1038.

Sutton, R. B., Fasshauer, D., Jahn, R. and Brunger, A. T. (1998) Crystal structure of a SNARE complex involved in synaptic exocytosis at 2.4 A resolution. *Nature* **395**, 347–353.

Suzuki, K., Kirisako, T., Kamada, Y., Mizushima, N., Noda, T. and Ohsumi, Y. (2001) The pre-autophagosomal structure organized by concerted functions of *APG* genes is essential for autophagosome formation. *EMBO Journal* **20**, 5971–5981.

Swanson, R., Locher, M. and Hochstrasser, M. (2001) A conserved ubiquitin ligase of the nuclear envelope/endoplasmic reticulum that functions in both ER-associated and Matalpha2 repressor degradation. *Genes and Development* **15**, 2660–2674.

Sweet, D. J. and Pelham, H. R. (1992) *The Saccharomyces cerevisiae SEC20* gene encodes a membrane glycoprotein which is sorted by the HDEL retrieval system. *EMBO Journal* **11**, 423–432.

Tam, Y. Y. and Rachubinski, R. A. (2002) *Yarrowia lipolytica* cells mutant for the *PEX24* gene encoding a peroxisomal membrane peroxin mislocalize peroxisomal proteins and accumulate membrane structures containing both peroxisomal matrix and membrane proteins. *Molecular Biology of the Cell* **13**, 2681–2691.

Taxis, C., Vogel, F. and Wolf, D. H. (2002) ER-golgi traffic is a prerequisite for efficient ER degradation. *Molecular Biology of the Cell* **13**, 1806–1818.

Titorenko, V. I. and Rachubinski, R. A. (1998a) The endoplasmic reticulum plays an essential role in peroxisome biogenesis. *Trends in Biochemical Sciences* **23**, 231–233.

Titorenko, V. I. and Rachubinski, R. A. (1998b) Mutants of the yeast *Yarrowia lipolytica* defective in protein exit from the endoplasmic reticulum are also defective in peroxisome biogenesis. *Molecular and Cellular Biology* **18**, 2789–2803.

Titorenko, V. I., Nicaud, J. M., Wang, H., Chan, H. and Rachubinski, R. A. (2002) Acyl-CoA oxidase is imported as a heteropentameric, cofactor-containing complex into peroxisomes of *Yarrowia lipolytica*. *Journal of Cell Biology* **156**, 481–494.

Tokatlidis, K., Junne, T., Moes, S., Schatz, G., Glick, B. S. and Kronidou, N. (1996) Translocation arrest of an intramitochondrial sorting signal next to Tim11 at the inner-membrane import site. *Nature* **384**, 585–588.

Travers, K. J., Patil, C. K., Wodicka, L., Lockhart, D. J., Weissman, J. S. and Walter, P. (2000) Functional and genomic analyses reveal an essential coordination between the unfolded protein response and ER-associated degradation. *Cell* **101**, 249–258.

Trilla, J. A., Duran, A. and Roncero, C. (1999) Chs7p, a new protein involved in the control of protein export from the endoplasmic reticulum that is specifically engaged in the regulation of chitin synthesis in *Saccharomyces cerevisiae*. *Journal of Cell Biology* **145**, 1153–1163.

Truscott, K. N., Kovermann, P., Geissler, A., Merlin, A., Meijer, M., Driessen, A. J., Rassow, J., Pfanner, N. and Wagner, R. (2001) A presequence- and voltage-sensitive channel of the mitochondrial preprotein translocase formed by Tim23. *Nature Structural Biology* **8**, 1074–1082.

Tyson, J. R. and Stirling, C. J. (2000) LHS1 and SIL1 provide a lumenal function that is essential for protein translocation into the endoplasmic reticulum. *EMBO Journal* **19**, 6440–6452.

Ulbrandt, N. D., Newitt, J. A. and Bernstein, H. D. (1997) The *E-coli* signal recognition particle is required for the insertion of a subset of inner membrane proteins. *Cell* **88**, 187–196.

Ungermann, C., Guiard, B., Neupert, W. and Cyr, D. M. (1996) The delta psi- and Hsp70/MIM44-dependent reaction cycle driving early steps of protein import into mitochondria. *EMBO Journal* **15**, 735–744.

Urbanowski, J. L. and Piper, R. C. (2001) Ubiquitin sorts proteins into the intralumenal degradative compartment of the late-endosome/vacuole. *Traffic* **2**, 622–630.

Vale, R. D. (2000) AAA proteins. Lords of the ring. *Journal of Cell Biology* **150**, F13–F20.

van Wilpe, S., Ryan, M. T., Hill, K. Maarse, A. C., Meisinger, C., Brix, J., Dekker, P. J., Moczko, M., Wagner, R., Meijer, M., Guiard, B., Honlinger, A. and Pfanner, N. (1999) Tom22 is a multifunctional organizer of the mitochondrial preprotein translocase. *Nature* **401**, 485–489.

Vashist, S., Kim, W., Belden, W. J., Spear, E. D., Barlowe, C. and Ng, D. T. (2001) Distinct retrieval and retention mechanisms are required for the quality control of endoplasmic reticulum protein folding. *Journal of Cell Biology* **155**, 355–368.

Verner, K. (1993) Co-translational protein import into mitochondria: an alternative view. *Trends in Biochemical Sciences* **18**, 366–371.

Vida, T. A., Huyer, G. and Emr, S. D. (1993) Yeast vacuolar proenzymes are sorted in the late Golgi complex and transported to the vacuole via a prevacuolar endosome-like compartment. *Journal of Cell Biology* **121**, 1245–1256.

Voisine, C., Craig, E. A., Zufall, N., von Ahsen, O., Pfanner, N. and Voos, W. (1999) The protein import motor of mitochondria: unfolding and trapping of preproteins are distinct and separable functions of matrix Hsp70. *Cell* **97**, 565–574.

von Mollard, G. F., Nothwehr, S. F. and Stevens, T. H. (1997) The yeast v-SNARE Vti1p mediates two vesicle transport pathways through interactions with the t-SNAREs Sed5p and Pep12p. *Journal of Cell Biology* **137**, 1511–1524.

Voos, W. and Stevens, T. H. (1998) Retrieval of resident late-Golgi membrane proteins from the prevacuolar compartment of *Saccharomyces cerevisiae* is dependent on the function of Grd19p. *Journal of Cell Biology* **140**, 577–590.

Vowels, J. J. and Payne, G. S. (1998) A dileucine-like sorting signal directs transport into an AP-3-dependent, clathrin-independent pathway to the yeast vacuole. *EMBO Journal* **17**, 2482–2493.

Walton, P. A., Hill, P. E. and Subramani, S. (1995) Import of stably folded proteins into peroxisomes. *Molecular Biology of the Cell* **6**, 675–683.

Wang, W., Sacher, M. and Ferro-Novick, S. (2000) TRAPP stimulates guanine nucleotide exchange on Ypt1p. *Journal of Cell Biology* **151**, 289–296.

Weber, T., Zemelman, B. V., McNew, J. A., Westermann, B., Gmachl, M., Parlati, F., Sollner, T. H. and Rothman, J. E. (1998) SNAREpins: minimal machinery for membrane fusion. *Cell.* **92**, 759–772.

Westermann, B., Prip-Buus, C., Neupert, W. and Schwarz, E. (1995) The role of the GrpE homologue, Mge1p, in mediating protein import and protein folding in mitochondria. *EMBO Journal* **14**, 3452–3460.

Westphal, V., Marcusson, E. G., Winther, J. R., Emr, S. D. and van den Hazel, H. B. (1996) Multiple pathways for vacuolar sorting of yeast proteinase A. *Journal of Biological Chemistry* **271**, 11865–11870.

Whyte, J. R. and Munro, S. (2001) A yeast homolog of the mammalian mannose 6-phosphate receptors contributes to the sorting of vacuolar hydrolases. *Current Biology* **11**, 1074–1078.

Wiebel, F. F. and Kunau, W. H. (1992) The Pas2 protein essential for peroxisome biogenesis is related to ubiquitin-conjugating enzymes. *Nature* **359**, 73–76.

Wiedemann, N., Pfanner, N. and Ryan, M. T. (2001) The three modules of ADP/ATP carrier cooperate in receptor recruitment and translocation into mitochondria. *EMBO Journal* **20**, 951–960.

Wiedmann, B., Sakai, H., Davis, T. A. and Wiedmann, M. (1994) A protein complex required for signal-sequence-specific sorting and translocation. *Nature* **370**, 434–440.

Wild, K., Weichenrieder, O., Strub, K., Sinning, I. and Cusack, S. (2002) Towards the structure of the mammalian signal recognition particle. *Current Opinion in Structural Biology* **12**, 72–81.

Wilkinson, B. M., Critchley, A. J. and Stirling, C. J. (1996) Determination of the transmembrane topology of yeast Sec61p, an essential component of the endoplasmic-reticulum translocation complex. *Journal of Biological Chemistry* **271**, 25590–25597.

Wilkinson, B. M., Tyson, J. R. and Stirling, C. J. (2001) Ssh1p determines the translocation and dislocation capacities of the yeast endoplasmic reticulum. *Developmental Cell* **1**, 401–409.

Wilson, D. W., Lewis, M. J. and Pelham, H. R. (1993) pH-dependent binding of KDEL to its receptor in vitro. *Journal of Biological Chemistry* **268**, 7465–7468.

Wilson, D. W., Wilcox, C. A., Flynn, G. C., Chen, E., Kuang, W. J., Henzel, W. J., Block, M. R., Ullrich, A. and Rothman, J. E. (1989) A fusion protein required for vesicle-mediated transport in both mammalian cells and yeast. *Nature* **339**, 355–359.

Wittke, S., Dunnwald, M., Albertsen, M. and Johnsson, N. (2002) Recognition of a subset of signal sequences by Ssh1p, a Sec61p-related protein in the membrane of endoplasmic reticulum of yeast *Saccharomyces cerevisiae. Molecular Biology of the Cell* **13**, 2223–2232.

Wooding, S. and Pelham, H. R. (1998) The dynamics of golgi protein traffic visualized in living yeast cells. *Molecular Biology of the Cell* **9**, 2667–2680.

Wurmser, A. E. and Emr, S. D. (1998) Phosphoinositide signaling and turnover: PtdIns(3)P, a regulator of membrane traffic, is transported to the vacuole and degraded by a process that requires lumenal vacuolar hydrolase activities. *EMBO Journal* **17**, 4930–4942.

Wurmser, A. E. and Emr, S. D. (2002) Novel PtdIns(3)P-binding protein Etf1 functions as an effector of the Vps34 PtdIns 3-kinase in autophagy. *Journal of Cell Biology* **158**, 761–772.

Xu, Z., Mayer, A., Muller, E. and Wickner, W. (1997) A heterodimer of thioredoxin and I(B)2 cooperates with Sec18p (NSF) to promote yeast vacuole inheritance. *Journal of Cell Biology* **136**, 299–306.

Yaglom, J. A., Goldberg, A. L., Finley, D. and Sherman, M. Y. (1996) The molecular chaperone Ydj1 is required for the $p34^{CDC28}$-dependent phosphorylation of the cyclin Cln3 that signals its degradation. *Molecular and Cellular Biology* **16**, 3679–3684.

Yahara, N., Ueda, T., Sato, K. and Nakano, A. (2001) Multiple roles of Arf1 GTPase in the yeast exocytic and endocytic pathways. *Molecular Biology of the Cell* **12**, 221–238.

Yamaguchi, T., Dulubova, I., Min, S. W., Chen, X., Rizo, J. and Sudhof, T. C. (2002) Sly1 binds to Golgi and ER syntaxins via a conserved N-terminal peptide motif. *Developmental Cell* **2**, 295–305.

Yamamoto, H., Esaki, M., Kanamori, T., Tamura, Y., Nishikawa, S. and Endo, T. (2002) Tim50 is a subunit of the TIM23 complex that links protein translocation across the outer and inner mitochondrial membranes. *Cell* **111**, 519–528.

Ye, Y., Meyer, H. H. and Rapoport, T. A. (2001) The AAA ATPase Cdc48/p97 and its partners transport proteins from the ER into the cytosol. *Nature* **414**, 652–656.

Yeung, B. G., Phan, H. L. and Payne, G. S. (1999) Adaptor complex-independent clathrin function in yeast. *Molecular Biology of the Cell* **10**, 3643–3659.

Yoshihisa, T., Barlowe, C. and Schekman, R. (1993) Requirement for a GTPase-activating protein in vesicle budding from the endoplasmic reticulum. *Science* **259**, 1466–1468.

Young, B. P., Craven, R. A., Reid, P. J., Willer, M. and Stirling, C. J. (2001) Sec63p and Kar2p are required for the translocation of SRP-dependent precursors into the yeast endoplasmic reticulum in vivo. *EMBO Journal* **20**, 262–271.

Zhang, C. J., Cavenagh, M. M. and Kahn, R. A. (1998) A family of Arf effectors defined as suppressors of the loss of Arf function in the yeast *Saccharomyces cerevisiae*. *Journal of Biological Chemistry* **273**, 19792–19796.

Zhang, J. W. and Lazarow, P. B. (1996) Peb1p (Pas7p) is an intraperoxisomal receptor for the NH2-terminal, type 2, peroxisomal targeting sequence of thiolase: Peb1p itself is targeted to peroxisomes by an NH2-terminal peptide. *Journal of Cell Biology* **132**, 325–334.

Zhang, Y., Nijbroek, G., Sullivan, M. L., McCracken, A. A., Watkins, S. C., Michaelis, S. and Brodsky, J. L. (2001) Hsp70 molecular chaperone facilitates endoplasmic reticulum-associated protein degradation of cystic fibrosis transmembrane conductance regulator in yeast. *Molecular Biology of the Cell* **12**, 1303–1314.

Zhou, M. and Schekman, R. (1999) The engagement of Sec61p in the ER dislocation process. *Molecules to Cells* **4**, 925–934.

Ziegelhoffer, T., Lopez-Buesa, P. and Craig, E. A. (1995) The dissociation of ATP from hsp70 of *Saccharomyces cerevisiae* is stimulated by both Ydj1p and peptide substrates. *Journal of Biological Chemistry* **270**, 10412–10419.

Chapter 8

Protein phosphorylation and dephosphorylation

Michael J. R. Stark

8.1 Introduction

Protein phosphorylation is a universal regulatory mechanism and in eukaryotic cells is a pervasive phenomenon that permeates almost every aspect of cell physiology and biochemistry. It is therefore almost impossible to write a book, let alone a single book chapter, that does justice to the subject of protein phosphorylation even in a simple eukaryotic organism such as yeast. In the yeast *Saccharomyces cerevisiae* as in all eukaryotes, protein phosphorylation plays critical regulatory roles in a large number of cellular processes. This chapter will review selected examples of protein kinases and protein phosphatases where sufficient progress has been made to allow a meaningful overview to be taken, and particularly those instances where the molecular events that have been uncovered can be linked to the physiological responses of the cell. The essence of protein phosphorylation is that by adding a phosphate group to a polypeptide, a protein kinase can modify the biological function of its target. Introduction of a phosphate group can have a variety of effects, including steric blocking and either allosteric or conformational inactivation or activation. A good example of the latter is the activation of some protein kinases by phosphorylation on their activation or T-loop (between subdomains VII and VIII), leading to a conformational change that allows access of the protein kinase to its substrate (Russo *et al.*, 1996; Huse and Kuriyan, 2002). For phosphoregulation to be reversible, it is necessary to be able to remove as well as add the key phosphate groups. Thus is it the balance between the phosphorylation and dephosphorylation of specific residues that controls their phosphorylation state and so, unless there is very rapid protein turnover, phosphoprotein phosphatases are also critical players. Although protein phosphatases have been considered to be rather non-specific enzymes in the past, providing simply a constant background level of dephosphorylation to oppose the protein kinases, it is now accepted that they frequently have highly specific roles to play.

The yeast *S. cerevisiae* provides an excellent model system in which to explore the role of protein phosphorylation in cellular regulation. While lacking the higher levels of organisation and specialisation required in multicellular organisms, many key intracellular processes are nonetheless surprisingly well conserved in yeast. Thus work in yeast on the role of protein phosphorylation in processes as disparate as the control of cell division and nutrient signalling have demonstrated that yeast has much to teach us and that yeast and higher cells frequently show striking parallels. Since the completion of the yeast genome sequence in April 1996 it has been possible to identify with relative certainty almost all of the protein kinase and protein phosphatase genes in this simple eukaryote (Stark, 1996; Hunter and

Plowman, 1997) and for over three quarters of them there is now at least some evidence regarding their likely function (see e.g. *Saccharomyces* Genome Database: Dolinski *et al.*, 2002). Nonetheless, while the functions of many protein kinases and phosphatases are now fairly well understood, it will clearly be a long time before the roles of the majority of these components in controlling the physiological responses of yeast are uncovered.

Although what follows broadly divides the subject between consideration of the protein kinases and the protein phosphatases, the two are clearly intimately entwined. Thus wherever possible, the interplay between the kinases and the phosphatases will be discussed. First, though, these two groups of enzymes will be introduced in the context of *S. cerevisiae*.

Protein kinases in Saccharomyces cerevisiae

Throughout the eukaryotic world, the vast majority of protein kinases form a family of related sequences that can phosphorylate proteins on serine and threonine, on tyrosine or on both (Hanks *et al.*, 1988). In the yeast *S. cerevisiae* there are 115 clearly identifiable genes encoding these 'conventional' protein kinases (Hunter and Plowman, 1997). Yeast also encodes a number of phosphatidylinositol kinase-related sequences that, like their mammalian homologues, also function as protein kinases (Hunter, 1995). Although yeast doesn't encode members of the novel protein kinase family defined by *Dictyostelium* myosin heavy chain kinase (Futey *et al.*, 1995) and eEF2 kinase (Ryazanov *et al.*, 1997), it does have a sequence related to the atypical human tyrosine kinase A6 (Hunter and Plowman, 1997). Finally, there are two examples of histidine kinases in *S. cerevisiae* (Hunter and Plowman, 1997). One of these is Sln1p, a protein related to the 'two-component' signal transducers more commonly found in bacteria. Thus in total, *S. cerevisiae* has 122 protein kinases (Table 8.1), corresponding to about 2% of the genome. Some of these may have highly specific functions (e.g. Swe1p, which phosphorylates tyr-19 on Cdc28p: Booher *et al.*, 1993), while others such as cAMP-dependent protein kinase (PKA) may have a plethora of different roles, and many act in combination with regulatory subunits that are required for their activity.

S. cerevisiae has no true members of the tyrosine kinase family, perhaps reflecting the fact that tyrosine kinases are involved primarily in signalling events required for multicellular existence. However, tyrosine phosphorylation is clearly important in yeast. Firstly, phosphorylation of MAP kinases (MAPKs) on a conserved tyrosine residue in their activation loop is critical for their activity and yeast has several examples of MAPK kinases (MAPKKs: see later). Secondly, even excluding the MAPKKs there are several examples of protein kinases that have been shown to phosphorylate tyrosine residues: Swe1p (Booher *et al.*, 1993), Rim1p (Malathi *et al.*, 1999), Mps1p (Lauze *et al.*, 1995), Rad53p (Spk1p: Zheng *et al.*, 1993), Mck1p (Lim *et al.*, 1993), the yeast casein kinase I isoforms (Hoekstra *et al.*, 1994) and Yak1p

Table 8.1 Identifiable *S. cerevisiae* protein kinases

Type	*Number*	*Comments*
Conventional protein kinases	115	See Hunter and Plowman (1997) for a complete list
Atypical protein kinases	1	Twf1p (Similar to human tyrosine kinase A6)
Phosphatidylinositol kinase-related protein kinases	4	Mec1p, Tel1p, Tor1p, Tor2p
Histidine kinases	2	Sln1p, YIL042c

(Kassis *et al.*, 2000). Indeed, a recent survey of the activity of 119 of the yeast protein kinases using protein chip technology showed that at least 27 could phosphorylate tyrosine *in vitro*, and this ability correlated with the presence of certain preferred residues at four positions within the catalytic domain (Zhu *et al.*, 2000b). Finally, since nearly half the protein phosphatase genes of yeast encode members of the tyrosine/dual specificity class (see Table 8.3), this implies that the cell has a significant requirement for the dephosphorylation of phosphotyrosine residues.

Protein phosphatases in Saccharomyces cerevisiae

Several different classes of protein phosphatase have been defined in eukaryotes. Protein ser/thr phosphatases (PPases) were initially characterised according to their biochemical properties (Cohen, 1989) as shown in Table 8.2, and yeast were found to contain protein phosphatase activities with similar properties (Cohen *et al.*, 1989). However, molecular cloning has revealed that while PP1, PP2A and PP2B belong to a single family of proteins (the PPP family), PP2Cs (the PPM family) are unrelated in sequence. Furthermore, both of these families comprise more members than suggested by the original biochemical classification. Even in *S. cerevisiae* there are twelve PPP phosphatases, only half of which are true homologues of mammalian PP1, PP2A and PP2B, and seven examples of PPM sequences (Table 8.3). Thus while yeast has definite homologues of mammalian enzymes such as PP1, PP2A and PP2B that encode the major PPP phosphatases contributing to the activities summarised in

Table 8.2 Biochemical classification of protein phosphatase activity

Type	*Sensitivity to inhibitors 1 and 2*	*Phosphorylase subunit preference*	*ID_{50} for okadaic acid*	*Divalent cation dependence*
PP1	yes	β	10–15 nM	none
PP2A	no	α	1 nM	none
PP2B	no	α	>50 mM	Ca^{2+}
PP2C	no	α	refractory	Mg^{2+}

Table 8.3 *S. cerevisiae* protein phosphatases

Type	*Specificity*	*Number*	*Names*
PPP family	Ser/Thr	12	PP1: Glc7p PP2A: Pph21p, Pph22p PP2B: Cna1p, Cna2p Others: Ppg1p, Pph3p, Ppq1p, Ppt1p, Ppz1p, Ppz2p, Sit4p
PPM family (PP2C)	Ser/Thr	7	Ptc1p, Ptc2p, Ptc3p, Ptc4p, Ptc5p, Ptc7p, YCR079w
Protein tyrosine phosphatase	Tyr and dual specificity (Ser/Thr)	12	Cdc14p, Mih1p, Msg5p, Pps1p, Ptp1p, Ptp2p, Ptp3p, Sdp1p, Siw14p, Tep1p, Yvh1p, YJR110w
Low molecular weight phosphatase	Tyr	1	Ltp1p
DXDX(T/V) family	Ser (Thr?)	3	Fcp1p, Psr1p, Psr2p

Table 8.2, there are clearly several other related phosphatases, and many of these (e.g. Sit4p) have important cellular roles. A third major family of protein phosphatases are the tyrosine and dual-specificity phosphatases (PTPases), which constitute a completely different family of polypeptides (twelve examples in *S. cerevisiae*), while the low molecular weight PTPases form a fourth group (one example in *S. cerevisiae*), which is thought to have shown convergent evolution with the main PTPase family (Zhang *et al.*, 1994). Since the last major review of *S. cerevisiae* protein phosphatases (Stark, 1996), a fifth group has also emerged that includes Fcp1p, Psr1p and Psr2p (Table 8.3). Fcp1p is the yeast homologue of the RNA polymerase II (RNAPII) C-terminal phosphatase that dephosphorylates ser-2 within C-terminal domain (CTD) heptad repeat motifs (Kobor *et al.*, 1999; Cho *et al.*, 2001), while Psr1p and Psr2p are plasma membrane-associated protein phosphatases that are required during sodium ion stress and contribute towards the induction of the Pmr2p sodium extrusion pump under such conditions (Siniossoglou *et al.*, 2000). These latter protein phosphatases share with phosphomutases a D-X-D-X-(T/V) motif in their catalytic domain that is required for phosphatase activity (Collet *et al.*, 1998; Siniossoglou *et al.*, 2000). Since yeast also encodes two further proteins with strong similarity to Psr1/2p in this region, it is possible that there may be additional members of this group. However, for now the number of protein phosphatase genes in *S. cerevisiae* totals thirty-five (Table 8.3). Figure 8.1 shows the remarkable sequence conservation among the yeast PPP phosphatases and also with their mammalian counterparts, in contrast to the much weaker conservation shown by the other groups (not shown).

An important consideration is how the specificity of protein phosphatases is controlled, since cells contain a huge variety of phosphoproteins generated by the action of many different protein kinases. The emerging theme is that, in many cases, their specificity is governed by the interaction of the catalytic subunit with one of a range of regulatory polypeptides or targeting subunits, which narrow the substrate specificity towards specific substrates or localise the catalytic subunit to specific subcellular locations (Hubbard and Cohen, 1993). This is clearest in the case of PP1 and PP2A and will be discussed later on in the chapter.

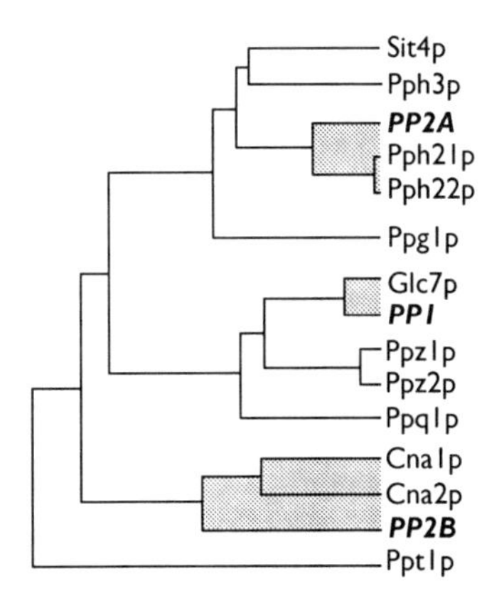

Figure 8.1 Phylogenetic relationships between the catalytic domains of yeast and mammalian PPP family members. The catalytic domains of each phosphatase sequence were first defined by pairwise comparison with the rabbit PP1 sequence and then multiple sequence alignment performed using Clustal (Higgins and Sharp, 1988). In addition to the yeast sequences, rabbit PP1 (***PP1***), rabbit PP2A (***PP2A***) and rat PP2B (***PP2B***) are included for comparison.

Thus yeast encodes homologues of mammalian PP1 and PP2A regulatory subunits and in yeast, as in higher cells, there may be as many as 30 different proteins that can interact with PP1 and modify its activity and function (see Bollen, 2001; Cohen, 2002; Walsh *et al.*, 2002). Thus even a single protein phosphatase such as PP1 should perhaps more properly be considered as a family of protein phosphatases that share a common catalytic subunit (Glc7p in this case), and the wider range of activities specified by different combinations of phosphatase regulatory and catalytic subunits may help narrow the apparent imbalance between the number of protein kinases (122) and phosphatases (35). However, even considering just the protein kinase and protein phosphatase catalytic subunits, up to 3% of the yeast genome appears to encode proteins directly involved in regulating the phosphorylation state of other proteins.

8.2 Cyclin-dependent protein kinases

S. cerevisiae encodes several examples of the cyclin-dependent protein kinase (CDK) subgroup, as summarised in Table 8.4. CDKs require the binding of a cyclin subunit via a conserved -A-I-R-E- motif in catalytic subdomain III (Jeffrey *et al.*, 1995), and generally also require an activatory phosphorylation on the conserved threonine in the activation loop region between subdomains VII and VIII. This latter modification is catalysed by a protein kinase activity known as CAK (CDK-activating kinase). CDKs are proline-directed protein kinases that phosphorylate S/T-P motifs. Classically, CDKs have been discovered because of their roles as regulators of cell division, but it is becoming clear that they can have other functions as well. Thus in mammalian cells Cdk7·cyclin H not only functions as CAK but is a component of TFIIH, a factor involved in RNAPII transcriptional activation (Nigg, 1996).

Table 8.4 Yeast cyclin-dependent protein kinases

Protein kinase	*Known cyclins*	*Function*
Cdc28p	Cln1p, Cln2p, Cln3p	Cell cycle: execution of 'Start'
	Clb5p, Clb6p	Cell cycle: initiation of DNA replication
	Clb1p, Clb2p, Clb3p, Clb4p	Cell cycle: mitosis
Pho85p	Pho80p	Regulation of *PHO5* expression by low phosphate
	Pcl1p[a], Pcl2p[b], Pcl9p	Cell cycle: G1 regulation
	Pcl5p, Pcl6p, Pcl7p, Clg1p	?
	Pcl8p, Pcl10p	Glycogen synthase kinase
Kin28p	Ccl1p	Transcriptional regulation; RNAPII CTD kinase (Hengartner *et al.*, 1998)
Srb10p[c]	Srb11p[d]	Transcriptional regulation; RNAPII CTD kinase (Hengartner *et al.*, 1998)
Ctk1p	Ctk2p	Transcriptional regulation; RNAPII CTD kinase (Sterner *et al.*, 1995)
Sgv1p[e]	Bur2p	Transcriptional elongation? (Yao *et al.*, 2000; Murray *et al.*, 2001)

Notes

a *PCL1* is also known as *HCS26*.

b *PCL2* is also known as *ORFD*.

c *SRB10* is also known as *SSN3* and *UME5*.

d *SRB11* is also known as *SSN8* and *UME3*.

e *SGV1* is also known as *BUR1*.

Of the *S. cerevisiae* CDKs, Cdc28p and Pho85p have been studied in particular detail as described later. Cdc28p plays a central role in cell cycle regulation and is the budding yeast homologue of the universally conserved $p34^{cdc2}$ kinase, while Pho85p is required for the cell's response to low levels of inorganic phosphate and complements the function of Cdc28p in G1. The other four yeast CDKs are involved in transcriptional regulation via interaction with RNAPII. Thus *S. cerevisiae* TFIIH also contains a CDK·cyclin pair (Kin28p and Ccl1p, respectively: Feaver *et al.*, 1994) and both these genes are essential, acting as positive regulators of RNAPII transcription by phosphorylating the CTD of the Rbp1p subunit of RNAPII (Cismowski *et al.*, 1995; Valay *et al.*, 1996). However, unlike Cdk7·cyclin H, Kin28p·Ccl1p appears not to provide CAK function (see later). Two other CDK-related polypeptides also share CTD kinase activity with Kin28p·Ccl1p. *SRB10* encodes a CDK-related kinase and with its cyclin subunit (Srb11p) is associated with the RNAPII holoenzyme (Liao *et al.*, 1995). While Srb10p·Srb11p is not essential, it is nonetheless important for transcriptional regulation in a number of instances and may be the yeast homologue of Cdk8·cyclin C (Leclerc *et al.*, 1996). The Ctk1p·Ctk2p divergent cdk·cyclin pair also phosphorylates the RNAPII CTD (Cho *et al.*, 2001) and appears either to promote the assembly of RNAPII elongation complexes or make them less prone to transcriptional arrest (Jona *et al.*, 2001). Finally, Sgv1p·Bur2p is most closely related to mammalian Cdk9·cyclin T and may play a role in transcriptional elongation (Yao *et al.*, 2000).

Cdc28p

As outlined in Chapter 1, the Cdc28p CDK in *S. cerevisiae* sits at the heart of the cell cycle regulatory machinery (Nasmyth, 1993; Mendenhall and Hodge, 1998), functioning in conjunction with nine cyclins that are required for its activity (Table 8.4). *cdc28* mutations generally arrest cells in G1, but some specific alleles cause a G2 block (Reed and Wittenberg, 1990) and these allele-specific phenotypes were important in the identification of many of the *CLN* and *CLB* genes (Nasmyth, 1993). Four of the Cdc28p cyclins are expressed periodically in G1 (*CLN1, 2* and *CLB5, 6*) while the remaining B-type cyclins (*CLB1–4*) are expressed in G2, the *CLB3, 4* transcripts appearing first. Conversely, *CLN3* transcription is relatively constant, but shows a small increase in late M/early G1 that depends on the function of the Mcm1p transcription factor (McInerny *et al.*, 1997). *CLN1, 2* expression requires Swi4p and Swi6p, which form a heterodimeric transcription factor termed SBF, while *CLB5, 6* expression requires a related factor (MBF) composed of Mbp1p and Swi6p (Andrews and Mason, 1993). Periodic expression of the *CLB1, 2* requires SFF (Swi Five Factor: Maher *et al.*, 1995), a transcription factor that involves a pair of forkhead proteins together with Mcm1p (Kumar *et al.*, 2000; Pic *et al.*, 2000; Zhu *et al.*, 2000a) and transcription appears to involve a positive feedback mechanism (Amon *et al.*, 1993). By comparison, *CLN3* levels are low and essentially constitutive throughout the cell division cycle. An understanding of the separate and specific roles of the different cyclin·Cdc28p CDKs was initially hampered by the redundancy shown by them, particularly given that some forms can carry out the functions of others in their absence. Early work suggested that the three *CLN* genes shared an essential function, since cells could survive with any one but not without all three, although it is now known that Cln3p has a quite distinct role from the other two Clns (Cross, 1995). Similarly, cells can survive with only one of the *CLB1–4* genes intact, although *CLB2* appears to be the most critical and some triple *clb* deletion strains were also inviable (Fitch *et al.*, 1992). Most work on the mitotic Clbs has therefore focussed on Clb2p. In fact *CLB1* and

CLB4 seem to be more important in meiosis and *CLB2* is specialised for mitosis (Dahmann and Futcher, 1995). Progression through the cell cycle follows the appearance of distinct cyclin·Cdc28p CDKs and the appearance and subsequently loss of particular Cdc28p CDK activities is subject to multiple controls. Clb·Cdc28p complexes are subject to inhibition early in the cell cycle by Sic1p (Schwob *et al.*, 1994), a CDK inhibitor (CKI), while another CKI (Far1p) mediates the arrest of cell division in late G1 in response to mating pheromone by inhibition of Cln·Cdc28p complexes (Peter *et al.*, 1993; see later). Proteolysis is also a critical element in removing specific cyclins (and hence particular forms of Cdc28p kinase). This involves ubiquitin-mediated protein turnover via the proteasome, and two types of ubiquitin-protein ligase or E3 enzyme (Hershko and Ciechanover, 1992) have been found to be important promoters of cyclin turnover. Turnover of Cln1p and Cln2p involves a complex of Cdc53p, Skp1p and Grr1p, which functions together with an E2 ubiquitin conjugating enzyme (either Ubc9p or Cdc34p: Deshaies *et al.*, 1995; Bai *et al.*, 1996; Blondel and Mann, 1996; Li and Johnston, 1997) in a complex termed SCF^{Grr1}. A similar complex (SCF^{Cdc4}), but in which Cdc4p replaces Grr1p and with the Cdc34p E2 enzyme, is required to destroy the Sic1p CKI (Feldman *et al.*, 1997; Li and Johnston, 1997). Destruction of the B-type cyclins requires a large and complex E3 function known as the anaphase-promotion complex or cyclosome (APC/C), which as well as destroying the Clbs is vital as a trigger for the anaphase separation of sister chromatids in mitosis (Zachariae and Nasmyth, 1999). Some cyclin·Cdc28p kinases also contain a small, conserved polypeptide called Cks1p, which has roles both at 'Start' (the point of commitment to a new cell cycle) and at the G2 $\rightarrow$ M transition (Tang and Reed, 1993). Recent work has shown that Cks1p is required for the activity of Cln·Cdc28p complexes (Reynard *et al.*, 2000), although the finding that its mammalian homologue is involved in ubiquitin-mediated destruction of the cell cycle regulatory protein p27 (Harper, 2001) suggests that it may have other roles in yeast as well.

The cyclin·Cdc28p cycle

Figure 8.2 shows the different forms of Cdc28p kinase, which form an orderly progression of events that orchestrate the cell division cycle in *S. cerevisiae*. Although the three *CLN* genes seemed initially to be redundant, careful examination of their roles (Dirick *et al.*, 1995) revealed that Cln3p·Cdc28p is specifically required to activate gene expression in response to the SBF transcription factor, setting in train a transcriptional programme in late G1 involving genes regulated by both SBF and MBF. The *CLN1, 2* genes are major targets of SBF and Cln3p·Cdc28p is therefore an upstream activator of Cln1,2p·Cdc28p (Tyers *et al.*, 1993). There is no evidence yet that activation of either SBF or MBF involves direct phosphorylation by Cln3p·Cdc28p, although all known roles of Cln3p require the common Swi6p subunit of these two transcription factors (Wijnen *et al.*, 2002). Recent work (Cosma *et al.*, 2001) has shown that SBF promotes the binding of Srb/mediator complex to SBF-dependent promoters and that Cdc28p function is required subsequently to recruit RNAPII, TFIIB and TFIIH so that transcription can initiate. Thus it is likely that the target of Cln·Cdc28p is not SBF itself. A truncation allele of *CLN3* (*CLN3-1*) that removes a destabilising 'PEST' sequence (Nash *et al.*, 1988; Tyers *et al.*, 1992) leads to a small cell phenotype due to advanced execution of 'Start', so the activity of Cln3p·Cdc28p is clearly important despite its lack of periodicity during the cell cycle. It is not yet completely clear how Cln3p·Cdc28p itself functions to integrate those signals (cell size and nutrient availability, for

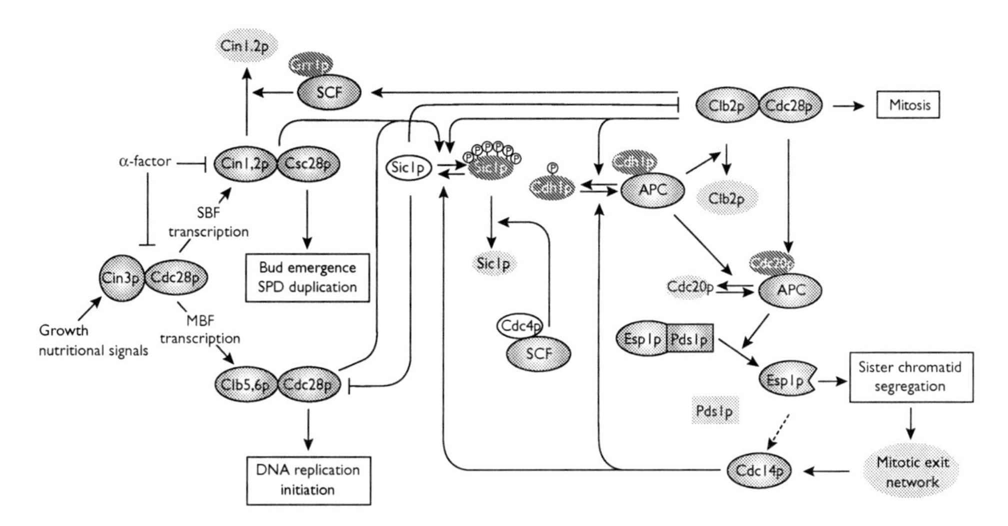

Figure 8.2 Control of the yeast cell division cycle by the Cdc28p CDK. The diagram depicts positive (→) and negative (⊣) interactions between the molecular components that regulate and are regulated by the Cdc28p CDK.

example) that ultimately trigger the cell to embark on a new cell division cycle, but it is likely to be under multiple controls. For example, *CLN3* expression is promoted by glucose (Hall *et al.*, 1998; Parviz *et al.*, 1998) and Cln3p function and expression is inhibited by nitrogen starvation (Gallego *et al.*, 1997) while conversely, inhibition of 'Start' by blocking the Target of Rapamycin (TOR) signalling pathway with rapamycin (see later) can be overcome by forced expression of *CLN3* (Barbet *et al.*, 1996).

The periodic accumulation of Cln1,2p·Cdc28p in late G1 as a result of SBF and MBF-mediated transcriptional activation triggers several key events that together constitute the 'Start' commitment point (Dirick *et al.*, 1995). The accumulation of *CLN1* and *CLN2* (but not *CLN3*) transcripts is negatively regulated by the cAMP signal (Baroni *et al.*, 1994; Tokiwa *et al.*, 1994) such that more growth is needed to generate levels of Cln1,2p·Cdc28p that can trigger the events of 'Start'. The cAMP pathway can therefore set the threshold size at which 'Start' occurs and this mechanism provides one means whereby cell growth and division can be coordinated. Although Cln1,2p·Cdc28p can be shown to stimulate their own synthesis by elevating SBF function (Dirick and Nasmyth, 1991), this has not been found to be involved normally in the execution of 'Start' (Dirick *et al.*, 1995). 'Start' involves the triggering of bud emergence and polarised growth, duplication of the spindle pole bodies (SPBs, which form the poles of the mitotic spindle) and initiation of DNA synthesis. There is evidence for a direct role of Cln1,2p·Cdc28p in the case of bud emergence (Cvrckova and Nasmyth, 1993), but the Cln1/2p requirement for DNA synthesis is probably indirect via triggering Clb5/6p·Cdc28p activity (Dirick *et al.*, 1995; see later). Thus Cln3p and Cln1, 2p have different functions and this is at least in part due to their different subcellular localisation: Cln3p is mainly nuclear while Cln1p and Cln2p are largely cytoplasmic (Miller and Cross, 2000; Edgington and Futcher, 2001).

Activation of Cln1,2p·Cdc28p ultimately induces expression of the six B-type cyclins. In the case of Clb5p and Clb6p, this involves activation of MBF (Andrews and Mason, 1993;

Schwob and Nasmyth, 1993), which also stimulates the expression of a large number of genes required for DNA synthesis. The mechanism by which *CLB1–4* transcription is activated involves Mcm1p and SFF as mentioned earlier. However, expression of *CLB* genes is on its own insufficient to ensure that functional Clb·Cdc28p CDKs appear. Firstly, the CKI Sic1p keeps any such complexes that form in G1 inactive, while APC/C-mediated cyclin B proteolysis system also ensures degradation of Clb proteins in G1. Threshold levels of Cln1,2p·Cdc28p are needed to inactivate the APC/C and to switch on Cdc34p-dependent proteolysis of Sic1p. The sole essential function of the Clns is to promote Cdc34p-dependent destruction of Sic1p, since deletion of *SIC1* suppressed the lethality of a triple *cln1Δ cln2Δ cln3Δ* strain (Schneider *et al.*, 1996). Sic1p is phosphorylated by Cln·Cdc28p and the Pho85p CDK, and phosphorylation is required for its degradation by SCF[Cdc4] (Verma *et al.*, 1997; Nishizawa *et al.*, 1998). Six CDK phosphorylation sites in Sic1p are needed for efficient degradation, and five or less are inadequate (Nash *et al.*, 2001). None of the Sic1p CDK sites is optimal. However, a single, optimal site supports efficient and precocious Sic1p destruction that leads to genome instability. This has lead to models whereby the requirement for multisite phosphorylation may both build a time lag into the destruction process and also make the transition more switch-like, setting a threshold for the level of Cln·Cdc28p activity needed as the trigger (Deshaies and Ferrell, 2001). In contrast to Sic1p degradation, destruction of Cln1p and Cln2p requires Clb·Cdc28p activity (Blondel and Mann, 1996). As the cell switches over to Clb·Cdc28p CDKs, not only are the Clns degraded but SBF is also inhibited, possibly by direct interaction with Clb2p·Cdc28p or additional Clb complexes (Amon *et al.*, 1993; Koch *et al.*, 1996).

Clb5p·Cdc28p and Clb6p·Cdc28p are required for initiation of DNA replication (Schwob and Nasmyth, 1993), but the appearance of these CDKs also marks the point at which Cdc6p, a key protein required for activation of replication origins, can no longer maintain the prereplicative state of origins that is needed for them to fire (Piatti *et al.*, 1996). Prereplicative complexes involving origin recognition complex (ORC), Mcm2p–Mcm7p and Cdc6p assemble at replication origins in G1 when Cdc28p activity is low, but cannot initiate replication without the action of Clb·Cdc28p; conversely, the high CDK activity associated with Clb·Cdc28p complexes is inhibitory to prereplicative complex formation (Diffley and Labib, 2002). It is these opposing roles of Clb·Cdc28p activity that help to ensure that replication can occur only once per cell division cycle. Clb·Cdc28p activity is needed to load Cdc45p into replication preinitiation complexes (Zou and Stillman, 1998), while another good candidate substrate of Clb·Cdc28p for mediating its positive role in promoting replication initiation is Sld2p (Masumoto *et al.*, 2002), which is part of a complex required for origin firing and which must be phosphorylated by S-phase CDKs to permit replication. Origin firing also requires DDK (Dbf4-Dependent Kinase: i.e. Cdc7p kinase in conjunction with its activator Dbf4p), and this has been shown to act after the Clb5,6p·Cdc28p-dependent step (Nougarede *et al.*, 2000). Clb·Cdc28p inhibits prereplicative complex formation via multiple, redundant mechanisms including ORC phosphorylation, Cdc6p downregulation and nuclear export of the Mcm proteins (Nguyen *et al.*, 2001). Cdc6p is destroyed following SCF[Cdc4] ubiquitination and Cdc28p phosphorylation promotes this (Perkins *et al.*, 2001). Remarkably, Cdc6p can also bind to and inhibit Clb·Cdc28p CDKs and thereby contributes to the mechanisms that are required for mitotic exit (Calzada *et al.*, 2001).

While Clb5·Cdc28p and Clb6p·Cdc28p are involved primarily with S-phase, the other Clb·Cdc28p complexes, principally Clb2p·Cdc28p, are needed for setting up the mitotic metaphase, being involved for example in production of a bipolar mitotic spindle (Lim *et al.*,

1996). Once chromosomes are attached in a bipolar manner to the mitotic spindle, the metaphase–anaphase transition is triggered by the multiprotein Anaphase Promotion Complex (APC/C), which ubiquitinates proteins that must be destroyed to permit anaphase and mitotic exit. The APC/C functions in combination with either Cdc20p or Cdh1p (Visintin *et al.*, 1997), which act as activators toward specific APC/C targets, and APC/C is phosphorylated by Cdc28p on its Cdc16p, Cdc23p and Cdc27p subunits (Rudner and Murray, 2000). This phosphorylation is required for the activity of APC/C^{Cdc20}, and so this is another way in which Clb·Cdc28p is likely to contribute towards setting up mitosis. However, it is clear that there are still many more Clb·Cdc28p substrates to be identified. The Cdh1p activator is also phosphorylated by Clb·Cdc28p but this modification is inhibitory, since it blocks the ability of Cdh1p to bind the APC/C (Jaspersen *et al.*, 1999). This mechanism explains how Clb·Cdc28p can switch off B cyclin proteolysis as cells traverse 'Start'. Activation of APC/C^{Cdc20} triggers the anaphase separation of sister chromatids by ubiquitination and degradation of Pds1p ('securin'), an inhibitor of the protease Esp1p ('separase') that cleaves Scc1p, a component of cohesin (Nasmyth *et al.*, 2000). Cohesin consists of a complex containing at least Scc1p, Scc3p, Smc1p and Smc3p, which holds sister chromatids firmly together until Scc1p is cleaved to initiate their separation, and cleavage of Scc1p is sufficient to trigger anaphase in metaphase cells (Uhlmann *et al.*, 2000). Phosphorylation of serine residues adjacent to the two separase cleavage sites in Scc1p by Cdc5p kinase strongly enhances their cleavage, explaining why cells lacking Pds1p still show a tightly regulated anaphase (Alexandru *et al.*, 2001). In addition to inhibiting separase prior to anaphase initiation, Pds1p is also required for the nuclear localisation of Esp1p, and this in turn depends on Clb·Cdc28p phosphorylation (Agarwal and Cohen-Fix, 2002).

Once cells have undergone anaphase, mitotic exit and G1 entry requires the destruction of Clb·Cdc28p activity to return to the low CDK state of G1, where cells can establish the prereplicative complexes needed for DNA replication in a subsequent cell cycle. At the end of mitosis and in G1, APC/C^{Cdh1} functions to destroy the Clbs and thereby switch off Clb·Cdc28p activity, although APC/C^{Cdc20} is clearly responsible for initiating Clb destruction (Shirayama *et al.*, 1999; Yeong *et al.*, 2000; Wasch and Cross, 2002). For APC/C^{Cdh1} to become functional Cdh1p must first be dephosphorylated, and the Cdc14p phosphatase, which sits at the end of a signal transduction pathway termed the Mitotic Exit Network (MEN), plays a critical role in this process (Visintin *et al.*, 1998). Cdc14p is normally bound in an inactive complex with Net1p in the nucleolus, but is released and activated at the end of mitosis to promote mitotic exit (Shou *et al.*, 1999). Cdc14p release occurs in two phases, the first of which requires Esp1p and occurs in early anaphase, the second and more extensive release occurring in response to MEN activation (Geymonat *et al.*, 2002). The MEN pathway includes Temp1p, a GTPase that acts upstream of the Cdc15p and Dbf2p protein kinases, and its components are localised to the SPB. Cdc14p released in the first phase also localises to the SPB and helps activate the MEN (Pereira *et al.*, 2002). Since the exchange factor Lte1p that activates Tem1p (and hence the MEN) is localised to the bud cortex, this provides a mechanism to coordinate mitotic exit with spindle elongation, and another GTPase (Cdc42p) assists Lte1p in this role (Hofken and Schiebel, 2002). As well as dephosphorylating Cdh1p, Cdc14p also negatively regulates Clb·Cdc28p activity by dephosphorylating and stabilising the Sic1p CKI and by dephosphorylating the Sic1p transcription factor Swi5p, which allows it to gain entry to the nucleus (Jans *et al.*, 1995; Visintin *et al.*, 1998). Cdc14p released in the second, MEN-dependent phase also functions to switch off the MEN by dephosphorylating and reactivating the Tem1p GTPase activating protein at the

SPB (Bub2p·Bfa1p: Pereira *et al.*, 2002), by dephosphorylating and delocalising the Lte1p exchange factor (Jensen *et al.*, 2002), and possibly also by dephosphorylating Net1p (Shou *et al.*, 1999).

The coordination and order of events of the yeast cell cycle is therefore maintained by the sequential activity of different forms of the Cdc28p CDKs in a complex network of regulator interactions. For example, Clb5/6p·Cdc28p CDK is needed to fire replication origins but blocks them from becoming competent to fire again, while the CDKs that promote mitotic metaphase are destroyed by the same process that triggers anaphase but which is carefully coordinated with the anaphase segregation of sister chromatids. The alternation of the different Cdc28p CDKs is ensured by both transcriptional and proteolytic regulation of the different cyclins, the Sic1p CKI and other proteins. Figure 8.2 summarises the main regulatory interactions discussed here. Some functions of Cdc28p are clearly unique to specific cyclin·Cdc28p complexes, for example, the Clb forms of the kinase are required for setting up a mitotic spindle (Fitch *et al.*, 1992) and for switching the pattern of bud growth from hyperpolarised to isotropic as the cell cycle progresses (Lew and Reed, 1993), while Clb5p and Clb6p are specialised for their roles at replication origins (Donaldson, 2000). However, some roles are at least partly dependent on localisation as mentioned above, while other functions can be adopted by cyclin·Cdc28p forms that are not usually involved if the normal cyclin is missing, showing that the substrate specificity of different complexes cannot be absolute. Thus the functions of Clb5p and Clb6p required for initiation of DNA replication can be taken on by other Clbs in their absence and cells can proliferate in the absence of Cln3p, albeit with changes to the normal timing of events (Dirick *et al.*, 1995). Although the Cdc28p CDKs clearly play a central role in the yeast cell cycle, it is important not to forget that there are many other protein kinases involved in addition to Cdc28p and that some of these also have a multiplicity of roles. Cdc5p kinase, the homologue of *polo* kinase in higher cells (Golsteyn *et al.*, 1996), is an example of such a kinase with many roles in the cell division cycle.

Cyclin·Cdc28p regulation by phosphorylation

Although Cdc28p must associate with one of its cognate cyclins for function, other modes of regulating its kinase activity must also be considered. As mentioned previously, CDKs require an activatory phosphorylation on a threonine in their activation loop, thr-169 in Cdc28p. In *S. cerevisiae*, this function is carried out by Cak1p (Kaldis *et al.*, 1996), which performs the equivalent role for at least some of the other CDKs shown in Table 8.3 (Espinoza *et al.*, 1998; Yao and Prelich, 2002). Cak1p is active as a monomer and its levels and activity are relatively constant through the cell cycle (Sutton and Freiman, 1997), while genetic evidence shows that Cak1p is needed for activation of both Cln and Clb forms of Cdc28p kinase. Thus while *cak1* conditional alleles lead to a G2 block and are partially suppressed by *CLB2* overexpression, they are also synthetically lethal with *sit4* mutations (Kaldis *et al.*, 1996; Sutton and Freiman, 1997). Synthetic lethality is a phenomenon where mutations in either of two genes alone are not lethal but the double mutant is inviable, that is, the two mutations are additive or synergistic, implying that they affect the same process. Sit4p is a protein phosphatase required for *CLN1, 2* expression and other events in late G1 as discussed later, so the *cak1 sit4* synthetic lethality strongly implies that Cak1p activates Cln·Cdc28p complexes too. In higher cells, PP2A may dephosphorylate the activation loop phosphothreonine and therefore acts as an inhibitor of CDKs *in vitro* (Lee *et al.*, 1991, 1994). However, since conditional alleles of PP2A delay mitotic entry in *S. cerevisiae* and PP2A is needed to maintain high

Clb·Cdc28p kinase activity (Lin and Arndt, 1995), it is not clear that PP2A performs this function *in vivo* in budding yeast. Mammalian cells also have a CDK-associated dual-specificity phosphatase (KAP) that dephosphorylates this residue (Hannon *et al.*, 1994), so it is conceivable that one of the as yet uncharacterised yeast enzymes in this class (Table 8.2) fulfils this role. Intriguingly, human KAP1 is most closely related to Cdc14p, the dual-specificity phosphatase required for mitotic exit (see earlier).

A second way in which CDKs can be regulated is by phosphorylation on a conserved tyrosine residue in the P-loop motif of the ATP binding site. In fact detailed studies on this phenomenon in *Schizosaccharomyces pombe* cdc2 and its homologues in higher eukaryotic cells have shown such regulation to be critical both in controlling mitotic entry (Nurse, 1990) and in the functioning of checkpoints that prevent mitosis in response to incomplete replication or DNA damage (see Boddy and Russell, 2001). Phosphorylation of this residue in *S. pombe* cdc2 (tyr-15) is inhibitory and is catalysed by wee1 and mik1, a pair of dual-specificity protein kinases (Lundgren *et al.*, 1991), while the cdc25 protein phosphatase catalyses the dephosphorylation reaction (Millar and Russell, 1992). In higher cells, the preceding threonine residue in cdc2 is also regulated by the homologous dual-specificity kinases and phosphatases (Norbury *et al.*, 1991), and evidence in fission yeast suggests that thr-14 is phosphorylated in this organism too (Denhaese *et al.*, 1995). Not surprisingly, *S. cerevisiae* has homologues both of *wee1*$^+$ (*SWE1*: Booher *et al.*, 1993) and *cdc25*$^+$ (*MIH1*: Russell *et al.*, 1989). However, neither of these activities seems to be important for regulation of the cell cycle under normal circumstances (Sia *et al.*, 1996). Thus *CDC28*-F19, encoding a non-phosphorylatable variant of Cdc28p, does not advance mitosis in *S. cerevisiae*, nor does it abrogate the checkpoints that block mitosis prior to completion of DNA replication or DNA damage repair (Amon *et al.*, 1992; Sorger and Murray, 1992). Cells blocked by either of these two checkpoints have high Clb·Cdc28p activity (Amon *et al.*, 1992; Sorger and Murray, 1992) and at least in the case of DNA damage, checkpoint-dependent arrest in mitosis is mediated through mechanisms controlling the stability of Pds1p (Wang *et al.*, 2001a). However, Cdc28p tyr-19 phosphorylation (Lew and Reed, 1995) and Swe1p kinase are required for a checkpoint that prevents mitosis in the absence of the morphological development of a bud or when the actin cytoskeleton is disassembled, and this morphology checkpoint operates by regulating the stability of Swe1p (Sia *et al.*, 1996; McMillan *et al.*, 2002). It has also been suggested that Cdc28p phosphorylation on tyr-19 constitutes an adaptive response to long-term arrest at the mitotic checkpoint by which the normally high Cdc28p activity is downregulated to permit mitotic exit (Minshull *et al.*, 1996).

A key difference between budding and fission yeasts is in the way they regulate cell size at division. In *S. cerevisiae*, this involves the timing of 'Start' and events in G1, whereas in *S. pombe* it is regulated by control of mitotic entry from G2 and the role of cdc2 tyr-15 phosphorylation is a key element in this (see Forsburg and Nurse, 1991). In budding yeast, mitosis occurs before the bud has reached the same size as the mother cell. Division is therefore asymmetric and the mother and daughter cells bud at different times in the next cell cycle. By comparison, when budding cells undergo pseudohyphal differentiation (see later), cell division is symmetrical and the data suggest that mitosis is restrained at the G2/M boundary to permit this (Kron *et al.*, 1994). Thus although it is not normally a critical part of cell cycle control in *S. cerevisiae*, the ability to inhibit Cdc28p by tyr-19 phosphorylation might become important under certain conditions.

PP2A-deficient *S. cerevisiae* cells, which are delayed for mitosis and cannot maintain high Clb·Cdc28p kinase activity, show sensitivity to loss of the Mih1p phosphatase. This sensitivity to loss of Mih1p is abolished by the non-phosphorylatable *CDC28*-F19 mutant (Lin and Arndt, 1995), which also elevates Clb-dependent kinase activity in checkpoint-arrested cells. This demonstrates additional situations in which the phosphorylation state of tyr-19 can become important. In higher cells, PP2A can dephosphorylate and inhibit the Mih1p homologue cdc25 (Clarke *et al.*, 1993) and may also activate wee1 by dephosphorylation. Since either of these dephosphorylations would reduce Clb·Cdc28p activity, such roles for PP2A again seem incompatible with the activatory role(s) for PP2A at mitotic entry in budding yeast deduced from genetic studies (Lin and Arndt, 1995).

Pho85p

Pho85p and the low phosphate response

When yeast experience conditions of limiting inorganic phosphate, the cells undergo a physiological response that leads to induction of genes required for scavenging phosphate, principally the *PHO5* gene encoding a repressible secreted acid phosphatase (see Lenburg and O'Shea, 1996). Initial studies sought to identify mutant genes that led either to constitutive *PHO5* expression or failure to induce under low phosphate conditions (see Kaffman *et al.*, 1994). *PHO85* was identified as a member of the first group, that is, Pho85p is needed to keep *PHO5* switched off when phosphate is plentiful. When *PHO85* was isolated it became clear that it encodes a CDK, having the canonical -P-S-T-A-I-R-E- motif found in true CDKs and showing 51% amino acid sequence identity with Cdc28p (Uesono *et al.*, 1987). *PHO80*, which genetically was defined as a negative regulator of *PHO5* expression, was found to encode a cyclin-like molecule (Kaffman *et al.*, 1994). *PHO81* (a positive regulator of *PHO5* that lies upstream of *PHO80* and *PHO85* genetically: Creasy *et al.*, 1993) encodes an inhibitor (CKI) of the Pho85p·Pho80p cdk·cyclin complex that interacts with the Pho80p cyclin (Schneider *et al.*, 1994; Ogawa *et al.*, 1995). Thus Pho85p and Pho80p act to inhibit *PHO5* induction, apparently working by blocking the activity of Pho4p, a bHLH transcriptional activator of *PHO5*. Pho4p forms a complex with Pho2p (a homeobox protein) in order to activate *PHO5* expression and this interaction is regulated by the phosphate switch (Hirst *et al.*, 1994; Barbaric *et al.*, 1996). Pho4p is a substrate of Pho80p·Pho85p *in vitro* and shows *PHO85*- and *PHO80*-dependent hyperphosphorylation in response to phosphate (Kaffman *et al.*, 1994). Five *in vivo* Pho80p·Pho85p phosphorylation sites (all -S-P- motifs) have been defined on Pho4p. Hyperphosphorylated Pho4p, which accumulates under high phosphate conditions, is excluded from the nucleus while the quintuple S→A *PHO4* mutant is constitutively nuclear and leads to 10% *PHO5* derepression (O'Neill *et al.*, 1996). This leads to a straightforward model whereby in high phosphate, the Pho85p·Pho80p complex is active and phosphorylates Pho4p, excluding it from the nucleus and preventing it from interacting with the *PHO5* promoter. Conversely, in low phosphate the Pho80p CKI inhibits Pho85p·Pho80p and allows hypophosphorylated Pho4p to enter the nucleus, bind Pho2p and activate *PHO5* expression (Figure 8.3). The phosphatase that removes the relevant phosphate groups from Pho4p has yet to be identified. Further analysis of the Pho4p phosphorylation sites has revealed that four are critical: phosphorylation of two sites promotes nuclear export, phosphorylation at a third inhibits nuclear import and at a fourth inhibits the interaction with Pho2p (Kaffman *et al.*, 1998a,b; Komeili and O'Shea, 1999). Pho2p also requires phosphorylation on ser-231 by Cdc28p kinase for transcriptional activation (Liu *et al.*, 2000). *PHO81* is induced on phosphate starvation in a *PHO4*- and *PHO2*-dependent manner

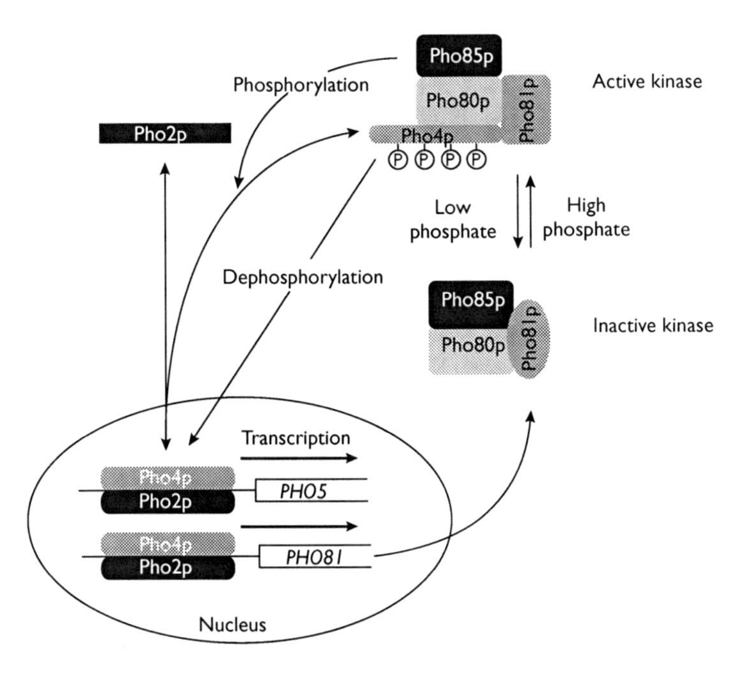

Figure 8.3 Control of *PHO5* expression by Pho85p CDK.

(Yoshida *et al.*, 1989), suggestive of a positive feedback mechanism to amplify the response. Pho81p is a key element in the phosphate-sensitive switch, and mutations in the N-terminal domain that lead to constitutive *PHO5* expression define a region needed to keep Pho81p inactive in high phosphate (Ogawa *et al.*, 1995). However, the Pho81p interaction with Pho85p·Pho80p is phosphate-independent (Hirst *et al.*, 1994), although studies with *pho85* mutants suggest that the Pho80p–Pho81p interaction is modulated by phosphate level (Huang *et al.*, 2001). Remarkably, a domain of just 80 amino acids within the 1178-residue Pho81p is sufficient to confer phosphate regulated *PHO5* expression (Huang *et al.*, 2001).

Both Pho85p·Pho80p and Pho81p show physical interactions with Pho4p that seem to be phosphate regulated (i.e. occur under high but not low phosphate conditions: Hirst *et al.*, 1994), and Pho80p can repress Pho4p function even in the absence of Pho85p (Madden *et al.*, 1990). Furthermore (and like the *PHO4* quintuple S→A mutant), deletion of *PHO85* only causes 15% *PHO5* derepression compared with deletion of *PHO80* (Hirst *et al.*, 1994). Taken together, this indicates that at least part of the regulation of Pho4p is via direct protein–protein interactions and that Pho80p is not only a cyclin for Pho85p but also a repressor of Pho4p function; while Pho80p is absolutely required for repression, Pho85p is only needed for full repression of *PHO5*. The model in Figure 8.3 encompasses the proposed role of the direct protein–protein interactions in Pho4p function. How the roles of the cyclin·cdk complex both as protein kinase and as direct repressor of Pho4p may be integrated is not yet fully resolved. Perhaps the regulation of Pho4p localisation by phosphorylation provides a coarse control, removing most of the protein from the nucleus, while the interactions of Pho4p with Pho80p and Pho81p function to inhibit any residual Pho4p present in the nucleus under conditions of high phosphate. Pho4p is also required for induction of *PHO8*, encoding an alkaline phosphatase. This induction does not require Pho2p, but is still regulated by Pho85p·Pho80p (Kaneko *et al.*, 1985).

Pho85p and the cell division cycle

Regulation of repressible acid phosphatase is only one of the functions of the Pho85p CDK and *pho85* mutants show a variety of additional phenotypes, including glycogen hyperaccumulation, reduced growth on non-fermentable carbon sources and slower growth and morphological abnormalities on glucose (see Measday *et al.*, 1997). In fact Pho80p is only one of ten Pho85p cyclins (Measday *et al.*, 1997; Carroll and O'Shea, 2002 see Table 8.4). Phylogenetically these fall into two groups: *PCL1, 2, 5, 9, CLG1* and *PCL6, 7, 8, 10, PHO80.* Although deletion of *PHO85* (unlike loss of *CDC28*) is not lethal, at least two of these cyclins (Pcl1p, Pcl2p) function in conjunction with Pho85p in the late G1 phase of the cell division cycle in parallel with Cdc28p·Cln1p and Cdc28p·Cln2p. Thus loss of *PCL1* is synthetically lethal with the normally viable *cln1Δ cln2Δ* double deletion in diploid cells and shows a severe defect in haploid cells (Espinoza *et al.*, 1994), while the *pcl1Δ pcl2Δ cln1Δ cln2Δ* combination is lethal in haploid cells (Measday *et al.*, 1994). Consistently, deletion of *pho85* is lethal in combination with *cln1Δ cln2Δ*. Pcl2p appears to be specific for Pho85p since it showed no interaction with Cdc28p using the two-hybrid system (Measday *et al.*, 1994), so it is unlikely that the Pcls function redundantly with the Clns via Cdc28p. However, the finding that Sic1p requires Pho85p phosphorylation for efficient destruction (Nishizawa *et al.*, 1998) provides a possible explanation for the functional overlap between *CLNs* and *PCLs*, although other genetic data are not entirely consistent with this model (see Carroll and O'Shea, 2002). This cell cycle role is emphasised by the finding of functional homology between Pho85p and mammalian Cdk5 (Nishizawa *et al.*, 1999). Like *CLN1* and *CLN2*, *PCL1, PCL2* and *PCL5* transcripts peak in late G1 and are SBF-dependent, and *PCL1* was first isolated as a high-copy suppressor of SBF deficiency (Espinoza *et al.*, 1994). *PCL9* also shows periodic expression but peaking at M/G1 (Tennyson *et al.*, 1998), and Pho85p is involved in the Swi5p-dependent expression of genes in late M/early G1 (Measday *et al.*, 2000). Deletion of all five members of the *PCL1* family caused a number of morphological abnormalities similar to those seen in the *pho85 Δ* strain (Measday *et al.*, 1997), and genetic evidence supports a role for at least Pcl1p and Pcl2p together with Pho85p in morphogenetic events in the cell cycle (Measday *et al.*, 2000). The *PCL1* family of cyclins therefore seem to form a family of Pho85p CDKs with multiple roles in the cell division cycle. The functions of the other Pho85p cyclins (apart from Pho80p) are not at all well understood yet, although Pcl8p·Pho85p and Pcl10p·Pho85p are known to regulate glycogen synthesis by phosphorylating and inactivating glycogen synthase (Huang *et al.*, 1998; see later) and Pho85p also plays a role in the stability of the transcription factor Gcn4p in partnership with Pcl5p (Shemer *et al.*, 2002). A more general role for Pho85p and its cyclins in the environmental stress response has been uncovered following genome transcript profiling in a strain where Pho85p had been mutated to become the sole kinase sensitive to a specific inhibitor (Carroll *et al.*, 2001). Thus inhibition of Pho85p, in addition to switching on the genes of the *PHO* regulon, also induces genes required for glycogen and trehalose synthesis, glycolysis, redox stress, protein folding, and protein degradation. A further study looking at the transcriptional profiles of strains lacking *PHO85* or different combinations of its cyclins (Huang *et al.*, 2002) has both underlined and extended the conclusion that Pho85p and its cyclins act more generally to control the cell's response to its environmental conditions.

8.3 Cyclic AMP-dependent protein kinase

Cyclic AMP-dependent protein kinase (PKA) lies at the heart of an important signal transduction pathway in *S. cerevisiae* (Figure 8.4). Although the core elements of the PKA pathway were identified principally in the 1980s, it has taken much longer to tease out both the molecular mechanisms that activate the pathway and the critical PKA substrates. As outlined in Chapters 1 and 3, PKA activity influences many of the physiological responses of the cell including glucose regulation of gene expression, the cell division cycle, polarised growth, glycogen synthesis and many more. In yeast, the sole essential function of cAMP is to activate PKA (Toda *et al.*, 1985) and the cAMP requirement for proliferation became evident from the identification of both adenylyl cyclase (Cyr1p/Cdc35p) and the guanine nucleotide exchange factor (Cdc25p) for its activatory Ras GTPase in a screen for cell division cycle mutants. Loss of PKA pathway function causes cells to arrest in G1 (see before) and the PKA pathway functions to stimulate Cln3p levels (Hall *et al.*, 1998), providing a possible explanation for this. On the other hand, PKA activation in response to glucose makes it harder for cells to express *CLN1* and *CLN2* (Baroni *et al.*, 1994; Tokiwa *et al.*, 1994), inhibiting 'Start' until cells have achieved a larger size than when glucose is limiting. The PKA pathway therefore plays both positive and negative roles in proliferation. PKA is essential to the cell and promotes growth as well as proliferation, but when hyperactivated it inhibits the cell's ability to respond properly to nutrient starvation and other stresses, leading to the notion that PKA signalling is linked to nutrient sensing. While some aspects of the PKA pathway were well understood over a decade ago, it is only more recently that the mechanisms by which the PKA pathway is activated and the identity of some of the key targets have been uncovered (see Thevelein and de Winde, 1999).

The pathway

Like its mammalian counterpart, *S. cerevisiae* PKA consists of a heterotetramer containing two regulatory (Bcy1p) and two catalytic (Tpk) subunits, which are encoded by triply redundant genes (*TPK1–TPK3*). However, recent work has shown that while at least one *TPK* gene is needed for viability, not all functions of these three PKA isoforms are shared. Thus only *TPK2* is required for the function of PKA in pseudohyphal growth (Robertson and Fink, 1998), while transcriptional profiling of strains lacking just one of the three *TPK* genes has revealed a unique signature for loss of each that demonstrates further diversity of function (Robertson *et al.*, 2000). On binding cAMP, the Bcy1p subunits dissociate from the complex, releasing active PKA that can go on to phosphorylate its many target proteins. cAMP is produced by the *S. cerevisiae* adenylyl cyclase (Cyr1p), which in turn requires two mammalian ras-related guanine-nucleotide binding proteins (Ras1p, Ras2p) for its activity. Cyr1p encodes a large protein of over 200 kDa. The catalytic domain is located in the C-terminal 20%, but this domain functions independently of the Ras proteins. The central region of the protein is made up of leucine-rich repeats and is probably involved in membrane localisation. This region, together with the C-terminal catalytic domain, is required for modulation by Ras1p and Ras2p. The two Ras proteins are extensively modified post-translationally like their mammalian counterparts. The GTP-bound form of the Ras proteins activates Cyr1p and the GTP-bound state is controlled by the opposing action of the Cdc25p guanine nucleotide exchange factor (that promotes Ras·GTP formation) and the GTPase activating proteins (GAPs) Ira1p and Ira2p, which stimulate hydrolysis of GTP by Ras1, 2p. For Cyr1p to be activated by Ras, association with another protein (Srv2p) is required (Fedor-Chaiken *et al.*, 1990). Srv2p appears to have a second, independent function relating to control of cell morphology and the actin cytoskeleton and in fact Srv2p interacts directly with the actin cytoskeleton via Abp1p, an actin-binding protein (Lila and Drubin, 1997; Yu *et al.*, 1999). Activation of PKA was found to be dependent on the balance between Ras·GDP and Ras·GTP in the cell and stimulation of the pathway could in principal have resulted from positive signals channelled to Cdc25p or by inhibition of Ira1/2p function. While this has been proved likely for one signal (intracellular acidification), which increases the ratio of GTP to GDP-bound Ras by a mechanism that operates via the Ira proteins (Colombo *et al.*, 1998), more recent work has shown that the effect of glucose in activating the pathway is mediated through the G-protein coupled receptor Gpr1p and its associated Gα subunit Gpa2p (Kraakman *et al.*, 1999). The synthetic lethality shown between *gpa2* and *ras2* (Kubler *et al.*, 1997) is consistent with a requirement for both Gpa2p and the Ras proteins for activation of adenylate cyclase. Rgs2p, a homologue of mammalian RGS proteins, is a negative regulator of signalling through Gpa2p and its deletion and overexpression phenotypes mirror those of other mutations that respectively elevate or compromise PKA signalling (Versele *et al.*, 1999). cAMP can be removed by the action of cAMP phosphodiesterase and in *S. cerevisiae* there are two such enzymes (Pde1p and Pde2p). Of these, Pde2p has the higher affinity for cAMP.

The genetics of the PKA pathway has been extensively studied (see Broach, 1991; Thevelein and de Winde, 1999). Deletion of *BCK1* leads to constitutive, cAMP-independent activation of PKA and dominant *RAS2* mutations such as *RAS2*$^{\text{val-19}}$, which inhibit GTP hydrolysis, are also activatory. Constitutive activation is very deleterious to the cell and leads to the rapid acquisition of suppressor mutations that alleviate the defects. High PKA activity induces breakdown of storage carbohydrates such as glycogen and trehalose, induces various growth-specific genes such as ribosomal protein genes and generally stimulates proliferation. This leads to a situation where cells apparently fail to respond to the normal nutrient cues that block proliferation in G1 when nutrients are limiting. In the case of nitrogen starvation, failure of cells with a hyperactivated PKA pathway to arrest in G1 may in part be due to overriding TOR pathway mediated responses (see later), in addition to which such cells may fail to store sufficient amino acids to buffer them until they reach the next G1 phase once external supplies run out (Markwardt *et al.*, 1995). Cells with elevated PKA function are also sensitive to heat shock, failing to acquire the normal thermotolerance when nutrient limited.

Conversely, loss of function of the PKA pathway is also highly detrimental, blocking growth and proliferation, increasing gluconeogenesis and reducing the transcription of growth-specific genes. In diploid cells, PKA pathway loss of function also triggers sporulation without the need for the normal nutrient starvation signal. Single *RAS* deletions are viable, but the double deletion is lethal, emphasising the redundant nature of these two GTPases. However, *RAS1* transcription is repressed on non-fermentable carbon sources, so *ras2* mutants fail to grow on carbon sources other than glucose. The order and function of the various components in the pathway has been extensively explored by genetic analysis, showing for example that *bck1* mutations can suppress defects in adenylyl cyclase, while *srv2* mutations block the deleterious effects of the dominant *RAS2*$^{\text{val-19}}$ mutation.

Thus generally speaking, the idea developed that the PKA pathway in some sense acts to transduce nutrient signals (principally glucose), promoting growth, glycolysis and proliferation in the presence of glucose and allowing growth arrest in G1, sporulation (in diploid cells), gluconeogenesis and the accumulation of storage carbohydrates on glucose deprivation. However, strains in which *BCK1* was deleted and in which a single, functional but mutant *TPK* allele was present (termed *tpk*^{w} strains) contained weakly active, constitutive PKA function that was unregulatable by cAMP and yet showed essentially normal regulation of glycogen content, heat shock resistance and sporulation in response to glucose (Cameron *et al.*, 1988). This showed that other mechanisms must exist through which glucose can regulate these aspects of cell physiology. The SNF1 kinase (involved in glucose repression, see Chapter 3 and also Section 8.4), the Snf3p and Rgt1p glucose sensors and their downstream signalling pathway and hexokinases are now each known to provide PKA-independent glucose signalling mechanisms (see Rolland *et al.*, 2001) that may account for this.

cAMP levels in *S. cerevisiae* are subject to stringent feedback regulation by PKA (Nikawa *et al.*, 1987), such that quite major perturbations of the pathway cause only modest changes in cAMP level. Thus combined deletion of the two cAMP phosphodiesterase genes or presence of *RAS2*$^{\text{val-19}}$ leads to only a few-fold increase in cAMP level. However, *tpk*^{w} strains have cAMP levels that are elevated by 1000-fold, whereas *bck1* mutants have very low cAMP levels. This demonstrates both a capacity for up to a 10,000-fold dynamic response range in the pathway and also clearly shows that PKA exerts a strict control over the cell's ability to synthesise cAMP. This feedback might play a role in enabling the cell to see small signals against a large background 'noise', and could be exerted by PKA-mediated phosphorylation of adenylyl cyclase. Pde1p has also emerged recently as a PKA substrate and probable target for PKA-mediated feedback on cAMP levels (Ma *et al.*, 1999).

Inputs and targets of the PKA pathway

A major input into the PKA pathway is the response to glucose and the presence of glucose supports a basal level of signalling through the PKA pathway (Figure 8.4). However, on addition of glucose or other rapidly fermentable sugar to yeast which are either starved for glucose or growing non-fermentatively, an extremely rapid rise in intracellular cAMP concentration is triggered, although this declines quickly to about twice the basal level (Broach, 1991). This transient activation of the PKA pathway by glucose requires the Ras proteins but does not result in an elevated proportion of GTP-bound Ras, leading to the idea that while the Ras proteins are required for this glucose induced PKA signalling, they do not transmit the glucose signal itself (Colombo *et al.*, 1998). Interpretation of the relative involvement of Ras and Gpa2p in promoting this sharp rise in cAMP level is complicated by the possible role of Ras in plasma membrane localisation of adenylate cyclase, by the feedback mechanisms that operate on the pathway and by the role of Ras in signalling intracellular acidification that is also dependent on glucose and the cell's ability to maintain ATP levels. However, it is likely that Gpa2p is a major mediator of signalling to adenylate cyclase when glucose is added to glucose-derepressed cells (Thevelein and de Winde, 1999). As yet there is no firm evidence that Gpr1p is a glucose sensor, but it certainly behaves like one (Rolland *et al.*, 2000). Glucose phosphorylation is also required for glucose activation PKA, but this most likely operates by making adenylate cyclase more responsive to the Gpr1p–Gpa2p signal (see Rolland *et al.*, 2001). The activation of PKA via Ras in response to intracellular acidification (Thevelein, 1991) is more sustained, providing a potential mechanism for the cell to replenish failing ATP levels due to carbon starvation (Thevelein and de Winde, 1999). Reduced ATP generation would compromise the cell's ability to extrude protons, leading to intracellular acidification. Activation of PKA in response to this

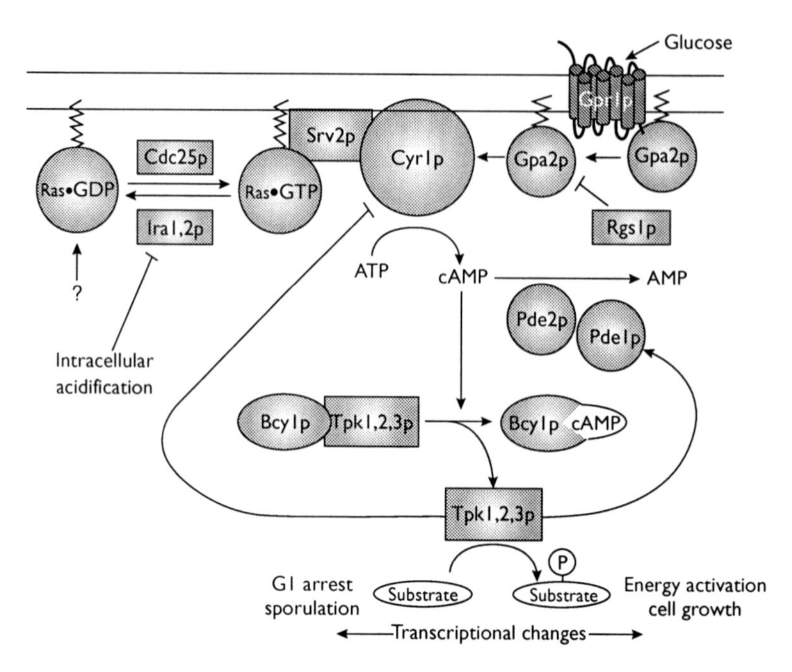

Figure 8.4 The PKA pathway in *S. cerevisiae*. The relationship between the different components of the PKA pathway is depicted showing stimulatory (→) and inhibitory (⊣) interactions.

could activate fermentation and breakdown of storage carbohydrates that could replenish ATP levels.

A number of proteins have been identified as potential targets of PKA, but while it may be clear that their activity is regulated in response to PKA, it is often difficult to be sure that activation is direct. Glycogen synthesis is inhibited when PKA is activated and part of this effect is mediated through regulation of glycogen synthase (see later). However, although PKA plays a role in the phosphorylation and hence inactivation of glycogen synthase, the key sites whose phosphorylation controls glycogen synthase activity do not seem to be directly modified by PKA itself. Glycogen phosphorylase, required for glycogen breakdown, may also be a substrate of PKA (see Broach and Deschenes, 1990), in this case phosphorylation resulting in activation. Activation of PKA by *bck1* deletion also reduces the mRNA level of glycogen synthase (Hardy *et al.*, 1994), suggesting transcriptional effects on glycogen accumulation as well. PKA also modulates expression of glycogen phosphorylase (*GPH1*) and *GAC1*, a gene required for activation of glycogen synthesis as discussed later, providing further inputs into the control of glycogen metabolism (François *et al.*, 1992). Synthesis of trehalose, another storage carbohydrate, also seems to be directly regulated by PKA. Thus yeast trehalase is activated by PKA *in vitro* (Uno *et al.*, 1983), while trehalose-6-phosphate synthase is inhibited by PKA phosphorylation and its activity *in vivo* mirrors the activation state of the PKA pathway in a variety of mutants in the predicted manner (Panek *et al.*, 1987).

PKA also targets key enzymes that regulate glycolytic flux, playing a role in the balance between glycolysis and gluconeogenesis. PKA influences the equilibrium between fructose-6-phosphate (F6P) and fructose-1,6-bisphosphate (F1,6P2) by stimulating the synthesis of fructose-2,6-bisphosphate (F2,6P2), an allosteric activator of phosphofructokinase-1 (catalysing the forward reaction for glycolysis) and an inhibitor of fructose-1,6-bisphosphatase (F1,6P2ase: catalysing the reverse reaction for gluconeogenesis). These effects are probably mediated by PKA phosphorylation and activation of Pfk2p (François *et al.*, 1988), which generates F2,6P2 from F6P, and by inactivation of F2,6P2 bisphosphatase (Kretschmer *et al.*, 1987), which converts F2,6P2 back to F6P. F1,6P2ase is also phosphorylated by PKA (Rittenhouse *et al.*, 1987), a modification which makes it much more sensitive to F1,6P2 inhibition (Marcus *et al.*, 1988). Another way in which PKA may affect carbon metabolism is via Adr1p. This protein is a transcription factor that is required for expression of the inducible alcohol dehydrogenase *ADH2*, but whose activity is inhibited during growth on glucose. This latter phenomenon seems to be mediated at least in part by PKA-dependent phosphorylation and inactivation of Adr1p on ser-230 (Cherry *et al.*, 1989; Taylor and Young, 1990).

In mammalian systems, many of the effects of PKA on gene expression are mediated via modulation of CREBs, bHLH transcription factors that bind to cAMP-responsive elements or CREs in the promoter sequences of PKA-regulated genes. In *S. cerevisiae*, Sko1p has been shown to bind to a CRE-like sequence in the promoters of various genes that are induced by high extracellular osmolarity, where it acts negatively in conjunction with the Ssn6p·Tup1p repressor under conditions of normal osmolarity, and recruitment of Sko1p to these promoters is dependent on PKA activity (Pascual-Ahuir *et al.*, 2001). However, Msn2p and Msn4p, two zinc finger transcription factors that bind to stress responsive promoter elements (STREs: 5′-CCCCT-3′) have been shown to mediate many of the effects of the PKA pathway on gene expression, and deletion of both *MSN2*and *MSN4* together suppresses many of the phenotypes associated with loss of PKA function (Smith *et al.*, 1998). STRE-regulated genes (Moskvina *et al.*, 1998) include *CTT1* (catalase), *HSP12* (heat-shock protein) and *TPS2* (trehalose-6-phosphate phosphatase) and show low expression during growth on glucose, but are induced when cells are grown on non-fermentative carbon sources and on entry into stationary phase. This regulation by glucose is mediated via the PKA pathway, which controls the localisation of the two transcription factors (Görner *et al.*, 1998), and in the case of Msn2p this has been shown to result from direct phosphorylation of the nuclear localisation sequence by PKA in response to glucose (Görner *et al.*, 2002). Cells with high PKA activity prevent the nuclear localisation of Msn2p and Msn4p even in the presence of other stress stimuli that normally activate STRE-regulated gene expression, whereas mutants with low PKA activity show constitutive nuclear localisation of Msn2p and Msn4p (Görner *et al.*, 1998). The 14-3-3 proteins Bmh1p and Bmh2p act as cytoplasmic anchors for Msn2p and Msn4p when they are excluded from the nucleus by phosphorylation (Beck and Hall, 1999). Since loss of Msn2/4p counteracts the growth arrest of PKA pathway mutants, a corollary of this is that PKA activity may be required to inhibit the expression of growth-inhibitory genes. Consistent with this notion, expression of the Yak1p kinase that antagonises PKA-dependent growth is Msn2/4p-dependent (Smith *et al.*, 1998). Interestingly, the localisation of the Bcy1p PKA regulatory subunit is dependent on Yak1p: Bcy1p is normally mainly nuclear, but on glucose withdrawal it shows rapid, Yak1p-dependent phosphorylation and is translocated to the cytoplasm. Although the significance of this is not yet clear, it may form part of an autoregulatory circuit.

Another downstream target of PKA is Rim15p, a kinase required for entry into meiosis (Vidan and Mitchell, 1997). The PKA pathway appears to regulate Rim15p in a negative manner, since deletion of *RIM15* suppressed a temperature-sensitive adenylate cyclase mutant whereas overexpression of Rim15p counteracted the effects of PKA pathway hyperactivity, and phosphorylation of Rim15p *in vitro* by PKA was strongly inhibitory to its activity (Reinders *et al.*, 1998). Loss of Rim15p also causes defects in trehalose and glycogen synthesis as cells enter stationary phase, and this is mediated via the Gis1p transcription factor that activates gene expression through the post-diauxic shift element (PDE1: Pedruzzi *et al.*, 2000). Thus PKA can suppress the expression of genes involved at the diauxic shift by at least two mechanisms, firstly by phosphorylating Msn2p and Msn4p to confine them to the cytoplasm and secondly by phosphorylating and inactivating Rim15p, thereby down-regulating Gis1p. Since Rim15p activity is required for proper induction of *IME1* (Vidan and Mitchell, 1997), a gene required for early meiotic induction, this explains at least in part the inhibitory effect of PKA on sporulation. The transcriptional regulator Sok2p functions as a repressor at the *IME1* promoter when phosphorylated by PKA, providing a second level of control (Shenhar and Kassir, 2001). Sok2p most likely regulates other genes downstream of PKA that are involved during mitotic growth and division as well (Ward *et al.*, 1995; see later).

8.4 SNF1 protein kinase and glucose repression

As described in Chapter 3, a major physiological response of yeast involves the control of metabolism in response to glucose availability. This becomes important both at the diauxic shift, when cultures become limiting for glucose and cells begin to utilise the ethanol generated during fermentative growth, and when fermentable carbon sources other than glucose are available. As discussed earlier, the PKA pathway is one mechanism by which yeast cells can respond to glucose. Glucose is the favoured carbon source and in its presence, the expression of many genes involved in utilisation of alternative carbon sources, respiration, gluconeogenesis and peroxisome biogenesis is repressed. During glucose starvation, derepression of these genes is a key adaptive response of *S. cerevisiae* and another protein kinase, Snf1p, is central to this process.

SNF1 was first identified in a genetic screen for *S. cerevisiae* mutants that failed to derepress the *SUC2* gene (encoding invertase) during growth in the absence of glucose (in fact *SNF* stands for sucrose non-fermenting). Thus *snf1* mutants cannot ferment sucrose because *SUC2* expression is repressed irrespective of the presence or absence of glucose (Neigeborn and Carlson, 1984). However, *SNF1* shows a more general involvement in glucose derepression: it has also been defined by mutations that prevent growth on glycerol (*cat1*) or ethanol (*ccr1*) and it is also needed for glycogen synthesis, peroxisome biogenesis and meiosis (which is also glucose-repressed; see Hardie *et al.*, (1998) for a review). Snf1p is also involved in invasive growth (see later), control of cell lifespan (Ashrafi *et al.*, 2000) and autophagy (Wang *et al.*, 2001b). Snf1p kinase is related in sequence to the α-subunit of mammalian AMP-activated protein kinase (AMPK), an important element in the stress response of higher cells (Hardie and Carling, 1997). However, while Snf1p activity correlates with cellular AMP levels (Wilson *et al.*, 1996), there is as yet no firm evidence for direct regulation of Snf1p by AMP in the way that has been clearly shown for the mammalian enzyme.

Snf1p is found in high molecular weight complexes that also contain Snf4p (Estruch *et al.*, 1992), homologous to the γ subunit of AMPK, and one of a family of AMPK β subunit-related polypeptides defined genetically (Sip1p, Sip2p and Gal83p: Yang *et al.*, 1994). The two *SIP* genes were identified by two-hybrid screening using Snf1p as 'bait' (Yang *et al.*, 1992, 1994), while *GAL83* was defined by a mutation changing the effect of glucose on the expression of the *GAL* genes (needed for galactose utilisation: see Erickson and Johnston, 1993). The physical interactions between Snf1p, Snf4p and the Sip/Gal83 proteins have been extensively studied by two-hybrid analysis, leading to a model whereby the Sip/Gal83 proteins interact with Snf1p and Snf4p via separate domains (termed KIS and ASC, respectively: see Hardie *et al.*, 1998 and Figure 8.5). Like *snf1* and *snf4* mutants, concurrent loss of all three β subunit genes confers a Snf$^-$ phenotype (Schmidt and McCartney, 2000), confirming that they constitute essential components of the SNF1 complex. The Sip/Gal83 proteins appear to act as a scaffold for Snf1p and Snf4p, although in the triple deletion mutant there was still some detectable Snf1p–Snf4p interaction (Jiang and Carlson, 1997). The β subunits show different subcellular localisations during growth under conditions of glucose derepression, with Gal83p largely nuclear, Sip1p in the vacuole and Sip2p in the cytoplasm and excluded from the nucleus, although in glucose-grown cells all three polypeptides are cytoplasmic (Vincent *et al.*, 2001). Nuclear localisation of Snf1p under derepressing conditions is Gal83p dependent, but nuclear exclusion of Gal83p on glucose is independent of either Snf1p or PKA. The β subunits may therefore play a major role in Snf1p substrate specificity by differential subcellular localisation, and the finding that loss of different β subunits uncovers different subsets of the *snf1* mutant phenotype supports this notion (Schmidt and McCartney, 2000).

The protein kinase activity of Snf1p is required for derepression and its activity increases dramatically on glucose starvation (Woods *et al.*, 1994; Wilson *et al.*, 1996). Although glucose availability seems not to change the overall composition of the Snf1 complex, examination of the inter-subunit interactions of the kinase has indicated that glucose availability influences its conformation (Jiang and Carlson, 1996). Thus Snf1p and Snf4p only interact strongly under derepressing conditions (when the kinase is active) via the C-terminal regulatory domain of Snf1p, while the kinase domain of Snf1p interacts with a distinct region of the regulatory domain in the presence of glucose. This leads to a model whereby the glucose

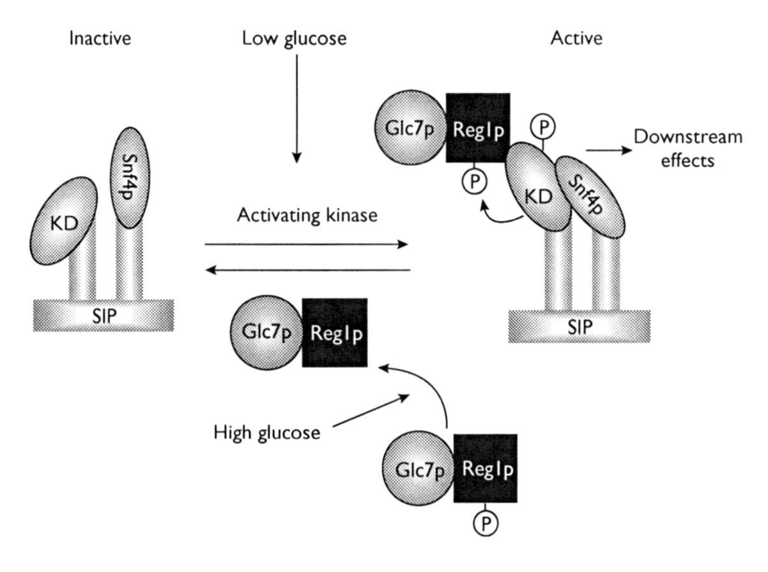

Figure 8.5 Regulation of the SNF1 protein kinase complex. Interconversion of the SNF1 complex between the closed (inactive) and open (active) conformations is shown based on the model proposed by Ludin *et al.* (1998). SIP refers to the Gal83p/Sip1p/Sip2p β subunit of the SNF1 complex.

derepression signal activates the Snf1 kinase by a Snf4p-mediated change in the conformation of the complex, releasing the catalytic domain of Snf1p to permit phosphorylation of key substrates (Figure 8.5).

A key question is how glucose starvation signals activation of Snf1 kinase. Mutations in *HXK2*, *GLC7* and *REG1* have each been found to relieve glucose repression in a *SNF1*-dependent manner (Neigeborn and Carlson, 1987), suggesting that they act upstream of Snf1 kinase. *HXK2* encodes the major isoform of hexokinase, required for glucose metabolism, while *GLC7* and *REG1* encode the catalytic and a regulatory subunit of PP1, respectively. Deletion of *REG1* causes constitutive invertase derepression and a specific mutant allele of *GLC7* (*glc7*-T152K: Tu and Carlson, 1994) confers a similar phenotype. Reg1p and Glc7p interact by several criteria and the *glc7*-T152K mutation specifically blocks the Reg1p-Glc7p interaction (Tu and Carlson, 1995). The *hxk2, glc7-T152K* and *reg1* mutations affect the glucose regulation of protein interactions within the Snf1 complex (Jiang and Carlson, 1996), since in each instance the mutation relieves the inhibition of the Snf1p–Snf4p two-hybrid interaction in the presence of glucose. In fact Reg1p interacts with the Snf1p kinase domain independently of Snf4p (Ludin *et al.*, 1998), demonstrating that it is likely to act on the Snf1 kinase directly, and use of a *reg1* mutant specifically defective for binding of Reg1p to Glc7p demonstrated that Reg1p functions by targeting Glc7p to the SNF1 complex (Sanz *et al.*, 2000a). Another protein (Sip5p) appears to mediate the Snf1p–Reg1p interaction (Sanz *et al.*, 2000b). The role of Reg1p involves modulating the conformation of the SNF1 complex, since a *reg1* deletion failed to derepress a strain dependent on just the Snf1p kinase domain. Like mammalian AMPK, Snf1p requires phosphorylation of its activation loop threonine (thr-210) for activity, although the upstream kinase has yet to be identified (McCartney and Schmidt, 2001). A T210→A mutation impairs the interaction between Snf1p and Snf4p in limiting glucose, consistent with inability to form the active SNF1 complex conformation (Ludin *et al.*, 1998). The T210→A mutation also impaired the interaction with Reg1p. Reg1p is phosphorylated in a Snf1p-dependent manner in the absence of glucose and shows Glc7p-dependent dephosphorylation on glucose readdition, while a 'kinase-dead' mutant Snf1p showed elevated interaction with Reg1p (Ludin *et al.*, 1998; Sanz *et al.*, 2000a). This suggests that Snf1p may negatively regulate its interaction with Reg1p·Glc7p. Phosphorylation of Reg1p appears to stimulate the ability of the Reg1p·Glc7p complex to return SNF1 to its inactive conformation, and an attractive model is that Reg1p·Glc7p is responsible for converting Snf1p to its inactive form by removing the activatory phosphorylation on thr-210. This is consistent with the finding that thr-210 is constitutively phosphorylated in a *reg1* deletion strain (McCartney and Schmidt, 2001). Hxk2p promotes the accumulation of phosphorylated Reg1p, consistent with the role of Hxk2p in glucose repression being mediated through Reg1p. It is not easy to integrate the role of Hxk2p in Reg1p phosphorylation with the roles of Reg1 phosphorylation in SNF1 inactivation and dissociation of Reg1p·Glc7p from Snf1p into a simple, unified model, while the fact that Reg1p·Glc7p can dephosphorylate Hxk2p (which is phosphorylated in response to glucose: Alms *et al.*, 1999) may provide yet an additional layer of complexity. In addition to Reg1p, *S. cerevisiae* encodes a related polypeptide (Reg2p) that is also associated with Glc7p. However, while the severe growth defect of the *reg1 reg2* double mutant shows that the two Reg proteins share an important function, Reg2p does not appear to be involved at all in glucose derepression (Frederick and Tatchell, 1996). Deletion of *SNF1* strongly suppresses the *reg1 reg2* growth defect (Frederick and Tatchell, 1996), showing that both the Snf1 protein kinase and the Reg·Glc7p protein phosphatase have additional, critical functions in the cell that are unrelated to glucose derepression.

Once Snf1 kinase is activated by glucose starvation, what substrates does it phosphorylate in order to mediate derepression? *In vitro* biochemical studies (see Hardie *et al.*, 1998) have shown that Snf1 kinase phosphorylates serine in a specific motif in its substrates, which can be defined as ϕ-X-R-X_2-S-X_3-ϕ (where ϕ is hydrophobic). In terms of glucose derepression, Mig1p is clearly a key substrate. Mig1p encodes a zinc finger protein that binds to the promoters of glucose-repressed genes such as *SUC2* (Nehlin and Ronne, 1990; Nehlin *et al.*, 1991), where it recruits a co-repressor (Ssn6p·Tup1p) that is also needed for glucose repression (Keleher *et al.*, 1992). Mig1p is rapidly phosphorylated on glucose deprivation and dephosphorylated in response to glucose addition and its nuclear localisation is also glucose dependent (DeVit *et al.*, 1997), but in an *snf1* mutant it is constitutively nuclear. Genetically, *mig1* mutations suppress the *snf1* glucose derepression defect (Johnston *et al.*, 1994; Vallier and Carlson, 1994) and the effects of Snf1 kinase on Mig1p are likely to be direct since Snf1p can efficiently phosphorylate Mig1p at four sites *in vitro* (Smith *et al.*, 1999). Mig1p nuclear export in response to glucose withdrawal is mediated by the exportin Msn5p, and mutation of the Snf1p phosphorylation sites in Mig1p rendered it constitutively nuclear (DeVit and Johnston, 1999). Thus Snf1p-regulated nuclear export of the Mig1p repressor is the mechanism of derepression for many glucose-repressed genes. There is also evidence of Mig1p-independent repression of *SUC2* on glucose (Vallier and Carlson, 1994) and this is mediated through the related repressor, Mig2p (Lutfiyya *et al.*, 1998). Surprisingly, Mig2p function is not regulated by SNF1, nor does its localisation show glucose-regulated changes like that of Mig1p (Lutfiyya *et al.*, 1998).

However, Mig1p is not the only downstream effector of the SNF1 kinase. Sip4p is a Cys_6 zinc cluster transcription factor that interacts with Snf1p and is phosphorylated in a Snf1p-dependent manner in glucose-derepressed cells (Lesage *et al.*, 1996). Sip4p binds to the carbon source-responsive element (CSRE) found upstream of gluconeogenic genes such as *FBP1* and is likely to be activated by SNF1 to promote their expression, and Gal83p is required both for binding to Snf1p and for the Snf1p-dependent phosphorylation (Vincent and Carlson, 1998, 1999). Expression of *SIP4* on glucose limitation requires Cat8p, which is itself induced upon SNF1 activation by relief of Mig1p-mediated repression, and *SIP4* can also activate its own expression via a CSRE in the *SIP4* promoter (Randez-Gil *et al.*, 1997; Vincent and Carlson, 1998).

Another manner in which SNF1 is emerging as a regulator of gene expression is as a histone kinase. Snf1p can phosphorylate ser-10 of histone H3 and this modification, followed by acetylation on lys-14 by Gcn5p, is required for induction of *INO1* on inositol starvation (Lo *et al.*, 2001). A similar mechanism operating via histone H3 phosphorylation-dependent chromatin remodelling may explain how SNF1 promotes *ADH2* expression by modulating the binding of Adr1p to the *ADH2* promoter under glucose-limiting conditions (Young *et al.*, 2002). However, since *ADH2* expression is also dependent on Cat8p (Walther and Schuller, 2001), SNF1 appears to act at multiple levels to upregulate *ADH2* expression.

In addition to its involvement in glucose repression, Snf1p also plays a role in haploid invasive growth (see Section 1.5 of Chapter 1), a phenomenon whereby yeast cells change their budding pattern and morphology on glucose limitation to allow penetration of the agar on which they are growing (Cullen and Sprague, 2000; Kuchin *et al.*, 2002). This regulation seems to be mediated through Snf1p-dependent inhibition of two repressors, Nrg1p and Nrg2p, which bind to the *FLO11* promoter during growth on glucose and are antagonised by Snf1p when it is activated. *FLO11* encodes an adhesin that is required for haploid invasive growth. Surprisingly, Snf1p is also required for pseudohyphal differentiation, a similar phenomenon that occurs in diploid cells when starved for nitrogen rather than carbon. Glucose limitation also causes Snf1p-dependent phosphorylation and nuclear import of Gln3p (Bertram *et al.*, 2002), a GATA transcription factor that has previously been shown to play a role in gene expression triggered by nitrogen starvation via the TOR pathway (Beck and Hall, 1999; see later). This effect is associated with elevated expression of several Gln3p target genes and is quite distinct from the regulation of Gln3p nuclear localisation by TOR-dependent phosphorylation discussed in the next section, which involves the phosphorylation of different sites on Gln3p (Bertram *et al.*, 2002).

8.5 The TOR signalling pathway

The TOR signalling pathway was first identified in yeast but represents a conserved mechanism by which cells can respond to nutrients. TOR stands for Target Of Rapamycin, and the two yeast kinases (Tor1p and Tor2p) that play a critical role in this pathway are the target of the immunosuppressant and anti-cancer drug rapamycin. Tor1p and Tor2p together with their mammalian homologue mTOR and similar proteins in other organisms

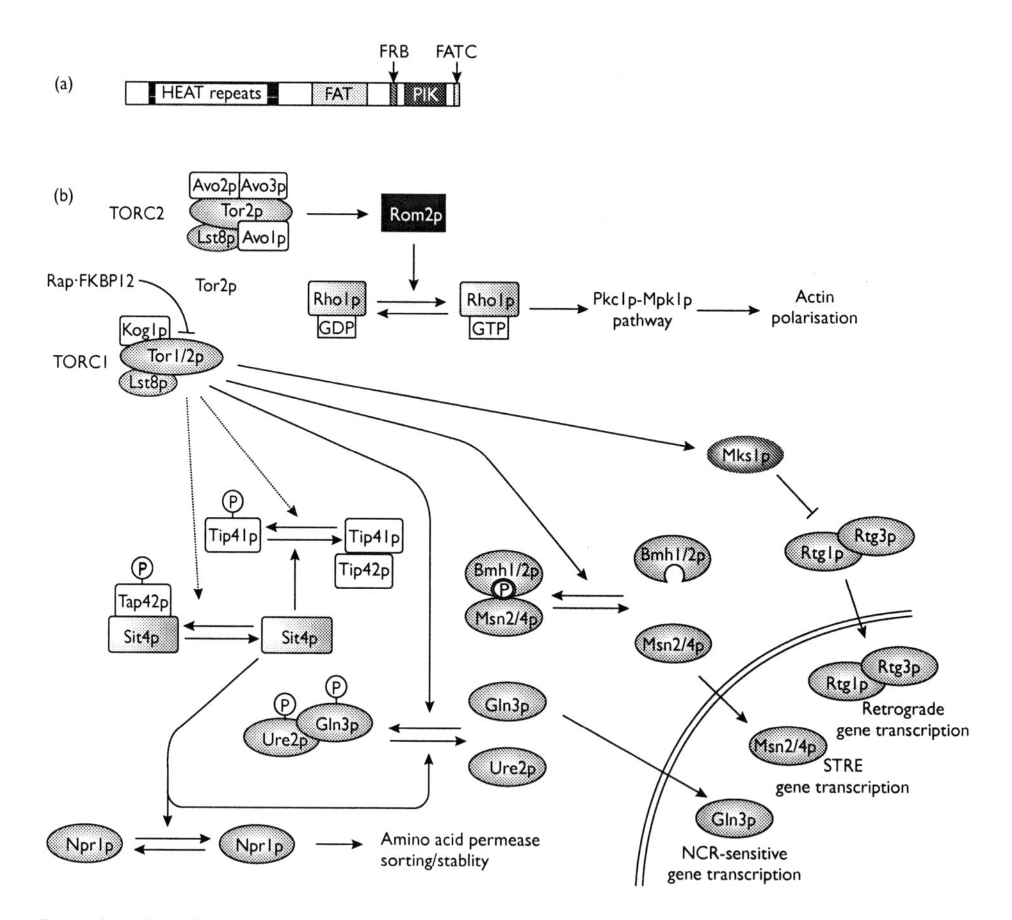

Figure 8.6 The TOR pathway. (a) Domain structure of Tor1p and Tor2p. (b) The TOR pathway, showing the shared and Tor2p unique functions.

belong to the PIK (phosphatidylinositol kinase) family, but act as protein kinases (Kunz *et al.*, 1993; Helliwell *et al.*, 1994; Alarcon *et al.*, 1999). They are large proteins with a C-terminal catalytic domain and their function is blocked by rapamycin, which binds to the TOR kinases in a complex with FKBP12, a prolyl *cis–trans* isomerase that also acts as the cellular receptor for the related drug FK506 (Heitman *et al.*, 1991). Figure 8.6(a) shows the structure of the TOR kinases. The rapamycin–FKBP12 complex binds to the FRB domain upstream of the C-terminal PIK domain and point mutations in the FRB region are responsible for conferring dominant, rapamycin resistance on yeast cells (Helliwell *et al.*, 1994; Lorenz and Heitman, 1995). Additional domains conserved with other PIK-related kinases (FAT, FATC: Bosotti *et al.*, 2000) bracket the FRB-PIK region while the N-terminal half of the protein consists of 20 HEAT repeats (see Groves *et al.*, 1999), a motif shared with the regulatory A subunit of PP2A (see later). The FAT domain mediates Tor1p toxicity when overexpressed and may represent a TOR effector domain (Alarcon *et al.*, 1999), while in the case of Tor2p the HEAT repeats have been shown to anchor the protein to the plasma membrane (Kunz *et al.*, 2000). *TOR2* is essential whereas *TOR1* is not, but in fact the two kinases share an essential, rapamycin-sensitive function involved in the regulation of translation, G1 progression and the response to nutrients (Kunz *et al.*, 1993; Helliwell *et al.*, 1994). Tor2p is essential because it has a unique, rapamycin-insensitive function concerned with the regulation of the actin cytoskeleton (Zheng *et al.*, 1995; Schmidt *et al.*, 1996). These two different functions are reflected by the existence of two distinct, TOR kinase-containing multiprotein complexes in yeast cells that appear to mediate them: TORC1 contains either Tor1p or Tor2p together with Kog1p and Lst8p and mediates the shared function, whereas TORC2 contains Tor2p together with Avo1p, Avo2p, Avo3p and Lst8p and mediates the Tor2p-unique function (Loewith *et al.*, 2002). Figure 8.6(b) summarises the mechanisms that underlie the shared and unique functions of TOR that are discussed next.

Rapamycin-sensitive shared functions of TOR

Genome-wide analysis of transcriptional changes in response to rapamycin treatment have demonstrated that the TOR pathway modulates the expression of a large number of genes including those encoding ribosomal proteins, glycolytic enzymes, citric acid cycle enzymes and genes subject to nitrogen catabolite repression (Cardenas *et al.*, 1999; Hardwick *et al.*, 1999), emphasising the role of TOR in coordinating the expression of many genes involved in growth in response to nutrients. The TOR pathway in yeast clearly responds to the quality of nitrogen source, and many of the outputs discussed next are triggered by growth on poor nitrogen sources such as urea or proline. Consistently, glutamine is the preferred nitrogen source in yeast and starvation for glutamine by the glutamine synthase inhibitor methionine sulfoxime (MSX) triggers a rapid, TOR-pathway-mediated response (Crespo *et al.*, 2002). How the nitrogen signal is transmitted to or detected by the TOR kinases is currently not known, but clearly one possibility is that TOR can sense the intracellular glutamine level.

Many of the shared functions of Tor1p and Tor2p involve Tap42p, a protein that associates with both PP2A and the Sit4p protein phosphatase in yeast (Di Como and Arndt, 1996). Tap42p is related in sequence to the mouse $\alpha4$ protein, which is phosphorylated in response to stimulation of the immunoglobulin receptor in B cells and which is also associated with PP2A and PP2A-related protein phosphatases (Murata *et al.*, 1997; Chen *et al.*, 1998). Deletion of *TAP42* is lethal, while a temperature-sensitive *tap42* mutant (*tap42-11*) causes reduced translational initiation and G1 arrest at the restrictive temperature as well as dominant rapamycin resistance. In the presence of nutrients, Tap42p is phosphorylated and remains associated with Sit4p, but on starvation or rapamycin treatment the two proteins dissociate (Di Como and Arndt, 1996). Since Sit4p activity is required for a number of the downstream signalling outputs of the TOR pathway (see later), this has led to the generally accepted notion that Tap42p functions as a negative regulator of Sit4p and hence probably also of PP2A. TOR kinase activity therefore keeps Tap42p phosphorylated and bound to Sit4p, whereas TOR inactivation (either by lack of nutrients or rapamycin) causes Tap42p to become dephosphorylated and to dissociate from Sit4p such that Sit4p can then dephosphorylate downstream effectors. The TOR kinases may phosphorylate Tap42p directly, while PP2A in combination with its Tpd3p and Cdc55p subunits (see later) is involved in Tap42p dephosphorylation (Jiang and Broach, 1999). However, the rate of Tap42p dephosphorylation seen on rapamycin treatment seems rather slow compared with the rate of dissociation of Tap42p from Sit4p and the rate at which Sit4p appears to be activated (see Crespo and Hall, 2002). Tip41p, a protein that binds to Tap42p and may relieve its inhibition of Sit4p, also shows TOR-dependent phosphorylation and rapamycin treatment stimulates the binding of Tip41p to Tap42p by a mechanism that involves dephosphorylation of Tip41p by Sit4p (Jacinto *et al.*, 2001). This may function to promote rapid amplification of phosphatase activity when TOR signalling is inhibited and it is possible that phosphorylation of Tip41p by TOR is the primary manner in which TOR controls Sit4p activity. Less is known about how PP2A is regulated by TOR and it is also possible that while Tap42p is inhibitory to Sit4p (and PP2A) for some TOR pathway functions, Tap42p might activate the dephosphorylation of other phosphatase substrates to mediate other functions. The finding that phosphatase regulatory subunits can stimulate activity towards some substrates while inhibiting dephosphorylation of others has certainly been a recurrent theme in the case of PP1. The ability of high-copy *TAP42* to suppress loss of function mutations in either *PPH21* (encoding a PP2A catalytic subunit isoform) or *SIT4*, plus the observations that excess Tap42p has almost no effect on growth but is strongly inhibitory to growth in cells with excess Sit4p or Pph21p (Di Como and Arndt, 1996), are both findings that do not fit comfortably with models in which Tap42p acts as a phosphatase inhibitor.

The TOR shared function in yeast controls a number of processes required for adaptation to poor nitrogen sources, and one of these concerns the spectrum of amino acid permeases exhibited on the cell surface (see Chapter 4, Section 4.3). Some permeases such as the general amino acid permease Gap1p are downregulated in the presence of a good nitrogen source while other, more specific permeases such as the tryptophan permease Tat2p are utilised during growth on good nitrogen sources but downregulated on nitrogen starvation. The Npr1p kinase is activated on poor nitrogen sources and acts to stabilise Gap1p by preventing its degradation, while Npr1p kinase is a negative regulator of the stability of permeases such as Tat2p (Schmidt *et al.*, 1998). Npr1p is kept in a phosphorylated and inactive state by the TOR pathway, but when TOR activity is inhibited it becomes dephosphorylated and active (Schmidt *et al.*, 1998). Activation of Npr1p requires both Sit4p phosphatase and Tip41p, showing that in this case at least, TOR signalling operates broadly in the manner discussed earlier (Jacinto *et al.*, 2001). Active Npr1p prevents cell surface Gap1p from being internalised and also prevents the direct targeting of newly synthesised Gap1p to the vacuole, most likely by antagonising Gap1p ubiquitination involving the Npi1p ubiquitin ligase (De Craene *et al.*, 2001). The inverse regulation of permeases such as Tat2p also involves similar effects on both protein stability and sorting (Beck *et al.*, 1999) and both the Gap1p and Tat2p modes of regulation may involve direct phosphorylation of the permeases by Npr1p kinase.

The TOR pathway also adapts cells to nitrogen limitation by regulating both nitrogen catabolite repression (NCR)-sensitive gene expression (e.g. *GAP1*) and retrograde gene expression (e.g. *CIT2*). The GATA transcription factors Gln3p and Gat1p are involved in control of the NCR-sensitive genes (see Cooper, 2002). On good nitrogen sources, Gat1p and Gln3p are localised to the cytoplasm where they are bound to Ure2p, which acts as their cytoplasmic receptor. On nitrogen limitation they are rapidly relocalised to the nucleus, where they can activate the expression of their target genes (Beck and Hall, 1999). The interaction between Gln3p and Ure2p requires Gln3p to be phosphorylated, and this phosphorylation was lost on rapamycin treatment in a Sit4p-dependent manner (Beck and Hall, 1999). Thus again it seems that activation of Sit4p in response to reduced TOR kinase signalling is responsible for this particular response to nitrogen limitation. TOR forms a complex with Gln3p, and Ure2p and Gln3p is phosphorylated directly by TOR *in vitro* (Bertram *et al.*, 2000), so it is likely that TOR directly phosphorylates the GATA transcription factors. Whether Sit4p dephosphorylates them directly is, however, not yet known. Ure2p also shows TOR-regulated phosphorylation and dephosphorylation and this may therefore form an additional aspect to the regulation of GATA factor function (Cardenas *et al.*, 1999; Bertram *et al.*, 2000). Although glutamine starvation using MSX treatment drives Gln3p into the nucleus, Gat1p remains cytoplasmic under such conditions (Crespo *et al.*, 2002). Thus although both transcription factors are involved in the regulation of NCR-sensitive genes they appear to be activated via TOR by different signals. The retrograde gene regulation involves the Rtg1p·Rtg3p transcription factor, whose nuclear localisation and function are negatively regulated by TOR and Rtg3p has been shown to become hyperphosphorylated in rapamycin-treated cells (Komeili *et al.*, 2000). Mks1p, a negative regulator of the Rtg1·Rtg3p heterodimer appears to mediate at least part of the signal by which TOR activity constrains Rtg1·Rtg3p to the cytoplasm, and this may involve antagonistic interactions with Rtg2p, required for nuclear localisation of the transcription factor (Dilova *et al.*, 2002; Tate *et al.*, 2002). A second possible role for Mks1p as a regulator of the NCR-sensitive genes via interactions with Ure2p is more controversial (see Tate *et al.*, 2002). Recently, it has been proposed that RTG target gene expression is regulated not by the nitrogen source *per se* but by the metabolic products of that source. Thus nitrogen sources that yield glutamate down-regulate RTG target genes whereas they are expressed on those yielding ammonia (Tate *et al.*, 2002). Nonetheless, MSX inhibition of glutamine synthase still drives the Rtg1p·Rtg3p complex into the nucleus of glutamate-grown cells, showing that glutamine is likely to be an important signal (Crespo *et al.*, 2002). The TOR pathway also regulates the localisation of Msn2p and Msn4p STRE transcription factors (see earlier text) in a similar manner to the GATA transcription factors, although in this case their cytoplasmic anchors are the redundant 14-3-3 proteins Bmh1p and Bmh2p (Beck and Hall, 1999). Rapamycin treatment drives Msn2p and Msn4p into the nucleus, although in this case the process does not require Sit4p phosphatase. Thus the PKA and TOR pathways intersect in the control of STRE gene expression by regulating Msn2p and Msn4p localisation.

The TOR pathway has recently been shown to regulate autophagy, another starvation response in which cytoplasmic components are delivered to the vacuole for degradation. Autophagy requires the Apg1p kinase as a positive effector and under good growth conditions, Apg1p activity is held in check because its activatory subunit, Apg13p, is inactivated by phosphorylation. Rapamycin treatment or nutrient starvation block the phosphorylation of Apg13p such that it can bind to and activate Apg1p (Kamada *et al.*, 2000). The TOR pathway is also required for pseudohyphal differentiation in diploid cells in response to nitrogen limitation although, surprisingly, TOR signalling appears necessary to permit pseudohyphal growth rather than mediating a nitrogen limitation signal, and low concentrations of rapamycin block the pseudohyphal response (Cutler *et al.*, 2001).

In mammalian cells, the TOR pathway has been shown to regulate translation in at least two different ways (see Schmelzle and Hall, 2000). Firstly, the TOR pathway promotes 5′-TOP mRNA translation by stimulating phosphorylation of ribosomal protein S6 (Jefferies *et al.*, 1997). Since this class of mRNA mainly encode proteins of the translation machinery, this leads to an upregulation of translation. Secondly, the TOR pathway promotes the phosphorylation of eIF4E-BP, thereby dissociating it from eIF4E and promoting translational initiation by allowing eIF4E to bind eIF4G. There is no evidence for the former mechanism operating in yeast, although TOR is clearly a positive regulator of production of the translational machinery at the transcriptional level (see e.g. Zaragoza *et al.*, 1998; Powers and Walter, 1999; Shamji *et al.*, 2000). However, it is quite possible that TOR regulates eIF4E function in a related manner to that just described. Thus Eap1p has been shown recently to play an eIF4E-BP-like role in that it can compete with yeast eIF4G for binding to eIF4E (Cdc33p) and an *eap1Δ* deletion confers partial rapamycin resistance (Cosentino *et al.*, 2000). Consistent with regulation of this step in translational initiation, rapamycin treatment or starvation also confer reduced stability of eIF4G (Berset *et al.*, 1998). As discussed in Section 8.2, expression of Cln3p is very dependent on Cdc33p function (Danaie *et al.*, 1999) and it is the ability of rapamycin to inhibit translational initiation that is responsible for the G1 arrest caused by rapamycin treatment (Barbet *et al.*, 1996). Consistent with this, expression of Cln3p using a 5′ leader region derived from a gene normally expressed during starvation (*UBI4*) suppressed the rapamycin-induced G1 arrest but conferred starvation sensitivity (Barbet *et al.*, 1996).

The Tor2p unique function

The unique, essential function of Tor2p concerns the polarisation of the actin cytoskeleton (Schmidt *et al.*, 1996). Like the shared roles of TOR in promoting translation initiation, ribosome biosynthesis and in antagonising autophagy, regulation of the actin cytoskeleton is also an important growth process in that it is required for polarised secretion into the bud. Tor2p signals to the actin cytoskeleton through the Rho1p and Rho2p GTPases, which it activates via the Rom2p guanine nucleotide exchange factor (GEF). Thus either deletion of the Sac7p Rho GAP or overexpression of the Rom2p GEF suppresses *tor2* mutations, while Rom2p GEF activity is reduced in a *tor2* mutant. Rho1p has a number of possible downstream effectors including β-glucan synthase, the formin Bni1p, Skn7p and the Pkc1p–Mpk1p MAP kinase pathway (see later) but the latter is probably the most critical for mediating the effect of Tor2p on the actin cytoskeleton (Helliwell *et al.*, 1998). As discussed next, a major output of this pathway is the transcriptional regulation of genes involved in cell integrity and cell wall synthesis, but it is also possible that more direct effects of Pkc1p and/or Mpk1p are involved in Tor2p-promoted actin polarisation.

8.6 MAP kinase pathways

The eponymous enzymes of MAP kinase (MAPK) pathways constitute a major family of proline-directed protein kinases that require activatory phosphorylations on both a threonine and a tyrosine in the sequence -thr-X-tyr- in their activation loop. These activatory phosphorylations are carried out by a specific dual-specificity upstream kinase or MEK (MAPK/ERK kinase), which in turn is also activated by phosphorylation on a pair of closely spaced serine residues by a MEKK (MEK kinase). The MEKK–MEK–MAPK cascades constitute a conserved signalling module found in many eukaryotic signal transduction pathways and may function both to amplify a signal and to convert graded inputs into a more switch-like output (Ferrell, 1996). Frequently, MAPK pathways function to modulate the activity of transcription pathways (Treisman, 1996) and budding yeast is no exception to this general rule. The genome of *S. cerevisiae* encodes five potential MAPK sequences, four MEKs and five MEKKs (Herskowitz, 1995; Hunter and Plowman, 1997). At each level in the MEKK–MEK–MAPK cassette the protein kinases identified form a family of related sequences. The yeast kinases are organised into four MEKK–MEK–MAPK cassettes (Figure 8.7: see also Gustin *et al.*, 1998). To date, no upstream MEK or MEKK enzymes have been assigned to the sporulation-specific Smk1p (see Hunter and Plowman, 1997). However, *SPS1* encodes a protein kinase which is required for sporulation-specific gene expression late in the sporulation developmental programme and whose catalytic domain is 44% identical to Ste20p, an activator of the pheromone and nitrogen starvation pathways shown in Figure 8.7 (Friesen *et al.*, 1994). Sps1p might therefore function upstream of Smk1p.

Protein kinase C and the MPK1 MAP kinase pathway

The Pkc1p–Mpk1p pathway protein kinases

Many of the elements in the Mpk1p pathway were identified because they show genetic interactions with Pkc1p, the budding yeast homologue of mammalian protein kinase C, and in fact Pkc1p is a positive activator of the Mpk1p pathway. The Pkc1p–Mpk1p pathway has been reviewed in detail elsewhere (Gustin *et al.*, 1998; Heinisch *et al.*, 1999) and so only a basic overview will be given here; it is often termed the cell integrity pathway because of its role and the properties of cells in which it is defective. Loss of Pkc1p function causes a cell lysis defect and sensitivity to low osmolarity, reduced cell wall thickness and cell lysis at the tips of

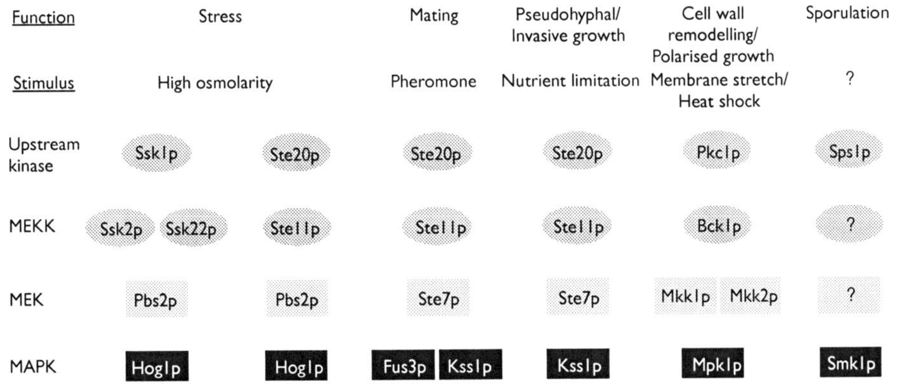

Function	Stress		Mating	Pseudohyphal/ Invasive growth	Cell wall remodelling/ Polarised growth	Sporulation
Stimulus	High osmolarity		Pheromone	Nutrient limitation	Membrane stretch/ Heat shock	?
Upstream kinase	Ssk1p	Ste20p	Ste20p	Ste20p	Pkc1p	Sps1p
MEKK	Ssk2p Ssk22p	Ste11p	Ste11p	Ste11p	Bck1p	?
MEK	Pbs2p	Pbs2p	Ste7p	Ste7p	Mkk1p Mkk2p	?
MAPK	Hog1p	Hog1p	Fus3p Kss1p	Kss1p	Mpk1p	Smk1p

Figure 8.7 The MAP kinase pathways of *S. cerevisiae*. The pathways into which each of the five *bona fide* yeast MAP kinases fit are indicated, together with the proposed function and input into each protein kinase cascade. MAPK, MAP kinase; MEK, MAPK/ERK kinase; MEKK, MEK kinase.

small buds as they undergo polarised growth (see Levin and Bartlett-Heubusch, 1992; Levin *et al.*, 1994). These defects can be suppressed by growth on osmotically stabilised media and suggest defects in cell wall metabolism. Bck1p was identified through the dominant *BCK1–20* allele, encoding a constitutively activated MEKK that suppresses the lysis defect of a *pkc1Δ* mutant (Lee and Levin, 1992). The *bck1Δ* strains themselves have a temperature-sensitive lysis defect and *MPK1* was identified as a dosage suppressor of this (Lee *et al.*, 1993b), while *MKK1* was isolated as dosage suppressors of another mutant with an osmoremedial lysis phenotype (Irie *et al.*, 1993). Deletion of *BCK1*, *MPK1* or the double *mkk1Δ mkk2Δ* all show a temperature-sensitive cell lysis defect and epistatic relationships indicated the order of the pathway depicted in Figure 8.7: high-copy *MPK1* is a dosage suppressor of *mkk1Δ mkk2Δ* or *bck1Δ*, high-copy *MKK1* is a dosage suppressor of *bck1Δ* but not *mpk1Δ* and *BCK1–20* was unable to suppress the *mkk1Δ mkk2Δ* double mutation (Irie *et al.*, 1993). As might be expected, phosphorylation of Mpk1p on the -T-E-Y- sequence in its activation loop is essential for activity, which was blocked by the double T190→A Y192→F mutation (Lee *et al.*, 1993b). Mpk1p tyr-192 phosphorylation is Bck1p- and Mkk1/2p-dependent (Davenport *et al.*, 1995; Zarzov *et al.*, 1996), consistent with their roles as the Mpk1p MEKK and MEKs, respectively. Of the two MEKs, Mkk1p plays the more important role in transmitting signals to Mpk1p (Martin *et al.*, 2000). Bck1p is probably activated directly by Pkc1p phosphorylation since it is phosphorylated on four sites by Pkc1p *in vitro* (see Watanabe *et al.*, 1994).

The mammalian PKC superfamily includes both PKCs and the PRK/ROCK kinases and yeast Pkc1p contains, in addition to its protein kinase domain, the C1, C2 and V5 domains found in mammalian PKCs together with two N-terminal HR1 domains found in the PRKs (Mellor and Parker, 1998). The PRKs are Rho-activated protein kinases (Aspenstrom, 1999) and the HR1 motifs provide a site of interaction for Rho GTPases. Unlike mammalian PKCs, Pkc1p failed to show activation by Ca^{2+}, correlating with the absence of key aspartate residues in its C2 domain (Ponting and Parker, 1996). Pkc1p also failed to show activation by phosphatidylserine or diacylglycerol (DAG), despite the presence of the C1 domain that mediates phospholipid binding in the mammalian PKCs (Antonsson *et al.*, 1994; Watanabe *et al.*, 1994). However, mutation of key cysteine residues in the presumptive DAG binding site clearly affects at least some functions of Pkc1p (Jacoby *et al.*, 1997a; Nanduri and Tartakoff, 2001). Like the mammalian enzymes, it has a pseudosubstrate domain which, when deleted, leads to much higher intrinsic activity (Watanabe *et al.*, 1994). This implies that Pkc1p activity should be activated by a mechanism that displaces the pseudosubstrate region, and in fact the small GTPase Rho1p interacts with Pkc1p and confers on it the ability to be stimulated by phosphatidylserine (Kamada *et al.*, 1996; see later). On activation of the pathway Mpk1p becomes rapidly phosphorylated on thr-190 and tyr-192 (Martin *et al.*, 2000). Since loss of these modifications is also very rapid when the activating stimulus is removed (less than 10 min), protein phosphatases must be involved in downregulating Mpk1p function (Martin *et al.*, 2000). Two functionally redundant protein phosphatase genes (*PPZ1* and *PPZ2*) have been isolated as dosage suppressors of both *mpk1Δ* and *pck1Δ* mutations but are most likely not involved in regulating the Mpk1p pathway itself, acting in parallel with it as positive regulators of cell integrity (Lee *et al.*, 1993a). The dual-specificity phosphatase Msg5p, which also has a role in regulating Fus3p (see later), can clearly influence the phosphorylation state of Mpk1p (Martin *et al.*, 2000) and the PP2C phosphatase Ptc1p may also be involved in regulating flux through the Pkc1p–Mpk1p pathway, since *ptc1* recessive mutations suppress *pkc1* mutations (Huang and Symington, 1995). However, the best candidate Mpk1p phosphatase is Sdp1p, which can clearly bind to and dephosphorylate active Mpk1p directly (Collister *et al.*, 2002; Hahn and Thiele, 2002). Sdp1p is induced in an Msn2/4p-dependent manner upon heat shock, providing a feedback mechanism whereby activated Mpk1p can be subsequently downregulated. Induction of *PTP2* by Mpk1p activation may also provide a similar feedback mechanism (Mattison *et al.*, 1999; Hahn and Thiele, 2002).

Inputs, outputs and functions

The available evidence suggests that Pkc1p may serve to integrate a number of inputs, and activation of the Mpk1p pathway is also not the only function of Pkc1p. *pkc1* mutants have a much more severe defect than mutants in the downstream MAPK pathway, while some of the effects mediated by Pkc1p are clearly Mpk1p-independent, indicating that Pkc1p has additional roles. This notion is supported by the fact that a variety of other genes can reduce the severity of the *pkc1* mutant phenotype when present in high copy, but when mutated they show an additive defect with mutations in the MAPK module. One such example is *KRE6*, which is required for $\beta(1\rightarrow6)$glucan biosynthesis (Roemer *et al.*, 1994). However, activation of the Mpk1p MAPK pathway is clearly a critical Pkc1p function.

A principal requirement for the Pkc1p–Mpk1p pathway is polarised growth. Thus *mpk1* mutants are synthetically lethal with some conditional *cdc28* alleles that block cells in G1 (i.e. at the stage where Cln1,2·Cdc28p is required to initiate polarised cell growth: Lew and Reed, 1993; Mazzoni *et al.*, 1993). Co-lethality is not seen with alleles such as *cdc28-1N*, which preferentially affect Clb·Cdc28p function and block in G2 (Mazzoni *et al.*, 1993), while the lethality of the *cln1Δ cln2Δ mpk1Δ* triple mutant (Gray *et al.*, 1997) supports a cooperative involvement of Cln·Cdc28p and Pkc1p–Mpk1p. Cln·Cdc28p activity has also been proposed to act as one input into the Pkc1p–Mpk1p pathway (Marini *et al.*, 1996). Thus *pkc1* mutants were identified in a genetic screen designed to identify targets of Cdc28p and overexpression of *CLN2* (which elevates Cln·Cdc28p activity) partially suppressed *pkc1* temperature-sensitive mutations, and in *cdc28* temperature-sensitive mutants the normally rapid activation of Mpk1p at 37°C (see later), the restrictive temperature for the *cdc28* mutations, was not observed (Marini *et al.*, 1996; Zarzov *et al.*, 1996). How Cdc28p might activate the Pkc1p–Mpk1p pathway is not fully understood, but a proposal that Cdc28p-dependent generation of DAG is involved (Marini *et al.*, 1996) seems unlikely given that DAG activation of yeast Pkc1p has yet to be demonstrated. *mpk1* mutants show delocalisation of cortical actin patches and chitin deposition, and are also synthetically lethal with conditional mutations in actin and the nonconventional myosin Myo2p, both of which are needed for targeting secretory vesicles to sites of cell surface growth (Mazzoni *et al.*, 1993). Mpk1p is tyrosine phosphorylated and activated both during bud growth (corresponding to the stage when *pkc1* mutant cells lyse: Levin and Bartlett-Heubusch, 1992) and during morphogenesis in response to mating pheromone (Zarzov *et al.*, 1996). In fact Mpk1p activation peaks at the same time as formation of the mating projection, although Mpk1p activation is not required for projection morphogenesis *per se* (Buehrer and Errede, 1997).

Heat shock is also an activator of the Pkc1p–Mpk1p pathway (Kamada *et al.*, 1995), which is also needed for induced thermotolerance. In at least some yeast strains Mpk1p is strongly activated in a Pkc1p-dependent manner by growth even at 37°C. However, this response is distinct from the upregulation of heat-shock gene expressions, which occurs normally in *pkc1* or *mpk1* mutants (Kamada *et al.*, 1995). Shifting cells to 37°C leads to a transient depolarisation of the actin skeleton and of $\beta(1\rightarrow3)$glucan synthase that is mediated by

Pkc1p but which is independent of Mpk1p function, although subsequent actin repolarisation requires Mpk1p function as well (Delley and Hall, 1999). Other sources of cell wall stress such as zymolyase or mild treatment with SDS also induce a similar response (Delley and Hall, 1999). Conversely, depolarisation of the actin cytoskeleton also activates Mpk1p, which is required for a robust arrest of cells in G2 via the morphology checkpoint that blocks mitosis in response to loss of actin polarity (Harrison *et al.*, 2001). There is evidence that the mechanism by which Mpk1p functions to inhibit mitosis under these circumstances is by inhibition of Mih1p, the Cdc28p tyr-19 phosphatase (see before). Chlorpromazine, which has been proposed to induce membrane stretch, is a potent Mpk1p activator (Kamada *et al.*, 1995), and Mpk1p is phosphorylated and activated by hypo-osmotic shock (Davenport *et al.*, 1995). Increasing osmolarity reduces signalling through the MAPK pathway, but the constitutively activating *BCK1-20* mutation blocks the damping down of signalling as external osmotic strength is increased (Davenport *et al.*, 1995). Mpk1p activation also occurs for unknown reasons in response to caffeine and vanadate (Martin *et al.*, 2000). Pkc1p, Rho1p and $\beta(1\rightarrow3)$glucan synthase are localised to the cell surface at the sites of polarised growth (i.e. the bud tip and, in larger cells the bud neck), placing them in optimal locations both for detecting cell wall stress and possibly for effecting localised responses to it. Cell wall stress leads to their delocalisation, a response that might be required to repair cell wall damage (Delley and Hall, 1999; Andrews and Stark, 2000a).

Overall, it is clear that the Pkc1p pathway is activated in response to a variety of factors that lead to cell surface stress including high temperature, cell wall problems, polarised morphogenesis and lowered osmolarity. Recently, the Pkc1p–Mpk1p pathway has been found to be required for viability on entry into stationary phase or on rapamycin treatment, and under these circumstances Mpk1p is normally activated (Krause and Gray, 2002; Torres *et al.*, 2002). The Sit4p phosphatase is involved in this response (Torres *et al.*, 2002) but seems to act as a negative regulator of Mpk1p activation, since *sit4* deletion mutants show enhanced Mpk1p phosphorylation under a variety of conditions (Angeles de la Torre-Ruiz *et al.*, 2002). While Sit4p may affect the Pkc1p–Mpk1p pathway other than through its role in the TOR pathway (see earlier), this would be consistent with an active role for the Tap42p·Sit4p complex as a negative regulator of Pkc1p–Mpk1p signalling.

At the molecular level, an important activator of Pkc1p is the small GTPase Rho1p, a protein involved in bud formation and organisation of the actin cytoskeleton (Yamochi *et al.*, 1994). As well as being a component of $\beta(1\rightarrow3)$glucan synthase (Drgonova *et al.*, 1996; Mazur and Baginsky, 1996; Qadota *et al.*, 1996) and a potential activator of both Bni1p and Skn7p (Kohno *et al.*, 1996; Alberts *et al.*, 1998), Rho1p·GTP binds to and activates Pkc1p, and *rho1* mutants show a cell lysis defect similar to that resulting from Pkc1p–Mpk1p pathway loss-of-function (Nonaka *et al.*, 1995; Drgonova *et al.*, 1996; Kamada *et al.*, 1996). Thus Rho1p appears to stimulate cell wall metabolism in at least two separate and independent ways (directly by modulating β-glucan synthase and indirectly by signalling through Pkc1p), and it may influence the actin cytoskeleton both through Pkc1p and Bni1p (Fujiwara *et al.*, 1998; Delley and Hall, 1999). Rho1p interacts with the C1 and pseudosubstrate regions of Pkc1p, but surprisingly, apparently not with the HR1-containing N-terminal region. Activation required phosphatidylserine, since Rho1·GTP alone was unable to activate Pkc1p. DAG was not required, so Rho1p may take on the role of DAG in Pkc1p activation (Nonaka *et al.*, 1995; Kamada *et al.*, 1996). Both Pkc1p and Rho1p shows polarised localisation to the cell cortex as mentioned above and the localisation of Pkc1p to these sites is Rho1p-dependent (Andrews and Stark, 2000a). Recent work has proposed that Spa2p may act as a scaffold to bring Mkk1p and Mpk1p to the sites where Rho1p and Pkc1p are localised, although it was not possible to demonstrate that Spa2p played a similar role for Bck1p, which provides the MEKK activity in the Mpk1p MAPK module (van Drogen and Peter, 2002).

Rho1p is activated by Rom1p and Rom2p, a pair of Plekstrin Homology (PH) domain proteins that function as its GEFs (Bickle *et al.*, 1998). Like Rho1p and Pkc1p, Rom2p is localised to the plasma membrane at sites of polarised growth and this requires an interaction between its PH domain and phosphatidylinositol-4,5-bisphosphate (PI4,5P_2) in the plasma membrane (Audhya and Emr, 2002). The PI4,5P_2 is generated by the sequential activity of Stt4p, a phosphatidylinositol (PI) 4-kinase, and Mss4p, a PI4P 5-kinase, both of which are plasma membrane localised (Audhya and Emr, 2002). Consistent with the requirement for Stt4p as an activator of Pkc1p, *stt4* mutants are hypersensitive to the PKC inhibitor staurosporine (Yoshida *et al.*, 1994), fail to activate Mpk1p in response to heat shock but show enhanced Mpk1p activation in cells lacking Sjl1p, the major PI4,5P_2 phosphatase (Audhya and Emr, 2002) and like *mss4* mutants, they show defects in actin polarity and cell wall integrity (Desrivieres *et al.*, 1998; Audhya *et al.*, 2000). Another Rho GEF, Tus1p, has recently been shown to share a role in transmitting cell surface stress signals to Rho1p (Schmelzle *et al.*, 2002). In addition to the GEFs, a number of Rho GAPs have been identified that may downregulate Rho1p activation of Pkc1p and its other effectors. These include Sac7p, Bem2p, Bag7p and Lrg1p (Lorberg *et al.*, 2001; Schmidt *et al.*, 2002). These GAPs regulate distinct functions of Rho1p, with the first two being particularly important for Pkc1p regulation (Schmidt *et al.*, 2002).

Another family of proteins involved in activation of Pkc1p are the *WSC* proteins (Wsc1p–Wsc4p), Mid2p and its related polypeptide Mtl1p. All of these proteins have a similar architecture in that they are single-pass transmembrane proteins with large, ser–thr rich extracellular domains and a smaller cytoplasmic domain, and both Wsc1p and Mid2p have been shown to reside on the cell surface (Verna *et al.*, 1997; Ketela *et al.*, 1999; Rajavel *et al.*, 1999). The extracellular domains are likely to be heavily glycosylated, as demonstrated for Mid2p and Wsc1p (Lodder *et al.*, 1999; Rajavel *et al.*, 1999), and thus will probably have an extended conformation that may be able to sense and respond to changes in the extracellular environment such as physical stress. Genetically, Mid2p and the *WSC* proteins behave as activators of the Pkc1p–Mpk1p pathway: *wsc* mutants show an osmoremedial cell lysis defect at higher growth temperatures and a defect in Mpk1p activation, while *mid2p* mutants are impaired in Mpk1p activation in response to pheromone (Verna *et al.*, 1997; Rajavel *et al.*, 1999). Wsc1p (and probably Mid2p), interact with the Rom2p GEF via their C-terminal cytoplasmic domains (Philip and Levin, 2001). Consistent with this above, *MID2* was originally isolated as a gene required for viability in the presence of pheromone although it may also have additional functions that are not mediated through Pkc1p but which may rely on other Rho1p effectors such as Skn7p (Ketela *et al.*, 1999). However, the synthetic growth defect seen in *wsc1Δmid2Δ* double mutants shows that both genes have a role in vegetative cells (Ketela *et al.*, 1999; Rajavel *et al.*, 1999; Stirling and Stark, 2000). These putative stress receptors therefore appear to be coupled to Pkc1p via activation of Rho1p, and they have been proposed to function together with PIP4,5P_2 generation in the plasma membrane in Rho1p-mediated Pkc1p activation (Audhya and Emr, 2002).

Pkc1p is also regulated by phosphorylation involving Pkh1p and Pkh2p, two redundant homologues of mammalian PDK1. Cells lacking *PKH* activity show cell lysis, depolarised actin and failure to activate Mpk1p-mediated gene expression, and they have reduced Pkc1p activity (Inagaki *et al.*, 1999). Pkh2p can phosphorylate Pkc1p on its activation loop threonine *in vitro*, and this modification is required for Pkc1p function. Recent work has shown that sphingolipid depletion regulates the phosphorylation and localisation of Ypk1p and has suggested that the *PKH* and *YPK* kinases may be the downstream part of a sphingolipid-mediated signalling pathway (Sun *et al.*, 2000). Since sphingolipids have been implicated in signalling heat stress in yeast (Skrzypek *et al.*, 1999), this may provide one explanation of how such a signal can activate Pkc1p. The *PKH* kinases also phosphorylate and activate these yeast homologues of mammalian SGK: Pkh1p phosphorylates Ypk1p and Pkh2p phosphorylates Ypk2/Ykr2p (Roelants *et al.*, 2002). The *YPK* kinases are also required for actin polarity, cell integrity and full activation of Mpk1p (Schmelzle *et al.*, 2002), and they may be involved

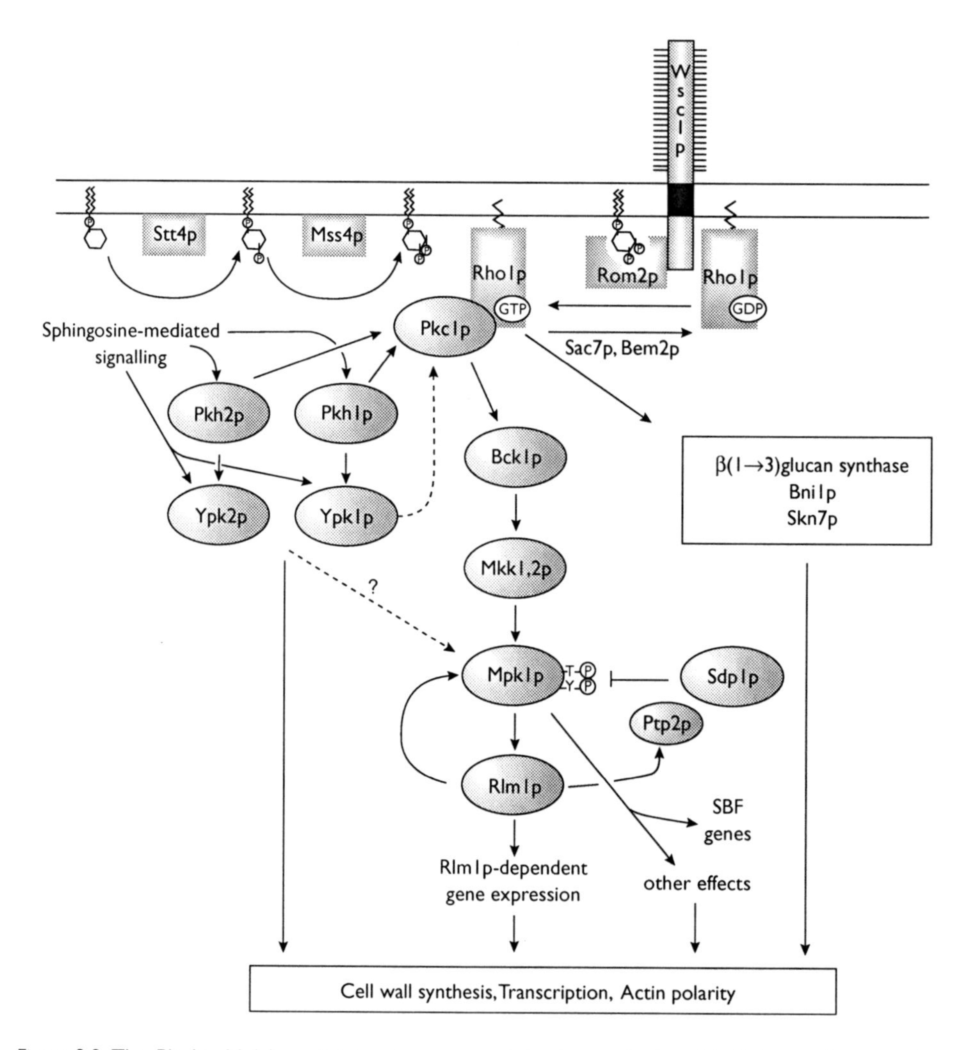

Figure 8.8 The Pkc1p–Mpk1p cell integrity pathway. Stimulatory (→) and inhibitory (⊣) interactions are indicated, and dashed lines indicate less certain links. Note that although only Wsc1p is shown as a representative cell surface stress receptor; Mid2p and the other *WSC* proteins are omitted for clarity.

through both Mpk1p-mediated and Mpk1p-independent routes (see Roelants *et al.*, 2002). Thus both protein phosphorylation and Rho GTPase binding are required for Pkc1p activations, and like the mammalian PRKs to which Pkc1p is closely related, binding of Rho1·GTP may lead to a conformational change that allows phosphorylation by a *PHK* kinase (Flynn *et al.*, 2000). Figure 8.8 summarises this organisation of the Pkc1p–Mpk1p pathway, including the downstream transcription factor Rlm1p discussed below.

Mpk1p targets

What are the targets of the Mpk1p MAPK pathway that mediate these effects? One clear downstream target is Rlm1p, an SRF-like transcription factor. *rlm1* mutations block the growth inhibition resulting from overproduction of an activated *MKK1* mutant allele (Watanabe *et al.*, 1995) and the C-terminal region of Rlm1p is directly phosphorylated by Mpk1p (Dodou and Treisman, 1997; Watanabe *et al.*, 1997). Modification on ser-427 and thr-439 is critical for activation by Mpk1p and affects transcriptional activation rather than DNA binding, while an Mpk1p docking domain in Rlm1p preceding the phosphorylation sites assists activation by Mpk1p (Jung *et al.*, 2002). Fusion of Rlm1p to the Gal4p transcriptional activation domain rendered Rlm1p independent of Mpk1p activation and suppressed the phenotypes associated with loss of the MAPK pathway (Watanabe *et al.*, 1995). Thus it is likely that Rlm1p mediates some of the responses to activation of Mpk1p, although there must be other downstream effectors since the phenotype of *rlm1Δ* mutations was not as severe as, for example, that of an *mpk1* deletion (Watanabe *et al.*, 1995). However, of the 18 genes showing greater than two fold induction in response to expression of a constitutively activated mutant Mkk1p, Rlm1p mediated the change in all but one instance and also mediated repression of five genes (Jung and Levin, 1999). Many of the Rlm1p-dependent genes encode cell surface proteins, consistent with a role for Mpk1p in cell integrity. Consistently, two of the Rlm1p-dependent genes were found to require a functional Pkc1p–Mpk1p pathway for optimal expression in a separate study (Igual *et al.*, 1996), although other cell wall biosynthesis genes that do not appear to be Rlm1p dependent were also identified. Further genes induced by activated Rho1p or Pkc1p have also been identified (Roberts *et al.*, 2000).

A second target of Mpk1p phosphorylation may be Swi6p, a component of the SBF and MBF transcription factors discussed before. Swi6p is a phosphoprotein whose phosphorylation state directly correlates with Mpk1p activity *in vivo* (Madden *et al.*, 1997), and both Swi6p and Swi4p (the other component of SBF) are substrates of Mpk1p *in vitro*. Phosphorylation of SBF is correlated with enhanced expression of a subset of SBF target genes (Madden *et al.*, 1997), while SBF mutants (*swi4, swi6*) show hypersensitivity to cell wall stress (Igual *et al.*, 1996). Although a number of genes involved in cell wall synthesis show periodic cell cycle expression mediated by SBF (Igual *et al.*, 1996), the contributions of SBF and Mpk1p to their expression appear to be independent, consistent with the synthetic defect obtained when SBF and Pkc1p–Mpk1p pathway mutants are combined (Igual *et al.*, 1996; Gray *et al.*, 1997). Thus while Mpk1p may directly regulate expression of some genes through SBF, this does not appear to be a general mechanism and this conclusion is supported by the fact that almost all gene expression induced by Mpk1p activation is mediated through Rlm1p (Jung and Levin, 1999). Finally, two HMG1-related proteins that had also been thought to play a specific role downstream of Mpk1p (Nhp6Ap and Nhp6Bp: Costigan *et al.*, 1994) are probably more generally involved in transcription through chromatin remodelling (Brewster *et al.*, 2001; Formosa *et al.*, 2001). Interestingly, another chromatin remodelling complex (RSC) has also recently been linked functionally with Pkc1p, but in this case with an Mpk1p-independent role (Chai *et al.*, 2002; Romeo *et al.*, 2002).

Mpk1p MAPK pathway-independent targets of Pkc1p

As discussed already, there are likely to be several outputs from Pkc1p that are not mediated via Mpk1p MAPK signalling, although the nature of these remains relatively obscure. However, one MAPK-independent role that has emerged recently is the involvement of Pkc1p in the 'arrest of secretion response' (ASR: see Ng, 2001). When the secretion pathway is blocked, transcription of ribosomal protein genes, ribosomal RNA and transfer RNA is downregulated and the organisation of the nucleus profoundly affected, with strong inhibition of nuclear import and relocation of many nuclear proteins and nucleoporins to the cytoplasm (Li *et al.*, 2000; Nanduri and Tartakoff, 2001). The ASR requires the Wsc1p and Wsc2p membrane proteins and Pkc1p, but not the downstream MAPK pathway, and while Wsc1p is more important for the ribosomal regulation, Wsc2p plays the predominant role for the nuclear reorganisation (Li *et al.*, 2000; Nanduri and Tartakoff, 2001). Intriguingly, the role of Wsc2p in the ASR pathway requires it to be located in an intracellular compartment, since cell surface Wsc2p could not activate the ASR. Since plasma membrane *WSC* proteins are also involved in signalling to the actin cytoskeleton through Pkc1p, this implies differential localisation of the downstream pathways involved or, alternatively, may also indicate that the membrane stress that most likely activates the ASR via the *WSC* proteins does not occur at the cell surface.

Pkc1p–Mpk1p summary

The Pkc1p–Mpk1p cell integrity pathway constitutes a signal transduction cassette that is required for polarised cell growth and may be activated by membrane stretch, a stimulus that could explain activation by low osmolarity, heat-shock and cell wall stress and that may also be triggered during polarised growth as the cell wall is remodelled. The direct and independent involvement of Rho1p in β-glucan synthase activation and Pkc1p function provides a point of coordination between the activity and expression level of cell wall biosynthetic enzymes. Changed patterns of gene expression in response to such signals are clearly an important consequence of activating the Pkc1p–Mpk1p pathway, but more direct effects (e.g. by direct regulation of cell wall biosynthetic enzymes or components of the actin cytoskeleton) cannot be excluded, and further work is still needed to clarify all the inputs and the outputs of the Pkc1p and the Mpk1p MAPK pathway for a complete picture to emerge.

The mating pheromone and filamentous growth response signal transduction pathways

Saccharomyces cerevisiae can secrete and respond to peptide mating pheromones, resulting in morphological changes, cell cycle arrest and induction of gene expression required for conjugation between cells of the two mating types. **a** cells secrete **a**-factor and synthesise α-factor receptor, while α cells secrete α-factor and synthesise **a**-factor receptor. The pheromones and their receptors are the only parts of the pheromone response pathway that are cell type specific (Bender and Sprague, 1986). The receptors for α and **a**-factor belong to the seven-transmembrane class and are encoded by *STE2* and *STE3* respectively (Burkholder and Hartwell, 1985; Hagen *et al.*, 1986). These are coupled via a heterotrimeric G-protein (Gpa1p·Ste4p·Ste18p) to a second MAP kinase module indicated in Figure 8.7 (see Herskowitz, 1995). In addition to being required for pheromone signal transduction, several of the components of the mating pheromone signalling pathway, including the MEKK Ste11p, the MEK Ste7p and the downstream transcription factor Ste12p, are also required for pseudohyphal differentiation and invasive growth (Liu *et al.*, 1993; Roberts and Fink, 1994). Pseudohyphal differentiation is a mode of growth in diploid cells triggered by nitrogen starvation and leading to delayed mitosis, elongated morphology, a unipolar budding pattern, incomplete cell separation such that cells remain connected, and expression of proteins that confer the ability to grow invasively beneath the agar surface (Gimeno *et al.*, 1992; Kron *et al.*, 1994). In haploids, a similar response occurring under colonies on rich medium is triggered by glucose depletion (Cullen and Sprague, 2000) and leads to haploid invasive growth. This also involves a change from an axial to unipolar budding pattern, allowing cells to form shorter filaments that can penetrate their substratum (Roberts and Fink, 1994). A similar phenomenon in diploid cells in response to glucose depletion has also been discovered recently (Cullen and Sprague, 2002). Filamentous growth has been observed in haploid cells in response to low pheromone concentrations and may represent a response designed to help yeast cells locate a mating partner (Erdman and Snyder, 2001). Only some laboratory strains show pseudohyphal differentiation in response to nitrogen starvation, in particular those derived from strain Σ1278. The two overlapping MAPK pathways which are needed for the mating pheromone response and for pseudohyphal or invasive growth will now be examined, focussing on the role of the protein kinases and protein phosphorylation in the signal transduction events.

The pheromone response pathway

When two haploid cells of opposite mating types come into close proximity, pheromone secreted by one cell is sensed by binding to a specific, cell surface receptor on the other, leading to a number of consequences. Firstly, both cells arrest in the G1 phase of the cell division cycle, allowing them to synchronise their cell division cycles prior to fusion and subsequent karyogamy. Secondly, they develop mating projections or 'shmoos' (shown in Figure 1.1), leading to polarised growth towards each other that occurs in response to the gradient of pheromone presented to each cell. Finally, they induce expression of a subset of genes required for conjugation such as *FUS1* and *FUS2*, which encode surface O-glycoproteins required for cell fusion (Trueheart *et al.*, 1987; Elion *et al.*, 1995), and *AGα1*, *AGA1* and *AGA2*, which encode cell surface agglutinins (Cappellaro *et al.*, 1991). These physiological responses involve a complex series of signal transduction events and the full pathway from receptor inwards into the cell is shown in Figure 8.9.

Binding of pheromone to the receptor leads to activation of the heterotrimeric G-protein, such that the G_α and $G_{\beta\gamma}$ components dissociate in response to exchange of GDP for GTP on G_α(Gpa1p), but both components remain localised to the plasma membrane by virtue of their lipid modifications. Signalling to the MAPK cascade is mediated by the Ste4p·Ste18p $G_{\beta\gamma}$ complex, since deletion of either *STE4* or *STE18* blocks signalling, whereas deletion of *GPA1* leads to a constitutive pheromone response, which can be blocked by mutations in the downstream components (Dietzel and Kurjan, 1987). Thus in unstimulated cells, G_α is a negative regulator of signalling by the $G_{\beta\gamma}$ complex, and inhibition is

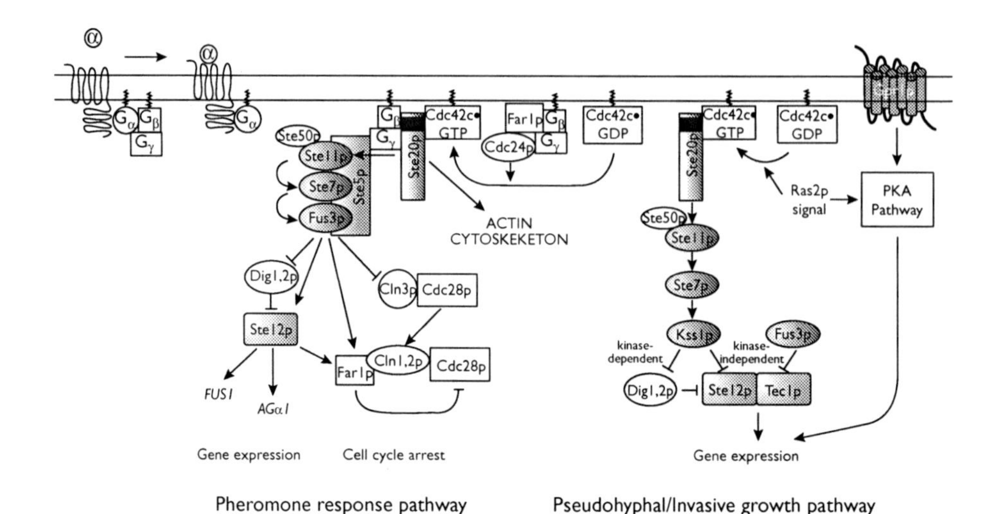

Figure 8.9 The pheromone response and filamentous growth response MAPK pathways. The pheromone and filamentous growth response MAPK pathways share common components but function differently, in particular because pheromone signal transduction involves the Ste5p scaffold protein and activation by the pheromone receptor $G_{\beta\gamma}$ complex. Although not shown, Kss1p and Fus3p both form complexes with Ste5p in the pheromone pathway and both contribute to gene expression in response to pheromone (see text). The independent, parallel role of the PKA pathway (see Figure 8.4) in promoting a filamentous growth response is also shown. Stimulatory (→) and inhibitory (⊣) interactions are indicated.

released by receptor-coupled guanine nucleotide exchange on G_α when pheromone binds, leading to $G_{\beta\gamma}$ release. However, Gpa1p also has a positive role to play in adaptation to pheromone (Metodiev *et al.*, 2002). With the exception of Ste5p, the order of the remaining components in the pathway was determined by genetic epistasis (see e.g. Cairns *et al.*, 1992) and suggested that the $G_{\beta\gamma}$ complex signals to Ste20p (Leberer *et al.*, 1992). Ste20p belongs to the PAK family of protein kinases and contains a domain that binds the GTP-loaded form of Cdc42p, a GTPase of the rho family (see Symons, 1996). Ste20p can phosphorylate Ste11p *in vitro* (Wu *et al.*, 1995) and appears to activate the MAPK cascade by direct phosphorylation of Ste11p (Van Drogen *et al.*, 2000). The role of Ste5p in the mating pathway was initially unclear, since it could not be ordered unambiguously relative to the other components shown in Figure 8.9 (see Herskowitz, 1995). However, it is now clear that it functions as a scaffold to anchor the MAPK–MEK–MEKK elements of the pathway into a functional complex. Thus different regions of Ste5p show specific interactions with Fus3p, Kss1p, Ste7p and Ste11p (Choi *et al.*, 1994; Kranz *et al.*, 1994; Marcus *et al.*, 1994; Printen and Sprague, 1994). Since *ste5* cells are by definition sterile, the formation of this complex must be essential for efficient signalling. Ste4p (G_β) can bind directly to Ste5p (Whiteway *et al.*, 1995) and this interaction recruits Ste5p to the plasma membrane in pheromone-treated cells (Pryciak and Huntress, 1998). Since artificially tethering Ste5p to the plasma membrane can activate the pathway independently of $G_{\beta\gamma}$, Ste5p recruitment is likely to be an important step in the signal transduction process. The role of Ste20p and its requirement for binding the Cdc42p GTPase with which it interacts was unclear for a while, in part because CRIB domain in PAK kinases (through which Cdc42p binds to Ste20p) is autoinhibitory (Hoffman and Cerione, 2000), so derivatives of Ste20p lacking this domain showed excessive activity (Moskow *et al.*, 2000). Recent work has clarified that Cdc42p is required and that it activates Ste20p (Moskow *et al.*, 2000; Lamson *et al.*, 2002). Although Cdc42p and Ste20p are both required for signalling they can function independently of regulation by pheromone of the $G_{\beta\gamma}$ complex, so the most economical model is that pheromone stimulation recruits the Ste5p complex to a location (the Shmoo tip) where the Ste11p that it carries can be phosphorylated by Ste20p, which is independently localised there by its association with Cdc42p. However, Ste4p (G_{β}) and Ste20p can interact directly and so this may enhance pathway activation (Leeuw *et al.*, 1998). Another protein (Ste50p) binds to Ste11p and is required for its full activation but is not absolutely essential (Wu *et al.*, 1999). Like other MEKKs, Ste11p has an N-terminal negative regulatory domain that can be deleted to yield a constitutively active variant, and Ste50p interacts with this region.

Ste11p phosphorylates and activates Ste7p on two closely spaced serine residues that are conserved with mammalian MEKs (Zheng and Guan, 1994). Activated Ste7p is the dual-specificity MEK that can activate (by activation loop phosphorylation) Fus3p and Kss1p, two MAPKs that function to transmit the pheromone signal (Elion *et al.*, 1991b; Gartner *et al.*, 1992; Errede *et al.*, 1993; Zhou *et al.*, 1993; Ma *et al.*, 1995). At first, Fus3p and Kss1p were thought to be partially redundant for the pheromone response although they clearly did not perform identical functions. Thus *kss1* mutants arrest their cell cycle in response to pheromone but show reduced transcriptional induction, while *fus3* mutants are normal for transcriptional induction but fail to block in G1 (Elion *et al.*, 1991a,b). The conclusion from this was that while either MAPK can transmit the pheromone response, only Fus3p was effective at arresting the cell division cycle. Many yeast strains naturally lack a functional *KSS1* gene due to a widespread polymorphism but are perfectly capable of mating, so clearly Fus3p alone is sufficient to carry out all the critical pheromone-activated responses (Elion *et al.*, 1991b). Conversely, the Kss1p MAPK is required for the filamentous growth response discussed in the next section (Madhani *et al.*, 1997). The idea therefore emerged that Fus3p, bound to its Ste5p scaffold, was specific for the mating pathway whereas Kss1p was dedicated to filamentous growth, but that in the absence of Fus3p (for example in a *fus3* deletion mutant) Kss1p could gain access to the MAPK binding site on Ste5p, leading to erroneous crosstalk with the filamentous growth pathway (Madhani *et al.*, 1997). However, it is now clear that Fus3p and Kss1p are both normally activated in response to pheromone in wild-type cells, both form complexes with Ste5p and both are required to obtain a full transcriptional response (see later), but Fus3p activation leads to downregulation of Kss1p and thus limits activation of filamentous growth (Choi *et al.*, 1994; Breitkreutz *et al.*, 2001; Sabbagh *et al.*, 2001).

Transcriptional induction is a major response to pheromone and involves the transcription factor Ste12p. Overproduction of Ste12p activates transcription of pheromone-inducible genes (Dolan and Fields, 1990), suggesting that it is limiting for their expression, and Ste12p becomes phosphorylated under conditions where the responsive genes are induced (Song *et al.*, 1991). Thus the simplest model is that the MAPKs directly phosphorylate and activate Ste12p. Consistent with a role in gene expression, Fus3p has been shown to shuttle between the nucleus and cytoplasm and thus can be translocated to the nucleus once activated (van Drogen *et al.*, 2001). However, Ste12p activation cannot be mediated solely by phosphorylation, and release of inhibition by Dig1p and Dig2p (also called Rst1p and Rst2p) is a major link in the pathway (Tedford *et al.*, 1997). These two proteins each bind to Ste12p and inhibit its activity by non-redundant mechanisms (Olson *et al.*, 2000). Release of Dig-mediated inhibition requires the MAPKs, which almost certainly phosphorylate Dig1p and Dig2p to achieve this (Cook *et al.*, 1996; Tedford *et al.*, 1997). Release of this inhibition is sufficient to activate Ste12p since a *dig1Δdig2Δ* double mutant showed constitutive activation of pheromone-responsive genes (Tedford *et al.*, 1997). However, activation of a mutant Ste12p that fails to bind Dig1p and Dig2p still required a signal from the MAPKs (Pi *et al.*, 1997). Thus activation may involve two independent steps mediated by the MAPKs, direct phosphorylation of Ste12p and phosphorylation and inactivation of Dig1p and Dig2p, although the latter event is certainly the most critical for conferring inducibility. The Dig proteins show strong interactions with the MAPKs as well as Ste12p, suggesting that they may also mediate localisation of the kinases to Ste12p complexes bound to DNA (Cook *et al.*, 1996; Tedford *et al.*, 1997). Ste12p phosphorylation might also be involved in downregulation of Ste12p function.

Ste12p binds to a specific DNA sequence in the promoters of pheromone-inducible genes termed the PRE (pheromone response element: 5′-TGAAACA-3′). Two or more of these elements together are sufficient to confer inducibility and often more PREs are present as in the case of *FUS1*, which has four (Hagen *et al.*, 1991). However, some inducible genes contain a single element. In this case, it is always associated with a P-box, the binding site for the SRF-like general transcription factor Mcm1p (Errede and Ammerer, 1989; Jarvis *et al.*, 1989; Hwang-Shum *et al.*, 1991). These genes show lesser degrees of pheromone-stimulated expression but which is wholly dependent on the single PRE.

G1 cyclin function is critical for passage of cells past ‘Start’ and this is the point of regulation by mating pheromone that leads to cell cycle arrest. Mutations that truncate and thereby stabilise Cln3p (see earlier text) lead to pheromone resistance as well as execution of ‘Start’ at a smaller size than usual, but can still induce gene expression in response to pheromone (Cross, 1988; Nash *et al.*, 1988). This implies that downregulation of G1 cyclin function is part of the cell cycle arrest mechanism. Fus3p and Far1p are required to mediate the cell cycle arrest promoted by α-factor (Chang and Herskowitz, 1990; Elion *et al.*, 1991b). Far1p can associate with Cln·Cdc28p complexes in an α-factor-dependent manner (Peter *et al.*, 1993; Tyers and Futcher, 1993) and in fact Far1p is a specific inhibitor (CKI) of Cln1,2p·Cdc28p kinase activity (Tyers and Futcher, 1993; Valdivieso *et al.*, 1993; Peter and Herskowitz, 1994). *far1* mutants are thus blocked specifically for cell cycle arrest but not for pheromone-induced signalling (Chang and Herskowitz, 1990). *FAR1* expression is induced by pheromone (Chang and Herskowitz, 1990; Oehlen *et al.*, 1996) and Far1p, phosphorylation by Fus3p in response to pheromone is required for its function as a CKI (Gartner *et al.*, 1998). Thus MAPK phosphorylation controls both the transcriptional and cell cycle arrest responses by phosphorylating and activating Ste12p and Far1p, respectively. Although induction of *FAR1* is not sufficient for arrest, it is clear that Ste12p-mediated *FAR1* expression is nonetheless important (Oehlen *et al.*, 1996), so in fact MAPK signalling also promotes cell cycle arrest by transcriptional mechanisms involving the Far1p CKI. Pheromone stimulation leads to a rapid loss of *CLN1* and *CLN2* mRNAs (Wittenberg *et al.*, 1990). However, pheromone can still arrest cells constitutively expressing *CLN2* because there is sufficient Far1p to inhibit the excess Cln2p·Cdc28p (Peter and Herskowitz, 1994) and so the decline in mRNA level is not an obligatory part of the mechanism. The loss of *CLN1, 2* transcripts on pheromone stimulation could in principal be mediated by Far1p inhibition of Cln3p·Cdc28p, since this CDK is required for *CLN1, 2* expression (see earlier) and Far1p associates well with Cln3p·Cdc28p. However, this interaction appears to occur too late after pheromone treatment to provide a workable mechanism (Tyers and Futcher, 1993). Fus3p may inhibit Cln3p function post-translationally (Elion *et al.*, 1991b) and so perhaps this provides an explanation. Exactly how Far1p functions as a CKI is also not yet clear since Cln·Cdc28p complexes containing Far1p retained significant kinase activity (Gartner *et al.*, 1998). Overexpression of *CLN2* represses pheromone signalling by interfering with the activity of Ste11p, so in fact there seems to be some reciprocal regulation between the pheromone response and vegetative growth: the former arrests the cell cycle by downregulating G1 cyclin function while the latter downregulates mating by antagonising pheromone signalling (Wassmann and Ammerer, 1997). Surprisingly, Far1p has a quite distinct role in mating projection formation in addition to its role as a CKI. Far1p is predominantly nuclear, where it functions to arrest the cell cycle in response to pheromone, but it also undergoes nuclear export after pheromone treatment (Blondel *et al.*, 1999). Far1p can bind to activated $G_{\beta\gamma}$, interacting at the cell cortex with polarity proteins that regulate the actin cytoskeleton (Butty *et al.*, 1998; Nern and Arkowitz, 1999), and this is required for the polarised growth of the cell towards its mating partner. Intriguingly, Far1p acts as a nuclear anchor for Cdc24p, the GEF for the Cdc42p rho GTPase discussed earlier, and nuclear export of Far1p is required for relocalisation of Cdc24p to the cell cortex where it is needed for the correct polarisation of the actin cytoskeleton (Shimada *et al.*, 2000). Far1p is normally phosphorylated by Cln1,2p·Cdc28p in late G1 and then degraded by the SCF^{Cdc4} complex (see earlier text), and a non-degradable mutant Far1p blocks normal development of cell polarity by sequestering Cdc24p in the nucleus throughout the cell cycle (Shimada *et al.*, 2000). Thus Far1p regulation of Cdc24p function is also a part of normal cell cycle regulation. In the absence of Far1p, mating pheromone stimulation leads to invasive growth (Roberts *et al.*, 2000), demonstrating a critical role of Far1p in defining the pheromone response.

So far, we have considered how the mating pheromone response is activated but it is important also to understand how the pathway is downregulated, including the involvement of protein phosphatases. A dual-specificity phosphatase (Msg5p) has been specifically implicated in the downregulation of Fus3p, the major MAPK discussed before. Overproduction of *MSG5* inhibits pheromone-induced cell cycle arrest whereas *msg5Δ* knockout strains show a poorer adaptive response and elevated Fus3p kinase activity. Thus Msg5p might oppose activation of Fus3p by Ste7p. Two other members of the tyrosine phosphatase family, Ptp2p and Ptp3p are also important in downregulation of stimulated Fus3p activity and Ptp3p is the major activity suppressing the basal activity of Fus3p (Zhan *et al.*, 1997). Ste5p (Kranz *et al.*, 1994) and Ste7p (Zhou *et al.*, 1993) are also substrates of Fus3p and these phosphorylations could also be part of the downregulation mechanism. Other elements of adaptation include pheromone-specific proteases (MacKay *et al.*, 1991; Marcus *et al.*, 1991), receptor endocytosis, ligand-induced receptor phosphorylation (Konopka *et al.*, 1988; Reneke *et al.*, 1988; Chen and Konopka, 1996) and both G_β; (Grishin *et al.*, 1994) and G_α mediated effects (Stratton *et al.*, 1996). The G_α GAP Sst2p (Dohlman *et al.*, 1996; Chen and Kurjan, 1997) is also important in the pheromone adaptation process and may be protected from degradation by MAPK phosphorylation in response to pheromone (Garrison *et al.*, 1999; Hoffman *et al.*, 2000).

Finally, we saw previously that the Pkc1p–Mpk1p pathway is also required for the morphological development that follows exposure to mating pheromone. Thus it is of interest to ask how this pathway is coordinated with the pheromone-activated MAPK pathway. The two pathways appear to act sequentially, activation of the Pkc1p–Mpk1p pathway requiring pheromone-induced transcription and translation (Buehrer and Errede, 1997). The timing and extent of Mpk1p activation is determined by Spa2p (already known to be required for mating projection formation: Gehrung and Snyder, 1990) and Bni1p, a member of the formin family that may be another target of Cdc42p (Evangelista *et al.*, 1997).

Filamentous growth response

Haploid invasive growth and diploid pseudohyphal differentiation (which will here be collectively termed 'filamentous growth') appear to be responses of yeast to inadequate nutrient supply that allow them to forage for additional nutrients. In both cases, the developmental programme that is set in train requires several of the components of the mating pheromone transduction pathway: the Ste20p kinase, Ste11p MEKK and its binding protein Ste50p, Ste7p MEK and Ste12p transcription factor (Liu *et al.*, 1993; Roberts and Fink, 1994; Ramezani Rad *et al.*, 1998). Deletion of these genes causes greatly reduced filamentous growth in *S. cerevisiae* and the involvement of the pathway is further underlined by the finding that the constitutively active *STE11–4* allele induces filamentous growth. However, there are also inputs from other signalling pathways into the filamentous growth response, in particular the PKA pathway, and these will also be discussed later.

Despite some shared components, the mating and filamentous growth MAPK pathways are not by any means identical. Firstly, the Ste5p scaffold protein is dispensable for filamentous growth, implying that the Ste5p-organised MEKK–MEK–MAPK complex is required specifically for pheromone signalling, and secondly, neither a pheromone receptor nor its associated heterotrimeric G-protein are involved. At first it was not at all clear which MAPK mediated signalling in the filamentous growth pathway. In diploids, Kss1p appeared to have a positive but non-essential role in pseudohyphal differentiation and Fus3p was completely dispensable (Liu *et al.*, 1993), whereas invasive growth in haploids required Kss1p but was negatively regulated by Fus3p (Roberts and Fink, 1994). This fits with the fact that Kss1p is expressed in both haploid and diploid cells and is not pheromone inducible (Courchesne *et al.*, 1989), whereas Fus3p is haploid specific and is pheromone-inducible (Elion *et al.*, 1990). More recent work (Cook *et al.*, 1997; Madhani *et al.*, 1997) showed that in fact Kss1p is the filamentous growth MAPK, but its role in the pathway was initially obscured due to dual inhibitory and activatory roles. In the dephosphorylated form, Kss1p and Fus3p are potent inhibitors of filamentous growth independently of their kinase activity. In the case of Kss1p, the inhibitory function maps to a particular region of the MAPK that is specific to the MAPK family of protein kinases (the MAPK insertion) and is mediated by the strong interaction between the MAPK and the Ste12p transcription factor that lies downstream (Madhani *et al.*, 1997). The activatory role of Kss1p requires its protein kinase function and the presence of a functional Ste11p–Ste7p module upstream, and phosphorylation of Kss1p by Ste7p converts it from an inhibitor into an activator (Cook *et al.*, 1996). Once again the Dig proteins (see earlier text) are involved in the activatory signal, since deletion of both *DIG1* and *DIG2* promotes constitutive invasive growth in haploids and pseudohyphal differentiation in diploids, even in strain backgrounds that do not normally show either of these responses (Cook *et al.*, 1996; Tedford *et al.*, 1997). In haploids cells (which express Fus3p), at least one requirement for Kss1p kinase activity is to overcome the inhibition of filamentous growth by Fus3p, which normally limits Kss1p activation (see below).

The specificity for expression of filamentous growth-specific genes could in principal be explained by the requirement for Tec1p, a TEF-1-related transcription factor that functions with Ste12p during filamentous growth (Gavrias *et al.*, 1996). TEF-1 binds to the sequences

5′-CATTCC/T-3′ and genes that are upregulated during filamentous growth contain a PRE (Ste12p binding site) adjacent to one of these sequences (presumptive Tec1p binding site or TCS: Madhani and Fink, 1997; Mosch and Fink, 1997). The combination of the PRE and TCS has been termed a *filamentous growth and invasion response element* or FRE, and mutation of either PRE or TCS component severely reduced its function. Together with the requirement for both Tec1p and Ste12p, this was consistent with a combinatorial nature of gene expression induction during filamentous growth (Madhani and Fink, 1997). *TEC1* itself contains an FRE, suggesting that a positive feedback loop is involved in the transcriptional response. However, the distinction between FRE- and PRE-mediated gene expression has become less clear recently with the realisation that reporter constructs generated to monitor FRE-mediated gene expression do not necessarily accurately reflect the behaviour of the endogenous genes (the *TEC1* gene, for example, is induced by pheromone), and it has not been possible to define distinctive pheromone response and filamentous growth gene sets by transcriptional profiling (see Breitkreutz and Tyers, 2002). Thus clear-cut assignment of FRE genes to the filamentous response pathway and PRE genes to the pheromone response cannot be made.

Pheromone-inducible genes are not expressed during filamentous growth and filamentous growth is not normally induced during mating, and yet the proposed endpoint of each pathway is similar (inactivation of the Dig proteins to permit Ste12p function). Thus a critical question is how crosstalk between the pathways is blocked. Initially, it was thought that *fus3* deletion mutants could signal in response to pheromone only because Kss1p could erroneously gain access to pheromone signalling complexes defined by the Ste5p scaffold (Madhani *et al.*, 1997), but it is now accepted that both Fus3p and Kss1p are normally activated in response to pheromone and so this model does not explain specificity (see Breitkreutz and Tyers, 2002). However, Kss1p activity is downregulated by activated Fus3p and crosstalk with the filamentous growth response occurs in its the absence (Madhani *et al.*, 1997) because Kss1p activity remains high (Sabbagh *et al.*, 2001). Thus while Ste5p complexes had been proposed to impart specificity to the pheromone pathway by being selective for the Fus3p MAPK, in fact such a mechanism does not normally shield Kss1p from activation by pheromone. Instead, it seems that the magnitude and duration of MAPK activation is responsible for pathway specificity, and that prolonged activation of Kss1p is required to activate the filamentous growth response (Sabbagh *et al.*, 2001). Although PREs and FREs appear to use different transcription factors (Ste12p·Ste12p/Ste12p·Mcm1p versus Ste12p·Tec1p, respectively), either mating or filamentous growth responses can be promoted by dominant activatory mutations such as *STE11-4* or by deletion of Dig1p and Dig2p and the differences in the global gene expression patterns underlying the two responses are remarkably small (Fambrough *et al.*, 1999; Roberts *et al.*, 2000). This underlines the idea that it is the strength and longevity of MAPK activation that is critical for output specificity and that determines whether a pheromone response or filamentous growth response is obtained, and will be interesting to see how the filamentous growth triggered by low pheromone concentrations mentioned before (Erdman and Snyder, 2001) fits into this model. Another important factor is whether or not Far1p is activated, given that pheromone also induces haploid invasive growth when *FAR1* has been deleted (Roberts *et al.*, 2000). Other aspects to the specificity of the two responses concern the specificity of the Fus3p and Kss1p kinases, and Fus3p shows strong substrate selectivity for Far1p when compared with Kss1p (Breitkreutz *et al.*, 2001) that explains the requirement for Fus3p to permit robust pheromone-induced cell cycle arrest. Thus the substrate selectivity of the two MAPKs is clearly important, and it is likely that other differences in substrate specificity between Fus3p and Kss1p will become apparent in the future that will explain other aspects of pathway specificity. Finally, since inputs from other pathways normally contribute to the filamentous growth response (see later), this might provide another mechanism to limit spillover from pheromone signalling into the haploid invasive growth response.

Given that the filamentous growth pathway is not triggered by pheromone, what might be the inputs that activate the pathway? Cdc42p is required to activate the Ste20p kinase to signal pseudohyphal development. Thus deletion of the Cdc42p binding domain in Ste20p blocks filamentous growth (Peter *et al.*, 1996; Leberer *et al.*, 1997), while activated *CDC42* alleles (e.g. *CDC42*$^{\text{val-12}}$) can stimulate it (Mosch *et al.*, 1996). The dominant activatory allele of *RAS2* (*RAS2*$^{\text{val-19}}$) also activates filamentous growth but this effect can be blocked by a *CDC42*$^{\text{ala-118}}$ dominant negative allele, consistent with Ras2p as an upstream activator of Cdc42p for filamentous growth (Mösch *et al.*, 1996). *RAS2*$^{\text{val-19}}$ also activates the PKA pathway (see Section 8.3) and in cells lacking Ste7p and Kss1p, filamentous growth was still triggered in response to activated *RAS* alleles (Cook *et al.*, 1996). Conversely, *RAS2*$^{\text{val-19}}$ containing additional mutations that block its interaction with adenylate cyclase can still trigger filamentous growth (Mösch *et al.*, 1999). Thus Ras2p seems to activate filamentous growth both via the MAPK pathway and by PKA activation. Mammalian Ras is thought to act via Rho family GTPases such as Cdc42Hs in some instances (see Minden *et al.*, 1995), and Ras2p may therefore activate Cdc42p in the filamentous growth pathway in a similar manner. Pseudohyphal development also requires the 14-3-3 proteins Bmh1p and Bmh2p and while cells lacking these proteins show normal mating pheromone-induced signalling, they fail to undergo pseudohyphal growth in response to activated alleles of *RAS2* or *CDC42*. In fact Bmh1/2p associate with Ste20p, suggesting that they provide an input into the MAPK pathway at this level (Roberts *et al.*, 1997).

Having focussed so far on the MAPK pathway, now we will consider the role of the PKA pathway and how the two signalling pathways cooperate in promoting filamentous growth. Not only is Ras2p involved in stimulating filamentous growth through cAMP generation and PKA, but the entire pathway including the Gpr1p receptor and Gpa2p G_α is required (Lorenz and Heitman, 1997; Pan and Heitman, 1999; Lorenz *et al.*, 2000). *RAS2*$^{\text{val-19}}$ can overcome the filamentous growth defect in *gpa2* mutants and like *RAS2*$^{\text{val-19}}$, activated alleles of *GPA2* also trigger filamentous growth in the absence of nitrogen starvation. Although the PKA pathway is thought to respond primarily to monosaccharides such as glucose (see earlier text), the *GPR1* gene is strongly expressed during nitrogen limitation (Xue *et al.*, 1998) and may therefore provide a link between this stimulus and PKA pathway activation for pseudohyphal differentiation in diploid cells, which in addition to nitrogen limitation requires an abundant fermentable carbon source (Lorenz *et al.*, 2000). As mentioned earlier, the three PKA isoforms do not function equally; Tpk2p is critical for activating filamentous growth whereas Tpk3p and possibly Tpk1p are inhibitory (Robertson and Fink, 1998; Pan and Heitman, 1999). The adhesin Flo11p is important for filamentous growth and increasing its expression is sufficient to induce filamentous growth in both haploid and diploid cells (Palecek *et al.*, 2000). *FLO11* has a large and complex promoter that responds to several transcription factors including Ste12p and Tec1p (Rupp *et al.*, 1999; see earlier text). Tpk2p activation upregulates *FLO11* expression by phosphorylating and activating the *FLO11* transcription factor Flo8p, and by phosphorylating and inhibiting the transcriptional repressor Sfl1p (Pan and Heitman, 2002). These two proteins interact with a common sequence within the *FLO11* promoter and phosphorylation of Sfl1p inhibits its dimerisation and DNA binding, allowing Flo8p to activate *FLO11* transcription. Thus the PKA pathway and the MAPK pathway required for filamentous growth converge on the *FLO11* promoter, and hyperactivation of one pathway can compensate for a block in the other (Rupp *et al.*, 1999). Sok2p is a negative regulator of pseudohyphal differentiation and also acts by modulating *FLO11* expression, since loss of Sok2p elevates the expression of three other *FLO11* transcription factors (Pan and Heitman, 2000). Although this regulation is independent of that which operates through Flo8p and Sfl1p, it might provide yet another link to the PKA pathway since Sok2p is known to be a PKA target (Shenhar and Kassir, 2001). Many commonly used laboratory strains fail to undergo pseudohyphal development in response to nitrogen starvation, emphasising the importance of strain background when looking at physiological responses in yeast. In the case of S288C this is because of a mutation in the *FLO8* gene (Liu *et al.*, 1996), and the original selection of laboratory yeast strains may have involved selection for loss of genes such as *FLO8*.

In addition to the PKA and MAPK pathways just discussed, other regulatory mechanisms influence filamentous growth as well. As mentioned before, the SNF1 kinase is also required in and may also regulate *FLO11* expression by antagonising the repressors Nrg1p and Nrg2p on glucose limitation (Kuchin *et al.*, 2002). However, Snf1p has also been proposed to be more important as a regulator of the changes in bud site selection that are needed to permit invasive growth, at least in haploid cells (Cullen and Sprague, 2002; Palecek *et al.*, 2002). *ELM1* encodes a protein kinase that confers an elongated cell morphology when deleted in haploid strains and constitutive pseudohyphal development in homozygous *elm1Δ/elm1Δ* diploids (Blacketer *et al.*, 1993), and may therefore function as a negative regulator of cell elongation required for filamentous growth. Filamentous growth is also potentiated by activation of the Swe1p kinase, which phosphorylates tyr-19 on Cdc28p and delays the activation of the Cbl·Cdc28p kinases, resulting in hyperpolarised growth (La Valle and Wittenberg, 2001), while the transcriptional repressor Xbp1p mediates a reduction in Clb levels in nitrogen-limited diploid cells that also enhances filamentous morphology (Miled *et al.*, 2001). In fact genetic screens aimed at identifying other genes required for filamentous growth identified a range of functions including cytoskeletal elements and genes required for bud site selection (Mösch and Fink, 1997; Palecek *et al.*, 2000), confirming the importance of changed budding pattern, cell elongation and agar invasion as independent physiological responses that each contribute to filamentous growth.

In conclusion, multiple pathways regulate filamentous growth and although pseudohyphal differentiation and haploid invasive growth share many features and signalling components in common, it still remains to be discovered how nutrient limitation promotes signalling through these pathways and for the full complement of downstream effectors to be identified. Figure 8.9 summarises our current understanding of the filamentous growth MAPK pathway.

The Hog1p pathway

The last MAPK pathway to be considered here is the Hog1p pathway (Figure 8.10), whose function is required for growth on high osmolarity (Brewster *et al.*, 1993). Similar stress-activated MAP kinase pathways are found in both fission yeast and in mammalian cells although they seem to respond to a wider range of cellular stresses (Degols *et al.*, 1996; Shiozaki *et al.*, 1997). The whole of chapter 9 is devoted to stress responses. The Mpk1p and Hog1p pathways are therefore the yin and yang of osmoregulation: Mpk1p is activated

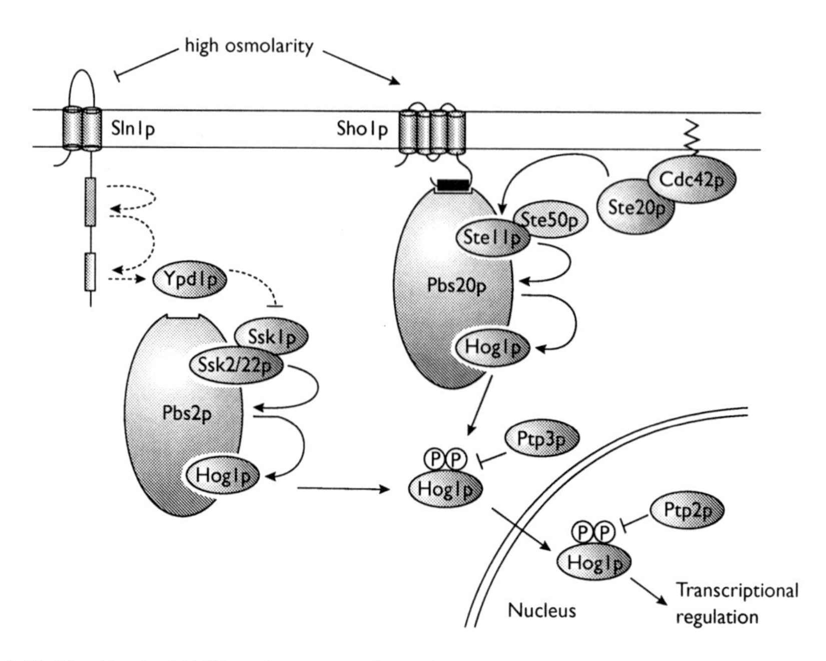

Figure 8.10 The Hog1p MAPK pathway transduces distinct osmolarity signals into an osmotic stress response. The Sln1p–Ypd1p–Ssk1p phosphorelay system that signals under low osmolarity conditions is indicated with dashed lines showing positive (→) and negative (⊣) effects. Interactions occurring in high osmolarity are similarly depicted using solid lines.

by low osmolarity and inhibited by high osmolarity, whereas Hog1p is activated by high osmolarity and inhibited by low osmolarity. There are two main inputs known to activate the Hog1p MAPK, both of which respond to high external osmolarity but which are specialised for responding to different osmotic conditions (O'Rourke *et al.*, 2002). One of these involves proteins related to the two-component systems that are mainly found in bacterial systems, but which have latterly been discovered in fungi and plants as well (see Morgan *et al.*, 1995). Sln1p is a plasma membrane protein that contains two intracellular domains, one of which is a 'histidine kinase' and the second a 'response regulator'. In conditions of low osmolarity, the kinase domain autophosphorylates on a conserved histidine residue and the phosphate group is then passed sequentially to a conserved aspartate in the response regulator domain, a histidine residue in the small protein Ypd1p and finally to the conserved aspartate in Ssk1p, which is also related in sequence to other response regulator domains (Posas *et al.*, 1996). Ssk1p is an activator of the redundant MEKKs Ssk2p and Ssk22p, but phosphorylation of Ssk1p by this phosphorelay is inhibitory. On high osmolarity medium, Sln1p no longer autophosphorylates and the phosphorelay is blocked, releasing Ssk1p to activate the MEKKs via its interaction with the noncatalytic inhibitory domains of the latter. In fact activation of the Hog1p pathway under conditions of low external osmolarity is lethal, so loss of either *SLN1* or *YPD1* causes inviability under normal conditions. It has been proposed that Sln1p responds to the general osmotic balance rather than specifically to high external osmolarity (Tao *et al.*, 1999), and osmosensing may require Sln1p dimerisation (Ostrander and Gorman, 1999). The second input into the pathway is via Sho1p, another membrane protein that can activate the Hog1p pathway at the level of the MEKK Pbs2p (Maeda *et al.*, 1995). Curiously, this activation is mediated by the Ste11p MEKK together with its binding protein Ste50p (Posas and Saito, 1997; Posas *et al.*, 1998), but since high osmolarity doesn't trigger the mating pathway, a separate pool of Ste11p must be involved. This pool may be defined by protein–protein interactions between Pbs2p and each of Ste11p, Sho1p and Hog1p (Posas and Saito, 1997), with Pbs2p performing a scaffold function analogous to that provided by Ste5p in the pheromone response MAPK pathway (see earlier text). Interestingly, high osmolarity can induce expression of pheromone-responsive genes in *hog1* mutants (Hall *et al.*, 1996; Davenport *et al.*, 1999), and this may reflect the necessity to form multiprotein complexes involving several components of the pathway in order to limit crosstalk (see O'Rourke *et al.*, 2002). However, such crosstalk may be important under some circumstances in normal cells (Lee and Elion, 1999; Cullen *et al.*, 2000). Although originally considered to be an osmosensor, the sole role of Sho1p appears to be as a membrane anchor for Pbs2p, with which it can interact via its cytoplasmic SH3 domain. A membrane-tethered form of Pbs2p can completely bypass the Sho1p requirement for osmolarity signalling through this arm of the HOG pathway (Raitt *et al.*, 2000), bringing into question the exact nature of the osmosensor for Sho1-dependent signalling. Pbs2p is localised throughout the cytoplasm in the absence of high osmolarity, but on hyperosmotic stress it shows Sho1p-dependent localisation to sites of polarised growth, although activation of Pbs2p appears to trigger rapid dissociation from the cell cortex (Reiser *et al.*, 2000). The PAK kinase Ste20p shows Cdc42p GTPase-dependent polarised localisation and both these proteins are required for Sho1p-dependent osmotic signalling. It is likely that Sho1p-dependent recruitment of Pbs2p to polarised growth sites in response to osmotic stress allows Ste20p to activate the HOG pathway via Ste11 (Raitt *et al.*, 2000; see Figure 8.10), in a manner analogous to the recruitment of the Ste5p complex by $G_{\beta\gamma}$ to allow activation of Ste11p by Ste20p during pheromone signalling. Another plasma membrane protein, Msb2p, has also been shown to function in parallel with Sho1p to provide and input to the HOG pathway (O'Rourke and Herskowitz, 2002). In addition to responding to high osmolarity, the Sho1p arm of the HOG pathway is also activated by heat stress (Winkler *et al.*, 2002).

Activation of the Pbs2p either via the Sln1p or Sho1p-dependent arms leads to phosphorylation and activation of Hog1p, which is then translocated into the nucleus where it can regulate the expression of many different genes (Posas *et al.*, 2000). Examples of these are stress response genes such as catalase (*CTT1*) and *HSP12* together with genes required for glycerol synthesis, including the key enzyme glycerol-3-phosphate dehydrogenase (*GPD1*: Albertyn *et al.*, 1994; Schuller *et al.*, 1994; Varela *et al.*, 1995). Elevated intracellular glycerol synthesis can then protect cells from high external osmolarity. The mechanisms by which Hog1p regulates gene expression are not well understood at present (de Nadal *et al.*, 2002). As discussed in Section 8.3, genes such as *CTT1* show Msn2/4p-dependent expression from STRE promoter elements and osmotic stress causes the Msn proteins to be translocated into the nucleus and bind to the STRE-containing promoters. Activated Hog1p shows Msn-dependent recruitment and might phosphorylate components of the general transcriptional machinery (Alepuz *et al.*, 2001). Hog1p phosphorylates Sko1p, which binds upstream of some osmotically inducible genes and recruits the Ssn6p·Tup1p repressor (Rep *et al.*, 2001). Ssn6·Tup1p remains bound to the promoter during activation, and Hog1p seems to act by converting it into an activator that can recruit the SAGA and SWI/SNF complexes required for transcriptional initiation (Proft and Struhl, 2002). Another relevant

Hog1p target is Hot1p, which is required for stress-induced expression of genes such as *GPD1*. Osmotic stress leads to Hot1p-dependent recruitment of Hog1p to the *GPD1* promoter and Hot1p itself becomes hyperphosphorylated, but how this promotes *GPD1* expression is not yet known (Alepuz *et al.*, 2001). In addition to transcriptional regulation, Hog1p also phosphorylates and activates Rck2p, which is required for the downregulation of translation that occurs on osmotic shock. This response may be mediated through the phosphorylation of elongation factor 2 (Teige *et al.*, 2001). There are also likely to be other targets of Hog1p since, for example, osmotic stress also causes a transient cell cycle arrest by an unknown mechanism (Alexander *et al.*, 2001).

Two tyrosine phosphatases, Ptp2p and Ptp3p, have been implicated in the downregulation of Hog1p activity. Thus overexpression of *PTP2* and *PTP3* can compensate for the deleterious effects of hyperactivating the Hog1p pathway on normal growth media (Jacoby *et al.*, 1997b; Wurgler-Murphy *et al.*, 1997). Furthermore, cells lacking both these activities show constitutive Hog1p tyrosine phosphorylation, and the two tyrosine phosphatases are essential for preventing Hog1p hyperactivation during growth at high temperatures (Winkler *et al.*, 2002). Interestingly, Hog1p catalytic activity is required for its own tyrosine dephosphorylation and in fact Hog1p may activate Ptp2p and Ptp3p by a post-transcriptional mechanism (Wurgler-Murphy *et al.*, 1997). Catalytically inactive variants of the two phosphatases form complexes with Hog1p (Jacoby *et al.*, 1997b; Wurgler-Murphy *et al.*, 1997). Ptp2p is nuclear but Ptp3p is cytoplasmic, and the complexes they form with Hog1p may help modulate the localisation of Hog1p: loss of Ptp2p leads to reduced nuclear accumulation of active Hog1p while loss of Ptp3p confers prolonged nuclear retention after activation (Mattison and Ota, 2000). Ptp2p and Ptp3p, therefore, act as Hog1p tethers as well as regulating its state of phosphorylation. In fission yeast, PP2C activities have also been strongly implicated in controlling signalling through the corresponding MAPK pathway (Shiozaki and Russell, 1995) and in budding yeast too there is evidence for their role in Hog1p pathway activity acting specifically at the level of Hog1p, probably by direct dephosphorylation of activated Hog1p (Warmka *et al.*, 2001).

Sln1p does not just function as a regulator of the Hog1p MAPK pathway as it is also a regulator of Mcm1p-dependent transcriptional events (Yu *et al.*, 1995). Thus cells defective for Sln1p show reduced Mcm1p-dependent transcription, while activatory mutations in *SLN1* enhance Mcm1p-dependent transcription but not surprisingly inhibit the Hog1p pathway (Fassler *et al.*, 1997). Thus the balance between phosphorylated and unphosphorylated Sln1p may change the balance of signal transduction by these two routes and given the fact that Mcm1p plays multiple roles in the physiological responses of the cell, this shifting balance could be an important element in the cell's stress response (Fassler *et al.*, 1997). Mcm1p is a phosphoprotein, and one particular phosphorylated variant is specific to cells treated with NaCl (Kuo *et al.*, 1997). Furthermore, mutations in Mcm1p can modulate the survival of cells in response to high osmolarity, showing that Mcm1p has a role to play in mediating the response of cells to osmotic stress (Kuo *et al.*, 1997). The regulation of Mcm1p by Sln1p is mediated by phosphotransfer to Skn7p, which contains a response regulator domain that can receive a phosphate group from Sln1p (Li *et al.*, 1998).

8.7 Protein phosphatase 1

The catalytic subunit of yeast PP1 was first defined by a mutation that blocked glycogen accumulation (*glc7*), demonstrating the key role that PP1 plays in this process. However, PP1 is involved in a wide range of functions other than glycogen biosynthesis and mutations in *GLC7* define a range of different phenotypes, specific mutational alterations appearing to target different functions of the phosphatase (Hisamoto *et al.*, 1994; MacKelvie *et al.*, 1995; Baker *et al.*, 1997). In performing these multiple roles, the interaction of Glc7p with regulatory subunits is critical and evidence from several groups suggests that there may be as many as twenty or thirty of these (Stark, 1996; Walsh *et al.*, 2002). By analogy with mammalian systems where a number of PP1 regulatory polypeptides have been well-characterised (Cohen, 2002), these will be responsible for directing the catalytic activity of PP1 to different locations or substrates within the cell. Thus in combination with specific regulators, PP1 may be able to adopt the high specificity required for tightly defined regulatory events. As in mammalian PP1, interaction of yeast Glc7p with its regulatory subunits is generally mediated by a short motif (V/I-X-F) in the latter polypeptides, which bind to a groove on the opposite side of the enzyme to its active site (Egloff *et al.*, 1997; Wu and Tatchell, 2001). We have already considered PP1 in the context of SNF1 function, and here the function of PP1 in a number of other cellular processes where there is significant knowledge about its role will be described.

Glc7p and glycogen biosynthesis

Glycogen biosynthesis is dependent on the activation of glycogen synthase, the enzyme that forms the main $\alpha(1\rightarrow4)$glycosidic linkages in the main glycogen chain. There are two isoforms of glycogen synthase encoded by *GSY1* and *GSY2*, the latter encoding the major isoform that is activated as cells approach stationary phase (Farkas *et al.*, 1991). Although several protein phosphatases have been identified as candidates for the dephosphorylation and activation of glycogen synthase (see Stark, 1996), PP1 (Glc7p) is the best candidate for the key enzyme functioning *in vivo*. A variety of *glc7* mutants show deficiency in glycogen accumulation, although most work has been done with the original *glc7-1* mutation (Cannon *et al.*, 1994), which leads to accumulation of glycogen synthase in the inactive (glucose-6-phosphate-dependent) form (Peng *et al.*, 1990). *glc7* strains accumulate glycogen synthase in a hyperphosphorylated state, phosphorylated in the case of the major isoform (Gsy2p) on three key residues (ser-650, ser-654, and thr-667) close to the C-terminus (Hardy and Roach, 1993). A *GSY2* allele in which these three residues are mutated to alanine or a C-terminal *GSY2* truncation suppressed the *glc7-1* glycogen phenotype (Hardy and Roach, 1993), leading to a model whereby PP1 dephosphorylates and activates glycogen synthase to promote glycogen synthesis.

In mammals, PP1's role in glycogen metabolism is mediated by $PP1_G$, a complex of the catalytic subunit with a G subunit that binds it to glycogen particles (Hubbard and Cohen, 1989; Doherty *et al.*, 1995). Association of G and catalytic subunits is regulated by phosphorylation of the G subunit (Hubbard and Cohen, 1989; Dent *et al.*, 1990). In *S. cerevisiae*, glycogen synthesis is inhibited by mutations that cause activation of PKA and *GAC1* was isolated as a high copy suppressor of this inhibition (François *et al.*, 1992). Gac1p is related in sequence to the G subunit of skeletal muscle. Increased *GAC1* dosage boosted glycogen synthesis in wild type or *glc7-1* mutant strains while a *gac1* deletion (*gac1Δ*) was viable but glycogen-deficient, causing glycogen synthase to accumulate in the inactive (i.e. phosphorylated) form (François *et al.*, 1992). The *gac1Δ* phenotype was again suppressed by the mutant *GSY2* alleles discussed (Hardy and Roach, 1993; Huang *et al.*, 1996b) and *GAC1* mRNA is strongly induced at the diauxic shift when glycogen synthesis normally commences. As in mammalian $PP1_G$, Gac1p and Glc7p form a complex and the *glc7-1* mutant is specifically defective in this interaction (Stuart *et al.*, 1994; Tu *et al.*, 1996). In fact the interaction between Glc7p and Gac1p may be stronger in the absence of glucose (Tu *et al.*, 1996), consistent with the commencement of glycogen synthesis when glucose becomes limiting. Thus dephosphorylation and activation of glycogen synthase by PP1 is mediated by its association with Gac1p. *PIG1* encodes a Gac1p-related polypeptide, which might provide weak G subunit function and that shows an additive glycogen defect with deletion of *GAC1*. Both Pig1p and Gac1p associate with Gsy2p, consistent with a role in the direct targeting PP1 dephosphorylation of glycogen synthase (Cheng *et al.*, 1997), and different regions within Gac1p direct its association with Glc7p and Gsy2p (Wu *et al.*, 2001). Interestingly, the Gac1p·Glc7p glycogen synthase phosphatase also plays a role in the heat shock induction of the *CUP1* metallothionein gene (Lin and Lis, 1999).

If dephosphorylation of glycogen synthase is required for activation of glycogen synthesis, then clearly there must be protein kinases that phosphorylate and inactivate the enzyme. Since PKA activity is inhibitory to glycogen synthesis it is an obvious candidate. However, while PKA does phosphorylate glycogen synthase at its C-terminus, the sites known to be inhibitory and those modified by PKA are not the same (Hardy and Roach, 1993). Thus the input of PKA into the regulatory network is via some other route. One such route may be transcriptional, since elevated PKA activity suppresses *GAC1* gene expression (François *et al.*, 1992). However, constitutive high expression of *GAC1* in a *bcy1* deletion strain did not restore glycogen synthesis whereas similar expression of the C-terminal truncated allele gave partial suppression, showing that PKA can influence the phosphorylation state of Gsk2p as well as its expression (Hardy *et al.*, 1994).

Snf1p protein kinase is also involved in glycogen synthesis and *snf1Δ* deletion strains fail to accumulate glycogen (Thompson-Jaeger *et al.*, 1991). This effect is accentuated by PKA pathway activatory mutations such as *RAS2*$^{\text{val-19}}$, while *snf1Δ* cells with lowered PKA activity show better glycogen accumulation. Thus the two protein kinases have antagonistic and at least partially independent effects on glycogen synthesis. The effects of Snf1p are mediated in part by affecting the phosphorylation state of the C-terminus of Gsy2p, since the truncated C-terminal Gsy2p mutant suppresses the glycogen defect of *snf1*-deleted cells (Hardy *et al.*, 1994). In *snf1Δ* cells glucose-6-phosphate (G6P) is reduced. G6P is an allosteric activator of glycogen synthase and also inhibits glycogen synthase kinase activity in cell extracts. Thus G6P can modulate glycogen synthase activity in two ways and the *snf1* defect may be mediated in part by reduced intracellular G6P levels (Huang *et al.*, 1997). A role for Snf1p kinase has also emerged from studies on Reg1p, which encodes the PP1 regulatory subunit discussed before. Deletion of *REG1* confers glycogen hyperaccumulation and *reg1* loss of function can correct the glycogen defect of *glc7-1* or *gac1Δ* (Tu and Carlson, 1995; Huang *et al.*, 1996b). This effect requires a functional Snf1p but is not due to derepression of some critical glucose-repressed genes. Recently, it has emerged that it is important to distinguish between the cell's ability to activate glycogen synthesis and accumulation. Glycogen accumulation is dependent on autophagy and this in turn requires active Snf1p kinase, so part of the role of Snf1p is clearly mediated in this way (Wang *et al.*, 2001c).

The nature of the glycogen synthase kinase is most likely to be the Pho85p. Mutations in Pho85p cause glycogen hyperaccumulation and elevated glycogen synthase activity, at least part of which is due to elevated *GSY2* expression (Timblin *et al.*, 1996). *pho85* mutations also restore glycogen synthesis to *snf1* mutants without elevating G6P levels and Pho85p is both required for glycogen synthase kinase activity in cell extracts and co-purifies with it

(Huang *et al.*, 1996a). By using ser→ala mutants at each of the three critical Gsy2p C-terminal phosphorylation sites, Pho85p was deduced to be required for phosphorylation of ser-654 and thr-667 but not ser-650 (Huang *et al.*, 1996a). Of the ten Pho85p cyclins, Pcl8p and Pcl10p are required for glycogen synthase kinase function *in vitro* and *in vivo* and Pcl10p·Pho85p can phosphorylate Gsy2p on ser-654 and thr-667 (Huang *et al.*, 1998; Wilson *et al.*, 1999). However, Pho85p has additional roles in glycogen accumulation in that it is inhibitory to autophagy, which is required for glycogen accumulation (Wang *et al.*, 2001c), and together with Pcl6p and Pcl6p, it can phosphorylate and activate Glc8p, a Glc7p-interacting protein that functions positively for glycogen accumulation (Tan *et al.*, 2003). Clearly, though, Pho85p is not the only glycogen synthase kinase (since ser-650 is phosphorylated in a *pho85* mutant: Huang *et al.*, 1996a) and so further protein kinases remain to be identified. Since activation of the PKA pathway can suppress glycogen accumulation in a *pho85* mutant, perhaps this is mediated by indirectly promoting ser-650 phosphorylation.

Glc7p and the cell division cycle

An indication that yeast PP1 plays an important role in mitosis came with the realisation that *S. pombe dis2*$^{+}$, mutationally altered in cells deficient in sister chromatid disjunction, encodes a PP1 catalytic subunit (Ohkura *et al.*, 1989). In fact fission yeast encode two PP1 isoforms, *sds21*$^{+}$ encoding the second catalytic subunit gene but *dis2*$^{+}$ being responsible for most of the activity (Kinoshita *et al.*, 1990). In fission yeast cells deficient in both PP1 catalytic subunits, mitosis is blocked in metaphase (Ishii *et al.*, 1996).

In *S. cerevisiae*, the catalytic subunit of PP1 is also required for mitosis. The cold-sensitive *glc7*$^{Y\text{-}170}$ allele blocks mitosis at the restrictive temperature with vast majority of cells arrested with large buds, elevated Clb·Cdc28p kinase activity and a short metaphase spindle (Hisamoto *et al.*, 1994), while a number of other temperature-sensitive *glc7* mutants show a similar mitotic arrest phenotype (MacKelvie *et al.*, 1995; Bloecher and Tatchell, 1999; Sassoon *et al.*, 1999; Andrews and Stark, 2000b). While the mitotic block could be due to a direct effect on progression through mitosis, it might also result from a defect that activates one of the checkpoint controls inhibiting anaphase in response, for example, to an unattached kinetochore (Musacchio and Hardwick, 2002), and some of the *glc7* mutants that arrest in mitosis certainly do so in a mitotic checkpoint-dependent manner (Bloecher and Tatchell, 1999; Sassoon *et al.*, 1999; Andrews and Stark, 2000b). Some of these *glc7* mutants also show elevated chromosome loss (Hisamoto *et al.*, 1994; Sassoon *et al.*, 1999), consistent with a mitotic defect.

One role for PP1 in mitosis that might account for some of these phenotypes appears to be in a process that ensures correct attachment of kinetochores to the mitotic spindle. Ipl1p is an essential protein kinase required for chromosome segregation and genetic studies indicated that it might function antagonistically with Glc7p in this process (Chan and Botstein, 1993; Francisco *et al.*, 1994). Thus a high copy plasmid encoding an inactive fragment of Glc7p suppressed *ipl1* mutants (but not an *ipl1* deletion) by what is presumed to be a dominant negative effect of the inactive fragment titrating a regulatory protein away from active, full-length Glc7p in the cell. Conversely, the *ipl1-1* mutant showed enhanced sensitivity to wild-type Glc7p. In fact Ipl1p activity is inhibitory to the binding of kinetochores to microtubules in an *in vitro* assay, while Glc7p function is required to reverse this inhibition (Biggins *et al.*, 1999; Sassoon *et al.*, 1999). Consistent with opposing roles *in vivo*, Ipl1p and Glc7p have also been shown to regulate the phosphorylation state of ser-10 on histone H3 in budding yeast (Hsu *et al.*, 2000). *ipl1* mutants lose viability at the restrictive temperature as they enter mitosis because they separate their DNA into unequal masses, indicative of severe chromosome non-disjunction. Recent work has revealed that this occurs because for most replicated chromosomes, both sister chromatids attached to a single SPB. In the majority of cases this is the old SPB, which is segregated to the bud. Ipl1p kinase functions to promote detachment of one sister kinetochore from its microtubule followed by reattachment to a microtubule from the opposite pole (Tanaka *et al.*, 2002). In double *glc7–10 ipl1–321* mutants, the severe *ipl1* defect in bipolar attachment of sister chromatids and in the subsequent non-disjunction of chromosomes is largely corrected, but surprisingly the *glc7–10* mutant itself does not show a defect of the same magnitude, as might be expected if Ipl1p and Glc7p function sequentially in a pathway that is obligatory for correct chromosome segregation (N. Rachidi and M. J. R. Stark, unpublished data). This may be because the *glc7–10* mutant activates the mitotic checkpoint whereas the defect in *ipl1* mutants does not, or it might imply that the suppression reflects negative regulation of Ipl1p activity by Glc7p (and hence partial rescue of Ipl1p function in the *glc7–10* mutant). If Glc7p does oppose Ipl1p by dephosphorylation of a key substrate, a good candidate for this is Dam1p, a protein of the yeast outer kinetochore. Thus Ipl1p phosphorylates Dam1p both *in vitro* and *in vivo* and mutation of all four of the Ipl1p phosphorylation sites is lethal, while mutation of two or three of the sites in combination with a point mutation in *SPC34* (encoding another outer kinetochore component) confers a defect that is closely similar to that of *ipl1* mutants (Cheeseman *et al.*, 2002).

If PP1's interaction with regulatory polypeptides is critical for mediating its effects, then which protein might regulate its mitotic function? A potential candidate is encoded by the *SDS22* in budding yeast and *sds22*$^+$ in fission yeast, both of which are essential genes (Ohkura and Yanagida, 1991; Hisamoto *et al.*, 1995; MacKelvie *et al.*, 1995). These genes encode highly related conserved proteins consisting largely of 11 degenerate ~22-residue leucine-rich repeats. In fission yeast, *sds22* mutants cause a mid-mitotic arrest (Ohkura and Yanagida, 1991; Stone *et al.*, 1993) and the protein physically associates with the PP1 catalytic subunit, changing the substrate specificity (Stone *et al.*, 1993). Like its fission yeast counterpart, *S. cerevisiae* Sds22p is a largely nuclear protein that shows direct interaction with Glc7p (Hisamoto *et al.*, 1995; MacKelvie *et al.*, 1995; Peggie *et al.*, 2002) and extra copies of *SDS22* efficiently suppressed the mitotic arrest phenotype of the *glc7–12* mutant, providing evidence that Sds22p plays a role specifically in the mitotic function of PP1 (MacKelvie *et al.*, 1995). However, a range of *sds22* temperature-sensitive mutants does not show a convincing mitotic arrest phenotype although one (*sds22–6*) can suppress the conditional growth defect of an *ipl1* mutant (Peggie *et al.*, 2002). Analysis of *sds22–6* revealed that at the restrictive temperature, the normal nuclear localisation of Glc7p was rapidly lost: instead of being uniformly distributed throughout the nucleus (but enriched in the nucleolus), the distribution of Glc7p instead collapsed to a small number of foci (Peggie *et al.*, 2002). Thus Sds22p may have a more general role in the function of PP1 in the nucleus by promoting its proper distribution, and the putative regulator of PP1's role in kinetochore function probably awaits discovery.

Other roles of PP1 in yeast

In yeast cells, amino acid starvation induces expression of *GCN4*, encoding a transcriptional activator for a wide range of amino acid biosynthetic genes that mediates general amino acid control (See Chapter 4 and Hinnebusch, 1992). *GCN2* encodes an eIF-2α protein kinase that is required for *GCN4* induction, phosphorylating the initiation factor on ser-51 (Dever *et al.*, 1992) and inhibiting its function. Paradoxically, this elevates *GCN4* translation, which is normally blocked in an eIF-2-dependent manner by the presence of four short open reading frames that precede the *GCN4* start codon. An inactive Glc7p fragment was isolated as a high copy suppressor of a *gcn2* mutation, but suppression was blocked by an S51→A mutation in eIF-2α (Sui2p: Wek *et al.*, 1992). As in the case of the *ipl1* mutant discussed before, the truncated phosphatase must somehow exert a dominant negative effect. A *gcn2* deletion was not suppressed (showing that residual Gcn2p function was needed), while extra copies of full-length *GLC7* partially reduced *GCN4*-dependent gene expression. This supports opposing roles for PP1 and Gcn2p in controlling the phosphorylation state of eIF-2α.

glc7 mutations that do not block the mitotic cell division cycle have been found that confer a sporulation defect in homozygous diploids (Tu and Carlson, 1994; Baker *et al.*, 1997; Ramaswamy *et al.*, 1998), supporting one or more distinct roles for PP1 specifically in meiosis. In fact, Glc7p interacts in the two-hybrid assay with two proteins (Red1p and Gip1p) required for meiosis (Tu *et al.*, 1996). *RED1* is an early meiotic gene required for synaptonemal complex formation and successful execution of meiosis I (Rockmill and Roeder, 1990), while Gip1p is required for late meiotic gene expression and is expressed under the control of the early meiotic gene transcription factor Ime1p (Tu and Carlson, 1994). Red1p, together with the protein kinase Mek1p and Glc7p, is specifically required for the pachytene checkpoint (Roeder and Bailis, 2000), a regulatory mechanism that ensures meiotic recombination and chromosome synapsis have occurred. Mek1p-dependent phosphorylation of Red1p is needed for normal levels of sister chromatid cohesion and synapsis and the two proteins localise to the chromosomes early in meiotic prophase (Bailis and Roeder, 1998). Glc7p is present in a complex with Red1p and Mek1p, dephosphorylation of Red1p by Glc7p is required for pachytene exit and overproduction of Glc7p bypasses the checkpoint arrest observed in a *zip1* mutant, which is defective in synapsis (Bailis and Roeder, 2000). Glc7p localises to chromosomes in late pachytene consistent with its role in triggering pachytene exit, and this localisation depends on Gip1p. The Glc7p encoded by the sporulation-defective *glc7-T152K* failed to associate with Gip1p in the two-hybrid assay, supporting the notion that Gip1p-dependent PP1 dephosphorylation is required for meiosis. Recent work has shown that the Gip1p·Glc7p phosphatase also plays a specific role in spore wall development, which is defective both in *gip1* mutants and in *glc7* mutants that are defective in their association with Gip1p (Tachikawa *et al.*, 2001). Gip1p and Glc7p co-localise with the meiotic septin structures and also seem to be required for normal meiotic septin organisation.

Finally, PP1 may also have an involvement in polarised cell growth. Some *glc7* mutants show hyperpolarisation of the actin cytoskeleton, elongated bud morphology and a temperature-sensitive cell lysis defect reminiscent of mutations in the Pkc1p–Mpk1p MAPK pathway discussed earlier (Andrews and Stark, 2000b). In fact *PKC1* in high copy compensates for the defects in this *glc7* mutant, while loss of function of the MAPK pathway or reduced Pkc1p function has an additive effect. This suggests that PP1 may be required in parallel with the MAPK pathway for promoting cell wall remodelling during growth. Glc7p also functions at a late stage in the fusion of vesicles with their target membrane during protein trafficking, although the precise role that it plays is not yet known (Peters *et al.*, 1999). It also appears to regulate ion homeostasis (Williams-Hart *et al.*, 2002).

8.8 Protein phosphatase 2A

Saccharomyces cerevisiae has two PP2A catalytic subunit genes (*PPH21* and *PPH22*), the products of which are each 74% identical to mammalian PP2A (Sneddon *et al.*, 1990). Deletion of either gene alone is without phenotype, although individual loss of *PPH21* and *PPH22* reduced measurable PP2A activity, and the double deletion is viable but has a severe slow growth phenotype (Ronne *et al.*, 1991; Lin and Arndt, 1995; Evans and Stark, 1997). In combination with deletion of *PPH3*, encoding a non-essential protein phosphatase related to PP2A (Figure 8.1), the double *pph21Δ pph22Δ* knockout is lethal. Thus Pph3p can provide residual overlapping function with PP2A (Ronne *et al.*, 1991). Fission yeast also encodes redundant PP2A catalytic subunits (genes $ppa1^+$ and $ppa2^+$). While the double deletion is inviable, deletion of just $ppa2^+$ (*ppa2Δ*) conferred a small cell phenotype that suggested premature entry into mitosis (Kinoshita *et al.*, 1990), a conclusion strengthened by genetic interactions of the *ppa2* mutation with the components that modulate mitotic activation of fission yeast cdc2 CDK (Kinoshita *et al.*, 1993). However, as we shall see, budding yeast PP2A seems to be an activator rather than an inhibitor of mitotic entry. This section will consider the roles of PP2A in the cell cycle and in morphogenesis but, like many other yeast protein phosphatases, PP2A is likely to have a multiplicity of other functions.

Mammalian PP2A consists of the catalytic subunit in combination with A and B regulatory subunits, the latter consisting of a small number of families of unrelated polypeptides (Cohen, 1989). Homologues of these polypeptides are conserved in yeast and interact with the PP2A catalytic subunits. The A subunit of *S. cerevisiae* PP2A is encoded by *TPD3*, a gene originally identified as required for tRNA production (van Zyl *et al.*, 1989). Like its mammalian homologues, it largely consists of 15 ~39-residue imperfect HEAT repeats. Deletion of *TPD3* conferred growth defects at both low and high growth temperatures. The tRNA production defect *in vivo* at high temperatures was reproducible *in vitro* and could be repaired by addition of TFIIIB, despite normal levels of TFIIIB and RNA polymerase III in the mutant extracts (van Zyl *et al.*, 1992). This implied the presence of an inhibitor in the mutant and suggests that dephosphorylation of some element of the transcriptional machinery by PP2A is important for regulating RNA polymerase III transcription.

Two B-type subunits of yeast PP2A have been identified. Cdc55p is a member of the mammalian B family (Healy *et al.*, 1991), while Rts1p is related in sequence to mammalian B′ polypeptides (Shu *et al.*, 1997; Zhao *et al.*, 1997). On the basis of several observations, Cdc55p and Rts1p mediate different functions of PP2A. *rts1* mutants are temperature-sensitive, arrest in G2 and are defective in certain stress responses (Evangelista *et al.*, 1996; Shu *et al.*, 1997), while *cdc55* mutants are cold-sensitive and exhibit a bud morphology defect (Healy *et al.*, 1991). The A subunit is limiting for assembly of PP2A heterotrimers, since the catalytic subunits are in approximately eightfold excess and the B family subunits in fourfold excess (Gentry and Hallberg, 2002). Interestingly, Rts1p is far more abundant than Cdc55p by a factor of at least tenfold (Gentry and Hallberg, 2002). The C-terminus of all PP2As consists of the highly conserved sequence -DYFL and the terminal leucine residue is carboxymethylated by the Ppm1p methyl transferase and demethylated by the Ppe1p methyl esterase. This modification is needed for the stability of heterotrimeric complexes and is of particular importance for the assembly of complexes containing Cdc55p (Evans and Hemmings, 2000; Wu *et al.*, 2000; Kalhor *et al.*, 2001; Wei *et al.*, 2001). Using green fluorescent protein (GFP) fusions to Rts1p, Cdc55p and Pph21p, it has been shown recently that yeast PP2A is enriched in the nucleus, but Cdc55p appears to localise PP2A to the bud tip and Rts1p to target PP2A to the kinetochore, a conclusion confirmed by the use of chromatin immuno-precipitation. In addition, both Cdc55p and Rts1p appear to localise PP2A to the bud neck in large-budded cells (Gentry and Hallberg, 2002).

PP2A is thought to be one of the major activities that dephosphorylates phospho-ser/thr-pro motifs, which in turn are favoured by MAPKs and CDKs. An added twist to regulation at such sites has come with the realisation that the rate of dephosphorylation is greatly influenced by the state of the proline residue, which can exist in a peptide as either a *cis* or *trans* isomer. PP2A effectively dephosphorylates sites in which the proline is in the *trans* state and the Pin1 proline *cis–trans* isomerase (Ess1p) facilitates dephosphorylation by catalysing the *cis–trans* isomerisation reaction (Zhou *et al.*, 2000). This type of regulation is likely to be important *in vivo* because extra Ess1p can partially suppress the growth defect of Ts^- *pph22* mutants, while excess PP2A suppresses the mitotic arrest phenotype of an *ess1* mutant, and up to 20% of phospho-ser/thr-pro bonds in proteins may be in the *cis* configuration (Zhou *et al.*, 2000).

PP2A, the cell division cycle and cell morphology

Although it is possible to grow cells lacking both PP2A catalytic subunit genes, it is nonetheless possible to generate strains temperature-sensitive for PP2A function by deleting one gene and selecting for suitable mutations in the other (Lin and Arndt, 1995; Evans and Stark, 1997), while the additional deletion of *PPH3* confers a stronger temperature-sensitive phenotype (Evans and Stark, 1997). Temperature-sensitive *pph21* mutants fail to enter mitosis at the restrictive temperature, arresting as budded cells with replicated DNA, a short spindle and an undivided nucleus (Lin and Arndt, 1995). The mutant is unable to maintain high levels of Clb·Cdc28p protein kinase (required for entry into mitosis) at the restrictive temperature and the level of *CLB2* mRNA was reduced. *CLB2* overexpression partially suppressed the block to M-phase in some cells but did not restore normal levels of Clb2p·Cdc28p activity (Lin and Arndt, 1995). Thus PP2A is required for the normal activation of Clb·Cdc28p kinase. This function of PP2A seems to be mediated by a form of the enzyme containing Rts1p, since *rts1* mutants are temperature-sensitive for growth and arrest in G2 like the *pph21* mutant (Shu *et al.*, 1997). That this block is due to reduced Clb·Cdc28p function is underlined by the ability of *CLB2* in high copy to suppress the *rts1* mutant phenotype as well. In addition to playing a role in promoting mitotic entry, Rts1p may also mediate roles of PP2A in the general stress response (Evangelista *et al.*, 1996).

While temperature-sensitive *pph22* strains also appear to block entry into mitosis, significant fractions of cells become binucleate subsequently, indicating an aberrant division in which the one of the daughter nuclei failed to segregate into the bud (Evans and Stark, 1997). *pph22* mutant strains show aberrant microtubule organisation with multiple and/or misoriented microtubule structures in many cells. This implies that the block to mitosis in the mutant cells is not irrevocable, but that when cells do eventually try to divide they are impeded from properly segregating the two daughter nuclei by defects in microtubule organisation. Since actin has been shown to play a role in spindle orientation and function (Palmer *et al.*, 1992; Theesfeld *et al.*, 1999; Beach *et al.*, 2000), it is possible that these defects result from changes in actin organisation in the *pph* mutants noted below.

An indication that *S. cerevisiae* PP2A might be involved in morphogenesis came from examining the phenotypes of *cdc55* and *tpd3* mutant strains at low growth temperatures. *cdc55* was identified as a cold-sensitive mutant affecting cell morphology. Unlike Tpd3p,

Cdc55p is not also required at higher growth temperatures. The phenotype of both *cdc55* and *tpd3* mutants at low temperatures is similar: cells become highly elongated and multiply budded, indicative of delayed cytokinesis, forming large clusters of unseparated cells in which septa are not present between every cell compartment. Mutations in *BEM2* (a Rho1p GAP required for bud emergence: see Kim *et al.*, 1994) can suppress *cdc55Δ*, while *bem2* deletion is synthetically lethal with the *pph21Δ pph22Δ* double knockout (Lin and Arndt, 1995), strengthening the morphogenesis link. The morphology defects in *cdc55* mutants can be largely corrected by mutation of tyr-19 in Cdc28p to a non-phosphorylatable phenylalanine (Wang and Burke, 1997; Yang *et al.*, 2000), showing that the abnormal morphology results from misregulation of Cdc28p activity. While *cdc55* knockout strains seem normal for Mih1p phosphatase activity and function, they do, however, accumulate excess Swe1p kinase. Loss of Cdc55p results in the stabilisation of Swe1p from degradation and hence Cdc28p activity is reduced (Yang *et al.*, 2000). Since loss of both PP2A catalytic subunit genes also removes much of the morphological abnormality of a *cdc55* deletion strain and reduced the Swe1p level, the effect seems to be mediated by inappropriate PP2A activity rather than by loss of a Cdc55p-dependent phosphatase function (Yang *et al.*, 2000). *cdc55* mutants also show a mitotic checkpoint defect in that they are hypersensitive to nocodazole and cannot maintain a mitotic metaphase arrest in response to impaired kinetochore function (Minshull *et al.*, 1996; Wang and Burke, 1997). This effect might be mediated by Cdc28p tyr-19 phosphorylation due to the elevated Swe1p levels in *cdc55* mutant strains, but there is currently some disagreement regarding whether this is the case and whether their nocodazole sensitivity is suppressible by nonphosphorylatable *CDC28* mutants.

Strains temperature-sensitive for PP2A catalytic subunit function have yielded evidence for roles of PP2A in the actin cytoskeleton. At the restrictive temperature bud morphology was perturbed; in around 40% of cells the buds were angular, tube-like or hooked (Lin and Arndt, 1995; Evans and Stark, 1997). This was associated with delocalisation of cortical actin patches into the mother cell at stages of the cell cycle when they are normally restricted to the bud, with concentrations of actin at the sites of abnormal growth in those buds showing the heterogeneous abnormal morphologies just described (Lin and Arndt, 1995; Evans and Stark, 1997). Actin cables were also perturbed and chitin deposition completely delocalised. Thus PP2A function is required for proper organisation of the actin cytoskeleton and this may determine the bud morphology and chitin defects.

Temperature-sensitive *pph22* mutants are largely suppressed by the inclusion of 1 M sorbitol in the growth medium and *pph21Δ pph22Δ* strains, which also show temperature sensitivity, are similarly partially suppressed (Evans and Stark, 1997). PP2A-deficient cells have a temperature-sensitive cell lysis phenotype, supporting the notion that they develop defective cell walls. However, to date it has not been possible to identify any strong genetic interactions between this phenotype and mutations in the Pkc1p–Mpk1p MAPK pathway, unlike the situation with PP1 (see before). Thus while PP2A may also play a role in cell wall integrity, it appears to act in a different manner to PP1.

8.9 Sit4p protein phosphatase

Sit4p phosphatase is related to PP2A (Figure 8.1) and was first defined in a search for mutations that restored *HIS4* expression in the absence of the normal transcriptional activators (Arndt *et al.*, 1989). However, *sit4* mutations lead to alterations in the transcription of many other genes indicating that Sit4p may be important for transcriptional regulation despite its predominantly cytoplasmic localisation (Sutton *et al.*, 1991). Interpretation of the functions of Sit4p is complicated because the effects of *sit4* mutations are very strain-dependent. *SIT4* is in principal an essential gene but the phenotype of *sit4* deleted (*sit4Δ*) and temperature-sensitive (Ts$^-$) *sit4* mutant strains is dependent on a polymorphic locus termed *SSD1*. In *ssd1-d* strains such as W303, *sit4Δ* is lethal and *sit4–102* is Ts$^-$; in *SSD1-v* strains such as S288C *sit4Δ* is viable, conferring a slow growth phenotype, while *sit4–102* is Ts$^+$ (Sutton *et al.*, 1991). *SSD1-v* alleles are dominant to *ssd1-d* and Ssd1p is a 1250-residue protein of unknown function but which may bind RNA (Uesono *et al.*, 1997). The *SSD1* status of cells affects many different pathways (Stettler *et al.*, 1993; Kikuchi *et al.*, 1994; Uesono *et al.*, 1994) and *SSD1-v* alleles ameliorate the defects of cells with a hyperactivated PKA pathway (Sutton *et al.*, 1991; Wilson *et al.*, 1991), deficient in the *PKC1–MPK1* MAPK pathway or in *ins1* mutants (Wilson *et al.*, 1991) to mention just a few examples. The temperature sensitivity of cells deficient in chitin synthase III is also suppressed by *SSD1-v* alleles (Bulawa, 1993), as is the slow growth defect of G1 cyclin-deficient strains (Cvrckova and Nasmyth, 1993). Although the common connection between these effects is undoubtedly protein phosphorylation, how Ssd1p comes to have such pleiotropic effects remains to be established.

Drosophila and mammalian PP6 (PPV: Mann *et al.*, 1993; Becker *et al.*, 1994) are the most closely related metazoan protein phosphatases to Sit4p and can functionally complement *sit4* mutants when expressed in yeast (Mann *et al.*, 1993; Bastians and Ponstingl, 1996). The unique character of PP6 in providing Sit4p function resides in the first 52 codons, since a chimeric phosphatase in which this region was used to replace the equivalent section of a *Drosophila* PP1 cDNA was as effective as the complete PP6 cDNA, whereas the unmodified PP1 cDNA was unable to complement (Mann *et al.*, 1993). The distinctive function(s) of Sit4p are thus inferred to lie within the N-terminal domain and may be mediated by the specific binding of proteins such as the SAPs (discussed later). Sit4p and Ppz1p, a PP1-related phosphatase (Table 8.1), appear to act antagonistically for one or more critical Sit4p functions. Thus the slow growth defect in viable *sit4* deletion strains is greatly improved by deletion of *PPZ1*, overexpression of Hal3p/Sis2p (a negative regulator of Ppz1p) rescues *sit4* mutants and deletion of *HAL3* is lethal in combination with otherwise viable *sit4* mutants (Di Como *et al.*, 1995; de Nadal *et al.*, 1998; Clotet *et al.*, 1999). The exact basis for this antagonism remains to be uncovered.

We have already considered one of the most important roles of Sit4p in the context of the TOR pathway (Section 8.5), where it forms part of the mechanism whereby TOR inhibition (by rapamycin or nutrient starvation) communicates with TOR pathway effectors. Here we will consider some other aspects of Sit4p function.

Sit4p, CLN expression and bud emergence

In *ssd1-d* strains, *sit4^{ts}* mutants are temperature-sensitive and block cells in G1 at 'Start'. This effect is due to independent effects of Sit4p on *CLN* expression and on bud emergence. *SIT4* is needed for the accumulation of *CLN1*, *CLN2* and *PCL1* G1 cyclin mRNAs and *sit4Δ* strains are sensitive to reduced G1 cyclin gene dosage (Fernandez-Sarabia *et al.*, 1992). Since Cln1,2·Cdc28p is needed for execution of Start and entry into a new cell cycle, this correlates nicely with the arrest phenotype of *sit4* mutants. Since Sit4p is required for *SWI4* transcription and Swi4p (a component of SBF) is needed for *CLN1* and *CLN2* expression (see before), the effect of *sit4* mutation on G1 cyclin mRNAs appears to be mediated largely by modulation of *SWI4* function. *CLN3* levels were not *SIT4*-dependent and *SIT4* and *CLN3* appear to provide additive pathways for G1 cyclin expression (Fernandez-Sarabia *et al.*, 1992), in keeping with the distinct role of *CLN3* as an activator of SBF-mediated transcriptional activation of *CLN1* and *CLN2* (Dirick *et al.*, 1995; see before). Problems with G1 cyclin expression are not the only deficiency in *sit4* mutants, though, because ectopic *CLN2* expression could not bypass loss of Sit4p (Fernandez-Sarabia *et al.*, 1992), although it did allow DNA replication in the absence of budding. This demonstrates an additional, *CLN2*-independent function of Sit4p concerned with bud emergence and morphogenesis (Fernandez-Sarabia *et al.*, 1992).

SAPs may regulate Sit4p function

Sit4p was found initially to associate in separate complexes with two high molecular weight proteins (Sap155p and Sap190p). These complexes were absent in G1 but appeared around Start, persisting until the end of mitosis (Sutton *et al.*, 1991). There are four members of the *SAP* gene family; *SAP185* also encodes a Sit4-associated protein (SAP) and *SAP4* is related in sequence to the other three but has not yet been shown to complex with Sit4p (Luke *et al.*, 1996). Any of the four *SAP* genes can partially suppress *sit4* mutants in high copy, while *sap* deletions are additive and loss of all four genes confers the same phenotype as deletion of *SIT4*. The SAPs form two distinct subgroups; Sap185p and Sap190p are more closely related to each other than to the other SAPs, as are Sap4p and Sap155p, and association of the SAPs with Sit4p is specific. The SAPs are probably regulators of Sit4p, although it is formally possible that they could be substrates that act as effectors of Sit4p function.

The two SAP subgroups may mediate distinct functions, since the synthetic lethality between *sit4* mutations in *SIT4* and *BEM2* (which encodes a Rho1p GAP: Kim *et al.*, 1994) is a property of the *SAP185/SAP190* subgroup, while increased dosage of a *SAP* gene from one sub group failed to correct the growth defect caused by loss of the other *SAP* subgroup (Luke *et al.*, 1996). In addition, high copy *SAP155* confers resistance to *Kluyveromyces lactis* zymocin whereas high copy *SAP185* and *SAP190* do not (Jablonowski *et al.*, 2001a). In fact this phenotype is mediated by competition between the SAPs for Sit4p and so toxin sensitivity is also a property of the *SAP185/SAP190* subgroup (Jablonowski *et al.*, 2001a). Like loss of Sit4p, *K. lactis* zymocin causes cells to arrest in the G1 phase of the cell division cycle at Start and a major target of this toxin appears to be the RNA polymerase II Elongator complex (Otero *et al.*, 1999; Winkler *et al.*, 2001), since mutations in many of its known components confer resistance to the *K. lactis* zymocin (Frohloff *et al.*, 2001; Jablonowski *et al.*, 2001b). *sit4* deletion mutants are also resistant to the zymocin and share with Elongator mutants a number of phenotypes including 6-azauracil sensitivity, although it is not yet completely clear how Sit4p-dependent processes are involved in sensitivity to the zymocin. Sit4p may regulate the function of RNA polymerase II through Elongator, and *SIT4* was first identified through its effects on transcription on transcription of a number of genes including *HIS4* as mentioned above (Arndt *et al.*, 1989). Apart from Tap42p with which only a small fraction of Sit4p associates (see above), the SAPs represent the major Sit4p-interacting proteins and are therefore likely to mediate many of its functions.

8.10 PP2B

PP2B, otherwise known as calcineurin, consists of a heterodimeric complex, the activity of which is regulated by interaction with Ca^{2+}-calmodulin. In budding yeast the catalytic

A subunit is encoded by the redundant genes *CNA1* and *CNA2*, while the B subunit, an EF-hand-containing protein related to calmodulin, is encoded by *CNB1* (see Cyert, 2001). The A subunit contains the calmodulin binding site and an autoinhibitory region that is displaced by Ca^{2+}-calmodulin binding. Yeast PP2B plays roles in regulating gene expression in response to stress, Ca^{2+} homeostasis and the cell division cycle. PP2B is inhibited by the drug FK506, which binds to PP2B in a complex with its cytoplasmic receptor FKBP12 (Foor *et al.*, 1992), and the isolation of FK506-hypersensitive mutants that mapped to one of the genes for $\beta(1\rightarrow3)$glucan synthase (*FKS1*) reflect a role for PP2B in regulating expression of the second isoform (*FKS2*: Foor *et al.*, 1992).

PP2B regulates gene expression in response to cell wall damage, mating pheromone treatment and high extracellular concentrations of ions such as Ca^{2+} and Na^{+} through a transcription factor dubbed Crz1p (Matheos *et al.*, 1997; Stathopoulos and Cyert, 1997). PP2B can dephosphorylate Crz1p directly, leading to its translocation into the nucleus where it binds to the CDRE (calcineurin dependent response elements) upstream of genes such *FKS2* (Stathopoulos-Gerontides *et al.*, 1999), thereby inducing their transcription. Some 160 yeast genes show PP2B-dependent transcriptional activation in response to high extracellular Ca^{2+} or Na^{+} and Crz1p is largely responsible for this (Yoshimoto *et al.*, 2002). Activation of Crz1p by PP2B requires a Ca^{2+} signal to activate PP2B itself, and at least in the case of the response to mating pheromone this is furnished by Ca^{2+} entry through the Mid1p·Cch1p Ca^{2+} channel in the plasma membrane (Paidhungat and Garrett, 1997) and a second unidentified influx system (Muller *et al.*, 2001). Ca^{2+} entry is essential for viability following pheromone stimulation (Iida *et al.*, 1990), and this may reflect the need to induce PP2B-dependent gene expression to support activities such as cell wall synthesis during the polarised morphogenesis that occurs during mating.

The effects of PP2B on Ca^{2+} homeostasis are less well understood. Ca^{2+} homeostasis involves the V-ATPase present in the vacuolar membrane, which creates a proton gradient that is used to drive entry of Ca^{2+} from the cytosol into the vacuole via the Vcx1p Ca^{2+}/H^{+} exchanger, while the Pmc1p P-type ATPase can also pump Ca^{2+} into the vacuole against its concentration gradient (reviewed in Cyert, 2001). PP2B induces the expression of Pmc1p through the Crz1p pathway described before, but it is clear that there are other PP2B targets and one of these may be the Vcx1p exchanger. Finally, PP2B acts in a pathway that can destabilise Hsl1p, a protein kinase that inhibits the activity of Swe1p. Thus PP2B can activate Swe1p, which phosphorylates Cdc28p on tyr-19 leading to a G2-M delay (Mizunuma *et al.*, 1998, 2001). PP2B therefore may provide a link between stress-induced Ca^{2+} signalling and the cell cycle.

8.11 Concluding remarks

Protein phosphorylation is a pervasive phenomenon that affects just about every imaginable aspect of yeast cell physiology. Wherever the cell has to respond to changes in the internal or external environment, it is very likely that protein kinases and protein phosphatases will be involved somewhere to mediate the necessary physiological changes. The topics discussed here are but a few examples demonstrating the roles of yeast protein kinases and phosphatases in cellular regulation but are far from being an exhaustive list. With a significant fraction of yeast protein kinases and protein phosphatases yet to be properly analysed it is clear that there is still much to learn and the task is certainly not simple. Even with protein kinases, it can be difficult to define precisely the *in vivo* substrates that are mediating the key regulatory events, while with the protein phosphatases this is even more difficult. New technologies such as the use of protein chips will greatly facilitate the analysis of protein kinase substrates (Zhu *et al.*, 2000b, 2001), while other recent approaches, for example the fractionation of cell extracts for substrate identification with either recombinant or purified kinases of interest, might compensate for such difficulties (see Knebel *et al.*, 2001). Such approaches will need to take cognisance of the potential importance of protein kinase regulatory subunits. The ability to engineer protein kinases to make them the sole enzyme in a cell (or extract) that is susceptible to an inhibitor, or able to use a modified ATP substrate, is also likely to have a major impact on the analysis of protein phosphorylation, both in yeast and in many other systems (see Shogren-Knaak *et al.*, 2001). In the case of protein phosphatases the problem of identifying the real substrates is even greater, and while the use of substrate-trapping mutants looked promising initially (Zhang *et al.*, 1999), it is far from clear that this can provide a general route to substrate identification for all phosphatase families. For phosphatases such as PP1, identification of the regulatory polypeptides should be useful in allowing subsets of the functions to be analysed in mutants whose effects are restricted to the process of interest. Perhaps the greatest promise for the future lies with the rapid advances in mass spectrometry that are starting to enable meaningful analysis of the phosphoproteome (Zhu *et al.*, 2000b; Ficarro *et al.*, 2002). Coupled with the use of specific kinase and phosphatase mutants and the further likely advances in the application of this type of technology, this should begin to allow an *in vivo* analysis of exactly which sites on which proteins are being modified and how these changes relate to the inner workings of the yeast cell.

References

Agarwal, R. and Cohen-Fix, O. (2002) Phosphorylation of the mitotic regulator Pds1/securin by Cdc28 is required for efficient nuclear localization of Esp1/separase. *Genes and Development* **16**, 1371–1382.

Alarcon, C. M., Heitman, J. and Cardenas, M. E. (1999) Protein kinase activity and identification of a toxic effector domain of the target of rapamycin TOR proteins in yeast. *Molecular Biology of the Cell* **10**, 2531–2546.

Alberts, A. S., Bouquin, N., Johnston, L. H. and Treisman, R. (1998) Analysis of RhoA-binding proteins reveals an interaction domain conserved in heterotrimeric G protein beta subunits and the yeast response regulator protein Skn7. *Journal of Biological Chemistry* **273**, 8616–8622.

Albertyn, J., Hohmann, S., Thevelein, J. M. and Prior, B. A. (1994) *GPD1*, which encodes glycerol-3-phosphate dehydrogenase, is essential for growth under osmotic stress in *Saccharomyces cerevisiae*, and its expression is regulated by the high-osmolarity glycerol response pathway. *Molecular and Cellular Biology* **14**, 4135–4144.

Alepuz, P. M., Jovanovic, A., Reiser, V. and Ammerer, G. (2001) Stress-induced map kinase Hog1 is part of transcription activation complexes. *Molecular Cell* **7**, 767–777.

Alexander, M. R., Tyers, M., Perret, M., Craig, B. M., Fang, K. S. and Gustin, M. C. (2001) Regulation of cell cycle progression by Swe1p and Hog1p following hypertonic stress. *Molecular Biology of the Cell* **12**, 53–62.

Alexandru, G., Uhlmann, F., Mechtler, K., Poupart, M. A. and Nasmyth, K. (2001) Phosphorylation of the cohesin subunit Scc1 by Polo/Cdc5 kinase regulates sister chromatid separation in yeast. *Cell* **105**, 459–472.

Alms, G. R., Sanz, P., Carlson, M. and Haystead, T. A. (1999) Reg1p targets protein phosphatase 1 to dephosphorylate hexokinase II in *Saccharomyces cerevisiae*: characterizing the effects of a phosphatase subunit on the yeast proteome. *EMBO Journal* **18**, 4157–4168.

Amon, A., Surana, U., Muroff, I. and Nasmyth, K. (1992) Regulation of p34^{CDC28} tyrosine phosphorylation is not required for entry into mitosis in *Saccharomyces cerevisiae*. *Nature* **355**, 368–371.

Amon, A., Tyers, M., Futcher, B. and Nasmyth, K. (1993) Mechanisms that help the yeast cell cycle clock tick: G2 cyclins transcriptionally activate G2 cyclins and repress G1 cyclins. *Cell* **74**, 993–1007.

Andrews, B. J. and Mason, S. W. (1993) Gene expression and the cell cycle: a family affair. *Science* **261**, 1543–1544.

Andrews, P. D. and Stark, M. J. (2000a) Dynamic, Rho1p-dependent localization of Pkc1p to sites of polarized growth. *Journal of Cell Science* **113**, 2685–2693.

Andrews, P. D. and Stark, M. J. (2000b) Type 1 protein phosphatase is required for maintenance of cell wall integrity, morphogenesis and cell cycle progression in *Saccharomyces cerevisiae*. *Journal of Cell Science* **113**, 507–520.

Angeles de la Torre-Ruiz, M., Torres, J., Arino, J. and Herrero, E. (2002) Sit4 is required for proper modulation of the biological functions mediated by Pkc1 and the cell integrity pathway in *Saccharomyces cerevisiae*. *Journal of Biological Chemistry* **277**, 33468–33476.

Antonsson, B., Montessuit, S., Friedli, L., Payton, M. A. and Paravicini, G. (1994) Protein kinase C in yeast: characteristics of the *Saccharomyces cerevisiae PKC1* gene product. *Journal of Biological Chemistry* **269**, 16821–16828.

Arndt, K. T., Styles, C. A. and Fink, G. R. (1989) A suppressor of a *HIS4* transcriptional defect encodes a protein with homology to the catalytic subunit of protein phosphatases. *Cell* **56**, 527–537.

Ashrafi, K., Lin, S. S., Manchester, J. K. and Gordon, J. I. (2000) Sip2p and its partner Snf1p kinase affect aging in *Saccharomyces cerevisiae*. *Genes and Development* **14**, 1872–1885.

Aspenstrom, P. (1999) Effectors for the Rho GTPases. *Current Opinion in Cell Biology* **11**, 95–102.

Audhya, A. and Emr, S. D. (2002) Stt4 PI 4-kinase localizes to the plasma membrane and functions in the Pkc1-mediated MAP kinase cascade. *Developmental Cell* **2**, 593–605.

Audhya, A., Foti, M. and Emr, S. D. (2000) Distinct roles for the yeast phosphatidylinositol 4-kinases, Stt4p and Pik1p, in secretion, cell growth, and organelle membrane dynamics. *Molecular Biology of the Cell* **11**, 2673–2689.

Bai, C., Sen, P., Hofmann, K., Ma, L., Goebl, M., Harper, J. W. *et al.* (1996) SKP1 connects cell cycle regulators to the ubiquitin proteolysis machinery through a novel motif, the F-box. *Cell* **86**, 263–274.

Bailis, J. M. and Roeder, G. S. (1998) Synaptonemal complex morphogenesis and sister-chromatid cohesion require Mek1-dependent phosphorylation of a meiotic chromosomal protein. *Genes and Development* **12**, 3551–3563.

Bailis, J. M. and Roeder, G. S. (2000) Pachytene exit controlled by reversal of Mek1-dependent phosphorylation. *Cell* **101**, 211–221.

Baker, S. H., Frederick, D. L., Bloecher, A. and Tatchell, K. (1997) Alanine-scanning mutagenesis of protein phosphatase type I in the yeast *Saccharomyces cerevisiae*. *Genetics* **145**, 615–626.

Barbaric, S., Munsterkotter, M., Svaren, J. and Horz, W. (1996) The homeodomain protein Pho2 and the basic-helix-loop-helix protein Pho4 bind DNA cooperatively at the yeast *PHO5* promoter. *Nucleic Acids Research* **24**, 4479–4486.

Barbet, N. C., Schneider, U., Helliwell, S. B., Stansfield, I., Tuite, M. F. and Hall, M. N. (1996) TOR controls translation initiation and early G1 progression in yeast. *Molecular Biology of the Cell* **7**, 25–42.

Baroni, M. D., Monti, P. and Alberghina, L. (1994) Repression of growth-regulated G1 cyclin expression by cyclic AMP in budding yeast. *Nature* **371**, 339–342.

Bastians, H. and Ponstingl, H. (1996) The novel human protein serine/threonine phosphatase 6 is a functional homologue of budding yeast Sit4p and fission yeast ppe1, which are involved in cell cycle regulation. *Journal of Cell Science* **109**, 2865–2874.

Beach, D. L., Thibodeaux, J., Maddox, P., Yeh, E. and Bloom, K. (2000) The role of the proteins Kar9 and Myo2 in orienting the mitotic spindle of budding yeast. *Current Biology* **10**, 1497–1506.

Beck, T. and Hall, M. N. (1999) The TOR signalling pathway controls nuclear localization of nutrient-regulated transcription factors. *Nature* **402**, 689–692.

Beck, T., Schmidt, A. and Hall, M. N. (1999) Starvation induces vacuolar targeting and degradation of the tryptophan permease in yeast. *Journal of Cell Biology* **146**, 1227–1238.

Becker, W., Kentrup, H., Klumpp, S., Schultz, J. E. and Joost, H. G. (1994) Molecular cloning of a protein serine/threonine phosphatase containing a putative regulatory tetratricopeptide repeat domain. *Journal of Biological Chemistry* **269**, 22586–22592.

Bender, A. and Sprague, G. F., Jr. (1986) Yeast peptide pheromones, **a**-factor and α-factor, activate a common response mechanism in their target cells. *Cell* **47**, 929–937.

Berset, C., Trachsel, H. and Altmann, M. (1998) The TOR (target of rapamycin) signal transduction pathway regulates the stability of translation initiation factor eIF4G in the yeast *Saccharomyces cerevisiae. Proceedings of the National Academy of Sciences USA* **95**, 4264–4269.

Bertram, P. G., Choi, J. H., Carvalho, J., Ai, W., Zeng, C., Chan, T. F. *et al.* (2000) Tripartite regulation of Gln3p by TOR, Ure2p, and phosphatases. *Journal of Biological Chemistry* **275**, 35727–35733.

Bertram, P. G., Choi, J. H., Carvalho, J., Chan, T. F., Ai, W. and Zheng, X. F. (2002) Convergence of TOR-nitrogen and Snf1-glucose signaling pathways onto Gln3. *Molecular and Cellular Biology* **22**, 1246–1252.

Bickle, M., Delley, P. A., Schmidt, A. and Hall, M. N. (1998) Cell wall integrity modulates RHO1 activity via the exchange factor ROM2. *EMBO Journal* **17**, 2235–2245.

Biggins, S., Severin, F. F., Bhalla, N., Sassoon, I., Hyman, A. A. and Murray, A. W. (1999) The conserved protein kinase Ipl1 regulates microtubule binding to kinetochores in budding yeast. *Genes and Development* **13**, 532–544.

Blacketer, M. J., Koehler, C. M., Coats, S. G., Myers, A. M. and Madaule, P. (1993) Regulation of dimorphism in *Saccharomyces cerevisiae*: involvement of the novel protein kinase homolog Elm1p and protein phosphatase 2A. *Molecular and Cellular Biology* **13**, 5567–5581.

Bloecher, A. and Tatchell, K. (1999) Defects in *Saccharomyces cerevisiae* protein phosphatase type I activate the spindle/kinetochore checkpoint. *Genes and Development* **13**, 517–522.

Blondel, M. and Mann, C. (1996) G2 cyclins are required for the degradation of G1 cyclins in yeast. *Nature* **384**, 279–282.

Blondel, M., Alepuz, P. M., Huang, L. S., Shaham, S., Ammerer, G. and Peter, M. (1999) Nuclear export of Far1p in response to pheromones requires the export receptor Msn5p/Ste21p. *Genes and Development* **13**, 2284–2300.

Boddy, M. N. and Russell, P. (2001) DNA replication checkpoint. *Current Biology* **11**, R953–R956.

Bollen, M. (2001) Combinatorial control of protein phosphatase-1. *Trends in Biochemical Sciences* **26**, 426–431.

Booher, R. N., Deshaies, R. J. and Kirschner, M. W. (1993) Properties of *Saccharomyces cerevisiae* wee1 and its differential regulation of $p34^{CDC28}$ in response to G1 and G2 cyclins. *EMBO Journal* **12**, 3417–3426.

Bosotti, R., Isacchi, A. and Sonnhammer, E. L. (2000) FAT: a novel domain in PIK-related kinases. *Trends in Biochemical Sciences* **25**, 225–227.

Breitkreutz, A. and Tyers, M. (2002) MAPK signaling specificity: it takes two to tango. *Trends in Cell Biology* **12**, 254–257.

Breitkreutz, A., Boucher, L. and Tyers, M. (2001) MAPK specificity in the yeast pheromone response independent of transcriptional activation. *Current Biology* **11**, 1266–1271.

Brewster, J. L., de Valoir, T., Dwyer, N. D., Winter, E. and Gustin, M. C. (1993) An osmosensing signal transduction pathway in yeast. *Science* **259**, 1760–1763.

Brewster, N. K., Johnston, G. C. and Singer, R. A. (2001) A bipartite yeast SSRP1 analog comprised of Pob3 and Nhp6 proteins modulates transcription. *Molecular and Cellular Biology* **21**, 3491–3502.

Broach, J. R. (1991) *RAS* genes in *Saccharomyces cerevisiae*: signal transduction in search of a pathway. *Trends in Genetics* **7**, 28–33.

Broach, J. R. and Deschenes, R. J. (1990) The function of *RAS* genes in *Saccharomyces cerevisiae. Advances in Cancer Research* **54**, 79–139.

Buehrer, B. M. and Errede, B. (1997) Coordination of the mating and cell integrity mitogen-activated protein kinase pathways in *Saccharomyces cerevisiae. Molecular and Cellular Biology* **17**, 6517–6525.

Bulawa, C. E. (1993) Genetics and molecular biology of chitin synthesis in fungi. *Annual Review of Microbiology* **47**, 505–534.

Burkholder, A. C. and Hartwell, L. H. (1985) The yeast alpha-factor receptor: structural properties deduced from the sequence of the *STE2* gene. *Nucleic Acids Research* **13**, 8463–8475.

Butty, A. C., Pryciak, P. M., Huang, L. S., Herskowitz, I. and Peter, M. (1998) The role of Far1p in linking the heterotrimeric G protein to polarity establishment proteins during yeast mating. *Science* **282**, 1511–1516.

Cairns, B. R., Ramer, S. W. and Kornberg, R. D. (1992) Order of action of components in the yeast pheromone response pathway revealed with a dominant allele of the *STE11* kinase and the multiple phosphorylation of the *STE7* kinase. *Genes and Development* **6**, 1305–1318.

Calzada, A., Sacristan, M., Sanchez, E. and Bueno, A. (2001) Cdc6 cooperates with Sic1 and Hct1 to inactivate mitotic cyclin-dependent kinases. *Nature* **412**, 355–358.

Cameron, S., Levin, L., Zoller, M. and Wigler, M. (1988) cAMP-independent control of sporulation, glycogen metabolism and heat shock resistance in *Saccharomyces cerevisiae*. *Cell* **53**, 555–566.

Cannon, J. F., Pringle, J. R., Fiechter, A. and Khalil, M. (1994) Characterization of glycogen-deficient *glc* mutants of *Saccharomyces cerevisiae*. *Genetics* **136**, 485–503.

Cappellaro, C., Hauser, K., Mrsa, V., Watzele, M., Watzele, G., Gruber, C. *et al.* (1991) *Saccharomyces cerevisiae* **a**- and α-agglutinin: characterisation of their molecular interaction. *EMBO Journal* **10**, 4081–4088.

Cardenas, M. E., Cutler, N. S., Lorenz, M. C., Di Como, C. J. and Heitman, J. (1999) The TOR signaling cascade regulates gene expression in response to nutrients. *Genes and Development* **13**, 3271–3279.

Carroll, A. S. and O'Shea, E. K. (2002) Pho85 and signaling environmental conditions. *Trends in Biochemical Sciences* **27**, 87–93.

Carroll, A. S., Bishop, A. C., DeRisi, J. L., Shokat, K. M. and O'Shea, E. K. (2001) Chemical inhibition of the Pho85 cyclin-dependent kinase reveals a role in the environmental stress response. *Proceedings of the National Academy of Sciences USA* **98**, 12578–12583.

Chai, B., Hsu, J. M., Du, J. and Laurent, B. C. (2002) Yeast RSC function is required for organization of the cellular cytoskeleton via an alternative *PKC1* pathway. *Genetics* **161**, 575–584.

Chan, C. S. M. and Botstein, D. (1993) Isolation and characterization of chromosome-gain and increase-in-ploidy mutants in yeast. *Genetics* **135**, 677–691.

Chang, F. and Herskowitz, I. (1990) Identification of a gene necessary for cell cycle arrest by a negative growth factor of yeast: FAR1 is an inhibitor of a G1 cyclin, CLN2. *Cell* **63**, 999–1011.

Cheeseman, I. M., Anderson, S., Jwa, M., Green, E. M., Kang, J., Yates, J. R., III *et al.* (2002) Phosphoregulation of kinetochore-microtubule attachments by the Aurora kinase Ipl1p. *Cell* **111**, 163–172.

Chen, Q. and Konopka, J. B. (1996) Regulation of the G-protein-coupled alpha-factor pheromone receptor by phosphorylation. *Molecular and Cellular Biology* **16**, 247–257.

Chen, T. and Kurjan, J. (1997) *Saccharomyces cerevisiae* Mpt5p interacts with Sst2p and plays roles in pheromone sensitivity and recovery from pheromone arrest. *Molecular and Cellular Biology* **17**, 3429–3439.

Chen, J., Peterson, R. T. and Schreiber, S. L. (1998) Alpha 4 associates with protein phosphatases 2A, 4, and 6. *Biochemical and Biophysical Research Communications* **247**, 827–832.

Cheng, C., Huang, D. Q. and Roach, P. J. (1997) Yeast *PIG* genes: *PIG1* encodes a putative type 1 phosphatase subunit that interacts with the yeast glycogen synthase Gsy2p. *Yeast* **13**, 1–8.

Cherry, J. R., Johnson, T. R., Dollard, C., Shuster, J. R. and Denis, C. L. (1989) Cyclic AMP-dependent protein kinase phosphorylates and inactivates the yeast transcriptional activator ADR1. *Cell* **56**, 409–419.

Cho, E. J., Kobor, M. S., Kim, M., Greenblatt, J. and Buratowski, S. (2001) Opposing effects of Ctk1 kinase and Fcp1 phosphatase at Ser 2 of the RNA polymerase II C-terminal domain. *Genes and Development* **15**, 3319–3329.

Choi, K. Y., Satterberg, B., Lyons, D. M. and Elion, E. A. (1994) Ste5 tethers multiple protein kinases in the MAP kinase cascade required for mating in *Saccharomyces cerevisiae*. *Cell* **78**, 499–512.

Cismowski, M. J., Laff, G. M., Solomon, M. J. and Reed, S. I. (1995) *KIN28* encodes a C-terminal domain kinase that controls mRNA transcription in *Saccharomyces cerevisiae* but lacks cyclin-dependent kinase-activating kinase (CAK) activity. *Molecular and Cellular Biology* **15**, 2983–2992.

Clarke, P. R., Hoffmann, I., Draetta, G. and Karsenti, E. (1993) Dephosphorylation of cdc25-C by a type-2A protein phosphatase-specific regulation during the cell cycle in *Xenopus* egg extracts. *Molecular Biology of the Cell* **4**, 397–411.

Clotet, J., Gari, E., Aldea, M. and Arino, J. (1999) The yeast ser/thr phosphatases Sit4 and Ppz1 play opposite roles in regulation of the cell cycle. *Molecular and Cellular Biology* **19**, 2408–2415.

Cohen, P. (1989) The structure and regulation of protein phosphatases. *Annual Review of Biochemistry* **58**, 453–508.

Cohen, P., Schelling, D. L. and Stark, M. J. R. (1989) Remarkable similarities between yeast and mammalian protein phosphatases. *FEBS Letters* **250**, 601–606.

Cohen, P. T. (2002) Protein phosphatase 1-targeted in many directions. *Journal of Cell Science* **115**, 241–256.

Collet, J. F., Stroobant, V., Pirard, M., Delpierre, G. and Van Schaftingen, E. (1998) A new class of phosphotransferases phosphorylated on an aspartate residue in an amino-terminal DXDX(T/V) motif. *Journal of Biological Chemistry* **273**, 14107–14112.

Collister, M., Didmon, M. P., MacIsaac, F., Stark, M. J., MacDonald, N. Q. and Keyse, S. M. (2002) YIL113w encodes a functional dual-specificity protein phosphatase which specifically interacts with and inactivates the Slt2/Mpk1p MAP kinase in *Saccharomyces cerevisiae*. *FEBS Letters* **527**, 186–192.

Colombo, S., Ma, P., Cauwenberg, L., Winderickx, J., Crauwels, M., Teunissen, A. *et al.* (1998) Involvement of distinct G-proteins, Gpa2 and Ras, in glucose- and intracellular acidification-induced cAMP signalling in the yeast *Saccharomyces cerevisiae*. *EMBO Journal* **17**, 3326–3341.

Cook, J. G., Bardwell, L., Kron, S. J. and Thorner, J. (1996) Two novel targets of the MAP kinase Kss1 are negative regulators of invasive growth in the yeast *Saccharomyces cerevisiae*. *Genes and Development* **10**, 2831–2848.

Cook, J. G., Bardwell, L. and Thorner, J. (1997) Inhibitory and activating functions for MAPK Kss1 in the *Saccharomyces cerevisiae* filamentous growth signalling pathway. *Nature* **390**, 85–88.

Cooper, T. G. (2002) Transmitting the signal of excess nitrogen in *Saccharomyces cerevisiae* from the Tor proteins to the GATA factors: connecting the dots. *FEMS Microbiological Reviews* **26**, 223–238.

Cosentino, G. P., Schmelzle, T., Haghighat, A., Helliwell, S. B., Hall, M. N. and Sonenberg, N. (2000) Eap1p, a novel eukaryotic translation initiation factor 4E-associated protein in *Saccharomyces cerevisiae*. *Molecular and Cellular Biology* **20**, 4604–4613.

Cosma, M. P., Panizza, S. and Nasmyth, K. (2001) Cdk1 triggers association of RNA polymerase to cell cycle promoters only after recruitment of the mediator by SBF. *Molecular Cell* **7**, 1213–1220.

Costigan, C., Kolodrubetz, D. and Snyder, M. (1994) *NHP6A* and *NHP6B*, which encode HMG1-like proteins, are candidates for downstream components of the yeast SLT2 mitogen-activated protein kinase pathway. *Molecular and Cellular Biology* **14**, 2391–2403.

Courchesne, W. E., Kunisawa, R. and Thorner, J. (1989) A putative protein kinase overcomes pheromone-induced arrest of cell cycling in *Saccharomyces cerevisiae*. *Cell* **58**, 1107–1119.

Creasy, C. L., Madden, S. L. and Bergman, L. W. (1993) Molecular analysis of the *PHO81* gene of *Saccharomyces cerevisiae*. *Nucleic Acids Research* **21**, 1975–1982.

Crespo, J. L. and Hall, M. N. (2002) Elucidating TOR signaling and rapamycin action: lessons from *Saccharomyces cerevisiae*. *Microbiology and Molecular Biology Reviews* **66**, 579–591.

Crespo, J. L., Powers, T., Fowler, B. and Hall, M. N. (2002) The TOR-controlled transcription activators GLN3, RTG1, and RTG3 are regulated in response to intracellular levels of glutamine. *Proceedings of the National Academy of Sciences USA* **99**, 6784–6789.

Cross, F. R. (1988) *DAF1* a mutant gene affecting size control, pheromone arrest, and cell cycle kinetics of *Saccharomyces cerevisiae*. *Molecular and Cellular Biology* **8**, 4675–4684.

Cross, F. R. (1995) Starting the cell cycle: what's the point? *Current Opinion in Cell Biology* **7**, 790–797.

Cullen, P. J. and Sprague, G. F., Jr. (2000) Glucose depletion causes haploid invasive growth in yeast. *Proceedings of the National Academy of Sciences USA* **97**, 13619–13624.

Cullen, P. J. and Sprague, G. F., Jr. (2002) The roles of bud-site-selection proteins during haploid invasive growth in yeast. *Molecular Biology of the Cell* **13**, 2990–3004.

Cullen, P. J., Schultz, J., Horecka, J., Stevenson, B. J., Jigami, Y. and Sprague, G. F., Jr. (2000) Defects in protein glycosylation cause SHO1-dependent activation of a STE12 signaling pathway in yeast. *Genetics* **155**, 1005–1018.

Cutler, N. S., Pan, X., Heitman, J. and Cardenas, M. E. (2001) The TOR signal transduction cascade controls cellular differentiation in response to nutrients. *Molecular Biology of the Cell* **12**, 4103–4113.

Cvrckova, F. and Nasmyth, K. (1993) Yeast G1 cyclins CLN1 and CLN2 and a GAP-like protein have a role in bud formation. *EMBO Journal* **12**, 5277–5286.

Cyert, M. S. (2001) Genetic analysis of calmodulin and its targets in *Saccharomyces cerevisiae. Annual Review of Genetics* **35**, 647–672.

Dahmann, C. and Futcher, B. (1995) Specialization of B-type cyclins for mitosis or meiosis in *S. cerevisiae. Genetics* **140**, 957–963.

Danaie, P., Altmann, M., Hall, M. N., Trachsel, H. and Helliwell, S. B. (1999) *CLN3* expression is sufficient to restore G1-to-S-phase progression in *Saccharomyces cerevisiae* mutants defective in translation initiation factor eIF4E. *Biochemical Journal* **340**, 135–141.

Davenport, K. D., Williams, K. E., Ullmann, B. D. and Gustin, M. C. (1999) Activation of the *Saccharomyces cerevisiae* filamentation/invasion pathway by osmotic stress in high-osmolarity glycogen pathway mutants. *Genetics* **153**, 1091–1103.

Davenport, K. R., Sohaskey, M., Kamada, Y., Levin, D. E. and Gustin, M. C. (1995) A second osmosensing signal transduction pathway in yeast – hypotonic shock activates the PKC1 protein kinase-regulated cell integrity pathway. *Journal of Biological Chemistry* **270**, 30157–30161.

De Craene, J. O., Soetens, O. and Andre, B. (2001) The Npr1 kinase controls biosynthetic and endocytic sorting of the yeast Gap1 permease. *Journal of Biological Chemistry* **276**, 43939–43948.

de Nadal, E., Clotet, J., Posas, F., Serrano, R., Gomez, N. and Arino, J. (1998) The yeast halotolerance determinant Hal3p is an inhibitory subunit of the Ppz1p Ser/Thr protein phosphatase. *Proceedings of the National Academy of Sciences USA* **95**, 7357–7362.

de Nadal, E., Alepuz, P. M. and Posas, F. (2002) Dealing with osmostress through MAP kinase activation. *EMBO Reports* **3**, 735–740.

Degols, G., Shiozaki, K. and Russell, P. (1996) Activation and regulation of the Spc1 stress-activated protein kinase in *Schizosaccharomyces pombe. Molecular and Cellular Biology* **16**, 2870–2877.

Delley, P. A. and Hall, M. N. (1999) Cell wall stress depolarizes cell growth via hyperactivation of RHO1. *Journal of Cell Biology* **147**, 163–174.

Denhaese, G. J., Walworth, N., Carr, A. M. and Gould, K. L. (1995) The wee1 protein kinase regulates T14 phosphorylation of fission yeast cdc2. *Molecular Biology of the Cell* **6**, 371–385.

Dent, P., Lavoinne, A., Nakielny, S., Caudwell, F. B., Watt, P. and Cohen, P. (1990) The molecular mechanism by which insulin stimulates glycogen synthesis in mammalian skeletal muscle. *Nature* **348**, 302–308.

Deshaies, R. J., Chau, V. and Kirschner, M. (1995) Ubiquitination of the G1 cyclin Cln2p by a Cdc34p-dependent pathway. *EMBO Journal* **14**, 303–312.

Deshaies, R. J. and Ferrell, J. E., Jr. (2001) Multisite phosphorylation and the countdown to S phase. *Cell* **107**, 819–822.

Desrivieres, S., Cooke, F. T., Parker, P. J. and Hall, M. N. (1998) MSS4, a phosphatidylinositol-4-phosphate 5-kinase required for organization of the actin cytoskeleton in *Saccharomyces cerevisiae. Journal of Biological Chemistry* **273**, 15787–15793.

Dever, T. E., Feng, L., Wek, R. C., Cigan, A. M., Donahue, T. F. and Hinnebusch, A. G. (1992) Phosphorylation of initiation factor-2α by protein kinase GCN2 mediates gene-specific translational control of GCN4 in yeast. *Cell* **68**, 585–596.

DeVit, M. J. and Johnston, M. (1999) The nuclear exportin Msn5 is required for nuclear export of the Mig1 glucose repressor of *Saccharomyces cerevisiae*. *Current Biology* **9**, 1231–1241.

DeVit, M. J., Waddle, J. A. and Johnston, M. (1997) Regulated nuclear translocation of the Mig1 glucose repressor. *Molecular Biology of the Cell* **8**, 1603–1618.

Di Como, C. J. and Arndt, K. T. (1996) Nutrients, via the Tor proteins, stimulate the association of Tap42 with type 2A phosphatases. *Genes and Development* **10**, 1904–1916.

Di Como, C. J., Bose, R. and Arndt, K. T. (1995) Overexpression of *SIS2* which contains an extremely acidic region, increases the expression of *SWI4 CLN1* and *CLN2* in *sit4* mutants. *Genetics* **139**, 95–107.

Dietzel, C. and Kurjan, J. (1987) The yeast *SCG1* gene: a G_α;-like protein implicated in the a- and alpha-factor response pathway. *Cell* **50**, 1001–1010.

Diffley, J. F. and Labib, K. (2002) The chromosome replication cycle. *Journal of Cell Science* **115**, 869–872.

Dilova, I., Chen, C. Y. and Powers, T. (2002) Mks1 in concert with TOR signaling negatively regulates RTG target gene expression in *Saccharomyces cerevisiae*. *Current Biology* **12**, 389–395.

Dirick, L. and Nasmyth, K. (1991) Positive feedback in the activation of G1 cyclins in yeast. *Nature* **351**, 754–757.

Dirick, L., Bohm, T. and Nasmyth, K. (1995) Roles and regulation of Cln-Cdc28 kinases at the start of the cell cycle of *Saccharomyces cerevisiae*. *EMBO Journal* **14**, 4803–4813.

Dodou, E. and Treisman, R. (1997) The *Saccharomyces cerevisiae* MADS-box transcription factor Rlm1 is a target for the Mpk1 mitogen-activated protein kinase pathway. *Molecular and Cellular Biology* **17**, 1848–1859.

Doherty, M. J., Moorhead, G., Morrice, N., Cohen, P. and Cohen, P. T. W. (1995) Amino acid sequence and expression of the hepatic glycogen-binding (G_L)-subunit of protein phosphatase-1. *FEBS Letters* **375**, 294–298.

Dohlman, H. G., Song, J., Ma, D., Courchesne, W. E. and Thorner, J. (1996) Sst2, a negative regulator of pheromone signaling in the yeast *Saccharomyces cerevisiae*: expression, localization, and genetic interaction and physical association with Gpa1 (the G-protein alpha subunit). *Molecular and Cellular Biology* **16**, 5194–5209.

Dolan, J. W. and Fields, S. (1990) Overproduction of the yeast STE12 protein leads to constitutive transcriptional induction. *Genes and Development* **4**, 492–502.

Dolinski, K., Balakrishnan, R., Christie, K. R., Costanzo, M. C., Dwight, S. S., Engel, S. R. *et al.* *Saccharomyces Genome Database*, Online. Available HTTP: <http://genome-www.stanford.edu/Saccharomyces/> (accessed 28/11/2002).

Donaldson, A. D. (2000) The yeast mitotic cyclin Clb2 cannot substitute for S phase cyclins in replication origin firing. *EMBO Reports* **1**, 507–512.

Drgonova, J., Drgon, T., Tanaka, K., Kollar, R., Chen, G. C., Ford, R. A. *et al.* (1996) Rho1p, a yeast protein at the interface between cell polarization and morphogenesis. *Science* **272**, 277–279.

Edgington, N. P. and Futcher, B. (2001) Relationship between the function and the location of G1 cyclins in *Saccharomyces cerevisiae*. *Journal of Cell Science* **114**, 4599–4611.

Egloff, M. P., Johnson, D. F., Moorhead, G., Cohen, P. T., Cohen, P. and Barford, D. (1997) Structural basis for the recognition of regulatory subunits by the catalytic subunit of protein phosphatase 1. *EMBO Journal* **16**, 1876–1887.

Elion, E. A., Grisafi, P. L. and Fink, G. R. (1990) *FUS3* encodes a $cdc2^+$/CDC28-related kinase required for the transition from mitosis into conjugation. *Cell* **60**, 649–664.

Elion, E., Brill, J. A. and Fink, G. R. (1991a) Functional redundancy in the yeast cell cycle: *FUS3* and *KSS1* have both overlapping and unique functions. *Cold Spring Harbor Symposium on Quantitative Biology* **56**, 41–49.

Elion, E. A., Brill, J. A. and Fink, G. R. (1991b) FUS3 represses *CLN1* and *CLN2* and in concert with KSS1 promotes signal transduction. *Proceedings of the National Academy of Sciences USA* **88**, 9392–9396.

Elion, E. A., Trueheart, J. and Fink, G. R. (1995) Fus2 localizes near the site of cell fusion and is required for both cell fusion and nuclear alignment during zygote formation. *Journal of Cell Biology* **130**, 1283–1296.

Erdman, S. and Snyder, M. (2001) A filamentous growth response mediated by the yeast mating pathway. *Genetics* **159**, 919–928.

Erickson, J. R. and Johnston, M. (1993) Genetic and molecular characterization of *GAL83*: its interaction and similarities with other genes involved in glucose repression in *Saccharomyces cerevisiae. Genetics* **135**, 655–664.

Errede, B. and Ammerer, G. (1989) STE12, a protein involved in cell-type-specific transcription and signal transduction in yeast, is part of protein-DNA complexes. *Genes and Development* **3**, 1349–1361.

Errede, B., Gartner, A., Zhou, Z. Q., Nasmyth, K. and Ammerer, G. (1993) MAP kinase-related FUS3 from *Saccharomyces cerevisiae* is activated by STE7 *in vitro. Nature* **362**, 261–264.

Espinoza, F. H., Ogas, J., Herskowitz, I. and Morgan, D. O. (1994) Cell cycle control by a complex of the cyclin HCS26 (PCL1) and the kinase PHO85. *Science* **266**, 1388–1391.

Espinoza, F. H., Farrell, A., Nourse, J. L., Chamberlin, H. M., Gileadi, O. and Morgan, D. O. (1998) Cak1 is required for Kin28 phosphorylation and activation *in vivo. Molecular and Cellular Biology* **18**, 6365–6373.

Estruch, F., Treitel, M. A., Yang, X. and Carlson, M. (1992) N-terminal mutations modulate yeast SNF1 protein kinase function. *Genetics* **132**, 639–650.

Evangelista, C. C., Torres, A. M. R., Limbach, M. P. and Zitomer, R. S. (1996) *ROX3* and *RTS1* function in the global stress response pathway in baker's yeast. *Genetics* **142**, 1083–1093.

Evangelista, M., Blundell, K., Longtine, M. S., Chow, C. J., Adames, N., Pringle, J. R. *et al.* (1997) Bni1p, a yeast formin linking Cdc42p and the actin cytoskeleton during polarized morphogenesis. *Science* **276**, 118–122.

Evans, D. R. and Hemmings, B. A. (2000) Mutation of the C-terminal leucine residue of PP2Ac inhibits PR55/B subunit binding and confers supersensitivity to microtubule destabilization in *Saccharomyces cerevisiae. Molecular and General Genetics* **264**, 425–432.

Evans, D. R. H. and Stark, M. J. R. (1997) Mutations in the *Saccharomyces cerevisiae* type 2A protein phosphatase catalytic subunit reveal roles in cell wall integrity, actin cytoskeleton organization and mitosis. *Genetics* **145**, 227–241.

Fambrough, D., McClure, K., Kazlauskas, A. and Lander, E. S. (1999) Diverse signaling pathways activated by growth factor receptors induce broadly overlapping, rather than independent, sets of genes. *Cell* **97**, 727–741.

Farkas, I., Hardy, T. A., Goebl, M. G. and Roach, P. J. (1991) Two glycogen synthase isoforms in *Saccharomyces cerevisiae* are coded by distinct genes that are differentially controlled. *Journal of Biological Chemistry* **266**, 15602–15607.

Fassler, J. S., Gray, W. M., Malone, C. L., Tao, W., Lin, H. and Deschenes, R. J. (1997) Activated alleles of yeast *SLN1* increase Mcm1-dependent reporter gene expression and diminish signaling through the Hog1 osmosensing pathway. *Journal of Biological Chemistry* **272**, 13365–13371.

Feaver, W. J., Svejstrup, J. Q., Henry, N. L. and Kornberg, R. D. (1994) Relationship of CDK-activating kinase and RNA polymerase II CTD kinase TFIIH/TFIIK. *Cell* **79**, 1103–1109.

Fedor-Chaiken, M., Deschenes, R. J. and Broach, J. R. (1990) *SRV2* a gene required for RAS activation of adenylate cyclase in yeast. *Cell* **61**, 329–340.

Feldman, R. M. R., Correll, C. C., Kaplan, K. B. and Deshaies, R. J. (1997) A complex of Cdc4p, Skp1p, and Cdc53p/cullin catalyzes ubiquitination of the phosphorylated CDK inhibitor Sic1p. *Cell* **91**, 221–230.

Fernandez-Sarabia, M. J., Sutton, A., Zhong, T. and Arndt, K. T. (1992) SIT4 protein phosphatase is required for the normal accumulation of *SWI4 CLN1 CLN2*, and *HCS26* RNAs during late G_1. *Genes and Development* **6**, 2417–2428.

Ferrell, J. E. (1996) Tripping the switch fantastic: how a protein kinase cascade can convert graded inputs into switch-like outputs. *Trends in Biochemical Sciences* **21**, 460–466.

Ficarro, S. B., McCleland, M. L., Stukenberg, P. T., Burke, D. J., Ross, M. M., Shabanowitz, J. *et al.* (2002) Phosphoproteome analysis by mass spectrometry and its application to *Saccharomyces cerevisiae. Nature Biotechnology* **20**, 301–305.

Fitch, I., Dahmann, C., Surana, U., Amon, A., Nasmyth, K., Goetsch, L. *et al.* (1992) Characterization of four B-type cyclin genes of the budding yeast *Saccharomyces cerevisiae. Molecular Biology of the Cell* **3**, 805–818.

Flynn, P., Mellor, H., Casamassima, A. and Parker, P. J. (2000) Rho GTPase control of protein kinase C-related protein kinase activation by 3-phosphoinositide-dependent protein kinase. *Journal of Biological Chemistry* **275**, 11064–11070.

Foor, F., Parent, S. A., Morin, N., Dahl, A. M., Ramadan, N., Chrebet, G. *et al.* (1992) Calcineurin mediates inhibition by FK506 and cyclosporin of recovery from alpha-factor arrest in yeast. *Nature* **360**, 682–684.

Formosa, T., Eriksson, P., Wittmeyer, J., Ginn, J., Yu, Y. and Stillman, D. J. (2001) Spt16-Pob3 and the HMG protein Nhp6 combine to form the nucleosome-binding factor SPN. *EMBO Journal* **20**, 3506–3517.

Forsburg, S. L. and Nurse, P. (1991) Cell cycle regulation in the yeasts *Saccharomyces cerevisiae* and *Schizosaccharomyces pombe. Annual Review of Cell Biology* **7**, 227–256.

Francisco, L., Wang, W. F. and Chan, C. S. M. (1994) Type 1 protein phosphatase acts in opposition to Ipl1 protein kinase in regulating yeast chromosome segregation. *Molecular and Cellular Biology* **14**, 4731–4740.

François, J., Van Schaftigen, E. and Hers, H. G. (1988) Characterization of phosphofructokinase 2 and of enzymes involved in the degradation of fructose 2,6-bisphosphate in yeast. *European Journal of Biochemistry* **171**, 599–608.

François, J. M., Thompson-Jaeger, S., Skroch, J., Zellenka, U., Spevak, W. and Tatchell, K. (1992) *GAC1* may encode a regulatory subunit for protein phosphatase type 1 in *Saccharomyces cerevisiae. EMBO Journal* **11**, 87–96.

Frederick, D. L. and Tatchell, K. (1996) The *REG2* gene of *Saccharomyces cerevisiae* encodes a type 1 protein phosphatase-binding protein that functions with Reg1p and the Snf1 protein kinase to regulate growth. *Molecular and Cellular Biology* **16**, 2922–2931.

Friesen, H., Lunz, R., Doyle, S. and Segall, J. (1994) Mutation of the SPS1-encoded protein kinase of *Saccharomyces cerevisiae* leads to defects in transcription and morphology during spore formation. *Genes and Development* **8**, 2162–2175.

Frohloff, F., Fichtner, L., Jablonowski, D., Breunig, K. D. and Schaffrath, R. (2001) *Saccharomyces cerevisiae* Elongator mutations confer resistance to the *Kluyveromyces lactis* zymocin. *EMBO Journal* **20**, 1993–2003.

Fujiwara, T., Tanaka, K., Mino, A., Kikyo, M., Takahashi, K., Shimizu, K. *et al.* (1998) Rho1p-Bni1p-Spa2p interactions: implication in localization of Bni1p at the bud site and regulation of the actin cytoskeleton in *Saccharomyces cerevisiae. Molecular Biology of the Cell* **9**, 1221–1233.

Futey, L. M., Medley, Q. G., Cote, G. P. and Egelhoff, T. T. (1995) Structural analysis of myosin heavy chain kinase A from *Dictyostelium*. Evidence for a highly divergent protein kinase domain, an amino-terminal coiled-coil domain, and a domain homologous to the β-subunit of heterotrimeric G proteins. *Journal of Biological Chemistry* **270**, 523–529.

Gallego, C., Gari, E., Colomina, N., Herrero, E. and Aldea, M. (1997) The Cln3 cyclin is down-regulated by translational repression and degradation during the G1 arrest caused by nitrogen deprivation in budding yeast. *EMBO Journal* **16**, 7196–7206.

Garrison, T. R., Zhang, Y., Pausch, M., Apanovitch, D., Aebersold, R. and Dohlman, H. G. (1999) Feedback phosphorylation of an RGS protein by MAP kinase in yeast. *Journal of Biological Chemistry* **274**, 36387–36391.

Gartner, A., Nasmyth, K. and Ammerer, G. (1992) Signal transduction in *Saccharomyces cerevisiae* requires tyrosine and threonine phosphorylation of FUS3 and KSS1. *Genes and Development* **6**, 1280–1292.

Gartner, A., Jovanovic, A., Jeoung, D. I., Bourlat, S., Cross, F. R. and Ammerer, G. (1998) Pheromone-dependent G1 cell cycle arrest requires Far1 phosphorylation, but may not involve inhibition of Cdc28-Cln2 kinase, in vivo. *Molecular and Cellular Biology* **18**, 3681–3691.

Gavrias, V., Andrianopoulos, A., Gimeno, C. J. and Timberlake, W. E. (1996) *Saccharomyces cerevisiae TEC1* is required for pseudohyphal growth. *Molecular Microbiology* **19**, 1255–1263.

Gehrung, S. and Snyder, M. (1990) The *SPA2* gene of *Saccharomyces cerevisiae* is important for pheromone-induced morphogenesis and efficient mating. *Journal of Cell Biology* **111**, 1451–1464.

Gentry, M. S. and Hallberg, R. L. (2002) Localization of *Saccharomyces cerevisiae* protein phosphatase 2A subunits throughout mitotic cell cycle. *Molecular Biology of the Cell* **13**, 3477–3492.

Geymonat, M., Jensen, S. and Johnston, L. H. (2002) Mitotic exit: the Cdc14 double cross. *Current Biology* **12**, R482–R484.

Gimeno, C. J., Ljungdahl, P. O., Styles, C. A. and Fink, G. R. (1992) Unipolar cell divisions in the yeast *Saccharomyces cerevisiae* lead to filamentous growth: regulation by starvation and RAS. *Cell* **68**, 1077–1090.

Golsteyn, R. M., Lane, H. A., Mundt, K. E., Arnaud, L. and Nigg, E. A. (1996) The family of polo-like kinases. *Progress in Cell Cycle Research* **2**, 107–114.

Görner, W., Durchschlag, E., Martinez-Pastor, M. T., Estruch, F., Ammerer, G., Hamilton, B. *et al.* (1998) Nuclear localization of the C_2H_2 zinc finger protein Msn2p is regulated by stress and protein kinase A activity. *Genes and Development* **12**, 586–597.

Görner, W., Durchschlag, E., Wolf, J., Brown, E. L., Ammerer, G., Ruis, H. *et al.* (2002) Acute glucose starvation activates the nuclear localization signal of a stress-specific yeast transcription factor. *EMBO Journal* **21**, 135–144.

Gray, J. V., Ogas, J. P., Kamada, Y., Stone, M., Levin, D. E. and Herskowitz, I. (1997) A role for the Pkc1 MAP kinase pathway of *Saccharomyces cerevisiae* in bud emergence and identification of a putative upstream regulator. *EMBO Journal* **16**, 4924–4937.

Grishin, A. V., Weiner, J. L. and Blumer, K. J. (1994) Control of adaptation to mating pheromone by G protein beta subunits of *Saccharomyces cerevisiae*. *Genetics* **138**, 1081–1092.

Groves, M. R., Hanlon, N., Turowski, P., Hemmings, B. A. and Barford, D. (1999) The structure of the protein phosphatase 2A PR65/A subunit reveals the conformation of its 15 tandemly repeated HEAT motifs. *Cell* **96**, 99–110.

Gustin, M. C., Albertyn, J., Alexander, M. and Davenport, K. (1998) MAP kinase pathways in the yeast *Saccharomyces cerevisiae*. *Microbiology and Molecular Biology Reviews* **62**, 1264–1300.

Hagen, D. C., McCaffrey, G. and Sprague, G. F., Jr. (1986) Evidence the yeast *STE3* gene encodes a receptor for the peptide pheromone a factor: gene sequence and implications for the structure of the presumed receptor. *Proceedings of the National Academy of Sciences USA* **83**, 1418–1422.

Hagen, D. C., McCaffrey, G. and Sprague, G. F., Jr. (1991) Pheromone response elements are necessary and sufficient for basal and pheromone-induced transcription of the *FUS1* gene of *Saccharomyces cerevisiae*. *Molecular and Cellular Biology* **11**, 2952–2961.

Hahn, J. S. and Thiele, D. J. (2002) Regulation of the *Saccharomyces cerevisiae* Slt2 kinase pathway by the stress-inducible Sdp1 dual specificity phosphatase. *Journal of Biological Chemistry* **277**, 21278–21284.

Hall, D. D., Markwardt, D. D., Parviz, F. and Heideman, W. (1998) Regulation of the Cln3-Cdc28 kinase by cAMP in *Saccharomyces cerevisiae*. *EMBO Journal* **17**, 4370–4378.

Hall, J. P., Cherkasova, V., Elion, E., Gustin, M. C. and Winter, E. (1996) The osmoregulatory pathway represses mating pathway activity in *Saccharomyces cerevisiae*: isolation of a *FUS3* mutant that is insensitive to the repression mechanism. *Molecular and Cellular Biology* **16**, 6715–6723.

Hanks, S. K., Quinn, A. M. and Hunter, T. (1988) The protein kinase family: conserved features and deduced phylogeny of the catalytic domain. *Science* **241**, 42–52.

Hannon, G. J., Casso, D. and Beach, D. (1994) KAP: a dual specificity phosphatase that interacts with cyclin-dependent kinases. *Proceedings of the National Academy of Sciences USA* **91**, 1731–1735.

Hardie, D. G. and Carling, D. (1997) The AMP-activated protein kinase: fuel gauge of the mammalian cell? *European Journal of Biochemistry* **246**, 259–273.

Hardie, D. G., Carling, D. G. and Carlson, M. (1998) The AMP-activated/SNF1 protein kinase subfamily: metabolic sensors of the eukaryotic cell? *Annual Review of Biochemistry* **67**, 821–855.

Hardwick, J. S., Kuruvilla, F. G., Tong, J. K., Shamji, A. F. and Schreiber, S. L. (1999) Rapamycin-modulated transcription defines the subset of nutrient-sensitive signaling pathways directly controlled by the Tor proteins. *Proceedings of the National Academy of Sciences USA* **96**, 14866–14870.

Hardy, T. A. and Roach, P. J. (1993) Control of yeast glycogen synthase 2 by COOH-terminal phosphorylation. *Journal of Biological Chemistry* **268**, 23799–23805.

Hardy, T. A., Huang, D. Q. and Roach, P. J. (1994) Interactions between cAMP-dependent and SNF1 protein kinases in the control of glycogen accumulation in *Saccharomyces cerevisiae*. *Journal of Biological Chemistry* **269**, 27907–27913.

Harper, J. W. (2001) Protein destruction: adapting roles for Cks proteins. *Current Biology* **11**, R431–R435.

Harrison, J. C., Bardes, E. S., Ohya, Y. and Lew, D. J. (2001) A role for the Pkc1p/Mpk1p kinase cascade in the morphogenesis checkpoint. *Nature Cell Biology* **3**, 417–420.

Healy, A. M., Zolnierowicz, S., Stapleton, A. E., Goebl, M., DePaoli-Roach, A. A. and Pringle, J. R. (1991) *CDC55* a *Saccharomyces cerevisiae* gene involved in cellular morphogenesis: identification, characterization, and homology to the B subunit of mammalian type 2A protein phosphatase. *Molecular and Cellular Biology* **11**, 5767–5780.

Heinisch, J. J., Lorberg, A., Schmitz, H. P. and Jacoby, J. J. (1999) The protein kinase C-mediated MAP kinase pathway involved in the maintenance of cellular integrity in *Saccharomyces cerevisiae*. *Molecular Microbiology* **32**, 671–680.

Heitman, J., Movva, N. R., Hiestand, P. C. and Hall, M. N. (1991) FK 506-binding protein proline rotamase is a target for the immunosuppressive agent FK 506 in *Saccharomyces cerevisiae*. *Proceedings of the National Academy of Sciences USA* **88**, 1948–1952.

Helliwell, S. B., Wagner, P., Kunz, J., Deuter-Reinhard, M., Henriquez, R. and Hall, M. N. (1994) TOR1 and TOR2 are structurally and functionally similar but not identical phosphatidylinositol kinase homologues in yeast. *Molecular Biology of the Cell* **5**, 105–118.

Helliwell, S. B., Schmidt, A., Ohya, Y. and Hall, M. N. (1998) The Rho1 effector Pkc1, but not Bni1, mediates signalling from Tor2 to the actin cytoskeleton. *Current Biology* **8**, 1211–1214.

Hengartner, C. J., Myer, V. E., Liao, S. M., Wilson, C. J., Koh, S. S. and Young, R. A. (1998) Temporal regulation of RNA polymerase II by Srb10 and Kin28 cyclin-dependent kinases. *Molecular Cell* **2**, 43–53.

Hershko, A. and Ciechanover, A. (1992) The ubiquitin system for protein degradation. *Annual Review of Biochemistry* **61**, 761–807.

Herskowitz, I. (1995) MAP kinase pathways in yeast: for mating and more. *Cell* **80**, 187–197.

Higgins, D. G. and Sharp, P. M. (1988) CLUSTAL: a package for performing multiple sequence alignment on a microcomputer. *Gene* **73**, 237–244.

Hinnebusch, A. G. (1992) General and pathway specific regulatory mechanisms controlling the synthesis of amino acid biosynthetic enzymes in *Saccharomyces cerevisiae*. In: E. W. Jones, J. R. Pringle and J. R. Broach (eds) *The Molecular and Cellular Biology of the Yeast Saccharomyces*, Vol. 2, pp. 319–414. New York: Cold Spring Harbor Laboratory Press.

Hirst, K., Fisher, F., McAndrew, P. C. and Goding, C. R. (1994) The transcription factor, the Cdk, its cyclin and their regulator: directing the transcriptional response to a nutritional signal. *EMBO Journal* **13**, 5410–5420.

Hisamoto, N., Sugimoto, K. and Matsumoto, K. (1994) The *GLC7* type 1 protein phosphatase of *Saccharomyces cerevisiae* is required for cell cycle progression in G2/M. *Molecular and Cellular Biology* **14**, 3158–3165.

Hisamoto, N., Frederick, D. L., Sugimoto, K., Tatchell, K. and Matsumoto, K. (1995) The *EGP1* gene may be a positive regulator of protein phosphatase type 1 in the growth control of *Saccharomyces cerevisiae*. *Molecular and Cellular Biology* **15**, 3767–3776.

Hoekstra, M. F., Dhillon, N., Carmel, G., DeMaggio, A. J., Lindberg, R. A., Hunter, T. *et al.* (1994) Budding and fission yeast casein kinase I isoforms have dual-specificity protein kinase activity. *Molecular Biology of the Cell* **5**, 877–886.

Hoffman, G. A., Garrison, T. R. and Dohlman, H. G. (2000) Endoproteolytic processing of Sst2, a multidomain regulator of G protein signaling in yeast. *Journal of Biological Chemistry* **275**, 37533–37541.

Hoffman, G. R. and Cerione, R. A. (2000) Flipping the switch: the structural basis for signaling through the CRIB motif. *Cell* **102**, 403–406.

Hofken, T. and Schiebel, E. (2002) A role for cell polarity proteins in mitotic exit. *EMBO Journal* **21**, 4851–4862.

Hsu, J. Y., Sun, Z. W., Li, X., Reuben, M., Tatchell, K., Bishop, D. K. *et al.* (2000) Mitotic phosphorylation of histone H3 is governed by Ipl1/aurora kinase and Glc7/PP1 phosphatase in budding yeast and nematodes. *Cell* **102**, 279–291.

Huang, D., Farkas, I. and Roach, P. J. (1996a) Pho85p, a cyclin-dependent protein kinase, and the Snf1p protein kinase act antagonistically to control glycogen accumulation in *Saccharomyces cerevisiae. Molecular and Cellular Biology* **16**, 4357–4365.

Huang, D. Q., Chun, K. T., Goebl, M. G. and Roach, P. J. (1996b) Genetic interactions between *REG1/HEX2* and *GLC7* the gene encoding the protein phosphatase type 1 catalytic subunit in *Saccharomyces cerevisiae. Genetics* **143**, 119–127.

Huang, D. Q., Wilson, W. A. and Roach, P. J. (1997) Glucose-6-P control of glycogen synthase phosphorylation in yeast. *Journal of Biological Chemistry* **272**, 22495–22501.

Huang, D., Moffat, J., Wilson, W. A., Moore, L., Cheng, C., Roach, P. J. *et al.* (1998) Cyclin partners determine Pho85 protein kinase substrate specificity in vitro and in vivo: control of glycogen biosynthesis by Pcl8 and Pcl10. *Molecular and Cellular Biology* **18**, 3289–3299.

Huang, D., Moffat, J. and Andrews, B. (2002) Dissection of a complex phenotype by functional genomics reveals roles for the yeast cyclin-dependent protein kinase Pho85 in stress adaptation and cell integrity. *Molecular and Cellular Biology* **22**, 5076–5088.

Huang, K. N. and Symington, L. S. (1995) Suppressors of a *Saccharomyces cerevisiae pkc1* mutation identify alleles of the phosphatase gene *PTC1* and of a novel gene encoding a putative basic leucine zipper protein. *Genetics* **141**, 1275–1285.

Huang, S., Jeffery, D. A., Anthony, M. D. and O'Shea, E. K. (2001) Functional analysis of the cyclin-dependent kinase inhibitor Pho81 identifies a novel inhibitory domain. *Molecular and Cellular Biology* **21**, 6695–6705.

Hubbard, M. J. and Cohen, P. (1989) Regulation of protein phosphatase-1G from rabbit skeletal muscle. 2. Catalytic subunit translocation is a mechanism for reversible inhibition of activity towards glycogen-bound substrates. *European Journal of Biochemistry* **186**, 711–716.

Hubbard, M. J. and Cohen, P. (1993) On target with a new mechanism for the regulation of protein phosphorylation. *Trends in Biochemical Sciences* **18**, 172–177.

Hunter, T. (1995) When is a lipid kinase not a lipid kinase? When it is a protein kinase. *Cell* **83**, 1–4.

Hunter, T. and Plowman, G. D. (1997) The protein kinases of budding yeast: six score and more. *Trends in Biochemical Sciences* **22**, 18–22.

Huse, M. and Kuriyan, J. (2002) The conformational plasticity of protein kinases. *Cell* **109**, 275–282.

Hwang-Shum, J. J., Hagen, D. C., Jarvis, E. E., Westby, C. A. and Sprague, G. F. J. (1991) Relative contributions of MCM1 and STE12 to transcriptional activation of **a**- and α-specific genes from *Saccharomyces cerevisiae. Molecular and General Genetics* **227**, 197–204.

Igual, J. C., Johnson, A. L. and Johnston, L. H. (1996) Coordinated regulation of gene expression by the cell cycle transcription factor SWI4 and the protein kinase C MAP kinase pathway for yeast cell integrity. *EMBO Journal* **15**, 5001–5013.

Iida, H., Yagawa, Y. and Anraku, Y. (1990) Essential role for induced Ca^{2+} influx followed by $[Ca^{2+}]_i$ rise in maintaining viability of yeast cells late in the mating pheromone response pathway. a study of $[Ca^{2+}]_i$ in single *Saccharomyces cerevisiae* cells with imaging of fura-2. *Journal of Biological Chemistry* **265**, 13391–13399.

Inagaki, M., Schmelzle, T., Yamaguchi, K., Irie, K., Hall, M. N. and Matsumoto, K. (1999) PDK1 homologs activate the Pkc1-mitogen-activated protein kinase pathway in yeast. *Molecular and Cellular Biology* **19**, 8344–8352.

Irie, K., Takase, M., Lee, K. S., Levin, D. E., Araki, H., Matsumoto, K. *et al.* (1993) MKK1 and MKK2, which encode *Saccharomyces cerevisiae* mitogen-activated protein kinase-kinase homologs, function in the pathway mediated by protein kinase C. *Molecular and Cellular Biology* **13**, 3076–3083.

Ishii, K., Kumada, K., Toda, T. and Yanagida, M. (1996) Requirement for PP1 phosphatase and 20S cyclosome/APC for the onset of anaphase is lessened by the dosage increase of a novel gene *sds23*$^{+}$. *EMBO Journal* **15**, 6629–6640.

Jablonowski, D., Butler, A. R., Fichtner, L., Gardiner, D., Schaffrath, R. and Stark, M. J. (2001a) Sit4p protein phosphatase is required for sensitivity of *Saccharomyces cerevisiae* to *Kluyveromyces lactis* zymocin. *Genetics* **159**, 1479–1489.

Jablonowski, D., Frohloff, F., Fichtner, L., Stark, M. J. and Schaffrath, R. (2001b) *Kluyveromyces lactis* zymocin mode of action is linked to RNA polymerase II function via Elongator. *Molecular Microbiology* **42**, 1095–1105.

Jacinto, E., Guo, B., Arndt, K. T., Schmelzle, T. and Hall, M. N. (2001) TIP41 interacts with TAP42 and negatively regulates the TOR signaling pathway. *Molecular Cell* **8**, 1017–1026.

Jacoby, J. J., Schmitz, H. P. and Heinisch, J. J. (1997a) Mutants affected in the putative diacylglycerol binding site of yeast protein kinase C. *FEBS Letters* **417**, 219–222.

Jacoby, T., Flanagan, H., Faykin, A., Seto, A. G., Mattison, C. and Ota, I. (1997b) Two protein-tyrosine phosphatases inactivate the osmotic stress response pathway in yeast by targeting the mitogen-activated protein kinase, Hog1. *Journal of Biological Chemistry* **272**, 17749–17755.

Jans, D. A., Moll, T., Nasmyth, K. and Jans, P. (1995) Cyclin-dependent kinase site-regulated signal-dependent nuclear localization of the SW15 yeast transcription factor in mammalian cells. *Journal of Biological Chemistry* **270**, 17064–17067.

Jarvis, E. E., Clark, K. L. and Sprague, G. J. (1989) The yeast transcription activator PRTF, a homolog of the mammalian serum response factor, is encoded by the *MCM1* gene. *Genes and Development* **3**, 936–945.

Jaspersen, S. L., Charles, J. F. and Morgan, D. O. (1999) Inhibitory phosphorylation of the APC regulator Hct1 is controlled by the kinase Cdc28 and the phosphatase Cdc14. *Current Biology* **9**, 227–236.

Jefferies, H. B., Fumagalli, S., Dennis, P. B., Reinhard, C., Pearson, R. B. and Thomas, G. (1997) Rapamycin suppresses 5′TOP mRNA translation through inhibition of p70^{s6k}. *EMBO Journal* **16**, 3693–3704.

Jeffrey, P. D., Ruso, A. A., Polyak, K., Gibbs, E., Hurwitz, J., Massague, J. *et al.* (1995) Mechanism of CDK activation revealed by the structure of a cyclinA-CDK2 complex. *Nature* **376**, 313–320.

Jensen, S., Geymonat, M., Johnson, A. L., Segal, M. and Johnston, L. H. (2002) Spatial regulation of the guanine nucleotide exchange factor Lte1 in *Saccharomyces cerevisiae. Journal of Cell Science* **115**, 4977–4991.

Jiang, R. and Carlson, M. (1996) Glucose regulates protein interactions within the yeast SNF1 protein kinase complex. *Genes and Development* **10**, 3105–3115.

Jiang, R. and Carlson, M. (1997) The Snf1 protein kinase and its activating subunit, Snf4, interact with distinct domains of the Sip1/Sip2/Gal83 component in the kinase complex. *Molecular and Cellular Biology* **17**, 2099–2106.

Jiang, Y. and Broach, J. R. (1999) Tor proteins and protein phosphatase 2A reciprocally regulate Tap42 in controlling cell growth in yeast. *EMBO Journal* **18**, 2782–2792.

Johnston, M., Flick, J. S. and Pexton, T. (1994) Multiple mechanisms provide rapid and stringent glucose repression of *GAL* gene expression in *Saccharomyces cerevisiae. Molecular and Cellular Biology* **14**, 3834–3841.

Jona, G., Wittschieben, B. O., Svejstrup, J. Q. and Gileadi, O. (2001) Involvement of yeast carboxy-terminal domain kinase I (CTDK-I) in transcription elongation in vivo. *Gene* **267**, 31–36.

Jung, U. S. and Levin, D. E. (1999) Genome-wide analysis of gene expression regulated by the yeast cell wall integrity signalling pathway. *Molecular Microbiology* **34**, 1049–1057.

Jung, U. S., Sobering, A. K., Romeo, M. J. and Levin, D. E. (2002) Regulation of the yeast Rlm1 transcription factor by the Mpk1 cell wall integrity MAP kinase. *Molecular Microbiology* **46**, 781–789.

Kaffman, A., Herskowitz, I., Tjian, R. and O'Shea, E. K. (1994) Phosphorylation of the transcription factor PHO4 by a cyclin-CDK complex, PHO80-PHO85. *Science* **263**, 1153–1156.

Kaffman, A., Rank, N. M., O'Neill, E. M., Huang, L. S. and O'Shea, E. K. (1998a) The receptor Msn5 exports the phosphorylated transcription factor Pho4 out of the nucleus. *Nature* **396**, 482–486.

Kaffman, A., Rank, N. M. and O'Shea, E. K. (1998b) Phosphorylation regulates association of the transcription factor Pho4 with its import receptor Pse1/Kap121. *Genes and Development* **12**, 2673–2683.

Kaldis, P., Sutton, A. and Solomon, M. J. (1996) The cdk-activating kinase (CAK) from budding yeast. *Cell* **86**, 553–564.

Kalhor, H. R., Luk, K., Ramos, A., Zobel-Thropp, P. and Clarke, S. (2001) Protein phosphatase methyltransferase 1 (Ppm1p) is the sole activity responsible for modification of the major forms of protein phosphatase 2A in yeast. *Archives of Biochemistry and Biophysics* **395**, 239–245.

Kamada, Y., Jung, U. S., Piotrowski, J. and Levin, D. E. (1995) The protein kinase C-activated MAP kinase pathway of *Saccharomyces cerevisiae* mediates a novel aspect of the heat shock response. *Genes and Development* **9**, 1559–1571.

Kamada, Y., Qadota, H., Python, C. P., Anraku, Y., Ohya, Y. and Levin, D. E. (1996) Activation of yeast protein kinase C by rho1 GTPase. *Journal of Biological Chemistry* **271**, 9193–9196.

Kamada, Y., Funakoshi, T., Shintani, T., Nagano, K., Ohsumi, M. and Ohsumi, Y. (2000) Tor-mediated induction of autophagy via an Apg1 protein kinase complex. *Journal of Cell Biology* **150**, 1507–1513.

Kaneko, Y., Tamai, Y., Tohe, A. and Oshima, Y. (1985) Transcriptional and post-transcriptional control of *PHO8* expression by *PHO* regulatory genes in *Saccharomyces cerevisiae. Molecular and Cellular Biology* **5**, 248–252.

Kassis, S., Melhuish, T., Annan, R. S., Chen, S. L., Lee, J. C., Livi, G. P. *et al.* (2000) *Saccharomyces cerevisiae* Yak1p protein kinase autophosphorylates on tyrosine residues and phosphorylates myelin basic protein on a C-terminal serine residue. *Biochemical Journal* **348**, 263–272.

Keleher, C. A., Redd, M. J., Schultz, J., Carlson, M. and Johnson, A. D. (1992) Ssn6-Tup1 is a general repressor of transcription in yeast. *Cell* **68**, 709–719.

Ketela, T., Green, R. and Bussey, H. (1999) *Saccharomyces cerevisiae* Mid2p is a potential cell wall stress sensor and upstream activator of the PKC1-MPK1 cell integrity pathway. *Journal of Bacteriology* **181**, 3330–3340.

Kikuchi, Y., Oka, Y., Kobayashi, M., Uesono, Y., Toh-e, A. and Kikuchi, A. (1994) A new yeast gene, *HTR1* required for growth at high temperature, is needed for recovery from mating pheromone-induced G1 arrest. *Molecular and General Genetics* **245**, 107–116.

Kim, Y. J., Francisco, L., Chen, G. C., Marcotte, E. and Chan, C. S. M. (1994) Control of cellular morphogenesis by the Ipl2/Bem2 GTPase-activating protein: possible role of protein phosphorylation. *Journal of Cell Biology* **127**, 1381–1394.

Kinoshita, N., Ohkura, H. and Yanagida, M. (1990) Distinct, essential roles of type 1 and 2A protein phosphatases in the control of the fission yeast cell division cycle. *Cell* **63**, 405–415.

Kinoshita, N., Yamano, H., Niwa, H., Yoshida, T. and Yanagida, M. (1993) Negative regulation of mitosis by the fission yeast protein phosphatase ppa2. *Genes and Development* **7**, 1059–1071.

Knebel, A., Morrice, N. and Cohen, P. (2001) A novel method to identify protein kinase substrates: eEF2 kinase is phosphorylated and inhibited by SAPK4/p38delta. *EMBO Journal* **20**, 4360–4369.

Kobor, M. S., Archambault, J., Lester, W., Holstege, F. C., Gileadi, O., Jansma, D. B. *et al.* (1999) An unusual eukaryotic protein phosphatase required for transcription by RNA polymerase II and CTD dephosphorylation in *Saccharomyces cerevisiae. Molecular Cell* **4**, 55–62.

Koch, C., Schleiffer, A., Ammerer, G. and Nasmyth, K. (1996) Switching transcription on and off during the yeast cell cycle: Cln/Cdc28 kinases activate bound transcription factor SBF (Swi4/Swi6) at Start, whereas Clb/Cdc28 kinases displace it from the promoter in G_2. *Genes and Development* **10**, 129–141.

Kohno, H., Tanaka, K., Mino, A., Umikawa, M., Imamura, H., Fujiwara, T. *et al.* (1996) Bni1p implicated in cytoskeletal control is a putative target of Rho1p small GTP binding protein in *Saccharomyces cerevisiae*. *EMBO Journal* **15**, 6060–6068.

Komeili, A. and O'Shea, E. K. (1999) Roles of phosphorylation sites in regulating activity of the transcription factor Pho4. *Science* **284**, 977–980.

Komeili, A., Wedaman, K. P., O'Shea, E. K. and Powers, T. (2000) Mechanism of metabolic control. Target of rapamycin signaling links nitrogen quality to the activity of the Rtg1 and Rtg3 transcription factors. *Journal of Cell Biology* **151**, 863–878.

Konopka, J. B., Jenness, D. D. and Hartwell, L. H. (1988) The C-terminus of the *S. cerevisiae* α-pheromone receptor mediates an adaptive response to pheromone. *Cell* **54**, 609–620.

Kraakman, L., Lemaire, K., Ma, P., Teunissen, A. W., Donaton, M. C., Van Dijck, P. *et al.* (1999) A *Saccharomyces cerevisiae* G-protein coupled receptor, Gpr1, is specifically required for glucose activation of the cAMP pathway during the transition to growth on glucose. *Molecular Microbiology* **32**, 1002–1012.

Kranz, J. E., Satterberg, B. and Elion, E. A. (1994) The map kinase Fus3 associates with and phosphorylates the upstream signaling component Ste5. *Genes and Development* **8**, 313–327.

Krause, S. A. and Gray, J. V. (2002) The protein kinase C pathway is required for viability in quiescence in *Saccharomyces cerevisiae*. *Current Biology* **12**, 588–593.

Kretschmer, M., Schellenberger, W., Otto, A., Kessler, R. and Hofmann, E. (1987) Fructose-2, 6-bisphosphatase and 6-phosphofructo-2-kinase are separable in yeast. *Biochemical Journal* **246**, 755–759.

Kron, S. J., Styles, C. A. and Fink, G. R. (1994) Symmetric cell division in pseudohyphae of the yeast *Saccharomyces cerevisiae*. *Molecular Biology of the Cell* **5**, 1003–1022.

Kubler, E., Mosch, H. U., Rupp, S. and Lisanti, M. P. (1997) Gpa2p, a G-protein α-subunit, regulates growth and pseudohyphal development in *Saccharomyces cerevisiae* via a cAMP-dependent mechanism. *Journal of Biological Chemistry* **272**, 20321–20323.

Kuchin, S., Vyas, V. K. and Carlson, M. (2002) Snf1 protein kinase and the repressors Nrg1 and Nrg2 regulate *FLO11* haploid invasive growth, and diploid pseudohyphal differentiation. *Molecular and Cellular Biology* **22**, 3994–4000.

Kumar, R., Reynolds, D. M., Shevchenko, A., Goldstone, S. D. and Dalton, S. (2000) Forkhead transcription factors, Fkh1p and Fkh2p, collaborate with Mcm1p to control transcription required for M-phase. *Current Biology* **10**, 896–906.

Kunz, J., Henriquez, R., Schneider, U., Deuter-Reinhard, M., Movva, N. R. and Hall, M. N. (1993) Target of rapamycin in yeast, TOR2, is an essential phosphatidylinositol kinase homolog required for G1 progression. *Cell* **73**, 585–596.

Kunz, J., Schneider, U., Howald, I., Schmidt, A. and Hall, M. N. (2000) HEAT repeats mediate plasma membrane localization of Tor2p in yeast. *Journal of Biological Chemistry* **275**, 37011–37020.

Kuo, M. H., Nadeau, E. T. and Grayhack, E. J. (1997) Multiple phosphorylated forms of the *Saccharomyces cerevisiae* Mcm1 protein include an isoform induced in response to high salt concentrations. *Molecular and Cellular Biology* **17**, 819–832.

La Valle, R. and Wittenberg, C. (2001) A role for the Swe1 checkpoint kinase during filamentous growth of *Saccharomyces cerevisiae*. *Genetics* **158**, 549–562.

Lamson, R. E., Winters, M. J. and Pryciak, P. M. (2002) Cdc42 regulation of kinase activity and signaling by the yeast p21-activated kinase Ste20. *Molecular and Cellular Biology* **22**, 2939–2951.

Lauze, E., Stoelcker, B., Luca, F. C., Weiss, E., Schutz, A. R. and Winey, M. (1995) Yeast spindle pole body duplication gene *MPS1* encodes an essential dual specificity protein kinase. *EMBO Journal* **14**, 1655–1663.

Leberer, E., Dignard, D., Harcus, D., Thomas, D. Y. and Whiteway, M. (1992) The protein kinase homologue Ste20p is required to link the yeast pheromone response G-protein $\beta\gamma$ subunits to downstream signalling components. *EMBO Journal* **11**, 4815–4824.

Leberer, E., Wu, C., Leeuw, T., Fourest-Lieuvin, A., Segall, J. E. and Thomas, D. Y. (1997) Functional characterization of the Cdc42p binding domain of yeast Ste20p protein kinase. *EMBO Journal* **16**, 83–97.

Leclerc, V., Tassan, J. P., O'Farrell, P. H., Nigg, E. A. and Leopold, P. (1996) *Drosophila* Cdk8, a kinase partner of cyclin C that interacts with the large subunit of RNA polymerase II. *Molecular Biology of the Cell* **7**, 505–513.

Lee, B. N. and Elion, E. A. (1999) The MAPKKK Ste11 regulates vegetative growth through a kinase cascade of shared signaling components. *Proceedings of the National Academy of Sciences USA* **96**, 12679–12684.

Lee, K. S. and Levin, D. E. (1992) Dominant mutations in a gene encoding a putative protein kinase (*BCK1*) bypass the requirement for a *Saccharomyces cerevisiae* protein kinase C homolog. *Molecular and Cellular Biology* **12**, 172–182.

Lee, K. S., Hines, L. K. and Levin, D. E. (1993a) A pair of functionally redundant yeast genes (*PPZ1* and *PPZ2*) encoding type 1-related protein phosphatases function within the *PKC1*-mediated pathway. *Molecular and Cellular Biology* **13**, 5843–5853.

Lee, K. S., Irie, K., Gotoh, Y., Watanabe, Y., Araki, H., Nishida, E. *et al.* (1993b) A yeast mitogen-activated protein kinase homolog (Mpk1p) mediates signalling by protein kinase C. *Molecular and Cellular Biology* **13**, 3067–3075.

Lee, T. H., Solomon, M. J., Mumby, M. C. and Kirschner, M. W. (1991) INH, a negative regulator of MPF, is a form of protein phosphatase 2A. *Cell* **64**, 415–423.

Lee, T. H., Turck, C. and Kirschner, M. W. (1994) Inhibition of cdc2 activation by INH/Pp2A. *Molecular Biology of the Cell* **5**, 323–338.

Leeuw, T., Wu, C., Schrag, J. D., Whiteway, M., Thomas, D. Y. and Leberer, E. (1998) Interaction of a G-protein beta-subunit with a conserved sequence in Ste20/PAK family protein kinases. *Nature* **391**, 191–195.

Lenburg, M. E. and O'Shea, E. K. (1996) Signaling phosphate starvation. *Trends in Biochemical Science* **21**, 383–387.

Lesage, P., Yang, X. and Carlson, M. (1996) Yeast SNF1 protein kinase interacts with SIP4, a C6 zinc cluster transcriptional activator: a new role for SNF1 in the glucose response. *Molecular and Cellular Biology* **16**, 1921–1928.

Levin, D. E. and Bartlett-Heubusch, E. (1992) Mutants in the *Saccharomyces cerevisiae PKC1* gene display a cell cycle-specific osmotic stability defect. *Journal of Cell Biology* **116**, 1221–1229.

Levin, D. E., Bowers, B., Chen, C. Y., Kamada, Y. and Watanabe, M. (1994) Dissecting the protein kinase C/MAP kinase signalling pathway of *Saccharomyces cerevisiae. Cellular and Molecular Biology Research* **40**, 229–239.

Lew, D. J. and Reed, S. I. (1993) Morphogenesis in the yeast cell cycle: regulation by Cdc28 and cyclins. *Journal of Cell Biology* **120**, 1305–1320.

Lew, D. J. and Reed, S. I. (1995) A cell cycle checkpoint monitors cell morphogenesis in budding yeast. *Journal of Cell Biology* **129**, 739–749.

Li, F. N. and Johnston, M. (1997) Grr1 of *Saccharomyces cerevisiae* is connected to the ubiquitin proteolysis machinery through Skp1: coupling glucose sensing to gene expression and the cell cycle. *EMBO Journal* **16**, 5629–5638.

Li, S., Ault, A., Malone, C. L., Raitt, D., Dean, S., Johnston, L. H. *et al.* (1998) The yeast histidine protein kinase, Sln1p, mediates phosphotransfer to two response regulators, Ssk1p and Skn7p. *EMBO Journal* **17**, 6952–6962.

Li, Y., Moir, R. D., Sethy-Coraci, I. K., Warner, J. R. and Willis, I. M. (2000) Repression of ribosome and tRNA synthesis in secretion-defective cells is signaled by a novel branch of the cell integrity pathway. *Molecular and Cellular Biology* **20**, 3843–3851.

Liao, S. M., Zhang, J., Jeffery, D. A., Koleske, A. J., Thompson, C. M., Chao, D. M. *et al.* (1995) A kinase-cyclin pair in the RNA polymerase II holoenzyme. *Nature* **374**, 193–196.

Lila, T. and Drubin, D. G. (1997) Evidence for physical and functional interactions among two *Saccharomyces cerevisiae* SH3 domain proteins, an adenylyl cyclase-associated protein and the actin cytoskeleton. *Molecular Biology of the Cell* **8**, 367–385.

Lim, M. Y., Dailey, D., Martin, G. S. and Thorner, J. (1993) Yeast MCK1 protein kinase autophosphorylates at tyrosine and serine but phosphorylates exogenous substrates at serine and threonine. *Journal of Biological Chemistry* **268**, 21155–21164.

Lim, H. H., Goh, P. Y. and Surana, U. (1996) Spindle pole body separation in *Saccharomyces cerevisiae* requires dephosphorylation of the tyrosine 19 residue of Cdc28. *Molecular and Cellular Biology* **16**, 6385–6397.

Lin, F. C. and Arndt, K. T. (1995) The role of *Saccharomyces cerevisiae* type 2A phosphatase in the actin cytoskeleton and in entry into mitosis. *EMBO Journal* **14**, 2745–2759.

Lin, J. T. and Lis, J. T. (1999) Glycogen synthase phosphatase interacts with heat shock factor to activate CUP1 gene transcription in *Saccharomyces cerevisiae. Molecular and Cellular Biology* **19**, 3237–3245.

Liu, C., Yang, Z., Yang, J., Xia, Z. and Ao, S. (2000) Regulation of the yeast transcriptional factor PHO2 activity by phosphorylation. *Journal of Biological Chemistry* **275**, 31972–31978.

Liu, H., Styles, C. A. and Fink, G. R. (1993) Elements of the yeast pheromone response pathway required for filamentous growth of diploids. *Science* **262**, 1741–1744.

Liu, H., Styles, C. A. and Fink, G. R. (1996) *Saccharomyces cerevisiae* S288C has a mutation in *FLO8*, a gene required for filamentous growth. *Genetics* **144**, 967–978.

Lo, W. S., Duggan, L., Tolga, N. C., Emre, Belotserkovskya, R., Lane, W. S. *et al.* (2001) Snf1 – a histone kinase that works in concert with the histone acetyltransferase Gcn5 to regulate transcription. *Science* **293**, 1142–1146.

Lodder, A. L., Lee, T. K. and Ballester, R. (1999) Characterization of the Wsc1 protein, a putative receptor in the stress response of *Saccharomyces cerevisiae. Genetics* **152**, 1487–1499.

Loewith, R., Jacinto, E., Wullschleger, S., Lorberg, A., Crespo, J. L., Bonenfant, D. *et al.* (2002) Two TOR complexes, only one of which is rapamycin sensitive, have distinct roles in cell growth control. *Molecular Cell* **10**, 457–468.

Lorberg, A., Schmitz, H. P., Jacoby, J. J. and Heinisch, J. J. (2001) Lrg1p functions as a putative GTPase-activating protein in the Pkc1p-mediated cell integrity pathway in *Saccharomyces cerevisiae. Molecular Genetics and Genomics* **266**, 514–526.

Lorenz, M. C. and Heitman, J. (1995) TOR mutations confer rapamycin resistance by preventing interaction with FKBP12-rapamycin. *Journal of Biological Chemistry* **270**, 27531–27537.

Lorenz, M. C. and Heitman, J. (1997) Yeast pseudohyphal growth is regulated by GPA2, a G protein alpha homolog. *EMBO Journal* **16**, 7008–7018.

Lorenz, M. C., Pan, X., Harashima, T., Cardenas, M. E., Xue, Y., Hirsch, J. P. *et al.* (2000) The G protein-coupled receptor Gpr1 is a nutrient sensor that regulates pseudohyphal differentiation in *Saccharomyces cerevisiae. Genetics* **154**, 609–622.

Ludin, K., Jiang, R. and Carlson, M. (1998) Glucose-regulated interaction of a regulatory subunit of protein phosphatase 1 with the Snf1 protein kinase in *Saccharomyces cerevisiae. Proceedings of the National Academy of Sciences USA* **95**, 6245–6250.

Luke, M. M., Dellaseta, F., Di Como, C. J., Sugimoto, H., Kobayashi, R. and Arndt, K. T. (1996) The SAPs, a new family of proteins, associate and function positively with the SIT4 phosphatase. *Molecular and Cellular Biology* **16**, 2744–2755.

Lundgren, K., Walworth, N., Booher, R., Dembski, M., Kirschner, M. and Beach, D. (1991) mik1 and wee1 cooperate in the inhibitory tyrosine phosphorylation of cdc2. *Cell* **64**, 1111–1122.

Lutfiyya, L. L., Iyer, V. R., DeRisi, J., DeVit, M. J., Brown, P. O. and Johnston, M. (1998) Characterization of three related glucose repressors and genes they regulate in *Saccharomyces cerevisiae. Genetics* **150**, 1377–1391.

Ma, D., Cook, J. G. and Thorner, J. (1995) Phosphorylation and localization of Kss1, a MAP kinase of the *Saccharomyces cerevisiae* pheromone response pathway. *Molecular Biology of the Cell* **6**, 889–909.

Ma, P., Wera, S., Van Dijck, P. and Thevelein, J. M. (1999) The *PDE1*-encoded low-affinity phosphodiesterase in the yeast *Saccharomyces cerevisiae* has a specific function in controlling agonist-induced cAMP signaling. *Molecular Biology of the Cell* **10**, 91–104.

MacKay, V. L., Armstrong, J., Yip, C., Welch, S., Walker, K., Osborn, S. *et al.* (1991) Characterization of the Bar proteinase, an extracellular enzyme from the yeast *Saccharomyces cerevisiae*. *Advances in Experimental Medicine and Biology* **306**, 161–172.

MacKelvie, S. H., Andrews, P. D. and Stark, M. J. R. (1995) The *Saccharomyces cerevisiae* gene *SDS22* encodes a potential regulator of the mitotic function of yeast type 1 protein phosphatase. *Molecular and Cellular Biology* **15**, 3777–3785.

Madden, K., Sheu, Y. J., Baetz, K., Andrews, B. and Snyder, M. (1997) SBF cell cycle regulator as a target of the yeast PKC-MAP kinase pathway. *Science* **275**, 1781–1784.

Madden, S. L., Johnson, D. L. and Bergman, L. W. (1990) Molecular and expression analysis of the negative regulators involved in the transcriptional regulation of acid phosphatase production in *Saccharomyces cerevisiae*. *Molecular and Cellular Biology* **10**, 5950–5957.

Madhani, H. D. and Fink, G. R. (1997) Combinatorial control required for the specificity of yeast MAPK signaling. *Science* **275**, 1314–1317.

Madhani, H. D., Styles, C. A. and Fink, G. R. (1997) MAP kinases with distinct inhibitory functions impart signalling specificity during yeast differentiation. *Cell* **91**, 673–684.

Maeda, T., Takekawa, M. and Saito, H. (1995) Activation of yeast PBS2 MAPKK by MAPKKKs or by binding of an SH3-containing osmosensor. *Science* **269**, 554–558.

Maher, M., Cong, F., Kindelberger, D., Nasmyth, K. and Dalton, S. (1995) Cell cycle-regulated transcription of the *CLB2* gene is dependent on Mcm1 and a ternary complex factor. *Molecular and Cellular Biology* **15**, 3129–3137.

Malathi, K., Xiao, Y. and Mitchell, A. P. (1999) Catalytic roles of yeast GSK3β/shaggy homolog Rim11p in meiotic activation. *Genetics* **153**, 1145–1152.

Mann, D. J., Dombradi, V. and Cohen, P. T. W. (1993) *Drosophila* protein phosphatase V functionally complements a *SIT4* mutant in *Saccharomyces cerevisiae* and its amino-terminal region can confer this complementation to a heterologous phosphatase catalytic domain. *EMBO Journal* **12**, 4833–4842.

Marcus, F., Rittenhouse, J., Moberly, L., Edelstein, I., Hiller, E. and Rogers, D. T. (1988) Yeast (*Saccharomyces cerevisiae*) fructose-1,6-bisphosphatase properties of phospho and dephospho forms and of two mutants in which serine 11 has been changed by site-directed mutagenesis. *Journal of Biological Chemistry* **263**, 6058–6062.

Marcus, S., Xue, C. B., Naider, F. and Becker, J. M. (1991) Degradation of **a**-factor by a *Saccharomyces cerevisiae* alpha mating type-specific endopeptidase: evidence for a role in recovery of cells from G1 arrest. *Molecular and Cellular Biology* **11**, 1030–1039.

Marcus, S., Polverino, A., Barr, M. and Wigler, M. (1994) Complexes between STE5 and components of the pheromone-responsive mitogen-activated protein kinase module. *Proceedings of the National Academy of Sciences of the United States of America* **91**, 7762–7766.

Marini, N. J., Meldrum, E., Buehrer, B., Hubberstey, A. V., Stone, D. E., Traynor-Kaplan, A. *et al.* (1996) A pathway in the yeast cell division cycle linking protein kinase C (Pkc1) to activation of Cdc28 at START. *EMBO Journal* **15**, 3040–3052.

Markwardt, D. D., Garrett, J. M., Eberhardy, S. and Heideman, W. (1995) Activation of the ras/cyclic AMP pathway in the yeast *Saccharomyces cerevisiae* does not prevent G1 arrest in response to nitrogen starvation. *Journal of Bacteriology* **177**, 6761–6765.

Martin, H., Rodriguez-Pachon, J. M., Ruiz, C., Nombela, C. and Molina, M. (2000) Regulatory mechanisms for modulation of signaling through the cell integrity Slt2-mediated pathway in *Saccharomyces cerevisiae*. *Journal of Biological Chemistry* **275**, 1511–1519.

Masumoto, H., Muramatsu, S., Kamimura, Y. and Araki, H. (2002) S-Cdk-dependent phosphorylation of Sld2 essential for chromosomal DNA replication in budding yeast. *Nature* **415**, 651–655.

Matheos, D. P., Kingsbury, T. J., Ahsan, U. S. and Cunningham, K. W. (1997) Tcn1p/Crz1p, a calcineurin-dependent transcription factor that differentially regulates gene expression in *Saccharomyces cerevisiae*. *Genes and Development* **11**, 3445–3458.

Mattison, C. P. and Ota, I. M. (2000) Two protein tyrosine phosphatases, Ptp2 and Ptp3, modulate the subcellular localization of the Hog1 MAP kinase in yeast. *Genes and Development* **14**, 1229–1235.

Mattison, C. P., Spencer, S. S., Kresge, K. A., Lee, J. and Ota, I. M. (1999) Differential regulation of the cell wall integrity mitogen-activated protein kinase pathway in budding yeast by the protein tyrosine phosphatases Ptp2 and Ptp3. *Molecular and Cellular Biology* **19**, 7651–7660.

Mazur, P. and Baginsky, W. (1996) *In vitro* activity of 1,3-β-D-glucan synthase requires the GTP-binding protein rho1. *Journal of Biological Chemistry* **271**, 14604–14609.

Mazzoni, C., Zarzov, P., Rambourg, A. and Mann, C. (1993) The SLT2 (MPK1) MAP kinase homolog is involved in polarized cell growth in *Saccharomyces cerevisiae*. *Journal of Cell Biology* **123**, 1821–1833.

McCartney, R. R. and Schmidt, M. C. (2001) Regulation of Snf1 kinase. Activation requires phosphorylation of threonine 210 by an upstream kinase as well as a distinct step mediated by the Snf4 subunit. *Journal of Biological Chemistry* **276**, 36460–36466.

McInerny, C. J., Partridge, J. F., Mikesell, G. E., Creemer, D. P. and Breeden, L. L. (1997) A novel Mcm1-dependent element in the *SWI4*, *CLN3*, *CDC6*, and *CDC47* promoters activates M/G1-specific transcription. *Genes and Development* **11**, 1277–1288.

McMillan, J. N., Theesfeld, C. L., Harrison, J. C., Bardes, E. S. and Lew, D. J. (2002) Determinants of Swe1p degradation in *Saccharomyces cerevisiae*. *Molecular Biology of the Cell* **13**, 3560–3575.

Measday, V., Moore, L., Ogas, J., Tyers, M. and Andrews, B. (1994) The PCL2 (ORFD)-PHO85 cyclin-dependent kinase complex: a cell cycle regulator in yeast. *Science* **266**, 1391–1395.

Measday, V., Moore, L., Retnakaran, R., Lee, J., Donoviel, M., Neiman, A. M. *et al.* (1997) A family of cyclin-like proteins that interact with the Pho85 cyclin-dependent kinase. *Molecular and Cellular Biology* **17**, 1212–1223.

Measday, V., McBride, H., Moffat, J., Stillman, D. and Andrews, B. (2000) Interactions between Pho85 cyclin-dependent kinase complexes and the Swi5 transcription factor in budding yeast. *Molecular Microbiology* **35**, 825–834.

Mellor, H. and Parker, P. J. (1998) The extended protein kinase C superfamily. *Biochemical Journal* **332**, 281–292.

Mendenhall, M. D. and Hodge, A. E. (1998) Regulation of Cdc28 cyclin-dependent protein kinase activity during the cell cycle of the yeast *Saccharomyces cerevisiae*. *Microbiology and Molecular Biology Reviews* **62**, 1191–1243.

Metodiev, M. V., Matheos, D., Rose, M. D. and Stone, D. E. (2002) Regulation of MAPK function by direct interaction with the mating-specific Gα in yeast. *Science* **296**, 1483–1486.

Miled, C., Mann, C. and Faye, G. (2001) Xbp1-mediated repression of *CLB* gene expression contributes to the modifications of yeast cell morphology and cell cycle seen during nitrogen-limited growth. *Molecular and Cellular Biology* **21**, 3714–3724.

Millar, J. B. A. and Russell, P. (1992) The cdc25 M-phase inducer: an unconventional protein phosphatase. *Cell* **68**, 407–410.

Miller, M. E. and Cross, F. R. (2000) Distinct subcellular localization patterns contribute to functional specificity of the Cln2 and Cln3 cyclins of *Saccharomyces cerevisiae*. *Molecular and Cellular Biology* **20**, 542–555.

Minden, A., Lin, A., Claret, F.-X., Abo, A. and Krin, M. (1995) Selective activation of the JNK signalling cascade and c-jun transcriptional activity by the small GTPase Rac and Cdc42Hs. *Cell* **81**, 1147–1157.

Minshull, J., Straight, A., Rudner, A. D., Dernburg, A. F., Belmont, A. and Murray, A. W. (1996) Protein phosphatase 2A regulates MPF activity and sister chromatid cohesion in budding yeast. *Current Biology* **6**, 1609–1620.

Mizunuma, M., Hirata, D., Miyahara, K., Tsuchiya, E. and Miyakawa, T. (1998) Role of calcineurin and Mpk1 in regulating the onset of mitosis in budding yeast. *Nature* **392**, 303–306.

Mizunuma, M., Hirata, D., Miyaoka, R. and Miyakawa, T. (2001) GSK-3 kinase Mck1 and calcineurin coordinately mediate Hsl1 down-regulation by Ca^{2+} in budding yeast. *EMBO Journal* **20**, 1074–1085.

Morgan, B. A., Bouquin, N. and Johnston, L. H. (1995) Two-component signal-transduction systems in budding yeast MAP a different pathway? *Trends in Cell Biology* **5**, 453–457.

Mösch, H. U. and Fink, G. R. (1997) Dissection of filamentous growth by transposon mutagenesis in *Saccharomyces cerevisiae*. *Genetics* **145**, 671–684.
Mösch, H. U., Roberts, R. L. and Fink, G. R. (1996) Ras2 signals via the Cdc42/Ste20/mitogen-activated protein kinase module to induce filamentous growth in *Saccharomyces cerevisiae*. *Proceedings of the National Academy of Sciences USA* **93**, 5352–5356.
Mösch, H. U., Kubler, E., Krappmann, S., Fink, G. R. and Braus, G. H. (1999) Crosstalk between the Ras2p-controlled mitogen-activated protein kinase and cAMP pathways during invasive growth of *Saccharomyces cerevisiae*. *Molecular Biology of the Cell* **10**, 1325–1335.
Moskow, J. J., Gladfelter, A. S., Lamson, R. E., Pryciak, P. M. and Lew, D. J. (2000) Role of Cdc42p in pheromone-stimulated signal transduction in *Saccharomyces cerevisiae*. *Molecular and Cellular Biology* **20**, 7559–7571.
Moskvina, E., Schuller, C., Maurer, C. T., Mager, W. H. and Ruis, H. (1998) A search in the genome of *Saccharomyces cerevisiae* for genes regulated via stress response elements. *Yeast* **14**, 1041–1050.
Muller, E. M., Locke, E. G. and Cunningham, K. W. (2001) Differential regulation of two Ca^{2+} influx systems by pheromone signaling in *Saccharomyces cerevisiae*. *Genetics* **159**, 1527–1538.
Murata, K., Wu, J. and Brautigan, D. L. (1997) B cell receptor-associated protein $\alpha 4$ displays rapamycin-sensitive binding directly to the catalytic subunit of protein phosphatase 2A. *Proceedings of the National Academy of Sciences USA* **94**, 10624–10629.
Murray, S., Udupa, R., Yao, S., Hartzog, G. and Prelich, G. (2001) Phosphorylation of the RNA polymerase II carboxy-terminal domain by the Bur1 cyclin-dependent kinase. *Molecular and Cellular Biology* **21**, 4089–4096.
Musacchio, A. and Hardwick, K. G. (2002) The spindle checkpoint: structural insights into dynamic signalling. *Nature Reviews Molecular Cell Biology* **3**, 731–741.
Nanduri, J. and Tartakoff, A. M. (2001) The arrest of secretion response in yeast: signaling from the secretory path to the nucleus via Wsc proteins and Pkc1p. *Molecular Cell* **8**, 281–289.
Nash, P., Tang, X., Orlicky, S., Chen, Q., Gertler, F. B., Mendenhall, M. D. *et al.* (2001) Multisite phosphorylation of a CDK inhibitor sets a threshold for the onset of DNA replication. *Nature* **414**, 514–521.
Nash, R., Tokiwa, G., Anand, S., Erickson, K. and Futcher, A. B. (1988) The *WHI1*$^+$ gene of *Saccharomyces cerevisiae* tethers cell division to cell size and is a cyclin homolog. *EMBO Journal* **7**, 4335–4346.
Nasmyth, K. (1993) Control of the yeast cell cycle by the Cdc28 protein kinase. *Current Opinion in Cell Biology* **5**, 166–179.
Nasmyth, K., Peters, J. M. and Uhlmann, F. (2000) Splitting the chromosome: cutting the ties that bind sister chromatids. *Science* **288**, 1379–1385.
Nehlin, J. O. and Ronne, H. (1990) Yeast MIG1 repressor is related to the mammalian early growth response and Wilms' tumour finger proteins. *EMBO Journal* **9**, 2891–2898.
Nehlin, J. O., Carlberg, M. and Ronne, H. (1991) Control of yeast *GAL* genes by MIG1 repressor: a transcriptional cascade in the glucose response. *EMBO Journal* **10**, 3373–3377.
Neigeborn, L. and Carlson, M. (1984) Genes affecting the regulation of *SUC2* gene expression by glucose repression in *Saccharomyces cerevisiae*. *Genetics* **108**, 845–858.
Neigeborn, L. and Carlson, M. (1987) Mutations causing constitutive invertase synthesis in yeast: genetic interactions with *snf* mutations. *Genetics* **115**, 247–253.
Nern, A. and Arkowitz, R. A. (1999) A Cdc24p-Far1p-$G_{\beta\gamma}$ protein complex required for yeast orientation during mating. *Journal of Cell Biology* **144**, 1187–1202.
Ng, D. T. (2001) Interorganellar signal transduction: the arrest of secretion response. *Developmental Cell* **1**, 319–320.
Nguyen, V. Q., Co, C. and Li, J. J. (2001) Cyclin-dependent kinases prevent DNA re-replication through multiple mechanisms. *Nature* **411**, 1068–1073.
Nigg, E. A. (1996) Cyclin-dependent kinase 7: at the cross-roads of transcription, DNA repair and cell cycle control? *Current Opinion in Cell Biology* **8**, 312–317.

Nikawa, J., Cameron, S., Toda, T., Ferguson, K. M. and Wigler, M. (1987) Rigorous feedback control of cAMP levels in *Saccharomyces cerevisiae. Genes and Development* **1**, 931–937.

Nishizawa, M., Kawasumi, M., Fujino, M. and Toh-e, A. (1998) Phosphorylation of sic1, a cyclin-dependent kinase (Cdk) inhibitor, by Cdk including Pho85 kinase is required for its prompt degradation. *Molecular Biology of the Cell* **9**, 2393–2405.

Nishizawa, M., Kanaya, Y. and Toh-e, A. (1999) Mouse cyclin-dependent kinase (Cdk) 5 is a functional homologue of a yeast Cdk, pho85 kinase. *Journal of Biological Chemistry* **274**, 33859–33862.

Nonaka, H., Tanaka, K., Hirano, H., Fujiwara, T., Kohno, H., Umikawa, M. *et al.* (1995) A downstream target of RHO1 small GTP-binding protein is PKC1, a homolog of protein kinase C, which leads to activation of the MAP kinase cascade in *Saccharomyces cerevisiae. EMBO Journal* **14**, 5931–5938.

Norbury, C., Blow, J. and Nurse, P. (1991) Regulatory phosphorylation of the $p34^{cdc2}$ protein kinase in vertebrates. *EMBO Journal* **10**, 3321–3329.

Nougarede, R., Della Seta, F., Zarzov, P. and Schwob, E. (2000) Hierarchy of S-phase-promoting factors: yeast Dbf4-Cdc7 kinase requires prior S-phase cyclin-dependent kinase activation. *Molecular and Cellular Biology* **20**, 3795–3806.

Nurse, P. (1990) Universal control mechanism regulating onset of M-phase. *Nature* **344**, 503–508.

O'Neill, E. M., Kaffman, A., Jolly, E. R. and O'Shea, E. K. (1996) Regulation of PHO4 nuclear localization by the PHO80-PHO85 cyclin-CDK complex. *Science* **271**, 209–212.

O'Rourke, S. M. and Herskowitz, I. (2002) A third osmosensing branch in *Saccharomyces cerevisiae* requires the Msb2 protein and functions in parallel with the Sho1 branch. *Molecular and Cellular Biology* **22**, 4739–4749.

O'Rourke, S. M., Herskowitz, I. and O'Shea, E. K. (2002) Yeast go the whole HOG for the hyperosmotic response. *Trends in Genetics* **18**, 405–412.

Oehlen, L. J., McKinney, J. D. and Cross, F. R. (1996) Ste12 and Mcm1 regulate cell cycle-dependent transcription of *FAR1*. *Molecular and Cellular Biology* **16**, 2830–2837.

Ogawa, N., Noguchi, K., Sawai, H., Yamashita, Y., Yompakdee, C. and Oshima, Y. (1995) Functional domains of Pho81p, an inhibitor of Pho85p protein kinase, in the transduction pathway of P_i signals in *Saccharomyces cerevisiae. Molecular and Cellular Biology* **15**, 997–1004.

Ohkura, H. and Yanagida, M. (1991) *S. pombe* gene *sds22*$^+$ essential for a midmitotic transition encodes a leucine-rich repeat protein that positively modulates protein phosphatase 1. *Cell* **64**, 149–157.

Ohkura, H., Kinoshita, N., Miyatani, S., Toda, T. and Yanagida, M. (1989) The fission yeast *dis2*$^+$ gene required for chromosome disjoining encodes one of two putative type 1 protein phosphatases. *Cell* **57**, 997–1007.

Olson, K. A., Nelson, C., Tai, G., Hung, W., Yong, C., Astell, C. *et al.* (2000) Two regulators of Ste12p inhibit pheromone-responsive transcription by separate mechanisms. *Molecular and Cellular Biology* **20**, 4199–4209.

Ostrander, D. B. and Gorman, J. A. (1999) The extracellular domain of the *Saccharomyces cerevisiae* Sln1p membrane osmolarity sensor is necessary for kinase activity. *Journal of Bacteriology* **181**, 2527–2534.

Otero, G., Fellows, J., Li, Y., de Bizemont, T., Dirac, A. M., Gustafsson, C. M. *et al.* (1999) Elongator, a multisubunit component of a novel RNA polymerase II holoenzyme for transcriptional elongation. *Molecular Cell* **3**, 109–118.

Paidhungat, M. and Garrett, S. (1997) A homolog of mammalian, voltage-gated calcium channels mediates yeast pheromone-stimulated Ca^{2+} uptake and exacerbates the *cdc1*^{Ts} growth defect. *Molecular and Cellular Biology* **17**, 6339–6347.

Palecek, S. P., Parikh, A. S. and Kron, S. J. (2000) Genetic analysis reveals that *FLO11* upregulation and cell polarization independently regulate invasive growth in *Saccharomyces cerevisiae. Genetics* **156**, 1005–1023.

Palecek, S. P., Parikh, A. S., Huh, J. H. and Kron, S. J. (2002) Depression of *Saccharomyces cerevisiae* invasive growth on non-glucose carbon sources requires the Snf1 kinase. *Molecular Microbiology* **45**, 453–469.

Palmer, R. E., Sullivan, D. S., Huffaker, T. and Koshland, D. (1992) Role of astral microtubules and actin in spindle orientation and migration in the budding yeast *Saccharomyces cerevisiae*. *Journal of Cell Biology* **119**, 583–593.

Pan, X. and Heitman, J. (1999) Cyclic AMP-dependent protein kinase regulates pseudohyphal differentiation in *Saccharomyces cerevisiae*. *Molecular and Cellular Biology* **19**, 4874–4887.

Pan, X. and Heitman, J. (2000) Sok2 regulates yeast pseudohyphal differentiation via a transcription factor cascade that regulates cell-cell adhesion. *Molecular and Cellular Biology* **20**, 8364–8372.

Pan, X. and Heitman, J. (2002) Protein kinase A operates a molecular switch that governs yeast pseudohyphal differentiation. *Molecular and Cellular Biology* **22**, 3981–3993.

Panek, A. C., De, A. P., Moura, N. V. and Panek, A. D. (1987) Regulation of the trehalose-6-phosphate synthase complex in *Saccharomyces*. I. Interconversion of forms by phosphorylation. *Current Genetics* **11**, 459–465.

Parviz, F., Hall, D. D., Markwardt, D. D. and Heideman, W. (1998) Transcriptional regulation of *CLN3* expression by glucose in *Saccharomyces cerevisiae*. *Journal of Bacteriology* **180**, 4508–4515.

Pascual-Ahuir, A., Posas, F., Serrano, R. and Proft, M. (2001) Multiple levels of control regulate the yeast cAMP-response element-binding protein repressor Sko1p in response to stress. *Journal of Biological Chemistry* **276**, 37373–37378.

Pedruzzi, I., Burckert, N., Egger, P. and de Virgilio, C. (2000) *Saccharomyces cerevisiae* Ras/cAMP pathway controls post-diauxic shift element-dependent transcription through the zinc finger protein Gis1. *EMBO Journal* **19**, 2569–2579.

Peggie, M. W., MacKelvie, S. H., Bloecher, A., Knatko, E. V., Tatchell, K. and Stark, M. J. (2002) Essential functions of Sds22p in chromosome stability and nuclear localization of PP1. *Journal of Cell Science* **115**, 195–206.

Peng, Z. Y., Trumbly, R. J. and Reimann, E. M. (1990) Purification and characterization of glycogen synthase from a glycogen-deficient strain of *Saccharomyces cerevisiae*. *Journal of Biological Chemistry* **265**, 13871–13877.

Pereira, G., Manson, C., Grindlay, J. and Schiebel, E. (2002) Regulation of the Bfa1p-Bub2p complex at spindle pole bodies by the cell cycle phosphatase Cdc14p. *Journal of Cell Biology* **157**, 367–379.

Perkins, G., Drury, L. S. and Diffley, J. F. (2001) Separate SCF^{CDC4} recognition elements target Cdc6 for proteolysis in S phase and mitosis. *EMBO Journal* **20**, 4836–4845.

Peter, M. and Herskowitz, I. (1994) Direct inhibition of the yeast cyclin-dependent kinase Cdc28-Cln by Far1. *Science* **265**, 1228–1231.

Peter, M., Gartner, A., Horecka, J., Ammerer, G. and Herskowitz, I. (1993) *FAR1* links the signal transduction pathway to the cell cycle machinery in yeast. *Cell* **73**, 747–760.

Peter, M., Neiman, A. M., Park, H. O., van Lohuizen, M. and Herskowitz, I. (1996) Functional analysis of the interaction between the small GTP binding protein Cdc42 and the Ste20 protein kinase in yeast. *EMBO Journal* **15**, 7046–7059.

Peters, C., Andrews, P. D., Stark, M. J., Cesaro-Tadic, S., Glatz, A., Podtelejnikov, A. *et al.* (1999) Control of the terminal step of intracellular membrane fusion by protein phosphatase 1. *Science* **285**, 1084–1087.

Philip, B. and Levin, D. E. (2001) Wsc1 and Mid2 are cell surface sensors for cell wall integrity signaling that act through Rom2, a guanine nucleotide exchange factor for Rho1. *Molecular and Cellular Biology* **21**, 271–280.

Pi, H., Chien, C.-T. and Fields, S. (1997) Transcriptional activation upon pheromone stimulation mediated by a small domain of *Saccharomyces cerevisiae* Ste12p. *Molecular and Cellular Biology* **17**, 6410–6418.

Piatti, S., Bohm, T., Cocker, J. H., Diffley, J. F. and Nasmyth, K. (1996) Activation of S-phase-promoting CDKs in late G1 defines a "point of no return" after which Cdc6 synthesis cannot promote DNA replication in yeast. *Genes and Development* **10**, 1516–1531.

Pic, A., Lim, F. L., Ross, S. J., Veal, E. A., Johnson, A. L., Sultan, M. R. *et al.* (2000) The forkhead protein Fkh2 is a component of the yeast cell cycle transcription factor SFF. *EMBO Journal* **19**, 3750–3761.

Ponting, C. P. and Parker, P. J. (1996) Extending the C2 domain family: C2s in PKCs delta, epsilon, eta, theta, phospholipases, GAPs, and perforin. *Protein Science* **5**, 162–166.

Posas, F. and Saito, H. (1997) Osmotic activation of the HOG MAPK pathway via Ste11p MAPKKK: scaffold role of Pbs2p MAPKK. *Science* **276**, 1702–1705.

Posas, F., Wurgler-Murphy, S. M., Maeda, T., Witten, E. A., Thai, T. C. and Saito, H. (1996) Yeast HOG1 MAP kinase cascade is regulated by a multistep phosphorelay mechanism in the SLN1-YPD1-SSK1 "two-component" osmosensor. *Cell* **86**, 865–875.

Posas, F., Witten, E. A. and Saito, H. (1998) Requirement of STE50 for osmostress-induced activation of the STE11 mitogen-activated protein kinase kinase kinase in the high-osmolarity glycerol response pathway. *Molecular and Cellular Biology* **18**, 5788–5796.

Posas, F., Chambers, J. R., Heyman, J. A., Hoeffler, J. P., de Nadal, E. and Arino, J. (2000) The transcriptional response of yeast to saline stress. *Journal of Biological Chemistry* **275**, 17249–17255.

Powers, T. and Walter, P. (1999) Regulation of ribosome biogenesis by the rapamycin-sensitive TOR-signaling pathway in *Saccharomyces cerevisiae*. *Molecular Biology of the Cell* **10**, 987–1000.

Printen, J. A. and Sprague, G. F., Jr. (1994) Protein-protein interactions in the yeast pheromone response pathway: Ste5p interacts with all members of the MAP kinase cascade. *Genetics* **138**, 609–619.

Proft, M. and Struhl, K. (2002) Hog1 kinase converts the Sko1-Cyc8-Tup1 repressor complex into an activator that recruits SAGA and SWI/SNF in response to osmotic stress. *Molecular Cell* **9**, 1307–1317.

Pryciak, P. M. and Huntress, F. A. (1998) Membrane recruitment of the kinase cascade scaffold protein Ste5 by the G$\beta\gamma$ complex underlies activation of the yeast pheromone response pathway. *Genes and Development* **12**, 2684–2697.

Qadota, H., Python, C. P., Inoue, S. B., Arisawa, M., Anraku, Y., Zheng, Y. *et al.* (1996) Identification of yeast rho1p GTPase as a regulatory subunit of 1,3-β-glucan synthase. *Science* **272**, 279–281.

Raitt, D. C., Posas, F. and Saito, H. (2000) Yeast Cdc42 GTPase and Ste20 PAK-like kinase regulate Sho1-dependent activation of the Hog1 MAPK pathway. *EMBO Journal* **19**, 4623–4631.

Rajavel, M., Philip, B., Buehrer, B. M., Errede, B. and Levin, D. E. (1999) Mid2 is a putative sensor for cell integrity signaling in *Saccharomyces cerevisiae*. *Molecular and Cellular Biology* **19**, 3969–3976.

Ramaswamy, N. T., Li, L., Khalil, M. and Cannon, J. F. (1998) Regulation of yeast glycogen metabolism and sporulation by Glc7p protein phosphatase. *Genetics* **149**, 57–72.

Ramezani Rad, M., Jansen, G., Buhring, F. and Hollenberg, C. P. (1998) Ste50p is involved in regulating filamentous growth in the yeast *Saccharomyces cerevisiae* and associates with Ste11p. *Molecular and General Genetics* **259**, 29–38.

Randez-Gil, F., Bojunga, N., Proft, M. and Entian, K. D. (1997) Glucose derepression of gluconeogenic enzymes in *Saccharomyces cerevisiae* correlates with phosphorylation of the gene activator Cat8p. *Molecular and Cellular Biology* **17**, 2502–2510.

Reed, S. I. and Wittenberg, C. (1990) Mitotic role for the Cdc28 protein kinase of *Saccharomyces cerevisiae*. *Proceedings of the National Academy of Sciences USA* **87**, 5697–5701.

Reinders, A., Burckert, N., Boller, T., Wiemken, A. and de Virgilio, C. (1998) *Saccharomyces cerevisiae* cAMP-dependent protein kinase controls entry into stationary phase through the Rim15p protein kinase. *Genes and Development* **12**, 2943–2955.

Reiser, V., Salah, S. M. and Ammerer, G. (2000) Polarized localization of yeast Pbs2 depends on osmostress, the membrane protein Sho1 and Cdc42. *Nature Cell Biology* **2**, 620–627.

Reneke, J. E., Blumer, K. J., Courchesne, W. E. and Thorner, J. (1988) The carboxy-terminal segment of the yeast alpha-factor receptor is a regulatory domain. *Cell* **55**, 221–234.

Rep, M., Proft, M., Remize, F., Tamas, M., Serrano, R., Thevelein, J. M. *et al.* (2001) The *Saccharomyces cerevisiae* Sko1p transcription factor mediates HOG pathway-dependent osmotic regulation of a set of genes encoding enzymes implicated in protection from oxidative damage. *Molecular Microbiology* **40**, 1067–1083.

Reynard, G. J., Reynolds, W., Verma, R. and Deshaies, R. J. (2000) Cks1 is required for G1 cyclin-cyclin-dependent kinase activity in budding yeast. *Molecular and Cellular Biology* **20**, 5858–5864.

Rittenhouse, J., Moberly, L. and Marcus, F. (1987) Phosphorylation *in vivo* of yeast (*Saccharomyces cerevisiae*) fructose-1,6-bisphosphatase at the cyclic AMP-dependent site. *Journal of Biological Chemistry* **262**, 10114–10119.

Roberts, C. J., Nelson, B., Marton, M. J., Stoughton, R., Meyer, M. R., Bennett, H. A. *et al.* (2000) Signaling and circuitry of multiple MAPK pathways revealed by a matrix of global gene expression profiles. *Science* **287**, 873–880.

Roberts, R. L. and Fink, G. R. (1994) Elements of a single MAP kinase cascade in *Saccharomyces cerevisiae* mediate two developmental programs in the same cell type: mating and invasive growth. *Genes and Development* **8**, 2974–2985.

Roberts, R. L., Mosch, H. U. and Fink, G. R. (1997) 14-3-3 proteins are essential for RAS/MAPK cascade signaling during pseudohyphal development in *Saccharomyces cerevisiae. Cell* **89**, 1055–1065.

Robertson, L. S. and Fink, G. R. (1998) The three yeast A kinases have specific signaling functions in pseudohyphal growth. *Proceedings of the National Academy of Sciences USA* **95**, 13783–13787.

Robertson, L. S., Causton, H. C., Young, R. A. and Fink, G. R. (2000) The yeast A kinases differentially regulate iron uptake and respiratory function. *Proceedings of the National Academy of Sciences USA* **97**, 5984–5988.

Rockmill, B. and Roeder, G. S. (1990) Meiosis in asynaptic yeast. *Genetics* **126**, 563–574.

Roeder, G. S. and Bailis, J. M. (2000) The pachytene checkpoint. *Trends in Genetics* **16**, 395–403.

Roelants, F. M., Torrance, P. D., Bezman, N. and Thorner, J. (2002) Pkh1 and Pkh2 differentially phosphorylate and activate Ypk1 and Ykr2 and define protein kinase modules required for maintenance of cell wall integrity. *Molecular Biology of the Cell* **13**, 3005–3028.

Roemer, T., Paravicini, G., Payton, M. A. and Bussey, H. (1994) Characterisation of the yeast (1→6)-β-glucan biosynthetic components, Kre6p and Skn1p, and genetic interactions between the PKC1 pathway and extracellular matrix assembly. *Journal of Cell Biology* **127**, 567–579.

Rolland, F., de Winde, J. H., Lemaire, K., Boles, E., Thevelein, J. M. and Winderickx, J. (2000) Glucose-induced cAMP signalling in yeast requires both a G-protein coupled receptor system for extracellular glucose detection and a separable hexose kinase-dependent sensing process. *Molecular Microbiology* **38**, 348–358.

Rolland, F., Winderickx, J. and Thevelein, J. M. (2001) Glucose-sensing mechanisms in eukaryotic cells. *Trends in Biochemical Sciences* **26**, 310–317.

Romeo, M. J., Angus-Hill, M. L., Sobering, A. K., Kamada, Y., Cairns, B. R. and Levin, D. E. (2002) HTL1 encodes a novel factor that interacts with the RSC chromatin remodelling complex in *Saccharomyces cerevisiae. Molecular and Cellular Biology* **22**, 8165–8174.

Ronne, H., Carlberg, M., Hu, G. Z. and Nehlin, J. O. (1991) Protein phosphatase 2A in *Saccharomyces cerevisiae*: effects on cell growth and bud morphogenesis. *Molecular and Cellular Biology* **11**, 4876–4884.

Rudner, A. D. and Murray, A. W. (2000) Phosphorylation by Cdc28 activates the Cdc20-dependent activity of the anaphase-promoting complex. *Journal of Cell Biology* **149**, 1377–1390.

Rupp, S., Summers, E., Lo, H. J., Madhani, H. and Fink, G. (1999) MAP kinase and cAMP filamentation signaling pathways converge on the unusually large promoter of the yeast *FLO11* gene. *EMBO Journal* **18**, 1257–1269.

Russell, P., Moreno, S. and Reed, S. I. (1989) Conservation of mitotic controls in fission and budding yeasts. *Cell* **57**, 295–303.

Russo, A. A., Jeffrey, P. D. and Pavletich, N. P. (1996) Structural basis of cyclin-dependent kinase activation by phosphorylation. *Nature Structural Biology* **3**, 696–700.

Ryazanov, A. G., Ward, M. D., Mendola, C. E., Pavur, K. S., Dorovkov, M. V., Wiedmann, M. *et al.* (1997) Identification of a new class of protein kinases represented by eukaryotic elongation factor-2 kinase. *Proceedings of the National Academy of Sciences USA* **94**, 4884–4889.

Sabbagh, W., Jr., Flatauer, L. J., Bardwell, A. J. and Bardwell, L. (2001) Specificity of MAP kinase signaling in yeast differentiation involves transient versus sustained MAPK activation. *Molecular Cell* **8**, 683–691.

Sanz, P., Alms, G. R., Haystead, T. A. and Carlson, M. (2000a) Regulatory interactions between the Reg1-Glc7 protein phosphatase and the Snf1 protein kinase. *Molecular and Cellular Biology* **20**, 1321–1328.

Sanz, P., Ludin, K. and Carlson, M. (2000b) Sip5 interacts with both the Reg1/Glc7 protein phosphatase and the Snf1 protein kinase of *Saccharomyces cerevisiae. Genetics* **154**, 99–107.

Sassoon, I., Severin, F. F., Andrews, P. D., Taba, M. R., Kaplan, K. B., Ashford, A. J. *et al.* (1999) Regulation of *Saccharomyces cerevisiae* kinetochores by the type 1 phosphatase Glc7p. *Genes and Development* **13**, 545–555.

Schmelzle, T. and Hall, M. N. (2000) TOR, a central controller of cell growth. *Cell* **103**, 253–262.

Schmelzle, T., Helliwell, S. B. and Hall, M. N. (2002) Yeast protein kinases and the RHO1 exchange factor TUS1 are novel components of the cell integrity pathway in yeast. *Molecular and Cellular Biology* **22**, 1329–1339.

Schmidt, A., Kunz, J. and Hall, M. N. (1996) TOR2 is required for organization of the actin cytoskeleton in yeast. *Proceedings of the National Academy of Sciences USA* **93**, 13780–13785.

Schmidt, A., Beck, T., Koller, A., Kunz, J. and Hall, M. N. (1998) The TOR nutrient signalling pathway phosphorylates NPR1 and inhibits turnover of the tryptophan permease. *EMBO Journal* **17**, 6924–6931.

Schmidt, A., Schmelzle, T. and Hall, M. N. (2002) The RHO1-GAPs SAC7, BEM2 and BAG7 control distinct RHO1 functions in *Saccharomyces cerevisiae. Molecular Microbiology* **45**, 1433–1441.

Schmidt, M. C. and McCartney, R. R. (2000) β-subunits of Snf1 kinase are required for kinase function and substrate definition. *EMBO Journal* **19**, 4936–4943.

Schneider, B. L., Yang, Q. H. and Futcher, A. B. (1996) Linkage of replication to start by the cdk inhibitor sic1. *Science* **272**, 560–562.

Schneider, K. R., Smith, R. L. and O'Shea, E. K. (1994) Phosphate-regulated inactivation of the kinase PHO80-PHO85 by the CDK inhibitor PHO81. *Science* **266**, 122–126.

Schuller, C., Brewster, J. L., Alexander, M. R., Gustin, M. C. and Ruis, H. (1994) The HOG pathway controls osmotic regulation of transcription via the stress response element (STRE) of the *Saccharomyces cerevisiae CTT1* gene. *EMBO Journal* **13**, 4382–4389.

Schwob, E. and Nasmyth, K. (1993) CLB5 and CLB6, a new pair of B-Cyclins involved in DNA replication in *Saccharomyces cerevisiae. Genes and Development* **7**, 1160–1175.

Schwob, E., Bohm, T., Mendenhall, M. D. and Nasmyth, K. (1994) The B-type cyclin kinase inhibitor $p40^{Sic1}$ controls the G1 to S transition in *Saccharomyces cerevisiae. Cell* **79**, 233–244.

Shamji, A. F., Kuruvilla, F. G. and Schreiber, S. L. (2000) Partitioning the transcriptional program induced by rapamycin among the effectors of the Tor proteins. *Current Biology* **10**, 1574–1581.

Shemer, R., Meimoun, A., Holtzman, T. and Kornitzer, D. (2002) Regulation of the transcription factor Gcn4 by Pho85 cyclin Pcl5. *Molecular and Cellular Biology* **22**, 5395–5404.

Shenhar, G. and Kassir, Y. (2001) A positive regulator of mitosis, Sok2, functions as a negative regulator of meiosis in *Saccharomyces cerevisiae. Molecular and Cellular Biology* **21**, 1603–1612.

Shimada, Y., Gulli, M. P. and Peter, M. (2000) Nuclear sequestration of the exchange factor Cdc24 by Far1 regulates cell polarity during yeast mating. *Nature Cell Biology* **2**, 117–124.

Shiozaki, K. and Russell, P. (1995) Counteractive roles of protein phosphatase 2C (PP2C) and a MAP kinase kinase homolog in the osmoregulation of fission yeast. *EMBO Journal* **14**, 492–502.

Shiozaki, K., Shiozaki, M. and Russell, P. (1997) Mcs4 mitotic catastrophe suppressor regulates the fission yeast cell cycle through the Wik1-Wis1-Spc1 kinase cascade. *Molecular Biology of the Cell* **8**, 409–419.

Shirayama, M., Toth, A., Galova, M. and Nasmyth, K. (1999) APC^{Cdc20} promotes exit from mitosis by destroying the anaphase inhibitor Pds1 and cyclin Clb5. *Nature* **402**, 203–207.

Shogren-Knaak, M. A., Alaimo, P. J. and Shokat, K. M. (2001) Recent advances in chemical approaches to the study of biological systems. *Annual Review of Cell and Developmental Biology* **17**, 405–433.

Shou, W., Seol, J. H., Shevchenko, A., Baskerville, C., Moazed, D., Chen, Z. W. *et al.* (1999) Exit from mitosis is triggered by Tem1-dependent release of the protein phosphatase Cdc14 from nucleolar RENT complex. *Cell* **97**, 233–244.

Shu, Y., Yang, H., Hallberg, E. and Hallberg, R. (1997) Molecular genetic analysis of Rts1p, a B′ regulatory subunit of *Saccharomyces cerevisiae* protein phosphatase 2A. *Molecular and Cellular Biology* **17**, 3242–3253.

Sia, R. A. L., Herald, H. A. and Lew, D. J. (1996) Cdc28 tyrosine phosphorylation and the morphogenesis checkpoint in budding yeast. *Molecular Biology of the Cell* **7**, 1657–1666.

Siniossoglou, S., Hurt, E. C. and Pelham, H. R. (2000) Psr1p/Psr2p, two plasma membrane phosphatases with an essential DXDX(T/V) motif required for sodium stress response in yeast. *Journal of Biological Chemistry* **275**, 19352–19360.

Skrzypek, M. S., Nagiec, M. M., Lester, R. L. and Dickson, R. C. (1999) Analysis of phosphorylated sphingolipid long-chain bases reveals potential roles in heat stress and growth control in *Saccharomyces*. *Journal of Bacteriology* **181**, 1134–1140.

Smith, A., Ward, M. P. and Garrett, S. (1998) Yeast PKA represses Msn2p/Msn4p-dependent gene expression to regulate growth, stress response and glycogen accumulation. *EMBO Journal* **17**, 3556–3564.

Smith, F. C., Davies, S. P., Wilson, W. A., Carling, D. and Hardie, D. G. (1999) The SNF1 kinase complex from *Saccharomyces cerevisiae* phosphorylates the transcriptional repressor protein Mig1p *in vitro* at four sites within or near regulatory domain 1. *FEBS Letters* **453**, 219–223.

Sneddon, A. A., Cohen, P. T. W. and Stark, M. J. R. (1990) *Saccharomyces cerevisiae* protein phosphatase 2A performs an essential cellular function and is encoded by two genes. *EMBO Journal* **9**, 4339–4346.

Song, D., Dolan, J. W., Yuan, Y. L. and Fields, S. (1991) Pheromone-dependent phosphorylation of the yeast STE12 protein correlates with transcriptional activation. *Genes and Development* **5**, 741–750.

Sorger, P. K. and Murray, A. W. (1992) S-phase feedback control in budding yeast independent of tyrosine phosphorylation of p34^{CDC28}. *Nature* **355**, 365–368.

Stark, M. J. R. (1996) Yeast protein serine/threonine phosphatases: multiple roles and diverse regulation. *Yeast* **12**, 1647–1675.

Stathopoulos, A. M. and Cyert, M. S. (1997) Calcineurin acts through the *CRZ1/TCN1*-encoded transcription factor to regulate gene expression in yeast. *Genes and Development* **11**, 3432–3444.

Stathopoulos-Gerontides, A., Guo, J. J. and Cyert, M. S. (1999) Yeast calcineurin regulates nuclear localization of the Crz1p transcription factor through dephosphorylation. *Genes and Development* **13**, 798–803.

Sterner, D. E., Lee, J. M., Hardin, S. E. and Greenleaf, A. L. (1995) The yeast carboxyl-terminal repeat domain kinase CTDK-I is a divergent cyclin-cyclin-dependent kinase complex. *Molecular and Cellular Biology* **15**, 5716–5724.

Stettler, S., Chiannil-Kulchai, N., Denmat, S. H. L., Lalo, D., Lacroute, F., Sentenac, A. *et al.* (1993) A general suppressor of RNA polymerase I, polymerase II and polymerase III mutations in *Saccharomyces cerevisiae*. *Molecular and General Genetics* **239**, 169–176.

Stirling, D. A. and Stark, M. J. (2000) Mutations in *SPC110*, encoding the yeast spindle pole body calmodulin-binding protein, cause defects in cell integrity as well as spindle formation. *Biochimica et Biophysica Acta* **1499**, 85–100.

Stone, E. M., Yamano, H., Kinoshita, N. and Yanagida, M. (1993) Mitotic regulation of protein phosphatases by the fission yeast sds22 protein. *Current Biology* **3**, 13–26.

Stratton, H. F., Zhou, J. L., Reed, S. I. and Stone, D. E. (1996) The mating-specific Gα protein of *Saccharomyces cerevisiae* downregulates the mating signal by a mechanism that is dependent on pheromone and independent of G$\beta\gamma$ sequestration. *Molecular and Cellular Biology* **16**, 6325–6337.

Stuart, J. S., Frederick, D. L., Varner, C. M. and Tatchell, K. (1994) The mutant type 1 protein phosphatase encoded by *glc7-1* from *Saccharomyces cerevisiae* fails to interact productively with the *GAC1*-encoded regulatory subunit. *Molecular and Cellular Biology* **14**, 896–905.

Sun, Y., Taniguchi, R., Tanoue, D., Yamaji, T., Takematsu, H., Mori, K. *et al.* (2000) Sli2 (Ypk1), a homologue of mammalian protein kinase SGK, is a downstream kinase in the sphingolipid-mediated signaling pathway of yeast. *Molecular and Cellular Biology* **20**, 4411–4419.

Sutton, A. and Freiman, R. (1997) The Cak1p protein kinase is required at G1/S and G2/M in the budding yeast cell cycle. *Genetics* **147**, 57–71.

Sutton, A., Immanuel, D. and Arndt, K. T. (1991) The SIT4 protein phosphatase functions in late G1 for progression into S phase. *Molecular and Cellular Biology* **11**, 2133–2148.

Symons, M. (1996) Rho family GTPases: the cytoskeleton and beyond. *Trends in Biochemical Sciences* **21**, 178–181.

Tachikawa, H., Bloecher, A., Tatchell, K. and Neiman, A. M. (2001) A Gip1p-Glc7p phosphatase complex regulates septin organization and spore wall formation. *Journal of Cell Biology* **155**, 797–808.

Tan, Y. S. H., Morcos, P. A. and Cannon, J. F. (2003) Pho85 phosphorylates the Glc7 protein phosphatase regulator Glc8 *in vivo*. *Journal of Biological Chemistry* **278**, 147–153.

Tanaka, T. U., Rachidi, N., Janke, C., Pereira, G., Galova, M., Schiebel, E. *et al.* (2002) Evidence that the Ipl1-Sli15 (Aurora kinase-INCENP) complex promotes chromosome bi-orientation by altering kinetochore-spindle pole connections. *Cell* **108**, 317–329.

Tang, Y. and Reed, S. I. (1993) The Cdk-associated protein Cks1 functions both in G1 and G2 in *Saccharomyces cerevisiae*. *Genes and Development* **7**, 822–832.

Tao, W., Deschenes, R. J. and Fassler, J. S. (1999) Intracellular glycerol levels modulate the activity of Sln1p, a *Saccharomyces cerevisiae* two-component regulator. *Journal of Biological Chemistry* **274**, 360–367.

Tate, J. J., Cox, K. H., Rai, R. and Cooper, T. G. (2002) Mks1p is required for negative regulation of retrograde gene expression in *Saccharomyces cerevisiae* but does not affect nitrogen catabolite repression-sensitive gene expression. *Journal of Biological Chemistry* **277**, 20477–20482.

Taylor, W. E. and Young, E. T. (1990) cAMP-dependent phosphorylation and inactivation of yeast transcription factor ADR1 does not affect DNA binding. *Proceedings of the National Academy of Sciences USA* **87**, 4098–4102.

Tedford, K., Kim, S., Sa, D., Stevens, K. and Tyers, M. (1997) Regulation of the mating pheromone and invasive growth responses in yeast by two MAP kinase substrates. *Current Biology* **7**, 228–238.

Teige, M., Scheikl, E., Reiser, V., Ruis, H. and Ammerer, G. (2001) Rck2, a member of the calmodulin-protein kinase family, links protein synthesis to high osmolarity MAP kinase signaling in budding yeast. *Proceedings of the National Academy of Sciences USA* **98**, 5625–5630.

Tennyson, C. N., Lee, J. and Andrews, B. J. (1998) A role for the Pcl9-Pho85 cyclin-cdk complex at the M/G1 boundary in *Saccharomyces cerevisiae*. *Molecular Microbiology* **28**, 69–79.

Theesfeld, C. L., Irazoqui, J. E., Bloom, K. and Lew, D. J. (1999) The role of actin in spindle orientation changes during the *Saccharomyces cerevisiae* cell cycle. *Journal of Cell Biology* **146**, 1019–1032.

Thevelein, J. M. (1991) Fermentable sugars and intracellular acidification as specific activators of the RAS-adenylate cyclase signalling pathway in yeast: the relationship to nutrient-induced cell cycle control. *Molecular Microbiology* **5**, 1301–1307.

Thevelein, J. M. and de Winde, J. H. (1999) Novel sensing mechanisms and targets for the cAMP-protein kinase A pathway in the yeast Saccharomyces cerevisiae. *Molecular Microbiology* **33**, 904–918.

Thompson-Jaeger, S., François, J., Gaughran, J. P. and Tatchell, K. (1991) Deletion of *SNF1* affects the nutrient response of yeast and resembles mutations which activate the adenylate cyclase pathway. *Genetics* **129**, 697–706.

Timblin, B. K., Tatchell, K. and Bergman, L. W. (1996) Deletion of the gene encoding the cyclin-dependent protein kinase Pho85 alters glycogen metabolism in *Saccharomyces cerevisiae*. *Genetics* **143**, 57–66.

Toda, T., Uno, I., Ishikawa, T., Powers, S., Kataoka, T., Broek, D. *et al.* (1985) In yeast, RAS proteins are controlling elements of adenylate cyclase. *Cell* **40**, 27–36.

Tokiwa, G., Tyers, M., Volpe, T. and Futcher, B. (1994) Inhibition of G1 cyclin activity by the Ras/cAMP pathway in yeast. *Nature* **371**, 342–345.

Torres, J., Di Como, C. J., Herrero, E. and de La Torre-Ruiz, M. A. (2002) Regulation of the cell integrity pathway by rapamycin-sensitive TOR function in budding yeast. *Journal of Biological Chemistry* **277**, 43495–43504.

Treisman, R. (1996) Regulation of transcription by MAP kinase cascades. *Current Opinion in Cell Biology* **8**, 205–215.

Trueheart, J., Boeke, J. D. and Fink, G. R. (1987) Two genes required for cell fusion during yeast conjugation: evidence for a pheromone-induced surface protein. *Molecular and Cellular Biology* **7**, 2316–2328.

Tu, J. and Carlson, M. (1994) The *GLC7* type 1 protein phosphatase is required for glucose repression in *Saccharomyces cerevisiae*. *Molecular and Cellular Biology* **14**, 6789–6796.

Tu, J. and Carlson, M. (1995) REG1 binds to protein phosphatase type 1 and regulates glucose repression in *Saccharomyces cerevisiae*. *EMBO Journal* **14**, 5939–5946.

Tu, J., Song, W. and Carlson, M. (1996) Protein phosphatase type 1 interacts with proteins required for meiosis and other cellular processes in *Saccharomyces cerevisiae*. *Molecular and Cellular Biology* **16**, 4199–4206.

Tyers, M. and Futcher, B. (1993) Far1 and Fus3 link the mating pheromone signal transduction pathway to three G1-phase Cdc28 kinase complexes. *Molecular and Cellular Biology* **13**, 5659–5669.

Tyers, M., Tokiwa, G., Nash, R. and Futcher, B. (1992) The Cln3-Cdc28 kinase complex of *S. cerevisiae* is regulated by proteolysis and phosphorylation. *EMBO Journal* **11**, 1773–1784.

Tyers, M., Tokiwa, G. and Futcher, B. (1993) Comparison of the *Saccharomyces cerevisiae* G1 cyclins: Cln3 may be an upstream activator of Cln1, Cln2 and other cyclins. *EMBO Journal* **12**, 1955–1968.

Uesono, Y., Tanaka, K. and Toh-e, A. (1987) Negative regulators of the PHO system in *Saccharomyces cerevisiae*: isolation and structural characterization of *PHO85*. *Nucleic Acids Research* **15**, 10299–10309.

Uesono, Y., Fujita, A., Toh-e, A. and Kikuchi, Y. (1994) The *MCS1/SSD1/SRK1/SSL1* gene is involved in stable maintenance of the chromosome in yeast. *Gene* **143**, 135–138.

Uesono, Y., Toh-e, A. and Kikuchi, Y. (1997) Ssd1p of *Saccharomyces cerevisiae* associates with RNA. *Journal of Biological Chemistry* **272**, 16103–16109.

Uhlmann, F., Wernic, D., Poupart, M. A., Koonin, E. V. and Nasmyth, K. (2000) Cleavage of cohesin by the CD clan protease separin triggers anaphase in yeast. *Cell* **103**, 375–386.

Uno, I., Matsumoto, K., Adachi, K. and Ishikawa, T. (1983) Genetic and biochemical evidence that trehalase is a substrate of cAMP-dependent protein kinase in yeast. *Journal of Biological Chemistry* **258**, 10867–10872.

Valay, J. G., Dubois, M. F., Bensaude, O. and Faye, G. (1996) Ccl1, a cyclin associated with protein kinase Kin28, controls the phosphorylation of RNA polymerase II largest subunit and mRNA transcription. *Comptes Rendu de L' Academis des Sciences Serie III – Sciences de la Vie* **319**, 183–189.

Valdivieso, M. H., Sugimoto, K., Jahng, K. Y., Fernandes, P. M. and Wittenberg, C. (1993) *FAR1* is required for posttranscriptional regulation of *CLN2* gene expression in response to mating pheromone. *Molecular and Cellular Biology* **13**, 1013–1022.

Vallier, L. G. and Carlson, M. (1994) Synergistic release from glucose repression by *mig1* and *ssn* mutations in *Saccharomyces cerevisiae*. *Genetics* **137**, 49–54.

van Drogen, F. and Peter, M. (2002) Spa2p functions as a scaffold-like protein to recruit the Mpk1p MAP kinase module to sites of polarized growth. *Current Biology* **12**, 1698–1703.

van Drogen, F., O'Rourke, S. M., Stucke, V. M., Jaquenoud, M., Neiman, A. M. and Peter, M. (2000) Phosphorylation of the MEKK Ste11p by the PAK-like kinase Ste20p is required for MAP kinase signaling in vivo. *Current Biology* **10**, 630–639.

van Drogen, F., Stucke, V. M., Jorritsma, G. and Peter, M. (2001) MAP kinase dynamics in response to pheromones in budding yeast. *Nature Cell Biology* **3**, 1051–1059.

van Zyl, W. H., Wills, N. and Broach, J. R. (1989) A general screen for mutant of *Saccharomyces cerevisiae* deficient in tRNA biosynthesis. *Genetics* **123**, 55–68.

van Zyl, W., Huang, W. D., Sneddon, A. A., Stark, M., Camier, S., Werner, M. *et al.* (1992) Inactivation of the protein phosphatase 2A regulatory subunit A results in morphological and transcriptional defects in *Saccharomyces cerevisiae. Molecular and Cellular Biology* **12**, 4946–4959.

Varela, J. C., Praekelt, U. M., Meacock, P. A., Planta, R. J. and Mager, W. H. (1995) The *Saccharomyces cerevisiae HSP12* gene is activated by the high-osmolarity glycerol pathway and negatively regulated by protein kinase A. *Molecular and Cellular Biology* **15**, 6232–6245.

Verma, R., Annan, R. S., Huddleston, M. J., Carr, S. A., Reynard, G. and Deshaies, R. J. (1997) Phosphorylation of Sic1 by G_1 cdk required for its degradation and entry into S phase. *Science* **278**, 455–460.

Verna, J., Lodder, A., Lee, K., Vagts, A. and Ballester, R. (1997) A family of genes required for maintenance of cell wall integrity and for the stress response in *Saccharomyces cerevisiae. Proceedings of the National Academy of Sciences USA* **94**, 13804–13809.

Versele, M., de Winde, J. H. and Thevelein, J. M. (1999) A novel regulator of G protein signalling in yeast, Rgs2, downregulates glucose-activation of the cAMP pathway through direct inhibition of Gpa2. *EMBO Journal* **18**, 5577–5591.

Vidan, S. and Mitchell, A. P. (1997) Stimulation of yeast meiotic gene expression by the glucose-repressible protein kinase Rim15p. *Molecular and Cellular Biology* **17**, 2688–2697.

Vincent, O. and Carlson, M. (1998) Sip4, a Snf1 kinase-dependent transcriptional activator, binds to the carbon source-responsive element of gluconeogenic genes. *EMBO Journal* **17**, 7002–7008.

Vincent, O. and Carlson, M. (1999) Gal83 mediates the interaction of the Snf1 kinase complex with the transcription activator Sip4. *EMBO Journal* **18**, 6672–6681.

Vincent, O., Townley, R., Kuchin, S. and Carlson, M. (2001) Subcellular localization of the Snf1 kinase is regulated by specific β subunits and a novel glucose signaling mechanism. *Genes and Development* **15**, 1104–1114.

Visintin, R., Prinz, S. and Amon, A. (1997) CDC20 and CDH1: a family of substrate-specific activators of APC-dependent proteolysis. *Science* **278**, 460–463.

Visintin, R., Craig, K., Hwang, E. S., Prinz, S., Tyers, M. and Amon, A. (1998) The phosphatase Cdc14 triggers mitotic exit by reversal of Cdk-dependent phosphorylation. *Molecular Cell* **2**, 709–718.

Walsh, E. P., Lamont, D. J., Beattie, K. A. and Stark, M. J. (2002) Novel interactions of *Saccharomyces cerevisiae* type 1 protein phosphatase identified by single-step affinity purification and mass spectrometry. *Biochemistry* **41**, 2409–2420.

Walther, K. and Schuller, H. J. (2001) Adr1 and Cat8 synergistically activate the glucose-regulated alcohol dehydrogenase gene *ADH2* of the yeast *Saccharomyces cerevisiae. Microbiology* **147**, 2037–2044.

Wang, Y. and Burke, D. J. (1997) Cdc55p, the B-type regulatory subunit of protein phosphatase 2A, has multiple functions in mitosis and is required for the kinetochore/spindle checkpoint in *Saccharomyces cerevisiae. Molecular and Cellular Biology* **17**, 620–626.

Wang, H., Liu, D., Wang, Y., Qin, J. and Elledge, S. J. (2001a) Pds1 phosphorylation in response to DNA damage is essential for its DNA damage checkpoint function. *Genes and Development* **15**, 1361–1372.

Wang, Z., Wilson, W. A., Fujino, M. A. and Roach, P. J. (2001b) Antagonistic controls of autophagy and glycogen accumulation by Snf1p, the yeast homolog of AMP-activated protein kinase, and the cyclin-dependent kinase Pho85p. *Molecular and Cellular Biology* **21**, 5742–5752.

Ward, M. P., Gimeno, C. J., Fink, G. R. and Garrett, S. (1995) *SOK2* may regulate cyclic AMP-dependent protein kinase-stimulated growth and pseudohyphal development by repressing transcription. *Molecular and Cellular Biology* **15**, 6854–6863.

Warmka, J., Hanneman, J., Lee, J., Amin, D. and Ota, I. (2001) Ptc1, a type 2C Ser/Thr phosphatase, inactivates the HOG pathway by dephosphorylating the mitogen-activated protein kinase Hog1. *Molecular and Cellular Biology* **21**, 51–60.

Wasch, R. and Cross, F. R. (2002) APC-dependent proteolysis of the mitotic cyclin Clb2 is essential for mitotic exit. *Nature* **418**, 556–562.

Wassmann, K. and Ammerer, G. (1997) Overexpression of the G1-cyclin gene *CLN2* represses the mating pathway in *Saccharomyces cerevisiae* at the level of the MEKK Ste11. *Journal of Biological Chemistry* **272**, 13180–13188.

Watanabe, M., Chen, C. Y. and Levin, D. E. (1994) *Saccharomyces cerevisiae PKC1* encodes a protein kinase C (PKC) homolog with a substrate specificity similar to that of mammalian PKC. *Journal of Biological Chemistry* **269**, 16829–16836.

Watanabe, Y., Irie, K. and Matsumoto, K. (1995) Yeast *RLM1* encodes a serum response factor-like protein that may function downstream of the Mpk1 (Slt2) mitogen-activated protein kinase pathway. *Molecular and Cellular Biology* **15**, 5740–5749.

Watanabe, Y., Takaesu, G., Hagiwara, M., Irie, K. and Matsumoto, K. (1997) Characterization of a serum response factor-like protein in *Saccharomyces cerevisiae*, Rlm1, which has transcriptional activity regulated by the Mpk1 (Slt2) mitogen-activated protein kinase pathway. *Molecular and Cellular Biology* **17**, 2615–2623.

Wei, H., Ashby, D. G., Moreno, C. S., Ogris, E., Yeong, F. M., Corbett, A. H. *et al.* (2001) Carboxymethylation of the PP2A catalytic subunit in *Saccharomyces cerevisiae* is required for efficient interaction with the B-type subunits Cdc55p and Rts1p. *Journal of Biological Chemistry* **276**, 1570–1577.

Wek, R. C., Cannon, J. F., Dever, T. E. and Hinnebusch, A. G. (1992) Truncated protein phosphatase GLC7 restores translational activation of GCN4 expression in yeast mutants defective for the eIF2α kinase GCN2. *Molecular and Cellular Biology* **12**, 5700–5710.

Whiteway, M. S., Wu, C., Leeuw, T., Clark, K., Fourest-Lieuvin, A., Thomas, D. Y. *et al.* (1995) Association of the yeast pheromone response G protein $\beta\gamma$ subunits with the MAP kinase scaffold Ste5p. *Science* **269**, 1572–1575.

Wijnen, H., Landman, A. and Futcher, B. (2002) The G1 cyclin Cln3 promotes cell cycle entry via the transcription factor Swi6. *Molecular and Cellular Biology* **22**, 4402–4418.

Williams-Hart, T., Wu, X. and Tatchell, K. (2002) Protein phosphatase type 1 regulates ion homeostasis in *Saccharomyces cerevisiae*. *Genetics* **160**, 1423–1437.

Wilson, R. B., Brenner, A. A., White, T. B., Engler, M. J., Gaughran, J. P. and Tatchell, K. (1991) The *Saccharomyces cerevisiae SRK1* gene, a suppressor of *bcy1* and *ins1*, may be involved in protein phosphatase function. *Molecular and Cellular Biology* **11**, 3369–3373.

Wilson, W. A., Hawley, S. A. and Hardie, D. G. (1996) Glucose repression/derepression in budding yeast: SNF1 protein kinase is activated by phosphorylation under derepressing conditions, and this correlates with a high AMP:ATP ratio. *Current Biology* **6**, 1426–1434.

Wilson, W. A., Mahrenholz, A. M. and Roach, P. J. (1999) Substrate targeting of the yeast cyclin-dependent kinase Pho85p by the cyclin Pcl10p. *Molecular and Cellular Biology* **19**, 7020–7030.

Winkler, A., Arkind, C., Mattison, C. P., Burkholder, A., Knoche, K. and Ota, I. (2002) Heat stress activates the yeast high-osmolarity glycerol mitogen-activated protein kinase pathway, and protein tyrosine phosphatases are essential under heat stress. *Eukaryotic Cell* **1**, 163–173.

Winkler, G. S., Petrakis, T. G., Ethelberg, S., Tokunaga, M., Erdjument-Bromage, H., Tempst, P. *et al.* (2001) RNA polymerase II elongator holoenzyme is composed of two discrete subcomplexes. *Journal of Biological Chemistry* **276**, 32743–32749.

Wittenberg, C., Sugimoto, K. and Reed, S. I. (1990) G1-specific cyclins of *S. cerevisiae*: cell cycle periodicity, regulation by mating pheromone, and association with the $p34^{CDC28}$ protein kinase. *Cell* **62**, 225–237.

Woods, A., Munday, M. R., Scott, J., Yang, X., Carlson, M. and Carling, D. (1994) Yeast SNF1 is functionally related to mammalian AMP-activated protein kinase and regulates acetyl-CoA carboxylase in vivo. *Journal of Biological Chemistry* **269**, 19509–19515.

Wu, C., Whiteway, M., Thomas, D. Y. and Leberer, E. (1995) Molecular characterization of Ste20p, a potential mitogen-activated protein or extracellular signal-regulated kinase kinase (MEK) kinase kinase from *Saccharomyces cerevisiae*. *Journal of Biological Chemistry* **270**, 15984–15992.

Wu, C., Leberer, E., Thomas, D. Y. and Whiteway, M. (1999) Functional characterization of the interaction of Ste50p with Ste11p MAPKKK in *Saccharomyces cerevisiae*. *Molecular Biology of the Cell* **10**, 2425–2440.

Wu, J., Tolstykh, T., Lee, J., Boyd, K., Stock, J. B. and Broach, J. R. (2000) Carboxyl methylation of the phosphoprotein phosphatase 2A catalytic subunit promotes its functional association with regulatory subunits in vivo. *EMBO Journal* **19**, 5672–5681.

Wu, X. and Tatchell, K. (2001) Mutations in yeast protein phosphatase type 1 that affect targeting subunit binding. *Biochemistry* **40**, 7410–7420.

Wu, X., Hart, H., Cheng, C., Roach, P. J. and Tatchell, K. (2001) Characterization of Gac1p, a regulatory subunit of protein phosphatase type I involved in glycogen accumulation in *Saccharomyces cerevisiae*. *Molecular Genetics and Genomics* **265**, 622–635.

Wurgler-Murphy, S. M., Maeda, T., Witten, E. A. and Saito, H. (1997) Regulation of the *Saccharomyces cerevisiae HOG1* mitogen-activated protein kinase by the *PTP2* and *PTP3* protein tyrosine phosphatases. *Molecular and Cellular Biology* **17**, 1289–1297.

Xue, Y., Batlle, M. and Hirsch, J. P. (1998) GPR1 encodes a putative G protein-coupled receptor that associates with the Gpa2p Gα subunit and functions in a Ras-independent pathway. *EMBO Journal* **17**, 1996–2007.

Yamochi, W., Tanaka, K., Nonaka, H., Maeda, A., Musha, T. and Takai, Y. (1994) Growth site localization of rho1 small GTP-binding protein and its involvement in bud formation in *Saccharomyces cerevisiae*. *Journal of Cell Biology* **125**, 1077–1093.

Yang, H., Jiang, W., Gentry, M. and Hallberg, R. L. (2000) Loss of a protein phosphatase 2A regulatory subunit (Cdc55p) elicits improper regulation of Swe1p degradation. *Molecular and Cellular Biology* **20**, 8143–8156.

Yang, X., Hubbard, E. J. and Carlson, M. (1992) A protein kinase substrate identified by the two-hybrid system. *Science* **257**, 680–682.

Yang, X., Jiang, R. and Carlson, M. (1994) A family of proteins containing a conserved domain that mediates interaction with the yeast SNF1 protein kinase complex. *EMBO Journal* **13**, 5878–5886.

Yao, S. and Prelich, G. (2002) Activation of the Bur1-Bur2 cyclin-dependent kinase complex by Cak1. *Molecular and Cellular Biology* **22**, 6750–6758.

Yao, S., Neiman, A. and Prelich, G. (2000) *BUR1* and *BUR2* encode a divergent cyclin-dependent kinase-cyclin complex important for transcription in vivo. *Molecular and Cellular Biology* **20**, 7080–7087.

Yeong, F. M., Lim, H. H., Padmashree, C. G. and Surana, U. (2000) Exit from mitosis in budding yeast: biphasic inactivation of the Cdc28-Clb2 mitotic kinase and the role of Cdc20. *Molecular Cell* **5**, 501–511.

Yoshida, K., Ogawa, N. and Oshima, Y. (1989) Function of the *PHO* regulatory genes for repressible acid phosphatase synthesis in *Saccharomyces cerevisiae*. *Molecular and General Genetics* **217**, 40–46.

Yoshida, S., Ohya, Y., Goebl, M., Nakano, A. and Anraku, Y. (1994) A novel gene, *STT4*, encodes a phosphatidylinositol 4-kinase in the PKC1 protein kinase pathway of *Saccharomyces cerevisiae*. *Journal of Biological Chemistry* **269**, 1166–1172.

Yoshimoto, H., Saltsman, K., Gasch, A. P., Li, H. X., Ogawa, N., Botstein, D. *et al.* (2002) Genome-wide analysis of gene expression regulated by the calcineurin/Crz1p signaling pathway in *Saccharomyces cerevisiae*. *Journal of Biological Chemistry* **277**, 31079–31088.

Young, E. T., Kacherovsky, N. and Van Riper, K. (2002) Snf1 protein kinase regulates Adr1 binding to chromatin but not transcription activation. *Journal of Biological Chemistry* **277**, 38095–38103.

Yu, G., Deschenes, R. J. and Fassler, J. S. (1995) The essential transcription factor, Mcm1, is a downstream target of Sln1, a yeast "two-component" regulator. *Journal of Biological Chemistry* **270**, 8739–8743.

Yu, J., Wang, C., Palmieri, S. J., Haarer, B. K. and Field, J. (1999) A cytoskeletal localizing domain in the cyclase-associated protein, CAP/Srv2p, regulates access to a distant SH3-binding site. *Journal of Biological Chemistry* **274**, 19985–19991.

Zachariae, W. and Nasmyth, K. (1999) Whose end is destruction: cell division and the anaphase-promoting complex. *Genes and Development* **13**, 2039–2058.

Zaragoza, D., Ghavidel, A., Heitman, J. and Schultz, M. C. (1998) Rapamycin induces the G_0 program of transcriptional repression in yeast by interfering with the TOR signaling pathway. *Molecular and Cellular Biology* **18**, 4463–4470.

Zarzov, P., Mazzoni, C. and Mann, C. (1996) The SLT2(MPK1) MAP kinase is activated during periods of polarized cell growth in yeast. *EMBO Journal* **15**, 83–91.

Zhan, X. L., Deschenes, R. J. and Guan, K. L. (1997) Differential regulation of FUS3 MAP kinase by tyrosine-specific phosphatases PTP2/PTP3 and dual-specificity phosphatase MSG5 in *Saccharomyces cerevisiae*. *Genes and Development* **11**, 1690–1702.

Zhang, M., Stauffacher, C. V. and van Etten, R. L. (1994) The three-dimensional structure, chemical mechanism and function of the low molecular weight protein tyrosine phosphatases. *Advances in Protein Phosphatases* **9**, 1–23.

Zhang, S. H., Liu, J., Kobayashi, R. and Tonks, N. K. (1999) Identification of the cell cycle regulator VCP (p97/CDC48) as a substrate of the band 4.1-related protein-tyrosine phosphatase PTPH1. *Journal of Biological Chemistry* **274**, 17806–17812.

Zhao, Y., Boguslawski, G., Zitomer, R. S. and DePaoli-Roach, A. A. (1997) *Saccharomyces cerevisiae* homologs of mammalian B and B′ subunits of protein phosphatase 2A direct the enzyme to distinct cellular functions. *Journal of Biological Chemistry* **272**, 8256–8262.

Zheng, C. F. and Guan, K. L. (1994) Activation of MEK family kinases requires phosphorylation of two conserved Ser/Thr residues. *EMBO Journal* **13**, 1123–1131.

Zheng, P., Fay, D. S., Burton, J., Xiao, H., Pinkham, J. L. and Stern, D. F. (1993) *SPK1* is an essential S phase-specific gene of *Saccharomyces cerevisiae* that encodes a nuclear serine/threonine/tyrosine kinase. *Molecular and Cellular Biology* **13**, 5829–5842.

Zheng, X. F., Fiorentino, D., Chen, J., Crabtree, G. R. and Schreiber, S. L. (1995) TOR kinase domains are required for two distinct functions, only one of which is inhibited by rapamycin. *Cell* **82**, 121–130.

Zhou, X. Z., Kops, O., Werner, A., Lu, P. J., Shen, M., Stoller, G. *et al.* (2000) Pin1-dependent prolyl isomerization regulates dephosphorylation of Cdc25C and tau proteins. *Molecular Cell* **6**, 873–883.

Zhou, Z., Gartner, A., Cade, R., Ammerer, G. and Errede, B. (1993) Pheromone-induced signal transduction in *Saccharomyces cerevisiae* requires the sequential function of three protein kinases. *Molecular and Cellular Biology* **13**, 2069–2080.

Zhu, G., Spellman, P. T., Volpe, T., Brown, P. O., Botstein, D., Davis, T. N. *et al.* (2000a) Two yeast forkhead genes regulate the cell cycle and pseudohyphal growth. *Nature* **406**, 90–94.

Zhu, H., Klemic, J. F., Chang, S., Bertone, P., Casamayor, A., Klemic, K. G. *et al.* (2000b) Analysis of yeast protein kinases using protein chips. *Nature Genetics* **26**, 283–289.

Zhu, H., Bilgin, M., Bangham, R., Hall, D., Casamayor, A., Bertone, P. *et al.* (2001) Global analysis of protein activities using proteome chips. *Science* **293**, 2101–2105.

Zou, L. and Stillman, B. (1998) Formation of a preinitiation complex by S-phase cyclin CDK-dependent loading of Cdc45p onto chromatin. *Science* **280**, 593–596.

Chapter 9

Stress responses

Ian W. Dawes

9.1 Introduction

As a unicellular organism yeast is exposed to many types of environmental insult, including sudden changes in temperature, desiccation, oxidative and osmotic stress, exposure to high salt or heavy metal ions or toxic chemicals. In order to combat these insults yeast cells have evolved a wide range of responses to many different types of stress, including osmotic, heat, oxidative, starvation, low pH and organic solvents (including the toxic effects of ethanol). Some of these defences are intrinsic (constitutive) and depend on the prior physiological or metabolic conditions under which the yeast has been grown, especially important are the growth phase and the stage of the organism in its life cycle. Other defence systems are inducible following stress. The inducible responses differ from one stress to another, but it is now clear that these can be overlapping, so that acquisition of resistance to one form of stress leads to cross resistance to others. The existence of regulatory pathways leading to induction of resistance comes from a range of sensing systems, identified by signal transduction pathways and the transcription factors they influence; these form an important part of the stress responses in yeast and are presently very actively under study. More recently, it has been recognised that complex processes such as cell division are sensitive to stress-induced damage to the cells. Often the stress response can lead to cell cycle arrest while the cell sorts out the damage induced.

The response of cells to different stresses turns out to be very complex and involves a relatively high proportion of the total genetic potential. Genome-wide technologies for transcript analysis using DNA microarrays (Iyer *et al.*, 1999) and for screening the entire set of yeast deletion mutants (Higgins *et al.*, 2002) has enabled a very detailed understanding of which genes are involved in or are activated in response to different forms of stress. Related proteome analysis has complemented these studies to track changes in proteins following heat shock (Boy-Marcotte *et al.*, 1999) and cadmium ion stress (Vido *et al.*, 2001). The extensive broad stress analyses of Gasch *et al.* (2000) and Causton *et al.* (2001), and more specific studies for oxidative stress (Gasch and Brown, 1999) saline stress (Posas *et al.*, 2000), osmotic stress (Rep *et al.*, 2000), have shown that a relatively high proportion (~15%) of the entire set of yeast genes is activated in response to particular stresses. Some genes were activated in a similar way across a range of stress conditions, while others were uniquely switched on for a particular stress. For every stress there was a unique programme of response, which indicates the degree of control and precision in the way yeast cells adjust to environmental change. This theme can be seen also in the results of screening the genome-wide deletion mutants for sensitivity to different oxidants – even here there are distinctive responses to different oxidants.

There have been a considerable number of reviews covering the effects of stress on yeast cells and how they respond to survive. These include a book (Hohmann and Mager, 1997) which covers many aspects of stress responses and more recent reviews of osmotic stress (Hohmann, 2002) and oxidative stress (Jamieson, 1998). Rather than try to be comprehensive this review covers only some of the major stresses (heat, oxidative stress, osmotic stress, salt stress and freeze–thaw injury) that yeast cells face. It concentrates on the nature of the damage caused by the stress, the cellular defence systems that can counter it and the general regulatory systems that control the responses.

9.2 Heat shock

Growth temperature and thermal injury

All organisms have a range of temperatures over which they can grow. For cells exposed to supra-optimal temperatures there is a multiplicity of possible target sites for heat-induced injury, including proteins which can aggregate or denature, cell membrane damage leading to permeability changes and ion leakage, ribosome breakdown, and DNA strand breakage. It has been speculated that the cell membrane is the target site for thermal damage ultimately leading to cell death, since those auxotrophic or temperature-sensitive mutants that lose viability (rather than stop growing) include ones that block membrane synthesis or function (Dawes, 1976). Incubation of *Saccharomyces* near the maximum growth temperature also leads to induction of respiratory-deficient petites (Sherman, 1959).

More recently it has been proposed that an important component of heat injury is the effect on membranes leading to increased fluidity and the permeability of the membrane to protons and other ions (Piper, 1997). These lead to delayed effects leading to alteration of the composition of membrane proteins as well as lipid saturation, and the activation of pathways stimulated by Ca^{2+}.

Defence systems

Almost all organisms tested have a system for responding to heat shock, and yeast is no exception. This is one of the earliest stress response systems to be studied in *S. cerevisiae* and the subject has been reviewed extensively from the physiological and molecular aspects (Piper, 1993, 1997; Lindquist and Craig, 1988). The sensitivity of cells to heat stress is dependent on their physiological state, those in the post-diauxic or stationary phase of culture are intrinsically much more resistant to a temperature shift, as they are to a number of other stress conditions.

The heat stress response of yeast cells to a shift in temperature (e.g. from 23 °C to 37 °C) leads to an increase in the thermotolerance of the cells (Miller *et al.*, 1979; McAlister and Holland, 1985). This initial phase is accompanied by an accumulation of trehalose which has been proposed as a marker event (Wiemken, 1990), by a transient arrest of cells in the G1 phase of the cell cycle and the induction of heat shock proteins. Subsequently the cells recover, resume growth at the elevated temperature and maintain thermotolerance. A major change seen in cells exposed to heat shock is the rapid shut-down in synthesis of most of the cellular proteins, with the concomitant up-regulation of expression of a set of heat shock genes encoding a limited number of proteins. Early studies with one-dimensional gel electrophoresis led to the identification of these as groups on the basis of their molecular mass.

These included in *Saccharomyces cerevisiae* Hsp104 (Hsp100 family), Hsp83 (Hsp 90 family) Hsp70, Hsp60 and the smaller Hsp30, Hsp26 and Hsp12. Subsequent two-dimensional gel electrophoretic analyses indicated that 52 to 80 of about 500 proteins were transiently induced by a relatively mild heat shock, the rest were produced at reduced rate (Boy-Marcotte *et al.*, 1999; Miller *et al.*, 1982). These proteins are produced at high rates for about 30 min, then the rates decline to steady-state levels that are greater than those before the shift. The regulation of this response is discussed later, however, it is clear from the (Boy-Marcotte *et al.*, 1999) study that there are two main, and largely non-overlapping sets of proteins that are induced following heat shock reflecting differences in their physiological role as well as their regulation. One set, encoded by genes subject to transcriptional activation by heat shock factor (Hsf1p) includes mostly chaperones and associated ubiquitin-related protein degradation systems, while the others (encoded by genes regulated by the general stress transcription factors Msn2p/Msn4p) included mostly antioxidant functions or enzymes involved in carbohydrate metabolism.

Heat shock-induced proteins and their functions

The functions of some of the heat shock proteins have been identified. Some play important roles in normal cellular functions, or are homologues of proteins produced in unstressed cells, for which there is an increased demand for their function following the cellular damage caused by heat. Many of the proteins are involved directly in the degradation or reactivation of damaged proteins (Parsell and Lindquist, 1993). Some that have been well characterised to date are discussed here.

Hsp70 family

These proteins belong to the group of ATP-binding chaperones that bind to proteins that are unfolded, or in a partially unfolded conformation to prevent their aggregation and assist in proper protein folding. Their action is required not only to assist in the refolding of damaged proteins, but also to assist in the folding of proteins that have been translocated across membranes. Their mode of action has been best elucidated in *E. coli* for the DnaK/DnaJ/GrpE and GroEL/GroES systems. These proteins act in a sequential way, for example during protein synthesis the DnaJ protein binds to the nascent polypeptide and recruits the ATP-bound form of DnaK to form a complex in which ATP hydrolysis occurs. This induces a conformational change and results in a tighter binding between DnaK and the folding polypeptide. This situation is resolved through the binding of the co-chaperone GrpE which acts as a nucleotide exchange factor exchanging ADP for ATP with the release of the polypeptide. This can then continue folding spontaneously, or participate in further folding and release cycles (Langer *et al.*, 1992; Hartl, 1996); some proteins appear to need the further participation of the GroEL and GroES chaperones to complete the folding.

Yeast has ten homologues of the DnaK protein distributed across most of the cellular compartments. These belong to the family of Hsp70 proteins which have been separated into at least four subfamilies on the basis of structural and functional similarities. Those that are involved in the translocation of proteins across the mitochondrial, vacuolar and endoplasmic reticulum membranes are encoded by the *SSA1* to *SSA4* genes, and three of these (not *SSA2*) are induced to varying extents relative to basal levels by heat shock. From analysis of the effects of single and multiple mutations in the *SSA* genes it is clear that each of the protein products can substitute at least in part for the absence of the other three. *ssa1 ssa2* double mutants are temperature sensitive for growth but curiously are more thermotolerant and produce constitutively high levels of many heat shock proteins. The *SSB1,2* gene products are largely associated with primary translation products at the polysome, and these are repressed by an upshift in temperature; the single null mutants are viable, but disruption of both leads to a cold-sensitive phenotype (Lindquist and Craig, 1988). *SSC1, SSD1* and *KAR2* are also members of the family, and these are transcribed abundantly during normal growth, and up-regulated three- to ten-fold following heat shock. *KAR2* and *SSC1* are both essential, and localise to the endoplasmic reticulum and the mitochondrion, respectively. In conjunction with the multitude of Hsp70 proteins in yeast there has been identified seven DnaJ homologues that are also distributed across compartments.

Since Hsp70 proteins bind to damaged proteins, it has been suggested that a depletion of free Hsp70 may lead to increased synthesis of these proteins, possibly through a direct interaction between Hsp70 and the heat shock transcription factor.

Hsp90

This abundant molecular chaperone is also found in evolutionarily diverse species, and can make up to 1% of soluble protein in the cell (Buchner, 1999). In yeast there are two highly homologous genes, *HSC83* and *HSP83*; the former of these is expressed normally at high levels and is moderately induced by heat, the latter is more strongly heat-inducible from a lower basal level. Genetic analysis using single and double mutants has shown that the genes encode proteins that serve identical functions, but neither gene alone is adequate for growth at temperatures above 37 °C.

The role of these proteins in yeast has yet to be clearly defined. In mammalian cells, the Hsp90 protein binds to several unrelated proteins including steroid receptors and tyrosine kinases, leading to the view that the binding of Hsp90 to a protein keeps it in an inactive state until it is released by receipt of an activating signal. For a steroid receptor this may require interaction with the DNA target sequence in the nucleus. For full chaperone activity, Hsp90 associates with several other proteins, including Hsp70 and its co-chaperone Ydj1, Sse1 (the yeast Hsp110 family member), the cyclophilin orthologues Cpr6 and Cpr7, Sti1, Cdc37 and Sba1. This complex is dynamic with distinct subunits being involved at different stages of the folding or chaperone process (Liu *et al.*, 1999).

Hsp60

These are ubiquitous proteins that are the homologues of the *E. coli* GroEL. They are not structurally related to the Hsp70 proteins, but have a similar function, binding to unfolded proteins and catalysing ATP-dependent structural changes that assist in formation of tertiary structure. The yeast heat-inducible Hsp60 is located in the mitochondrion and is encoded by the essential *MIF4* gene. The cytoplasmic equivalent is encoded by the *TCP1* gene which is essential for viability, but not induced by heat treatment.

Hsp104

Saccharomyces cerevisiae has one gene, *HSP104*, that encodes a protein of the Hsp100 family. Unlike the Hsp70 and Hsp90 proteins, which are essential for cell viability, but are required at higher concentrations for growth at higher temperatures, the *HSP104* gene product is critical for cell survival after heat shock. Hsp104 is expressed at very low levels at normal temperatures, but is induced very strongly by heat. *hsp104* mutants given a short heat pretreatment of 37 °C for 30 min show some initial induced thermotolerance when subjected to 50 °C, but then die at a rate two to three orders of magnitude faster than wild-type cells (Lindquist and Kim, 1996). By expressing the *HSP104* gene under the control of a hormone-regulated heterologous promoter they showed that only low levels of Hsp104 were required to provide full thermotolerance to a strain that had been pre-induced by heat, but that in the absence of pretreatment higher levels of Hsp104 were adequate. They speculate that manipulating the levels of Hsp100 proteins may provide a means of altering thermotolerance in other species.

Hsp104 is required for survival after heat shock to promote the resolubilisation and reactivation of proteins that have unfolded or aggregated as a result of exposure to heat (Parsell *et al.*, 1994). This activity is not seen in other chaperones, but is dependent on additional specific chaperones including *SSA1*-encoded Hsp70, a DnaK homologue and *YDJ1*-encoded Hsp40. This refolding has been demonstrated by *in vitro* reactivation of chemically denatured proteins that had previously formed high molecular mass aggregates (Glover and Lindquist, 1999). Using a temperature-sensitive luciferase protein as substrate, they showed that Hsp104 assists the solubilisation of heat-generated insoluble aggregates, and that this was independent of the Hsp70 function which may be more directed towards preventing aggregation. Following heat shock, Hsp104 has been shown to localise around protein aggregates appearing in both the cytoplasm and nucleus (Kawai *et al.*, 1999). Hsp104 may also interact with the Hsp90 cochaperones (listed before) in an interaction independent of Hsp90 (Abbas-Terki *et al.*, 2001).

Hsp26

The small heat shock proteins form a very diverse group, and some organisms produce a large number of them in the range of 16–40 kDa. *Saccharomyces cerevisiae* has only one main small heat shock protein encoded by the *HSP26* gene. This has no known function, and no differences have been seen in the phenotypes of isogenic wild-type and *hsp26* mutants, nor in strains overproducing the *HSP26* (Lindquist and Kim, 1996). Its role if any, in the heat shock response is therefore quite subtle.

Ubiquitin and ubiquitin-conjugating systems

Ubiquitin is a 76 amino acyl residue very highly conserved protein that is involved in the mechanism for the targeted destruction of damaged or unstable proteins in the cell (Finley and Chau, 1991). The protein is conjugated via its C-terminus to ε-amino groups of lysine residues in target proteins. Within ubiquitin is a critical Lys-48 residue which can serve as an acceptor for self-ubiquitination. Hence extended chains of ubiquitin groups can be attached to the target protein. The ligation of ubiquitin to proteins is catalysed by ubiquitin-conjugating enzymes and ubiquitin–protein ligase, and the degradation of ubiquitin–protein conjugates is catalysed by an ATP-dependent multienzyme complex called the proteasome. The ubiquitin catalysed degradation of proteins is illustrated in Figure 9.1.

In *S. cerevisiae* the four loci *UBI1* to *UBI4* encode ubiquitin (Ozkaynak *et al.*, 1987). All of these involve fusions of ubiquitin to other proteins, *UBI1–3* encode fusions of ubiquitin to a ribosomal protein while *UBI4* encodes five adjacent copies of the ubiquitin polypeptide, the locus is transcribed as a single transcript, and after translation as a polyubiquitin the ubiquitin polypeptides are released by proteolytic cleavage (Finley and Varshavsky, 1985; Ozkaynak *et al.*, 1987; Finley and Chau, 1991). Only the *UBI4* gene is responsive to stress. A variety of stress conditions result in induction of this polyubiquitin, including heat stress, starvation sporulation conditions, exposure to DNA-damaging agents and AMP depletion (Finley *et al.*, 1987; Treger *et al.*, 1988; Finley and Chau, 1991). Yeast has one ubiquitin-activating enzyme which covalently links with the C-terminal carboxyl group of ubiquitin in an ATP-dependent reaction. The ubiquitin is then transferred to the target protein catalysed by a ubiquitin-conjugating enzyme, of which there are eight in *S. cerevisiae*, encoded by *UBC1–7* and *PAS2* required for peroxisome biogenesis (Finley and Chau, 1991).

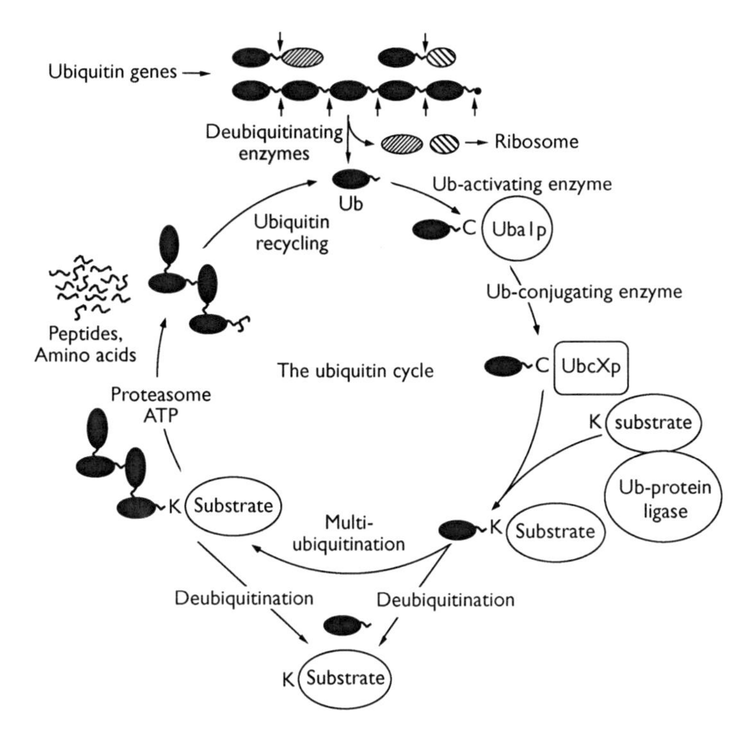

Figure 9.1 The ubiquitin cycle for degradation of damaged proteins. Ubiquitin precursors are processed by the action of deubiquitinating enzymes to release free ubiquitin (top). This is then activated as a thiolester by a ubiquitin-activating enzyme (Uba1p; C = cysteine residue) and then transferred to one of a family of ubiquitin conjugating enzymes (UbcXp; there are 13 in the yeast genome). The ubiquitin is then transferred to the side chain of an internal lysine (K) of a substrate protein with the assistance of a ubiquitin–protein ligase (E3) enzyme. Further rounds of ubiquitination build up a branched, multi-ubiquitin chain on the substrate. This chain is recognised by the proteasome, a large multi-subunit protease, which degrades the substrate protein to short peptides and amino acids. Ubiquitin can be recycled via the action of deubiquitinating enzymes. Deubiquitinating enzymes can also cleave ubiquitin from conjugates prior to proteolysis and thus prevent degradation. (This figure was kindly provided by Rohan Baker.)

Interestingly *UBC2* is *RAD6*, and *UBC3* is *CDC34* highlighting the role of ubiquitination in repair and cell cycle progression. *UBC4* and *UBC5* encode almost identical proteins and are also heat-shock inducible. Tongaonkar *et al.* (2000) have shown that the ubiquitin-conjugating enzymes Ubc1p, Ubc2p, Ubc4p and Ubc5p can interact with the 26S proteasome, and that the interaction between Ubc4p and the proteasome is strongly induced by heat stress. Curiously, while heat shock promotes this association it appears to lock 20S proteasomes in their latent inactive state, impairs further activation of the 26S proteasome by ATP, decreases the transcription of genes encoding proteasomal components and inhibits assembly of the complex (Kuckelkorn *et al.*, 2000).

The role of ubiquitination, and therefore of targeted destruction of damaged proteins following thermal injury is indicated by the phenotype of the *ubi4*, and *ubc4 ubc5* mutants. These grow at rates comparable to the wild type up to 37 °C, but are more sensitive than the wild type to prolonged incubation at 38 °C and are sensitive to amino acid analogues. The *ubi4 ubi5* mutant is resistant to heat stress at high temperatures (52 °C). This has been interpreted to indicate that an increased capacity for ubiquitination is needed for cells to survive a range of stresses, but ubiquitin is not necessary for resistance to acute exposure to high temperature (Finley and Chau, 1991).

Enzymes associated with carbon metabolism and antioxidant defence

Induction of the chaperone genes and *UBC4* depends on the activity of the heat shock transcription factor Hsf1p. As indicated before there is another set of heat shock-induced proteins whose genes are regulated by the Msn2p/Msn4p transcription factors instead (Boy-Marcotte *et al.*, 1999). These have a variety of cellular roles, but they include many involved in shifting carbon metabolism towards increased glycolysis and accumulation of trehalose. These genes included those encoding: phosphoglucomutase (*PGM2*), UDP-glucose pyrophosphorylase (*UGP1*), trehalose-6-phosphate synthase (*TPS1*), hexokinase (*HXK1*), glucokinase (*GLK1*), galactokinase (*GAL1*), trehalose-6-phosphate synthase (*TPS1*), transketolase 2 (*TKL2*), glycerol dehydrogenase (YBR149w) and malate dehydrogenase (*MDH1*).

Other proteins in this Msn2p/Msn4p dependent set included ones important for antioxidant defences and NADPH generation. These included the mitochondrial superoxide dismutase (Sod2p), the cytosolic catalase (Ctt1p) and the mitochondrial cytochrome-*c* peroxidase which would protect the cells from oxidative damage resulting from heat stress (Davidson *et al.*, 1996). This oxidative damage possibly results from heat damage to mitochondria leading to the generation of superoxide free radicals and possibly also the release of Fe^{2+} leading to Fenton chemistry. Other proteins induced included several that could contribute to regeneration of NADPH.

Compounds accumulating in heat-stressed cells

Trehalose

The non-reducing disaccharide trehalose (α-D-glucosyl-1$\rightarrow$1-α-D-glucose) accumulates in yeast cells under conditions that reduce their growth rate, and Wiemken (1990) proposed that its major role in the cell is as a protectant against heat, desiccation and freezing rather than as a reserve carbohydrate. This is supported by the fact that trehalose acts *in vitro* to protect enzymes from heat and drying (Hottinger *et al.*, 1994) and that heat shock causes a very rapid accumulation of the disaccharide in the cytoplasm. Mutations that block trehalose synthesis lead to a marked reduction in thermotolerance (Attfield *et al.*, 1994). In cultures growing on non-fermentable substrates, or starved cells, the level of trehalose in the cells correlates with the their thermotolerance, but not in cells growing on fermentable substrates (Van Dijck *et al.*, 1995). While trehalose plays an important role in thermotolerance, it cannot substitute for Hsp104, since mutants lacking Hsp104 are capable of synthesising trehalose, yet have 100- to 1000-fold less thermotolerance than wild-type cells (Lindquist and Kim, 1996). Trehalose appears to assist the Hsp104 chaperone to refold proteins in both the cytosol and endoplasmic reticulum since loss of Tsp1p leads to severe impairment of this folding activity (Simola *et al.*, 2000).

The balance between synthesis and degradation determines trehalose levels in the cell and the mechanisms of cellular regulation of trehalose levels have been reviewed by Winde *et al.* (1997). Synthesis is catalysed by trehalose synthase, a complex heteromeric protein formed from a trehalose-6-phosphate synthase subunit encoded by *TPS1* (*CIF1, GGS1*), a trehalose-6-phosphate phosphatase (encoded by *TPS2*), and a (probably) regulatory subunit encoded by either of the highly homologous *TPS3* and *TLS1* genes. The synthase activity is strongly activated by fructose-6-phosphate and inhibited by phosphate, whereas the phosphatase activity is enhanced by phosphate. Transcription of these genes is coordinately controlled by the Msn2p/Msn4p transcription factors in response to stress conditions, and also by growth phase, through stress responsive elements (STREs) identified in the promoter regions (Schmitt and McEntee, 1996; Winderickx *et al.*, 1996). There is additional control at the post-translational level since the synthase subunit is subject to catabolite inactivation due to proteolysis in cells growing on glucose (Francois *et al.*, 1991).

The mobilisation of trehalose occurs via the action of trehalases, mainly the neutral trehalase encoded by *NTH1*, although there appears to be a second neutral trehalase enzyme encoded by *NTH2* (Francois *et al.*, 1991); there is also an acidic trehalase located in the vacuole and encoded by *ATH1*. Neutral trehalase is the main enzyme degrading trehalose in proliferating, stationary phase or germinating yeast, and its activity is mainly controlled by phosphorylation (Winde *et al.*, 1997). The enzyme is activated by heat shock (De Virgilio *et al.*, 1991), or by restoring nutrients such as metabolisable nitrogen compounds, phosphate or sulphate to cells starved for the particular nutrient in the presence of glucose, or by adding fermentable sugars to cells in post-diauxic shift phase. This activation is interesting since it depends on protein kinase A (PKA), but does not require cAMP as a second messenger. It occurs under conditions in which cAMP levels do not change, and in mutants defective in the *BCY1* gene encoding the regulatory subunit of cAMP-dependent protein kinase (Hirimburegama *et al.*, 1992). The pathway concerned has been termed the Fermentable Growth Medium (FGM)-induced pathway (see Chapter 3, section 3.8) to indicate the dependence of its operation on a fermentable substrate such as glucose (Thevelein, 1994).

Parrou *et al.* (1997) have suggested that under conditions of heat stress there may be a recycling of trehalose and glycogen since both systems for synthesis and degradation of trehalose are activated by mild heat stress and salt shock. Trehalose does accumulate transiently following a heat shift from 33 °C to 37 °C, but the extent is probably limited by this recycling since mutants with defective trehalase accumulate more. Boy-Marcotte *et al.* (1999) have suggested that this apparently futile cycling may have a role in producing higher levels of trehalose-6-phosphate which is a potent inhibitor of the main hexokinase Hxk2p. Since they have shown that Glk1p and Hxk1p are induced following heat shock, and these hexokinases are not, or only poorly, inhibited by trehalose-6-phosphate this would have implications for the regulation of glycolytic flux and/or alternate pathways for glucose metabolism. As noted earlier, trehalose-6-phosphate synthase activity is modulated by temperature, such that higher temperature shifts to above 40°C trehalose accumulates to very high levels (Parrou *et al.*, 1997).

Sphingolipids

Recently a possible role for sphingolipids in yeast heat tolerance was proposed since addition of sphingolipids, especially C_{20} phytosphingosine and C_{20} di-hydroshingosine, to a suppressed mutant lacking these lipids led to correction of a heat-sensitive phenotype. The structure and biosynthesis of the sphingolipids are discussed in Chapter 6 (Lipids and Membranes). Following heat stress the concentrations of these sphingolipids increased by up to tenfold within 15 min, then declined over the next hour; two ceramides derived from the sphingolipids also increased markedly within one hour, and the changes in these lipids occurred by *de novo* synthesis (Jenkins *et al.*, 1997). These authors have speculated that these lipids may act in either of two ways. The first is based on the finding in yeast that about 30% of the phospholipid in the plasma membrane is composed of inositol phosphoceramides, and proposes that these are structural components of the membrane required for heat tolerance. On the other hand the yeast sphingolipids may play a more active role in the heat stress response by acting as second messengers or signal transducers, as sphingolipid long chain bases and their phosphorylated derivatives have been found to do during heat stress, apoptosis, senescence and differentiation in mammalian cells (Hannun, 1996).

On shifting cells from 25°C to 37°C phytosphingosine-1-phosphate (PHS-1-P) and dihydrosphingosine-1-phosphate (DHS-1-P) levels increase (Skrzypek *et al.*, 1999). Studies of mutations affecting synthesis and degradation of sphingosine (Mao *et al.*, 1999; Skrzypek *et al.*, 1999; Kim *et al.*, 2000) have indicated that either of these molecules may be involved in thermotolerance and control of cell growth. Cells unable to break down long chain base phosphates due to deletion of *DPL1* and *LCB3* show a 500-fold increase in PHS-1-P and DHS-1-P, grow slowly and survive heat stress 10-fold better than wild-type cells (Skrzypek *et al.*, 1999). The *YSR2* and *YSR3* genes encoding DHS-1-P phosphatases have been identified. Deletion of *YSR2* led to increased thermotolerance while overexpression of either gene led to decreased thermotolerance and cell cycle arrest in G1 (Mao *et al.*, 1999). Some of the effects on thermotolerance may be due to phytosphingosine metabolites activating ubiquitin-dependent proteolysis via the endocytic vacuolar degradation and 26S proteasome pathways (Chung *et al.*, 2000).

It remains to be shown whether these compounds added to wild-type cells can induce thermotolerance directly and if so whether they act via components of the heat shock response, or participate in the subsequent phase as the cells resume growth at the elevated temperature. Given their involvement in signal transduction in higher organisms this is an interesting area for future research.

9.3 Oxidative stress

Reactive oxygen species and oxidative damage

The evolution of aerobic metabolism made possible the efficient energy generation process of aerobic respiration, but brought with it attendant problems associated with the generation

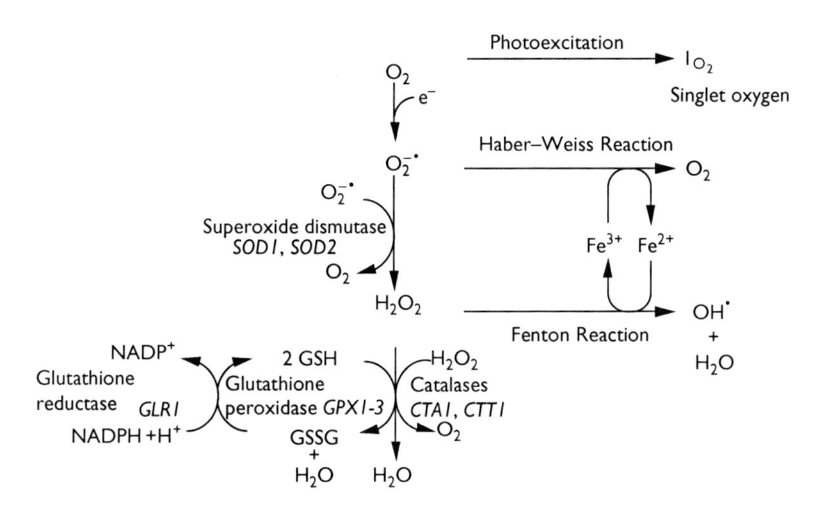

Figure 9.2 Reactive oxygen species and defence systems. The main reactive oxygen species are printed in large type, and include the superoxide anion, hydrogen peroxide and the hydroxyl free radical. Enzymic defences are indicated in italics, together with the yeast genes that encode the enzymes. The involvement of metal ions is shown using Fe ions as examples of the Fenton and Haber–Weiss reactions.

of reactive oxygen species that can react with, and destroy most constituents in the cell. These reactive species are indicated in Figure 9.2. In order to counteract this continuous threat, cells have evolved a variety of antioxidant defence systems that can cope with a broad range of oxidant types. Oxidative stress is the result of an imbalance that occurs when the survival mechanisms are unable to deal adequately with the reactive oxygen species present in the cell. It has been implicated in the process of ageing, and this aspect was covered in Chapter 2.

The chemistry of oxygen species has been reviewed by Halliwell and Gutteridge (1984). Oxygen, O_2, in its ground state is a radical with two unpaired electrons each located in a π^* antibonding orbital and having parallel spin. This makes it relatively unreactive in the context of biological macromolecules since if it were to oxidise another species the two electrons needed must have parallel spin to occupy these orbitals (Halliwell, 1987). The major reactive species derived from oxygen include:

- *Singlet oxygen ($^1O_2$)* This is a more reactive form of oxygen generated by the movement of one of the unpaired electrons to complete a shell, leaving the other empty. It is generated by photoexcited chlorophyll molecules in the triplet state, air pollutants and pathogenic fungi, and it can cause mutagenesis, membrane lipid peroxidation and photoxidation of amino acids (Scandalios, 1987).
- *Superoxide anion (O_2^-)* This moderately reactive radical is formed by the one-electron reduction of oxygen. The primary source of the superoxide anion is during respiration in mitochondria, with leakage of electrons from complex I (NADH:ubiquinone segment) and complex III (QH_2:cytochrome c). It is estimated from *in vitro* studies that more than 80% of the O_2^- in the cell is generated from complex III (Bouveris and Cadenas, 1982). Other sources of superoxide anion include the electron transport

events in microsomes, and the respiratory burst produced by phagocytic cells, including neutrophils, monocytes, macrophages and eosinophils as part of the process of killing invading bacteria. Superoxides, while abundant in the cell are not strongly reactive, although they can react directly with and damage some proteins (Kuo *et al.*, 1987; Gardner and Fridovich, 1991). They must, however, be removed efficiently from cells since they can give rise to hydrogen peroxide and other, more reactive, free radicals, including the highly damaging hydroxyl free radical.

- *Hydrogen peroxide (H_2O_2)* This compound is formed in the cell via the disproportionation of the superoxide anion via the action of superoxide dismutases (SOD), as well as being a product of several oxidases, including the glyoxosomal glycolate oxidase (Halliwell and Gutteridge, 1984, 1989). It can readily cross biological membranes, unlike the charged O_2^-. Hydrogen peroxide is also relatively unreactive, but it has deleterious effects mainly through its conversion to the highly reactive hydroxyl free radical via the Fenton reaction (Halliwell and Aruoma, 1991).
- *Hydroxyl radical ($^{\bullet}OH$)* The main toxic effects of both the superoxide anion and hydrogen peroxide result from their ready conversion to the extremely reactive $^{\bullet}OH$ species. These reactions are catalysed by metal ions, through the Haber–Weiss reaction which generates the reduced form of metal ions from Fe(III) or Cu(II) by reaction with superoxide:

$$Fe^{3+} + O_2^- \rightarrow Fe^{2+} + O_2$$

This makes available the reduced form of the metal ion for the Fenton reaction with H_2O_2:

$$Fe^{2+} + H_2O_2 \rightarrow Fe^{3+} + {}^{\bullet}OH + OH^-$$

The reduced forms of the redox active metals Fe(II), Cu(I) and Ti(III) can participate in this reaction. This brings in to focus the need for cells to maintain metal ion homeostasis, and indicates the overlap between metal ion toxicity and oxidative stress. The hydroxyl radical reacts in a diffusion-limited manner and it indiscriminately reacts with sugars, amino acids, phospholipids, nucleotides and organic acids (Lesko *et al.*, 1980; Halliwell, 1995). It can participate in hydrogen abstraction, addition and electron transfer reactions to generate other radicals that are generally less reactive than the $^{\bullet}OH$.

- *Reactive nitrogen species* This is a collective term for a group of radicals generated when the nitric oxide radical ($NO^{\bullet}$) reacts with the superoxide anion, and it includes among others the nitrogen dioxide radical ($NO_2^{\bullet}$) and peroxynitrite ($ONOO^-$). These species are formed in cells and have been shown to attack proteins by nitrating aromatic amino acid residues (Beckman *et al.*, 1994) and can cause lesions to DNA (Wiseman and Halliwell, 1996). In *S. cerevisiae* peroxynitrite has been found more potent than hydrogen peroxide in oxidising thiols, inducing heat shock proteins and enhancing protein ubiquitination, with glyceraldehyde-3-phosphate dehydrogenase as an especially sensitive target (Buchczyk *et al.*, 2000).

Cellular damage

Oxidants can damage a wide range of cellular molecules, including nucleic acids, proteins and lipids. Given the wide range of cellular targets, there is little knowledge of the events that lead to loss of cell viability following severe oxidative damage, and for cells in a given population the individuals may lose viability as a result of different types of damage. These presumably also differ depending on the definition of viability, which for micro-organisms is often taken to be loss of ability to reproduce by cell division, whereas with higher cell systems it is usually considered to be loss of cellular integrity. These processes could be caused by quite different types of cell lesion.

While DNA damage is known to occur following oxidant treatment, and reactive oxygen species have been implicated in mutagenesis and carcinogenesis (Ames, 1989; Ames and Gold, 1991; Joenje *et al.*, 1991) it is considered that only the hydroxyl radical ($^{\bullet}OH$) and singlet oxygen ($^1O_2$) are likely to affect DNA directly (Halliwell and Aruoma, 1991; Lafleur and Retel, 1993) and the superoxide radical O_2^- and H_2O_2 cause damage via the generation of these species. Brennan *et al.* (1994) have shown that five oxidative mutagens including, paraquat, cumene hydroperoxide, H_2O_2, the DNA cross-linking agent mitomycin C and phenylhydrazine also provoke significant levels of intrachromosomal recombination. Of these only H_2O_2 caused interchromosomal recombination at the doses used, although higher treatments of paraquat and mitomycin C have been reported to also cause interchromosomal recombination between heteroalleles in *S. cerevisiae* (Holliday, 1964; Parry, 1973). It is unlikely that DNA damage is one of the major contributors to cell death, since mutants affected in DNA repair functions are not prominent in the strains identified as being sensitive to oxidants in genome-wide surveys of deletion mutants (Higgins *et al.*, 2002; Thorpe, personal communication).

Free radical damage can lead to cross-linking of proteins, and in the presence of oxygen to protein fragmentation leading to enhanced proteolysis and tissue damage (Wolff and Dean, 1986). Hydroxyl free radical damage to proteins leads to the oxidation of amino acyl residues, especially tyrosine, phenylalanine, tryptophan and cysteine (Stadtman, 1992; Santoro and Thiele, 1997). Protein hydroperoxides are directly reactive and can decompose to free radicals which can lead to the formation of modified amino acids on the protein backbone, fragmentation, cross-linking and unfolding (Gebicki *et al.*, 2002). Hydroxylated derivatives can form from damage to aliphatic amino acids while oxidation of aromatic residues can also produce reactive phenoxyl radicals (Aeschbach *et al.*, 1976; Fu *et al.*, 1995). The generation of carbonyl derivatives may play a role in marking proteins for degradation, and these compounds are known to increase during ageing of cells. Proteins that are specifically susceptible to carbonylation following exposure of cells to hydrogen peroxide or a superoxide-generating agent include both isozymes (Tdh1p and Tdh2p) of glyceraldehyde-3-phosphate dehydrogenase, pyruvate dehydrogenase, 2-oxoglutarate dehydrogenase, aconitase, Hsp60 and fatty acid synthase as major targets. Cu/Zn SOD, cyclophilin I and phosphoglycerate mutase are others (Cabiscol *et al.*, 2000; Costa *et al.*, 2002). Inactivation of the Cu/Zn SOD may contribute to hydrogen peroroxide-induced cell death. Growth arrest caused by carbon or nitrogen starvation leads to increased protein carbonyl formation, and this appears to be linked to the degree of coupling of respiration and linked to the reactive oxygen species generated by semiquinones (Aguilaniu *et al.*, 2001).

Unsaturated lipids are a main target of attack by reactive oxygen species, leading to the process of autocatalytic lipid peroxidation (Gunstone, 1996). This process is illustrated in Figure 9.3, with the free radical species initially abstracting a hydrogen atom from a methylene group. Only certain reactive oxygen species are able to react in this way, including the hydroxyl radical and the protonated form of the superoxide anion, while the superoxide anion is not considered to be sufficiently reactive (Halliwell and Gutteridge, 1990). During

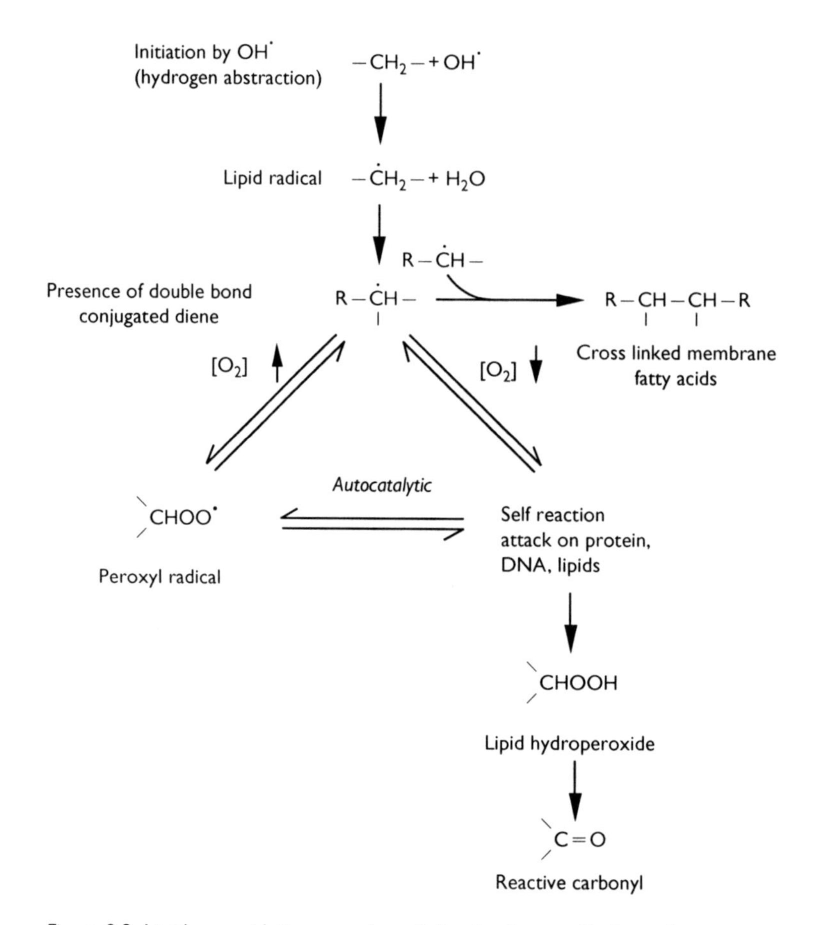

Figure 9.3 Lipid peroxidation reactions following free radical attack on unsaturated fatty acids. Lipid peroxidation is initiated by the abstraction of a hydrogen atom from a methylene group, the allylic hydrogens adjacent to double bonds in the unsaturated fatty acyl chains are highly susceptible. The lipid radical formed can form an unstable fatty acyl peroxy radical in the presence of oxygen, and this radical can initiate hydrogen abstraction from an adjacent lipid propagating the radical mechanism as well as producing a fatty acyl hydroperoxide. The fatty acyl hydroperoxide and the fatty acyl peroxy radical can undergo further reactions to generate aldehydes, including very reactive ones such as malondialdehyde and 4-hydroxynon-2-enal.

this process, reactive lipid radicals and lipid hydroperoxides are formed (Aoshima *et al.*, 1999) and these contribute to the autocatalytic nature of the reaction, and to further autooxidation which can involve radical coupling, fragmentation, rearrangement and cyclisation reactions (Porter *et al.*, 1995; Gunstone, 1996). Lipid hydroperoxides are amongst the most toxic hydroperoxides to yeast, affecting cells at concentrations as low as 10 μM, and killing above 80 μM (Alic *et al.*, 2001; Evans *et al.*, 1998). Lipid peroxidation leads to significant membrane damage, and is the basis of rancidity of foods. Highly reactive aldehydes such as malondialdehyde and 4-hydroxynonenal are formed in the subsequent breakdown of these lipid hydroperoxides and also lead to carbonylation of proteins (Levine, 2002).

Cellular responses to oxidants

For aerobically grown cells, there is continual exposure to some degree of oxidative stress, and this exposure is to a wide range of oxidant species. There has therefore presumably been strong evolutionary pressure since the earth's atmosphere became aerobic for organisms to evolve many efficient systems to deal with the range of reactive oxygen species and the types of damage that can occur. For facultatively aerobic yeast species such as *S. cerevisiae* this may apply more than for other forms of stress due to the continual nature of the exposure during aerobic growth.

Like the cellular response to other stresses, following the chemical and physical damage elicited by the stress, the cell sets in train defence mechanisms for protection against further damage, repair and recovery. This includes cell division cycle delay (Flattery-O'Brien and Dawes, 1998; Alic *et al.*, 2001) discussed later, and the triggering of several damage response pathways leading to activation of specific gene expression systems. Ultimately some of these lead to an adaptive response so that the cell becomes more resistant to subsequent damage. Since there is a strong overlap between oxidative and other stress responses (especially to heat and ethanol) in terms of the regulatory pathways involved, the adaptive and cell-cycle aspects are discussed later.

The defence systems that have evolved include ones that degrade or detoxify the reactive oxygen species, ones that maintain metal ion homeostasis to prevent free metal ions generating hydroxyl radicals, as well as those that repair damaged molecules in the cell. One of the interesting aspects of the cells' response to oxidative stress is that it often seems to involve relatively small changes in many systems. Induction levels for some of the antioxidant defence enzymes can be as low as twofold following damage that induces cells to adapt to an oxidant. This may be the result of the fact that some antioxidant systems can become pro-oxidants and therefore deleterious in excess, and there is therefore a need for the cell to balance its defence systems. The response of yeast to oxidative stress has been reviewed by Jamieson (1998).

Genome-wide surveys of transcription in response to oxidants have identified many genes that are activated and repressed. While these surveys have provided valuable insight in to the regulatory responses, they do not necessarily indicate the importance of the various genes identified in survival, repair of damage and cellular recovery. Analysis of mutants has been traditionally used to identify the role of functions in stress and other responses, and many antioxidant and repair functions have been identified by isolating mutants affecting known or suspected antioxidant functions, or screening for mutations altering resistance to oxidants, or genes that confer resistance when overexpressed. The availability of large genome-wide sets of deletion strains (Winzeler *et al.*, 1999) has enabled nearly comprehensive screening of the involvement of cellular functions in stress responses. In some cases genes that play a role may, when deleted, confer no specific phenotype due to gene redundancy or existence of compensatory parallel pathways. However, the power of this approach relies on the sheer number of genes under study and the fact that some components of a specific function or pathway will probably have a phenotype showing some degree of sensitivity or resistance even in cases where there is some redundancy.

Such screens have been performed for oxidative stress, initially on the large (>600) strains of the EUROFAN collection (Higgins *et al.*, 2002). Since this collection was largely of genes of unknown function (at the time they were chosen for deletion analysis) the results of this survey have particular relevance to the quest for identifying gene function. Subsequently, this laboratory has extended this screen to the entire genomic collection of deletion strains, to more oxidants/free radical-generating agents and to determine the degree of sensitivity/resistance (Thorpe, personal communication). From these studies several important principles have emerged. First, there were a substantial number of genes that led to sensitivity to at least one of the oxidants used – of nearly 5,000 gene deletants screened, about 15% showed sensitivity to at least one of the oxidants used. Secondly, different, but overlapping sets of genes (and cellular functions) were identified for the different oxidants, hydrogen peroxide, cumene hydroperoxide, linoleic acid hydroperoxide, diamide and the superoxide generating agent menadione. The different extents to which general functions are involved for each oxidant are indicated in a general way in Figure 9.4. A small core set of genes (~20) are needed for resistance to at least four of these conditions, representative genes and functions are listed in Table 9.1. These identified, in addition to known antioxidant functions such as the transcription factor Yap1p, those involved in translation, protein sorting and cell wall metabolism. The study has also highlighted functions that are unique to specific oxidants, and the differences between the cellular processes needed to provide resistance. More than half the genes required for full resistance to hydrogen peroxide replication were involved in mitochondrial genome, maintenance or respiratory metabolism (cf. the proportion to linoleate hydroperoxide in Figure 9.4). This points to the crucial requirement for

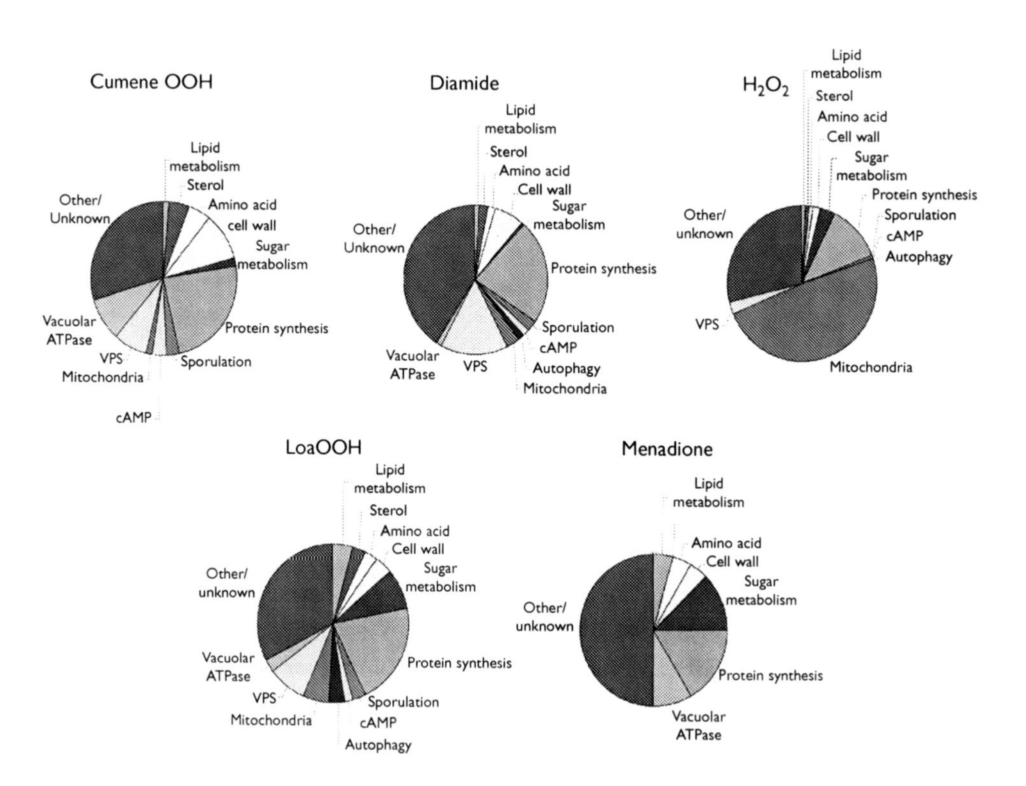

Figure 9.4 General functions involved in the response of *Saccharomyces cerevisiae* strains to oxidative stress from different oxidants/free radical generating agents. The data were obtained by Geoffrey Thorpe and Chii Fong from screening of approximately 5,000 yeast deletant strains.

Table 9.1 Core set of genes that are required for resistance to a range of oxidants. Genes in this set were identified by deletion mutations that conferred sensitivity to at least four of the reactive oxygen species generated by hydrogen peroxide, menadione, linoleate hydroperoxide, *t*-butyl hydroperoxide and diamide

Transcription	
YAP1	Yap1p transcription factor.
GAL11	Component of RNA polymerase II holoenzyme and Kornberg's mediator complex with positive and negative effects on transcription of individual genes.
SWI3	Component of SWI–SNF global transcription activator complex; acts to assist gene-specific activators through chromatin remodelling.
SNF6	Component of SWI–SNF global transcription activator complex; acts to assist gene-specific activators through chromatin remodelling.
Translation	
EAP1	Protein that inhibits <*cap*>-dependent translation, probable purine nucleotide-binding protein that associates with translation initiation factor eIF4E.
Protein sorting	
VID28	Protein involved in vacuolar import and degradation.
PEP3	Vacuolar peripheral membrane protein involved in vacuolar protein sorting and required for vacuole biogenesis.
VMA2	Vacuolar H(+)-ATPase (V-ATPase) regulatory subunit (subunit B) involved in nucleotide binding.
Cell wall	
CWH36	Protein involved in generation of the mannoprotein layer of the cell wall.

normal control of respiration to avoid or respond to hydrogen peroxide toxicity. For resistance to lipid hydroperoxides on the other hand, there was a broader spread of functions needed, and these included some that were not prominent in the other responses.

Specific defence systems

Glutathione (GSH)

The tripeptide glutathione (γ-glutamyl-cysteinyl-glycine) is the main low molecular mass thiol in yeast cells and is involved in a large number of cellular functions. These include synthesis of DNA precursors (via ribonucleotide reductase), protein synthesis, protein folding, amino acid transport, metabolism of various xenobiotics as well as in protecting cells from oxidative stress (Meister and Anderson, 1983; Penninckx, 2002). The oxidised form of glutathione, together with cystine may also play a role in redox homeostasis in the Golgi complex to counteract changes in reducing potential as protein disulphides are formed by the action of protein disulphide isomerase (Carelli *et al.*, 1997). This leads to secretion of these thiols from the cell.

Glutathione metabolism is summarised in Figure 9.5. The tripeptide is synthesised in two steps catalysed by γ-glutamyl-cysteine synthase which is encoded in yeast by the *GSH1* gene (Ohtake and Yabuuchi, 1991) and glutathione synthetase (encoded by *GSH2*: Grant *et al.*, 1997a). (For reviews of the synthesis of glutathione, and its role in protection against oxidative stress in yeast see Grant and Dawes, 1996; Penninckx, 2002.) When glutathione is used in the cell it is usually oxidised to the disulphide form (GSSG) from which the reduced form

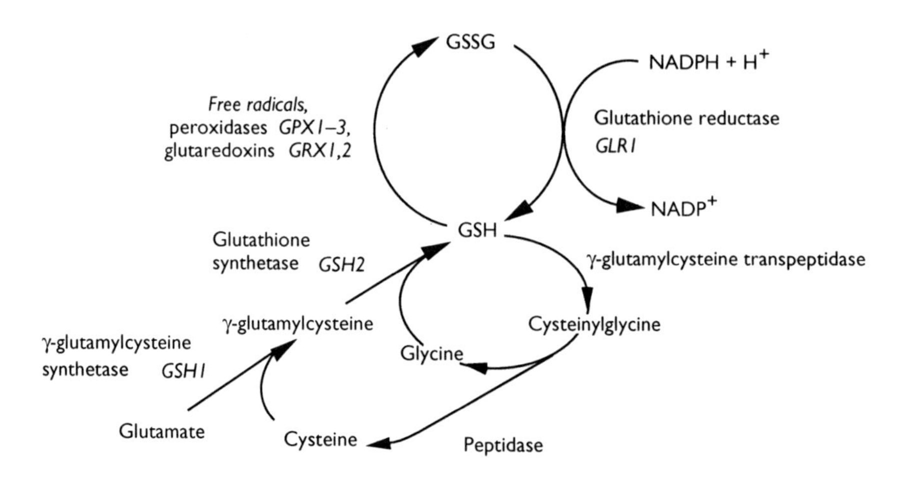

Figure 9.5 Glutathione metabolism in yeast. Genes involved are indicated in italics. The complete pathway for degradation of glutathione has not been conclusively demonstrated in yeast.

is regenerated by reaction with NADPH catalysed by glutathione reductase encoded by *GLR1* (Collinson and Dawes, 1995).

The role of glutathione in cellular defences against oxidants has been shown by isolation of mutants defective in synthesis of GSH. Null mutant *gsh1* mutants require a reductant for growth (although in some strains this can be satisfied by other thiols including 2-mercaptoethanol and dithiothreitol), but show increased sensitivity to oxidants including hydrogen peroxide and *tert*-butyl hydroperoxide (Wu and Moye-Rowley, 1994; Izawa *et al.*, 1995; Grant *et al.*, 1996b). This issue is complicated by the finding that the *gsh1* mutants isolated following disruption techniques were all phenotypically petite, and the fact that petites also exhibit increased sensitivity to oxidants (Collinson and Dawes, 1992; Grant *et al.*, 1997b). The *gsh1* mutants were, however, more sensitive than an isogenic ρ^0 petite, indicating the likely involvement of GSH in the protection against oxidative stress. This has been resolved by generating a null *gsh1* grande (by crossing a *gsh1* petite to a grande strain). This mutant also shows increased sensitivity to oxidants, as well as an elevated frequency of mutation of nuclear genes and petite formation relative to that seen in the isogenic wild type (Lee *et al.*, 2001). It is interesting that *gsh1* mutants are glutathione auxotrophs, but they have a very low requirement for the compound, about 100-fold below the concentration normally found in cells (Lee *et al.*, 2001). There is some essential role for glutathione in the cell that is as yet unidentified. It does not seem to involve antioxidant protection activity, ribonucleotide reduction or prevention of toxic accumulation of aberrant protein disulphides (Spector *et al.*, 2001). The concentration of glutathione in the cell is normally maintained at a very much higher level to fulfil the antioxidant buffering and other non-essential functions such as nitrogen storage. The *gsh1* mutant has been used extensively to probe the involvement of glutathione in a range of cellular processes, including the response to heat shock (Sugiyama *et al.*, 2000).

Comparison of the sensitivity to hydrogen peroxide of the *gsh1* mutant and the double *ctt1 cta1* mutant lacking catalases indicated that glutathione and the catalases provide overlapping defence systems. GSH appears to be the primary antioxidant for protection against hydrogen peroxide under the growth conditions used since mutants lacking GSH were sensitive to the oxidant whereas the catalase-deficient strains were not. Catalases do, however, provide some protection since the strains lacking catalase accumulate oxidised glutathione and the triple *gsh1 ctt1 cta1* mutant is more sensitive than the *gsh1* mutant (Grant *et al.*, 1998).

The outcome of disrupting the *GSH2* gene is interesting, since the mutants so generated have a reduced requirement for glutathione, and they are much less sensitive to oxidants than the *gsh1* mutant. This is probably the result of their accumulating the dipeptide γ-glutamyl-cysteine, which can act as an antioxidant in its own right (Grant *et al.*, 1997a). This may indicate that evolution of the pathway leading to synthesis of the tripeptide GSH occurred via an intermediate stage in which the dipeptide γ-glutamyl-cysteine served as an antioxidant. Interestingly, the *gsh2* null mutants overproduce the dipeptide, presumably because the tight control exerted over GSH synthesis is mediated via GSH feedback inhibition of γ-glutamyl cysteine synthetase and the dipeptide does not function as an inhibitor.

Disruption of the glutathione reductase (*GLR1*) gene does not lead to impairment of growth of yeast under normal aerobic conditions, but does cause them to become sensitive to a range of oxidants (Collinson and Dawes, 1995; Grant *et al.*, 1996a). Such mutants have a higher level of GSSG than the wild type, but only a slightly higher than normal level of total glutathione, indicating that the cells can synthesise reduced glutathione to maintain an appropriate concentration of GSH as it becomes oxidised.

Ubiquinol

Coenzyme Q, or ubiquinone (Figure 9.6), has a well-characterised function in the mitochondrial respiratory chain, but in its reduced form as ubiquinol also has an important role as a potent lipid-soluble antioxidant capable of inhibiting lipid peroxidation (Beyer, 1992; Ernster and Forsmark-Andree, 1993; Ernster and Dallner, 1995). It is also found in many intracellular membranes, including those of microsomes, lysosomes, peroxisomes and plasma membranes, and this is consistent with its role as an antioxidant (Takada *et al.*, 1982; Kalen *et al.*, 1987). Different organisms produce ubiquinones that differ in the length of the isoprenoid side chain; in *S. cerevisiae* this is six residues long ubiquinone-6 (Takada *et al.*, 1984), *E. coli* produces ubiquinone-8 and humans and *Schizosaccharomyces pombe* produce mainly ubiquinone-10 (Suzuki *et al.*, 1997). Apart from this, eukaryotes share the same coenzyme Q biosynthetic pathway (Olson and Rudney, 1983).

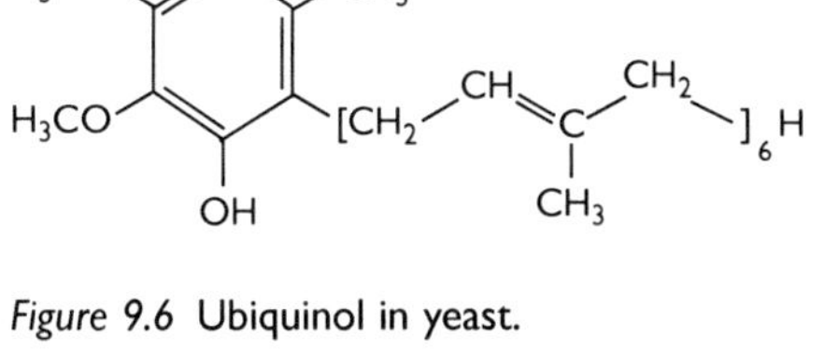

Figure 9.6 Ubiquinol in yeast.

Tzagoloff and his colleagues have described eight complementation groups (*coq1–coq8*) for yeast mutations that are defective in the synthesis of coenzyme Q (Tzagoloff *et al.*, 1975a,b; Tzagoloff and Dieckmann, 1990). Several of the genes, including *COQ1* (Ashby and Edwards, 1990); *COQ2* (Ashby and Edwards, 1990); *COQ3* (Clarke *et al.*, 1991); *COQ5* (Barkovich *et al.*, 1997; Dibrov *et al.*, 1997); and *COQ7* (Marbois and Clarke, 1996) encoding enzymes in the pathway have been identified and characterised (Schulz and Clark, 1999). Mutations in the *COQ* genes lead to a petite phenotype as a result of disruption of the respiratory chain. The *coq3* mutant is hypersensitive to treatment with polyunsaturated fatty acids compared to the wild-type strain. Since this sensitivity can be rescued by the addition of the synthetic antioxidant butylated hydroxytoluene or α-tocopherol, it was concluded that ubiquinol plays an important role *in vivo* in protecting cells against the products of lipid autoxidation (Do *et al.*, 1996).

Saccharomyces cerevisiae does not produce polyunsaturated fatty acids, and hence the extent to which lipid oxidation occurs is uncertain. The yeast does, however, produce significant intracellular levels of at least one product of fatty acid breakdown (malondialdehyde) and this is increased following exposure to oxidants such as H_2O_2; malondialdehyde can, however, be formed from other sources in the cell (Turton *et al.*, 1997). Yeast cells can also take up exogeneous polyunsaturated fatty acids and readily incorporate them into membrane lipid (Kohlwein and Paltauf, 1983; Bossie and Martin, 1989).

D-Erythroascorbic acid and L-ascorbate

L-Ascorbic acid (vitamin C) is found in many organisms and has strong *in vitro* antioxidant activity with the ability to scavenge a range of free radical species including the lipid peroxy radical ($RO_2^{\bullet}$), O_2^{-} and $^{\bullet}OH$. In the cell it appears to play an important role in recycling of vitamin E after its reaction with lipid radicals. It can, however, reduce ferric ions to the ferrous form and in the presence of H_2O_2 can stimulate $^{\bullet}OH$ formation via the Fenton reaction making it a pro-oxidant under certain conditions (Halliwell, 1994). *Saccharomyces cerevisiae* does not appear to synthesise significant amounts of L-ascorbic acid (5 μg/g dry wt), instead making ten times more (50 μg/g dry wt) of the closely related five-carbon compound D-erythroascorbic acid (Figure 9.7), indicating that ascorbate may play only a minor role, if any, in yeast metabolism (Leung and Loewus, 1985; Nick *et al.*, 1986; Kim *et al.*, 1993). Yeast can, however, convert exogenously added precursor L-galactono-1,4-lactone, or L-galactose to ascorbate via the D-erythroascorbate synthesis pathway, but the role of ascorbate (or erythroascorbate) in yeast metabolism remains to be fully resolved (Leung and Loewus, 1985; Hancock *et al.*, 2000).

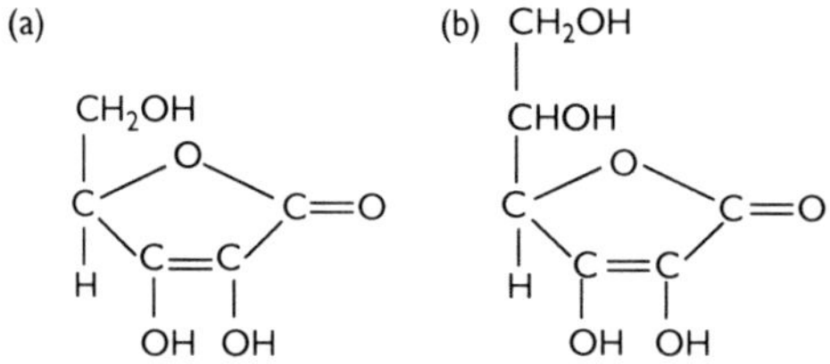

Figure 9.7 (a) D-erthrythroascorbic acid and (b) L-ascorbic acid.

The enzymes required for synthesis of D-erythroascorbate from D-galactose (D-arabinose dehydrogenase and D-arabino-1,4-lactone oxidase have been purified from yeast and shown to catalyse the conversion of D-arabinose to the antioxidant *in vitro* (Huh *et al.*, 1998; Kim *et al.*, 1998). Mutants that are unable to produce D-erythroascorbate have been generated by deletion of the *ALO1* gene encoding the last step in the biosynthetic pathway catalysed by D-arabino-1,4-lactone oxidase (Huh *et al.*, 1998). These mutants were more sensitive to oxidative stress, and overexpression of *ALO1* made them more resistant. This contradicts the view that D-erythroascorbate is of limited importance as an antioxidant in *S. cerevisiae*. This was proposed by (Spickett *et al.*, 2000) on the basis of observations of its oxidation in pre-loaded cells by external *t*-butyl hydroperoxide and hydrogen peroxide and its limited effect on glutathione levels in cells treated with oxidants.

Little is known about the control of erythroascorbate synthesis in yeast, although some progress has been made in *Neurospora crassa*. In this ascomycete, D-erythroascorbate levels are affected by the level of cAMP in the cell, as shown by the increased levels in the *cr-1* mutant that is defective in adenylate cyclase, and decreased amounts in the presence of exogenous cAMP. The level is also elevated on deprivation of nitrogen source (Dumbrava and Pall, 1987). If this is the case in *S. cerevisiae*, there are interesting parallels with other stress response systems, such as the induction of the *CTT1* and *SOD2* genes, as discussed later.

Since the development of dough in breadmaking depends on the cross-linking of gluten via oxidation of its cysteine residues, the antioxidants produced by yeast could have important effects by affecting the speed or extent of this reaction. This is important in frozen doughs since the freezing and thawing processes, as well as extended maintenance at low temperatures can lead to release of the yeast metabolites in to the dough. Release of glutathione is known to cause problems under such circumstances. Kim *et al.* (1993) have shown that D-erythroascorbate treatment slightly increased the flow of dough as rest time increased, and led to gluten that stretched at a faster rate than control gluten isolated from untreated dough. Like L-ascorbate, the compound did not change dough development time; it differed from L-ascorbate, however, since it did not show an oxidising effect on dough or gluten or an improving effect on bread.

D-Erythroascorbate is almost as readily oxidised as L-ascorbate in aqueous systems, and hence may have similar antioxidant activity. However, when this was tested in the tobacco hornworm *Manduca sexta* which requires vitamin C for growth and development, it was found that the D-erythroascorbate could support growth of the larvae, but not their ecdysis to pupae and adults, indicating a difference in its metabolic effects from those of L-ascorbate (Shao *et al.*, 1993).

Superoxide dismutases

Yeast cells contain two SOD genes: *SOD1* encodes the Cu/Zn SOD which in yeast, as in most other eukaryotes, is present in the cytoplasm; *SOD2* encodes the MnSOD which is located in the mitochondrion (Gralla and Kosman, 1992). Recently, it has been shown that a fraction of the Cu/Zn SOD and its metallochaperone CCS are co-localised to the mitochondrial intermembrane space, and that this form of the enzyme helps protect the mitochondrion against oxidative damage. Mutants with deletion of *SOD1* show elevated carbonylation of mitochondrial proteins and yeast cells enriched for the inner membrane space localised Cu/Zn SOD show extended survival in stationary phase (Sturtz *et al.*, 2001). Compared to the MnSOD, the Cu/Zn SOD is expressed at high levels in most organisms. (For a review of SODs and their role in cellular metabolism see Culotta, 2000.)

Oxygen, paraquat treatment, and growth on the non-fermentable carbon sources acetate and lactate lead to an increase in cellular levels of the Cu/Zn SOD (Lee and Hassan, 1985; Gralla and Kosman, 1992), and the enzyme is also increased in response to exposure to copper ion (Lee and Hassan, 1985; Greco *et al.*, 1990). Transcription of the *SOD2* gene encoding the mitochondrial MnSOD is induced by growth in oxygen, or on non-fermentable carbon sources, or by treatment of cells with superoxide-generating agents such as paraquat (Lowry and Zitomer, 1984; Westerbeek-Marres *et al.*, 1988; Galiazzo and Labbe-Bois, 1993; Pinkham *et al.*, 1997). In some strains it is also induced, to a much greater extent than by paraquat, as cells become starved, or enter the post-diauxic shift phase following growth on glucose (Lowry and Zitomer, 1984; Westerbeek-Marres *et al.*, 1988; Chang *et al.*, 1991; Galiazzo and Labbe-Bois, 1993; Flattery-O'Brien *et al.*, 1997; Pinkham *et al.*, 1997). The MnSOD provides a major defence against superoxide toxicity, and one of its roles is to protect mitochondrial Fe–S enzymes, such as aconitase, from superoxide-induced damage (Longo *et al.*, 1999). Interestingly, the lack of the cytoplasmic Cu/Zn SOD leads to an up-regulation of the *FET3* gene encoding the high affinity iron transporter and increased intracellular iron levels (De Freitas *et al.*, 2000), presumably as a response to assist reconstitution of Fe–S cluster-containing enzymes. Neither gene is absolutely essential for growth, but *sod1* mutants have reduced rates of aerobic growth in synthetic glucose medium and reduced ability to grow on non-fermentable substrates (De Freitas *et al.*, 2000). Mutants of *S. cerevisiae* with deletions of both *SOD* genes are unable to synthesise methionine and lysine, are defective in sporulation and have an increased mutation rate (Liu *et al.*, 1992). The auxotrophy for methionine and lysine is mainly the result of the *sod1* null mutation, but the *sod1 sod2* double mutant showed a slower growth rate than the *sod1* mutant in defined media lacking the auxotrophic requirements. Efficient sporulation requires at least one functional copy of both *SOD1* and *SOD2*, and *sod2* homozygous diploids do not show any ascus formation in the 10,000 tested. This presumably reflects the high demand for oxygen during sporulation (which can only occur in respiring cells) as well as the need to remove the superoxide anions generated in the mitochondrion during the process.

Both *sod1* and *sod2* deletion strains are hypersensitive to oxygen. When cells were incubated in a non-fermentable carbon source at low aeration rates both singly mutant strains and the double mutant grew, however, at high aeration rates growth of the single mutants was reduced, and the double mutant failed to grow (van Loon *et al.*, 1986; Longo *et al.*, 1996). The toxicity of oxygen to cells of *S. cerevisiae* lacking *SOD1* (encoding the cytoplasmic Cu/Zn SOD) can be suppressed by overexpression of the *ATX2* gene. This also reverses the lysine and methionine auxotrophies of the *sod1* deletion mutant, and enhances the resistance of these strains to paraquat and atmospheric oxygen. This gene appears to regulate Mn^{2+} levels, encoding a Mn^{2+}-trafficking protein that is located in Golgi-like vesicles (Lin and Culotta, 1996). These authors have identified several genes in which mutations can act as extragenic suppressors to bypass *SOD1* deficiency. These include mutations in *PMR1* (encoding a member of the P-type ATPase family which is another vesicle-located Mn^{2+}-trafficking protein involved in Ca^{2+} and Mn^{2+} homeostasis (Lapinskas *et al.*, 1995)) and *BSD2* which is a novel gene that probably functions in the homeostasis of heavy metal ions (Lin and Culotta, 1996).

Recently Cu/Zn SOD has been identified as a protein inhibitor of the PK60S kinase that phosphorylates acidic ribosomal proteins. The reversible phosphorylation of these proteins has a role in regulating the number of active ribosomes. This has led to the suggestion that the cytosolic SOD may participate in the regulation of 80S-ribosome activity under stress conditions (Zielinski *et al.*, 2002).

Metal ion uptake and homeostasis, in the context of oxidative stress has been reviewed in detail (Santoro and Thiele, 1997). Fe and Cu are the most important metal ions since they participate readily in the production of damaging free radical species, and the role of metallothionein as a scavenger of excess Cu ions is discussed later.

Catalases

Saccharomyces cerevisiae has two haemoprotein catalases which can dismutate hydrogen peroxide into water and oxygen. One, catalase A (encoded by the *CTA1* gene: Cohen *et al.*, 1985) is located in the peroxisome (Seah *et al.*, 1973), and the other is cytosolic (Seah and Kaplan, 1973) and is encoded by *CTT1* (Spevak *et al.*, 1986; Ruis and Hamilton, 1992). Neither of these catalases has the ability to react with larger hydroperoxides such as *tert*-butyl hydroperoxide whereas glutathione peroxidases can act on both hydrogen peroxide and the larger hydroperoxides. In animal cells the principal antioxidant enzymes for H_2O_2 detoxification has long been thought to be glutathione peroxidases rather than catalase, since catalase has a much lower affinity for H_2O_2 than glutathione peroxidase, and human erythrocytes lacking catalase activity are not susceptible to oxidant-induced haemolysis (Jacob *et al.*, 1965).

Disruption of the *CTT1* gene has been reported to lead to sensitivity to hydrogen peroxide and to exposure to elevated temperatures (Wieser *et al.*, 1991). However, Izawa *et al.* (1996) have shown that a mutant devoid of both catalases *cta1 ctt1*, as well as ones lacking either *cta1* or *ctt1* showed a similar growth rate to the wild type under non-oxidative stress conditions. Moreover, in the exponential phase of growth there was no difference in the susceptibility of the mutant strains to H_2O_2. During stationary phase the double mutant was much more sensitive to peroxide than the singly mutant or wild-type strains which showed a similar response. There is an increased resistance of cells to many forms of stress, including H_2O_2 as cells enter stationary phase (Jamieson *et al.*, 1994), which was seriously impaired in the acatalasaemic strain, but not in either singly mutant strain. This indicates that the catalases in yeast may not be important in growing cells in scavenging of endogenous hydrogen peroxide that is produced during respiration or β-oxidation, and that this function is performed by cytochrome *c* peroxidase, or glutathione peroxidases. This is consistent with the demonstration that in exponentially growing cells glutathione has a more important role than the catalases in the cellular response to hydrogen peroxide (Grant *et al.*, 1998). The catalases have, however, an important role in removal of H_2O_2 during stationary phase, and may play some part in the adaptive response to oxidative stress discussed (Izawa *et al.*, 1996). These results indicate that when catalases are functional, then the two forms, one located in the cytoplasm and the other in the peroxisome can each substitute for the absence of the other; this is presumably a reflection of the mobility of H_2O_2 in the cell.

Expression of the catalase genes is tightly regulated: *CTA1* is induced by oxygen, growth on non-fermentable carbon sources and fatty acids, and repressed by glucose (Ruis and Hamilton, 1992). The *CTT1* gene is dependent on haem for its expression and is controlled by oxygen, heat shock, osmotic stress, oxidative stress, copper ions and the availability of nitrogen, phosphorus and sulphur (Hörtner *et al.*, 1982; Bissinger *et al.*, 1989).

Thioredoxins and glutaredoxins

Many studies have indicated the key role played by sulphydryl moieties in the cellular response to oxidative stress, and in maintenance of redox homeostasis (Carmel-Harel and

Stortz, 2000; Grant, 2001). The major systems that are involved, the thioredoxins and glutaredoxins, are of considerable current interest. These are structurally very similar and low molecular mass proteins that have two conserved cysteine residues in their active site (Holmgren, 1989). These proteins can exist in either reduced or oxidised forms.

Thioredoxin and glutaredoxin are directly involved as hydrogen donors for ribonucleotide reductase in nucleic acid biosynthesis. They have, however, also been implicated in the mechanism and regulation of protein folding, antioxidant activity possibly by scavenging reactive oxygen species or reduction of dehydroascorbate or regenerating proteins that have been inactivated and in sulphur metabolism (Fernando *et al.*, 1992; Mitsui *et al.*, 1992; Trotter and Grant, 2003). The two systems and their overlap are illustrated in Figure 9.8. The oxidised forms of both proteins are ultimately reduced at the expense of NADPH. The reduced form of thioredoxin is regenerated directly by reaction with NADPH catalysed by thioredoxin reductase (Holmgren, 1985; Wollman *et al.*, 1988) encoded by the *TTR1* gene in yeast. Glutaredoxins differ in that the reduced form is regenerated from the oxidised form by reaction with GSH, which is then reduced in an NADPH-dependent reaction catalysed by glutathione reductase encoded by *GLR1*.

Two thioredoxin systems have been identified in *S. cerevisiae*. The most extensively characterised is cytoplasmically located and involves two thioredoxin genes, *TRX1* and *TRX2* (Gan, 1991; Muller, 1991) and one *TRR1* encoding thioredoxin reductase. Recently, an active mitochondrial system has been identified with *TRX3* encoding the thioredoxin and *TRR2*, the reductase (Pedrajas *et al.*, 1999). None of the three genes for the cytoplasmic system is essential since all double mutants and the triple deletant strain, *trx1 trx2 trr1*, can grow (Grant, 2001). Deletion of *TRX2* does, however, lead to increased sensitivity to H_2O_2 and other hydroperoxides during stationary phase growth (Garrido and Grant, 2002) and this gene is regulated by the yAP-1 transcription factor that mediates responses of many genes

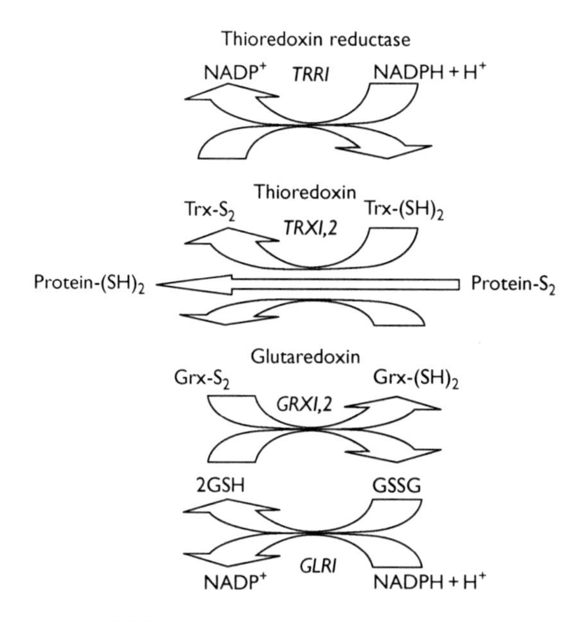

Figure 9.8 The glutaredoxin/thioredoxin systems.

to oxidants and xenobiotics (Kuge and Jones, 1994). This may, however, reflect the demands of DNA synthesis for the glutaredoxin system to provide reducing equivalents to ribonucleotide reductase when the thioredoxin system is impaired, and have less bearing on the response of cells to oxidants. The *trx1 trx2* double mutant is inhibited in vacuole inheritance and also shows a disturbance to the cell cycle progression in which S phase is prolonged and G1 is reduced (Gan, 1991; Muller, 1991); this is not due to changes in the levels of deoxyribonucleotides. Additionally, the double mutant cannot grow in the absence of cysteine and methionine presumably due to a defect in sulphate assimilation, indicating that thioredoxin under normal circumstances is the donor of hydrogen for 3-phosphoadenosine 5′-phosphosulphate reductase in yeast (Muller, 1991). It is interesting that a synthetic-lethality screen using the *trx1 trx2* double mutant led to the isolation of the *GLR1* gene, since the triple *trx1 trx2 glr1* mutant is unable to grow (Muller, 1996); the overlap between the thioredoxin and glutaredoxin systems is discussed next.

Deletion of either the mitochondrial thioredoxin or reductase did not affect viability or growth of yeast on fermentative or respiratory substrates. The *trx3* deletant was no more sensitive than the wild type to hydrogen peroxide, whereas the *trr2* deletant was (Pedrajas *et al.*, 1999). These authors suggest that there may be a thioredoxin-independent antioxidant activity of the reductase – in mammalian systems this may occur via its acting as an electron donor for glutathione peroxidase (Bjornstedt *et al.*, 1994) or direct reduction of lipid hydroperoxides (Bjornstedt *et al.*, 1995).

Glutaredoxins are heat-stable proteins that act as glutathione-dependent disulphide oxidoreductases (Holmgren and Aslund, 1995). They overlap, in their function with the thioredoxins, since they have an active pair of cysteine residues that can function in hydrogen donation to ribonucleotide reductase (Bushweller *et al.*, 1992). They also catalyse the reduction of low molecular mass mixed disulphides and some protein disulphides using GSH as co-factor (Holmgren, 1979). For this latter reaction, an initial mixed disulphide between GSH and the protein disulphide is formed non-enzymatically, followed by reaction with glutaredoxin to release the reduced protein and forming a mixed disulphide between GSH and glutaredoxin (Bushweller *et al.*, 1992). This is consistent with the finding that glutaredoxin is specific for GSH-containing mixed disulphides (Gravina and Mieyal, 1993). They have also been considered to be involved synergistically in GSH-dependent protein folding catalysed by protein disulphide isomerase (Lundström-Ljung and Holmgren, 1995) and in protection against oxidative stress.

Glutaredoxins catalyse the cleavage of mixed disulphides in the presence of low GSH concentrations, and they may therefore act to reduce any mixed disulphides formed following oxidative damage (Chrestensen *et al.*, 1995). *In vitro* glutaredoxins can reactivate a number of oxidised enzymes by reducing the mixed disulphides formed as a result of thiol oxidation (Terada, 1994). They may therefore function *in vivo* to repair oxidatively damaged proteins and serve as electron donors, for example for human plasma glutathione peroxidase (Bjornstedt *et al.*, 1994) and in the regeneration of ascorbic acid in a GSH-dependent reaction catalysed by glutaredoxin and protein disulphide isomerase (Wells *et al.*, 1990; Meister, 1994). Whether this reaction occurs with the erythroascorbic acid present in yeast remains to be determined.

There are five glutaredoxin homologues in yeast encoded by *GRX1–5*. Two show high homology to each other and to the family of glutaredoxins from other species and have the pair of cysteines at the catalytic site. The others have less similarity and lack one of the putative active site cysteines. The two main glutaredoxin genes have been termed *GRX1* and

GRX2 (Luikenhuis *et al.*, 1998), although *GRX2* gene was first identified by its thioltransferease activity and originally called *TTR1* (Gan, 1992). Recently, these enzymes have also been shown to have glutathione peroxidase activities. They can reduce peroxides directly using GSH to form the corresponding alcohol, which can be subsequently detoxified by conjugation to GSH. The conjugate can then be removed from the cytoplasm via the glutathione-conjugate Cf1p or the glutathione-*S*transferases Gtt1p and Gtt2p (Collinson and Dawes, 2002). Grx2p exists in two different isoforms, one of which is predominantly cytoplasmic and the other mitochondrially located (Pedrajas *et al.*, 2002).

The set of singly mutant, *grx1* and *grx2*, and double, *grx1grx2*, null mutant strains have been constructed (Luikenhuis *et al.*, 1998). The two genes appear to have somewhat different roles in the cell: one of them, *GRX2*, encodes the majority of the disulphide reductase activity, since a null *grx2* mutant has only 5% of the wild-type enzyme level. Moreover the *grx1* mutant is sensitive to superoxide-generating agents and not peroxides, whereas the *grx2* mutant is sensitive to H_2O_2. This provides further confirmation that there is a substantial difference in the spectrum of damage caused by the two different oxidants, despite the fact that the superoxide anion is converted to H_2O_2 by SODs. Detoxification of H_2O_2 is brought about both by catalase and glutathione peroxidases (which use GSH as reductant). In yeast, the GSH-mediated reactions appear to be the primary means of detoxifying H_2O_2, although catalases are (and induced strongly) during stationary phase (Izawa *et al.*, 1995). This is consistent with the finding that Grx2p was required for H_2O_2 resistance during exponential phase growth but not during the stationary phase (Luikenhuis *et al.*, 1998). Since Grx2p was the main source of mixed disulphide oxidoreductase activity in yeast, its primary function may be to detoxify any proteins damaged by reactive oxygen species. The genes are differentially regulated. Both are up-regulated by a variety of stress conditions, including exposure to oxidants, hyperosmotic conditions, heat and also during the stationary phase. Similar levels of induction of each genes were seen following treatment with H_2O_2, menadione and diamide, but *GRX1* was induced to a much greater extent in response to heat and osmotic stress (Luikenhuis *et al.*, 1998; Grant *et al.*, 2000), mediated by STREs in the promoters of both genes.

Of the monothiol glutaredoxins Grx3–5p, only the role of Grx5p has been studied in much detail. This protein is mitochondrially located and mutants lacking it show constitutive oxidative damage augmenting that caused by externally added oxidants, and (Rodriguez-Manzaneque *et al.*, 2002) have linked its function to the assembly of iron/sulphur clusters into proteins. Mutants lacking the enzyme are also unable to dethiolate a glyceraldehyde-3-phosphate isozyme that is glutathionylated following exposure of cells to hydrogen peroxide (Shenton *et al.*, 2002).

The extent of functional overlap between the thioredoxins and glutaredoxins has been investigated by analysis of multiple mutants affected in both systems. All combinations of *trx1, trx2, grx1, grx2* are viable with the exception of the *trx1trx2grx1grx2* quadruple mutant, hence the essential function of these enzymes can be met by any one of the four proteins. As note earlier, the *trx1trx2glr1* mutant is inviable, as is the *trr1glr1* double mutant which cannot reduce oxidised forms of either thioredoxin or glutaredoxin. Trr1p is required during normal growth, but Glr1p is not (Draculic *et al.*, 2000; Trotter and Grant, 2003). Despite this functional overlap, the redox states of the two systems are maintained separately. The redox state of the thioredoxins is maintained independently of the glutathione system, whereas there is a strong correlation between the redox states of the glutaredoxins and the GSH/GSSG couple (Trotter and Grant, 2003). This has important implications for the maintenance of normal redox homeostasis in cells.

Glutathione and Thioredoxin peroxidases

In addition to catalases, many organisms contain a range of peroxidases that can reduce hydroperoxides. Since catalases do not react with bulky hydroperoxides, other peroxidases are required to reduce lipid peroxides and probably also those formed in the amino acid side chains of proteins. Peroxidases can have a variety of reductants, including glutathione, thioredoxin, cytochrome *c* and ascorbate.

Saccharomyces cerevisiae has been known to contain glutathione peroxidase activity for some time and three genes encoding glutathione peroxidases GPX1–3 were identified from the genome sequencing data. The proteins they encode have been shown to possess peroxidase activity (Inoue *et al.*, 1999). Avery and Avery (2001) have shown that yeast is unusual in that these probably encode phospholipid hydroperoxide peroxidases which differ from more widespread glutathione peroxidases in being monomeric, partly membrane-associated, and capable of reducing lipid hydroperoxides esterified to membranes. Of the deletants only the *gpx3* was sensitive to peroxides, and the *GPX1* gene was induced by glucose starvation, the *GPX2* gene by oxidative stress and *GPX3* was reported to be constitutively expressed (Inoue *et al.*, 1999). The genes are expressed at different times during respiration. *GPX1* is controlled via the Msn2,4p regulated pathway, *GPX2* as part of the respiratory up-regulation and GPX3 under the control of the oleate-response system mediated by Pip1p and Oaf1p transcription factors. Gpx1p and Gpx2p contribute to the adaptive response to lipid peroxide stress by an inducible response controlled by the yAP-1 pathway, while Gpx3p provides the greatest protection to hydroperoxide stress (Israel, personal communication). Given the recent demonstration that Gpx3p acts as the sensor of hydrogen peroxide damage in cells (Delaunay *et al.*, 2002) this is not surprising.

Yeast cells have been reported to have up to five genes encoding thioredoxin peroxidases (peroxiredoxins) whose products are distributed in the cytoplasm, mitochondrion and nucleus (Park *et al.*, 2000; Pedrajas *et al.*, 2000). Deletion of the main cytosolic enzyme encoded by *TSA1*, or *TSA2* or the gene for the mitochondrial enzyme leads to some hypersensitivity to oxidants, and the *tpa1 tpa2* double mutant is more susceptible than either of the single mutations (Wong *et al.*, 2002). The *TSA1* gene is expressed to a greater extent than *TSA2*, although the latter is more highly induced by oxidatives stress.

Metallothioneins

These are small cysteine-rich metal-binding proteins that act as scavengers for metal ions, especially copper ions (Kagi and Schaffer, 1988). In yeast there are two metallothionein genes, *CUP1* and *CRS5*; the *CUP1* gene plays the more dominant role in protection against Cu toxicity since deletion of this gene makes the cell much more sensitive to Cu than deletion of *CRS5* (Culotta *et al.*, 1995). Metallothionein appears to play more than just a Cu-scavenging role in the protection against oxidative stress, acting as an antioxidant. Expression of *CUP1* suppresses the oxidative stress-induced defects of *sod1* mutants, restoring their ability to grow on lactate and suppressing the methionine and cysteine auxotrophy, and moreover it increases the cellular resistance to paraquat (Tamai *et al.*, 1994; Santoro and Thiele, 1997). The Cup1p protein shows a distinct antioxidant activity *in vivo*, and pulse radiolysis data indicate that the Cu(I)-bound form can scavenge both superoxide anion and hydroxyl radical (Santoro and Thiele, 1997). *CUP1* expression is regulated not only by the presence of metal ions, but also by heat shock, high oxygen, respiration, and in response to redox-cycling drugs (Tamai *et al.*, 1994; Liu and Thiele, 1996), highlighting its antioxidant role.

9.4 Osmotic stress (hypo- and hyperosmotic stress) and salt stress

Cellular responses to osmotic stress have been reviewed by Varela and Mager (1996), Hohmann (2002) and Mager and Siderius, (2002). There are several forms of osmotic stress; the more obvious differences are between those concerned with exposure of cells to hyperosmotic and hypo-osmotic conditions, for these cells respond differently, with different control mechanisms. For hyperosmotic stress there are also separable components depending on the nature of the solute causing the stress. This depends on whether accumulation of the solute in the medium causes high osmolarity *per se* (as in stress with an uncharged solute) or whether it results in the additional dimension of ionic stress due to the presence of charged ionic species.

Hyperosmolarity

Yeast cells shifted to a medium of higher osmotic pressure lose water from the cytoplasm almost instantaneously. This leads to a reduction of cell volume which depends on the osmolality of the resulting solution, for example, a reduction to 35% of the starting cell volume occurred after exposure of cells to a 5 osM NaCl solution, within 0.5 min (Morris *et al.*, 1983). This is partially offset by the transfer of water from the vacuole, but the whole process leads to a decrease or abolition of the turgor pressure of the membrane on the cell wall (Blomberg and Adler, 1992; Latterich and Watson, 1992). Following stress applied in the range from 1 to 5 osM NaCl there is a 50–75% loss in viability (Morris *et al.*, 1983). In contrast to plant cells in which the plasma membrane shrinks away from the cell wall during plasmolysis, the entire cell envelope shrinks when yeast cells are placed in hypertonic solutions. Morris *et al.* (1983) have shown by freeze-etch electron microscopy that the cell wall increases in thickness, buckling on the inner face to form irregular projections into the cytoplasm. The cytoplasmic membrane has often been considered the primary site of osmotic injury, however, following resuspension of cells from a hypertonic stress that would have caused a major loss of viability, all of the cells were osmotically responsive and recovered their original volume. The cell damage that occurs following osmotic stress has not been documented clearly, although there are indications that components of the plasma membrane may be affected (Hohmann, 1997), and that disassembly of the actin cytoskeleton occurs (Novick and Botstein, 1985; Chowdhury *et al.*, 1992).

As a result of these physicochemical changes, primary defence responses are triggered in the cells to provide protection, repair and initiate recovery. These include temporary growth and cell cycle arrest, closure of the Fsp1p glycerol channel and activation of the high-osmolarity glycerol (HOG) mitogen-activated protein (MAP) kinase and the general stress response (Msn2p–Msn4p-mediated) pathways (Mager and Siderius, 2002). In addition to the up-regulation of genes involved in glycerol metabolism there is a substantial number of other genes up-regulated as a result of these responses reflecting the overall adjustment of the cell's physiology (Gasch *et al.*, 2000; Rep *et al.*, 2000; Causton *et al.*, 2001).

Those yeast cells that survive, respond in the longer term to increased osmolarity of the growth medium by enhanced production and accumulation within the cell of the compatible solute glycerol at concentrations in the molar range (Albertyn *et al.*, 1994). They also extrude ions by the operation of membrane ATPases. Compatible solutes are ions or compounds that accumulate in cells to balance internal and external osmolarity without seriously affecting the physical and molecular processes in the cell, and for other organisms these can include sugars, sugar alcohols, K^+, amino acids and glycine betaine. For yeast, glycerol appears to be the sole compatible solute (Brown, 1978; Blomberg and Adler, 1992; Mager and Varela, 1993; Varela and Mager, 1996) and its intracellular concentration can increase 2-to 30-fold depending on the strain and degree of osmotic stress. Under high salt conditions other solutes such as trehalose may be important. This may be due to an overlap between the physiological effects of salt stress and other damaging conditions such as heat shock and oxidative stress and the putative ability of trehalose to protect cellular structures such as membranes (Wiemken, 1990; Andre *et al.*, 1991; Iwahashi *et al.*, 1995).

From genome-wide screens, more than 600 genes are activated in the cellular response to hyperosmolarity (Gasch *et al.*, 2000; Posas *et al.*, 2000). Activation of a gene does not necessarily mean that it has a significant involvement in the physiological processes that contribute to protection, repair or recovery from the stress. A more limited number of genes that function for survival or growth under high osmotic conditions have been identified. These include *GPD1* and *GPP2* needed for synthesis of the compatible solute glycerol, *FPS1* for maintenance of intracellular glycerol. It also includes genes involved in transport processes, the maintenance of cell structure and size, and the osmosensing signal transduction pathways discussed later (Brüning *et al.*, 1998). Brüning *et al.* (1998) isolated mutants by direct selection for osmosensitivity identifying eleven genes including *GPD1, PBS2* and *KAR2*. Subsequent transposon mutagenesis has identified 31 genes that when mutated led to salt sensitivity. This did not include *HOG1, GPD1* and *PBS2* but did identify several involved in the response to salt (Na^+ and Li^+) including the Na^+-ATPase encoded by the *ENA* gene cluster, and genes involved in the calcineurin pathway (see later). Other functions affected included signalling functions, energy generation, actin-related processes and vesicular transport (Ferreira *et al.*, 2001).

The HOG pathways of response to hyperosmotic stress

Hohmann (2002) has recently very comprehensively reviewed the systems by which yeast cells sense, respond and adapt to osmotic stress. So far, the mechanisms whereby cells sense osmotic stress have not been elucidated, although there has been considerable progress on the signalling pathways that transduce the signal to activate gene expression and activate other physiological changes. High osmolarity in yeast leads to the induction of two specific osmosensing signal systems (Figure 9.9), both of which activate the common HOG signal transduction pathway (Gustin *et al.*, 1998).

The Sln1 branch of this pathway acts as a negative regulator of the pathway, since deletion of either the sensor gene *SLN1* or *YDP1* is lethal due to overaction of the HOG pathway. The sensor Sln1p is therefore activated by cell swelling and inhibited by cell shrinkage (Maeda *et al.*, 1994; Hohmann, 2002). In this pathway the Sln1p–Ssk1p-(Ypd1p)-osmosensor activates a MAP kinase cascade composed of the Ssk2p and Ssk22p MAP kinase kinases (MAPKKK), the Pbs2p MAPKK and the Hog1p MAPK. *SLN1*, which is an essential gene for growth on rich medium has strong homology with the prokaryotic 'two-component' control systems that mediate the responses to a variety of extracellular stimuli (Ota and Varshavsky, 1993). These two-component systems are usually composed of

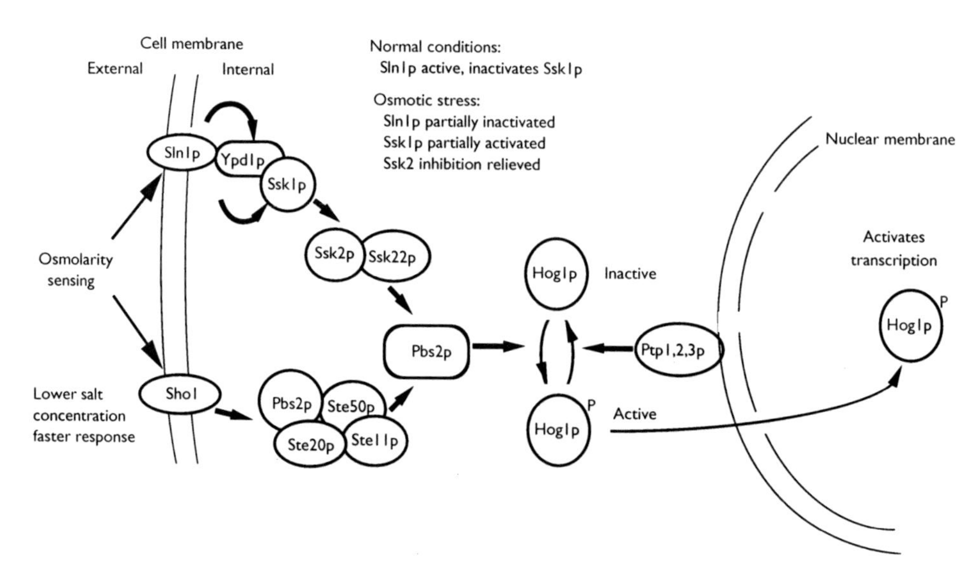

Figure 9.9 The pathways leading to HOG1 activation during osmosensing in yeast. Two different osmosensors converge to activate the MAP kinase signal transduction cascade identified by Ssk2p, Pbs2p and Hog1p. The pathway activates amongst others the genes for synthesis of glycerol in response to elevated osmotic stress.

a sensor protein in the membrane which is autophosphorylated on a histidyl residue on receiving the stimulus. This in turn has kinase activity and transfers the phosphate group to an aspartyl residue on the response regulator to alter its activity. In bacteria this regulator is typically a transcription factor, whereas in yeast the regulator (encoded by *SSK1*) initiates a protein kinase cascade whose components show homology to the MAP kinase pathway in mammalian cells. High osmolarity appears to partially inhibit the phosphorylation of Ssk1p by the Sln1p sensor (Maeda *et al.*, 1994).

The second HOG pathway osmosensory system involves the Sho1p transmembrane protein which has an SRC homology 3 (SH3) domain that interacts with a proline-rich motif in the NH_2-terminal region of Pbs2p (Maeda *et al.*, 1995). Sho1p appears not to be the direct sensor of osmolarity since deletion of its transmembrane domain does not affect signalling, and a membrane-targeted version of Pbs2p functions as well (Mager and Siderius, 2002). This signal transduction pathway requires the involvement of a signalling complex of Cdc42p, Ste20p, Ste50p and Ste11p proteins. The small GTPase Cdc42p appears to recruit this Pbs2p complex to the site of bud emergence where polarised growth occurs (Reiser *et al.*, 2000). Curiously, this pathway involves the Ste11p MAPKKK (Posas and Saito, 1997), which is an important part of the mating pheromone response pathway in yeast (Rhodes *et al.*, 1990). Posas and Saito (1997) propose that cross-talk between the two pathways is restricted by the formation of complexes; for the osmoregulation pathway Ste11p would be held in one complex by interaction with Pbs2p, preventing its phosphorylation of the Ste7p MAPKK of the pheromone receptor pathway.

Mager and Siderius (2002) have suggested that there is a third system that can activate Pbs2p since mutants defective in both of these branches still show Hog1p activation under severe osmotic conditions (>1.2 M sorbitol). Disruption of *BCK1* encoding the MAPKKK of the protein kinase C (PKC) cell integrity pathway abolished this activation, although it remains to be seen whether this is due to a direct effect of Bck1p-mediated phosphorylation. Mager and Siderius (2002) have also speculated that the PKC pathway and a recently proposed 'sterile vegetative growth pathway' that makes use of the main components of the filamentous growth MAP kinase system may be involved in cell wall sensing and signalling of osmotic stress.

Hog1p is phosphorylated transiently on residues Thr174 and Tyr176 in the cytoplasm, and once phosphorylated it is rapidly translocated to the nucleus to activate transcription. Phosphorylation is necessary and sufficent for this transport. This nuclear concentration can be detected within less than a minute of hyperosmotic shock and is transient with the timing dependent on the extent of osmotic shock. Nuclear localisation correlates in wild-type strains with the extent and duration of Hog1p phosphorylation. Activated Hog1p also exerts effects on functions external to the nucleus (Hohmann, 2002).

Several transcription factors, including Sko1p/Acr1p, Hot1p, Msn2p/4p and Smp1p appear to be targets of the HOG pathway. Sko1p can be a repressor or an activator depending on context, and its activity is regulated directly by Hog1p (Proft *et al.*, 2001). It has some role in regulating *ENA1* encoding a plasma membrane Na^+-exporter, *GRE2* and *HAL1*. Hot1p is a helix–loop–helix factor needed for the activation of the *GPD1* and *GPP2* genes involved in glycerol biosynthesis, and for some promoters its binding seems to involve association with the Hog1p kinase (Hohmann, 2002). The Msn2p, 4p factors are discussed later.

What do the sensors detect? Varela and Mager (1996) have pointed out that cells could respond to the sudden increase in the intracellular concentration of a solute such as Ca^{2+}, or through the extracellular activation of a plasma-membrane bound receptor that senses membrane extension. While the molecular mechanism of sensing remains to be determined, there is now some evidence that the latter of these models may occur for at least part of the response. Hohmann (2002) has speculated on why there are two branches to the HOG pathway, and has suggested that these may reflect different roles in osmosensing. Since components of the Sho1p branch are localised to places of active cell growth it is suggested that this branch is more involved in monitoring osmotic changes during cell growth and expansion while the Sln1p branch senses environmental changes.

The activity of the HOG pathway can be down-regulated by dephosphorylation of Hog1p by the serine–threonine phosphatases, Ptc1–3p (Warmka *et al.*, 2001) and the tyrosine phosphatase Ptp2p and Ptp3p (Jacoby *et al.*, 1997). Hohmann (2002) has discussed the physiological role of these phosphatases and their likely involvement in the osmotic response. He has suggested that while some are essential to prevent lethality due to overactivity of the phosphorylated Hog1p, they need not act as specific regulators of the overall process. He proposes this down-regulation on recovery may be achieved through reversal of the signalling functions of the MAP kinase pathway and that this takes place at the level of the membrane sensing (Hohmann, 2002).

The Ca^{2+}/calmodulin-dependent phosphoprotein phosphatase (calcineurin) pathway and salt stress

High salt concentrations can stress cells either through their osmotic activity, or through ion toxicity. For cells growing on glucose it appears that NaCl exerts its effects through Na^+ toxicity, since the cells apparently have a high capacity for osmotic adjustment and a reduced ability to pump out Na^+; on other substrates osmotic effects may predominate (Serrano, 1996; Serrano *et al.*, 1997; Proft *et al.*, 2001).

Saccharomyces cerevisiae can adapt to high salt stress by extrusion of the Na^+ ions via the action of a P-type ATPase encoded by the *ENA1/PMR2* gene. This gene is a member of a multi-gene complex consisting of four tandem arrays of highly conserved genes encoding a P-type ATPase (Rudolph and Fink, 1990; Haro *et al.*, 1993). The *ENA1* gene is regulated by the HOG pathway when the cells are subjected to low osmotic stress. It is, however, also specifically induced by high Na^+, mediated by the Ca^{2+}/calmodulin-dependent phosphoprotein phosphatase (calcineurin) pathway (Mendoza *et al.*, 1994; Marquez and Serrano, 1996); in mammalian cells calcineurin plays an important role in the T-cell receptor-mediated signalling pathway that leads to induction of interleukin-2 gene transcription. In *S. cerevisiae* signalling is triggered by an increase in extracellular Na^+, K^+ or Ca^{2+} ions, hypo-osmotic shock or an increase in temperature leading to an increase in intracellular Ca^{2+} (Cyert, 2001; Hohmann, 2002).

Calcineurin, or protein phosphatase 2B is composed of two non-identical subunits, one is a catalytic subunit (60 kDa) which has a C-terminal autoinhibition domain and a Ca^{2+}–calmodulin binding site, the other is a 20 kDa essential Ca^{2+}-binding regulatory subunit (see Chapter 8). Calcineurin is activated by the binding of Ca^{2+} to calmodulin and the regulatory subunit, displacing the autoinhibitory domain to activate the phosphatase (Figure 9.10). The *CNA1* (*CMP1*) and *CNA2* (*CMP2*) genes encode the calcineurin catalytic subunits and *CNB1* the regulatory subunit. Null mutations in any one of these three genes abolishes calcineurin activity. These mutants have increased sensitivity to high Na^+ or Li^+ in the growth medium; they also show slow recovery from α-factor pheromone cell division arrest (Nakamura *et al.*, 1993; Mendoza *et al.*, 1994; Cyert and Thorner, 1992). The Cna1p or Cna2p phosphatases activate gene expression by dephosphorylating the Crz1p transcription factor which is then localised to the nucleus (Stathopoulos and Cyert, 1997). Target genes include *ENA1*, *TRK1* and *TRK2* encoding K^+ uptake systems and the β-1,3-glucanase gene *FKS2* (Hohmann, 2002).

The *PDE1* gene encoding the low affinity cAMP phosphodiesterase was found to act as a multicopy suppressor of the Li^+ sensitivity of a *cmp1 cmp2* double disruptant, which led to the demonstration that the *ENA1/PMR2* gene was regulated antagonistically by calcineurin and the cAMP-dependent PKA pathway. This latter response could be mediated by a STRE in the promoter of the gene (Hirata *et al.*, 1995).

Hypo-osmotic stress

Less work has been done on the pathways involved in the yeast response to hypotonic shock, and there is some argument about the role of the signal transduction pathway mediated by the PKC-like protein encoded by *PKC1* (Levin *et al.*, 1990). This pathway is one of the four MAP kinase cascades in yeast (see section 8.6 of chapter 8), proceeding via Pkc1p to the MAPKKK Bck1p or Slk1p to two MAPKK (Mkk1p and Mkk2p) to MAPK (Mpk1p or Slt1p) (Davenport *et al.*, 1995). Mutations in the *MPK1* gene, or others in the pathway lead to a cell lysis defect (Levin *et al.*, 1994) and to a requirement for osmotic support in the growth medium (Torres *et al.*, 1991). Davenport *et al.* (1995) have shown that tyrosine phosphorylation of Mpk1p occurs rapidly after shifting cells to a medium of lower osmolarity, to a degree dependent on the extent of the shift, and this phosphorylation was dependent on action of protein kinases upstream in the pathway. While this indicates that the pathway

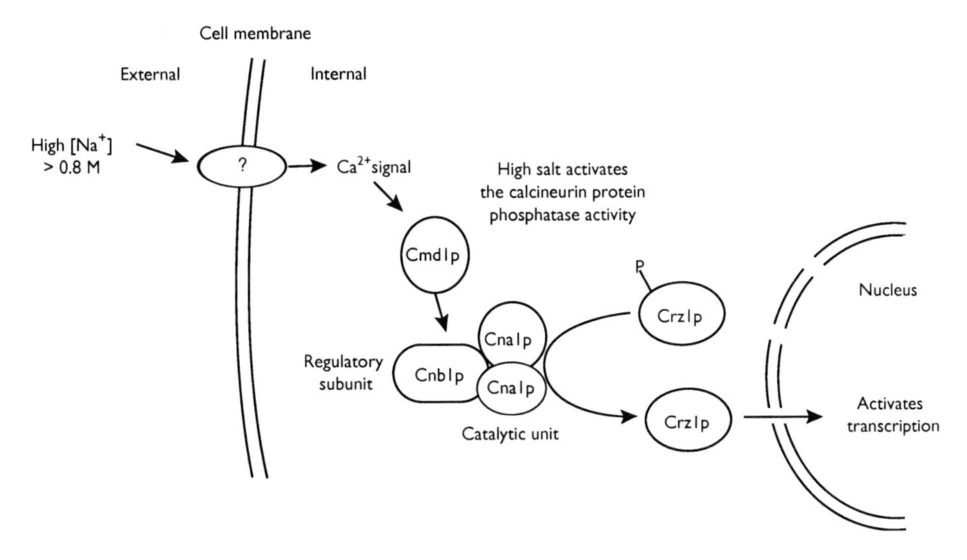

Figure 9.10 The calcineurin pathway of signalling high salt stress. By a mechanism that remains to be determined extracellular high salt concentrations lead to an increase in intracellular calcium ions, which activates calmodulin (Cmd1p). Calmodulin binds to the regulatory subunit of calcineurin (Cnb1p) activating the calcineurin catalytic subunit composed of the two polypeptides Cna1p and Cna2p. The activated calcineurin functions as a protein phosphatase to eventually activate transcription of the *ENA1* gene encoding the major ATPase exporting sodium ions from the cell.

signals a decrease in osmolarity, other signals including heat shock (Kamada *et al.*, 1995), polarized cell growth, cell cycle progression (Levin *et al.*, 1990) and nutrient sensing have been reported to activate it, and (Hohmann, 1997) has suggested that a unifying theme for these functions might be morphogenesis and growth which may involve a remodelling of the cell wall. This is along the lines suggested by (Kamada *et al.*, 1995) for the activation of the pathway in response to weakness in the cell wall created during growth under thermal stress. Hypotonic conditions would thus lead to cell expansion which Hohmann suggests may be sensed by plasma membrane sensors to signal a need for partial hydrolysis and remodelling of the wall. Since yeast cells are often growing in hypotonic conditions there may be no major requirement for a specific pathway to signal a decrease in osmotic pressure. Many of the changes in gene expression seen following a mild hypotonic shock appear to be a reversal of the changes seen on exposure to hypertonic conditions (Gasch *et al.*, 2000; Hohmann, 2002).

9.5 Freeze–thaw injury

Damage caused by freezing and thawing

Freezing has been used as a form of preservative for many years, but can also lead to loss of cell viability. This is true of *S. cerevisiae* which can undergo moderate to substantial loss of viability depending on the freezing parameters and the physiological state of the culture.

Cells can be damaged by physical factors such as ice crystal formation and dehydration. Rate of freezing is an important parameter determining the type of damage suffered by cells (Mazur, 1970; Toner *et al.*, 1993), at high freezing rates, intracellular freezing occurs, leading to damage mainly due to ice crystal formation. At low freezing rates, extracellular ice predominates leading to intracellular dehydration as the water equilibrates across the membrane. The type of freezing depends on the size and shape of cells as well as membrane permeability, and for yeast freezing rates below 7 °C min^{-1} result in slow freezing (Mazur, 1970).

During freeze–thaw cycles, cells can also suffer chemical damage, including oxidative damage caused by reactive oxygen species (Park *et al.*, 1998). Using mutants defective in a range of various physiological functions, including many affected in oxidative stress (including *glr1, gsh1, ctt1, cta1, yap1, sod1* and *sod2*) it was shown that the most sensitive strain by far was the *sod1* mutant. A major source of freeze–thaw injury in cells is therefore due to the appearance of superoxide radicals in the cytoplasm, probably as a result of an oxidative burst caused by mitochondrial respiration during the re-oxygenation process on thawing. The production of free radicals in cells following freezing and thawing was confirmed by the use of electron paramagnetic resonance spectroscopy. Wild-type cells produced a low level of free radical species, these were much more extensively produced in *sod1* mutants following freezing and thawing as indicated in Figure 9.11 (Park *et al.*, 1998).

Other factors also probably contribute to the ability of cells to survive freezing and thawing. Some clue to their nature has been gained from physiological and cross-protection studies. Prior heat shock, or treatment with H_2O_2, cycloheximide or NaCl at high concentration led to cross-protection to freeze–thaw injury (Lewis *et al.*, 1995; Park *et al.*, 1997). As with many other stresses, cells in the post-diauxic and stationary phases are much more resistant than those in exponential phase, and respiration and functional mitochondria are necessary for full resistance (Lewis *et al.*, 1993; Park *et al.*, 1997). The full extent of freeze–thaw injury depends on the operation of the RAS–cAMP signal transduction pathway since *ras2* mutants were more freeze–thaw tolerant than the wildtype; this effect was reversed in a *ras2 sra1* double mutant and other mutations in the pathway gave consistent results (Park *et al.*, 1997). This has been exploited in producing a yeast strain that maintains freeze–thaw and other stress resistance during fermentation by isolating a mutant with a point mutation in the adenylate cyclase gene *CYR1*. This reduces the enzyme activity tenfold and hence leads to a reduction in the glucose-induced increase in cAMP (Van Dijck *et al.*, 2000).

There appears to be a correlation between the expression of the *AQY1* and *AQY2* genes encoding acquaporins and freeze resistance which support the hypothesis that plasma membrane water transport activity has a role in determining freeze tolerance. It has been suggested that a rapid, osmotically driven efflux of water during the freezing process reduces intracellular ice formation and resulting cell damage (An *et al.*, 2002).

Despite speculation about the possible role of osmoprotectants in protection of cells against freeze–thaw damage, mutations in the pathways for glycerol and trehalose synthesis did not lead to increased sensitivity (Park, personal communication). Under slow-freezing conditions *hog1* mutants also showed similar levels of survival compared to the wild type, which indicated that while cellular dehydration may be occurring on slow freezing, the *HOG* pathway does not seem to play a significant role in the resistance of cells to freeze–thaw damage. The calcineurin pathway has been suggested to play a role since high salt concentrations are needed to stimulate resistance to freeze–thaw damage (Park *et al.*, 1997). Trehalose does however, appear to provide some protection against freeze–thaw stress since addition of trehalose to cells prior to freezing or freeze-drying provided protection to the

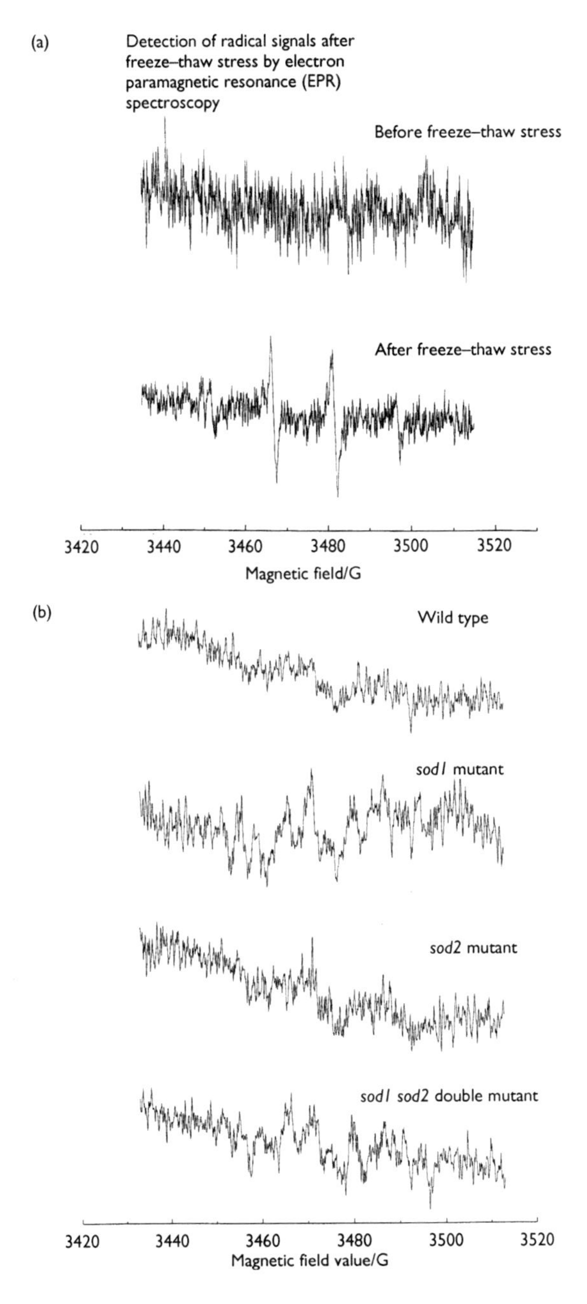

Figure 9.11 Free radicals generated in cells as a response to freeze–thaw stress. (a) Free radical signals detected by electron paramagnetic resonance spectroscopy (EPR) in wild-type cells before and after freezing and thawing. Signals due to the hydroxyl free radical were detected in the presence of the spin trapping agent α-phenyl-*tert*-butylnitrone. (b) Signal production following freezing and thawing of isogenic wild-type, *sod1* and *sod2* mutant strains using the spin trap agent 5,5-dimethylpyrroline-*N*-oxide. Figures provided by Jong-In Park.

cells (Diniz-Mendes *et al.*, 1999). Mutants with deletions of the neutral trehalase *NTH1* and acid trehalase *ATH1* genes were more tolerant of freezing (Shima *et al.*, 1999). Mutants that have increased intracellular proline are also more resistant to freeze–thaw and desiccation stresses, and this may relate to the fact that proline is used as an osmoprotectant in other organisms (Takagi *et al.*, 2000).

9.6 General regulatory mechanisms

Adaptive responses

It had been noted for some time that cells of most organisms that survive heat stress acquire an increased resistance to subsequent exposure. This adaptive response has now been found for many other types of stress, including oxidative, ethanol and hypertonic stress. In many cases these adaptive responses overlap, possibly as a result of the operation of stress response systems that are activated to differing extents by a range of stresses.

Thermotolerance

The response of cells to a sudden shift to an elevated temperature that is not lethal leads to the rapid induction of substantially increased thermotolerance. This heat shock response is a transient process and is induced most strongly by temperature shift to the maximum growth temperature or just above it, and by the extent of the shift: the greater the temperature difference the higher the response (Kirk and Piper, 1991). Coote *et al.* (1991) have shown that thermotolerance is most rapidly induced by temperatures up to 45°C, which is beyond that at which protein synthesis can be detected by pulse-labelling. The induction of thermotolerance is not, however, solely brought about by temperature increase. A wide range of conditions or treatments, including ethanol at concentrations above about 4% (v/v), compounds inducing aberrant protein synthesis (Grant *et al.*, 1989) or inhibiting transcription (Adams and Goss, 1991), protease inhibitors (Gropper and Rensing, 1993) or even sphaeroplasting (Piper *et al.*, 1994) can activate the heat shock response.

The mechanisms that lead up to the acquisition of thermotolerance are not clear. Activation of the heat shock factor can occur without cells becoming thermotolerant, as is seen following treatment of cells with a proline analogue that is incorporated into proteins and leads to misfolding (Trotter *et al.*, 2001). For some time it has been considered likely that the damage to proteins leading to aberrant, misfolded, or aggregated forms provides the stimulus leading to activation of the heat shock transcription factor (Craig and Gross, 1991; Parsell and Lindquist, 1993). Piper (1997) has put forward another possibility; that the cell membrane plays an important role, with either the proton motive force at the membrane, or its fluidity being sensitive potential monitors of heat stress via stress-responsive sensors located in the membrane. This is supported by data in which membrane potential was reduced by affecting the plasma membrane H^+-ATPase activity by inhibition or mutation, or by use of weak organic acids at low pH (Panaretou and Piper, 1990; Cheng and Piper, 1994). Kamada *et al.* (1995) have shown that thermotolerance depends on activation of the PKC–MAP kinase pathway and that the Mpk1p protein kinase in this pathway is strongly activated by mild heat shock. The process is relatively slow, taking up to 45 min, and requires protein synthesis. They propose that the activation of the PKC pathway is triggered in response to cell surface weakness during growth under stress, and involves the cell measuring stretch of the plasma membrane in some way. The cell membrane proteins Hcs77p(Wsc1p) and Mid2p are required for the activation of the PKC sytem, and these have been proposed to be sensors of the state of the cell surface. Heat shock also activates the HOG–MAPK pathway, but only through the Sho1p and not the Sln1p membrane sensor (Winkler *et al.*, 2002).

Oxidative stress

Cells can adapt to become more resistant to a subsequent exposure to oxidants such as hydrogen peroxide and the free radical-generating compounds menadione and paraquat. The adaptive responses to hydrogen peroxide differ from those to the O_2^--free radical generating agents (Collinson and Dawes, 1992; Jamieson, 1992, 1995; Flattery-O'Brien *et al.*, 1993) since there are differences in cross adaptation. There is an hierarchical response in the sense that heat shock makes cells more resistant to both hydrogen peroxide and menadione, hydrogen peroxide does not confer thermotolerance but does provide some protection against menadione, whilst cells adapted to menadione were neither hydrogen peroxide resistant nor more thermotolerant. Similar adaptive responses have been noted for the fission yeast *Schizosaccharomyces pombe* (Lee *et al.*, 1995), in this petite-negative yeast the extent of adaptation is much greater than for *S. cerevisiae*. These adaptive responses are fairly rapid since cells become resistant to H_2O_2 within 45 min of the challenge. They lose the protection completely within 4 h of removal of the cells from H_2O_2 but can mount a further adaptive response (Davies *et al.*, 1995). During adaptation the synthesis of a number of polypeptides is induced, some of these are also produced following heat shock, while others are unique to the peroxide treatment (Collinson and Dawes, 1992; Jamieson *et al.*, 1994; Davies *et al.*, 1995). A number of enzymes, including catalase T and glutathione reductase have been shown to be induced during oxidative stress, although the levels of induction are not great. Izawa *et al.* (1995, 1996) have reported that glutathione and catalases play a role in the adaptive response of cells to hydrogen peroxide although the role of glutathione is still open to question (Stephen and Jamieson, 1996).

Saccharomyces cerevisiae also has an adaptive response to damage caused by the toxic intermediates and products of oxidation of lipids. These include lipid peroxides such as linoleic acid hydroperoxide (Evans *et al.*, 1998) and to malondialdehyde which can arise in cells from a variety of sources, including the breakdown of fatty acid peroxides (Turton *et al.*, 1997). Malondialdehyde pretreatment leads to cross adaptation to linoleic hydroperoxide, but not to other peroxides or menadione. Linoleic acid hydroperoxide treatment does not lead to cross adaptation. This may be due to the accumulation of malondialdehyde in cells provoking detoxification systems that export the lipid hydroperoxide (Turton *et al.*, 1997). The lipid hydroperoxide is toxic at such low concentrations that a similar mechanism probably does not arise to confer cross-resistance to malondialdehyde.

The mechanisms involved in these adaptations are not clearly understood. Cells have a variety of responses to oxidants, and these are selectively brought into play to cope with a particular oxidant and the products as they are generated. The adaptive response to H_2O_2 is reduced in mutants deleted for either of the *YAP1* or *YAP2* genes encoding the oxidant response transcription factors, but these mutants are unaffected in their response to superoxide-generating agents such as menadione and paraquat (Stephen *et al.*, 1995). Glutathione is not necessary for the adaptive response to H_2O_2 since a *gsh1* deletion mutant is still capable of adaptation (Stephen and Jamieson, 1996). Depletion of glutathione by addition of 1-chloro-2,4-dinitrobenzene or buthionine sulphoxime during low dose pretreatment suppressed some of the ability of yeast cells to adapt (Izawa *et al.*, 1995); this may, however, have resulted from other effects of the compounds added in conjunction with the H_2O_2. From secondary screening analysis of the oxidant-sensitive mutants identified by genome-wide screening of deletion mutants in this laboratory, it is clear that most oxidant sensitive strains are not affected in their ability to adapt, and so far, from several hundred hydrogen peroxide-sensitive mutants, only six genes have been identified. These are involved in a range of functions, including the generation of NADPH and transcriptional coactivators (Ng, personal communication). The adaptive response to linoleic acid hydroperoxide is almost completely lost in mutants lacking both *GPX1* and *GPX2*, implicating these glutathione peroxidases, but not the product of *GPX3* in this response (Israel, personal communication).

Cells therefore have different control systems to cope with H_2O_2, free-radical damage brought about through electron leakage to oxygen from respiration and lipid peroxides. Adaptation to H_2O_2 stress, as well as the repair of the damage it causes depends on the effective operation of systems subject to Yap1p control, which include metal ion homeostasis and glutathione metabolism. Currently, less known about the mechanisms involved in the adaptation to free radical stress, although some indication may come from a consideration of the active elements in the *SOD2* promoter discussed later.

9.7 Regulation pathways

There are a number of transcriptional control systems that respond to stress. Some of these are more specific to one type of stress, or even subset of a stress response, while others are more general. The main transcription factors involved in stress responses in yeast are summarised in Table 9.2.

Heat shock factor (Hsf1p) control of transcription

Many, but not all yeast heat shock genes are controlled at the level of transcription by the binding to their promoters of a specific heat shock transcription factor (encoded by *HSF1* (Sorger and Pelham, 1987; Wiederrecht *et al.*, 1988). Hsf1p is essential for survival presumably since a degree of activation of some heat shock gene(s) is needed during normal growth in yeast (Sorger and Pelham, 1988). The transcription factor is a homotrimer and binds to a heat shock element (HSE) which was originally described in the *Drosophila* HSP70 promoter as CnnGAAnnTTCnnG (Pelham, 1982). For high affinity interaction three adjacent nGAAn units (possibly AGAAn; (Fernandes *et al.*, 1994)) arranged alternatively are required (e.g. nGAAnnTTCnnGAAn) and only slight induction is afforded by two tandem units (Yang *et al.*, 1991). In yeast, unlike other eukaryotes, Hsp1p is constitutively present as the trimer bound to the HSE, and this trimer undergoes a conformational change on heat shock that does not affect its affinity for the HSE but exposes an activation domain (Sorger *et al.*, 1987).

The molecular basis of activation of Hsf1p following heat shock is not clear. The protein can be phosphorylated, and the extent of phosphorylation of a number of serine and threonine residues is correlated with an increase in transcriptional activation over a range of temperatures (Sorger, 1991). The relevance of this hyper-phosphorylation to the control remains to be established (Mager and De Kruijff, 1995). One intriguing possibility is that the Hsp70 heat shock proteins act as regulators of Hsf1p. Reduction of Hsp70p leads to a high

Table 9.2 Consensus binding sites for stress response motifs

*Control site**	*Consensus motif*	*Representative genes regulated*
HSE (*Hsf1p*)	nGAAnnTTCnnGAAn (multimers of nGAAn)	*SSA1, SSA3, HSP82, UBI4*
STRE, TRS** (*Msn2p, Msn4p*)	AAGGGG	*CTT1, DDR2, HSP12, HSP26, HSP104, GSY2, UBI4, GPD1, HAL1, ENA1, GAC1, TPS1, TPS2, SOD2*
HAP2,3,4,5 (*Hap2p–5p*)	TNATTGGC/T	*CYC1, HEM1, LPD1, KGD1, KGD2, SOD2, ASN1, ASN2, GDH1, CIT1, COX4, COX5a, COX5b*
HAP1 (*Hap1p*)	CGGN$_3$TANCGGN$_3$TA	*CYC1, CYC7, CTT1, CTA1, SOD2*
AP-1 (*Yap1p, Yap2p, Yap3p?, Yap5p?*)	TTA(G/C)TAA	*GSH1, TRX2, YCF1, GLR1 SSA1, PDR5, SNQ2, ATR1*
CDRE (*Sko1p*)	TGACGTCA	*ENA1, GRE2, AHP1, GLR1, SFA1, HAL1*
? (*Hot1p*)	?	*GPD1, GPP2, CHA1, PHO84*
CRZ1 (*Crz1p*)	GNGGC(G/T))CA	*PMC1, PMR1, PMR2, ENA1 TRK1, TRK2, FKS2*
ACE1 (*Ace1p*)	TTTTGCTG	*CUP1, SOD1, CRS5*
MAC1 (*Mac1p*)	?	*CTT1, FRE1*

Note
* Transcription factors binding to the site are given in parentheses.
** See Kobayashi and McEntee (1993).

level of expression activated by Hsf1p even during normal growth (Boorstein and Craig, 1990b), and this requires Hsf1p since this constitutive expression can be eliminated by mutating the HSEs (Craig and Gross, 1991). One hypothesis put forward to explain this involves the binding of Hsp70 to the Hsf1p directly to inhibit its transcriptional activating function. On heat shock, or other induction of damage to cellular proteins, the level of available Hsp70 protein would fall as it bound to the damaged proteins, thereby leading to activation of Hsf1p (Craig and Gross, 1991). However, Stone and Craig (1990) have shown that while the *SSA1* gene is regulated by its own product and there are HSE elements in the *SSA1* promoter, there are different elements (proximal to, or overlapping HSE2) which are involved in the self-regulation mechanism. This overall hypothesis has been challenged by (Hjorth-Sorenson *et al.* (2001) since overexpression of heat shock protein genes, including *SSA2* did not inhibit heat shock, but did increase thermotolerance.

yAP1 and its role in oxidative stress responses

AP-1 proteins are a family of transcriptional activators that stimulate the expression of specific sets of genes in response to different extracellular stimuli. They are found in all eukaryotic organisms and contain a conserved bZIP DNA-binding domain consisting of a leucine zipper that participates in dimerisation, and an adjacent basic region that binds to specific AP-1 motifs in DNA (Fernandes *et al.*, 1997). Mammalian cells contain many AP-1 proteins that can interact among themselves to form heterodimers capable of binding to AP-1 sites; different members of the family play distinctive roles in the cell. The best known, and most extensively studied of the yeast AP-1 factors is Gcn4p (see Chapter 4) which coordinates transcription from 539 genes following a sudden starvation for amino acids, or in response to UV treatment (Engelberg *et al.*, 1994; Natarajan *et al.*, 2001). This protein binds to the 'optimal AP-1 site', TGACTCA. *Saccharomyces cerevisiae* contains eight additional members of the family (encoded by *YAP1–8*) that bind to related sequences, with TTACTAA as the likely consensus for at least four of these proteins (Yap1p, Yap2p, Yap3p and Yap5p).

Of these the most extensively studied is Yap1p (yAP-1) (Harshman *et al.*, 1988; Moye-Rowley *et al.*, 1989). The *YAP1* gene (also termed *SNQ3, PAR1, PDR4*) was found to confer pleiotropic drug resistance and cadmium resistance when present in high copy number (Hertle *et al.*, 1991; Hussein and Lenard, 1991; Kuge and Jones, 1994; Schnell *et al.*, 1992; Wu *et al.*, 1993). *YAP1*-defective strains have increased sensitivity to menadione, paraquat, the thiol oxidant diamide, H_2O_2 and cadmium (Schnell and Entian, 1991; Schnell *et al.*, 1992; Kuge and Jones, 1994). Yap1p is an essential transcription factor for the hydrogen peroxide adaptive response and controls the expression of many genes of the hydrogen peroxide stimulon (Lee *et al.*, 1999). These include those encoding most cellular antioxidants and enzymes of the glutathione and pentose phosphate pathways (Delauney *et al.*, 2000). It is also involved in the control of a number of genes concerned with metal ion resistance, heat shock protein synthesis and multidrug resistance. These include those encoding: the Ycf1p vacuolar transporter of glutathione–X conjugates which functions to detoxify cadmium ions (Wemmie *et al.*, 1994), the stress-inducible HSP70, Ssa1 (Stephen *et al.*, 1995), Pdr5p and Snq2 ABC transporter proteins involved in multidrug resistance (Miyahara *et al.*, 1996) and the Atr1p integral membrane protein involved in the efflux of toxic compounds (Coleman *et al.*, 1997). Based on its binding motif in these genes, the consensus sequence for its recognition has been given as TTA(G/C)TAA (Coleman *et al.*, 1997), which conforms well to the consensus given here.

Yap1p is differentially activated by hydrogen peroxide and the thiol oxidant, diamide. Considerable progress has been made in determining the mechanism of activation of Yap1p in response to hydrogen peroxide. (Wemmie *et al.* (1997) have shown that there are three repeated cysteine–serine–glutamate sequences located in the C-terminus that are required for regulation of Yap1p during oxidative stress. Replacement of these by alanines did not lead to similar effects on H_2O_2 and diamide regulation of the transcription factor's function, implying differences in the ways the protein responds to these two oxidants, and this has been confirmed by subsequent studies which have elucidated the mechanism of the hydrogen peroxide response which is illustrated in Figure 9.12. In unstressed cells the protein is restricted to the cytoplasm through rapid nuclear export via the export receptor Crm1p (Kuge *et al.*, 1998). When cells are exposed to hydrogen peroxide, Yap1p accumulates in the nucleus to act as a transcriptional activator (Kuge *et al.*, 1997) due to the loss of the Yap1p–Crm1p interaction through oxidation of critical cysteines Cys303, and Cys598 in the C-terminal domain carrying the nuclear export signal (Delauney *et al.*, 2000). Nuclear import of Yap1p depends on the receptor Pse1p, but import is not affected by stress (Isoyama *et al.*, 2001). The signal for activation and sensing of oxidative stress is not direct oxidation of Yap1p by hydrogen peroxide, but through the oxidised form of glutathione peroxidase Gpx3p. The Cys36 of Gpx3p, when oxidised, forms a Yap1p–Gpx3 disulphide through Cys598 of Yap1p, which is then resolved into an intramolecular Yap1p disulphide. Thioredoxin deactivates the system, by reducing both the sensor and Yap1p (Delaunay *et al.*, 2000, 2002), and since *TRX2* is regulated by Yap1p this system is elegantly autoregulated. Since nuclear localisation of Yap1p depends on export of the reduced form from the nucleus it remains to be resolved how oxidised Yap1p is reduced in the nucleus – by the mitochondrial thioredoxin system?

Fernandes *et al.* (1997) have shown that expression from an heterologous Yap-based promoter was subject to negative control by the Ras-pathway activated PKA, and this control was mediated in part by only a modest reduction in Yap1p levels as well as by inhibition of its activity (probably at the level of promoter occupancy). Yap1p exists in the phosphorylated form after its translocation to the nucleus following hydrogen peroxide stress (Delauney *et al.*, 2000). *SKN7* is also strongly negatively regulated by the PKA pathway, and high activity of PKA leads to complete loss of activation from a fusion protein with the GAL4 DNA-binding domain fused to Skn7p following oxidative stress (Charizanis *et al.*, 1999).

Of the other Yap proteins that activate transcription through a Yap site, Yap2p has also been shown to be involved in stress responses. Overexpression alleviates the toxic effects of metal ions and a range of drugs (Bossier *et al.*, 1993; Hirata *et al.*, 1994; Fernandes *et al.*, 1997), albeit at a lower level than for Yap1p. Yap2p-dependent transcription is strongly induced by 3-aminotriazole (AT; about five-fold) and Cd^{2+} (seven-fold), but not by H_2O_2 whereas the Yap1p-dependent transcription is strongly activated by H_2O_2 but not by AT or Cd^{2+}. Yap3p only responds to AT and not the other compounds tested. These results are consistent with the phenotypes of deletion mutants since *yap1* strains are sensitive to H_2O_2, whereas a *yap2 yap4 yap5* strain is not (Fernandes *et al.*, 1997). The role of Yap1p but not Yap2p in protection against H_2O_2-induced damage is highlighted by the demonstration that the *GSH1*, *TRX2* and *SSA1* genes are H_2O_2-inducible in *yap2*, but not *yap1* strains (Stephen *et al.*, 1995). Genome-wide transcript analysis has shown that Yap1p and Yap2p activate non-overlapping sets of genes and that the Yap2p regulon controls a set of genes over-represented for functions stabilising proteins (Cohen *et al.*, 2002).

Skn7p (Pos9p) is a transcription factor that cooperates with Yap1p to activate a subset of the genes activated by Yap1p, including *TRX2* and *TRR1* that are critical for redox homeostasis. Both proteins are needed for resistance to hydrogen peroxide, whereas Yap1p is

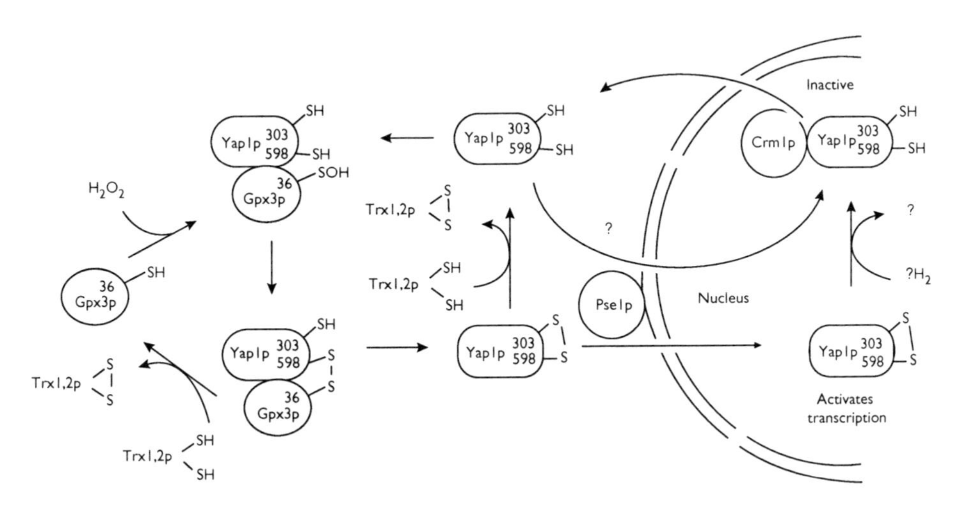

Figure 9.12 The involvement of glutathione peroxidase 3 (Gpx3p/Ors1p) in sensing and transduction of hydrogen peroxide-induced damage to the Yap1p transcription factor.

important for cadmium resistance whereas Skn7p has a negative effect on this response. Lee *et al.* (1999) have shown by two-dimensional gel electrophoresis of peroxide-induced proteins that about half of the 32 proteins induced required the presence of Skn7p for induction to occur. Some of these genes do not contain a consensus motif for recognition of Yap1p. There are, therefore, two subsets of genes that are activated by Yap1p, and these separate those genes involved in metabolic pathways regenerating glutathione and NADPH (which are Skn7p dependent) from those involved in antioxidant scavenging enzymes (Skn7p-dependent including *CTT1, SOD1, SOD2, TRR1, TRX2, TSA1 SSA1* and *AHP1*.

Skn7p seems to have the capacity to bind to (or interact with) other transcription factors, and this is reflected in differences in reported DNA sequences to which it binds. The partial overlap between heat shock and oxidative stress responses may be explained, at least in part, by the finding that Skn7p and the heat shock transcription factor Hsf1p interact *in vitro* and that Skn7p can bind to heat shock elements. From analysis of double mutants it is apparent that Skn7p and Hsf1p cooperate to achieve maximal induction of heat shock genes in response to oxidative stress (Raitt *et al.*, 2000). Skn7p can also cooperate with the calcineurin-dependent transcription factor Crz1p and the cell cycle transcription factor Mbp1p, and Hohmann (2002) has suggested that the transcription factor may mediate control of cell wall biogenesis under growth and stress conditions.

Msn2p and Msn4p and general stress responses via the STRE

Following identification of the Hsf1p-dependent heat shock element in genes responding to heat induction, another STRE was identified in the promoters of *CTT1* (Wieser *et al.*, 1991) and *DDR2* (DNA damage responsive gene, Kobayashi and McEntee, 1993) that were responsive to heat shock. This element, however, mediates responses to a much wider variety of stress conditions, including oxidative stress, osmotic stress, low pH, and the presence of weak organic acids and ethanol (Marchler *et al.*, 1993; Schüller *et al.*, 1994). It is responsible for the induction of a large block of genes seen as a 'general stress response' to many types of environmental insult (Gasch *et al.*, 2000). STREs have been identified in the promoters of a substantial number of stress responsive genes with functions in a range of shock responses. These include *HSP12, HSP26, HSP104, GSY2* (glycogen synthase), *UBI4, GPD1, HAL1, ENA1, GAC1, TPS1, TPS2* (Mager and De Kruijff, 1995; Ruis and Schüller, 1995) and *SOD2* (Flattery-O'Brien *et al.*, 1997). In *SOD2* the element functions mainly to activate transcription in starvation or the diauxic shift to respiration rather than as a response to oxidants (Flattery-O'Brien *et al.*, 1997). There is a related post-diauxic shift (PDS) element which in part mediates the response of cells to the shift from fermentative to respiratory metabolism when fermentable sugars are exhausted. This sequence has consensus T/AAGGGA, and is possibly a form of the STRE. It was first identified in the promoter of the Hsp70 gene *SSA3* (Boorstein and Craig, 1990a), and like the STRE, control through this element is negatively regulated by cAMP levels which decrease at the PDS. Control through this element is mediated by a separate zinc finger transcription factor, Gis1p (Pedruzzi *et al.*, 2000).

The STRE promoters are responsive to two stress signalling pathways:

- the HOG pathway specifically activates STRE-dependent genes in response to hyper-osmotic stress (Brewster *et al.*, 1993; Schüller *et al.*, 1994; Kamada *et al.*, 1995);
- the RAS–cAMP pathway which activates PKA negatively regulates gene expression via the STRE (Weiser *et al.*, 1991; Marchler *et al.*, 1993).

Two transcription factors that bind to the STRE have been identified (Kobayashi and McEntee, 1993; Marchler *et al.*, 1993; Piper, 1997). These are the highly homologous Msn2p and Msn4p that are functionally redundant proteins with Cys_2His_2 zinc finger DNA-binding domains. These proteins are required to activate *CTT1*, *DDR2* and *HSP12* via the STRE elements in response to a variety of stresses, including heat shock, low pH, sorbate and ethanol. It is of interest, however, that neither hydrogen peroxide nor osmotic stress induction of an STRE reporter construct was affected in an *msn2 msn4* mutant, indicating that there may be some other factor which mediates this control (Martinez-Pastor *et al.*, 1996).

The primary stress signal to the Msn2p/4p system is not known, although Moskvina *et al.* (1999) have suggested that some of the factors inducing it result in an increase in plasma membrane permeability leading to a decrease in membrane potential and that this may be a primary cellular stress signal. The activation of these factors, as with Yap1p, leads to their translocation from the cytoplasm to the nucleus (Gorner *et al.*, 1998). The signal and mechanism for import of Msn2p and Msn4p into the nucleus is not known. The proteins become hyper-phosphorylated following heat shock or at the diauxic transition in a process independent of the PKA pathway (Garreau *et al.*, 2000), and this phosphorylation may be important to the nuclear import and activation of the stress response genes. This is reversed by dephosphorylation involving a complex of the plasma membrane phosphatases Whi2p and Psr1p/Psr2p (Kaida *et al.*, 2002). The Tor1p and Tor2p protein kinases of the TOR pathway that signals nutrient limitation are candidates for this activity since the TOR pathway inhibitor rapamycin causes nuclear localisation of Msn2p as well as the regulator of nitrogen catabolite repression Gln3p (Beck and Hall, 1999). Nuclear localisation is inversely correlated with cAMP and PKA activity, which phosphorylates Msn2p (and presumably Msn4p) leading to its rapid export from the nucleus (Görner *et al.*, 1998). The general stress response mediated by Msn2p and Msn4p is therefore reduced under optimal growth conditions, and is sensitive to the nutritional status of the cells.

Haem activated factors: Hap1p and the Hap2p/Hap3p/Hap4p/Hap5p complex

The haem-activated transcription factors Hap1p and the Hap2/3/4/5p complex are mainly involved in the activation of genes involved in respiration, although Hap2/3/4/5 appears to have a broader role in the cell (Dang *et al.*, 1994). Given this association with respiratory metabolism, and the fact that increased respiration will lead to increased generation of superoxide and related free radicals due to leakage of electrons from the respiratory chain, it is not surprising that some antioxidant-defence genes respond to these activators. Hap1p is activated by haem which is produced in the mitochondrion and requires O_2 for its synthesis. This factor stimulates transcription of *CYC1* (Guarente *et al.*, 1984) and *CYC7* (Pfeifer *et al.*, 1987) genes encoding cytochrome c proteins as well as the cytosolic catalase T gene (*CTT1*) and seems to play a minor role in activation of *SOD2* (Winkler *et al.*, 1988; Pinkham *et al.*, 1997). The Hap1p protein recognises an extended motif in the promoters of genes with consensus $CGGN_3TANCGGN_3TA$ (Ha *et al.*, 1996).

The Hap2/3/4/5p complex is a heteromeric transcriptional activator of genes involved in respiration including *CYC1*, *COX4*, *COX5a*, *COX5b*, *HEM1*, *CIT1*, *KGD1*, *KGD2* and *LPD1* with the consensus binding motif TNATTGGT/C (Forsburg and Guarente, 1988; Bowman *et al.*, 1992). It is involved directly in part of the haem-dependent response of the *SOD2* gene to growth under respiratory conditions, and as cells undergo nutrient shift as they enter the post-diauxic shift phase (Flattery-O'Brien *et al.*, 1997; Pinkham *et al.*, 1997).

These haem-dependent systems may function in other genes involved in oxidative stress responses, currently few of the promoters of these genes have been studied in as much detail as *CTT1* and *SOD2*.

Transcription factors involved in metal ion homeostasis: Ace1p and Mac1p

Metal ion homeostasis is important for the cell to maintain its defences against oxidative stress, and hence the systems regulating the expression of genes involved in Cu and Fe entry into the cell, and their distribution between organelles, are important in the overall response of cells to stress. Cells respond to elevated levels of copper ions by inducing the *CUP1* and *CRS5* genes encoding metallothionein isoforms (Santoro and Thiele, 1997). These genes are regulated by the transcription factor encoded by *ACE1* (*CUP2*). This factor binds to the promoter of regulated genes at motifs with consensus TTTTGCTG (Thiele and Hamer, 1986). It has a cuprous ion-activated DNA-binding domain containing twelve cysteine residues. Eleven of these cysteines are involved in binding four to six Cu(I) ions, which brings about a conformational change that allows the protein to bind to the DNA control region and activate transcription; co-operative binding of the Cu(I) ions ensures a rapid and sensitive response to fluctuations in intracellular copper ion concentration (Fürst *et al.*, 1988).

The promoter of the *SOD1* gene encoding the cytosolic Cu/Zn SOD has an Ace1p-binding site which is active in the induction of the gene by copper (Gralla *et al.*, 1991). Santoro and Thiele (1997) point out that this control may have two advantages to the cell: control of the synthesis of the Cu/Zn SOD apoprotein when copper ion is limiting; and, up-regulation of the Cu/Zn SOD when copper is in excess to increase the protection against deleterious free radicals.

MAC1 encodes a transcriptional activator that has homology to *ACE1*, and which also binds copper. Strains with disruption of *MAC1* are hypersensitive to cadmium, zinc, lead, heat and H_2O_2. Mac1p regulates expression of *CTT1* in response to H_2O_2; it also controls the expression of the *FRE1* gene encoding the plasma membrane-located Fe reductase which reduces Fe(III) and Cu(II) ions prior to their transport into the cytoplasm (Jungmann *et al.*, 1994). Iron, while being an essential nutrient is also a potent toxin if its metabolism and storage is not properly regulated possibly through the ability of Fe(II) ions to catalyse the Fenton reaction. Genes involved in iron uptake are tightly regulated via the Aft1p transcription factor which has as its consensus motif PyPuCACCCPu (Yamaguchi-Iwai *et al.*, 1996).

9.8 Cell division cycle arrest: checkpoints

While this chapter is not directly concerned with DNA repair, the cell cycle response of cells to damage is of interest. Prokaryotes and eukaryotes have been known for some time to have a response to DNA damage caused by UV, X-, or γ-radiation in which cell division cycle progression is delayed to allow repair of damage to the DNA as well as to prevent replication of affected chromosomes. As outlined in Chapter 1, in *S. cerevisiae*, inhibition of DNA replication, or induction of DNA damage, causes the cells to arrest in the cell division cycle prior to mitosis until the damage has been repaired (Burns, 1956; Burnborg and Williamson,

1978). This type of control to ensure the order of cell cycle (or other developmental) events has been described as a checkpoint (Hartwell and Weinert, 1989). Weinert and Hartwell (1988) demonstrated the involvement of the *RAD9* gene which acts post-translationally as a negative regulator to arrest cells in the G2 phase after DNA damage by irradiation with X-rays (Weinert and Hartwell, 1990). *RAD9* has subsequently been shown to act in the G1 phase if cells have previously been arrested with α factor (Siede *et al.*, 1993), and *RAD17* has also been implicated in G1 control (Weinert and Hartwell, 1993). Slowing of the rate of cell cycle progression in S phase has also been reported, although there are conflicting reports concerning the involvement of *RAD9. POL2* (encoding DNA polymerase ε) but not *RAD9* is required in response to UV (Navas *et al.*, 1996), whereas *RAD9, RAD17* and *RAD24* are important for S-phase slowing of replication in response to methylmethanesulphonate alkylation of DNA (Paulovitch *et al.*, 1997). Additional genes involved in S and G2 phase checkpoint control were identified as mutations leading to lethality in a *cdc13* background. These mitosis entry checkpoint (*MEC*) genes included *MEC3* (which was G2 specific) while *MEC1* and *MEC2* (identical to *RAD53*) and *SPK1* encoding a protein kinase (Kim and Weinert, 1997) acted on S-phase cells to slow replication and G2 phase cells to block entry into mitosis (Weinert *et al.*, 1994). Kiser and Weinert (1996) have shown that DNA damage leads to transcriptional induction of four distinct pathways identified by activation of specific genes. *MEC1* mediates this response in three of the pathways, *RAD53* in two, and *RAD17* in only one pathway. The details of these responses and how they relate to different repair systems on the one hand, and physiological outcomes of cell cycle arrest and induction of DNA repair systems have been reviewed in (Lowndes and Murguia, 2000).

The target of inhibition by the *RAD9* checkpoint has been the subject of speculation. In fission yeast, the *cdc2* protein kinase (equivalent to *S. cerevisiae CDC28* protein kinase) is phosphorylated (on Tyr^{15}) and inactivated after DNA damage or when DNA replication is inhibited, and this has been shown to be mediated via the *cdc25* encoded protein kinase (Furnari *et al.*, 1997). In *S. cerevisiae*, however, *cdc28* mutations that prevent this phosphorylation were still capable of cell cycle arrest in early S phase or at the *RAD9* checkpoint in the late S/G_2 phase (Weinert and Hartwell, 1993). The control appears to be mediated via the Swi6p transcription factor involved in S-phase cell cycle regulation.

Cell division cycle arrest in response to stress is not limited to agents that cause DNA damage directly. Cells that have been subjected to an increase in temperature within their normal range of growth undergo a reduction in growth rate over several hours. In the initial stages they undergo a transient arrest in the cell cycle, accumulating as unbudded cells in the G1 phase at Start (Rowley *et al.*, 1993). This effect has been attributed to decreased transcription of the *CLN1* and *CLN2* genes that is needed for the cell to activate the $p34^{CDC28}$ protein kinase to promote progression of the cells into S phase (Wittenberg *et al.*, 1990; Rowley *et al.*, 1993).

Similarly, cell cycle arrest is induced by oxidative stress. Slow proliferation of a *sod1* mutant is due to an increase in the period spent in G1 phase, and in the presence of hyperbaric oxygen the mutant permanently arrests in G1 due to inhibition of transcription of the autoregulated Cln1p and Cln2p cyclins (Lee *et al.*, 1997). Treatment of cells with paraquat results in pronounced arrest in G1-synchronised cells independent of the *RAD9* gene (Nunes and Sieda, 1996); in contrast, hydrogen peroxide treatment at non-lethal doses led to arrest of cells in G2, in a *RAD9*-dependent way (Flattery-O'Brien and Dawes, 1998). Lipid hydroperoxides and their breakdown products also cause cell division delay (Wonisch *et al.*, 1997; Alic *et al.*, 2001). In this laboratory 11 genes required for the cell-cycle response to linoleate hydroperoxide have been identified by secondary screening of oxidant-sensitive mutants. These include *OCA1* (Alic *et al.*, 2001) and two of its homologues whose products may associate *in vivo* to form a protein phosphatase complex, and *SWI6* encoding the S-phase specific cell cycle transcription factor (Fong, personal communication).

These results raise the interesting question of what is being sensed in the cell to cause cell cycle arrest following oxidative damage, and how the signal is transduced. Clearly, there seem to be different mechanisms for responding to superoxide damage from those needed to cope with exposure to hydrogen peroxide. These pathways need to be dissected in more detail to determine whether the cells are responding to damage to DNA caused by the reactive oxygen species, or to some other effect of the oxidants. The results do indicate that there are several mechanisms operating to coordinate repair with cell division arrest in response to different oxidants.

Recently, it has also been shown that following hyperosmotic stress, cells also undergo temporary cell cycle arrest. There may be more than one process and stage of the cell cycle involved since Belli *et al.* (2001) have shown that a block occurs in G1 probably due to down-regulation of the Cln3p–Cdc28 kinase activity. On the other hand Alexander *et al.* (2001) have shown that there is G2/M arrest that seems to depend on the Hog1p pathway and inhibition of the Clb2p–Cdc28p kinase.

9.9 Overlapping regulatory networks and the future

Clearly, from the discussion made here there are multiple signal transduction pathways, and multiple sensors are involved in stress responses, and that there are many genes that are activated by different types of stress. For osmotic stress, the sensors and signal transduction pathways are fairly well defined, and how these activate transcription is under fairly intensive study. On the other hand, for heat stress the transcription factor(s) that mediate the response are known better than the pathways that sense and signal the stress.

For the overall integration of stress responses, the future challenges will be to understand how they overlap, integrating the vast amount of data that has very rapidly accumulated from different high-throughput genomic screens. This will be a complicated task since overlap could be at the level of cross-talk between signalling pathways, pathways acting on other systems than gene expression pathways, pathways activating multiple transcription factors, multiple recognition motifs within enhancers of specific genes, and interaction between specific transcription factors. Heat shock leads to an induction of cross protection to many other forms of stress, hence it may be that the type of damage response induced by heat stress (e.g. removal of damaged proteins) may be a common factor in the survival of the cell to other stress systems. For the response of cells to reactive oxygen species it is clear that there are several pathways particularly in terms of the responses to superoxide radicals and peroxide damage, and the subtlety of the overlap between these is yet to be explored in detail.

Many stress genes have in their upstream sequences an array of motifs that could bind transcription factors mediating stress responses such as STRE and HAP2/3/4 and yAP-1 sites in the *SOD2* gene and a similar extensive array of motifs in the catalase genes (as discussed earlier). The *HSP104* gene is under dual control from Htf1p through HSEs and from Msn2,4p through STREs in its promoter, thereby ensuring that this critical protein is activated under a wide range of stress conditions. Where promoters of stress response genes have been examined in detail it is clear that the genes are responsive to multiple stress transcription factors. However, some of these putative motifs identified *in silico* do not function to any great extent in cells exposed to stress conditions, as has been found for the yAP-1-binding site in *SOD2* (Flattery-O'Brien *et al.*, 1997). Some transcription factors respond directly or indirectly to multiple signal transduction pathways. The HOG1 and RAS2/cAMP/PKA pathways converge on genes that are regulated by STRE elements (Marchler *et al.*, 1993). Coleman *et al.* (1997) have also shown that the yAP-1 and Gcn4p transcriptional activators overlap in their control of the *ATR1* gene. This gene encodes a putative integral membrane protein that may mediate efflux of the inhibitor 3-amino-1,2,4-triazole (3-AT) and of the DNA-damaging agent 4-nitroquinoline oxide (4-NQO). This is an indication that the two transcription factors that were previously thought to be involved in the separate control of oxidative stress (yAP-1) and general control of amino acid biosynthesis (Gcn4p) can overlap in their regulatory networks. This may occur through their recognition of a common DNA motif related to the yAP-1 recognition element found upstream of other yAP-1 regulated genes since there is a fair degree of similarity between the recognition sequences for these bZip transcription factors. The action of Skn7p with a range of other transcription factors to assist in activating various subsets of genes has been discussed here as a way to integrate various response pathways.

One of the main challenges that highlights the excitement that lies ahead for yeast research is the integration of functional analyses of all yeast genes into a unified picture of how the cell responds to external challenges.

Acknowledgements

Thanks to Chris Grant, Jacinta Flattery-O'Brien, Jong-In Park, Marguerite Evans-Galea, Priyanka Bandara, Lisa Israel, Nazif Alic, Tamara Drakulic, Geoffrey Thorpe, Gabriel Perrone, Anthony Beckhouse, Ji-Chul Fong, Chong-Han Ng, Vince Higgins and Geoff Kornfeld and other members of the laboratory for their patience, help and provision of references and advice. Thanks also to Rohan Baker for providing material for illustrations. Research in the laboratory was supported by the Ramaciotti Centre for Gene Function Analysis and the Australian Research Council.

References

Abbas-Terki, T., Donze, O., Briand, P. A. and Picard, D. (2001) Hsp104 interacts with Hsp90 cochaperones in respiring yeast. *Molecular and Cellular Biology* **21**, 7569–7575.

Adams, C. C. and Goss, D. S. (1991) The yeast heat shock response is induced by conversion of cells to sphaeroplasts and by potent transcriptional inhibitors. *Journal of Bacteriology* **173**, 7429–7435.

Aeschbach, R., Amado, R. and Neukom, H. (1976) Formation of dityrosine cross-links in proteins by oxidation of tyrosine residues. *Biochimica et Biophysica Acta* **439**, 292–301.

Aguilaniu, H., Gustafsson, L., Rigoulet, M. and Nystrom T. (2001) Protein oxidation in G(0) cells of *Saccharomyces cerevisiae* depends on the state rather than the rate of respiration and is enhanced in *pos9* but not *yap1* mutants. *Journal of Biological Chemistry* **276**, 35396–35404.

Albertyn, J., Hohmann, S., Thevelein, J. M. and Prior, B. A. (1994) *GPD1*, which encodes glycerol-3-phosphate dehydrogenase, is essential for growth under osmotic stress in *Saccharomyces cerevisiae*, and its expression is regulated by the high-osmolarity glycerol response pathway. *Molecular and Cellular Biology* **14**, 4135–4144.

Alexander, M. R., Tyers, M., Perret, M., Craig, B. M., Fang, K. S. and Gustin, M. C. (2001) Regulation of cell cycle progression by Swe1p and Hog1p following hypertonic stress. *Molecular Biology of the Cell* **12**, 53–62.

Alic, N., Higgins, V. J., and Dawes, I. W. (2001) Identification of a *Saccharomyces cerevisiae* gene that is required for G1 arrest in response to the lipid oxidation product linoleic acid hydroperoxide. *Molecular Biology of the Cell* **12**, 1801–1810.

Ames, B. N. (1989) Endogenous DNA damage as related to cancer and aging. *Mutation Research* **250**, 3–16.

Ames, B. N. and Gold, L. S. (1991) Endogenous mutagens and the causes of aging and cancer. *Mutation Research* **214**, 41–46.

An, T. H., Van Dijck, P., Dumortier, F., Teunissen, A., Hohmann, S. and Thevelein, J. A. (2002) Aquaporin expression correlates with freeze tolerance in baker's yeast, and overexpression improves freeze tolerance in idustrial strains. *Applied and Environmental Microbiology* **68**, 5981–5989.

Andre, L., Hemming, A. and Adler, L. (1991) Osmoregulation in *Saccharomyces cerevisiae*. Studies on the osmotic induction of glycerol production and glycerol 3-phosphate dehydrogenase (NAD^+). *FEBS Letters* **286**, 13–17.

Aoshima, H., Kadoya, K., Taniguchi, H., Satoh, T. and Hatanaka, H. (1999) Generation of free radicals during the death of *Saccharomyces cerevisiae* caused by lipid hydroperoxide. *Bioscience Biotechnology and Biochemistry* **63**, 1025–1031.

Ashby, M. N. and Edwards, P. A. (1990) Elucidation of the deficiency in two yeast coenzyme Q mutants. Characterization of the structural gene encoding hexaprenyl pyrophosphate synthetase. *Journal of Biological Chemistry* **265**, 13157–13164.

Attfield, P. V., Kletsas, S. and Hazell, B. W. (1994) Concomitant appearance of intrinsic thermotolerance and storage of trehalose in *Saccharomyces cerevisiae* during early respiratory phase of batch culture is *CIF1*-dependent. *Microbiology* **140**, 2625–2632.

Avery, A. M. and Avery, S. V. (2001) *Saccharomyces cerevisiae* expresses three phospholipid hydroperoxide glutathione peroxidases. *Journal of Biological Chemistry* **276**, 33730–33735.

Barkovich, R. J., Shtankov, A., Shepherd, J. A., Lee, P. T., Myles, D. C., Tzagoloff, A. and Clarke, C. F. (1997) Characterization of the *COQ5* gene from *Saccharomyces cerevisiae*. Evidence for a *C*-methyltransferase in ubiquinone biosynthesis. *Journal of Biological Chemistry* **272**, 9182–9188.

Beck, T. and Hall, M. N. (1999) The TOR signalling pathway controls nuclear localization of nutrient-regulated transcription factors. *Nature* **402**, 689–692.

Beckman, J. S., Chen, J., Ischiropolous, H. and Crow, J. P. (1994) Oxidative chemistry of peroxynitrite. *Methods in Enzymology* **233**, 229–240.

Belli, G., Gari, E., Aldea, M. and Herrero, E. (2001) Osmotic stress causes a G1 cell cycle delay and downregulation of Cln3/Cdc28 activity in *Saccharomyces cerevisiae*. *Molecular Microbiology* **39**, 1022–1035.

Beyer, R. E. (1992) An analysis of the role of coenzyme Q in free radical generation and as an antioxidant. *Biochemistry and Cell Biology* **70**, 390–403.

Bissinger, P. H., Wieser, R., Hamilton, B. and Ruis, H. (1989) Control of *Saccharomyces cerevisiae* catalase T gene (*CTT1*) expression by nutrient supply via the RAS-cyclic AMP pathway. *Molecular and Cellular Biology* **9**, 1309–1315.

Bjornstedt, M., Xue, J., Huang, W., Akesson, B. and Holmgren, A. (1994) The thioredoxin and glutaredoxin systems are efficient electron donors to human plasma glutathione peroxidase. *Journal of Biological Chemistry* **269**, 29382–29384.

Bjornstedt, M., Hamberg, M., Kumar, S., Xue, J. and Holmgren, A. (1995) Human thioredoxin reductase directly reduces lipid hydroperoxides by NADPH and selenocystine strongly stimulates the reaction via catalytically generated selenols. *Journal of Biological Chemistry* **270**, 11761–11764.

Blomberg, A. and Adler, L. (1992) Physiology of osmotolerance in fungi. *Advances in Microbial Physiology* **33**, 145–212.

Boorstein, W. R. and Craig, E. A. (1990a) Regulation of a yeast HSP70 gene by a cAMP responsive transcriptional control element. *The EMBO Journal* **9**, 2543–2553.

Boorstein, W. R. and Craig, E. A. (1990b) Transcriptional regulation of *SSA3*, an Hsp70 gene from *Saccharomyces cerevisiae*. *Molecular and Cellular Biology* **10**, 3262–3269.

Bossie, M. A. and Martin, C. E. (1989) Nutritional regulation of yeast d9-fatty acid desaturase activity. *Journal of Bacteriology* **171**, 6409–6413.

Bossier, P., Fernandes, L., Rocha, D. and Rodrigues-Pousada, C. (1993) Overexpression of *YAP2*, coding for a new yAP protein, and *YAP1* in *Saccharomyces cerevisiae* alleviates growth inhibition caused by 1,10-phenathroline. *Journal of Biological Chemistry* **268**, 23640–23645.

Bouveris, A. and Cadenas, E. (1982) Production of superoxide radicals and hydrogen peroxide in mitochondria. In: L. W. Oberley. (ed.) *Superoxide dismutases*, pp. 15–30. Boca Raton, Florida: CRC Press.

Bowman, S. B., Zaman, Z., Collinson, L. P., Brown, A. J. P. and Dawes, I. W. (1992) Positive regulation of the *LPD1* gene of *Saccharomyces cerevisiae* by the HAP2/HAP3/HAP4 activation system. *Molecular and General Genetics* **231**, 296–303.

Boy-Marcotte, E., Lagniel, G, Perrot, M., Bussereau, F., Boudsocq, A., Jacquet, M. and Labarre, J. (1999) The heat shock response in yeast: differential regulations and contributions of the Msn2p/Msn4p and Hsf1p regulons. *Molecular Microbiology* **33**, 274–283.

Brennan, R. J., Swoboda, B. E. P. and Schiestl, R. H. (1994) Oxidative mutagens induce intrachromosomal recombination in yeast. *Mutation Research* **308**, 159–167.

Brewster, J. L., de-Valoir, T., Dwyer, N. D., Winter, E. and Gustin, M. C. (1993) An osmosensing signal transduction pathway in yeast. *Science* **259**, 1760–1763.

Brown, A. D. (1978) Compatible solute and extreme water stress in eukaryotic microorganisms. *Advances in Microbial Physiology* **17**, 181–242.

Brüning, A. R. N. E., Bauer, J., Krems, B., Entian, K.-D. and Prior, B. A. (1998) Physiological and genetic characterisation of osmosensitive mutants of *Saccharomyces cerevisiae. Archives of Microbiology* **170**, 99–105.

Buchczyk, D. P., Briviba, K., Harti, F. U. and Sies, H. (2000) Responses to peroxynitrite in yeast: glyceraldehyde-3-phosphate dehydrogenase (GAPDH) as a sensitive intracellular target for nitration and enhancement of chaperone expression and ubiquitination. *Biological Chemistry* **381**, 121–126.

Buchner, J. (1999) Hsp90 & Co. – a holding for folding. *Trends in Biochemical Sciences* **24**, 136–141.

Burnborg, G. and Williamson, D. H. (1978) The relevance of the nuclear division cycle to radiosensitivity in yeast. *Molecular and General Genetics* **162**, 277–285.

Burns, V. W. (1956) X-ray induced division delay of individual yeast cells. *Radiation Research* **4**, 394–412.

Bushweller, J. H., Aslund, F., Wuethrich, K. and Holmgren, A. (1992) Structural and functional characterization of the mutant *Escherichia coli* glutaredoxin (C14→S) and its mixed disulfide with glutathione. *Biochemistry* **31**, 9288–9293.

Cabiscol, E., Piulats, E., Echave, P., Herrero, E. and Ros, J. (2000) Oxidative stress promotes specific protein damage in *Saccharomyces cerevisiae. Journal of Biological Chemistry* **275**, 27393–27398.

Carelli, S., Ceriotti, A., Cabibbo, A., Fassina, G., Ruvo, M. and Sitia, R. (1997) Cysteine and glutathione secretion in response to disulfide bond formation in the ER. *Science* **277**, 1681–1684.

Carmel-Harel, O. and Stortz, G. (2000) Roles of glutathione- and thioredoxin-dependent reduction systems in the *Escherichia coli* and *Saccharomyces cerevisiae* responses to oxidative stress. *Annual Reviews of Microbiology* **54**, 439–461.

Causton, H. C., Ren, B., Koh, S. S., Harbison, C. T., Kanin, E., Jennings, E. G., Lee, T. I., True, H. L., Lander, E. S. and Young, R. A. (2001) Remodelling of yeast genome expression in response to environmental change. *Molecular Biology of the Cell* **12**, 323–337.

Chang, E. C., Crawford, B. F., Hong, Z. Bilinksi, T. and Kosman, D. J. (1991) Genetic and biochemical characterization of Cu,Zn superoxide dismutase mutants in *Saccharomyces cerevisiae. Journal of Biological Chemistry* **266**, 4417–4424.

Charizanis, C. J. H., Krems, B. and Entian, K. D. (1999) The oxidative stress response mediated via Pos9/Skn7 is negatively regulated by the Ras PKA pathway in *Saccharomyces cerevisae. Molecular and General Genetics* **261**, 740–752.

Cheng, L. and Piper, P. W. (1994) Weak acid preservatives block the heat shock response and heat shock element directed *lacZ* expression of low pH *Saccharomyces cerevisiae* cultures. *Microbiology* **140**, 1085–1096.

Chowdhury, S., Smith, K. W., Gustin, M. C. (1992) Osmotic stress and the yeast cytoskeleton: phenotype-specific suppression of an actin mutation. *Journal of Cell Biology* **118**, 561–571.

Chrestensen, C. A., Eckman, C. B., Starke, D. W. and Mieyal, J. J. (1995) Cloning, expression and characterization of the human thioltransferase (glutaredoxin) in *E. coli*. *FEBS Letters* **374**, 25–28.

Chung, N., Jenkins, G., Hannun, Y. A., Heitman, J. and Abeid, L. M. (2000) Sphingolipids signal heat stress-induced ubiquitin-dependent proteolysis. *Journal of Biological Chemistry* **275**, 17229–17232.

Clarke, C. F., Williams, W. and Teruya, J. H. (1991) Ubiquinone biosynthesis in *Saccharomyces cerevisae*. Isolation and sequence of *COQ3*, the 3,4-dihydroxy-5-hexaprenylbenzoate methyltransferase gene. *Journal of Biological Chemistry* **266**, 16636–16644.

Cohen, B. A., Pilpel, Y., Mitra, R. D. and Church, G. M. (2002) Discrimination between paralogs using microarray analysis: application to the Yap1p and Yap2p transcription networks. *Molecular Biology of the Cell* **13**, 1608–1614.

Cohen, G., Fessl, F., Traczyk, A., Rytka, J. and Ruis, H. (1985) Isolation of the catalase A gene of *Saccharomyces cerevisiae* by complementation of the *cta1* mutation. *Molecular and General Genetics* **200**, 74–79.

Coleman, S. T., Tseng, E. and Moye-Rowley, W. S. (1997) *Saccharomyces cerevisiae* basic region-leucine zipper protein regulatory networks converge at the *ATR1* structural gene. *Journal of Biological Chemistry* **272**, 23224–23230.

Collinson, E. J., Wheeler, G. L., Garrido, E. O., Avery, A. M., Avery, S. V. and Grant, C. M. (2002) The yeast glutaredoxins are active as glutathione peroxidases. *Journal of Biological Chemistry* **277**, 16712–16717.

Collinson, L. P. and Dawes, I. W. (1992) Inducibility of the response of yeast cells to peroxide stress. *Journal of General Microbiology* **138**, 329–335.

Collinson, L. P. and Dawes, I. W. (1995) Isolation, characterization and overexpression of the yeast gene, *GLR1*, encoding glutahione reductase. *Gene* **156**, 123–127.

Coote, P. J., Cole, M. B. and Jones, M. V. (1991) Induction of increased thermotolerance in *Saccharomyces cerevisiae* may be triggered by a mechanism involving intracellular pH. *Journal of General Microbiology* **137**, 1701–1708.

Costa, W. M. V., Amorim, M. A., Quintanilha, A. and Moradas-Ferreira, P. (2002) Hydrogen geroxide-induced carbonylation of key metabolic enzymes in *Saccharomyces cerevisiae*: the involvement of the oxidative stress regulators Yap1 and Skn7. *Free Radical Biology and Medicine* **33**, 1507–1515.

Craig, E. A. and Gross, C. A. (1991) Is hsp70 the cellular thermometer? *Trends in Biochemical Sciences* **16**, 135–139.

Culotta, V. C. (2000) Superoxide disumtase, oxidative stress, and cell metabolism. *Current Topics in Cellular Regulation* **36**, 117–132.

Culotta, V. C., Howard, W. R. and Liu, X. F. (1995) *CRS5* encodes a metallothionein-like protein in *Saccharomyces cerevisiae*. *Journal of Biological Chemistry* **269**, 25295–25302.

Cyert, M. S. (2001) Genetic analysis of clamodulin and its targets in *Saccharomyces cerevisiae*. *Annual Reviews of Genetics* **35**, 647–672.

Cyert, M. S. and Thorner, J. (1992) Regulatory subunit (*CNB1* gene product) of yeast Ca^{2+}/calmodulin-dependent phosphoprotein phosphatases is required for adaptation to pheromone. *Molecular and Cellular Biology* **12**, 3460–3469.

Dang, V. D., Valens, M., Bolotin-Fukahara, M. and Daignan-Fornier, B. (1994) A genetic screen to isolate genes regulated by the yeast CCAAT-box binding protein Hap2p. *Yeast* **10**, 1273–1283.

Davenport, K. R., Sohaskey, M., Kamada, Y., Levin, D. E. and Gustin, M. C. (1995) A second osmosensing signal transduction pathway in yeast. Hypotonic shock activates the PKC1 protein kinase-regulated cell integrity pathway. *Journal of Biological Chemistry* **270**, 30157–30161.

Davidson, J. F., Whyte, B., Bissinger, P. H. and Schiestl, R. H. (1996) Oxidative stress is involved in heat-induced cell death in *Saccharomyces cerevisiae*. *Proceedings of the National Academy of Sciences, USA* **93**, 5116–5121.

Davies, J. M. S., Lowry, C. V. and Davies, K. J. A. (1995) Transient adaptation to oxidative stress in yeast. *Archives of Biochemistry and Biophysics* **317**, 1–6.

Dawes, I. W. (1976) Inactivation of yeasts. In: J. A. Skinner and W. B. Hugo (eds) *Inactivation and Inhibition of Vegetative Organisms*, pp. 279–304. London: Academic Press.

De Freitas, J. M., Liba, A., Meneghini, R., Valentine, J. S. and Gralla, E. B. (2000) Yeast lacking Cu-Zn superoxide dismutase show altered iron homeostasis – role of oxidative stress in iron metabolism. *Journal of Biological Chemistry* **275**, 11645–11649.

De Virgilio, C., Bürckett, N., Boller, T. and Wiemken, A. (1991) A method to study the rapid phosphorylation-related modulation of neutral trehalase activity by temperature shifts in yeast. *FEBS Letters* **291**, 355–358.

Delauney, A., Isnard A-D. and Toledano, M. B. (2000) H_2O_2 sensing through oxidation of the Yap1 transcription factor. *EMBO Journal* **19**, 5157–5166.

Delaunay, A., Pflieger, D., Barrault, M. B., Vinh, J. and Toledano, M. B. (2002) A thiol peroxidase is an H_2O_2 receptor and redox-transducer in gene activation. *Cell* **111**, 471–481.

Dibrov, E., Robinson, K. M. and Lemire, B. D. (1997) The *COQ5* gene encodes a yeast mitochondrial protein necessary for ubiquinone biosynthesis and the assembly of the respiratory chain. *Journal of Biological Chemistry* **272**, 9175–9181.

Diniz-Mendes, L., Bernades, E., de Araujo, P. S., Panek, A. D. and Paschoalin, V. M. F. (1999) Preservation of frozen yeast cells by trehalose. *Biotechnology and Bioengineering* **65**, 572–578.

Do, T. Q., Schultz, J. R. and Clarke, C. F. (1996) Enhanced sensitivity of ubiquinone-deficient mutants of *Saccharomyces cerevisiae* to products of autoxidised polyunsaturated fatty acids. *Proceedings of the National Academy of the USA* **93**, 7534–7539.

Draculic, T., Dawes, I. W. and Grant, C. M. (2000) A single glutaredoxin or thioredoxin is essential for viability in the yeast *Saccharomyces cerevisiae. Molecular Microbiology* **36**, 1167–1174.

Dumbrava, V.-A. and Pall, M. L. (1987) Control of nucleotide and erythroascorbic acid pools by cyclic AMP in *Neurospora crassa. Biochimica et Biophysica Acta* **926**, 331–338.

Engelberg, D., Klein, C., Martinetto, H., Struhl, K. and Karin, M. (1994) The UV response involving the Ras signalling pathway and AP-1 transcription factors is conserved between yeast and mammals. *Cell* **77**, 381–390.

Ernster, L. and Dallner, G. (1995) Biochemical, physiological and medical aspects of ubiquinone function. *Biochimica et Biophysica Acta* **1271**, 195–204.

Ernster, L. and Forsmark-Andree, P. (1993) Ubiquinol: an endogenous antioxidant in aerobic organisms. *Clinical Investigation* **71**, S60–S65.

Evans, M., Turton, H. E., Grant, C. M. and Dawes, I. W. (1998) Toxicity of linoleic acid hydroperoxide to *Saccharomyces cerevisiae*: involvement of a respiration-related process for maximal sensitivity and adaptive response. *Journal of Bacteriology* **180**, 483–490.

Fernandes, L., Rodrigues-Pousada, C. and Struhl, K. (1997) Yap, a novel family of eight bZIP proteins in *Saccharomyces cerevisiae* with distinct biological functions. *Molecular and Cellular Biology* **17**, 6982–6993.

Fernandes, M., Xiao, H. and Lis, J. T. (1994) Fine structure analyses of the Drosophila and Saccharomyces heat-shock factor heat-shock element interactions. *Nucleic Acids Research* **22**, 167–173.

Fernando, M. R., Nanri, H., Yoshitkae, S., Ngata-Kuno, K. and Minakami, S. (1992) Thioredoxin regenerates proteins inactivated by oxidative stress in endothelial cells. *European Journal of Biochemistry* **209**, 917–922.

Ferreira, M. C., Bao, X., Laizé, V. and Hohmann, S. (2001) Transposon mutagenesis reveals novel loci affecting tolerance to salt stress and growth at low temperature. *Current Genetics* **40**, 27–39.

Finley, D. and Chau, V. (1991) Ubiquitination. *Annual Reviews of Cell Biology* **7**, 25–69.

Finley, D., Ozkaynak, E. and Varshavsky, A. (1987) The yeast polyubiquitin gene is essential for resistance to high temperatures, starvation and other stresses. *Cell* **48**, 1035–1046.

Finley, D. and Varshavsky, A. (1985) The ubiquitin system: functions and mechamisms. *Trends in Biochemical Sciences* **10**, 343–347.

Flattery-O'Brien, J., Collinson, L. P. and Dawes, I. W. (1993) *Saccharomyces cerevisiae* has an inducible response to menadione which differs from that to hydrogen peroxide. *Journal of General Microbiology* **139**, 501–507.

Flattery-O'Brien, J. A. and Dawes, I. W. (1998) Hydrogen peroxide causes *RAD9*-dependent arrest in G2 in *Saccahromyces cerevisiae* whereas menadione causes G1 arrest independent of *RAD9* function. *Journal of Biological Chemistry* **273**, 8564–8571.

Flattery-O'Brien, J. A., Grant, C. M. and Dawes, I. W. (1997) Stationary phase regulation of the *Saccharomyces cerevisiae SOD2* gene is dependent on additive effects of HAP2,3,4,5- and STRE-binding elements. *Molecular Microbiology* **23**, 303–312.

Forsburg, S. and Guarente, L. (1988) Mutational analysis of upstream activation sequence 2 of the *CYC1* gene of *Saccharomyces cerevisiae*: a HAP2-HAP3-responsive site. *Molecular and Cellular Biology* **8**, 647–654.

Francois, J., Neves, M. J. and Hers, H. G. (1991) The control of trehalose biosynthesis in *Saccharomyces cerevisiae* – evidence for a catabolite inactivation and repression of trehalose-6-phosphate synthase and trehalose-6-phosphate phosphatase. *Yeast* **7**, 575–587.

Fu, S., Gebicki, S., Jessup, W., Gebicki, J. and Dean, R. T. (1995) Biological fate of amino acid, peptide and protein hydroperoxides. *Biochemical Journal* **311**, 821–827.

Furnari, B., Rhind, N. and Russell, P. (1997) Cdc25 mitotic inducer tergeted by Chk1 DNA damage checkpoint kinase. *Science* **277**, 1495–1497.

Fürst, P., Hu, S., Hackett, R. and Hamer, D. (1988) Copper acitvates metallothionein gene transcription by altering the conformation of a specific DNA binding protein. *Cell* **55**, 705–717.

Galiazzo, F. and Labbe-Bois, R. (1993) Regulation of Cu,Zn- and Mn-superoxide dismutase transcription in *Saccharomyces cerevisiae. FEBS Letters* **315**, 197–200.

Gan, Z. R. (1991) Yeast thioredoxin genes. *Journal of Biological Chemistry* **266**, 1692–1696.

Gan, Z. R. (1992) Cloning and sequencing of the gene encoding yeast thioltransferease. *Biochemical and Biophysical Research Communications* **187**, 949–955.

Gardner, P. R. and Fridovich, I. (1991) Superoxide sensitivity of the *Escherichia coli* 6-phosphogluconate dehydratase. *Journal of Biological Chemistry* **266**, 1478–1483.

Garreau, H., Hasan, R. N., Renault, G., Estruch, F., Boy-Marcotte, E. and Jacquet, M. (2000) Hyperphosphorylation of Msn2p and Msn4p in response to heat shock and the diauxic shift is inhibited by cAMP in *Saccharomyces cerevisiae. Microbiology* **146**, 2113–2120.

Garrido, E. O. and Grant, C. M. (2002) Role of thioredoxins in the response of *Saccharomyces cerevisiae* to oxidative stress induced by hydroperoxides. *Molecular Microbiology* **43**, 993–1003.

Gasch, A. P. and Brown, P. O. (1999) Gene expression patterns of *Saccharomyces cerevisiae* in response to oxidative stress analyzed using cDNA microarrays. *Current Genetics* **35**, 189.

Gasch, A. P., Spellman, P. T., Kao, C. M., Carmen-Harel, O., Eisen, M. B., Storz, G., Botstein, D. and Brown, P. O. (2000) Genomic expression programs in the response of yeast cells to environmenal changes. *Molecular Biology of the Cell* **11**, 4241–4257.

Gebicki, S., Gill, K. H., Dean, R. T. and Gebicki, J. M. (2002) Action of peroxidases on protein hydroperoxides. *Redox Report* **7**, 235–242.

Glover, J. R. and Lindquist, S. (1999) Hsp104, Hsp70, and Hsp40: a novel chaperone system that rescues previously aggregated proteins. *Cell* **94**, 73–82.

Görner, W., Durchschlag, E., Martinez-Pastor, M. T., Estruch, F., Ammerer, G., Hamilton, B., Ruis, H. and Schuller, C. (1998) Nuclear localization of the C_2H_2 zinc finger protein Msn2p is regulated by stress and protein kinase A activity. *Genes and Development* **15**, 586–597.

Gralla, E. B. and Kosman, D. J. (1992) Molecular genetics of superoxide dismutases. *Advances in Genetics* **30**, 251–319.

Gralla, E. B., Thiele, D. J., Silar, P. and Valentine, J. S. (1991) ACE1, a copper-dependent transcription factor, activates expression of the yeast copper, zinc superoxide dismutase gene. *Proceedings of the National Academy of Sciences, USA* **88**, 8558–8562.

Grant, C. M. (2001) Role of the glutathione/glutaredoxin and thioredoxin systems in yeast growth and response to stress conditions. *Molecular Microbiology* **39**, 533–541.

Grant, C. M. and Dawes, I. W. (1996) Synthesis and role of glutathione in protection against oxidative stress in yeast. *Redox Report* **2**, 223–229.

Grant, C. M., Firoozan, M. and Tuite, M. F. (1989) Mistranslation induces the heat shock response in yeast. *Molecular Microbiology* **3**, 215–220.

Grant, C. M., Collinson, L. P., Roe, J.-H. and Dawes, I. W. (1996a) Yeast glutathione reductase is required for protection against oxidative stress and is a target gene for yAP-1 transcriptional regulation. *Molecular Microbiology* **21**, 739–746.

Grant, C. M., MacIver, F. H. and Dawes, I. W. (1996b) Glutathione is an essential metabolite required for resistance to oxidative stress in the yeast *Saccharomyces cerevisiae. Current Genetics* **29**, 511–515.

Grant, C. M., MacIver, F. H. and Dawes, I. W. (1997a) Glutathione synthetase is dispensable for growth under both normal and oxidative stress conditions in the yeast *Saccharomyces cerevisiae* due to an accumulation of the dipeptide γ-glutamylcysteine. *Molecular Biology of the Cell* **8**, 1699–1707.

Grant, C. M., MacIver, F. H. and Dawes, I. W. (1997b) Mitochondrial function is required for resistance to oxidative stress in the yeast *Saccharomyces cerevisiae. FEBS Letters* **410**, 219–222.

Grant, C. M., Perrone, G. and Dawes, I. W. (1998) Glutathione and catalase provide overlapping defenses for protection against hydrogen peroxide in the yeast *Saccharomyces cerevisiae. Biochemical and Biophysical Research Communications* **253**, 893–898.

Grant, C. M., Luikenhuis, S., Beckhouse, A., Soderbergh, M. and Dawes, I. W. (2000) Differential regulation of glutaredoxin gene expression in response to stress conditions in the yeast *Saccharomyces cerevisiae. Biochimica et Biophysica Acta* **1490**, 33–42.

Gravina, S. A. and Mieyal, J. (1993) Thioltransferase is a specific glutathionyl mixed disulfide oxidoreductase. *Biochemistry* **32**, 3368–3376.

Greco, M. A., Hrab, D. I., Magner, W. and Kosman, D. J. (1990) Cu,Zn superoxide dismutase and copper deprivation and toxicity in *Saccharomyces cerevisiae. Journal of Bacteriology* **172**, 317–325.

Gropper, T. and Rensing, L. (1993) Inhibitors of proteases and other stressors induce low molecular weight heat shock proteins in *Saccharomyces cerevisiae. Experimental Mycology* **17**, 46–54.

Guarente, L., Lalonde, B., Gifford, P. and Alani, E. (1984) Distinctly regulated tandem upstream activation sites mediate catabolite repression of the *CYC1* gene in *S. cerevisiae. Cell* **36**, 503–511.

Gunstone, F. D. (1996) *Fatty Acid and Lipid Chemistry*. London: Blackie Academic and Professional.

Gustin, M. C., Albertyn, J., Alexander, M. and Davenport, K. (1998) MAP kinase pathways in the yeast *Saccharomyces cerevisiae. Microbiology and Molecular Biology Reviews* **62**, 1264–1300.

Ha, N., Hellauer, K. and Turcotte, B. (1996) Mutations in target DNA elements of yeast HAP1 modulate its transcriptional activity without affecting DNA binding. *Nucleic Acids Research* **24**, 1453–1459.

Halliwell, B. (1987) Oxidants and human disease: some new concepts. *FASEB Journal* **1**, 358–364.

Halliwell, B. (1994) Vitamin C: the key to health or a slow-acting carcinogen? *Redox Reports* **1**, 5–9.

Halliwell, B. (1995) The biological significance of oxygen-derived species. In: J. S. Valentine, C. S., Foote, A., Greenberg, and J. F. Liebman, (eds) *Active Oxygen in Biochemistry*, pp. 313–335. Glasgow: Blackie Academic and Professional.

Halliwell, B. and Gutteridge, J. M. C. (1984) Oxygen toxicity, oxygen radicals, transition metals and disease. *Biochemical Journal* **219**, 1–14.

Halliwell, B. and Gutteridge, J. M. C. (1989) *Free Radicals in Biology and Medicine*. Oxford: Clarendon Press.

Halliwell, B. and Gutteridge, J. M. C. (1990) Role of free radicals and catalytic metal ions in human disease: an overview. *Methods in Enzymology* **186**, 1–85.

Halliwell, B. and Aruoma, O. I. (1991) DNA damage by oxygen-derived species: its mechanism and measurement in mammalian systems. *FEBS Letters* **281**, 9–19.

Hancock, R. D., Galpin, J. R. and Viola, R. (2000) Biosynthesis of L-ascorbic acid (vitamin C) by *Saccharomyces cerevisiae. FEMS Microbiology Letters* **186**, 245–250.

Hannun, Y. A. (1996) Functions of ceramide in coordinating cellular responses to stress. *Science* **274**, 1855–1859.

Haro, R., Banuelos, M. A., Quintero, F. J., Rubio, F. and Rodriguez-Navarro, A. (1993) Genetic basis of sodium exclusion and sodium tolerance in yeast. A model for plants. *Physiologia Plantarum* **89**, 868–874.

Harshman, K. D., Moye-Rowley, W. S. and Parker, C. S. (1988) Transcriptional activation by the SV40 AP-1 recognition element in yeast is mediated by a factor similar to AP1-1 that is distinct from GCN4. *Cell* **53**, 321–330.

Hartl, F.-U. (1996) Molecular chaperones in cellular protein folding. *Nature* **381**, 571–580.

Hartwell, L. H. and Weinert, T. A. (1989) Checkpoints: controls that ensure the order of cell cycle events. *Science* **246**, 629–634.

Hertle, K., Haase, E. and Brendel, M. (1991) The *SNQ3* gene of *Saccharomyces cerevisiae* confers hyper-resistance to several functionally unrelated chemicals. *Current Genetics* **19**, 429–433.

Higgins, V. J., Alic, N., Thorpe, G. W., Breitenbach, M, Larsson, V. and Dawes, I. W. (2002) Phenotypic analysis of gene deletant strains for sensitivity to oxidative stress. *Yeast* **19**, 203–214.

Hirata, D., Harada, S.-I., Namba, H. and Miyakawa, T. (1995) Adaptation to high-salt stress in *Saccharomyces cerevisiae* is regulated by Ca^{2+}/calmodulin-dependent phosphoprotein phosphatase (calcineurin) and cAMP-dependent protein kinase. *Molecular and General Genetics* **249**, 257–264.

Hirata, D., Yano, K. and Miyakawa, T. (1994) Stress-induced transcriptional activation mediated by *YAP1* and *YAP2* genes that encode the Jun family of transcriptional activators in *Saccharomyces cerevisiae*. *Molecular and General Genetics* **242**, 250–256.

Hirimburegama, K., Durnez, P., Keleman, J., Oris, E., Vergauwen, R., Mergelsberg, H. and Thevelein, J. M. (1992) Nutrient-induced activation of trehalase in nutrient-starved cells of the yeast *Saccharomyces cerevisiae*: cAMP is not involved as second messenger. *Journal of General Microbiology* **138**, 2035–2043.

Hjorth-Sorenson, B., Hoffman, E. R., Lissin, N. M., Sewell, A. K. and Jakobsen, B. K. (2001) Activation of heat shock transcription factor in yeast is not influenced by the levels of expression of heat shock proteins. *Molecular Microbiology* **39**, 914–923.

Hohmann, S. (1997) Shaping up: the response of yeast to osmotic stress. In: S. Hohmann and W. H. Mager (eds) *Yeast Stress Responses*, pp. 101–143. Austin: R.G. Landes Co.

Hohmann, S. (2002) Osmotic stress signaling and osmoadaptation in yeasts. *Microbiology and Molecular Biology Reviews* **66**, 300–372.

Hohmann, S. Mager, W.H. (eds) (1997) *Yeast Stress Responses*, Austin: R.G. Landes Co.

Holliday, R. (1964) The induction of mitotic recombination by mitomycin C in *Ustilago* and *Saccharomyces*. *Genetics* **50**, 323–335.

Holmgren, A. (1979) Glutathione-dependent synthesis of deoxyribonucleotides. *Journal of Biological Chemistry* **254**, 3664–3671.

Holmgren, A. (1985) Thioredoxin. *Annual Reviews of Biochemistry* **54**, 237–271.

Holmgren, A. (1989) Thioredoxin and glutaredoxin systems. *Journal of Biological Chemistry* **264**, 13963–13966.

Holmgren, A. and Aslund, F. (1995) Glutaredoxin. *Methods in Enzymology* **252**, 283–292.

Hörtner, H., Ammreer, G., Hartter, E., Hamilton, B., Rytka, J., Bilinski, T., and Ruis, H. (1982) Regulation of synthesis of catalases and iso-1-cytochrome c in *Saccharomyces cerevisiae. European Journal of Biochemistry* **128**, 179–184.

Hottinger, T., de Virgilio, C., Hall, M. N., Boller, T. and Wiemken, A. (1994) The role of trehalose synthesis for the acquisition of thermotolerance in yeast. II. Physiological concentrations of trehalose increase the thermal stability of proteins in vitro. *European Journal of Biochemistry* **219**, 187–193.

Huh, W. K., Lee, B. H., Kim, S. T., Kim, Y. R., Rhie, G. E., Back, Y. W., Hwang, C. S., Lee, J. S. and Kang, S. O. (1998) D-erythroascorbic acid is an important antioxidant molecule in *Saccharomyces cerevisiae*. *Molecular Microbiology* **30**, 895–903.

Hussain, M. and Lenard, J. (1991) Characterization of *PDR4*, a *Saccharomyces cerevisiae* gene that confers pleiotropic drug resistance in high-copy number: identity with *YAP1*, encoding a transcriptional activator. *Gene* **101**, 149–152.

Inoue, Y., Matsuda, T., Sugiyama, K.-I., Izawa, S. and Kimura, A. (1999) Genetic analysis of glutathione peroxidase in oxidative stress response of *Saccharomyces cerevisiae*. *Journal of Biological Chemistry* **274**, 27002–27009.

Isoyama, T., Murayama, A., Nomoto, A. and Kuge, S. (2001) Nuclear import of the yeast AP-1-like transcription factor Yap1p is mediated by transport receptor Pse1p, and this import step is not affected by oxidative stress. *Journal of Biological Chemistry* **276**, 21863–21869.

Iwahashi, H., Obuchi, K., Fujii, S. and Komatsu, Y. (1995) The correlative evidence suggests that trehalose stabilizes membrane structure in the yeast *Saccharomyces cerevisiae. Cellular and Molecular Biology* **41**, 763–769.

Iyer, V. R. Eisen, M. B., Ross, D. T., Schuler, G., Moore, T., Lee, J. C. F., Trent, J. M., Staudt, L. M., Hudson, J., Boguski, M. S., Lashkari, D., Shalon, D., Botstein, D. and Brown, P. O. (1999) The transcriptional program in the response of human fibroblasts to serum. *Science* **283**, 83–87.

Izawa, S., Inoue, Y. and Kimura, A. (1995) Oxidative stress response in yeast: effect of glutathione on adaptation to hydrogen peroxide stress in *Saccharomyces cerevisiae. FEBS Letters* **368**, 73–76.

Izawa, S., Inoue, Y. and Kimura, A. (1996) Importance of catalase in the adaptive response to hydrogen peroxide: analysis of acatalasaemic *Saccharomyces cerevisiae. Biochemical Journal* **320**, 61–67.

Jacob, H. S., Ingbar, S. H. and Jandl, J. H. (1965) Oxidative hemolysis and erythrocyte metabolism in heredity acatalasia. *Journal of Clinical Investigation* **44**, 1187–1199.

Jacoby, T., Flanagan, H., Faykin, A., Seto, A. G., Mattison, C. and Ota, I. (1997) Two protein-tyrosine phosphatases inactivate the osmotic stress response pathway in yeast by targeting the mitogen-activated protein kinase Hog1. *Journal of Biological Chemistry* **272**, 17749–17755.

Jamieson, D. J. (1992) *Saccharomyces cerevisiae* has distinct adaptive responses to both hydrogen peroxide and menadione. *Journal of Bacteriology* **174**, 6678–6681.

Jamieson, D. J. (1995) Oxidative stress responses of *Saccharomyces cerevisiae. Redox Reports* **1**, 89–95.

Jamieson, D. J. (1998) Oxidative stress responses of the yeast *Saccharomyces cerevisiae. Yeast* **14**, 1511–1527.

Jamieson, D. J., Rivers, S. T. and Stephen, D. W. S. (1994) Analysis of *Saccharomyces cerevisiae* proteins induced by peroxide and superoxide stress. *Microbiology* **140**, 3277–3283.

Jenkins, G. M., Richards, A., Wahl, T., Mao, C., Obeid, L. and Hannun, T. (1997) Involvement of yeast sphingolipids in the heat stress response of *Saccharomyces cerevisiae. Journal of Biological Chemistry* **272**, 32566–32572.

Joenje, H., Lafleur, M. V. M. and Retèl, J. (1991) Biological consequences of oxidative DNA damage. In: C. Vigo-Pelfrey (ed.) *Membrane Lipid Oxidation III*, pp. 87–113. Boca Raton, FL: CRC Press.

Jungmann, J., Reins, H. A., Lee, J., Romeo, A., Hassett, R., Kosman, D. and Jentsch, S. (1994) MAC1, a nuclear regulatory protein related to Cu-dependent transcription factors is involved in Cu/Fe utilization and stress resistance in yeast. *EMBO Journal* **13**, 5051–5056.

Kagi, J. H. R. and Schaffer, A. (1988) Biochemistry of metallothionein. *Biochemistry* **27**, 8509–8515.

Kaida, D., Yashiroda, H., Toh-e, A. and Kikuchi, Y. (2002) Yeast Whi2 and Psr-1 phosphatase form a complex and regulate STRE-mediated gene expression. *Genes to Cells* **7**, 543–552.

Kalen, A., Norling, B., Appelkvist, E. L. and Dallner, G. (1987) Ubiquinone biosynthesis by the microaomal fraction from rat liver. *Biochimica et Biophysica Acta* **926**, 70–78.

Kamada, Y., Jung, U. S., Piotrowski, J. and Levin, D. E. (1995) The protein kinase C-activated MAP kinase pathway of *Saccharomyces cerevisiae* mediates a novel heat shock response. *Genes and Development* **9**, 1559–1571.

Kawai, R. F., K., Iwahashi, H. and Komatsu, Y. (1999) Direct evidence for the intracellular localization of Hsp104 in *Saccharomyces cerevisiae* by immunoelectron microscopy. *Cell Stress and Chaperones* **4**, 46–53.

Kim, H. S., Seib, P. A. and Chung, O. K. (1993) D-erythroascorbic acid in bakers' yeast and effects on wheat dough. *Journal of Food Science* **58**, 845–862.

Kim, S. and Weinert, T. A. (1997) Characterization of the checkpoint gene *RAD53/MEC2* in *Saccharomyces cerevisiae. Yeast* **13**, 735–745.

Kim, S., Fyrst, H. and Saba, J. (2000) Accumulation of phosphorylated sphingoid long chain bases results in cell growth inhibition in *Saccharomyces cerevisiae. Genetics* **156**, 1519–1529.

Kim, S. T., Huh, W. K., Lee, B. H. and Kang, S. O. (1998) D-arabinose dehydrogenase and its gene from *Saccharomyces cerevisiae. Biochimica et Biophysica Acta* **1429**, 29–39.

Kirk, N. and Piper, P. W. (1991) The determinants of heat shock element-directed *lacZ* expression in *Saccharomyces cerevisiae*. *Yeast* **7**, 539–546.

Kiser, G. L. and Weinert, T. A. (1996) Distinct roles of yeast *MEC* and *RAD* checkpoint genes in transcriptional induction after DNA damage and implications for function. *Molecular Biology of the Cell* **7**, 703–718.

Kobayashi, N. and McEntee, K. (1993) Identification of *cis* and *trans* components of a novel heat shock stress regulatory pathway in *Saccharomyces cerevisiae*. *Molecular and Cellular Biology* **13**, 248–256.

Kohlwein, S. D. and Paltauf, F. (1983) Uptake of fatty acids by yeasts *Saccharomyces uvarum* and *Saccharomycopsis lipolytica*. *Biochimica et Biophysica Acta* **792**, 310–317.

Kuckelkorn, U., Knuehl, C., Boes-Fabian, B., Drung, I. and Kloetzel, P. M. (2000) The effect of heat shock on 20/26S proteasomes. *Biological Chemistry* **381**, 1017–1023.

Kuge, S. and Jones, N. (1994) *YAP1*-dependent of activation of TRX2 is essential for the response of *Saccharomyces cerevisiae* to oxidative stress by hydroperoxides. *EMBO Journal* **13**, 655–664.

Kuge, S., Jones, N. and Nomoto, A. (1997) Regulation of yAP-1 nuclear localization in response to oxidative stress. *EMBO Journal* **16**, 1710–1720.

Kuge, S., Toda, T., Lizuka, N. and Nomoto, A. (1998) Crm1 (Xpo1) dependent nuclear export of the budding yeasy transcription factor yAP-1 is sensitive to oxidative stress. *Genes to Cells* **3**, 521–532.

Kuo, C. F., Masino, T. and Fridovich, I. (1987) α–β-dihydroxyisovalerate dehydratase: a superoxide-sensitive enzyme. *Journal of Biological Chemistry* **262**, 4724–4727.

Lafleur, M. V. M. and Retèl, J. (1993) Contrasting effects of SH-compounds on oxidative DNA damage: repair and increase of damage. *Mutation Research* **295**, 1–10.

Langer, T., Lu, C., Echols, H., Flanagan, J., Hayer, M. K. and Hartl, F. H. (1992) Successive action of DnaK, DnaJ and GroEL along the pathway of chaperone-mediated protein folding. *Nature* **356**, 683–689.

Lapinskas, P. J., Cunningham, K. W., Liu, X. F., Fink, G. R. and Culotta, V. C. (1995) Mutations in *PMR1* suppress oxidative damage in yeast cells lacking superoxide dismutase. *Molecular and Cellular Biology* **15**, 1382–1388.

Latterich, M. and Watson, M. D. (1992) Isolation and characterization of osmosensitive vacuolar mutants of *Saccharomyces cerevisiae*. *Molecular Microbiology* **5**, 2417–2426.

Lee, F. J. and Hassan, H. M. (1985) Biosynthesis of superoxide dismutase in *Saccharomyces cerevisiae*: effects of paraquat and copper. *Free Radicals in Biology and Medicine* **1**, 319–325.

Lee, J., Romeo, A. and Kosman, D. J. (1997) Transcriptional remodelling and G_1 arrest in dioxygen stress in *Saccharomyces cerevisiae*. *Journal of Biological Chemistry* **271**, 24885–24893.

Lee, J., Godon, C., Lagniel, G., Spector, D., Garin, J., Labarre, J. and Toledano, M. B. (1999) Yap1 and Skn7 control two specialized oxidative stress response regulons in yeast. *Journal of Biological Chemistry* **274**, 16040–16046.

Lee, J.-C., Straffon, M. J., Jang, T-Y., Higgins, V. J., Grant C. M. and Dawes, I. W. (2001) The essential and ancillary role of glutathione in *Saccharomyces cerevisiae* analysed using a grande *gsh1* disruptant strain. *FEMS Yeast Research* **1**, 57–65.

Lee, J.-S., Dawes, I. W. and Roe J-H. (1995) Adaptive response of *Schizosaccharomyces pombe* to hydrogen peroxide and menadione. *Microbiology* **141**, 3127–3132.

Lesko, S., Lorentarn, A. and Tso, R. (1980) Involvement of the polycyclic aromatic hydrocarbons and reduced oxygen radicals in carcinogenesis. *Biochemistry* **19**, 3023–3028.

Leung, C. Y. and Loewus, F. A. (1985) Concerning the presence and formation of ascorbic acid in yeasts. *Plant Science* **38**, 65–69.

Levin, D. E., Fields, F., Kunisawa, R., Bishop, J. M. and Thorner, J. (1990) A candidate protein kinase C gene, *PKC1*, is required for the *Saccharomyces cerevisiae* cell cycle. *Cell* **62**, 213–224.

Levin, D. E., Bowers, B., Chen, C-Y., Kamada, Y. and Watanabe, M. (1994) Dissecting the protein kinase C/MAP kinase signalling pathway of *Saccharomyces cerevisiae*. *Cellular and Molecular Biology Research* **40**, 229–239.

Levine, R. L. (2002) Carbonyl modified proteins in cellular regulation, aging, and disease. *Free Radical Biology and Medicine* **32**, 790–796.
Lewis, J. G., Learmonth, R. P. and Watson, K. (1993) Role of growth phase and ethanol infreeze-thaw stress resistance of *Saccharomyces cerevisiae*. *Applied and Environmental Microbiology* **59**, 1065–1071.
Lewis, J. G., Learmonth, R. P. and Watson, K. (1995) Induction of heat, freezing and salt tolerance by heat and salt shock in *Saccharomyces cerevisiae*. *Microbiology* **141**, 687–694.
Lin, S.-J. and Culotta, V. C. (1996) Suppression of oxidative damage by *Saccharomyces cerevisiae ATX2*, which encodes a manganese-trafficking protein that localizes to Golgi-like vesicles. *Molecular and Cellular Biology* **16**, 6303–6312.
Lindquist, S. and Craig, E. A. (1988) The heat shock proteins. *Annual Reviews of Genetics* **55**, 631–677.
Lindquist, S. and Kim, G. (1996) Heat-shock protein 104 expression is sufficient for thermotolerance in yeast. *Proceedings of the National Academy of Sciences USA* **93**, 5301–5306.
Liu, X. D. and Thiele, D. J. (1996) Oxidative stress induced heat shock factor phosphorylation and HSF-dependent activation of yeast metallothionein gene transcription. *Genes and Development* **10**, 592–603.
Liu, X. D., Morano, K. A. and Thiele, D. J. (1999) The yeast Hsp110 family member, Sse1, is an Hsp90 cochaperone. *Journal of Biological Chemistry* **274**, 26654–26660.
Liu, X. F., Elashivili, I, Gralla, E. B., Valentine, J. S., Lapinskas, P. and Culotta, V. C. (1992) Yeast lacking superoxide dismutase. Isolation of genetic suppressors. *Journal of Biological Chemistry* **267**, 18298–18302.
Longo, V. D., Gralla, E. B. and Valentine, J. S. (1996) Superoxide dismutase is essential for stationary phase survival in *Saccharomyces cerevisiae*. *Journal of Biological Chemistry* **271**, 12275–12280.
Longo, V. D., Liou, L. L., Valentine, J. S. and Gralla, E. B. (1999) Mitochondrial superoxide decreases yeast survival in stationary phase. *Archives of Biochemistry and Biophysics* **365**, 131–142.
Lowndes, N. F. and Murguia, J. R. (2000) Sensing and responding to DNA damage. *Current Opinions in Genetics and Development* **10**, 17–25.
Lowry, C. V. and Zitomer, R. S. (1984) Oxygen regulation of anaerobic and aerobic genes mediated by a common factor in yeast. *Proceedings of the National Academy of Sciences USA* **81**, 6129–6133.
Luikenhuis, S. (1998) The yeast *Saccharomyces cerevisiae* contains two glutaredoxin genes that are required for protection against reactive oxygen species. *Molecular Biology of the Cell* **9**, 1081–1091.
Lundström-Ljung, J. and Holmgren, A. (1995) Glutaredoxin accelerates glutathione-dependent folding of reduced ribonuclease A together with protein disulfide-isomerase. *Journal of Biological Chemistry* **270**, 7822–7828.
Maeda, T., Wurgler-Murphy, S. M. and Saito, H. (1994) A two-component system that regulates an osmosensing MAP kinase cascade in yeast. *Nature* **369**, 242–245.
Maeda, T., Takekawa, M. and Saito, H. (1995) Activation of yeast PBS2 MAPKK by MAPKKKs or by binding of an SH3-containing osmosensor. *Science* **269**, 554–558.
Mager, W. H. and De Kruijff, A. J. J. (1995) Stress-induced transcriptional activation. *Microbiological Reviews* **1995**, 506–531.
Mager, W. H. and Siderius, M. (2002) Novel insights into the osmotic stress response of yeast. *FEMS Yeast Research* **2**, 251–257.
Mager, W. H. and Varela, J. C. S. (1993) Osmostress response of the yeast *Saccharomyces*. *Molecular Microbiology* **10**, 253–258.
Mao, G. C., Saba, J. D. and Obeid, L. M. (1999) The dihydrosphingosine-1-phospate phosphatases of *Saccharomyces cerevisiae* are important regulators of cell proliferation and heat stress responses. *Biochemical Journal*, **342,** 667–675.
Marbois, B. N. and Clarke, C. F. (1996) The *COQ7* gene encodes a protein in *Saccharomyces cerevisiae* necessary for ubiquinone biosynthesis. *Journal of Biological Chemistry* **271**, 2995–3004.
Marchler, G., Schüller, C., Adam, G. and Ruis, H. (1993) A *Saccharomyces cerevisiae* UAS element controlled by protein kinase A activates transcription in response to a variety of stress conditions. *EMBO Journal* **12**, 1997–2003.
Marquez, J. A. and Serrano, R. (1996) Multiple transduction pathways regulate the sodium-extrusion gene *PMR2/ENA1* during salt stress in yeast. *FEBS Letters* **382**, 89–92.

Martinez-Pastor, M., T., Marchler, G., Schuller, C., Marchler-Bauer, A., Ruis, H. and Estruch, F. (1996) The *Saccharomyces cerevisiae* zinc finger proteins Msn2p and Msn4p are required for transcriptional induction through the stress response element (STRE). *EMBO Journal* **15**, 2227–2235.

Mazur, P. (1970) Cryobiology: the freezing of biological systems. *Science* **168**, 939–949.

McAlister, L. and Holland, M. J. (1985) Differential expression of the three yeast glyceraldehyde-3-phosphate dehydrogenase genes. *Journal of Biological Chemistry* **260**, 15019–15027.

Meister, A. (1994) Glutathione-ascorbic acid antioxidant systems in animals. *Journal of Biological Chemistry* **269**, 9397–9400.

Meister, A. and Anderson, M. E. (1983) Glutathione. *Annual Reviews of Biochemistry* **52**, 711–760.

Mendoza, I., Rubio, F., Rodriguez-Navarro and Pardo, J. M. (1994) The protein phosphatase calcineurin is essential for NaCl tolerance of *Saccharomyces cerevisiae*. *Journal of Biological Chemistry* **269**, 8792–8796.

Miller, M. J., Xuong, N. H. and Geiduschek, E. P. (1979) A response of protein synthesis to temperature shift in the yeast *Saccharomyces cerevisiae*. *Proceedings of the National Academy of Sciences USA* **76**, 5222–5225.

Miller, M. J., Xuong, N. H. and Geiduschek, E. P. (1982) Quantitative analysis of the heat shock response of *Saccharomyces cerevisiae*. *Journal of Bacteriology* **151**, 311–327.

Mitsui, A., Hirakawa, T. and Yodoi, J. (1992) Reactive oxygen-reducing and protein-folding activities of adult T cell leukemia-derived factor/human thioredoxin. *Biochemical and Biophysical Research Communications* **186**, 1220–1226.

Miyahara, K., Hirata, D. and Miyakawa, T. (1996) yAP-1 and yAP-2-mediated, heat shock-induced transcriptional activation of the multidrug resistance ABC transporter genes in *Saccharomyces cerevisiae*. *Current Genetics* **29**, 103–105.

Morris, G. J., Winters, L., Coulson, G. E. and Clarke, K. J. (1983) Effect of osmotic stress on the ultrastructure and viability of the yeast *Saccharomyces cerevisiae*. *Journal of General Microbiology* **129**, 2023–2034.

Moskvina, E., Imre, E. M. and Ruis, H. (1999) Stress factors acting at the level of the plasma membrane induce transcription via the stress response element (STRE) of the yeast *Saccharomyces cerevisiae*. *Molecular Microbiology* **32**, 1263–1272.

Moye-Rowley, W. S., Harshman, K. D. and Parker, C. S. (1989) Yeast *YAP1* encodes a novel form of the jun family of transcriptional activator proteins. *Genes and Development* **3**, 283–292.

Muller, E. G. D. (1991) Thioredoxin deficiency in yeast prolongs S phase and shortens the G1 interval of the cell cycle. *Journal of Biological Chemistry* **266**, 9194–9202.

Muller, E. G. D. (1996) A glutathione reductase mutant of yeast accumulates high levels of oxidized glutathione and requires thioredoxin for growth. *Molecular Biology of the Cell* **7**, 1805–1813.

Nakamura, T., Liu, Y., Hirata, D., Namba, H., Harada, S., Hirokawa, T. and Miyakawa, T. (1993) Protein phosphatase type 2B (calcineurin)-mediated, FK506-sensitive regulation of intracellular ions in yeast is an important determinant for adaptation to high salt stress conditions. *EMBO Journal* **11**, 4063–4071.

Natarajan, J., Meyer, M. R., Jackson, B. M., Slade, D., Roberts, C., Hinnebusch, A. G. and Marton, M. J. (2001) Transcriptional profiling shows that Gcn4 is a master regulator of gene expression during amino acid starvation. *Molecular and Cellular Biology* **21**, 4347–4368.

Navas, T. A., Sanchez, Y. and Elledge, S. J. (1996) *Rad9* and DNA polymerase ε form parallel sensory branches for transducing the DNA damage checkpoint signal in *Saccharomyces cerevisiae*. *Radiation and Environmental Biophysics* **35**, 55–57.

Nick, J. A., Leung, C. T. and Loewus, F. A. (1986) Isolation and identification of erythroascorbic acid in *Saccharomyces cerevisiae* and *Lypomyces starkeyi*. *Plant Science* **46**, 181–187.

Novick, P. and Botstein, D. (1985) Phenotypic analysis of temperature-sensitive yeast actin mutants. *Cell* **40**, 415–426.

Nunes, E. and Siede, W. (1996) Hyperthermia and paraquat-induced G1 arrest in the yeast *Saccharomyces cerevisiae* is independent of the *RAD9* gene. *Radiation and Environmental Biophysics* **35**, 55–57.

Ohtake, Y. and Yabuuchi, S. (1991) Molecular cloning of the γ-glutamyl cysteine synthetase gene of *Saccharomyces cerevisiae. Yeast* **7**, 953–961.

Olson, R. E. and Rudney, H. (1983) Biosynthesis of ubiquinone. *Vitamins and Hormones* **40**, 1–43.

Ota, I. M. and Varshavsky, A. (1993) A yeast protein similar to bacterial two-component regulators. *Science* **262**, 566–569.

Ozkaynak, E., Finley, D. and Varshavsky, A. (1987) The yeast ubiquitin genes: a family of natural gene fusions. *EMBO Journal* **6**, 1429–1439.

Panaretou, B. and Piper, P. W. (1990) Plasma membrane ATPase action affects several stress tolerances of *Saccharomyces cerevisiae* and *Schizosaccharomyces pombe* as well as the extent and duration of the heat shock response. *Journal of General Microbiology* **136**, 1763–1770.

Park, J.-I., Grant, C. M., Attfield, P. V. and Dawes, I. W. (1997) The freeze–thaw stress response of the yeast *Saccharomyces cerevisiae* is growth phase specific and is controlled by nutritional state via the *RAS*-cyclic AMP signal transduction pathway. *Applied and Environmental Biology* **63**, 3818–3824.

Park, J.-I., Davies, M. J, Grant, C. M. and Dawes, I. W. (1998) The cytoplasmic Cu, Zn superoxide dismutase of *Saccharomyces cerevisiae* is required for resistance to freeze–thaw stress: generation of free radicals during freezing and thawing. *Journal of Biological Chemistry* **273**, 22921–22928.

Park, S. G., Cha, M. K., Jeong, W., and Kim, I. H. (2000) Distinct physiological functions of thiol peroxidase isoenzymes in *Saccharomyces cerevisiae. Journal of Biological Chemistry* **275**, 5723–5732.

Parrou, J. L., Teste, M-A. and Francois, J. (1997) Effects of various types of stress on the metabolism of reserve carbohydrates in *Saccharomyces cerevisiae*: genetic evidence for a stress-induced recycling of glycogen and trehalose. *Microbiology* **143**, 1891–1900.

Parry, J. M. (1973) The induction of gene conversion in yeast by herbicide preparations. *Mutation Research* **21**, 83–91.

Parsell, D. A. and Lindquist, S. (1993) The function of heat shock proteins in stress tolerance: degradation and reactivation of damaged proteins. *Annual Review of Genetics* **27**, 437–496.

Parsell, D. A., Kowal, A. S., Singer, M. A. and Lindquist, S. (1994) Protein disaggregation mediated by heat-shock protein Hsp104. *Nature* **372**, 475–478.

Paulovitch, A. G., Margulies, R. U., Garvik, B. M. and Hartwell, L. H. (1997) *RAD9*, *RAD17*, and *RAD24* are required for S phase regulation in *Saccharomyces cerevisiae* in response to DNA damage. *Genetics* **145**, 45–62.

Pedrajas, J. R., Kosmidou, E., Miranda-Vizuete, A., Gustafsson, J. A., Wright, A. P. H. and Spyrou, G. (1999) Identification and functional characterization of a novel mitochondrial thioredoxin system in *Saccharomyces cerevisiae. Journal of Biological Chemistry* **274**, 6366–6373.

Pedrajas, J. R., Miranda-Vizuete, A., Javanmardy, N., Gustaffson, J. A. and Spyrou, G. (2000) Mitochondria of *Saccharomyces cerevisiae* contain one-conserved cysteine type peroxiredoxin with thioredon peroxidase activity. *Journal of Biological Chemistry* **275**, 16296–16301.

Pedrajas, J. R., Porras, P., Nartinez-Galisteo, E., Padilla, C. A., Miranda-Vizuete, A. and Barcena, J. A. (2002) Two isoforms of *Saccharomyces cerevisiae* glutaredoxin 2 are expressed in vivo and localize to different subcellular compartments. *Biochemical Journal* **364**, 617–623.

Pedruzzi, L., Burckert, N., Egger, P. and De Virgilio, C. (2000) *Saccharomyces cerevisiae* Ras/cAMP pathway controls post-diauxic shift element-dependent transcription through the zinc finger protein Gis1. *EMBO Journal* **19**, 2569–2579.

Pelham, H. R. B. (1982) A regulatory upstream promoter element in the Drosophila Hsp70 gene. *Cell* **30**, 517–528.

Penninckx, M. J. (2002) An overview on glutathione in *Saccharomyces* versus non-conventional yeasts. *FEMS Yeast Research* **2**, 295–305.

Pfeifer, K., Prezant, T. and Guarente, L. (1987) Yeast HAP1 activator binds to two upstream activation sites of different sequence. *Cell* **49**, 19–27.

Pinkham, J. L., Wang, Z. and Alsina, J. (1997) Heme regulates *SOD2* transcription by activation and repression in *Saccahromyces cerevisiae. Current Genetics* **31**, 281–291.

Piper, P. (1997) The yeast heat shock response. In: S. Hohmann and W.H. Mager (eds) *Yeast Stress Responses*, pp. 75–99. Austin: R.G. Landes Co.

Piper, P. W. (1993) Molecular events associated with acquisition of heat tolerance by the yeast *Saccharomyces cerevisiae. FEMS Microbiology Reviews* **11**, 339–356.

Piper, P. W., Talreja, K., Panaretou, B., Moradas-Ferreira, P., Byrne, K., Praekelt, U. M., Meacock, P., Regnacq, M. and Boucherie, H. (1994) Induction of major heat shock proteins of *Saccharomyces cerevisiae*, including plasma membrane Hsp30, by ethanol levels above a critical threshold. *Microbiology* **140**, 3031–3038.

Porter, N. A., Caldwell, S. A. and Mills, K. A. (1995) Mechanisms of free radical oxidation of unsaturated lipids. *Lipids* **30**, 277–290.

Posas, F. and Saito, H. (1997) Osmotic activation of the HOG MAPK pathway via Ste11p MAPKKK: scaffold role of Pbs2p MAPKK. *Science* **276**, 1702–1705.

Posas, F., Chambers, J. R., Heyman, J. A., Hoeffler, J. P., de Nadal, E. and Ariño, J. (2000) The transcriptional response of yeast to saline stress. *Journal of Biological Chemistry* **275**, 17249–17255.

Proft, M., Pascual-Ahuir, A., de Nadal, E., Arino, J., Serrano, R. and Posas, F. (2001) Regulation of the Sko1 transcriptional repressor by the Hog1 MAP kinase in response to osmotic stress. *EMBO Journal* **20**, 1123–1133.

Raitt, D. C., Johnson, A. L., Erkine, A. M., Makino, K., Morgan, B., Gross, D. S. and Johnston, L. H. (2000) The Skn7 response regulator of *Saccharomyces cerevisiae* interacts with Hsf1 in vivo and is required for the induction of heat shock genes by oxidative stress. *Molecular Biology of the Cell* **11**, 2335–2347.

Reiser, V., Salah, S. M. and Ammerer, G. (2000) Polarized localization of yeast Pbs2 depends on osmostress, the membrane Sho1 and Cdc42. *Nature Cell Biology* **2**, 620–627.

Rep, M., Krantz, M., Thevelein, J. M., Prior, B. A. and Hohmann, S. (2000) The transcriptional response of *Saccharomyces cerevisiae* to osmotic shock. Hot1p and Msn2p/Msn4p are required for the induction of subsets of high osmolarity glycerol pathway-dependent genes. *Journal of Biological Chemistry* **275**, 8290–8300.

Rhodes, N., Connell, L. and Errede, B. (1990) STE11 is a protein kinase required for cell-type-specific transcription and signal transduction in yeast. *Genes and Development* **4**, 1862–1874.

Rodriguez-Manzaneque, M. T., Tamarit, J., Belli, G., Ros, J. and Herrero, E. (2002) Grx5 is a mitochondrial glutaredoxin required for the activity of iron'sulfur enzymes. *Molecular Biology of the Cell* **13**, 1109–1121.

Rowley, A., Johnston, G. C., Butler, B., Werner-Washburne, M. and Singer, R. A. (1993) Heat shock-mediated cell cycle blockage and G1 cyclin expression in the yeast *Saccharomyces cerevisiae. Molecular and Cellular Biology* **13**, 1034–1041.

Rudolph, H. K. and Fink, G. R. (1990) Multiple plasma membrane $Ca2^+$-pumps in yeast. *Yeast* **6**, S561.

Ruis, H. and Hamilton, B. (1992) *Regulation of Yeast Catalase Genes.* Cold Spring Harbor, New York: Cold Spring Harbor Laboratory Press.

Ruis, H. and Schüller, C. (1995) Stress signalling in yeast. *BioEssays* **17**, 959–965.

Santoro, N. and Thiele, D. J. (1997) Oxidative stress responses in the yeast *Saccharomyces cerevisiae.* In: S. Hohmann and W. H. Mager (eds) *Yeast Stress Responses*, pp. 171–211. Austin: R.G. Landes Co.

Scandalios, J. G. (1987) The antioxidant enzyme genes *Cat* ans *Sod* of maize: regulation, functional significance, and molecular biology. *Isozymes: Current Topics in Biological and Medical Research* **14**, 19–44.

Schmitt, A. P. and McEntee, K. (1996) Msn2p, a zinc finger DNA-binding protein, is the transcriptional activator of the multistress response in *Saccharomyces cerevisiae. Proceedings of the National Academy of Sciences USA* **93**, 5777–5782.

Schnell, N. and Entian, K-D. (1991) Identification and characterization of a *Saccharomyces cerevisiae* gene (*PAR1*) conferring resistance to iron chelators. *European Journal of Biochemistry* **200**, 487–493.

Schnell, N., Krems, B. and Entian, K-D. (1992) The *PAR1* (*YAP1/SNQ3*) gene of *Saccharomyces cerevisiae*, a c-*jun* homologue, is involved in oxygen metabolism. *Current Genetics* **21**, 269–273.

Schüller, C., Brewster, J. L., Alexander, M. R., Gustin, M. C. and Ruis, H. (1994) The HOG pathway controls osmotic regulation of transcription via the stress response element (STRE) of the *Saccharomyces cerevisiae CTT1* gene. *EMBO Journal* **13**, 4382–4389.

Schulz, J. R. and Clark, C. F. (1999) Characterization of *Saccharomyces cerevisiae* ubiquinone-deficient mutants. *Biofactors* **9**, 121–129.

Seah, T. C. M. and Kaplan, J. G. (1973) Purification and properties of the catalase of bakers' yeast. *Journal of Biological Chemistry* **248**, 2889–2893.

Seah, T. C. M., Bhatti, A. R. and Kaplan, J. G. (1973) Novel catalytic proteins of bakers' yeast. *Canadian Journal of Biochemistry* **51**, 1551–1555.

Serrano, R. (1996) Salt tolerance in plants and microorganisms: toxicity targets and defense responses. *International Reviews of Cytology* **165**, 1–52.

Serrano, R., Márquez, J. A. and Rios, G. (1997) Crucial factors in salt stress tolerance. In: S. Hohmann W. H. Mager (eds) *Yeast Stress Responses*, pp. 147–169. Austin: R.G. Landes Company (Springer).

Shao, Y.-Y., Seib, P. A., Kramer, K. J. and Van Galen, D. A. (1993) Synthesis and properties of D-erythroascorbic acid and its vitamin C activity in the tobacco hornworm (*Manduca sexta*). *Journal of Agricultural and Food Chemistry* **41**, 1391–1396.

Shenton, D., Perrone, G., Quinn, K. A., Dawes, I. W. and Grant, C. M. (2002) Regulation of prtotein S-thiolation by glutaredoxin 5 in the yeast *Saccharomyces cerevisiae*. *Journal of Biological Chemistry* **277**, 16853–16859.

Sherman, F. (1959) The effects of elevated temperatures on yeast. II. Induction of respiratory-deficient mutants. *Journal of Cellular and Comparative Physiology* **54**, 37–52.

Shima, J., Hino, A., Yamada-Iyo, C., Suzki, Y., Nakajima, R., Watanabe, H., Mori, K. and Takano, H. (1999) Stress tolerance in doughs of *Saccharomyces cerevisiae* trehalase mutants derived from commercial baker's yeast. *Applied and Environmental Microbiology* **65**, 2841–2846.

Siede, W., Friedberg, A. S. and Friedberg, E. C. (1993) *RAD9*-dependent G1 arrest defines a second checkpoint for damaged DNA in the cell cycle of *Saccharomyces cerevisiae*. *Proceedings of the National Academy of Sciences USA* **90**, 7985–7989.

Simola, M., Hänninen, A. L., Stranius, S. M. and Makarow, M. (2000) Trehalose is required for conformational repair of heat-denatured proteins in the endoplasmic reticulum but not for maintenance of membrane traffic functions after severe heat stress. *Molecular Microbiology* **37**, 42–53.

Skrzypek, M. S., Nagiee, M. M., Lester, R. L., and Dickson, R. C. (1999) Analysis of phosphorylated sphingolipid long-chain bases reveals potential roles in heat stress and growth control in *Saccharomyces*. *Journal of Bacteriology* **181**, 1134–1140.

Sorger, P. K. (1991) Heat shock factor and the heat shock response. *Cell* **65**, 363–366.

Sorger, P. K. and Pelham, H. R. B. (1987) Purification and characterization of a heat shock element binding protein from yeast. *EMBO Journal* **6**, 3035–3041.

Sorger, P. K. and Pelham, H. R. B. (1988) Yeast heat shock factor is an essential DNA-binding protein that exhibits temperature dependent phosphorylation. *Cell* **54**, 855–864.

Sorger, P. K., Lewis, M. J. and Pelham, H. R. B. (1987) Heat shock factor is regulated differently in yeast and HeLa cells. *Nature* **329**, 81–84.

Spector, D., Labarre, J. and Toledano, M. B. (2001) A genetic investigation of the essential role of glutathione. Mutations in the proline biosynthesis pathway are the only suppressors of glutathione auxotrophy in yeast. *Journal of Biological Chemistry* **276**, 7011–7016.

Spevak, W., Hartig, A., Meindl, P. and Ruis, H. (1986) Heme control region of the catalase T gene of the yeast *Saccharomyces cerevisiae*. *Molecular and General Genetics* **203**, 73–78.

Spickett, C. M., Smirnoff, N. and Pitt, A. R. (2000) The biosynthesis of erythroascorbate in *Saccharomyces cerevisiae* and its role as an antioxidant. *Free Radical Biology and Medicine* **28**, 183–192.

Stadtman, E. R. (1992) Protein oxidation and aging. *Science* **257**, 1220–1224.

Stathopoulos, A. M. and Cyert, M. S. (1997) Calcineurin acts through the *CRZ1/TCN1*-encoded transcription factor to regulate gene expression in yeast. *Genes and Development* **11**, 3432–3444.

Stephen, D. W. S. and Jamieson, D. J. (1996) Glutathione is an important antioxidant molecule in the yeast *Saccharomyces cerevisiae*. *FEMS Microbiology Letters* **141**, 207–212.

Stephen, D. W. S., Rivers, S. L. and Jamieson, D. J. (1995) The role of *YAP1* and *YAP2* genes in the regulation of the adaptive stress responses of *Saccharomyces cerevisiae*. *Molecular Microbiology* **16**, 415–423.

Stone, D. E. and Craig, E. A. (1990) Self-regulation of 70-kilodalton heat shock proteins in *Saccharomyces cerevisiae*. *Molecular and Cellular Biology* **10**, 1622–1632.

Sturtz, L. A., Diekert, K., Jensen, L. T., Lill, R. and Culotta, V. C. (2001) A fraction of Cu,Zn-superoxide dismutase and its metallochaperone, CCS, localize to the intermembrane space of mitochondria – a physiological role for SOD1 in guarding against mitochondrial oxidative damage. *Journal of Biological Chemistry* **276**, 38084–38089.

Sugiyama, K., Izawa, S. and Inoue, Y. (2000) The Yap1p-dependent induction of glutathione synthesis in heat shock response of *Saccharomyces cerevisiae*. *Journal of Biological Chemistry* **275**, 15535–15540.

Suzuki, K., Okada, K., Kamiya, Y., Zhu, X. F., Nakagawa, T., Kawamukai, M. and Matsuda, H. (1997) Analysis of the decaprenyl diphosphate synthase (*dps*) gene in fission yeast suggests a role of ubiquinone as an antioxidant. *Journal of Biochemistry* **121**, 496–505.

Takada, M., Ikenoya, S., Yuzuriha, T. and Katayama, K. (1982) Studies on reduced and oxidised coenzyme Q (ubiquinones). II. The determination of oxidation–reduction levels of coenzyme Q in mitochondria, microsomes and plasma by high performance liquid chromatography. *Biochimica et Biophysica Acta* **679**, 308–314.

Takada, M., Ikenoya, S., Yuzuriha, T. and Katayama, K. (1984) Simultaneous determination of reduced and oxidised ubiquinones. *Methods in Enzymology* **105**, 147–155.

Takagi, H., Sakai, K., Morida, K. and Nakamori, S. (2000) Proline accumulation by mutation or disruption of the proline oxidase gene improves resistance to freezing and desiccation stresses in *Saccharomyces cerevisiae*. *FEMS Microbiology Letters* **184**, 103–108.

Tamai, K. T., Liu, X., Silar, P., Sosinowski, T. and Thiele, D. J. (1994) Heat shock transcription factor activates yeast metallothionein gene expression in response to heat and glucose starvation via distinct signalling systems. *Molecular and Cellular Biology* **14**, 8155–8165.

Terada, T. (1994) Thioltransferase can utilize cysteamide as same as glutathione as a reductant during the restoration of cysteamide-treated glucose 6-phosphate dehydrogenase. *Biochemistry and Molecular Biology International* **34**, 723–727.

Thevelein, J. M. (1994) Signal transduction in yeast. *Yeast* **10**, 1753–1790.

Thiele, D. J. and Hamer, D. H. (1986) Tandemly duplicated upstream control sequences mediate copper-induced transcription of the *Saccharomyces cerevisiae* copper-metallothionein gene. *Molecular and Cellular Biology* **6**, 1158–1163.

Toner, M., Cravalho, E. G. and Karel, M. (1993) Cellular response of mouse oocytes to freezing stress: prediction of intracellular ice formation. *Journal of Biomechanical Engineering* **115**, 169–174.

Tongaonkar, P., Chen, L., Lambertson, D., Ko, B. and Madura, K. (2000) Evidence for an interaction between ubiquitin-conjugating enzymes and the 26S proteasome. *Molecular and Cellular Biology* **20**, 4691–4698.

Torres, L., Martin, H., Garcia-Saez, M. I., Arroyo, J., Molina, M., Sanchez, M. and Nombela, C. (1991) A protein kinase gene complements the lytic phenotype of *Saccharomyces cerevisiae lyt2* mutants. *Molecular Microbiology* **5**, 2845–2854.

Treger, J., Hiechman, K. A. and McEntee, K. (1988) Expression of the yeast *UBI4* gene increases in response to DNA-damaging agents and in meiosis. *Molecular and Cellular Biology* **8**, 1132–1136.

Trotter, E. W. and Grant, C. M. (2003) Non-reciprocal regulation of the redox state of the glutathione-glutaredoxin and thioredoxin systems. *EMBO Reports* **4**, 184–188.

Trotter, E. W., Berenfeld, L., Krause, S. A., Petsko, G. A. and Gray, J. V. (2001) Protein misfolding and temperature up-shift cause G(1) arrest via a common mechanism dependent on heat shock factor in *Saccharomyces cerevisiae*. *Proceedings of the National Academy of Sciences USA* **98**, 7313–7318.

Turton, H. E., Dawes, I. W. and Grant, C. M. (1997) *Saccharomyces cerevisiae* exhibits an adaptive response to malondialdehyde, a product formed by oxidative stress, and this response is mediated via the yAP-1 transcriptional regulator. *Journal of Bacteriology* **179**, 1096–1011.

Tzagoloff, A. and Dieckmann, C. L. (1990) *PET* genes of *Saccharomyces cerevisiae. Microbiological Reviews* **54**, 211–225.

Tzagoloff, A., Akai, A. and Needleman, R. B. (1975a) Assembly of the mitochondrial membrane system. Characterization of nuclear mutants of *Saccharomyces cerevisiae* with defects in mitochondrial ATPase and respiratory enzymes. *Journal of Biological Chemistry* **250**, 8228–8235.

Tzagoloff, A., Akai, A. and Needleman, R. B. (1975b) Assembly of the mitochondrial membrane system: isolation of nuclear and cytoplasmic mutants of *Saccharomyces cerevisiae* with specific defects in mitochondrial functions. *Journal of Bacteriology* **122**, 826–831.

Van Dijck, P., Colavizza, D., Smet, P. and Thevelein, J. M. (1995) Differential importance of trehalose in stress resistance in fermenting and nonfermenting *Saccharomyces cerevisiae* cells. *Applied and Environmental Microbiology* **61**, 109–115.

Van Dijck, P., Ma, P. S., Versele, M., Gorwa, M. F., Colombo, S., Lemaire, K., Bossi, D., Loiez, A. and Thevelein, J. M. (2000) A baker's yeast mutant (*fil1*) with a specific, partially inactivating mutation in adenylate cyclase maintains a high stress resistance during active fermentation and growth. *Journal of Molecular Microbiology and Biotechnology* **2**, 521–530.

van Loon, A. P. G. M., Pesold-Hurt, B. and Schatz, G. (1986) A yeast mutant lacking mitochondrial manganese-superoxide dismutase is hypersensitive to oxygen. *Proceedings of the National Academy of Sciences USA* **83**, 3820–3824.

Varela, J. C. S. and Mager, W. H. (1996) Response of *Saccharomyces cerevisiae* to changes in osmolarity. *Microbiology* **142**, 721–731.

Vido, K., Spector, D., Lagniel, G., Lopez, S. Toledano, M. B. and Labarre, J. (2001) A proteome analysis of the cadmium response in *Saccharomyces cerevisiae. Journal of Biological Chemistry* **276**, 8469–8474.

Warmka, J., Hanneman, J., Lee, J., Amin, D. and Ota, I. (2001) Ptc1, a type 2C Ser/Thr phosphatse, inactivates the HOG pathway by dephosphorylating the mitogen-activated protein kinase Hog1. *Molecular and Cellular Biology* **21**, 51–60.

Weinert, T. A. and Hartwell, L. H. (1988) The *RAD9* gene controls the cell cycle response to DNA damage in *Saccharomyces cerevisiae. Science* **241**, 317–322.

Weinert, T. A., Hartwell, L. H. (1990) Characterization of *RAD9* of *Saccharomyces cerevisiae* and evidence that its function acts post-translationally in cell cycle arrest after DNA damage. *Molecular and Cellular Biology* **10**, 6554–6564.

Weinert, T. A. and Hartwell, L. H. (1993) Cell cycle arrest of *cdc* mutants and specificity of the *RAD9* checkpoint. *Genetics* **134**, 63–80.

Weinert, T. A., Kiser, G. L. and Hartwell, L. H. (1994) Mitotic checkpoint genes in budding yeast and the dependence of mitosis on DNA replication and repair. *Genes and Development* **8**, 652–665.

Weiser, R., Adam, G., Wagner, A., Schüller, C., Marchler, G., Ruis, H., Krawiec, A. and Bilinski, T. (1991) Heat shock factor-independent heat control of transcription of the *CTT1* gene encoding the cytosolic catalse T of *Saccharomyces cerevisiae. Journal of Biological Chemistry* **266**, 12406–12411.

Wells, W. W., Xu, D. P., Yang, Y. and Rocque, P. A. (1990) Mammalian thioltransferase (glutaredoxin) and protein disulfide isomerase have dehydroascorbate reductase activity. *Journal of Biological Chemistry* **265**, 15361–15364.

Wemmie, J. A., Szczypka, M. S., Thiele, D. J. and Moye-Rowley, W. S. (1994) Cadmium tolerance mediated by the yeast AP-1 protein requires the presence of an ATP-binding cassette transporter-encoding gene, *YCF1*. *Journal of Biological Chemistry* **269**, 32592–32597.

Wemmie, J. A., Steggarda, S. M. and Moye-Rowley, W. S. (1997) The *Saccharomyces cerevisiae* AP-1 protein discriminates between oxidative stress elicited by the oxidants H_2O_2 and diamide. *Journal of Biological Chemistry* **272**, 7908–7914.

Westerbeek-Marres, C. A. M., Moore, M. M. and Autor, A. P. (1988) Regulation of manganese superoxide dismutase in *Saccharomyces cerevisiae. European Journal of Biochemistry* **174**, 611–620.

Wiederrecht, G., Seto, D. and Parker, C. S. (1988) Isolation of the gene encoding the *S. cerevisiae* heat shock factor. *Cell* **48**, 507–515.

Wiemken, A. (1990) Trehalose in yeast, stress protectant rather than reserve carbohydrate. *Antonie Leeuwenhoek* **58**, 209–217.

Wieser, R., Adam, G., Wagner, A., Schuller, C., Marchler, G. Ruis, H., Krawiec, Z. and Bilinski, T. (1991) Heat-shock factor-independent heat control of transcription of the *CTT1* gene encoding the cytosolic catalase T of *Saccharomyces cerevisiae*. *Journal of Biological Chemistry* **266**, 12406–12411.

Winde, J. H., Thevelein, J. M. and Winderickx, J. (1997) From feast to famine: adaptation to nutrient depletion in yeast. In: S. Hohmann and W. H. Mager (eds) *Yeast Stress Responses*, pp. 7–52. Austin: R.G. Landes Co.

Winderickx, J., de Winde, J. H., Crauwels, M., Hino, A, Hohmann, S. Van Dijck, P. and Thevelein, J. M. (1996) Regulation of genes encoding the subunits of the trehalose synthase complex in *Saccharomyces cerevisiae*. Novel variations of STRE-mediated transcriptional control? *Molecular and General Genetics* **252**, 470–482.

Winkler, A., Arkind, C., Mattison, C. P., Burkholder, A., Knoche, K. and Ota, I. (2002) Heat stress activates the yeast high-osmolarity glycerol mitogen-activated protein kinase pathway, and protein tyrosine phosphatases are essential under heat stress. *Eukaryotic Cell* **1**, 163–173.

Winkler, H., Adam, G., Mattes, E., Schanz, M., Hartig, A. and Ruis, H. (1988) Co-ordinate control of synthesis of mitochondrial and non-mitochondrial hemoproteins, a binding site for the HAP1 (CYP1) protein in the UAS region of the yeast catalase T gene (*CTT1*). *EMBO Journal* **7**, 1799–1804.

Winzeler, E. A. *et al.* (1999) Functional characterization of the *S. cerevisiae* genome by gene deletion and parallel analysis. *Science* **285**, 901–906.

Wiseman, H. and Halliwell, B. (1996) Damage to DNA by reactive oxygen and nitrogen species: role in inflammatory disease and progression to cancer. *Biochemical Journal* **313**, 17–29.

Wittenberg, C., Sugimoto, K. and Reed, S. I. (1990) G1-specific cyclins of *S. cerevisiae*: cell cycle periodicity, regulation by mating pheromone, and association with the $p34^{CDC28}$ protein kinase. *Cell* **62**, 225–237.

Wolff, S. P. and Dean, R. T. (1986) Fragmentation of proteins by free radicals and its effect on their susceptibility to enzymic hydrolysis. *Biochemical Journal* **234**, 399–403.

Wollman, E. E., D'Auriol, L., Rimsky, L., Shaw, A., Jacquot, J. P., Wingfield, P. *et al.* (1988) Cloning and expression of a cDNA for human thioredoxin. *Journal of Biological Chemistry* **263**, 15506–15512.

Wong, C. M., Zhou, Y., Ng, R. W. M., Kung, H. F. and Jin, D. Y. (2002) Cooperation of yeast peroxiredoxins Tsa1p and Tsa2p in the cellular defense against oxidative stress. *Journal of Biological Chemistry* **277**, 5385–5394.

Wonisch, W., Tatzber, F., Schaur, J. R., Larkovic, N., Guttenberger, H. and Esterbauer, H. (1997) Cell cycle inhibition by the lipid peroxidation product 4-hydroxynonenal in the yeast *Saccharomyces cerevisiae*. *Naunyn-Schmiedebergs Archives of Pharmacology* **356** (Supplement) **1**, 72.

Wu, A. and Moye-Rowley, W. S. (1994) *GSH1*, which encodes γ-glutamylcysteine synthetase, is a target gene for YAP-1 transcriptional regulation. *Molecular and Cellular Biology* **14**, 5832–5839.

Wu, A., Wemmie, J. A., Edgington, N. P., Goebl, M., Guevara, J. L. and Moye-Rowley, W. S. (1993) Yeast bZIPproteins mediate pleiotropic drug and metal resistance. *Journal of Biological Chemistry* **268**, 18850–18858.

Yamaguchi-Iwai, Y., Stearman, R., Dancis, A. and Klausner, R. D. (1996) Iron-regulated DNA binding by the AFT1 protein controls the iron regulon in yeast. *EMBO Journal* **15**, 3377–3384.

Yang, W., Gahl, W. and Hamer, D. (1991) Role of heat shock transcription factor in yeast metallothionein gene expression. *Molecular and Cellular Biology* **11**, 3676–3681.

Zielinski, R., Pilecki, M., Kubinski, K., Zien, P., Hellman, U. and Szyszka, R. (2002) Inhibition of yeast ribosomal stalk phosphorylation by Cu-Zn superoxide dismutase. *Biochemical and Biophysical Research Communications* **296**, 1310–1316.

Index

Numbers in **bold face** type refer to occurrences in figures or tables.

Printed in the United States
129230LV00006B/7/A

9 780415 299008